## Geophysical Monograph Series

Including

**IUGG Volumes**
**Maurice Ewing Volumes**
**Mineral Physics Volumes**

85 **The Polar Oceans and Their Role in Shaping the Global Environment** *O. M. Johannessen, R. D. Muench, and J. E. Overland (Eds.)*

86 **Space Plasmas: Coupling Between Small and Medium Scale Processes** *Maha Ashour-Abdalla, Tom Chang, and Paul Dusenbery (Eds.)*

87 **The Upper Mesosphere and Lower Thermosphere: A Review of Experiment and Theory** *R. M. Johnson and T. L. Killeen (Eds.)*

88 **Active Margins and Marginal Basins of the Western Pacific** *Brian Taylor and James Natland (Eds.)*

89 **Natural and Anthropogenic Influences in Fluvial Geomorphology** *John E. Costa, Andrew J. Miller, Kenneth W. Potter, and Peter R. Wilcock (Eds.)*

90 **Physics of the Magnetopause** *Paul Song, B.U.Ö. Sonnerup, and M.F. Thomsen (Eds.)*

91 **Seafloor Hydrothermal Systems: Physical, Chemical, Biological, and Geological Interactions** *Susan E. Humphris, Robert A. Zierenberg, Lauren S. Mullineaux, and Richard E. Thomson (Eds.)*

92 **Mauna Loa Revealed: Structure, Composition, History, and Hazards** *J. M. Rhodes and John P. Lockwood (Eds.)*

93 **Cross-Scale Coupling in Space Plasmas** *James L. Horowitz, Nagendra Singh, and James L. Burch (Eds.)*

94 **Double-Diffusive Convection** *Alan Brandt and H. J. S. Fernando (Eds.)*

95 **Earth Processes: Reading the Isotopic Code** *Asish Basu and Stan Hart (Eds.)*

96 **Subduction: Top to Bottom** *Gray E. Bebout, David Scholl, Stephen Kirby, and John Platt (Eds.)*

97 **Radiation Belts: Models and Standards** *J. F. Lemaire, D. Heynderickx, and D. N. Baker (Eds.)*

98 **Magnetic Storms** *Bruce T. Tsurutani, Walter D. Gonzalez, Yohsuke Kamide, and John K. Arballo (Eds.)*

99 **Coronal Mass Ejections** *Nancy Crooker, Jo Ann Joselyn, and Joan Feynman (Eds.)*

100 **Large Igneous Provinces** *John J. Mahoney and Millard F. Coffin (Eds.)*

101 **Properties of Earth and Planetary Materials at High Pressure and Temperature** *Murli Manghnani and Takehiki Yagi (Eds.)*

102 **Measurement Techniques in Space Plasmas: Particles** *Robert F. Pfaff, Joseph E. Borovsky, and David T. Young (Eds.)*

103 **Measurement Techniques in Space Plasmas: Fields** *Robert F. Pfaff, Joseph E. Borovsky, and David T. Young (Eds.)*

104 **Geospace Mass and Energy Flow: Results From the International Solar-Terrestrial Physics Program** *James L. Horwitz, Dennis L. Gallagher, and William K. Peterson (Eds.)*

105 **New Perspectives on the Earth's Magnetotail** *A. Nishida, D. N. Baker, and S. W. H. Cowley (Eds.)*

106 **Faulting and Magmatism at Mid-Ocean Ridges** *W. Roger Buck, Paul T. Delaney, Jeffrey A. Karson, and Yves Lagabrielle (Eds.)*

107 **Rivers Over Rock: Fluvial Processes in Bedrock Channels** *Keith J. Tinkler and Ellen E. Wohl (Eds.)*

108 **Assessment of Non-Point Source Pollution in the Vadose Zone** *Dennis L. Corwin, Keith Loague, and Timothy R. Ellsworth (Eds.)*

109 **Sun-Earth Plasma Interactions** *J. L. Burch, R. L. Carovillano, and S. K. Antiochos (Eds.)*

110 **The Controlled Flood in Grand Canyon** *Robert H. Webb, John C. Schmidt, G. Richard Marzolf, and Richard A. Valdez (Eds.)*

111 **Magnetic Helicity in Space and Laboratory Plasmas** *Michael R. Brown, Richard C. Canfield, and Alexei A. Pevtsov (Eds.)*

112 **Mechanisms of Global Climate Change at Millennial Time Scales** *Peter U. Clark, Robert S. Webb, and Lloyd D. Keigwin (Eds.)*

113 **Faults and Subsurface Fluid Flow in the Shallow Crust** *William C. Haneberg, Peter S. Mozley, J. Casey Moore, and Laurel B. Goodwin (Eds.)*

114 **Inverse Methods in Global Biogeochemical Cycles** *Prasad Kasibhatla, Martin Heimann, Peter Rayner, Natalie Mahowald, Ronald G. Prinn, and Dana E. Hartley (Eds.)*

115 **Atlantic Rifts and Continental Margins** *Webster Mohriak and Manik Talwani (Eds.)*

116 **Remote Sensing of Active Volcanism** *Peter J. Mouginis-Mark, Joy A. Crisp, and Jonathan H. Fink (Eds.)*

117 **Earth's Deep Interior: Mineral Physics and Tomography From the Atomic to the Global Scale** *Shun-ichiro Karato, Alessandro Forte, Robert Liebermann, Guy Masters, Lars Stixrude (Eds.)*

118 **Magnetospheric Current Systems** *Shin-ichi Ohtani, Ryoichi Fujii, Michael Hesse, and Robert L. Lysak (Eds.)*

119 **Radio Astronomy at Long Wavelengths** *Robert G. Stone, Kurt W. Weiler, Melvyn L. Goldstein, and Jean-Louis Bougeret (Eds.)*

120 **GeoComplexity and the Physics of Earthquakes** *John B. Rundle, Donald L. Turcotte, and William Klein (Eds.)*

Geophysical Monograph 121

# The History and Dynamics of Global Plate Motions

Mark A. Richards
Richard G. Gordon
Rob D. van der Hilst
*Editors*

American Geophysical Union
Washington, DC

## Published under the aegis of the AGU Books Board



**Library of Congress Cataloging-in-Publication Data**
The history and dynamics of global plate motions / Mark A. Richards, Richard G. Gordon and Rob D. van der Hilst, editors.
p. cm. -- (Geophysical monograph ; 121)
Includes bibliographical references.
ISBN 0-87590-979-5
1. Plate tectonics. I. Richards, Mark A., 1955- II. Gordon, Richard G., 1953- III. Hilst, Robert Dirk van der, 1961- IV. Series.

QE511.4 .H57 2000
551.1'36--dc21 00-059361

ISBN 0-87590-979-5
ISSN 0065-8448



Printed in the United States of America.

# CONTENTS

# CONTENTS

## Section 4: Seismic Tomography and Global Mantle Flow



## Section 5: Hotspots and Plate Motions

# PREFACE

The new global tectonics that emerged three decades ago profoundly changed our view of the Earth and its evolution. Although the theory of plate tectonics gives a kinematic description of much of Earth's surface motions, our dynamical understanding remains incomplete and unsatisfactory in many ways. Key remaining issues include the mechanics of plate boundaries and intraplate deformation, the relation between plate-scale dynamics and hotspot volcanism, vertical surface motions (dynamic topography) associated with subduction, and the origin of the plate-tectonic style of convection itself. The past two decades have seen fairly steady progress on a number of these problems, spurred in large part by the advent of 3-D seismic imaging of the Earth's interior. More recently, advances in high performance computing have provided long-needed 3-D geodynamic modeling tools that are yielding new insights into the relation between mantle convection and plate tectonics, and the interpretation of seismic heterogeneity structure.

Plate tectonics emerged as a synthesis of previously disparate disciplines of geoscience (e.g., seismology, paleomagnetism, marine geology and geophysics, petrology, and paleontology), and the field continues to advance through multi-disciplinary collaboration. An AGU Chapman conference on the *History and Dynamics of Global Plate Motions* was convened in June 1997 at Pt. Reyes, California, with the purpose of bringing together geodynamicists, tectonicists, and seismologists to improve our understanding of the dynamics of global plate motions. Following this conference, a group of participants and a few selected non-participants were invited to contribute to this volume. Contributors were encouraged to write for a general audience of Earth scientists, to provide a modern view of the field, and to present a combination of tutorial, review, and new research. The volume is organized along scientific themes rather than disciplines. Topics include the basics of mantle convection and its relation to plate motions, continental dynamics and intraplate deformation, detailed application of modern reconstruction and modeling techniques to the Australian region, the relation between 3-D seismic structure and mantle dynamics, and the relation between hotspots and plate motions, the latter emerging from the Chapman conference as a particularly challenging subject.

It is hoped that this grouping of topics and the tutorial and review material contained in the papers will promote a wider appreciation for our current state of knowledge of the dynamics of plate motions, and for the major questions that remain. The editors wish to thank the authors for producing an excellent group of papers and, in particular, their attention to writing for a more general audience than would be normal for journal articles. We thank the many reviewers of the manuscripts, and we are also grateful to the participants at the AGU Chapman Conference, who made for a most enjoyable and exciting gathering of minds. Steven Bell of AGU did a wonderful job of coordinating the conference.

Mark A. Richards
*University of California*

Richard G. Gordon
*Rice University*

Rob D. van der Hilst
*Massachusetts Institute of Technology*

# Introduction: Plate Tectonics and Mantle Convection Three Decades Later

Mark A. Richards

*Department of Geology and Geophysics, University of California, Berkeley, California*

Richard G. Gordon

*Department of Geology and Geophysics, Rice University, Houston, Texas*

Rob D. van der Hilst

*Department of Earth, Atmospheric, and Planetary Sciences, MIT, Cambridge, Massachusetts*

The advent of the theory of plate tectonics and its portrait of a mobile solid Earth marked the most profound intellectual advance in the twentieth century in the Earth sciences. Plate tectonics is fundamentally a kinematic theory: it gives the framework for describing the main features of Earth's horizontal surface motions, but does not quantify the forces that drive (and resist) plate motions. Spreading ridges and subduction zones are sites of upwelling and downwelling flow in the mantle, with subducted oceanic lithosphere sinking to at least 670 km depth, where deep earthquakes cease. The thickness, depth, and heat flow of oceanic lithosphere as a function of age are described well by a cooling thermal boundary layer model, making clear the role of the lithosphere as the top boundary to thermal convection in the mantle. However, three decades after the plate tectonics revolution geophysicists, geochemists, and geologists agree on remarkably little else about the dynamics of plate tectonics and mantle convection.

Volcanic hotspots, such as Hawaii, Iceland, and Yellowstone, do not fit at all into the theory of plate tectonics, except as mysterious tracers of plate motions over hotspot sources. Whatever the nature of these sources may be, they appear to have relative motions much slower than between the fastest-moving plates, but possibly comparable to motions between the slower moving plates. The leading theory, developed along with plate tectonics, is that sites of major intraplate volcanism are due to concentrated hot upwelling plumes from the deep mantle, but this theory remains controversial and has so far evaded definitive tests.

During the past decade, missions to Mars, Venus, and the moons of the outer planets have greatly increased our perspective on Earth's dynamic style. In our solar system, plate tectonics appears to be unique to Earth, but plume-related volcanism appears to occur on Venus, Mars, and perhaps several moons. Otherwise, Mars appears to be a "one-plate" planet, i.e., its lithosphere exhibits little mobility at present. Venus's surface appears to experience a more spatially continuous style of deformation instead of plate tectonics, and there is evidence for a catastrophic resurfacing of Venus about 500 million years ago. Thus Mars appears to operate in a "frozen lid" mode of convection, whereas Venus exists perhaps in an unstable, non-platelike regime. Both of these modes have straightforward explanations in terms of mantle convection theory when the effects of temperature dependent rheology and lithospheric failure are included.

The History and Dynamics of Global Plate Motions
Geophysical Monograph 121

It is plate tectonics on Earth that is hardest to explain in terms of dynamical models — to date none can be considered very successful. Moreover, no convection theory accounts satisfactorily for the coexistence of mantle plumes (hotspots) and plate motions. Simple thermal convection models, with or without plates, overpredict Earth's observed dynamic topography by at least a factor of two, and fail to account for the isolation of geochemical reservoirs necessary to explain the diversity of isotopic and trace element signatures observed in mantle-derived lavas.

Have geodynamicists failed, or are dynamic processes in the Earth's interior so well concealed that they cannot be discerned from surface observations?

We believe the answer is "no" on both counts: During the past two decades, major advances in our understanding of the complex Earth system have been made, with computational geodynamics connecting aspects of mantle convection theory with geophysical, geochemical, and geological observations. The most significant of these advances has been the advent and subsequent development of seismic tomography, a class of imaging that now provides increasingly detailed, 3-D maps of seismic velocity variations that can be related to thermal (and chemical) structures in the mantle. For example, despite observed complexity in the upper mantle transition zone, at least some subducted slabs appear to sink essentially to the bottom of the mantle, and there is a huge hot (low velocity) upwelling structure, not predicted by the theory of plate tectonics or computer simulations of thermal convection, which accounts for the anomalous elevation of much of Africa. Geodynamic models of flow associated with these structures successfully predict the observed geoid, and most of the cold (high seismic velocity) features are associated with the history of subduction of oceanic lithosphere. Vertical motions of continental plates also appear at least partly explainable in terms of the evolution of these dynamic structures in the mantle.

As to the possible "unknowability" of mantle dynamics, it appears instead that we are beset by too much information, at least for the simple geodynamic models developed so far. Geochemical data indicate a multiplicity, or continuum, of internal reservoirs that cannot be explained in terms of thermal convection alone. Indeed, the most peculiar geochemical signatures are associated with hotspot-type volcanism, further suggesting the isolation of this process from the main large-scale convection pattern associated with plate motions. Seismic imaging indicates unexpectedly strong velocity anomalies at the core-mantle boundary, equal in magnitude to those of the Earth's near-surface lithospheric layers and hinting at exotic dynamics yet to be discovered.

New information is also arriving from higher resolution studies of the history of the ocean floor, especially via new satellite altimetry data, high-resolution swath mapping, marine magnetic profiles, and refinements to the geomagnetic reversal time scale. These data have increased the precision and reliability of plate motion reconstructions in many parts of the globe, especially in the southern oceans, creating a stronger basis for testing geodynamic theories. Our knowledge of the structure and dynamics of midocean ridges has matured greatly in the past decade, due in no small part to concentrated geophysical studies such as the RIDGE program. In particular, the geometry of magma production and movement at ridges has been illuminated by high-resolution 3-D seismic studies.

New data, especially space geodetic (VLBI, SLR, GPS, DORIS) measurements, have also vastly increased our understanding of plate boundaries characterized by distributed deformation. These areas do not behave in plate tectonic fashion at all and constitute about 15 percent of the Earth's surface, with the continental area of diffuse plate boundaries being two to three times larger than the oceanic area. Some active continental areas, such as the Himalayas and Tibetan Plateau, Andes, and Basin and Range (USA), exhibit instead multiply block-faulted, tectonic styles near the surface, which may overlie broad deep regions of continuous deformation. Hence, the theory of plate tectonics itself has undergone a gradual but highly significant transformation since it's inception, although most geologists and geophysicists still think (and teach) in the simpler terms of a global surface velocity field described by relatively few plates, separated by narrow boundaries and undergoing rigid body motions described by angular velocities.

We find it surprising after three decades of research following the acceptance of plate tectonics that there are no agreed-upon models for the forces driving plate motions, or for the stresses within plates. Models developed in the 1970's were cast in terms of prescribed forces acting on the edges of plates ("ridge push," "slab pull"), or acting in arbitrary fashion along the bases of plates ("asthenosphere drag"). However, more recent large-scale mantle convection models, in which these forces arise naturally from numerical solutions of the full dynamical equations, are no better at predicting intraplate stresses and plate motions, and may in fact be worse. The essential problem is that we have only a very poor understanding of how the forces that act on and within the lithosphere are coupled to the dynamics of flow in the mantle, so that even a very good model for thermo-chemical convection in the Earth's interior (hardly a reality at present) might not meaningfully constrain inter- and intra-plate forces.

Several developments in geophysics, however, are facilitating progress in understanding the underlying dynamics of convection, plate tectonics, and hotspots on Earth. The most obvious advances are the continual improvements being made in documenting the record of present and past plate motions, much of this coming from improved data from the southern oceans. Much work remains, however. The documentation in the past three years of two previously unrecognized active diffuse plate boundaries, one intersecting the Southwest Indian Ridge and the other intersecting the Southeast Indian Ridge, raises a note of caution about the reliability of global plate reconstructions over the past 40 million years, not to mention attempts to reconstruct plate motions back to 80 million years ago, or further back into Mesozoic time, for which fundamental ambiguities remain. There remains much room for improvement, especially in regard to documenting punctuated plate motion rearrangements, or "stage boundaries."

Striking improvements have recently been made in the resolution and reliability of seismic imaging of the Earth's deep interior. Subducted slabs can now be seen with startling clarity, and new experiments employing ocean bottom seismic arrays should soon allow direct imaging of mantle plumes, if indeed they exist.

Less obvious, perhaps, is the likely impact of improvements in geodynamic modeling. During the past five years, we have crossed a threshold in high performance computing that allows modeling of mantle convection in three dimensions and at the resolution necessary to capture the full vigor (high Rayleigh number) of mantle convection. Furthermore, rapid advances are being made in modeling various candidate rheologies that probably account for the great diversity of styles of planetary dynamics, especially surface expressions such as faulting and volcanism. In other words, computational geodynamics now appears up to the task of hypothesis testing using the full range of geophysical, geochemical, and geological data bearing on the dynamics of the solid Earth. It is becoming increasingly possible for models of mantle convection to make some useful predictions, albeit determined with many adjustable parameters. These models can then be compared with a broader class of constraints, including more traditional geologic observations, and lead to investigations with a more hypothetico-deductive flavor.

These developments hold great potential for inter- and multi-disciplinary research in plate tectonics and mantle convection. In recent years, geodynamicists have begun working directly with plate tectonicists to use plate motion models to constrain regional and global models of the Earth's interior. Most recently, seismic tomography has been used by geodynamicists and plate tectonicists to better understand the evolution of subduction zones, both spatially and temporally, effectively using seismic imaging as an aid in mapping the history of surface plate motions.

In concert with these developments, an AGU Chapman conference was convened in June 1997 at Pt. Reyes, California, with the purpose of bringing together geodynamicists, tectonicists, and seismologists to improve our understanding of the dynamics of global plate motions. One theme that emerged at the conference was an emphasis on the importance and difficulties of the so-called "hotspot reference frame," both from the standpoint of geodynamics and the practice of plate motion reconstruction.

A group of authors, mainly conference participants plus a few selected nonparticipants, were invited to contribute to this volume, with the purpose of providing an accurate picture of where we currently stand in our understanding of the plate tectonic and mantle convection system. The volume is organized along scientific themes rather than disciplines, with an emphasis on dynamics that perhaps distinguishes this volume from others on plate tectonics. We begin with geodynamic analyses of the key importance of rheology in determining the dynamics of convection on Earth. After an extensive review paper, two distinct approaches to implementation of plate-like rheology follow. Nowhere is the richness of rheologically induced styles of tectonics more evident than on continents, and two more papers are devoted, respectively, to the problem of heat flow and deep continental structure, and to quantifying and modeling continuous deformation in an active continental region. These papers are followed by an analysis of intraplate deformation globally, both on continental and oceanic plates. The emphasis then becomes more regional with three papers focussing on the Australian region: The first analyzes geological events in the Cenozoic and Mesozoic, the second addresses the relations of plate motions, mantle flow, and volcanism, and the third develops a model for Cenozoic dynamics, including both horizontal and vertical motions of the Australian continent.

Seismic tomography is represented first by two studies employing the global basis function approach (e.g., using spherical harmonics), the first paper emphasizing lateral structure and anisotropy in the uppermost mantle, which may help improve our understanding of the plate cooling process, and the second paper comparing whole mantle structure with that predicted by geodynamic models. These are followed by a paper summarizing progress using the local basis function approach (e.g., non-overlapping constant-slowness volumes, or cubic splines that interpolate between grid values), which, in regions of dense data coverage, provides the highest resolution of features in the mantle related to plate motions. The seismic tomography papers are followed by a discussion of fledgling "global circulation models" for mantle convection which draw on both the history of plate motions and on seismic tomography as constraints on mantle dynamics and structure through Mesozoic and Cenozoic time.

The final section of the volume includes five papers on hotspots. A review of large igneous provinces (e.g., continental flood basalts, oceanic plateaux) is followed by three papers addressing the issue of hotspot fixity: The first of these suggests that the Pacific hotspots remain fixed with respect to one another. The next two papers conclude that Pacific hotspots (Hawaii) are not stationary relative to the Atlantic-Indian Ocean hotspot cluster, and suggest that the great Hawaiian-Emperor bend may not be due to a sudden change in overlying Pacific plate motion, as has long been assumed by many geophysicists. A final paper discusses geodynamic models that attempt to account for some observed inter-hotspot motions.

The papers in this volume are aimed at a general audience of Earth scientists, especially beginning graduate students who want to gain a modern view of the field. Authors were encouraged to include considerable review and tutorial material to serve this purpose, and it is possible that even the experts will benefit from this style of presentation. The volume is not entirely balanced, as more contributions on global plate reconstructions would have given a more complete picture of the current state of the field. We hope, however, that the papers contained herein give at least a sense of the excitement and rapid rate of progress in answering long-standing questions about plate tectonics and mantle convection, and that they also give a strong flavor of the interdisciplinary nature of this progress.

# The Relation Between Mantle Dynamics and Plate Tectonics: A Primer

David Bercovici[1], Yanick Ricard

*Laboratoire des Sciences de la Terre, Ecole Normale Supérieure de Lyon, France*

Mark A. Richards

*Department of Geology and Geophysics, University of California, Berkeley*

We present an overview of the relation between mantle dynamics and plate tectonics, adopting the perspective that the plates are the surface manifestation, i.e., the top thermal boundary layer, of mantle convection. We review how simple convection pertains to plate formation, regarding the aspect ratio of convection cells; the forces that drive convection; and how internal heating and temperature-dependent viscosity affect convection. We examine how well basic convection explains plate tectonics, arguing that basic plate forces, slab pull and ridge push, are convective forces; that sea-floor structure is characteristic of thermal boundary layers; that slab-like downwellings are common in simple convective flow; and that slab and plume fluxes agree with models of internally heated convection. Temperature-dependent viscosity, or an internal resistive boundary (e.g., a viscosity jump and/or phase transition at 660km depth) can also lead to large, plate sized convection cells. Finally, we survey the aspects of plate tectonics that are poorly explained by simple convection theory, and the progress being made in accounting for them. We examine non-convective plate forces; dynamic topography; the deviations of seafloor structure from that of a thermal boundary layer; and abrupt plate-motion changes. Plate-like strength distributions and plate boundary formation are addressed by considering complex lithospheric rheological mechanisms. We examine the formation of convergent, divergent and strike-slip margins, which are all uniquely enigmatic. Strike-slip shear, which is highly significant in plate motions but extremely weak or entirely

[1]Permanently at Department of Geology and Geophysics, School of Ocean and Earth Science and Technology, University of Hawaii

The History and Dynamics of Global Plate Motions
Geophysical Monograph 121

absent in simple viscous convection, is given ample discussion. Many of the problems of plate boundary formation remain unanswered, and thus a great deal of work remains in understanding the relation between plate tectonics and mantle convection.

## 1. INTRODUCTION

In the late 1930's, following the introduction of Alfred Wegener's theory of Continental Drift [Wegener, 1924], several driving mechanisms for the Earth's apparent surface motions were proposed. While Wegener himself favored tidal and *pole fleeing* (i.e., centrifugal) forces, Arthur Holmes and others hypothesized that thermal convection in the Earth's mantle provided the necessary force to drive continental motions [Holmes, 1931; see Hallam, 1987]. Even with the spurning and demise of the theory Continental Drift, and it's resurrection and revision 30 years later in the form of "Plate Tectonics" [e.g., Morgan, 1968], mantle convection is still widely believed to be the engine of surface motions, and, indeed, for many good reasons as we shall see in this review [see also review by Oxburgh and Turcotte, 1978]. However, there is still no complete physical theory which predicts how plate tectonics in its entirety is driven, or, more appropriately, caused by mantle convection.

There is little doubt that the direct energy source for plate tectonics and all its attendant features (mountain building, earthquakes, volcanoes, etc.) is the release of the mantle's gravitational potential energy through convective overturn (and of course that radiogenic heating and core cooling continue to replenish the mantle's gravitational potential energy). However, the precise picture of how plate motions are caused by convection is far from complete. With the recognition of the importance of slab pull, and that subducting slabs are essentially cold downwellings, it is becoming more widely accepted that the plates are an integral part of mantle convection, or more to the point they *are* mantle convection [e.g., Davies and Richards, 1992]. However, the idea that the plates arise from or are generated by mantle convection yields new complications and questions. Perhaps most difficult of these questions concerns the property of the mantle-lithosphere system which permits a form of convective flow that is unlike most forms of thermal convection in fluids (save perhaps lava lakes; see Duffield, 1972), i.e., a convection which looks like plate tectonics at the surface.

In this paper we review progress on the problem of how plate tectonics itself is explained by mantle convection theory. We begin by surveying the aspects of basic convection in simple fluid systems which are applicable to plate tectonics. We next discuss those features of plate tectonics which are reasonably well explained by basic convection theory. We then examine the numerous features of plate tectonics which are poorly (or not at all) explained by simple convection, and, in that regard, we discuss the possible new phyiscs which is needed to improve (and in some cases radically alter) present theories of mantle convection. This paper is not only a review and tutorial on the aspects of convection relevant to plate tectonics; it is also an exposition of how little we know about the causes of tectonic motions, and how much work remains in the important problem of unifying the theories of plate tectonics and mantle convection.

## 2. BASIC CONVECTION

Basic thermal convection in fluids is perhaps the most fundamental paradigm of self-organization in nonlinear systems [see Nicolis, 1995]. Such self-organization is characterized by the forcing of simple homogeneous yet perturbable or mobile systems (i.e., whose particles can move) far from equilibrium (e.g. by heating or imposition of chemical disequilibrium); in many cases, the systems can develop complex patterns and oscillatory or chaotic temporal behavior. In basic thermal convection, for example, a layer of fluid is heated uniformly and subsequently develops organized polygonal patterns and cells of cold and hot thermals. Indeed, the formation of tectonic plates is invariably a form of convective self-organization.

Although the phenomenon of convection was first recognized by Count Rumford [1870] and James Thomson [1882] (the brother of William Thomson, Lord Kelvin), the first systematic experimental study of basic convection was carried out by Henri Bénard on thin layers of spermaceti (whale fat) and paraffin [Bénard, 1900, 1901]. Bénard's experiments yielded striking images of honey-combed patterns and vertical cellular structure. After John William Strutt, Lord Rayleigh, failed to theoretically explain the experiments with hydrodynamic stability theory [Rayleigh, 1916], it was eventually deduced that Bénard had witnessed not thermal convection but (because his thin layers of spermaceti were exposed to air) surface-tension driven convection, also known as Marangoni convection [Pearson, 1958; see Berg et al., 1966]. When the thermal convection experiments were carried out correctly, they were found to be very well predicted by Rayleigh's theory [see Chandrasekhar, 1961, ch.2, §18]. Nevertheless, convection in a thin layer of fluid heated along its base is still called Rayleigh-Bénard (and often just Bénard) convection.

### 2.1. *Convective instability and the Rayleigh number*

Rayleigh's stability analysis of convection predicted the conditions necessary for the onset of convection, as well as the expected size of convection cells (relative to the layer thickness). Without delving into the mathematics of this analysis, suffice it to say that at the heart of the theory is a search for the infinitesimal thermal disturbance or ***perturbation*** most likely to trigger convective overturn in a fluid layer that is gravitationally unstable (i.e., hotter, and thus less dense, on the bottom than the top). This most destabilizing perturbation is often called the least-stable or most-unstable mode; its prediction and analysis is fundamental to the study of pattern-selection in convection, as well as other nonlinear systems.

Depending on the nature of the binding top and bottom surfaces of the layer (i.e., whether they are freely mobile, or adjacent to immobile walls), the most unstable mode leads to convection cells which have a width approximately equal to (though as much as 50% larger than) the depth of the fluid layer. This result is quite relevant to plate tectonics since it is the first basic step in predicting the size of convection cells, and thus the size of plates.

The perturbation which leads to this form of convection is deemed most unstable because, of all the possible perturbations, it will induce overturn with the least amount of heating. This heating however is measured relative to the properties of the system and as such defines the Rayleigh number. For example, say the layer is heated by basal heating only; i.e., the bottom boundary is kept isothermal at temperature $\Delta T$ hotter than the top boundary (which is also kept isothermal). The vigor of convection caused by this temperature drop $\Delta T$ depends on the properties of the system, in particular the fluid's density $\rho$, thermal expansivity $\alpha$, dynamic viscosity $\mu$, thermal diffusivity $\kappa$, the layer's thickness $d$, and the gravitational field strength $g$ (acting normal to the layer and downward). These properties reflect how much the system facilitates convection (e.g., larger $\rho$, $g$, and $\alpha$ allow more buoyancy) or impedes convection (e.g., larger $\mu$ and $\kappa$ imply that the fluid more readily resists motion or diffuses thermal anomalies away). The combination of these properties in the ratio $\frac{\mu\kappa}{\rho g \alpha d^3}$ gives a temperature that $\Delta T$ must exceed in order to cause convection. The Rayleigh number is the dimensionless ratio of these temperatures

$$Ra = \frac{\rho g \alpha \Delta T d^3}{\mu\kappa}$$

and therefore indicates how well the heating represented by $\Delta T$ will drive convection in this system.

Stability analysis predicts that the onset of convection, triggered by the least stable mode, will occur when $Ra$ is of the order of 1000; this varies between $Ra \approx 600$ (for the most readily induced convection with freely-slipping – called ***free-slip***– top and bottom boundaries) to $Ra \approx 2000$ (with the more resistive rigid or ***no-slip*** top and bottom boundaries). This $Ra$ for the onset of convection is usually referred to as the ***minimum critical Rayleigh number***, or $Ra_{cr}$. The analysis of the onset of convection and the evaluation of $Ra_{cr}$ with various added complications (e.g., rotation, magnetic fields, spherical geometry) is elegantly covered in the classic treatise by Chandrasekhar [1961].

In the end, the amount that the Rayleigh number of a convecting fluid exceeds $Ra_{cr}$ is a measure of its convective vigor. While $Ra_{cr}$ for any fluid layer is of order $10^3$, the present Rayleigh number of Earth's mantle is between $10^7$ and $10^9$. Thus, even though the mantle convects very slowly, it does so over great distances (thousands of kilometers) and against an extreme viscosity ($10^{21} Pa\ s$, as compared to, say, $10^2 Pa\ s$ for cold molasses), and is therefore actually undergoing highly vigorous convection.

### 2.2. *Vertical structure of simple convection*

*2.2.1. Symmetry.* Once convection initiates and becomes fully developed it is called ***finite amplitude*** convection, as opposed to the infinitesimal convection discussed above. The simplest form of such convection involves two-dimensional flow of constant-viscosity fluid in a plane layer that is heated on the bottom and cooled on the top (but whose upper and lower boundaries are otherwise mechanically identical, i.e. both free-slip or both no-slip). Such convection is a highly symmetric system. This symmetry is prescribed by the fact that when convection reaches its final state (i.e., a steady state or time-averaged steady state) the top boundary is cooling the fluid by the same amount that the bottom boundary is heating the fluid, otherwise the layer would heat up or cool down and is thus not in its final state. Thus, cold currents sinking from the top boundary will generally be equal and opposite in temperature

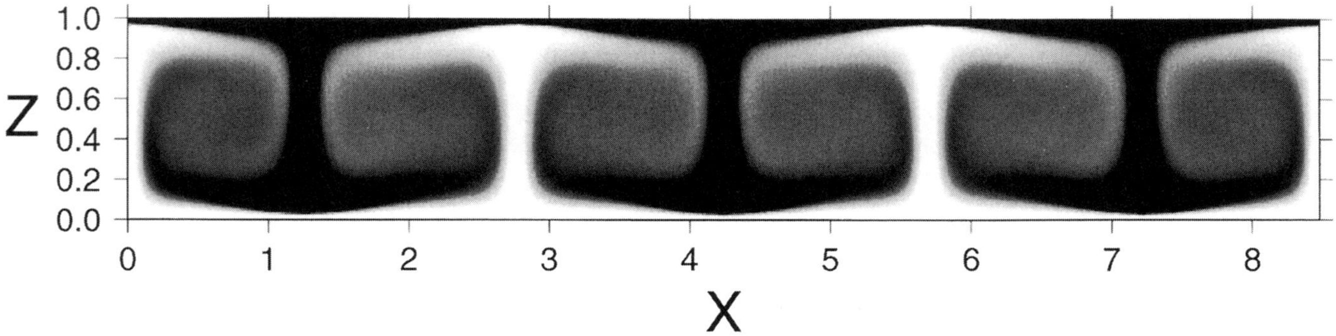


**Figure 1.** A vertical cross section of the temperature field of a numerical simulation of two-dimensional plane-layer, basally heated, isoviscous (Bénard), convection, with free-slip top and bottom boundaries and $Ra = 10^5$. Black represents cold fluid, light gray is hot fluid. The temperature field shows symmetric convection cells, upwellings, downwellings and thermal boundary layers thickening in the direction of motion (at the top and bottom of the layer, in between the upwellings and downwellings).

and velocity to the hot currents emanating from the bottom boundary (Fig. 1). The vertical cross-section of such 2-D steady convection therefore shows sequences of mirror image pairs of counter-rotating convection cells; each cell's vertical wall is either a hot upwelling or cold downwelling current and the upwelling and downwellings are themselves symmetric mirror images of each other through a $180^o$ rotation (Fig. 1). The breaking of this symmetry by various effects is vital to understanding the plate tectonic style of mantle convection.

*2.2.2. Thermal boundary layers.* When basally heated convection is reasonably vigorous (i.e. has moderate to high Rayleigh number *Ra*) it tends to stir the fluid until most of it is as homogeneous as possible, i.e., most of the fluid is at or near the average temperature of the top and bottom boundaries, thereby mitigating a gravitationally unstable situation for the greater part of the layer (Fig. 2). However, directly adjacent to the top and bottom boundaries the fluid temperature must make a transition from the largely homogeneous interior value to the colder (at the top) or hotter (at the bottom) boundary value (again see Fig. 2). These transition regions are called the ***thermal boundary layers.*** In effect, convection confines the gravitationally unstable parts of the fluid to these relatively thin boundary layers, keeping most of the rest of the fluid gravitationally stable. The Earth's lithosphere and tectonic plates are essentially the horizontal thermal boundary layer along the top surface of the Earth's convecting mantle, thus we will discuss thermal boundary layers in some detail and refer to them repeatedly.

Because of their large vertical temperature gradients, the thermal boundary layers control, via thermal conduction, the influx (through the bottom) and efflux (through the top) of heat. If the fluid is more vigorously stirred by convection then more of the fluid will be homogeneous, and thus the boundary layers will become thinner; although this leads to yet more gravitationally stable material it also causes sharper thermal gradients in the boundary layers and thus even larger heat fluxes in to and/or out of the layer. Therefore, the more vigorous the convective stirring, the greater the heatflow through the layer. (For basally heated convection in which the top and bottom boundaries are isothermal, the heatflow through the layer is, in theory, unbounded; i.e., convection can be so vigorous as to give the fluid layer the appearance of being a material with infinite thermal conductivity. However, the same is not true for convection with only internal heating (see below) since the heatflow is limited by the net rate of heat production in the layer.)

As mentioned, the thermal boundary layers are where the gravitationally unstable material is confined, and thus fluid in these layers must eventually either sink (if in the cold top boundary layer) or rise (if in the hot bottom boundary layer), thereby feeding the vertical convective currents, i.e., downwellings and upwellings, respectively. The feeding of vertical currents induces motion of the boundary layers toward convergent zones (e.g., a downwelling for the top boundary layer) and away from divergent zones (e.g., over an upwelling). Such flow naturally causes the boundary layers to thicken in the direction of motion (see Fig. 1). For example, in the top cold thermal boundary layer, fluid from hotter depths newly arrived at the divergent zone heats and thus thins the boundary layer; but as the fluid moves toward the convergent zone it cools against the surface, and the boundary layer gradually thickens before eventually growing heavy enough to sink into the downwelling. We will revisit boundary layer thickening again when discussing subduction zones and seafloor topography.

*2.2.3. The size of convection cells.* As mentioned previously, stability theory predicts that convection initiates with cells that are approximately as wide as the layer is thick. For the most part, this also holds true in fully developed basic convection. One of the things determining the lateral extent of a single convection cell is the length of the horizontal boundary layer currents. For example, as material in the top boundary layer current flows laterally it loses heat and buoyancy (being adjacent to the cold surface) and thus can travel only a certain distance before it becomes so cold and negatively buoyant that it sinks and feeds the cold downwelling. In the end, the lateral distance that material in the top (or bottom) boundary layer current can traverse before sinking (or rising) essentially determines the size of convection cells, and by inference (in a naive sense) the size of the tectonic plates themselves.

*2.2.4. Thermal boundary layer forces.* What force makes the thermal boundary layer currents flow laterally? While it may seem intuitively obvious that boundary layer fluid feeding a vertical current must flow horizontally (such as fluid going toward a drain), the forces behind these currents are of significance in regard to plate forces. It should be clearly understood that buoyancy does not drive the boundary layer currents directly; buoyancy only acts vertically while boundary layer currents move horizontally (buoyancy or gravity does eventually deflect these currents back into the convecting layer, but it cannot drive their lateral flow). Indeed, horizontal pressure gradients are the primary driving force for thermal boundary layers. For example, when hot upwelling fluid impinges on the top surface, it is forced to move horizontally away from a high pressure region which is centered above the upwelling itself; the high pressure region results from a force exerted by the surface on the fluid to stop the vertical motion of the upwelling thermal. As the top cold thermal boundary layer moves away from an upwelling to its own downwelling it thickens, gets heavier and acts to pull away from the surface; this induces a suction effect and thus, because of the boundary layer's growing weight, increasingly lower pressures in the direction of motion, eventually culminating in a concentrated low pressure zone where the downwelling separates from the surface. Thus the horizontal boundary layer current flows from the induced high pressure over the upwelling to the low pressure over the downwelling, i.e., it flows down the pressure gradient. Invariably plate driving forces are related to these pressure highs and lows, i.e., pressure gradient effects. Ridge push is the gradual pressure gradient going from the pressure high at a ridge outward; slab pull is effectively due to a concentrated pressure low caused by slabs pulling away from the surface (and the concept of a slab stress-guide simply means that the pressure low is kept concentrated thus leading to sharp pressure gradients). We will discuss plate forces in greater detail later.

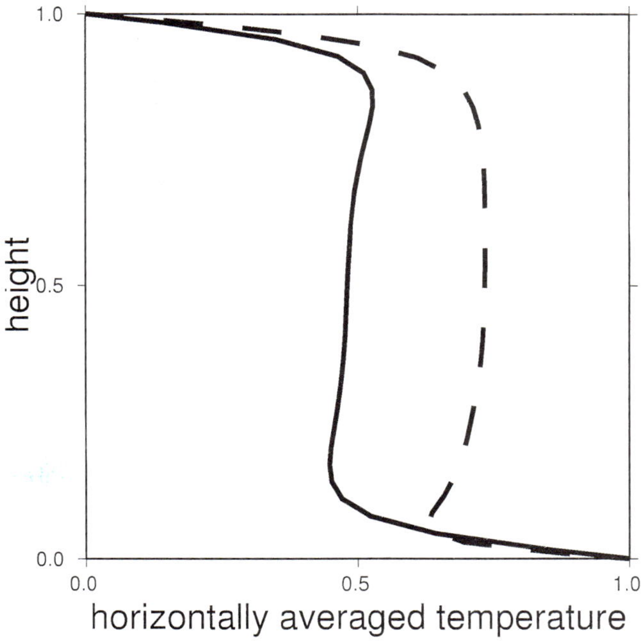


**Figure 2.** Temperature profiles (i.e., horizontally averaged temperature versus depth) for a basally heated, plane layer of fluid undergoing thermal convection when its viscosity is constant (solid curve) and temperature-dependent (dashed). The profiles show that most of the temperature change across the fluid occurs in relatively narrow *thermal boundary layers* near the top and bottom surfaces. In between the two boundary layers, most of the fluid is stably stratified or (if very well mixed) homogeneous. The fluid with temperature-dependent viscosity develops a stiffer upper thermal boundary layer which acts as a heat plug (i.e., it reduces convection's ability to eliminate heat), causing most of the rest of the fluid to heat up to a larger average temperature. (After Tackley [1996a].)

Though in this paper we are mainly concerned with the top boundary layer current of the mantle (i.e., the plates), suffice it to say that identical dynamics occurs at the bottom boundary layer current: the cold downwelling induces a high pressure region when it hits the bottom boundary, thereby forcing fluid away from it, etc.

### *2.3. Patterns of convection*

The two-dimensional convection described above is a special type of convective flow also known as convection with a roll planform; i.e., the 2D convection cells – when extended into three-dimensions – are infinitely long counter-rotating cylinders or rolls. This pattern of convection is typically unstable at moderately high Rayleigh numbers (i.e., somewhat greater than the critical $Ra$) at which point convection becomes three dimensional (3D) [Busse and Whitehead, 1971]; i.e., roll-like convection cells break down into more complicated geometrical shapes. The study of convective planforms and pattern selection is a very rich and fundamental field in itself, although a full quantitative discussion is beyond the scope of this paper; the interested reader should read Busse [1978]. However, we will attempt a brief qualitative discussion of convective patterns.

Much of the work on convective planforms was motivated by laboratory experiments in convection in thin layers [e.g., Busse and Whithead, 1971; Whitehead, 1976; Whitehead and Parsons, 1978; see Busse, 1978]. The planforms are most readily (albeit qualitatively) observed through a simple experimental technique called the shadowgraph [Busse and Whitehead, 1971]. Most laboratory observations of convective planforms are done through this method, so it deserves a brief mention. In the laboratory model of convection, the convecting fluid is bounded between two glass plates; the bottom is heated from below by hot water flushing along its outer surface, while the top is cooled by cold water flushed along the top. Collimated light (i.e., with parallel rays) is shown vertically up through the tank. In the convecting fluid, the index of refraction of light is dependent on temperature such that light rays diverge away from hot regions and converge toward colder regions. The projection of the resulting rays on a white screen thus show hot zones as dark shadows, and cold zones as concentrations of brightness.

In the end, the convection patterns observed with experimental (and computational) techniques are quite varied, though we will attempt to provide some order and logic to their presentation. As discussed in the previous section, purely basally-heated, plane-layer isoviscous convection is naturally a very symmetric system. Thus, even when convection is three-dimensional, the upwelling and downwelling currents retain some basic symmetry; e.g., in *bimodal* convection, the upwellings assume the geometry of two adjacent sides of a square, while the downwellings form the other two sides of the square (Fig. 3). However, the convection planform can also become highly irregular as with the *spoke* pattern which shows several nearly linear upwellings joined at a common vertex (and likewise with downwellings) (Fig. 3); even in this irregular planform the upwellings and downwellings are symmetric in that, apart from the sign of their thermal anomalies, their general structure is indistinguishable, i.e., they have the same irregular spoke shapes.

Other patterns of convection are also possible, but are mostly seen with non-constant viscosity fluids. Even in this case, regular polyhedral patterns like squares and hexagons are common (as are irregular patterns like the spoke pattern). One of the basic rules of the formation of regular patterns is that if the fluid layer is to be filled with identical convection cells, then the cells must all fit together exactly (otherwise, a misfit would constitute a cell with an aberrant shape). However, there are only a few geometries that will allow identical cells to fit together exactly, and these are simply infinitely long rolls, rectangles and squares, and hexagons (and triangles). However, even regular cells with, say, hexagonal planform involve signifcant asymmetry between upwellings and downwellings as we will discuss below. Indeed, the breaking of the symmetry of basally heated, isoviscous convection by various effects, most notably internal heating and variable viscosity, is fundamental to understanding the nature of mantle convection and its relation to plate tectonics.

### *2.4. Influence of internal heating*

As we have discussed above, purely basally heated convection in plane layers leads to upwellings and downwellings of equal intensity (equal and oppposite thermal anomalies and velocities). However, one of the defining characteristics of convection in the Earth's mantle is that it is very likley powered by radiogenic heating from the decay of uranium, thorium and potassium distributed throughout the mantle [see Turcotte and Schubert, 1982]. Thus, the mantle is not only heated along the core-mantle boundary by the hotter molten iron outer core, it is also heated throughout its interior. This component of internal heating leads to a very significant breaking of the symmetry between upwellings and downwellings.

The simplest form of convection driven by internal heating is one in which the heat sources are distributed uniformly, the top boundary is kept cold and isothermal, and the bottom boundary is thermally insulated, i.e., no heat passes through it. This is called purely internally heated convection. In this case, the bottom boundary cannot develop any thermal boundary layers on it, as these would carry heat-conducting thermal gradients which are disallowed by the insulating condition of the boundary. Since there is no relatively hot bottom thermal boundary layer to provide concentrations

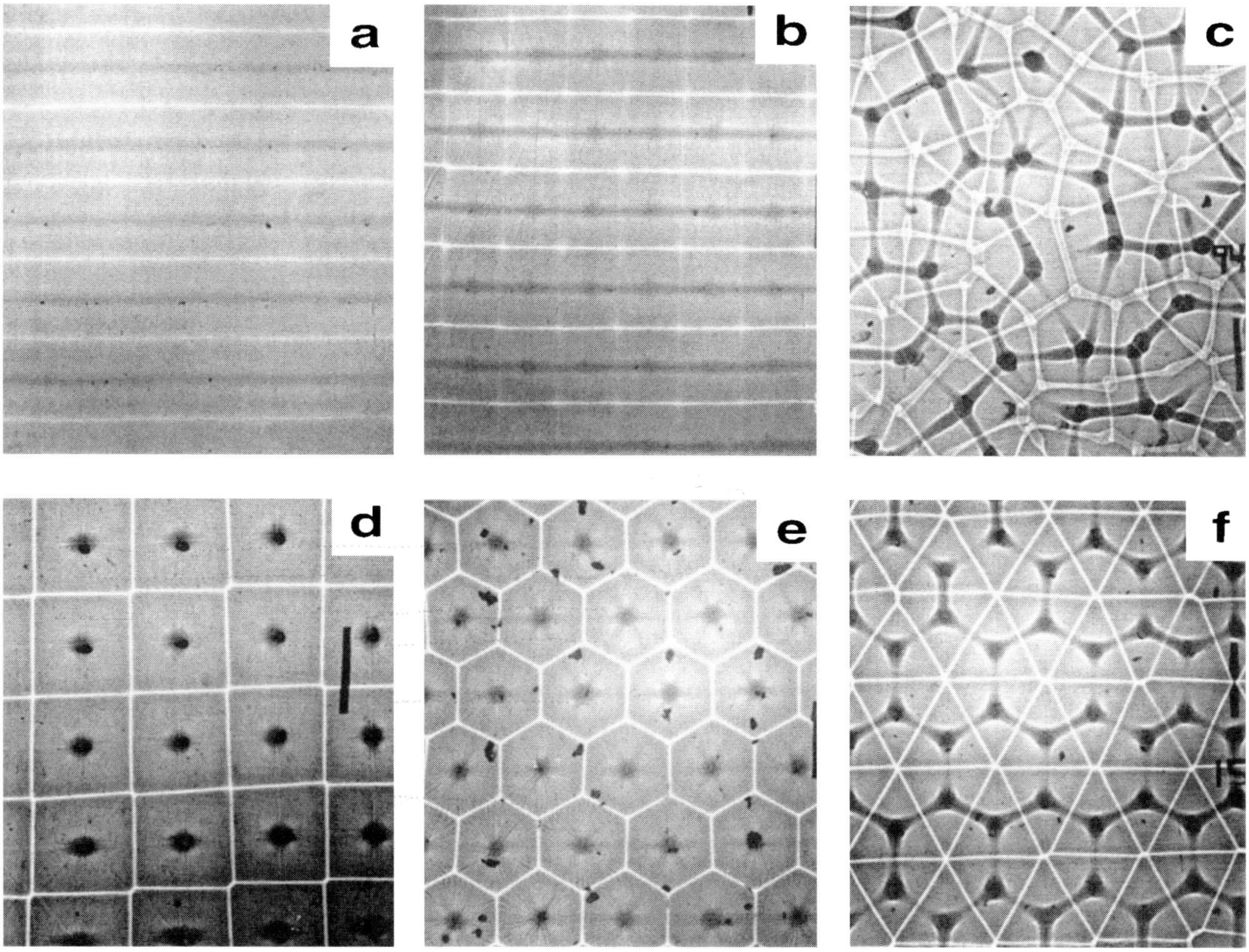


**Figure 3.** Laboratory experiments on convection in viscous liquids in a plane layer, bounded between two horizontal rigid glass plates, and heated from below. All images shown employ the shadowgraph techniqe (see text for discussion) and thus dark regions are relatively hot, while light regions are colder. The roll pattern (a) is also called 2-D convection; this pattern becomes unstable to the bimodal pattern (b) at moderate Rayleigh numbers, which eventually gives way to the spoke pattern (c) at yet higher Rayleigh numbers (around $Ra = 10^5$). The roll, bimodal and spoke pattern are typical for convection in isoviscous fluids, although they also occur in fluids with temperature-dependent viscosity (the photographs shown for figures (a)-(c) are in fact for weakly temperature-dependent viscosity fluids). More unique to convection in temperature-dependent viscosity fluids are squares (d), hexagons (e) and even triangles (f). (After White [1988].)

of positive buoyancy then there are no focussed active upwellings emanating from the base of the layer. However, the top boundary is cold and thus does develop a thermal boundary layer whose temperature gradients are responsible for conducting out all of the internally generated heat. Material in this boundary layer is all relatively cold and heavy, and is pulled horizontally (again, via pressure gradients) toward a convergent zone where it eventually sinks to form cold downwellings. The downwellings cool the interior of the fluid layer. Thus, in purely internally heated convection, there are only concentrated downwelling currents; to compensate for the resulting downward mass flux, upwelling motion occurs, but it tends to be a broad background of diffuse flow, rising passively in response to the downward injection of cold material (Plate 1). Therefore, since the downwellings descend under their own negative buoyancy they are typically called *active* currents, while the backgrond upwelling is deemed *passive*.

If we now allow for convection with both internal and basal heating (i.e., the bottom boundary is hot and isothermal – instead of insulating – and thus permits the passage of heat) we have a form of convection a bit more complex than with either purely basal heating or purely internal heating. However, the nature of the resulting convection can be understood if one realizes that the bottom thermal boundary layer must conduct in the heat injected through the bottom while the top thermal boundary layer must conduct out both the heat injected through the bottom as well as the heat generated internally. Thus to accomodate this extra heat flux, the top thermal boundary layer develops a larger temperature drop (to sustain a larger thermal gradient) than does the bottom boundary layer. In this way, the top boundary layer has a larger thermal anomaly than the bottom one, leading to larger, more numerous and/or more intense cold downwellings than hot upwellings. Invariably, internal heating breaks the symmetry between upwellings and downwellings by leading to a preponderance of downwellings driving convective flow (Fig. 4). In the case of the Earth, whose net surface heat flux is thought to be 80% or more due to radiogenic sources, the mantle is predominantly heated internally, and only a small amount basally. Thus, one can expect a top thermal boundary layer with a large temperature drop across it feeding downwellings which dominate the overall convective circulation; active upwellings from the heated bottom boundary provide only a small amount of the net outward heat flux and circulatory work. Although we have yet to discuss many of the complexities leading to the plate-tectonic style of mantle convection, this simple picture of internally heated convection is in keeping with the idea that the large scale circulation is driven by downwellings (slabs) fed by an intense thermal boundary layer (the lithosphere and plates), while active upwellings (mantle plumes) are relatively weak and/or few in number [see Bercovici et al., 1989a; Davies and Richards, 1992].

### *2.5. Influence of temperature-dependent viscosity*

Mantle material is known to have to have a highly temperature dependent viscosity for subsolidus, or solid-state, flow. Whether such subsolidus flow occurs by diffusion creep (deformation through the diffusion of molecules away from compressive stresses toward tensile stresses) or dislocation power-law creep (dislocations in the crystal lattice propagate to relieve compression and tension), the mobility of the molecules under applied stresses depends strongly on thermal activation; i.e., the atom's thermal kinetic energy determines the probability that it will jump out of a lattice site [Weertman and Weertman, 1975; Evans and Kohlstedt, 1995; Ranalli, 1995]. The viscosity law for silicates therefore contains a quantum-mechanical probablity distribution in the form of the Arrhenius factor $e^{H_a/RT}$ where $H_a$ is the activation enthalpy (basically the height of the energy potential-well of the lattice site out of which the mobilized atom must jump), $R$ is the gas constant, $T$ is temperature, and thus $RT$ represents the average kinetic energy of the atoms in the lattices sites. Because of this factor, a few hundred degree changes in temperature can cause many orders of magnitude changes in viscosity. Moreover, with the inverse dependence on $T$ in the Arrhenius exponent, viscosity is highly sensitive to temperature fluctuations at lower temperatures (i.e., the viscosity versus temperature curve is steepest at low $T$). Thus, in the coldest region of the mantle, i.e., the lithosphere, viscosity undergoes drastic changes: mantle viscosity goes from $10^{21} Pa\ s$ in the lower part of the upper mantle, to as low as $10^{18} Pa\ s$ in the asthenosphere [see King, 1995] to $10^{25} Pa\ s$ [Beaumont, 1976; Watts et al., 1982] or potentially higher in the lithosphere. Thus the viscosity may change by as much as 7 orders or magnitude in the top 200 hundred kilometers of the mantle. In the end, the effect of temperature-dependent viscosity on mantle convection is to make the top colder thermal boundary – i.e. the lithosphere – much stronger than the rest of the mantle. This phenomenon helps make thermal convection in the mantle plate-like at the surface in some respects, but it can also make convection *less* plate-like in other respects.

A strongly temperature dependent viscosity can break the symmetry between upwellings and downwellings in much the same way as internal heating. This occurs because the top cold thermal boundary layer is mechanically much stronger and less mobile than the hotter bottom thermal boundary layer. The less mobile upper boundary induces something of a heat plug that forces the fluid interior to warm up; this in turn causes the temperature contrast between the fluid interior and the surface to increase, and the contrast between the fluid interior and underlying medium (the core in the Earth's case) to decrease. This effect leads to a larger temperature jump across the top boundary layer (which partially reduces the boundary layer's stiffness by increasing its average temperature), and a smaller one across the bottom boundary layer (see Fig. 2).

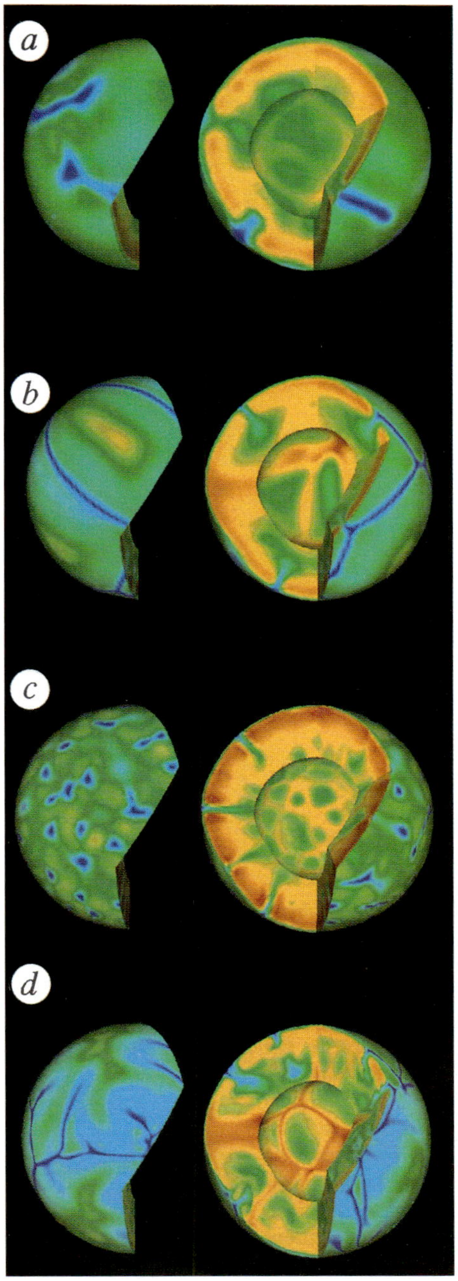


**Plate 1.** Purely internally heated convection in a spherical-shell model of the mantle with free-slip top and bottom surfaces, at moderate and high $Ra$, with and without a viscosity jump at 660km depth. Frame (a): Isoviscous mantle at $Ra = 1.6 \times 10^5$ Frame (b): Same as (a), but with the viscosity of the upper mantle (above 660km) reduced by a factor of 30. Frame (c): Same as (a) except the entire mantle viscosity is reduced by 30 (thus $Ra$ is larger by 30). Frame (d): Same as (b), except the viscosity of both layers is reduced by 10. Note that even while an increase in $Ra$ causes more columnar downwellings (from frame (a) to frame (c)), a viscosity jump tends to re-inforce or restore the tendency for sheet-like downwellings (frame (d)). (After Bunge et al. [1996].)

Thus, the temperature-dependent viscosity can lead to an asymmetry in the thermal anomalies of the top and bottom boundary layers much as we see in the Earth.

Temperature dependent viscosity can also cause a significant change in the lateral extent of convection cells. Because material in the top thermal boundary must cool a great deal and thus for a long time to become negatively buoyant enough to sink against its cold, stiff surroundings, it must travel horizontally a long distance while waiting to cool sufficiently, assuming it travels at a reasonable convective velocity. This can cause the upper thermal boundary layer and thus its convection cell to have excessively large lateral extents relative to the layer depth (i.e., large *aspect ratios*). This effect has been verified in laboratory and numerical experiments (Fig. 5) [e.g., Weinstein and Christensen, 1991; Giannandrea and Christensen, 1993; Tackley, 1996a; Ratcliff et al., 1997] As we will discuss later (§3.6), large aspect ratio convection cells are considered to occur in the Earth, especially if one assumes that the Pacific plate and its subduction zones reflect the dominant convection cell in the mantle. Thus, temperature-dependent viscosity can be used to explain the large aspect convection cells of mantle convection, but with some serious qualifications.

With very strongly temperature dependent viscosity, the top thermal boundary layer can also become completely immobile and the large aspect ratio effect vanishes. The immobile boundary layer happens simply because it is so strong that it cannot move. As a result, the top boundary layer effectively imposes a rigid lid on the rest of the underlying fluid, which then convects much as if it were nearly in isoviscous convection with a no-slip top boundary condition. Convection then has cells which are again about as wide as they are deep (i.e., with nearly unit aspect ratio). Moreover, in this form of convection the planform can assume various simple geometries such as squares and hexagons (Fig. 3); however, the upwellings and downwellings are not symmetric in that the downwellings form the edges of the hexagons (or squares), and the hot, low-viscosity upwellings tend to rise as pipes or plumes through the center of the hexagons (or squares) [see Busse, 1978],

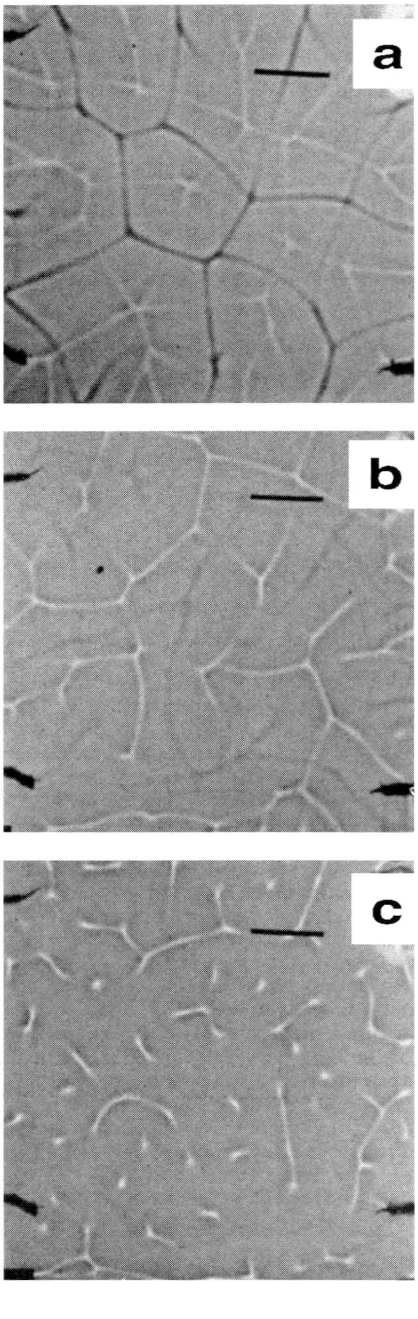


**Figure 4.** Laboratory experiment for internally heated convection in a plane layer. Images shown employ the shadowgraph technique (see Fig. 3 and text for discussion). Apparent internal heating is accomplished by initiating basally heated convection and then steadily decreasing the temperature of the top and bottom surfaces; this causes the average fluid temperature to be greater than the average temperature of the two boundaries. The fluid thus loses its net heat and the rate of this bulk cooling is a proxy for internal heating. The frames show the convective pattern when the layer is all basally heated (a) and thus upwellings (dark zones) and downwellings (light zones) are comparable; when bulk cooling (internal heating) is stronger than basal heating, the downwellings (light) are dominant (b); when bulk cooling is much stronger than basal heating the upwellings are not distinct enough to register in the shadowgraph (c). (After Weinstein and Olson [1990].)

similar in many respects to plumes in the Earth's mantle. Regardless of the pattern of convection, the immobilization of the top thermal boundary layer leads to convection that is unlike the Earth: unit aspect ratio cells and an immobile lithosphere.

Overall, three different styles or regimes of convection with temperature depedent viscosity were first noted by Christensen [1984a] and elucidated and summarized by Solomatov [1995] with a basic scaling analysis (Fig. 6). These regimes are as follows: 1) for weakly temperature-dependent viscosity, convection tends to appear as if it were nearly isoviscous (unit aspect-ratio cells and a mobile top boundary); 2) with moderately temperature-dependent viscosity, convection develops a sluggish cold top boundary layer that is mobile but with large horizontal dimensions i.e. large aspect-ratio cells; 3) with strongly temperature-dependent viscosity the top boundary layer becomes rigid and immobile and convection assumes much of the appearance (apart from detailed patterns) of isoviscous convection beneath a rigid lid. The size of these regimes depends on the Rayleigh number since vigorous convection is more capable of mobilizing a cold top boundary layer than is gentle convection. These regimes have been deemed the nearly-isoviscous or low-viscosity-contrast regime, the sluggish convection regime, and the stagnant-lid regime. Although the sluggish regime is characterized by large aspect ratio cells suggestive of the Earth's plates, it is most likely that the Earth's Rayleigh number and viscosity variability would put the Earth's mantle in the staganant-lid regime, i.e., if the mantle were a fluid with only temperature dependent viscosity. Indeed, both Venus and Mars, lacking obvious signs of plate tectonics or even universally-mobile lithospheres, appear to operate in the stagnant-lid mode, at least to some extent. However, the very presence of mobile plates on Earth shows that its lithosphere-mantle system has some extremely important effects which mitigate the demobilization of the top thermal boundary layer caused by temperature-dependent viscosity. In fact, it is not at all clear that the extreme Arrhenius-type temperature dependence of mantle or lithosphere viscosity actually occurs from a practical standpoint; i.e., as we shall discuss later, with multiple competing microscopic deformation mechanisms, there is always an effective competition to be the weakest. That is, thermal stiffening caused by one mechanism can be superceded by the onset of new deformation mechanisms at higher stresses. Evidence for this is the fact that stress releases in deep earthquakes do not appear elevated from those of near-surface earthquakes, suggesting that deep, cold and thus apparently stiff slabs do not support inordinately high shear stresses [Tao and O'Connell, 1993].

A final word on the effects of temperature dependent viscosity is necessary with regard to the broad concept of "self-regulation" in solid-state convection, both in Earth and other terrestrial planets. Tozer [1972] has elegantly made the case that the internal thermodynamic

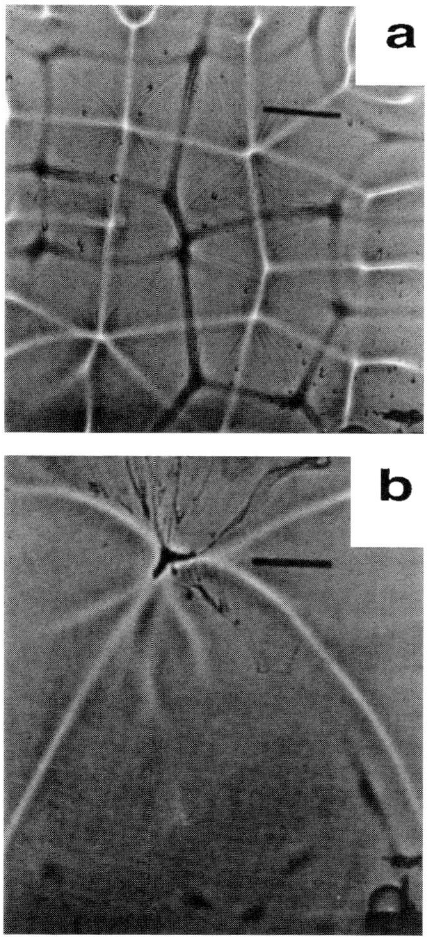


**Figure 5.** Lab experiment for convection with temperature-dependent viscosity fluid and mobile top thermal boundary layer. The mobilitity of the top boundary layer is facilitated by a free-slip upper boundary, accomplished by inserting a layer of low-viscosity silicon oil between the working fluid (corn syrup) and the top, cold glass plate. Without the oil, the rigidity of the top glass plate tends to help immobilize the cold and stiff top boundary layer. Frame (a) shows, for comparison, the experiment when the top boundary is rigid, yielding a predominantly spoke-like pattern. Frame (b) shows the planform when the oil is used to make the top boundary free-slip, causing a dramatic increase in the convection cell size to what is deemed the *spider* planform. (After Weinstein and Christensen [1991].)

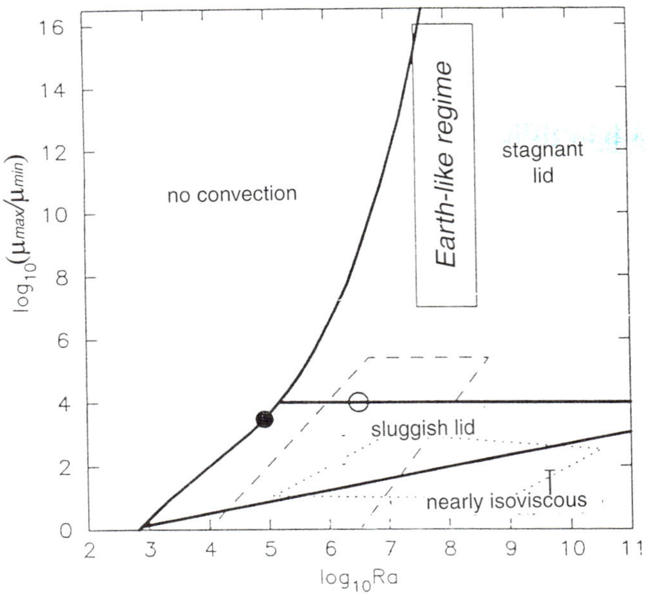


**Figure 6.** Diagram showing the different convective regimes in "*Ra* versus viscosity ratio" space for convection in fluid with temperature-dependent viscosity; $\mu_{max}$ and $\mu_{min}$ are the maximum and minimum allowable viscosities of the fluid, respectively. Dashed and dotted boxes show the regime of various numerical convection experiments. The box with the solid boundary shows the likely regime for the Earth. See text for discussion. (After Solomatov [1995].)

state of the mantle must be controlled to a large extent by thermal activation of creep. Put crudely, if mantle viscosity is too high for convection to be vigorous enough to remove the heat generated internally, then the mantle will simply heat up until the viscosity is reduced sufficiently. Thus there is a more profound role for temperature-dependent viscosity than just the effects on style or planform discussed here, and considerations of long-term evolution of the plate-mantle system must account for the extreme sensitivity of heat flow to internal temperature via viscosity variability [Davies, 1980; Schubert et al., 1980].

### *2.6. A brief note on the influence of sphericity*

Clearly because the Earth is primarily a sphere, models of convection in thick spherical fluid shells would appear to be most realistic [e.g., Bercovici et al., 1989a; Tackley et al., 1993; Bunge et al., 1996]. However, the effects of sphericity are possibly not so fundamental to the generation of plates. In particular, sphericity tends to break the symmetry between the top and bottom boundary layers, and upwellings and downwellings in the opposite sense of what we think is relevant for the Earth. For purely basally heated spherical shells, conservation of energy prescribes that for a convective solution to be stable (i.e. steady or statistically steady) the total heat input through the bottom boundary must be equal to the heat output through the top. However, because of sphericity, the bottom boundary has significantly less surface area than the top boundary through which heat passes; to compensate for this smaller area, the bottom thermal boundary layer generally has a larger temperature drop (and thus larger thermal gradients) across it than does the top boundary layer. This asymmetry leads to upwellings with larger temperature anomalies and velocities than the downwellings, which is opposite to the asymmetry between upwellings and downwellings thought to exist in the Earth. Thus, effects like internal heating and temperature dependent viscosity are even more important in order to overcome the asymmetry imposed by sphericity, and to give a more Earth-like asymmetry.

### *2.7. Poloidal and toroidal flow*

Convective motion, with its upwelling and downwelling currents, and the associated divergent and convergent zones at the surface and lower boundary, is also called *poloidal* flow. Basic convection in highly viscous fluids essentially has only poloidal flow. However, while a great deal of the Earth's plate motion is also poloidal (speading centers and subduction zone), much of it – with perhaps as much as 50% of the total kinetic energy – also involves strike-slip motion and spin of plates, which is called *toroidal* motion [Hager and O'Connell, 1978, 1979, 1981; Kaula, 1980; Forte and Peltier, 1987; O'Connell et al., 1991; Olson and Bercovici, 1991; Gable et al., 1991; Cadek and Ricard, 1992; Lithgow-Bertelloni et al., 1993; Bercovici and Wessel, 1994; Dumoulin et al., 1998]. The existence of toroidal flow in the Earth's plate-tectonic style of mantle convection is a major quandary for geodynamicists and is at the heart of a unified theory of mantle convection and plate tectonics. Therefore, the reason that toroidal motion does not exist in basic convection deserves some explanation.

Most models of mantle convection treat the mantle as nearly incompressible, i.e., as if it has constant density. In fact, convection models must allow for thermal buoyancy, thus they really treat the mantle as a *Boussinesq* fluid; this means that while density is a function of temperature (and thus actually not constant), the density fluctuations are so small that the fluid is still essentially incompressible *except* when the density fluctuations are acted on by gravity. Thus the fluid acts incompressible, but can still be driven by buoyancy. Moreover, even without thermal density anomalies, the mantle is still not really incompressible; due to increases in pressure its density changes by almost a factor of 2 from the top to the bottom of the mantle (e.g., from the IASP91 reference Earth model [Kennett and Engdah, 1991]). However, mantle flow occurs on a much slower time scale than compressional phenomena – in particular acoustic waves – and thus the mantle is really considered *anelastic* for which the separation of flow into poloidal and toroidal parts still applies [Jarvis and McKenzie, 1980; Glatzmaier, 1988; Bercovici et al., 1992]. However, even the influence of this anelastic component of compressibility is rather small compared to other effects in thermal convection [Bercovici et al., 1992; Bunge et al., 1997].

The incompressibility or Boussinesq condition requires that the rate at which mass is injected into a fixed volume must equal the rate at which the mass is ejected, since no mass can be compressed into the volume if it is incompressible; i.e., what goes in must equal what goes out. Mathematically, when considering infinitesimal fixed volumes – or, equivalently, individual points in space (since a point is essentially an infinitesimal volume) – this condition is called the continuity equation and is written as

$$\nabla \cdot \underline{v} = 0 \tag{1}$$

where $\underline{v}$ is the velocity vector and $\nabla$ is the gradient operator. This equation says that the net divergence of fluid away from or toward a point is zero; i.e., if some fluid diverges away from the point, an equal amount must converge into it in order to make up for the loss of fluid: again, what goes in must balance what goes out. The most general velocity field that automatically satisfies this equation has the form (in Cartesian coordinates)

$$\underline{v} = \nabla \times \nabla \times (\phi \hat{z}) + \nabla \times (\psi \hat{z}) \tag{2}$$

(where $\hat{z}$ is the unit vector in the vertical direction) since $\nabla \cdot \nabla \times$ of any vector is zero. (Note that the velocity field in (2) uses only two independent quantities, namely $\phi$ and $\psi$, to account for three independent velocity components $v_x, v_y$ and $v_z$. This is allowed because equation (1) enforces a dependence of one of the velocity components on the other two and thus there are really only two independent quantitities; i.e., in the parlance of elementary algebra, one equation with three unknowns means that there are really only two unknowns.) The quantity $\phi$ is called the poloidal potential which represents upwellings, downwellings, surface divergence and convergence and (as we will show below) is typical of convective motion. The variable $\psi$ is the toroidal potential and involves horizontal rotational or vortex-type motion, such as strike-slip motion and spin about a vertical axis; toroidal flow is not typical of convective motion (Fig. 7).

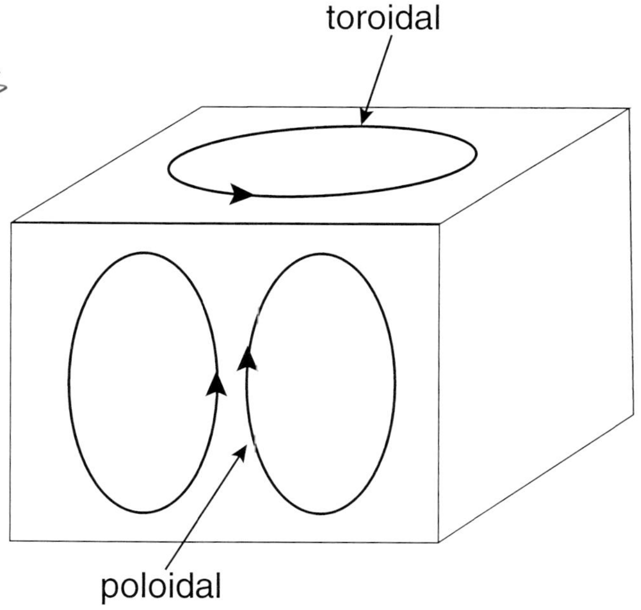


**Figure 7.** Cartoon illustrating simple flow lines associated with toroidal and poloidal motion.

What actually forces the velocity fields in a highly viscous fluid (where fluid particles are always at terminal velocity, i.e., acceleration is negligible) is determined by the balance between buoyancy (i.e., gravitational) forces, pressure gradients and viscous resistance; in simple convection with a constant viscosity fluid, this force balance is expressed as

$$0 = -\nabla P + \mu \nabla^2 \underline{v} - \rho g \hat{z} \tag{3}$$

where $P$ is pressure, $\mu$ is viscosity, $\rho$ is density, and $g$ is gravitational attraction ($9.8 m/s^2$). (The viscous resistance term, proportional to $\mu$, in (3) expresses that viscous forces are due to imbalances or gradients in stress, while stress is due to gradients in velocity – i.e., shearing, stretching and squeezing – imposed on a fluid with a certain stiffness or viscosity.) Taking $\hat{z} \cdot \nabla \times \nabla \times$ of (3) leads to

$$\mu \nabla^4 \nabla_H^2 \phi = -\nabla_H^2 (\rho g) \tag{4}$$

where $\nabla_H^2 = \frac{\partial^2}{\partial x^2} + \frac{\partial^2}{\partial y^2}$ is the 2D horizontal Laplacian. Equation (4) shows that the *poloidal* motion (left side of the equation) is directly driven by buoyancy forces (right side). However, if we take $\hat{z} \cdot \nabla\times$ of (3) we only obtain

$$\nabla^2 \nabla_H^2 \psi = 0 \tag{5}$$

which shows that there are no internal driving forces for toroidal flow. (Note, an identical analysis also exists for spherical geometry [e.g., see Chandrasekhar, 1961; Busse, 1975; Bercovici et al., 1989b].) Thus, for isoviscous convection, toroidal motion does not occur naturally and the only way to generate it is to excite it from the top and/or bottom boundaries; otherwise, toroidal motion does not exist. (The same argument holds even if viscosity $\mu$ is a function of height $z$; see §4.8.2.) To have convective flow drive toroidal motion itself, i.e., from within the medium, requires a forcing term on the right side of (5); as discussed later (§4.8.2), this can only be done if the viscosity varies laterally, i.e., $\mu = \mu(x, y)$ (see Plate 2 for a qualitative explanation). The problem of how to get buoyancy driven poloidal motion to induce toroidal flow, either through boundary conditions or through horizontal viscosity variations, is called the ***poloidal-toroidal coupling*** problem. This will be discussed in more detail later in the paper (see §4.8).

## 3. WHERE DOES BASIC CONVECTION THEORY SUCCEED IN EXPLAINING PLATE TECTONICS?

Having reviewed some of the essential aspects of simple viscous convection, we now examine those features of plate tectonics which are reasonably well explained as being characteristic of basic convective flow. Subsequently (§4), we discuss those aspects of plate tectonics which are *poorly* explained by basic convection.

### 3.1. *Convective forces and plate driving forces*

*3.1.1. Slab pull and convective downwellings.* If we are to compare simple viscous convection to mantle flow and plate tectonics, then we wish to identify subducting slabs with the downwelling of a cold upper thermal boundary layer. As demonstrated by Forsyth and Uyeda [1975], the correlation between the connectivity of a plate to a slab (i.e., the percent of its perimeter taken by subduction zones) and the plate's velocity argues rather conclusively for the dominance of slab pull as a plate driving force. Therefore, if slabs are simply cold downwellings, then the descent velocity of a downwelling in simple convection should be comparable to that of a slab and by inference the velocity of a plate (in particular a fast slab-connected or *active* plate). (As mentioned earlier (§2.2), the pull of a slab on a plate is in fact a horizontal pressure gradient acting in the cold upper thermal boundary layer and caused by the low pressure associated with a slab pulling away from the surface; invariably the pressure gradient is established so that the boundary layer or plate feeds the slab steadily and thus leads to the appearance that the slab is pulling the plate.) We can estimate the force and velocity of such a cold downwelling with simple scaling analysis, and compare the calculated velocity with tectonic plate velocities.

Generalizing the scaling analysis and boundary layer theory of Davies and Richards [1992] we consider a cold top thermal boundary layer which is $L$ long (from its creation at a divergent zone to its destruction at a convergent zone) and $W$ wide (Fig. 8). As it moves from the divergent to convergent zone, the boundary layer thickens by vertical heat loss which, in the horizontally moving boundary layer, is only due to thermal diffusion.

To estimate the boundary layer thickness $\delta(t)$ we apply dimensional homogeneity [see Furbish, 1997]. Assuming that only thermal diffusivity $\kappa$ (with units of $m^2/s$) and age $t$ (where $t = 0$ at the divergent zone where the boundary layer is initiated) control the growth of the boundary layer, then we write $\delta(t) = A\kappa^a t^b$ where $A$ is a dimensionless constant ($A$ is dimensionless since we assume no other dimensional properties of the system have any influence on the cooling process). The constants $a$ and $b$ are then determined to match the dimensions of either side of the equation for $\delta$, leading to $a = b = 1/2$. Indeed, a more careful analysis [see Turcotte and Schubert, 1982] shows that $\delta \approx \sqrt{\kappa t}$, within a factor of order unity.

When it reaches the site of downwelling the thickness of the boundary layer is thus $\delta = \sqrt{\kappa L/v}$, where $v$ is the as yet undetermined velocity of the boundary layer. After downwelling, the descending vertical current extends to depth $D$. Moreover, because the boundary layer involves a temperature contrast of $\Delta T$ (from the deeper fluid interior to the surface), then it and the vertical current it feeds have an average thermal anomaly of $\Delta T/2$. The buoyancy force of the downwelling current is therefore

$$F_B = [\rho\alpha\Delta T/2][\delta W D]g \tag{6}$$

where the first factor in the square brackets represents the density anomaly of the downwelling, and the second factor in brackets is the volume of the downwelling. The drag force on the downwelling current is the viscous stress (which, in this example, is approximately

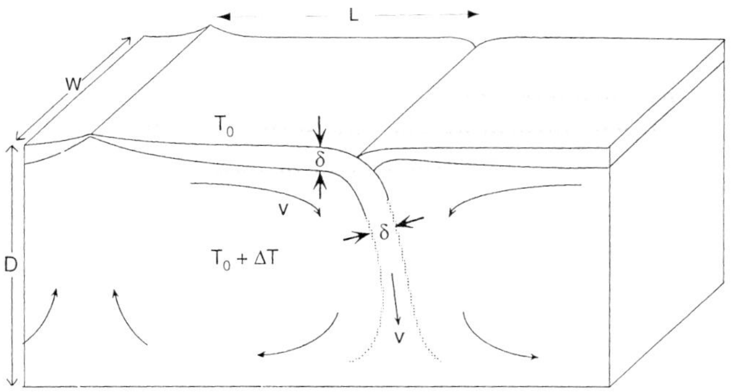


**Figure 8.** Cartoon illustrating the slab-pull force problem discussed in §3.1.1. For definition of labeled variables refer to text. (Cartoon after Davies and Richards [1992].)

the viscosity times the strain-rate of the fluid as it is sheared vertically between the downwelling and a stagnation point roughly half-way back to the divergent zone) acting on the current times the area of the current on which the stress acts, i.e.,

$$F_D = \left[\mu \frac{v}{L/2}\right][DW] = 2\mu v DW/L \qquad (7)$$

The force balance $F_B - F_D = 0$ yields the downwelling velocity

$$v = \frac{\rho\alpha\Delta T\delta L g}{4\mu} \qquad (8)$$

Substituting $\delta = \sqrt{\kappa L/v}$ to eliminate $v$ we obtain

$$v = \left[\left(\frac{\rho\alpha\Delta T g}{4\mu}\right)^2 L^3\kappa\right]^{1/3} \qquad (9)$$

As we have seen for most forms of basic convection $L \approx D$ (i.e., convection cells are roughly unit aspect ratio). Thus using $D = 3\times 10^6 m$, $\kappa = 10^{-6} m^2/s$, $\rho = 4000 kg/m^3$, $\alpha = 3\times 10^{-5} K^{-1}$, $\Delta T = 1400K$, $g = 10m/s^2$, $\mu = 10^{22} Pa\ s$ (this viscosity represents the effective whole mantle viscosity which is dominated by the lower mantle) we obtain a velocity of $v \approx 10cm/yr$. (We can adjust less well constrained parameters, e.g., use, within reason, a lower or higher viscosity, or a value of $L > D$; even so, however, the velocity remains of the order of magnitude of $10cm/yr$.) This velocity is indeed the correct order of magnitude for velocities of active (i.e., slab-connected) plates, strongly suggesting that the plate force deemed "slab pull" is well modelled by a simple cold downwelling current in basic convection.

*3.1.2. Ridge push is a convective lateral pressure gradient.* The ridge-push plate force was initially conceived of as an edge force (i.e. acting on the edge of a plate at a midocean ridge) due to the weight of a topographically high ridge [e.g., Forsyth and Uyeda, 1975]. It has since been recognized as a force that is distributed across the area of the plate [Hager and O'Connell, 1981]; since plate area seems to have little correlation with plate velocity [Forsyth and Uyeda, 1975] ridge push is typically assumed secondary to slab pull, (or it is almost exactly cancelled by mantle drag along the base of plates [Hager and O'Connell, 1981]). Indeed, Lithgow-Bertelloni and Richards [1998] estimate that ridge push constitutes only 5-10% the driving force due to subducted slabs. Nevertheless, after slab pull, ridge push may be one of the more significant forces and thus deserves some discussion from the perspective of basic convection.

Ridges are topographically high simply because they are younger and composed of hotter more buoyant material than the surrounding material, which has cooled and grown heavier since moving away from the ridge. The subsidence of material away from a ridge is slow enough that the entire ridge-lithosphere system is essentially in isostatic balance, at least far from subduction zones. If topography is isostatic, then there is a *compensation depth* beneath the lithosphere at which every overlying, infinitely high, column of material with equal cross-sectional area weighs the same; this also means that hydrostatic pressure (i.e., weight per area) is the same at every horizontal position along the compensation level. However, columns that do not extend to the compensation depth do not weigh the same; in particular the column centered on the ridge axis weighs the most since off-axis columns still include too much light material (like water) and not enough heavy material (like lithosphere) to equal the weight of the column of hot asthenosphere beneath the ridge. Thus, for depths less than the isostatic compensation depth, pressure is not horizontally uniform and is in fact highest beneath the ridge axis, thereby causing a pressure gradient pushing outward from the ridge [see Turcotte and Schubert, 1982]; however, this pressure gradient exists everywhere within the lithosphere (i.e., where ever sea floor depth changes laterally), and not as a line-force at the ridge axis.

How does this ridge-force pressure gradient relate to convective forces? As discussed previously (§2.2), the horizontal pressure gradient in the convective thermal boundary layers is what drives the lateral motion of the boundary layers. If we permitted the upper surface to be deformable we would find that pressure highs at divergent zones would push up the surface and the pressure lows at convergent zones would pull down on the surface. Thus the pressure gradient in the thermal

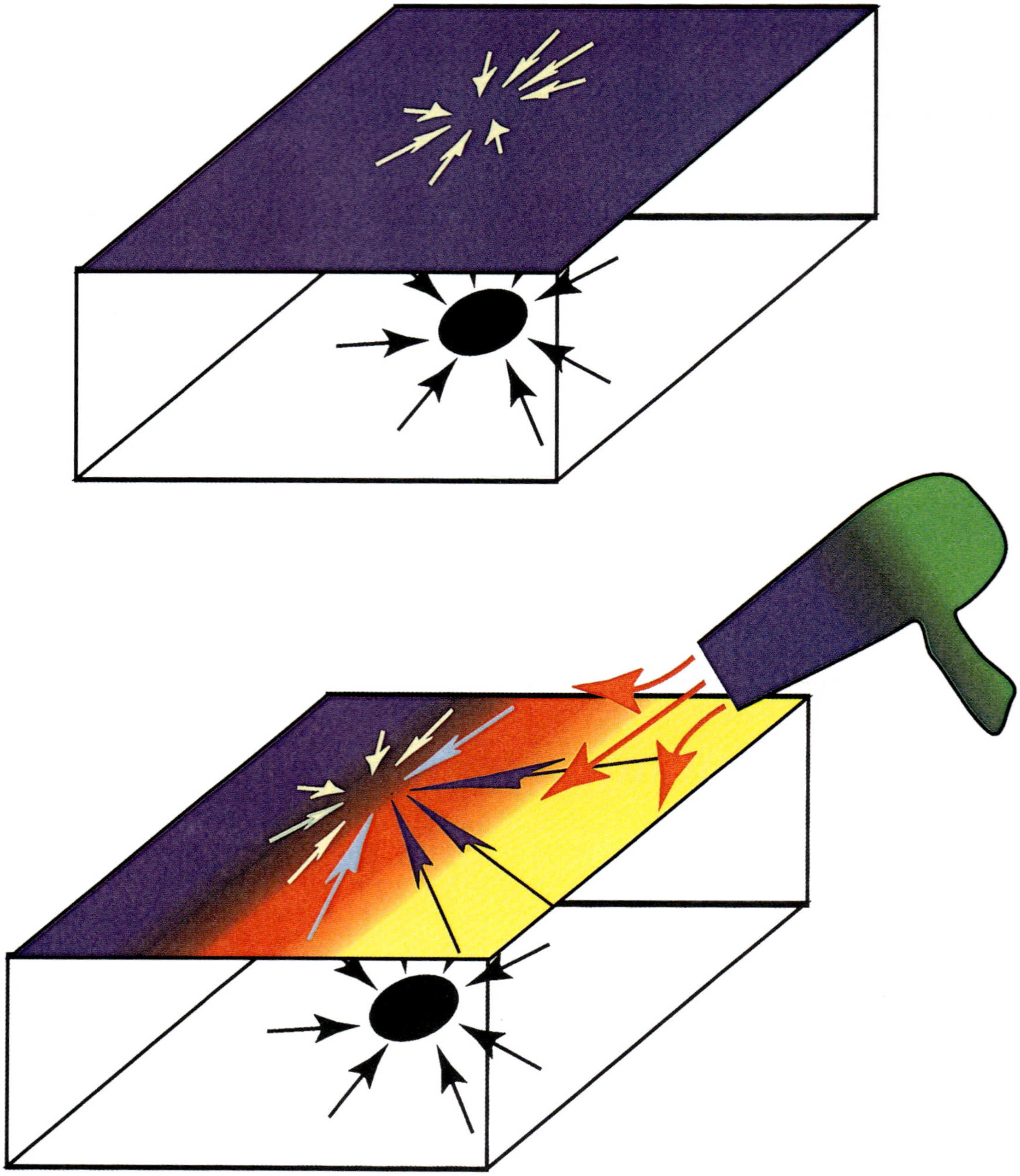

**Plate 2.** Cartoon illustrating the need for variable viscosity to obtain toroidal flow. Clear viscous fluid is drained through a hole in the bottom of a tank; on the clear fluid floats a uniformly cold congealed oil (e.g., butter or lard), shown in blue. The flow of clear fluid down the drain is predominanty poloidal, i.e., it only involves vertical motion and convergence toward the drain. The overlying oil is forced by the motion of the clear fluid to converge over the drain, and as long as the oil remains at one temperature it converges over the drain symmetrically, thus its flow is also poloidal. However, once one side of the oil is heated up (bottom frame), the warmer and softer side of the oil converges over the drain more readily. The differential motion between the warm, soft oil and the colder, stiffer oil appears as shear or toroidal motion (e.g., compare the light-blue and adjacent yellow flow vectors). Note that toroidal shear concentrates over the largest viscosity gradient in the oil (i.e., in the red region, at the transition between the warmest (yellow) and coldest (blue) oil); moreover, the toroidal motion only occurs because the viscosity gradient is perpendicular to the convergent poloidal flow.

boundary layer would be manifested as surface subsidence from a divergent to convergent zone. In fact, as we shall see in the next section, the resulting surface subsidence of a simple convective boundary layer is nearly identical to that observed for oceanic lithosphere. Thus, the ridge-push pressure gradient is essentially indistinguishable from the pressure gradient in a convective thermal boundary layer.

### *3.2. Structure of ocean basins*

Large scale variations in bathymetry, or sea-floor topoparphy, can be related to simple convective processes, in particular the cooling of the top thermal boundary layer. Although we have already discussed the cooling-boundary-layer concept in the previous section, we will re-iterate the arguments in slightly more detail in order to examine the validity of the underlying assumptions.

As mentioned above, when the top thermal boundary layer of a convection cell moves horizontally, it cools by heat loss to the colder surface. In simple convection, the surface is assumed impermeable, i.e., no mass crosses it and thus heat loss out of the top thermal boundary layer is assumed entirely due to thermal diffusion; i.e., since the boundary layer abuts the impermeable surface, vertical heat loss due to ejection of hot mass or ingestion of cold mass across the surface is assumed negligible. While this assumption is reasonable for most of the extent of the thermal boundary layer, we will later see that it causes convective theory to mispredict some details of ocean-basin topography and heat flow (§4.2).

As discussed above, the assumption of diffusive heat loss means that thickening of the boundary layer is controlled by thermal diffusivity $\kappa$, with units of $m^2/s$; thus, again by dimensional homogeneity (see §3.1.1 and Furbish [1997]), the boundary layer thickness $\delta \sim \sqrt{\kappa t}$. As the thermal boundary layer thickens it weighs more and thereby pulls down on the surface with increasing force; if the surface is deformable, it will be deflected downward (until isostasy is established). The surface subsidence thus mirrors the increasing weight of the boundary layer; the weight increases only because $\delta$ grows, and thus surface subsidence goes as $\sqrt{t}$. Therefore, sea floor subsidence is predicted by convective theory (more precisely convective boundary layer theory [Turcotte and Oxburgh, 1967]) to increase as the square-root of lithospheric age; this is called the $\sqrt{age}$ law. Moreover, this type of boundary layer theory predicts that the heatflow (units of $Watts/m^2$) out of the ocean floor should obey a $1/\sqrt{age}$ law, which can be inferred by the fact that the heatflow across the boundary layer is mostly conductive and thus goes as $k\Delta T/\delta$

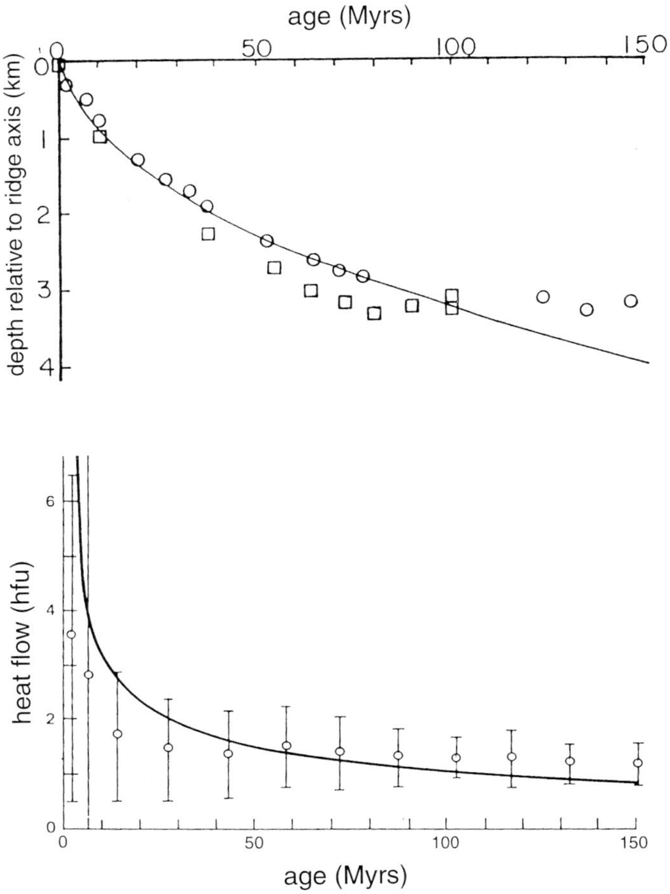


**Figure 9.** Bathymetry – i.e., seafloor topography – (top frame) and heatflow (bottom frame) versus age for typical ocean floor. Bathymetry is for the North Pacific (circles) and North Atlantic (squares) oceans; heatflow (in $hfu = 41.84mW/m^2$) shows average global values (circles), and associated error bars. Both frames show the predictions from convective boundary layer theory (solid curves); i.e., a $\sqrt{age}$ law for bathymetry and a $1/\sqrt{age}$ law for heatflow. The theoretical curves fit the observations reasonably well, except near and far from the ridge axis. Adapted from Turcotte and Schubert [1982] after Parsons and Sclater [1977] and Sclater et al., [1980]. See Stein and Stein [1992] for more recent analysis.

(where $k$ is thermal conductivity and $\Delta T$ is the temperature drop across the thickening thermal boundary layer, as also defined in §3.1.1). Profiles, perpendicular to the spreading center, of sea floor bathymetry and heatflow versus age (Fig. 9) show that, to first order, sea floor subsidence and heatflow do indeed follow boundary layer theory (i.e., follow $\sqrt{age}$ and $1/\sqrt{age}$ laws, respectively) [Parsons and Sclater, 1977; Sclater et al., 1980; see also Turcotte and Schubert, 1982, ch.4; and Stein and Stein, 1992], further emphasizing that the oceanic lithosphere is primarily a convective thermal boundary layer. However, at a greater level of detail, the $\sqrt{age}$ laws fail; this will be discussed later in the paper.

### *3.3. Slab-like downwellings are characteristic of 3D convection*

In simple, purely basally heated plane-layer convection, the convective pattern is often characterized by interconnected sheets of downwellings. For example, the hexagonal pattern is characterized by one upwelling at the center of each hexagon, surrounded by a hexagonal arrangement of downwelling walls (Fig. 3). A similar situation occurs for spherical systems [Busse, 1975; Busse and Riahi, 1982, 1988; Bercovici et al., 1989b]. When internal heating of the fluid is included, the upwellings and downwellings can be so imbalanced as to destroy any trace of symmetry; even so, the downwellings can persist as linear, albeit no longer interconnected, sheet-like downwellings at low to moderate Rayleigh numbers [Bercovici et al., 1989a]. The linear downwellings can desist and give way to downwelling cylinders and blobs at high Ra [Glatzmaier, et al., 1990; Bunge et al., 1996]; however, even at high Ra, the linearity of the downwellings is recovered by allowing viscosity to increase with depth, as is thought to occur in the Earth [Bunge et al., 1996] (see §3.6 and Plate 1). Thus, although sheet-like downwellings do not always occur in simple models of thermal convection, they are a very common feature. In that sense, the occurence of slab-like downwellings in mantle convection is characteristic of basic thermal convection. However, as we will discuss later, there are many other characteristics of subducting slabs that are unlike simple convective flow.

*3.3.1. Cylindrical upwellings and mantle plumes are also characteristic of 3D convection.* Although mantle plumes and hotspots are not, strictly speaking, part of plate tectonics, they play an important role in our ability to measure absolute plate motions (e.g., because of assumed hotspot fixity), and in understanding the nature of mantle convection and the relative sizes of its thermal boundary layers.

As with sheet-like downwellings, cylindrical or plume like upwellings are prevalent in simple thermal convection. (It is possible to get different shaped – e.g. sheet-like – upwellings, but they tend to break down into more columnar features before reaching the surface.) However, such upwellings always require a significant bottom thermal boundary layer heated from below by a relatively hot reservoir, such as the Earth's outer core; this boundary layer provides a source region for plumes and its thickness therefore determines their size [Loper and Eltayeb, 1986; Christensen, 1984b; Bercovici and Kelly, 1997]. In basic convection with significant internal heating, the bottom thermal boundary layer typically has a much smaller temperature drop than does the top thermal boundary layer. In this case, upwellings still retain some plume-like quality though they are fewer in number and/or quite weak. Nevertheless, the existence of cylindrical upwelling plumes in the mantle – indirectly inferred from hotspot volcanism, geochemical analyses [see Olson, 1990; Duncan and Richards, 1991; White and McKenzie, 1995] and perhaps most directly in seismic analysis [Nataf and VanDecar, 1993; Wolfe et al., 1997; Chen et al., 1998] – is consistent with basic models of thermal convection. However, it should be understood that even the weak plumes seen in complete 3D models of internally heated convection are an integral part of the convective flow [see Bercovici et al., 1989a]; i.e., while they may not have much influence on the dynamics of the top cold boundary layer, they will position themselves in accordance with the whole convective pattern, typically as far from downwellings as possible [Weinstein and Olson, 1989]. Plumes well integrated into the global convective circulation are very unlikely to establish themselves where they must pass through adverse conditions (e.g., beneath a downwelling or across a strong "mantle wind"), and even if they do, they can only exist as evanescent features [c.f. Steinberger and O'Connell, 1998].

### *3.4. Relative fluxes of plumes and slabs are in agreement with the mantle's presumed heating mode*

The heat flux transported by mantle plumes, relative to that transported by the cooling lithosphere and slabs, is also well predicted by convection models using the Earth's presumed proportion of internal heating. In basic convection, the net plume heat flux is approximately the same as the heating injected through the bottom boundary; i.e., plumes essentially carry only heat input from below [see Davies and Richards, 1992]. The Earth's mantle is thought to be between 80 and 90% heated internally by radiogenic sources [see Turcotte and Schubert, 1982]. Temperatures at the core-mantle boundary are not well known enough [see Boehler, 1996; also discussion by Davies and Richards, 1992] to definitively constrain the basal heat flux (i.e., to determine the temperature drop across the bottom thermal boundary layer, which is possibly the D" seismic layer; see Loper and Lay, 1995). However, the energy output necessary to power the geodynamo (i.e., the Earth's magnetic field) places the heat flux out of the core at approximately 10% of the net terrestrial flux [see Gubbins,

1977; Verhoogen, 1981, ch.4]. Thus, by the convective picture, plumes should transport on the whole between 10-20% of the total heat flux. Plume heat flux can be estimated by calculating the buoyancy flux necessary to keep hotspot swells inflated [Davies, 1988b; Sleep, 1990]. These calculations suggest that plumes transport roughly 10% of the net heat flow which is very consistent with our picture of basic, predominantly internally heated convection.

### *3.5. Mantle heterogeneity and the history of subduction*

So far we can construct a simple view of the connection between plate tectonics and mantle convection in which subduction zones are associated with convective downwellings and plumes with upwellings; whether ridges are also associated with active convective upwellings is less certain, if not doubtful (see §4.7). If heating is largely internal, then downwelling of the upper boundary layer, or lithosphere, at subduction zones would then be expected to dominate the structure of mantle temperature and density heterogeneity. Indeed, this seems to be the case, but the picture is complicated by the fact that plate motions, and hence subduction zones, are ephemeral on timescales of tens to hundreds of million years. Thus one does not expect a perfect correspondence between present-day subduction zones and mantle downwellings, and it is necessary to invoke plate motion history in addressing this issue.

Averaged over the past 120-150 million years, the locations of subduction zones (in a hotspot reference frame) correspond remarkably well to the location of high seismic velocity anomalies (cold material) in the mantle, at least at very long-wavelengths [Richards and Engebretson, 1992], a view supported in some detail by higher-resolution tomographic studies of mantle structure [Grand et al., 1997]. In fact, both the geoid (the Earth's gravitational equipotential surface) [Ricard et al., 1993] and global plate motions [Lithgow-Bertelloni and Richards, 1998] are well predicted using a mantle density model derived from slabs subducted during Cenozoic and Mesozoic times; again, however, there appears to be no evidence of deeply-seated upwellings beneath mid-ocean ridges that would provide plate driving forces of similar magnitude. A detailed discussion of this type of model is given in the paper by Richards et al. in this volume. For the present purpose, we note mainly that the expected relationships among downwelling flow, mantle density heterogeneity, and plate motion (via subduction) have been confirmed observationally, so that this constitutes a clear success of concepts derived from standard convection theory applied to plate motions. This, of course, bears little on the question of how the plates themselves are generated in the system.

### *3.6. Temperature-dependent viscosity, internal resistive boundaries, and the aspect ratio of convection cells*

The Pacific plate, with most of the kinetic energy and subduction zones of the plate tectonic system, can in many ways be considered the top of the Earth's dominant convection cell. As such, it is a very large convection cell with an excessively large aspect ratio (much longer than it is possibly deep). (The size of several other massive plates could also be used to argue for large aspect ratio cells; however, except for the Indo-Australian plate – which in itself may be composed of several other plates [Gordon et al., 1998] – no other super-large plate is considered *active*, i.e. slab-connected and "fast", and therefore does not really qualify as a possible convection cell.) Large aspect-ratio or long-wavelength convection cells in the mantle are also inferred from seismic tomography [Su and Dziewonski, 1992]. Because of the dominance of such long wavelength features, tomography and by inferrence convection are deemed to have a *reddened* spectrum.

Therefore, in many ways mantle convection in the Earth is thought to be characterized by very large aspect ratio convection cells. As discussed previously (§2.5), convection with temperature-dependent viscosity can allow for large aspect ratio cells [e.g., Weinstein and Christensen, 1991], in particular in the *sluggish* convection regime (with moderately temperature-dependent viscosity and a sluggish but mobile upper thermal boundary layer; see Solomatov, 1995). This effect can theoretically allow for a Pacific-sized convection cell; however, if the Earth's mantle were only a temperature-dependent-viscsosity fluid, estimates of the its convective regime places it more in the *stagnant lid* regime where the top thermal boundary layer is essentially frozen [Solomatov, 1995] and thus would be more characterized by unit aspect ratio convection cells [White, 1988].

However, other relatively simple effects may also cause long wavelength convection cells; these effects primarily involve an internal resistive boundary which, while not a feature of the fundamental fluid dynamics of convection, can be readily added to simple convection models to improve their match with the Earth.

As is well known, the mantle is nominally divided into the upper and lower mantle at the 660km seismic discontinuity [see review by Silver et al., 1988]. This discontinuity is most likely a phase change from spinel to perovskite plus magnesiowüstite, [Liu, 1979; Ito and Takahashi, 1989; Ito et al., 1990] and also involves a viscosity jump with depth by most likely a factor of 30-100 [e.g., Ricard et al., 1984; Richards and Hager, 1984; Hager, 1984; King and Masters, 1992; Ricard et al., 1993; Corrieu et al., 1994; see King, 1995]. Although the nature of convective mass transfer across this 660km boundary has been the subject of extensive debate, seismic tomography has given very compelling evidence for transfer of both slabs [Creager and Jordan, 1984; Grand, 1994; Grand et al. 1997; van der Hilst et al., 1991, 1997] and plumes [Shen et al., 1998] across the phase change. However, because of both the phase change and the viscosity jump, this boundary can still provide resistance to downwellings as they attempt to penetrate into the lower mantle. The phase change resists mass transfer because it is inferred to have a negative Clapeyron slope [Ito and Takahashi, 1989; Ito et al., 1990; Wicks and Richards, 1993]; this means that in relatively cold material the phase transition will occur at a higher pressure than in hotter material. Thus a cold downwelling moves the phase transition to greater pressures, causing a downward deflection of the phase boundary that, because of buoyancy, acts to rebound up and thus resist the motion of the downwelling [Schubert, Turcotte and Yuen, 1975; see also Christensen and Yuen, 1985; Machetel and Weber, 1991; Tackley et al., 1993, 1994; Honda et al., 1993; Weinstein, 1993]. A viscosity jump retards the motion of downwellings because of the greater resistance of the material beneath the boundary. In either case, convection tends to develop large-volume downwellings – and thus large wavelength convection cells – in order to gather enough negative buoyancy to penetrate a moderately resistive boundary. This "reddening" of the convective spectrum (thus creating large-aspect ratio, nearly-Pacific size cells) has been well documented in three-dimensional models of spherical convection at reasonably high Rayleigh number for both the phase change [Tackley et al., 1993, 1994] and the viscosity-jump effects [Bunge et al., 1996; see also Tackley, 1996b].

In the end we see that with temperature-dependent viscosity, or an internal boundary resistive to downwellings, basic convection can generate Earth-like, large-wavelength (or long aspect ratio) plate-sized convection cells. However, while these effects move us closer to obtaining plate-like-scales in convection models, none actually generate plates.

## 4. WHERE DOES BASIC CONVECTION THEORY FAIL IN EXPLAINING PLATE TECTONICS, AND WHAT ARE WE DOING (OR MIGHT WE BE DOING) TO FIX IT?

Although models of basic thermal convection in the Earth's mantle have made significant progress in explaining many features of plate tectonics, there remain a great number of unsolved problems. While we make no claim to present an exhaustive list of such unresolved issues, here we survey some of the large-scale problems remaining as well as progress being made to solve them.

### *4.1. Plate forces not well explained by basic convection*

While slab pull and to a large extent ridge push (really, distributed ridge push) are manifestations of simple convection, other forces are not. Most of the forces unaccounted for by convection are related to some feature of plate tectonics that convection does not (or does not easily) generate. One such force is transform resistance, as this involves strike-slip motion which cannot be readily generated, if generated at all, in simple convection. The lack of excitation of strike-slip motion is a major short-coming of basic convection theory, thus we will defer discussion of this until later (see §4.8). The precise distribution of the *ridge push* force also cannot be predicted in simple convection models since they do not generate narrow passive divergent zones, i.e. thin ridges fed by shallow nonbuoyant upwellings. The lack of formation of passive but focussed upwellings is also a signficant shortcoming of basic convection theory and will be discussed later as well (see §4.7). Finally, collision resistance does not explicitly occur in basic convection since most convection models do not include continents that resist each other's motion at their contact points. Collision resistance must in the end be a manifestation of chemical segregation, i.e., placement of chemically light continental material at the surface which retains a buoyancy that resists being drawn into a downwelling [e.g., Lenardic and Kaula, 1996].

### *4.2. Structure of ocean basins (deviations from the $\sqrt{age}$ law)*

As discussed previously, the classic bathymetry and heatflow profiles (Fig. 9) show a predominant $\sqrt{age}$ dependence. However, bathymetry and especially heatflow deviate significantly from the $\sqrt{age}$ curve near the ridge axis itself and far from the ridge axis. One must recall that the $\sqrt{age}$ prediction assumes that only diffusive transport of heat occurs in the upper thermal boundary layer. This assumption likely becomes invalid at and far from the ridge axis. At the ridge axis, heat transport by magma migration [see Spiegelman, 1996 and references therein] and cooling by hydrothermal circulation are highly significant [see Turcotte and Schubert, 1982, ch.4; Stein and Stein, 1992, and references therein]. It is likely that supression of the bathymetric and, in particular, the heatflow highs at the ridge is due to cooling not predicted by the diffusive thermal boundary layer theory; the most likely candidate for this extra cooling is indeed hydrothermal circulation. Incorporation of realistic or more complete transport phenonema relevant to ridges (e.g., magmatism and/or water ingestion) into convection models is also related to the problem of how to generate focussed but passive ridges (see §4.7).

The flattening of bathymetry, relative to the $\sqrt{age}$ law, far from the ridge axis has been the subject of some concern. It has been proposed that such flattening is the result of extra heating of the lithosphere (causing effective rejuvenation) brought on by secondary small-scale convection [Parsons and McKenzie, 1978], viscous heating [Schubert et al., 1976] and mantle plumes [Davies, 1988c]. This flattening has also been attributed to active, pressure-driven, asthenospheric flow that lifts up the lithosphere as the flow is forced into an increasingly constricted asthenospheric channel (the constriction being due to the cooling, thickening lithosphere) [Phipps Morgan and Smith, 1992]. However, since three-dimensional convection itself differs significantly from the two-dimensional boundary-layer theory from which the $\sqrt{age}$ law is derived [Turcotte and Oxburgh, 1967], such flattening and other deviations in the far-axis bathymetry are perhaps not altogether surprising from a mantle convection perspective.

### *4.3. Dynamic topography*

Convective currents impinging on or separating from the top boundary of the mantle induce vertical stresses that deflect the Earth's surface. In particular, upwellings should be associated with topographic highs and downwellings with depressions. Of course the most obvious topographic signal is not related to these thermal anomalies but to the chemical differences between the thick, light, continental crust and the denser, thinner oceanic crust. However, one of the more distinct features of the Earth's topography, namely the mid-oceanic ridges, is a direct consequence of mantle convection (see §3.2).

What is referred to as *dynamic topography* is what remains of the observed topography when the topography due to shallow, isostatic mass anomalies in the crust and lithosphere (i.e., thickness variations in the crust and lithosphere) are removed. This remaining topography is thus sometimes called non-isostatic topography. In fact, the distinction between isostatic and dynamic or non-isostatic topography is somewhat nebulous, e.g., while ridges and subsiding seafloor are isostatically compensated, they are also related to dynamic convective processes and plate motion. However, even in this case strict removal of isostatic topography is informative. In particular, the residual (i.e, dynamic or non-isostatic) topography of the sea floor (observed topography minus $\sqrt{age}$ subsidence) reveals, in theory, the presence of thermal anomalies below the lithosphere, especially far from the ridge axis since deviations from the $\sqrt{age}$ law near ridges are presumably due to hydrothermal circulation (see §4.2).

The dynamic topography associated with long-wavelength, tomographically inferred, deep-seated heterogeneities should be theoretically quite large, i.e., $\pm 1000$ meters [e.g., Hager and Clayton, 1987] in order to explain undulations in the geoid (the gravitational equipotential surface) of the Earth [e.g., Ricard et al., 1984; Richards and Hager, 1984]. However attempts to detect this dynamic topography have been unsuccessful, or have concluded that it cannot exceed $\pm 300$ meters, which is in contradiction with theroretical expectations [Colin and Fleitout, 1990; Cazenave and Lago, 1991; Kido and Seno, 1994; Le Stunff and Ricard, 1995]. On a more regional scale, Lithgow-Bertelloni and Silver [1998] have concluded that South-East Africa does indeed have anomalously high topography which correlates, as predicted by theory, with both lower-mantle heterogeneity and geoid undulation. However, many other lower-mantle heterogeneities mapped by tomography are not reflected in the surface topography. The theoretical, convective dynamic topography and the observed topography might be reconciled by assuming that part of the vertical stresses of convective currents are balanced by deflection of internal boundaries (e.g., the phase transitions at 400 and 670 km depth) [Thoraval et al., 1995; Le Stunff and Ricard, 1997; Thoraval and Richards, 1997].

Since mantle convection is likely quite vigorous and time-dependent then convectively generated dynamic topography is also a time-dependent phenomenon. This means that two continents should have significantly different relative sea-level histories. Thus, the sedimentological records of continental platforms should provide rigorous constraints on time-dependent, convection based models of dynamic topography [Gurnis, 1990, 1993; Mitrovica et al., 1989]. That sea-level changes possibly reflect large-scale convective processes is still somewhat alien to the conventional sedimentological view that such changes reflect only the global eustatic signal [Vail et al., 1977]. However, Gurnis [1993] has shown that a qualitative agreement exists between possible variations of dynamic topography during the Phanerozoic and the inundation history of continents. Clearly, a quantitative solution to the dynamic topography problem would be facilitated by further interactions between sedimentologists and geodynamicists.

### *4.4. Changes in plate motion*

Plate motions evolve with various time-scales. Some are clearly related to mantle convection, such as those associated with the so-called Wilson cycle [Wilson, 1966], i.e. the periodic formation and breakup of Pangea, approximately every 500 Myrs). Various hypotheses have been proposed to explain the dispersal of supercontinents. It is generally accepted that the presence of a supercontinent tends to insulate and thus, with radiogenic heating, warm the underlying mantle, eventually inducing a hot upwelling which weakens and breaks up the overlying lithosphere [Gurnis, 1988]. The fact that Pangea was surrounded by subduction also facilitates the thermal insulation of that portion of mantle trapped under the continent. Although this mechanism is physically sound, the present-day continents do not suggest that they stand above hotter than normal mantle. On the contrary, their basal heat flux seems very low [e.g., Guillou et al. 1994]. The effect of a super-continent on the mantle, however, may have been different from that of the present-day normal-sized continents.

Plate motion changes that have occurred on shorter time-scales are much more difficult to understand. The plate tectonic history presumably recorded in hotspot tracks consists of long stages of quasi-steady motions separated by abrupt reorganizations. During a stage of quasi-steady motion, our understanding of force balances on the plates, either in terms of plate or boundary forces (slab pull, ridge push, mantle drag, etc.) [e.g. Forsyth and Uyeda, 1975], or in terms of the buoyancy of large-scale, tomographically-inferred mantle heterogeneities [e.g., Ricard and Vigny, 1989; Lithgow-Bertelloni and Richards, 1995], is certainly one of the obvious quantitative successes of geodynamics. This success is reinforced by the fact that the parameters entering the theory (e.g., mantle viscosity, lithospheric thickness) are in agreement with independent observations (e.g., postglacial rebound, sea-floor topography). However, the success in predicting the direction of plate velocities is much less obvious, or is based on a tautology. For constant plate motion, the pattern of sea floor age and the location of slabs are obviously such that their associated forces push and pull the plates in the right direction. Thus, it is not surprising that, when starting from the observed positions of ridges and slabs, we predict the correct directions of plate motions. As the seismic tomography is well correlated, at least in the upper mantle, with the distribution of ridges and trenches, the ability to predict the surface motion from mantle heterogeneity simply confirms that surface tectonics and mantle dynamics belong to the same unique convecting system.

Abrupt changes in plate motion, however, are not easily related to convective processes. The most dramatic plate motion change is recorded in the Hawaiian-Emperor bend, dated at 43Ma; this bend suggests a velocity change of a major plate of approximately $45^o$ during a period no longer than 5 Myrs, as inferred from the sharpness of the bend. It is only because other hotspots were further away from the rotation pole of the Pacific that they did not record such a sharp bend. However all the Pacific hotspots have moved coherently and the various Pacific hotspot chains are indeed in agreement with a sudden change in the Pacific plate velocity [Fleitout & Moriceau, 1992]. Recent papers have challenged this conventional interpretation and suggest that the kink in the hotspot lines reflects, on the contrary, the motion of hostpots [Norton, 1995; Tarduno and Cottrell, 1997; see also Christensen, 1998]. Nevertheless, we believe, for the time being, that an abrupt change in Pacific plate motion is more physically plausible than the simultaneous motion of all the Pacific hotspots and intevening mantle over such short time scales. All convective type plate-driving forces, either those existing in the buoyancy of tomographically-inferred mantle heterogeneity or those associated with the thermal boundary layer (e.g., slabs, thickening of the oceanic lithosphere) can only evolve with convective time-scales (> 100 Myrs, which is based on a mantle overturn at a typical convective velocity; see §3.1.1). Therefore, changing convective motions in less than ~ 5 Myrs seems physically impossible. Perhaps the only way to explain short time-scale changes would be to invoke mechanisms that pertain to fracture or some other rapid rheological response (not controlled by the convective temperature field), or to changes in plate boundaries and geometries (which are possibly also controlled by rheology; see below). Plate reorganizations due to 1) the anihilation of a subduction boundary by continental collision, 2) loss of a ridge and/or trench by subduction of a ridge, or 3) the initiation of a new subduction zone, [e.g. Hilde et al., 1977] possibly occur on relatively rapid time scales, although such mechanisms have never been quantified or formulated rigorously. An attempt to quantitatively test the idea that the change in the Pacific plate motion was related to the Himalayan collision has simply shown that our present understanding of plate driving forces fails to explain plate velocity changes [Richards and Lithgow-Bertelloni, 1996]. Offsets in plate age and thickness along already-weak oceanic transform boundaries may provide excellent sites for initiation of subduction [e.g., Stern and Bloomer, 1992 and references therein], yielding a possible mechanism for rapid global plate motion changes. This, coupled with the notion of transform faults as long-lived "motion guides" (see §4.8.4), emphasizes the possible key role of transform boundaries in the plate-mantle system. This clearly points to the necessity of future studies to understand the interactions between the long time-scale convective processes, and the short time-scale effects, such as those likely to occur in the lithosphere's rheological response, e.g., fracture, fault sliding and various strain-localization mechanisms (see below).

### *4.5. Plate-like strength distributions: weak boundaries and strong interiors*

One of the predominant plate-like characteristics of the Earth's top thermal boundary layer is its strength or viscosity distribution. This distribution is marked by most of the boundary layer, in particular that which is undergoing horizontal motion, as being strong or of high viscosity; these strong regions are separated by narrow zones of intense deformation (i.e., the boundaries) which are necessarily weak to permit the observed levels of strain. The cause for such a strength distribution is very much at the heart of understanding how plate tectonics arises from convection. and is thus naturally the subject of much debate.

*4.5.1. Do plates form because the lithosphere breaks?.* It is often assumed that plates and their strength distribution form because the entire oceanic lithosphere is brittle and breaks under convective stresses. Thus, once the lithosphere is broken it is not easily mended; the broken edges remain – more or less permanently – as sites of weakness and low-frictional sliding [e.g., Davies and Richards, 1992]. However, the depth of brittle failure in oceanic lithosphere has undergone considerable revision in the rheological literature and is no longer thought to be a simple and deep transition to ductile motion. Indeed, brittle failure and frictional sliding cease at quite shallow depths, on the order of 10km [Kohlstedt et al., 1995], and give way to semi-ductile/semi-brittle behavior which involves ductile cracking and void growth (involving noncontiguous, distributed populations of voids and microcracks; Kohlstedt et al., 1995; Evans and Kohlstedt, 1995) and a variety of complex rheological phenomena. This semi-ductile behavior, and not fracture and frictional sliding, persists for most of the thickness of the lithosphere, before giving way to ductile/viscous flow at greater depths. The model of the broken lithosphere is therefore no longer valid. The lithospheric deformation mechanisms leading to plates likely contain both brittle discontinuous and ductile continuous (as well as other highly complex) processes.

*4.5.2. Continuum models and plateness.* Given the above considerations, many convection models seeking to generate plates employ a continuum approach; i.e., they do not break the top cold boundary layer into pieces, but instead keep it whole while employing rheologies that allow parts of it to become weaker than other parts. In doing so, they try to obtain the property of a plate-like strength (or viscosity) distribution for their top thermal boundary layer. This property is quantifiable and has been called *plateness* [Weinstein and Olson, 1992]. A boundary layer with broad, strong, slowly deforming regions (representing strong plate interiors) separated by narrow zones of weak, rapidly deforming material (representing plate boundaries) have high plateness; clearly convection in the Earth has a top thermal boundary layer with high plateness. A boundary layer with no strength variations, as occurs in basic convection with constant or only depth-dependent viscosity, has zero plateness.

Since plateness involves lateral variations in strength or viscosity, convection models can obviously only generate sufficient plateness if they employ a fully variable viscosity law. Generally viscosity is variable by virtue

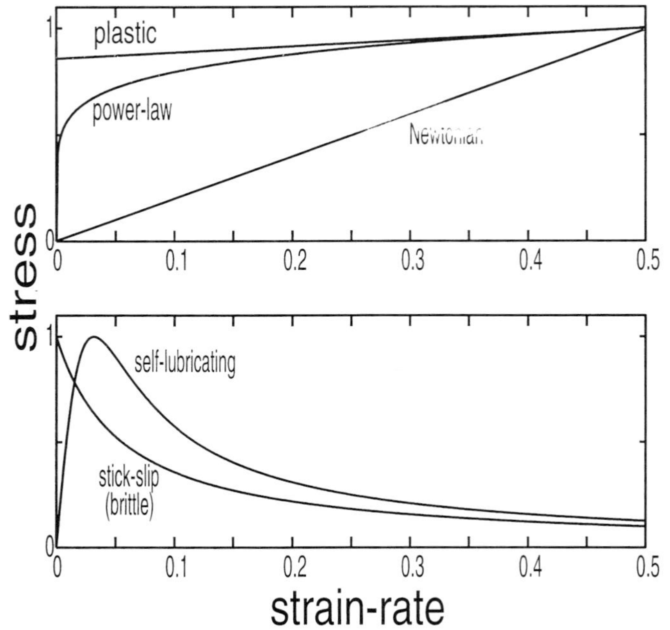


**Figure 10.** Constitutive (stress versus strain-rate) laws for (1) plastic, (2) Newtonian, (3) non-Newtonian power-law, (4) brittle stick-slip and (5) self-lubricating (also called continuous stick-slip) rheologies.

of being a function of the thermodynamic state (e.g. temperature and pressure), state of stress and/or composition of the medium.

If viscosity is temperature-dependent, the top boundary layer will tend to be weakest at the divergent zone (where the boundary layer is thinnest) which approaches plate-like behavior. However, the viscosity will be highest over the cold convergent zone, tending to immobilize it and eventually the entire boundary layer, which is of course one of the root causes for the subduction initiation problem (see §4.6); in the end, this also leads to a thermal boundary layer with little plateness. Therefore, it is a reasonable assumption that plate-like strength distributions – or high plateness – require a lithospheric rheology that permits boundaries to be weak, regardless of temperature.

At high stresses mantle rocks undergo non-Newtonian creep wherin deformation responds nonlinearly to the applied stress; i.e., the stress–strain-rate constitutive law is nonlinear through a power-law relation such that

$$\dot{e} \sim \sigma^n \tag{10}$$

where $\dot{e}$ is strain-rate, $\sigma$ is stress, and $n$ is the *power-law index* (see Fig. 10); for mantle rocks, $n = 3$, typically [Weertman and Weertman, 1975; Evans and Kohlstedt, 1995]. The effective viscosity is

$$\mu \sim \sigma/\dot{e} \sim \dot{e}^{\frac{1}{n}-1} \tag{11}$$

which means that for $n > 1$ viscosity decreases with increasing strain-rate. The power-law rheology causes regions of the material that are rapidly deformed to become softer, while slowly-deforming regions become relatively stiff. This rheology not only yields reasonable plateness, but also leads to a feedback effect: as the deformed parts of the fluid soften they become more readily deformed, causing further strain to concentrate there, thereby inducing further softening, etc. (Comparable effects can be caused with other rheologies such as Bingham plastics and bi-viscous laws; however, these are just further mathematical models for essentially the same strain-softening effect represented by a power-law rheology.) Such non-Newtonian rheologies have been incorporated into numerous 2-D convection models for $n = 3$ and have found little plate like behavior [e.g., Parmentier et al., 1976; Christensen, 1984c]. Other models, concentrating on plate formation, placed a thin non-Newtonian "lithospheric" fluid layer atop a thicker convecting layer, or mathematically confined the non-Newtonian behavior to a near-surface layer [Weinstein and Olson, 1992; Weinstein, 1996; Moresi and Solomatov, 1998; see also Schmeling and Jacoby, 1981] (Fig. 11). In these latter models, the non-Newtonian effect does indeed yield an upper layer with reasonably high *plateness* for basally heated convection: the layer becomes weaker over both divergent zones (i.e. over upwellings) and convergent zones (downwellings), and is relatively immobile and strong in between the these two zones. Moreover, as noted by King et al. [1992], 2D models with non-Newtonian rheology are often indistinguishable in some respects from those with imposed plate geometries [e.g., Olson and Corcos, 1980; Davies, 1988a] and lithospheric weak zones [e.g., King and Hager, 1990; Zhong and Gurnis, 1995a]. However, to obtain sufficient plateness with the non-Newtonian models, it is necessary to use a power-law index $n$ considerably higher than inferred from rheological experiments (typically $n > 7$ is needed). Thus, it seems that a general strain-weakening type rheology is a plausible and simple solution to generating high *plateness*, i.e. weak boundaries and strong plate interiors, in a convection model. However, when the convective flow is driven by internally heated convection, which has little or no concentrated upwellings, then the divergent zones in the lithospheric layer tend to be broad and diffuse,

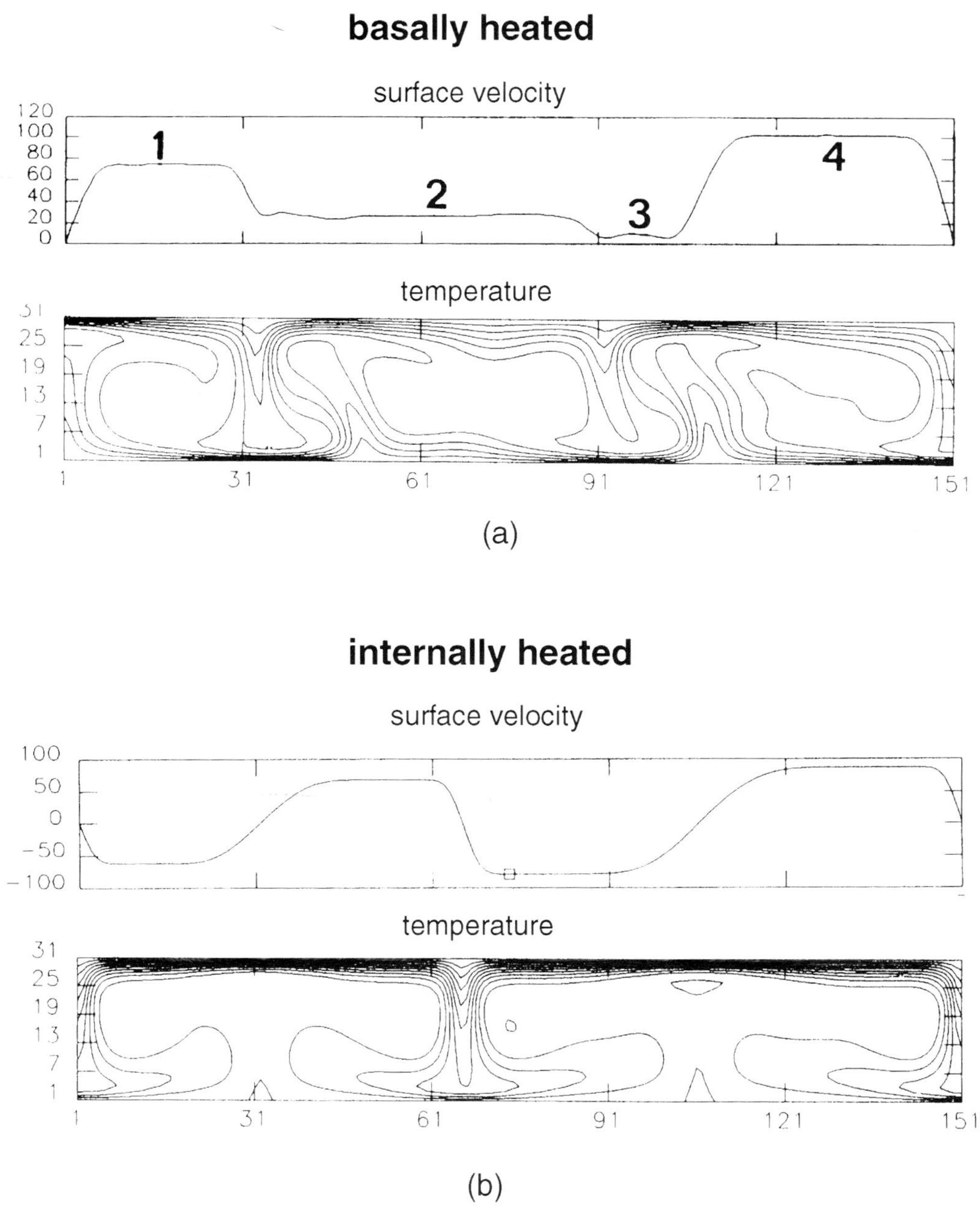


**Figure 11.** A lithospheric layer with non-Newtonian (power-law) rheology is driven by underlying 2D convection for both basal (a) and internal (b) heating. For both cases the convecting layers have $Ra = 50,000$ and the overlying lithosphere has a rheology with a very large power-law index of $n = 15$. The bottom of segments of (a) and (b) show a cross section of a temperature field in the convecting layer. Above each of these is shown the horizontal velocity for the non-Newtonian lithospheric layer. The step-like transitions in velocities suggest plate-like motion (high plateness) since a segment of nearly constant velocity is moving as a contiguous block with little internal deformation (the plate-like segments are numbered in the top frame). The step-like character of the velocity field is much more distinct for the basally heated convection where there are both active upwellings and downwellings to define narrow divergent and convergent margins. The internal heating has broad passive upwellings which cause the overlying divergent zones to have smooth, wide and thus unplate-like changes in velocity. (After Weinstein and Olson [1992].)

Subduction model.

i.e., not at all like narrow, passively spreading ridges [Weinstein and Olson, 1992] (Fig. 11).

It is undoubtedly naive to assert that plate-like behavior simply requires the generation of weak boundaries; in fact each type of boundary, i.e., convergent, divergent and strike-slip, develops under very different deformational and thermal environments. Obtaining sufficient plateness is invariably related to the the nature of how the different boundaries form. As each type of boundary is uniquely enigmatic, they warrant individual discussion. These will be the focus of the following sections.

*4.6. Convergent margins: initiation and asymmetry of subduction*

The downwelling currents in simple convection are for the most part very symmetric; i.e., one downwelling is composed of two cold thermal boundary layers converging on each other (Plate 1). In the Earth, the downwellings associated with plate motions – i.e., subducting slabs – are highly asymmetric, i.e., only one side of the convergent zone – only one plate – subducts. The cause for this asymmetry remains one of the greater mysteries in geodynamics. A common hypothesis is that the asymmetry is caused by chemically light continents counteracting the negative buoyancy of one of the boundary layers approaching the convergent zone [e.g., see Lenardic and Kaula, 1996]; however this hypothesis is not universely valid since there are subduction zones entirely in the ocean.

The asymmetry may also reflect an inequality of pressure on either side of the subducting slab if it deviates slightly out of vertical; i.e., a more acute corner flow on one side of the subduction zone induces lower pressure than on the opposite side, thereby causing the slab to be torqued to the more acute side, i.e., to enhance its obliquity [see Turcotte and Schubert, 1982]. The oblique downwelling may then act to impede the motion and descent of the boundary layer approaching the acute angle, since it would have considerable resistance to making a greater than $90^o$ downward turn, eventually leading to asymmetric convergence. This effect is most significant for slabs that are effectively rigid relative to the surrounding mantle, in order that they act as stiff paddles while being lifted up; thus if it occurs, the effect would theoretically be visible in basic convection calculations with strongly temperature-dependent viscosity. However, if the viscosity is so temperature-dependent that it induces such strong slabs then it will also likely place convection itself into the *stagnant* lid regime; i.e., to make the slabs strong enough will also lock up the lithosphere (see §2.5 and §3.6). To then adjust convection models to mitigate this extra problem requires proper initiation of subduction from cold, thick and strong lithosphere [see Mueller and Phillips, 1991; Kemp and Stevenson, 1996; Schubert and Zhang, 1997]. Models with imposed weak zones in the lithosphere show that cold rigid lithosphere subducts readily and can even assume a fairly oblique slab-like angle [Gurnis and Hager, 1988; King and Hager, 1990; Zhong and Gurnis, 1995a]. However, it is important to note that, based on seismically inferred deformation of slabs, it is not clear whether actual slabs are excessively stiff, or even stronger than the surrounding mantle [Tao and O'Connell, 1993].

The generation of the weak zone which permits subduction is itself enigmatic. Recent models have suggested that the necessary weak zone occurs as the result of faulting and rifting [Kemp and Stevenson, 1996; Schubert and Zhang, 1997]. However, brittle failure of the entire lithosphere at its thickest point is problematic (see §4.5.1). Other weakening mechanisms at the subduction zones may be necessary [e.g., King and Hager, 1990], although the nature of these mechanisms is still unclear, showing that the problem of downwelling asymmetry and subduction initiation remains one of the more elusive yet fruitful fields of geodynamical research.

*4.7. Divergent margins: ridges, and narrow, passive upwellings*

Basic convection predicts that upwellings occur as either columnar plumes rising actively under their own buoyancy, or as a very broad background of upwelling ascending passively in response to the downward flux of concentrated cold thermals. No where in basic convection theory or modelling does there occur concentrated but shallow and passive upwellings (i.e., which rise in response to lithospheric spreading motion, not because of any buoyancy of their own) analagous to ridges [Lachenbruch, 1976; see Bercovici et al., 1989a]. That all mid-ocean ridges involve passive upwelling is not necessarily universal [Lithgow-Bertelloni and Silver, 1998]. However, the fastest and arguably the most signficant ridge, the East Pacific Rise, is almost entirely devoid of a gravity anomaly, implying shallow, isostatic support and thus that there is no deep upwelling current lifting it up [see Davies, 1988b]. (Although the interpretation of gravity by itself is nonunique, the lack of a signficant free-air gravity anomaly over a topographic feature suggests that the gravity field of the topographic mass excess is being cancelled by the field of a nearby mass deficit; this deficit is most readily associated with a buoyant and shallow isostatic root on which the topographic feature is floating.) Moreover, 90% of the structure of the Earth's geoid (the equipotenial surface) can be explained by the gravitational attraction of the mass anomalies due only to slabs, implying that other mass anomalies, e.g., those due to active buoyant upwellings, are much less significant [Ricard et al., 1993].

The cause and initiation of narrow ridges is easily as enigmatic as the problems with subduction. The orientation of ridges more or less mirrors the subduction zones that they eventually feed, thus they may initiate as a strain localization, such as a self-focussing necking instability [e.g., Ricard et al., 1986], or quite simply an effective tear in the lithosphere. This would suggest that the stresses in the lithosphere due to the pull of slabs is guided considerable distances such that they can be concentrated on particular regions (otherwise, continuous release and diffusion of stresses becomes manifested as distributed – and thus very unridge-like – deformation). A cold lithosphere, strong by virtue of its temperature-dependent viscosity, makes a plausible stress guide. Strain-localization may be due to a variety of nonlinear feedback mechanisms, e.g., necking in a non-Newtonian strain-softening medium, or by weakening due to the presence of magma (induced by pressure release melting within the passive upwelling) or ingestion of volatiles (e.g., water, which both provides lubrication and facilitates melting by reducing the the melting temperature; see Hirth and Kohlstedt [1996]). That strain-localization occurs because of brittle failure across the entire lithosphere is, again, probably not likely (see §4.5.1). Finally, while stress guides and strain-localization are viable mechanisms for ridge formation, the physics that determines the distances to which the stresses are guided before causing a ridge – distances which also determine plate size – is not clear and remains a fruitful area of research.

### *4.8. Strike-slip margins: generation of toroidal motion*

While the upwellings, downwellings, divergent and convergent zones in basic convection may or may not appear exactly in the plate-tectonic form of convection seen on the Earth, they at least do exist in both forms of convection. However, as mentioned earlier in this paper, while strike-slip motion (i.e., toroidal flow, which also includes plate spin) is very significant in the Earth's plate-tectonic style of convection – comprising possibly as much as 50% of the plates net kinetic energy, depending on the choice of reference frame [Hager and O'Connell, 1978, 1979, 1981; O'Connell et al, 1991] (see Plate 3) – it does not exist at all in basic convection with simple rheologies (i.e., with constant or depth dependent viscosity) and is very weak in convection with slightly more complex rheologies (e.g., with temperature-dependent viscosity). Thus, the correction to basic convection theory necessary to obtain strike-slip motion is nontrivial; this is because it is not merely a matter of adjusting the form of the existing convective (poloidal) flow field, but entails obtaining an entirely new flow field that would not otherwise exist.

A very basic but important aspect of toroidal motion is that it cannot occur in two-dimensional (2-D) models of convection (i.e. models with only $x$ and $z$ coordinates). Since toroidal flow occurs as vortex type motion in the $x-y$ plane (see Fig. 7), and since convection is driven by buoyancy forces which point in the $z$ direction, a flow model with convection and toroidal motion needs all three directions; thus only 3-D convection models can obtain toroidal motion. Two-dimensional models of lithospheric motion in the $x-y$ plane are able to generate toroidal motion [Bercovici, 1993, 1995a], as discussed below, however these do not explicitly involve thermal convection. Another important criterion for plate-like toroidal motion is that it be manifest in narrow but intense bands of strike-slip shear which is what deformation on a fault would look like.

*4.8.1. Imposing plates or faults.* One way of inducing toroidal motion in mantle flow is to apply plate motion on the surface of an isoviscous mantle; the imposed strike-slip motion at the surface forces toroidal motion in the fluid, or mathematically speaking, it provides a boundary condition to (5) that yields non-null values of $\psi$ [e.g., Hager and O'Connell, 1978, 1979; Bunge and Richards, 1996]. This method of plate-driven mantle flow has yielded a variety of successful predictions, e.g., the observation that the flow lines of the forced motion correlates with slab dip [Hager and O'Connell, 1978]. A possibly more self-consistent approach is to drive the plate motion with buoyancy driven (poloidal) flow; the tractions on the base of a plate are first provided by the poloidal flow and then the resulting plate motion subsequently drives toroidal flow. This method can be used to infer the mantle's density and viscosity structure by solving for the internal mass anomalies and viscosity stratification that gives the observed plate motions, as well as the observed geoid and topography [Hager and O'Connell, 1981; Ricard et al., 1988, 1989, 1993; Ricard and Vigny, 1989; Ricard and Bai, 1991; Ricard and Froidevaux, 1991; Vigny and Ricard, 1991; Lithgow-Bertelloni and Richards, 1998]. It is also possible to use this technique to drive plates and

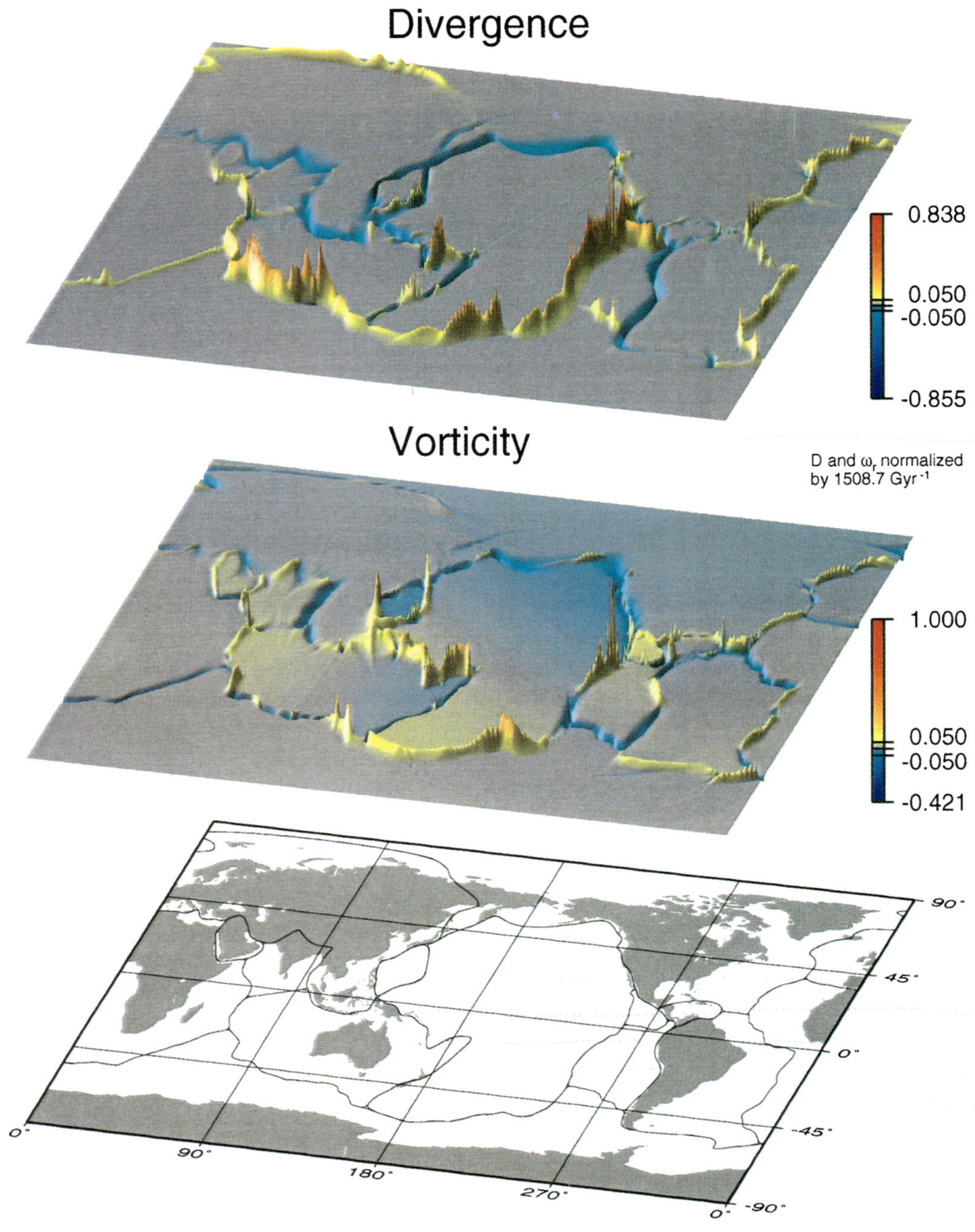


**Plate 3.** Approximate horizontal divergence (top) and vertical vorticity (middle) fields of the Earth (plate and continental outlines are shown for reference in the bottom frame). Horizontal divergence is $\nabla_h \cdot \underline{v} = \nabla_h^2 \phi$ (where $\underline{v}$ is the surface velocity); it is thus a fine-scaled representation of poloidal flow at the Earth's surface and represents the so-called rate of creation (divergence) and destruction (convergence or negative divergence) of the plates at ridges and subduction zones, respectively. Vertical vorticity is $\hat{z} \cdot \nabla \times \underline{v} = -\nabla_h^2 \psi$; it is thus a fine-scaled representation of toroidal flow and represents the rate of strike-slip shear between plates as well as plate spin (vorticity is essentially a measure of the local angular velocity, i.e., angular velocity of a point in the fluid). The divergence and vorticity fields shown are estimated directly from a model that uses finite-width plate margins based on seismicity distributions [Dumoulin et al., 1998]; for the standard plate model with infinitesimally thin margins, these fields are singularities and thus cannot be reliably estimated.

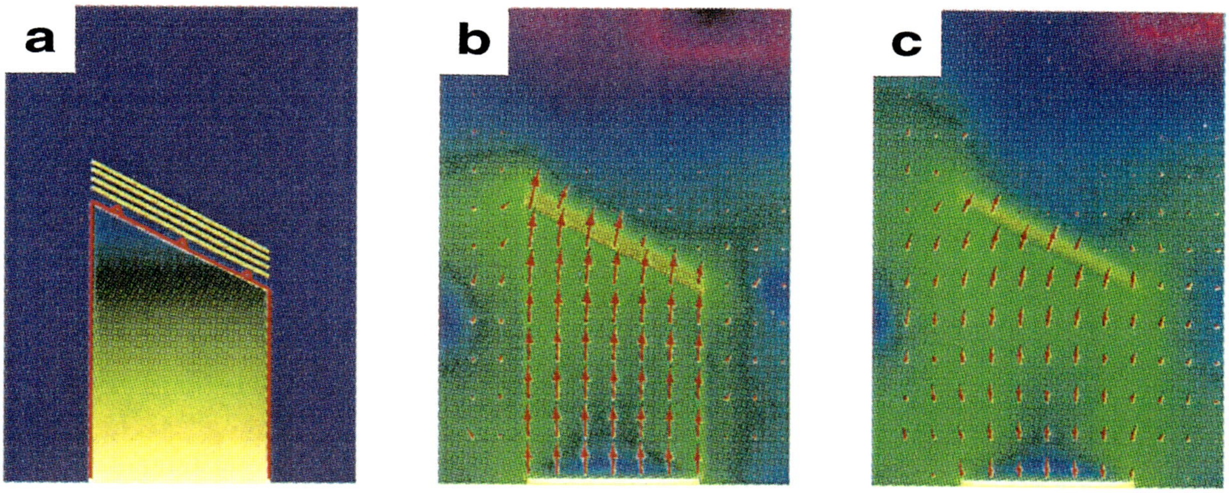


**Plate 4.** A model of lithospheric flow driven by "ridge push" (i.e., a lateral pressure gradient due to aging and thickening of the lithosphere) and slab pull with both non-Newtonian rheology and weak faults. Frame (a): the color shading shows lithospheric age (proportional to thickening), slab dip (yellow contours indicate slab depth) and imposed weak faults (red lines). Frame (b): arrows show velocity field and color shows viscosity for the case with weak faults. Frame (c): same as middle frame, but for the case without faults. The weak faults provide lubricated tracks that facilitate much more plate-like motion (middle frame) than without the faults (right frame). (After Zhong and Gurnis [1996]; see also Zhong et al. [1998].)

toroidal motion with thermal convection [Gable et al., 1991]; models employing this method give insight into how toroidal and convective/poloidal flow co-evolve and co-exist. However, methods employing prescribed plate geometries cannot address the problem of how plates are generated.

A still further step toward self-consistent generation of plates is to impose faults or narrow weak zones in a lithosphere which either overlies a convecting mantle [Zhong and Gurnis, 1995a,b], or moves under its own pressure gradients [Zhong and Gurnis, 1996; Zhong et al., 1998; see also Gurnis et al., 1999, this volume]. This approach is largely motivated by the fact that many weak zones persist within the lithosphere for long times and are always available to be reactivated by convective forces; this concept is addressed below in §4.8.6 and in more detail in Gurnis et al. [1999, this volume]. In these models, plate-like motion is best generated by low-friction faults, and the direction of motion is neither solely determined by the convective (or plate) forces nor the fault orientation; lithospheric motion instead assumes some optimum configuration that takes best advantage of the orientation of both the forces and the faults (Plate 4). Generally, the motion will arrange itself so that strike-slip shear occurs along the low friction faults (see discussion below §4.8.4 on the purpose of toroidal flow) and thus the models generate significant toroidal motion.

However, the methods described above permanently prescribe the plate or fault geometries and thus these methods do not permit the plates or plate boundaries themselves to evolve from the convective flow. As with the 2-D convection models which strive to generate plateness, 3-D models that attempt to generate toroidal motion (as well as plateness) out of convective flow use a continuum approach with variable viscosity [e.g., Kaula, 1980; Christensen and Harder, 1991; Ribe, 1992; Cadek et al., 1993; Balanchandar and Yuen, 1995a,b; Bercovici, 1993, 1995a,b, 1996, 1998; Zhang and Yuen, 1995; Tackley, 1998, 1999 this volume; Weinstein, 1998; Trompert and Hansen, 1998].

*4.8.2. Why is variable viscosity necessary for toroidal motion?.* The reason that the generation of toroidal motion in continuum models requires variable viscosity is somewhat mathematical and thus warrants some theoretical development. As shown earlier (§2.7), if viscosity is constant in viscous buoyancy-driven flows (with homogeneous boundaries), the toroidal motion is zero (see equation (5)). If viscosity $\mu$ is variable, then instead of the equation for toroidal flow being homogeneous as in (5) we have

$$\mu\nabla^2\nabla_H^2\psi = \hat{z}\cdot\nabla\mu\times\nabla^2\underline{v} + \hat{z}\cdot\nabla\times\left(\nabla\mu\cdot(\nabla\underline{v}+[\nabla\underline{v}]^t)\right) \tag{12}$$

(where $[..]^t$ implies transpose of a tensor). Thus the toroidal field is forced internally (i.e. within the fluid) if $\nabla\mu \neq 0$; if this condition is not met, then the right side of (12) is zero leading to a null solution for $\psi$, as in (5). This might lead one to believe that toroidal motion can be generated for any variable viscosity, including one that is dependent only on depth (which might occur because of the natural stratification and compression of mantle rock). However this is not the case. In order to guarantee that the toroidal field obtains some energy from convection, even indirectly, we require that the convective poloidal field $\phi$ appears on the right side of (12); i.e., the poloidal field gets energy from convective buoyancy (see (4)) and then passes on some of this energy to the toroidal field via viscosity gradients. However, the poloidal field $\phi$ does not appear at all in (12) if $\mu$ is only a function of height $z$. Therefore, the toroidal field only couples to thermal convection through the poloidal field if $\nabla_H\mu \neq 0$, i.e. if viscosity is at least horizontally variable.

The necessity for laterally variable viscosity can be understood qualitatively as well. If convective forces pull or push the upper thermal boundary layer, and part of this boundary layer is weak, then the weak area tends to get moved or deformed more easily than the neighboring stronger zones. The differential motion between the strong and the weak zones can lead to strike-slip type shear and toroidal motion (see Plate 2).

*4.8.3. What variable viscosities give both plateness and toroidal flow?.* As noted already in this paper, viscosity is variable by virtue of being dependent on the thermodynamic state, state of stress and composition of the fluid. Nevertheless, a temperature-dependent viscosity by itself is probably not capable of generating the requisite toroidal flow and plateness. Numerous studies of basic 3-D convection with temperature-dependent viscosity yielded little plate-like motion, usually with negligible plateness and toroidal kinetic energies on the order of 10% or less of the total [Christensen and Harder, 1991; Ogawa et al., 1991]. (Recall, toroidal motion on the Earth has possibly as much as 50% of the net kinetic energy of the plates.) The reason these models yielded little toroidal motion is likely due to the upper thermal boundary layer being too cold and stiff to permit any additional motion such as toroidal flow. It may also reflect the fact that since both viscosity and buoyancy are dependent on temperature, the viscosity and poloidal flow fields are in phase, whereas viscosity gradients orthogonal to the poloidal flow are probably necessary to generate toroidal flow (e.g., note in Plate 2 that strike-slip shear concentrates over viscosity gradients only because these gradients are perpendicular to the convergent flow in the upper buoyant layer). Thus, the necessary heterogeneities in viscosity likely arise from other nonlinear rheological mechanisms.

Given the success of creating respectable *plateness* from non-Newtonian power-law or plastic rheologies in 2-D convection models, it stands to reason that similar rheologies would generate significant toroidal motion in 3-D convection models. However, this appears not to be the case. Models using power-law rheologies (even with an excessively large power-law index $n > 3$) or plastic rheologies yield little plate-like toroidal flow (i.e., the strike-slip shear zones tend to be broad and diffuse) [Christensen and Harder, 1991; Bercovici 1993, 1995a; Cadek et al., 1993; Tackley, 1998; c.f. Trompert and Hansen, 1998].

It has been proposed that strain-softening type power-law rheologies are not sufficient to excite plate-like flow and instead that *self-lubricating* rheologies are necessary to generate such motion [Bercovici, 1993, 1995a]. A self-lubrication rheology involves a constitutive law that acts like a normal, highly viscous fluid at low strain-rates, but at larger strain-rates the viscosity weakens so rapidly with increasing strain-rate that the very resistance to flow itself – i.e., the stress – decreases [Whitehead and Gans, 1974]; this rheology is most simply represented by the constitutive law

$$\sigma \sim \frac{\dot{e}}{1 + \dot{e}^2/\gamma^2} \tag{13}$$

where $\gamma$ is the strain-rate at which the stress $\sigma$ reaches a maximum and begins to decrease with increasing strain-rate (Fig. 10). When employed in various flow models (which permit motion in the $x-y$ plane and thus toroidal flow) the self-lubricating rheology yields much more plate-like toroidal motion than do the power-law rheologies [Bercovici, 1993, 1995a; Tackley, 1998] (see Plate 5, Plate 6). In addition to generating sufficient toroidal motion, these models with self-lubricating rheologies also generate high plateness [Bercovici, 1993, 1995a; Tackley, 1998] and even suggest formation of relatively narrow passive upwellings [Tackley, 1998], something theretofore never accomplished in a fluid model. However, none of the models which have employed the self-lubricating rheology [Bercovici, 1993, 1995a; Tackley, 1998; see also Tackley, 1999 this volume] explicitly involve thermal convection, but instead prescribe the poloidal or buoyancy fields. Moreover, a technical difficulty with the self-lubricating rheology is that it can cause self-focussing of weak zones to sub-grid scale in numerical calculations [Tackley, 1998], thus the practical implementation of this rheology may require some arbitrary "clipping" of the effective viscosity field at low values. Nevertheless, these studies show that fluid mechanical models can produce plate-like toroidal motion, and that the self-lubricating rheology is reasonably successful at obtaining plate-like flows.

*4.8.4. Why does toroidal motion form at all?.* Continuum models which generate toroidal motion from first principles allow us to address an important question: Why does toroidal motion form in the first place, and what role does it play? Since poloidal motion involves upwellings and downwellings, it can enhance its own energy uptake by increasing the rate at which gravitational potential energy is released; this feedback effect is of course the reason that heated layers go convectively unstable in the first place. However, toroidal flow only dissipates the energy it is given: since it involves purely horizontal motion it does not explicitly enhance gravitational energy release, nor does it, by the same token, transport heat and directly facilitate cooling of the fluid. By all appearances, the toroidal field is superfluous. And yet it is omnipresent in the Earth and is readily generated in fluid models with non-Newtonian (especially self-lubricating) rheologies.

The reason that toroidal motion is generated appears to relate to the thermodynamic efficiency of surface flow. Examination of the total viscous dissipation in the continuum models of Bercovici [1993, 1995a] show that even though the toroidal flow field is added on to the convectively driven poloidal field, it in fact makes the system *less* dissipative than if it were not present [Bercovici 1995b]; i.e., generation of toroidal motion *enhances* the efficiency of surface motion. This result can be simply interpretted to mean that the toroidal field establishes itself in focussed zones of strike-slip shear that act as lubricated motion guides. The bulk of the fluid outside of these zones of intense toroidal motion remain relatively undeformed and thus entail little dissipation. Within these zones deformation may be intense, but the

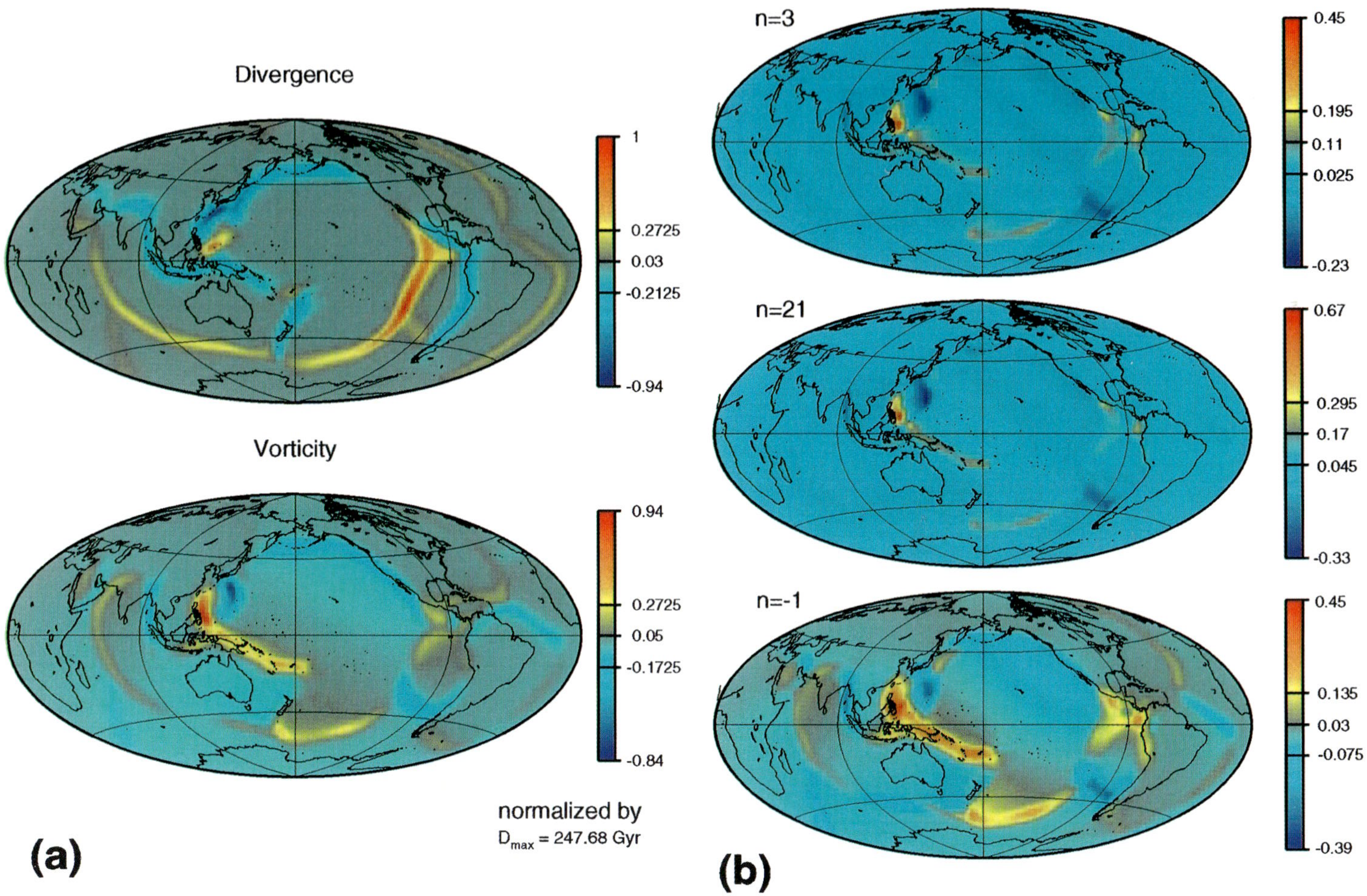


**Plate 5.** Generation of the Earth's toroidal motion via a source-sink model. The Earth's horizontal divergence (Fig. a – top) (a simpler and smoother version of that shown in Plate 3) is used as a source-sink field to drive flow in a non-Newtonian thin (lithospheric) fluid shell. The divergent zones (yellow and red) are used as sources of material, while the convergent zones (blue) are used as sinks. Ideally, the interaction of the source-sink field and non-Newtonian rheology in the lithospheric fluid shell would recover the Earth's vertical vorticity field (Fig. (a) – bottom) (again, simpler and smoother than that shown in Plate 3). Flow in the models with power-law rheologies (see equation (10)) do not generate Earth-like vorticity (Fig. (b) – top and middle frames) for power-law indices that are mantle-like, $n = 3$, and extreme, $n = 21$. Flow with the continuous stick-slip or self-lubricating rheologies (labelled as $n = -1$; see equation (13)) (Fig. (b) – bottom) generates a much more Earth-like vorticity field. See Bercovici [1995a] for further discussion.

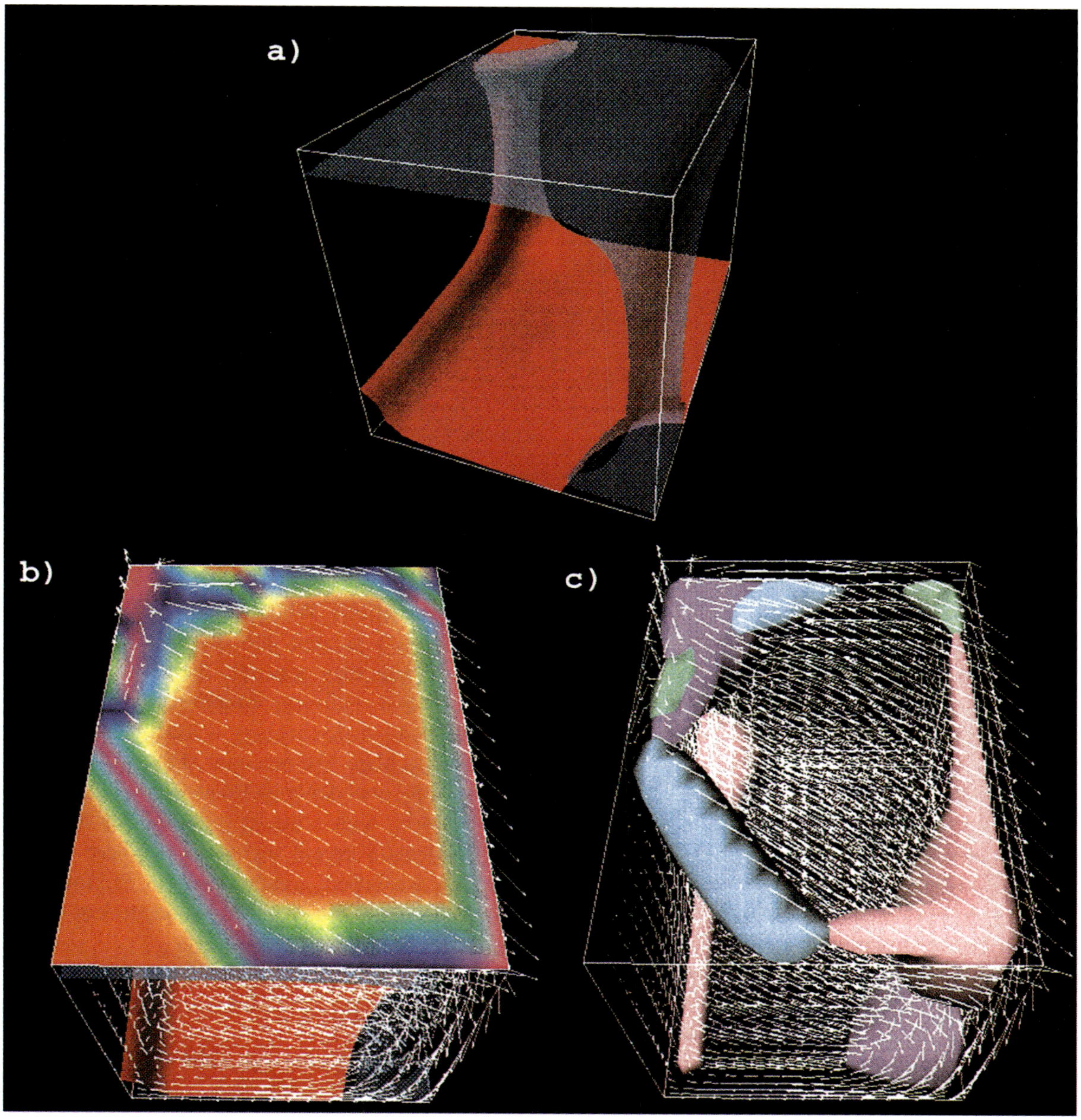


**Plate 6.** Generation of plate-like motion in lithospheric flow driven from below by a convective-type buoyancy field. Frame (a) shows the temperature field from simple basally heated convection. This is used as a static source of buoyancy to drive flow that then drags an overyling layer of fluid with stick-slip rheology (equation (13)). The sharp changes in viscosity (b) surrounding nearly isoviscous (red) blocks, along with nearly uniform velocity fields in these blocks, is suggestive of highly plate like flow. Frame (c) shows isosurfaces of horizontal divergence (poloidal flow) in the lithospheric layer (pink and purple), and vertical vorticity or toroidal flow (light blue and green). The elongated blue surface corresponds to a long, fault-like concentration of strike-slip shear. (After Tackley [1998]; see also Tackley [1999], this volume.)

viscosity is so low that the dissipation is small. Moreover, the zones of deformation are ideally so narrow and involve so little volume that they make little contribution to the total dissipation generated throughout the whole medium. This correlates with studies by Zhong and Gurnis [1996] and Zhong et al. [1998] which show that optimum convective flows occur when strike-slip motion concentrates on frictionless faults.

*4.8.5. Plateness and toroidal motion are not completely related.* While variable viscosity is a requirement for both plateness and toroidal flow, it does not mean that they are manifestations of the same phenomenon. For example, mantle flow models with imposed faults may yield significant toroidal motion in the lithosphere if the faults are very weak; however, the plateness may become poor if the viscosity of the underlying asthenosphere is too high [Zhong et al., 1998]. This occurs because highly plate-like flows (i.e., with large plateness and toroidal motion) seem to occur if a weak zone or fault can decouple the lithosphere on one side of the weak zone/fault from the other. However, a viscous asthenosphere transmits stresses (that were normally decoupled across the fault) from one side of the fault to the other, causing the lithosphere to deform, weaken (via non-Newtonian behavior) and reach low plateness in response, even while the overall strike-slip motion remains significant. (A similar but less illustrative decorrelation of plateness and toroidal motion was found by Bercovici [1995a].)

*4.8.6. Instantaneous and time-dependent rheologies and "true" self-lubricating mechanisms.* Up to now we have discussed the use of strain-rate- (or equivalently stress-) dependent rheologies, and amongst such rheologies the most successful one to date at generating plate-like motion appears to be the self-lubricating mechanism. However, there are two problems in this approach so far: 1) as pointed out by Zhong et al. [1998] and Gurnis et al. [1999, this volume], these strain-rate dependent rheologies are ***instantaneous rheologies***, even though weak plate boundaries are not instantaneously formed; and 2) the self-lubricating rheology is an ad hoc parameterization of more complex physics. In fact, these two problems and their possible solution are closely related.

The explicitly strain-rate–dependent rheologies are called ***instantaneous*** because viscosity responds only to the instantaneous strain-rate field. Thus weak zones (what we would like to call plate boundaries) exist only as long as they are being deformed; once deformation stops, the boundaries cease. However, it is known that while new plate boundaries continue to be formed, old inactive ones persist as fabric in the lithospere, and even provide intrinsic weak zones that are preferred sites at which to re-activate deformation. Thus, boundaries have memory: they can remain weak even without being deformed, and can be moved around with the material in which they are embedded. This implies that the weakness is an actual transportable property, like temperature or chemical concentration. Moreover, such weak zones can persist for extremely long times, possibly much longer than the typical convective time scale of $10^8$ years (a typical convective overturn time) [see Gurnis et al., 1998, this volume]; this suggests that the property which determines plate boundary weakness obeys a much longer time scale than does convection [see Bercovici, 1998].

That plate boundaries are caused by some weakening material property rather than an instantaneous rheological response in fact correlates with the fundamental origins of the the self-lubricating rheology. The rheological law expressed in (13) is actually a simplified, one-dimensional (1-D), steady-state representation [Whitehead and Gans, 1974; Bercovici, 1993] of the classical self-lubrication mechanism involving the coupling of viscous heating with temperature-dependent viscosity [Schubert and Turcotte, 1972]. In this self-lubrication mechanism, deformation causes frictional heating, which warms and weakens the material; the weak zone is more readily deformed causing a concentration of deformation, and therefore more heating, weakening, etc. In this case, a weak zone corresponds to a temperature anomaly, and thus temperature acts as the transportable time-dependent property whose history can be retained for a finite period of time.

However, when this viscous-heating-based, time-dependent self-lubrication mechanism is incorporated into flow models, it is not altogether successful at generating large-scale plate-like motions [Balachandar et al., 1995a,b; Zhang and Yuen, 1995; Bercovici, 1998]. This lack of plate motion likely occurs because temperature-anomalies due to viscous dissipation are generally not as important as the temperature anomalies due to the heating which is powering convection (e.g., heating from the core, or more importantly radiogenic heating). Thus, the temperature drop across the cold upper thermal boundary layer is usually much larger than any temperature anomaly due to viscous heating, causing this boundary layer to continue to be stiff and immobile. In some cases, viscous heating can be significant on small scales in convection, and, with temperature-dependent viscosity, can lead to shear-focussing and modest concentrations of toroidal motion on a local scale [Balachandar et al., 1995a,b; Zhang and Yuen, 1995]; however, the surface motions and strength distributions created are not distinctly plate-like and the global toroidal energy is typically very small. Moreover, if the Earth's weak zones were simply temperature anomalies (and were fortuitously not overwhelmed by convective temperature anomalies), they would diffuse away on the order of a few million years, at most a few hundred million years (assuming plate boundaries are $10 - 100km$ wide and thermal diffusivity is $10^{-6}m^2/s$); i.e., their lifetime would be of the same order or shorter than the convective time scale, which is not observed [Gurnis et al., 1998, this volume]. Finally, if plate boundaries, in particular strike-slip margins, were caused by viscous heating then they should be perceptible with heat-flow measurements; however, this is generally though not to be the case [Lachenbruch and Sass, 1980; c.f., Thatcher and England, 1998].

Self-lubricating mechanisms involving other weakening properties are of course also possible. One well documented shear-localization mechanism involves grain-size dependent rheology and dynamic recrystallization [Karato, 1989]. Although this is a fairly complicated mechanism, the possible self-lubricating effect relies on the fact that at low stresses silicate rheology is largely controlled by diffusion creep where the viscosity depends on grain size (such that viscosity decreases with decreasing grain-size). In the process of diffusion creep, crystals metamorphose in response to tensile or compressive stresses; however, recrystallization and grain growth occurs up to a maximum crystal size which is itself inversely dependent on stress to some power. Thus in a region of higher stress, new crystals form with a smaller size causing the viscosity in the region to diminish. This zone of decreased viscosity represents a weakness on which deformation concentrates, causing higher stresses, thus even smaller recrystallized grain sizes, smaller viscosities, more stress concentration, and so on. (See Karato [1989] and Jin et al. [1998] for more rigorous explanations; see also Kameyama et al., [1997]; and Tackley [1999] this volume.) Shear localization due to this feedback mechanism is well documented in field evidence in continental lithospheric shear zones [e.g., Jin et al. 1998]. Theoretical models of shear zones incorporting this mechanism have yielded strain-localization [Kameyama et al., 1997], although the localization only occurs transiently and requires the addition of viscous heating and thermally activated viscosity. However, the time-scales controlling this mechanism, i.e., for grain growth, are relatively short [Karato, 1989] and thus probably cannot by themselves yield long-lived plate boundaries.

An attractive yet somewhat speculative alternative mechanism involves using water as the weakening agent; this of course recalls the oft-stated hypothesis that plate tectonics only exists with the lubricating effects of water and water-laden sediments (and which also partially explains the lack of plate tectonics on the terrestrial planets devoid of water) [Tozer, 1985; Lenardic and Kaula, 1994, 1996]. Moreover, if the longevity of weak zones generated by ingestion of water are controlled by water's chemical diffusivity in rocks (which is several orders of magnitude slower than thermal diffusivity [Brady, 1995]), then such weak zones would survive for extremely long times, much longer than convective thermal anomalies [Bercovici, 1998]. However, in order for water to induce self-lubrication it must not only weaken the material into which it is ingested, but the water content must be affected by the deformation itself. Only in this way does one obtain the necessary shear-focussing feedback mechanism (similar to the viscous heating mechanism) whereby water content causes weakening, deformation concentrates on the weak zone, the enhanced deformation causes greater ingestion of water which causes further weakening, etc. The physics by which ingestion of water is controlled by rate of deformation is a necessary (albeit poorly understood) link and may occur when voids are created by damaging of the material [see Bercovici, 1998; Regenauer-Lieb, 1998 and references therein]; water (or any volatile) is then ingested into the voids thereby weakening the material (e.g. through pore pressure; Kohlstedt et al., 1995). This idea also draws on the fact that much of the lithosphere exists in a transitional region between brittle and ductile failure involving ductile cracking and void growth [Kohlstedt et al., 1995; Evans and Kohlstedt, 1995]. This *void-volatile* model of self-lubrication has, to date, only been used (with signficant simplifying assumptions) in an idealized mantle flow model; nevertheless, this model yielded extremely plate-like behavior with narrow, concentrated shear zones highly suggestive of discontinuous strike-slip faults [Bercovici 1998] (Plate 7). However, this simple void-volatile model requires considerably more work and rigorous analysis. Indeed, while water likely plays a role in facilitating the formation and maintenance of plate tectonics, it is unclear over what depths within the lithosphere it is available to

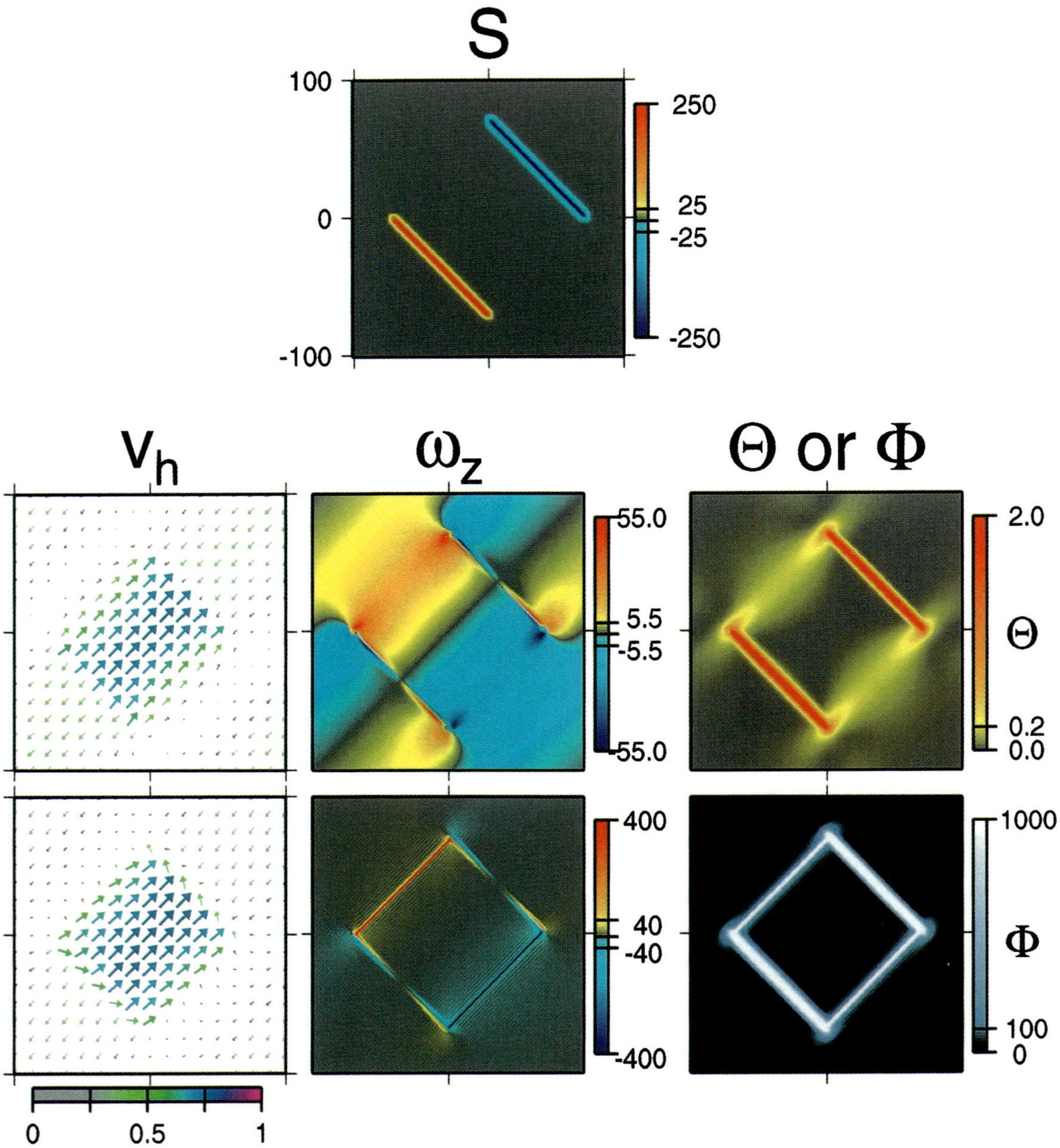


**Plate 7.** A simple model of a thin lithospheric-type layer of fluid driven by a basic source-sink field $S$ (top row; yellow and red represent a source of material, blues represent sinks). The viscosity of the fluid layer is variable. In one model the viscosity is a function of temperature $\Theta$ and heat is generated by viscous dissipation (middle row showing, left to right, dimensionless horizontal velocity $\underline{v}_h$, vertical vorticity or strike-slip shear $\omega_z$ [see also Plate 5], and temperature $\Theta$). In the second model viscosity is a function of volatile (e.g., water) content measured in terms of volume fraction of voids or porosity $\Phi$; the voids are generated according to the amount that the fluid is stressed and damaged (bottom row, showing left to right, velocity, vorticity and porosity). The fields $S$, $\omega_z$, $\Theta$ and $\Phi$ as shown are multiplied by a factor of 1000. The interaction of the source-sink field and the variable viscosity leads to toroidal flow. Toroidal motion (represented with vertical vorticity) and plate strength or *plateness* (measured by temperature or porosity distribution, where hot or porous is weak, cold and unporous is strong) remains diffuse and unplate-like in the viscous-heating based mechanism even with very temperature-dependent viscosity. However, toroidal motion in the void-volatile model can allow extremely focussed, nearly fault-like concentrations of strike-slip shear or toroidal motion (bottom middle frame) and plate-like strength distributions (bottom right frame). (After Bercovici [1998].)

cause weakening [e.g., see Hirth and Kohlstedt, 1996].

In the end, it is unlikely that any one shear-focussing or self-lubrication mechanism will be the so-called magic bullet of plate boundary formation and plate generation. A distinct possibility may be that a combination of rheological mechanisms, such as brittle failure at shallow depths, ductile void growth and water/volatile ingestion at intermediate lithospheric depths, and grain-size dependent rheologies at greater lithospheric depths, yields the requisite shear-localizing behavior [see also Tackley, 1999, this issue]. Indeed, in this case, the longevity of plate boundaries may really only reflect the diffusion and healing rates of failed zones and volatile-filled voids at shallow depths; while these long-lived weak zones may not, therefore, penetrate the entire lithosphere, they may serve as initiation sites for shear-localization due to grain-size variations at greater depths. Unfortunately, the depth and material properties of weak zones associated with plate boundaries is poorly constrained. Many of these questions, from what shear-localizing mechanisms dominate within the lithosphere, to the depth of plate boundaries must be addressed observationally (e.g., with detailed, short-wavelength seismic studies).

## 5. FUTURE DIRECTIONS

Although the theory of plate tectonics has provided one of the greatest predictive tools in the solid-Earth sciences, it is by no means a complete theory. Plate tectonics is a kinematic model in that it describes motions, but it does not involve dynamics, i.e., either the forces or energy behind plate motions; nor does it explain the cause and generation of the plates themselves. In contrast, mantle convection theory describes and incorporates the essential dynamics behind plate motions, but is not yet sophisticated enough to predict from ab initio models the formation of the plates – with weak boundaries, strong plate interiors, asymmetric subuction, passive spreading centers, sudden and sometimes global changes in plate motions, and toroidal motion. A theory of mantle convection which generates plates – in essence a unified theory of mantle dynamics and plate tectonics – remains a lofty goal of geodynamics.

Considerable progress has been made in the last decade on obtaining fluid mantle flow models that yield plate-like behavior. In many ways, much of this success has come about by prescribing lithospheric weak zones and ad hoc rheologies (or rheologies like brittle failure that, while empirically based, are not applicable to the entire lithosphere) that yield more or less the correct behavior. While ad hoc formulations give important clues about what basic stress-strain-rate laws will yield plate-like behavior, they do not elucidate the underlying physics leading to these rheologies or to plate generation [see also Tackley, 1999, this issue]. In this sense, future progress will not necessarily come about by yet more convection modelling, but instead by an examination of the physical and microscopic mechanisms behind shear-localization and the depths and conditions at which these mechanisms occur. The areas of physics and chemistry involving melting and continent formation, the state and fate of water in the mantle, grain-growth, recrystallization, microcracking and void nucleation are but a few examples mentioned here that will play important roles in understanding the dynamics of plate generation. Moreover, observational, in particlar seismic data on the material properties and 3-D structure of plate boundaries (for boundaries of many ages, from ancient and inactive to incipient ones) is of vital importance in constraining dynamic models and microscopic theories of plate boundary formation. As seismic coverage of the Earth continues to improve with the deployment of ocean-bottom seismometers, the amount of such valuable information will increase tremendously.

In the end, it is clear that far from being tried, true and completed models, the theories of plate tectonics and mantle convection are in their youth, requiring years (if not decades) of research, as well as advances in microscopic physics, continuum mechanics, and observational geophysics of which we have yet to conceive.

*Acknowledgments.* This work was supported by NSF grant EAR-9458405, NASA HPCC Grant NCCS5-147, and the Centre National de la Recherche Scientifique (CNRS) of France. Figure 1 was made by Nicolas Coltice using ConMan (King et al., 1990) and GMT (Wessel and Smith, 1998).

## REFERENCES

Balachandar,S., D.A. Yuen, and D.M. Reuteler, Localization of toroidal motion and shear heating in 3-D high Rayleigh number convection with temperature-dependent viscosity, *Geophys. Res. Lett., 22*, 477-480, 1995a.

Balachandar,S., D.A. Yuen, D.M. Reuteler and G.S. Lauer, Viscous dissipation in 3-dimensional convection with temperature-dependent viscosity, *Science, 267*, 1150-1152, 1995b.

Beaumont, C., The evolution of sedimentary basins on a viscoelastic lithosphere, *Geophys. J.R. Astron. Soc., 55* 471–497, 1976.

Bénard, H., Les Tourbillons cellulaires dans une nappe liquide, *Revue générale des Sciences pures et appliquées, 11*, 1261–1271 and 1309–1328, 1900.

Bénard, H., Les Tourbillons cellulaires dans une nappe liquide transportant de la chaleur par convection en régime permanent, *Annales de Chimie et de Physique, 23*, 62-144, 1901.

Bercovici, D., A simple model of plate generation from mantle flow, *Geophys. J. Int., 114*, 635–650, 1993.

Bercovici, D., A source-sink model of the generation of plate tectonics from non-Newtonian mantle flow, *J. Geophys. Res., 100*, 2013–2030, 1995a.

Bercovici, D., On the purpose of toroidal flow in a convecting mantle, *Geophys. Res. Lett., 22*, 3107–3110, 1995b.

Bercovici, D., Plate generation in a simple model of lithosphere-mantle flow with dynamic self-lubrication, *Earth Planet. Sci. Lett., 144*, 41–51, 1996.

Bercovici, D., Generation of plate tectonics from lithosphere-mantle flow and void-volatile self-lubrication, *Earth Planet. Sci. Lett., 154*, 139- 151, 1998.

Bercovici, D., G. Schubert and G. A. Glatzmaier, Three-dimensional, spherical models of convection in the Earth's mantle, *Science, 244*, 950–955, 1989a.

Bercovici, D., G. Schubert, G. A. Glatzmaier, and A. Zebib, Three-dimensional thermal convection in a spherical shell, *J. Fluid Mech., 206*, 75–104, 1989b.

Bercovici, D., G. Schubert and G. A. Glatzmaier, Three-dimensional, infinite Prandtl number, compressible convection in a basally heated spherical shell, *J. Fluid Mech., 239*, 683–719, 1992.

Bercovici, D. and P. Wessel, A continuous kinematic model of plate tectonic motions, *Geophys. J. Int., 119*, 595–610, 1994.

Bercovici, D. and A. Kelly, Nonlinear initiation of diapirs and plume heads, *Phys. Earth Planet. Int., 101*, 119-130, 1997.

Berg, J.C., A. Acrivos, and M. Boudart, Evaporative convection, *Advances Chem. Eng, 6*, 61-123, 1966.

Boehler, Melting temperature of the Earth's mantle and core: Earth's thermal structure, *Ann. Rev. Earth Planet. Sci., 24*, 15–40, 1996.

Brady, J.B., Diffusion data for silicate minerals, glasses, and liquids, in *Mineral Physics and Crystallography: A Handbook of Physical Constants, AGU Reference Shelf 2*, American Geophysical Union, 269–290, 1995.

Bunge, H.-P. and M.A. Richards, The origin of large scale structure in mantle convection: effects of plate motions and viscosity stratification, *Geophys. Res. Lett., 23*, 2987–2990, 1996.

Bunge, H.-P., M.A. Richards, and J.R. Baumgardner, Effect of depth-dependent viscosity on the planform of mantle-convection, *Nature, 379*, 436–438, 1996.

Bunge, H.-P., M.A. Richards, and J.R. Baumgardner, A sensitivity study of 3-D spherical mantle convection at $10^8$ Rayleigh number: Effects of depth-dependent viscosity, heating mode, and an endothermic phase change, *J. Geophys. Res., 102*, 11991-12007, 1997.

Busse, F. H., Patterns of convection in spherical shells, *J. Fluid Mech., 72*, 67–85, 1975.

Busse, F. H., Non-linear properties of thermal convection, *Rep. Prog. Phys., 41*, 1929, 1978.

Busse, F. H. and Riahi, N., Patterns of convection in spherical shells. Part 2, *J. Fluid Mech., 123*, 283–301, 1982.

Busse, F. H. and Riahi, N., Mixed-mode patterns of bifurcations from spherically symmetric basic states, *Nonlinearity, 1*, 379–388, 1988.

Busse, F.H. and J.A. Whitehead, Instabilities of convection rolls in high Prandtl number fluid, *J. Fluid Mech., 47* 305–320, 1971.

Cadek, O. and Y. Ricard, Toroidal/poloidal energy partitioning and global lithospheric rotation during Cenozoic time, *Earth Planet. Sci. Lett., 109*, 621-632, 1992.

Cadek, O., Y. Ricard, Z. Martinec and C. Matyska, Comparison between Newtonian and non-Newtonian flow driven by internal loads, *Geophys. J. Int., 112*, 103–114, 1993.

Cazenave, A. and B. Lago, Long wavelength topography, sea floor subsidence and flattening, *Geophys. Res. Lett., 19*, 1257-1260, 1991.

Chandrasekhar, S., *Hydrodynamic and Hydromagnetic Stability*, Oxford University Press, New York, 1961.

Christensen, U.R., Heat transport by variable-viscosity convection and implications for the Earth's thermal evolution, *Phys. Earth Planet Int., 35*, 264–282, 1984a.

Christensen, U.R., Instability of a hot thermal boundary layer and initiation of thermo-chemical plumes, *Anal. Geophys., 2*, 311-314, 1984b.

Christensen, U.R., Convection with pressure- and temperature-dependent non-Newtonian rheology, *Geophys. J.R. Astron. Soc, 77*, 343–384, 1984c.

Christensen, U.R., Fixed hotspots gone with the wind, *Nature, 391*, 739-740, 1998.

Christensen, U.R., and H. Harder, Three-dimensional convection with variable viscosity, *Geophys. J. Int., 104*, 213-226, 1991.

Christensen, U.R., and D.A. Yuen, Layered convection induced by phase changes, *J. Geophys. Res., 90*, 291–300, 1985.

Creager, K.C. and T.H. Jordan, Slab penetration into the lower mantle, *J. Geophys. Res., 89* 3031–3049, 1984.

Colin, P., and L. Fleitout, Topography of the ocean floor: thermal convection of the lithosphere and interaction of the deep mantle heterogeneity with the lithosphere, *Geophys. Res. Letters, 11*, 1961-1964, 1990.

Corrieu, V., Y. Ricard and C. Froidevaux, Radial viscosity of the Earth deduced from mantle tomography, *Phys. Earth Planet. Inter., 84*, 3-13, 1994.

Davies, G.F., Thermal histories of convective Earth models and constraints on radiogenic heat production in the Earth, *J. Geophys. Res., 85*, 2517–2530, 1980.

Davies, G.F., The role of the lithosphere in mantle convection, *J. Geophys. Res., 93*, 10451–10466, 1988a

Davies, G.F., Ocean bathymetry and mantle convection, 1. Large-scale flow and hotspots, *J. Geophys. Res., 93*, 10467–10480, 1988b.

Davies, G.F., Ocean bathymetry and mantle convection, 2. Small scale flow, *J. Geophys. Res., 93*, 10481–10488, 1988c.

Davies, G.F. and M.A. Richards, Mantle convection, *J. Geology, 100* 151–206, 1992.

Duffield, W.A., A naturally occurring model of global plate tectonics, *J. Geophys. Res.,* 77, 2543–2555, 1972.

Dumoulin, C., D. Bercovici, and P. Wessel, A continuous plate-tectonic model using geophysical data to estimate plate margin widths, with a seismicity based example, *Geophys. J. Int., 133*, 379–389, 1998.

Duncan, R.A. and M.A. Richards, Hotspots, mantle plumes, flood basalts and true polar wander, *Rev. Geophys., 29*, 31–50, 1991.

Evans, B. and D.L. Kohlstedt, Rheology of rocks, in *Rock*

*Physics and Phase Relations: A Handbook of Physical Constants, AGU Reference Shelf 3*, American Geophysical Union, 148–165, 1995.

Fleitout, L. and C. Moriceau, Short-wavelength geoid, bathymetry and the convective pattern beneath the Pacific Ocean, *Geophys. J. Int., 110*, 6–28, 1992.

Forsyth, D., and S. Uyeda, On the relative importance of the driving forces of plate motion, *Geophys. J. R. Astr. Soc., 43*, 163–200, 1975.

Forte, A.M. and W.R. Peltier, Plate tectonics and aspherical earth structure: The importance of poloidal-toroidal coupling, *J. Geophys. Res., 92*, 3645-3679, 1987.

Furbish, D.J., *Fluid Physics in Geology*, Oxford Univ. Press, 1997.

Gable, C.W., R.J. O'Connell, and B.J. Travis, Convection in three dimensions with surface plates: Generation of toroidal flow, *J. Geophys. Res., 96*, 8391–8405, 1991.

Giannandrea, E. and U. Christensen, Variable viscosity convection experiments with a stress-free upper boundary and implications for the heat transport in the Earth's mantle, *Phys. Earth. Planet. Int., 78*, 139–152, 1993.

Glatzmaier, G.A., Numerical simulations of mantle convection: Time-dependent, three-dimensional, compressible, spherical shell, *Geophys. Astrophys. Fluid Dyn., 43*, 223–264, 1988.

Glatzmaier, G.A., G. Schubert and D. Bercovici, Chaotic subduction-like downflows in a spherical model of convection in the Earth's mantle, *Nature, 347*, 274–277, 1990.

Grand, S.P., R.D. van der Hilst, and S. Widiyantoro, Global seismic tomography: A snapshot of convection in the Earth, *GSA Today, 7*, 1-7, 1997.

Gordon, R.G., C. DeMets, and J.-Y. Royer, Evidence for long-term diffuse deformation of the lithosphere of the equatorial Indian Ocean, *Nature, 395*, 370–374, 1998.

Gubbins, D. Energetics of the Earth's core, *J. Geophys., 43*, 453–464, 1977.

Guillou, L., J.-C. Mareschal, C. Jaupart, C. Gariépy, G. Bienfait, and R. Lapointe, Heat flow, gravity, and structure of the Abitibi belt, Superior Province, Canada: Implications for mantle heat flow, *Earth Planet. Sci. Lett., 122*, 103-123, 1994.

Gurnis, M., Large-scale mantle convection and the aggregation and dispersal of supercontinents, *Nature, 335*, 695–699, 1988.

Gurnis, M., Bounds on global dynamic topography from Phanerozoic flooding of continental platforms, *Nature, 344*, 754-756, 1990.

Gurnis, M., Phanerozoic marine inundation of continents driven by dynamic topography above subducting slabs, *Nature, 364*, 589–593, 1993.

Gurnis, M., and B.H. Hager, Controls of the structure of subducted slabs, *Nature, 335*, 317–321, 1988.

Gurnis, M., S. Zhong, and J. Toth, On the competing roles of fault reactivation and brittle failure in generating plate tectonics from mantle convection, *AGU Monograph on the History and Dynamics of Global Plate Motions*, edited by M. Richards, R. Gordon and R. van der Hilst, American Geophysical Union, 1999.

Hager, B.H. and R.W. Clayton, Constraints on the structure of mantle convection using seismic observations, flow models and the geoid, in *Mantle Convection*, ed. by W.R. Peltier, pp.657–763, Gordon and Breach, Newark, N.J., 1989.

Hager, B.H., and R.J. O'Connell, Subduction zone dip angles and flow driven by plate motion, *Tectonophysics, 50*, 111–133, 1978.

Hager, B.H., and R.J. O'Connell, Kinematic models of large-scale flow in the Earth's mantle, *J. Geophys. Res., 84*, 1031–1048, 1979.

Hager, B.H., and R.J. O'Connell, A simple global model of plate dynamics and mantle convection, *J. Geophys. Res., 86*, 4843–4867, 1981.

Hager, B.H., Subducted slabs and the geoid: constraints on mantle rheology and flow, *J. Geophys. Res., 89*, 6003-6015, 1984

Hallam, A., Alfred Wegener and the hypothesis of continental drift, in *Scientific Genius and Creativity: Readings from 'Scientific American'*, W.H. Freeman & Co., pp77-85. (Reprinted from *Scientific American*, February, 1975.) 1987.

Hilde, T.W.C., S. Uyeda, and L. Kroenke, Evolution of the western Pacific and its margins, *Tectonophysics, 38*, 145–165, 1977.

Hirth, G. and D.L. Kohlstedt, Water in the oceanic upper mantle: implications for rheology, melt extraction and the evolution of the lithosphere, *Earth Planet. Sci. Lett., 144*, 93–108, 1996.

Holmes, A., Radioactivity and earth movements, *Geol. Soc. Glasgow Trans., 18*, 559–606, 1931.

Honda, S., D.A. Yuen, S. Balachandar, and D. Reuteler, Three-dimensional instabilities of mantle convection with multiple phase transitions, *Science, 259*, 1308–1311, 1993.

Ito, E., and E. Takahashi, Postspinel transformation in the system $Mg_2SiO_4 - Fe_2SiO_4$ and some geophysical implications, *J. Geophys. Res., 94*, 10,637–10,646, 1989.

Ito, E., M. Akaogi, L. Topor, and A. Navrotsky, Negative pressure–temperature slopes for reactions forming $MgSiO_3$ perovskite from calorimetry, *Science, 249*, 1275–1278, 1990.

Jarvis, G.T. and D.P. McKenzie, Convection in a compressible fluid with infinite Prandtl number, *J. Fluid Mech., 96* 515–583, 1980.

Jin, D., S.-I. Karato, and M. Obata, Mechanisms of shear localization in the continental lithosphere: inference from the deformation microstructures of peridotites from the Ivrea zone, northwestern Italy, *J. Struct. Geol., 20*, 195–209, 1998.

Kameyama, M., D.A. Yuen and H. Fujimoto, The interaction of viscous heating with grain-size dependent rheology in the formation of localized slip zones, *Geophys. Res. Lett., 24*, 2523–2526, 1997.

Karato, S.-I., Grain growth kinetics in olivine aggregates, *Tectonophysics, 168*, 255–273, 1989.

Kaula, W.M., Material properties for mantle convection consistent with observed surface fields, *J. Geopys. Res., 85*, 7031-7044, 1980.

Kennet, B.L.N., and E.R. Engdahl, Travel times for global earthquake location and phase identification, *Geophys. J. Int., 105*, 429–466, 1991.

Kido M. and T. Seno, Dynamic topography compared with residual depth anomalies in oceans and implications for age-depth curves, *Geophys. Res. Lett., 21*, 717-720, 1994.

King, S. D., The viscosity structure of the mantle, *Reviews of Geophysics (Supplement): U.S. Quadrennial Report to the IUGG 1991-1994*, pp. 11-17, American Geophysical Union, Washington, DC., 1995.

King, S.D., A. Raefsky, and B.H. Hager, ConMan: Vectorizing a finite element code for incompressible two-dimensional convection in the Earth's mantle, *Phys. Earth Planet. Int., 59*, 196–208, 1990.

King, S.D. and B.H. Hager, The relationship between plate velocity and trench viscosity in Newtonian and power-law subduction calculations, *Geophys. Res. Lett., 17*, 2409–2412, 1990.

King, S.D., C.W. Gable and S. Weinstein, Models of convection driven plates: A comparison of methods and results, *Geophys. J. Int., 109*, 481-487, 1991.

King, S.D. and G.T. Masters, An inversion for radial viscosity structure using seismic tomography, *Geophys. Res. Lett., 19*, 1551–1554, 1992.

Kemp, D.V. and D.J. Stevenson, A tensile flexural model for the initiation of subduction, *Geophys. J. Int., 125*, 73–94, 1996

Kohlstedt, D.L., B. Evans and S.J. Mackwell, Strength of the lithosphere: Constraints imposed by laboratory experiments, *J. Geophys. Res., 100*, 17587–17602, 1995.

Lachenbruch, A.H., Dynamics of a passive spreading center, *J. Geophys. Res., 81*, 1883-1902, 1976.

Lachenbruch, A.H. and J.H. Sass, Heat flow and energetics of the San Andreas fault zone, USGS Open File Rep. 80-625, pp.47-146, 1980.

Lenardic, A., and W.M. Kaula, Self-lubricated mantle convection: Two-dimensional models, *Geophys. Res. Lett., 21*, 1707–1710, 1994.

Lenardic, A., and W.M. Kaula, Near surface thermal/chemical boundary layer convection at infinite Prandtl number: Two-dimensional numerical experiments, *Geophys. J. Int., 126*, 689–711, 1996.

Le Stunff Y., and Y. Ricard, Topography and geoid due to lithospheric mass anomalies, *Geophys. J. Int., 122*, 982-990, 1995.

Le Stunff, Y., and Y. Ricard, Partial advection of equidensity surfaces: a solution for the dynamic topography problem? *J. Geophys. Res., 102*, 24655-24667, 1997.

Lithgow-Bertelloni, C., M.A. Richards, Y. Ricard, R.J. O'Connell, and D.C. Engebretson, Toroidal-poloidal partitioning of plate motions since 120 Ma, *Geophys. Res. Lett., 20*, 375–378, 1993.

Lithgow-Bertelloni, C. and M.A. Richards, Cenozoic plate driving forces, *Geophys. Res. Lett., 22*, 1317-1320, 1995.

Lithgow-Bertelloni, C. and P.G. Silver, Dynamic topography, plate driving forces and the African superswell, *Nature, 395*, 269–272, 1998.

Liu, L.-G., in *Earth: Its Origins, Structure and Evolution*, M.W. McElhinny, ed., 177-202, Academic Press, San Diego, CA, 1979.

Loper, D.E. and I.A. Eltayeb, On the stability of D" layer, *Geophys. Astrophys. Fluid Dyn., 36*, 229–255, 1986.

Loper, D.E. and T. Lay, The core-mantle boundary region, *J. Geophys. Res., 100* 6397–6420, 1995.

Machetel and Weber, Intermittent layered convection in a model mantle with an endothermic phase change at 670km, *Nature, 350*, 55-57, 1991.

Mitrovica, J. X., C. Beaumont, and, G. T. Jarvis, Tilting of continental interiors by the dynamical effects of subduction, *Tectonics, 8*, 1079-1094, 1989.

Moresi, L. and V. Solomatov, Mantle convection with a brittle lithosphere: Thoughts on the global tectonic style of the Earth and Venus, *Geophys. J. Int., 133*, 669–682, 1998.

Morgan, W.J., Rises, trenches, great faults and crustal blocks, *J. Geophys. Res., 73*, 1959–1982, 1968.

Mueller, S. and R.J. Phillips, On the initiation of subduction, *J. Geophys. Res., 96* 651–665, 1991

Nataf, H.-C. and J. VanDecar, Seismological detection of a mantle plume? *Nature, 364*, 115–120, 1993.

Nicolis, G., *Introduction to Nonlinear Science*, Cambridge Univ. Press, 1995.

Norton, I.O., The 43Ma nonevent, *Tectonics, 14*, 1080-1094, 1995.

O'Connell, R.J., C.W. Gable, and B.H. Hager, Toroidal-poloidal partitioning of lithospheric plate motion, in *Glacial Isostasy, Sea Level and Mantle Rheology*, edited by R. Sabadini et al., pp. 535–551, Kluwer Academic, Norwell, Mass., 1991.

Ogawa, M., G. Schubert and A. Zebib, Numerical simulations of three-dimensional thermal convection in a fluid with strongly temperature-dependent viscosity, *J. Fluid Mech., 233*, 299–328, 1991.

Olson, P. and G.M. Corcos, A boundary layer model for mantle convection with surface plates, *Geophys. J.R. Astr. Soc., 62* 195–219, 1980.

Olson, P., Hotspots, swells and mantle plumes, in *Magma Transport and Storage*, M.P. Ryan, ed., John Wiley, New York, pp33-51, 1990.

Olson, P. and D. Bercovici, On the equipartitioning of kinetic energy in plate tectonics, *Geophys. Res. Lett., 18*, 1751-1754, 1991.

Oxburgh, E.R. and D.L. Turcotte, Mechanisms of continental drift, *Reports on Progress in Physics, 41*, 1249–1312, 1978.

Parmentier, E. M., Turcotte, D. L., & Torrance, K. E., Studies of finite amplitude non-Newtonian thermal convection with applications to convection in the Earth's mantle, *J. Geophys. Res. 81*, 1839–1846, 1976.

Parsons, B and J.G. Sclater, An analysis of the variation of ocean floor bathymetry with age, *J. Geophys. Res., 82*, 803–827, 1977.

Parsons, B. and D.P. McKenzie, Mantle convection and the thermal structure of plates, *J. Geophys. Res., 83*, 4485–4496, 1978.

Pearson, J.R.A, On convection cells induced by surface tension, *J. Fluid Mech., 4*, 489–500, 1958.

Phipps Morgan, J. and W.H.F. Smith, Flattening of the seafloor depth-age curve as a response to asthenospheric flow? *Nature, 359*, 524–527, 1992.

Ranalli, G., *Rheology of the Earth*, Chapman and Hall Publishers, London, 1995.

Ratcliff, J.T., P.J. Tackley, G. Schubert and A. Zebib, Transitions in thermal convection with strongly variable viscosity, *Phys. Earth. Planet. Int., 102*, 201–212, 1997.

Rayleigh, Lord, On convective currents in a horizontal layer of fluid when the higher temperature is on the under side, *Phil. Mag., 32*, 529–546, 1916.

Regenauer-Lieb, K., Dilatant plasticity applied to Alpine collision: Ductile void growth in the intraplate area beneath the Eifel volcanic field, *J. Geodynamics, 27*, 1-21, 1998.

Ribe, N.M., The dynamics of thin shells with variable viscosity and the origin of toroidal flow in the mantle, *Geophys. J. Int., 110*, 537–552, 1992.

Ricard, Y., L. Fleitout, and C. Froidevaux, Geoid heights and lithospheric stresses for a dynamic Earth, *Annales Geophysicae, 2*, 267-286, 1984.

Ricard, Y., and C. Froidevaux, Stretching instabilities and lithospheric boudinage, *J. Geophys. Res., 91*, 8314-8324, 1986.

Ricard, Y., C. Froidevaux and L. Fleitout, Global plate motion and the geoid: A physical model, *Geophys. J., 93*, 477-484, 1988.

Ricard, Y., and C. Vigny, Mantle dynamics with induced plate tectonics, *J. Geophys. Res., 94*, 17,543-17,559, 1989.

Ricard, Y., C. Vigny, and C. Froidevaux, Mantle heterogeneities, geoid and plate motion: a Monte Carlo inversion, *J. Geophys. Res., 94*, 13,739-13,754, 1989.

Ricard, Y. and Bai Wuming, Inferring the mantle viscosity and its three dimensional structure from geoid, topography and plate velocities, *Geophys. J., 105*, 561-571, 1991.

Ricard, Y. and C. Froidevaux, Seismic imaging, plate velocites and geoid, the direct and inverse problem, "Glacial isostasy, Sea-level and Mantle rheology", Sabadini and Lambeck eds, Kluwer Academic Publishers, 553-569, 1991.

Ricard Y., M. A. Richards, C. Lithgow-Berteloni and Y. Le Stunf, A geodynamic model of mantle mass heterogeneities, *J. Geophys. Res., 98*, 21895-21909, 1993.

Richards, M.A. and B.H. Hager, Geoid anomalies in a dynamic Earth, *J. Geophys. Res., 89*, 5487–6002, 1984.

Richards, M.A. and C. Lithgow-Bertelloni, Plate motion changes, the Hawaiian-Emperor bend, and the apparent success and failure of geodynamical models, *Earth Planet. Sci. Lett., 137*, 19–27, 1996.

Rumford, Count, Of the propagation of heat in fluids, *Complete Works, 1*, 239, American Academy of Arts and Sciences, Boston, 1870.

Schmeling, S., and Jacoby, W.R., On modeling the lithosphere in mantle convection with non-linear rheology, *J. Geophys., 50*, 89–100, 1981.

Schubert, G., C. Froidevaux and D.A. Yuen, Oceanic lithosphere and asthenosphere: thermal and mechanical structure, *J. Geophys. Res., 81*, 3525–3540, 1976.

Schubert, G., and D.L. Turcotte, One-dimensional model of shallow mantle convection, *J. Geophys. Res., 77*, 945–951, 1972.

Schubert, G., D.L Turcotte, and D.A. Yuen, Role of phase transitions in a dynamic mantle, *Geophys. J.R. Astron. Soc., 42*, 705–735, 1975.

Schubert, G., D. Stevenson, and P. Cassen, Whole planet cooling and the radiogenic heat source contents of the Earth and Moon, *J. Geophys. Res., 85*, 2531-2538, 1980.

Schubert, G. and Zhang, K., Foundering of the lithosphere at the onset of subduction, *Geophys. Res. Lett., 24*, 1527–1529, 1997

Sclater, J.G., C. Jaupart, and D. Galson, The heat flow through the oceanic and continental crust and the heat loss of the earth, *Rev. Geophys. and Space Phys., 18* 269–311, 1980.

Shen, Y., S.C. Solomon, I.Th. Bjarnason, and C.J. Wolfe, Seismic evidence for a lower mantle origin of the Iceland plume, *Nature, 395*, 62–65, 1998.

Silver, P.G., R.W. Carlson, and P. Olson, Deep slabs, geochemical heterogeneity, and the large-scale structure of mantle convection: an investigation of an enduring paradox, *Ann. Rev. Earth Planet. Sci., 16*, 477–541, 1988.

Sleep, N., Hotspots and mantle plumes: Some phenomenology, *J. Geophys. Res., 95*, 6715–6736, 1990.

Solomatov, V.S., Scaling of temperature dependent and stress dependent viscosity convection, *Phys. Fluids, 7* 266–274, 1995.

Spiegelman, M. Geochemical consequences of melt transport in 2-D: The sensitivity of trace elements to mantle dynamics, *Earth Planet. Sci. Lett., 139* 115–132, 1996.

Stein, C.A. and S. Stein, A model for the global variation in oceanic depth and heat flow with lithospheric age, *Nature, 359*, 123–129, 1992.

Steinberger B., and R.J. O'Connell, Advection of plumes in mantle flow; implications for hotspot motion, mantle viscosity and plume distribution, *Geophys. J. Int., 132*, 412-434, 1998.

Stern, R.J., and S.H. Bloomer, Subduction zone infancy: Examples from the Eocene Izu-Bonin-Mariana and Jurassic California arcs, *Geol. Soc. Am. Bull., 104* 1621–1636, 1992.

Su, W.J. and A.M. Dziewonski, On the scale of mantle heterogeneity, *Phys. Earth Planet. Int., 74*, 29–54, 1992.

Tackley, P.J., D.J. Stevenson, G. A. Glatzmaier, and G. Schubert, Effects of an endothermic phase transition at 670 km depth in a spherical model of convection in the Earth's mantle, *Nature, 361*, 699–704, 1993.

Tackley, P.J., D.J. Stevenson, G. A. Glatzmaier, and G. Schubert, Effects of a multiple phase transitions in a 3-dimensional spherical model of convection in Earth's mantle, *J. Geophys. Res., 99*, 15877–15901, 1994.

Tackley, P.J., Effects of strongly variable viscosity on three-dimensional compressible convection in planetary mantles, *J. Geophys. Res., 101*, 3311-3332, 1996a.

Tackley, P.J., On the ability of phase transitions and viscosity layering to induce long-wavelength heterogeneity in the mantle, *Geophys. Res. Lett., 23*, 1985–1988, 1996b.

Tackley, P.J., Self-consistent generation of tectonic plates in three-dimensional mantle convection, *Earth Planet. Sci. Lett., 157*, 9-22, 1998.

Tackley, P.J., The quest for self-consistent generation of plate tectonics in mantle convection models, *AGU Monograph on the History and Dynamics of Global Plate Motions*, edited by M. Richards, R. Gordon and R. van der Hilst, American Geophysical Union, 1999.

Tao, W.C. and R.J. O'Connell, Deformation of a weak subducted slab and variation of seismicity with depth, *Nature, 361*, 626–628, 1993.

Tarduno J.A. and R.D. Cottrell, Paleomagnetic evidence for motion of the Hawaiian hotspot during formation of the Emperor seamounts, *Earth Planet. Sci. Lett., 153*, 171-180, 1997.

Thatcher, W. and P.C. England, Ductile shear zones beneath strike-slip faults - implications for the thermomechanics of the San Andreas fault zone, *J. Geophys. Res, 103*, 891–905, 1998.

Thomson, J., On a changing tesselated structure in certain

liquids, *Proc. Phil. Soc. Glasgow, 13*, 464–468, 1882.

Thoraval, C., P. Machetel, and A. Cazenave, Locally layered convection inferred from dynamic models of the Earth's mantle, *Nature, 375*, 777–780, 1995.

Thoraval, C. and M.A. Richards, The geoid constraint in global geodynamics: Viscosity structure, mantle heterogeneity models, and boundary conditions, *Geophys. J. Int., 131,* 1-8, 1997.

Tozer, D.C., The present thermal state of the terrestrial planets, *Phys. Earth Planet. Int., 6*, 182–197, 1972.

Tozer, D.C., Heat transfer and planetary evolution, *Geophys. Surv., 7*, 213–246, 1985.

Trompert, R. and U. Hansen, Mantle convection simulations with rheologies that generate plate-like behavior, *Nature*, in press, 1998.

Turcotte, D.L. and E.R. Oxburgh, Finite amplitude convection cells and continental drift, *J. Fluid Mech., 28* 29–42, 1967.

Turcotte, D.L. and G. Schubert, *Geodynamics*, John Wiley & Sons, New York, 1982.

Vail, P.R., R.M. Mitchum, Jr., and S. Thompson III, Global cycles of relative changes in sea level, *Amer. Assoc. Petroleum Geologists Memoir, 26*, 83-97, 1977.

van der Hilst, R., R. Engdahl, W. Spakman and G. Nolet, Tomographic imaging of subducted lithosphere below northwest Pacific island arcs, *Nature, 353* 37–43, 1991.

van der Hilst, R., S. Widiyantaro, and R. Engdahl, Evidence for deep mantle circulation from global tomography, *Nature, 386* 578–584, 1997.

Verhoogen, J., *Energetics of the Earth*, National Academy of Sciences, 1981.

Vigny, C., Y. Ricard, and C. Froidevaux, The driving mechanism of plate tectonics, *Tectonophysics, 187*, 345-360, 1991.

Watts, A.B., G.D. Karner, and M.S. Steckler, Lithosphere flexure and the evolution of sedimentary basins, *Philos. Trans. R. Soc. London Ser. A, 305*, 249–281, 1982.

Weertman, J. and J.R. Weertman, High temperature creep of rock and mantle viscosity, *Ann. Rev. Earth. Planet. Sci., 3* 293–315, 1975.

Wegener, A., *The Origin of Continents and Oceans*, Methuen & Co., Ltd., 1924.

Weinstein, S.A., Catastrophic overturn of the Earth's mantle driven by multiple phase changes and internal heating, *Geophys. Res. Lett., 20*, 101–104, 1993.

Weinstein, S.A., Thermal convection in a cylindrical annulus with a non-Newtonian outer surface, *PAGEOPH, 146*, 551–572, 1996.

Weinstein, S.A., The effect of convection planform on the toroidal-poloidal energy ratio, *Earth Planet. Sci. Lett., 155*, 87-95, 1998.

Weinstein, S.A. and U. Christensen, The effects of a stress-free upper boundary on the convection planform of a fluid with a temperature-dependent viscosity, *Geophys. Res. Lett., 18*, 2035–2038, 1991

Weinstein, S.A. and P. Olson, The proximity of hotspots to convergent and divergent plate boundaries, *Geophys. Res. Lett., 16*, 433-436, 1989.

Weinstein, S.A. and P. Olson, Planforms in thermal convection with internal heat sources at large Rayleigh and Prandtl numbers, *Geophys. Res. Lett., 18*, 239-242, 1990

Weinstein, S.A., and P. Olson, Thermal convection with non-Newtonian plates, *Geophys. J. Int., 111*, 515–530, 1992.

Wessel, P., and W.H.F. Smith, New, improved version of the Generic Mapping Tools Released, *EOS Trans. AGU, 79*, 579, 1998.

White, D.B. The planforms and onset of convection with temperature-dependent viscosity, *J. Fluid Mech., 191*, 247–286, 1988.

White, R.S. and D.P. McKenzie, Mantle plumes and flood basalts, *J. Geophys. Res., 100*, 17543–17585, 1995.

Whitehead, J.A., Convection models: Laboratory versus mantle, *Tectonophysics, 35*, 215–228, 1976.

Whitehead, J.A. and B. Parsons, Observations of convection at Rayleigh numbers up to 760,000 in a fluid with a large Prandtl number, *Geophys. Astrophys. Fluid Dyn., 19*, 201–217, 1978.

Whitehead, J.A., and R.F. Gans, A new, theoretically tractable earthquake model, *Geophys. J. R. Astron. Soc., 39*, 11–28, 1974.

Wicks, C.W., and M.A. Richards, Depression of the seismic 670 km discontinuity beneath the Izu-Bonin and Marianas subduction zones, *Science, 261*, 1424–1427, 1993.

Wilson, J.T., Did the Atlantic close and then re-open? *Nature, 211*, 676-681, 1966.

Wolfe, C.J., I.Th. Bjarnason, J.C. VanDecar and S.C. Solomon, Seismic structure of the Iceland mantle plume, *Nature, 385*, 245–247, 1997.

Zhang, S. and D.A. Yuen, Intense local toroidal motion generated by variable viscosity compressible convection in 3-D spherical shell, *Geophys. Res. Lett., 23* 3135–3138, 1995.

Zhong S., and M. Gurnis, Mantle convection with plates and mobile, faulted plate margins, *Science, 267*, 838–843, 1995a.

Zhong S., and M. Gurnis, Towards a realistic simulation of plate margins in mantle convection, *Geophys. Res. Lett., 22*, 981–984, 1995b.

Zhong S., and M. Gurnis, Interaction of weak faults and non-Newtonian rheology produces plate tectonics in a 3D model of mantle flow, *Nature, 383*, 245–247, 1996.

Zhong, S., M. Gurnis, and L. Moresi, Role of faults, nonlinear rheology, and viscosity structure in generating plates from instantaneous mantle flow models, *J. Geophys. Res., 103*, 15255–15268, 1998.

---

D. Bercovici, Department of Geology and Geophysics, SOEST, University of Hawaii, Honolulu, HI 96822. (email: dberco@soest.hawaii.edu)

Y. Ricard, Laboratoire des Sciences de la Terre, Ecole Normale Supérieure de Lyon, 46 allée d'Italie, 69364 Lyon, Cedex 07, France. (email: Yanick.Ricard@ens-lyon.fr)

M.A. Richards, Department of Geology and Geophysics, University of California, Berkeley, CA 94720. (email: markr@seismo.berkeley.edu)

# The Quest for Self-Consistent Generation of Plate Tectonics in Mantle Convection Models

Paul J. Tackley

*Department of Earth and Space Sciences, University of California, Los Angeles*

Plate tectonics and mantle convection are different aspects of the same, coupled system, yet mantle convection calculations do not exhibit plate tectonic behavior unless it is imposed by the modeler. This paper explores the quest for self-consistent incorporation of plate tectonics into mantle convection models, and presents new results and parameterizations. Simulations of convection with buoyant, deformable continents but no oceanic plates show that supercontinent cycling is a possible, though not required, aspect of the dynamics, and that the presence of even non-episodic continents has a profound effect on the spectrum of mantle heterogeneity. Ductile shear localization may be the key process for generating weak plate boundaries. Possible localization mechanisms can be described using time-dependent evolution equations, allowing evolving lithospheric strength heterogeneity with long-term memory. Steady-state versions of these rheologies are used to demonstrate the formation of weak transform boundaries in a 2-D model, and convergent, divergent and strike-slip boundaries in 3-D models. However, the present models are far from realistic; future directions and some philosophical issues are discussed.

## INTRODUCTION

Plate tectonics, with attendant continental motions, is arguably the most important manifestation of mantle convection. However, despite the sophistication and scale of present-day mantle convection models (e.g., [*Balachandar et al.*, 1996; *Bunge et al.*, 1997; *Tackley et al.*, 1994]), they do not exhibit this fundamental mode of behavior unless constrained to do so by the modeler. Plate-like behavior does not emerge from the type of viscous flow formulation which has commonly been used to model mantle convection, because it does not include the modes of lithospheric behavior responsible for strain localization, weakening and thus plate boundary formation. The quest for self-consistent plate tectonics, in which plates arise naturally from the material description rather than being imposed by the modeler, is regarded as a type of "Holy Grail" by mantle convection modelers. We are now at an exciting time, when mantle convection modelers are beginning to incorporate more realistic lithospheric rheologies into their models, and the first "self-consistent" models are emerging [*Bercovici*, 1996; *Tackley*, 1998a; *Trompert and Hansen*, 1998], although such models are still in their infancy and far from realistic. The transition from models with imposed or kinematic plates to models with self-consistent plates is somewhat analagous to (although not as fundamental as) the recent transition in geodynamo modeling between kinematic dynamo models and fully self-consistent dynamo models [*Glatzmaier and Roberts*, 1995; *Kuang and Bloxham*, 1997]. Questions about the nature and evolution of the terrestrial planets Earth, Venus and Mars, including possible transitions between plate tectonic and non-plate-tectonic regimes [*Davies*, 1995; *Moresi and Solomatov*, 1998; *Turcotte*, 1993], could then be answered in a unified and self-consistent framework rather than piecemeal changing of boundary conditions, etc.

This paper aims to assess the current state of research in the field. One component of plate tectonics is continents, so the paper begins by reviewing models which include continents but not oceanic plates, and presents some new

The History and Dynamics of Global Plate Motions
Geophysical Monograph 121

3-D models to investigate whether supercontinent cycles are an inherent characteristic of the dynamics. This is followed by an examination of the problems inherent in incorporating plate tectonics in convection models, a review of our (presently incomplete) knowledge of lithospheric rheology and mechanisms responsible for localization, a presentation of some simple parameterization and models to illustrate how this may lead to plate boundary formation, and finally a discussion of fundamental issues and problems that must be addressed in future research. In keeping with the spirit of this volume a large fraction of the discussion is review and synthesis, although the majority of the presented results are new and previously unpublished, as is much of the parameterization. This paper does not attempt to cover brittle processes in any detail, but rather focuses on ductile processes.

There has often been a tendency to regard plates and mantle convection as somehow separate and distinct entities, with their own sets of driving and dissipative forces, and usually with one 'driving' the other. However, in reality plates and convection are *one coupled system*, since to first order the plates (particularly oceanic plates [*Davies*, 1989]) are simply the cold upper thermal boundary layer of the convection, with the addition of some compositional complexity. Thus, the forces acting on/in the plates are *the same* as the forces acting on/in the mantle, the dominant driving forces being the negative buoyancy associated with subducted oceanic slabs. To put this succinctly: the plates *are* the convection, and this view provides the starting point for plate modeling in this paper.

*Different Timescales, Different Modeling Approaches*

Memory plays an important role in the evolution of plate tectonics, since 'new' plate boundaries usually form along long-lived 'pre-existing' zones of weakness, and plate boundaries, once formed, may persist for hundreds of millions of years, possibly changing from one form to another [*Scotese*, 1991; *Scotese et al.*, 1988]. The present state of the lithosphere is thus related to history spanning billions of years.

It is clear that a self-consistent treatment of plate tectonics must include such memory in the form of evolving lithospheric heterogeneity, a point which is discussed in some depth later in this paper. However, even a perfect treatment will not be useful for the purposes of performing transient (i.e., of order 100 million years or less) or instantaneous flow calculations related to a specific present-day or historical plate configuration, since the lithospheric heterogeneity will not have time to develop in such calculations. Thus, for transient or instantaneous calculations it is necessary to insert the plate boundaries and other strength heterogeneity by hand, and much has been learned by so doing, e.g., [*Ricard and Vigny*, 1989; *Zhong and Gurnis*, 1995; *Zhong and Gurnis*, 1996].

Self-consistent plate treatments are most applicable to long-term calculations, which run for billions of years (simulated time)– long enough that the character of the final result is independent of initial conditions. The goal is not to reproduce the exact configuration of plates, mantle heterogeneity, etc in the present-day Earth, but to obtain a "planet X" which resembles Earth in its statistical characteristics (e.g., spectrum of mantle heterogeneity, range of plate sizes).

Which type (timescale) of model is most valuable? The answer is that instantaneous, transient and long-term modeling approaches are complementary, giving different information, and thus all three are necessary and useful. A successful description of Earth's material properties and behavior must be able to satisfy constraints from all three types of modeling approaches. For example, a model which can produce an Earth-like planet X in a long-term simulation but does not work in modeling specific events in real Earth's history is clearly an incorrect model. Similarly, a model which successfully reproduces an event in Earth's history or the present pattern of geoid anomalies is clearly incomplete if, when run for billions of years, it produces a planet X that looks like Venus or Mars rather than Earth. These different approaches are all necessary and complementary.

*Goals and Philosophy*

The goal of self-consistent generation of plate tectonics is to identify the correct physical description which allows plates to form and evolve naturally in the convective system, with nothing being imposed by the modeler. After billions of years of evolution from an arbitrary initial condition the system should resemble, in a statistical way, Earth. The same material description should apply everywhere in the domain (except for differences due to compositional variations), with different modes of behavior coming in to play as a function of the temperature, composition, pressure, stress, and history.

This paper reviews and extends two aspects of plate tectonic modeling that are linked by the common thread of attempting to introduce more realistic rheology into mantle convection models, thus bringing them closer to the goal of self-consistency. The first aspect is continents: their treatment as compositionally- and rheologically-distinct material and the resulting behavior, while the second and main aspect is the formation of plate boundaries by strain localization and weakening. These two aspects are treated individually both because previous studies have done so, and in order to simplify the models for better understanding. A full model should clearly incorporate both aspects since they interact– for example, there is much debate about the relative roles of localized vs. distributed deformation in the deformation of continents at plate boundaries [*England and McKenzie*, 1982; *Houseman and England*, 1993; *Kong et al.*, 1997].

## CONTINENTS AND SUPERCONTINENTS

Clearly, there are two types of lithosphere: oceanic and continental, and a plate may contain one or both types. A common modeling strategy is to treat only the continental parts, examining the dynamics of the coupled mantle-continents system without attempting a full treatment of plate tectonics (e.g., [*Gurnis*, 1988; *Zhong and Gurnis*, 1993]). Implicit in ignoring oceanic plates is the assumption that they passively adapt to the behavior inherent in the coupled continent-mantle system. This assumption is undoubtedly imperfect, but it is good experimental procedure to isolate and investigate one component of the system at a time. Pragmatic concerns also mandate this approach: we do not yet know how to include plate tectonics in models, as discussed later. In any case, such modeling can be performed using presently-available approaches, and can yield some important insights.

'Continents' contain buoyant, non-subductible continental crust with an average thickness of 45 km [*Mooney et al.*, 1998], and continental lithosphere. The lithosphere is of quite heterogeneous thickness, with very thin, tectonically-active regions and old, stable cratonic regions with 'roots' that may extend 300 km into the mantle [*Jordan*, 1975]. Continental mantle lithosphere is generally thought to be more highly depleted in crustal components (Ca and Al rich minerals) than oceanic lithosphere, rendering it buoyant [*Abbott et al.*, 1997; *Durrheim and Mooney*, 1994; *Pollack*, 1986]. This probably means that it is also more depleted with water, rendering it more highly viscous by as much as 2-3 orders of magnitude [*Hirth and Kohlstedt*, 1996; *Karato et al.*, 1986]. The hypothesized continental tectosphere [*Jordan*, 1978] may consist of such buoyant, viscous material.

The general modeling procedure is to insert continents, then let them move around self-consistently for long times. The inclusion of continents into mantle convection models was pioneered by [*Gurnis*, 1988; *Gurnis and Torsvik*, 1994; *Gurnis and Zhong*, 1991; *Zhong and Gurnis*, 1993], who included continents as rigid 'rafts' in two-dimensional geometry. Their models identified some basic cyclicity in continental motions with one or two continents. A similar rigid-block approach was used to model continents in 3-D [*Trubitsyn and Rykov*, 1995]. In 2-D, this approach was extended to include a parameterization of oceanic lithosphere, with rules governing the behavior and continental breakup [*Lowman and Jarvis*, 1993; *Lowman and Jarvis*, 1995; *Lowman and Jarvis*, 1996; *Lowman and Jarvis*, 1999].

The above models treat continents as idealized rigid blocks. [*Lenardic and Kaula*, 1995; *Lenardic and Kaula*, 1996] pioneered the treatment of continents in a more self-consistent manner: as fully dynamic buoyant, compositionally-distinct material, which satisfies the same equations of motion as the rest of the domain, closer to what they are in reality. Their models included continents using lithospheric and crustal material, in a two-dimensional domain with moderately temperature-dependent viscosity– variable enough to give the lithosphere some stiffness, but not enough to allow it to form a rigid lid (for a discussion of these see later). They found that the heat flow out of continents is relatively constant, and that the surface heat flux out of the continents scaled differently with convective vigor (Ra) than the heat flux out of oceanic regions, which helps to explain the Archean heat flux paradox [*Lenardic*, 1997].

### *New 3-D Results*

Here, new results pertaining to the effect of fully dynamic continents in three-dimensional mantle convection are presented. The goal of these is to explore how buoyant, rheologically-distinct continent-like masses interact with and modify the 3-D thermal convection.

*Continents.* As with the models of [*Lenardic and Kaula*, 1996], the continents are represented as buoyant, compositionally distinct material which is self-consistently governed by the same equations as the mantle. Thus, they can deform, break up or aggregate without any additional 'rules' being necessary. The representation is different from that of [*Lenardic and Kaula*, 1996] in that we choose to make viscosity dependent on composition only, rather than including temperature-dependence, and that we treat only the continental lithosphere, not crust. These are simply idealizations made to simplify the system to its most basic level. Due to the lack of temperature-dependent viscosity, the continental material is made more viscous than the rest of the mantle. Pragmatically this is necessary so that the continents move around as somewhat contiguous units, as observed; realistically it is justified because the continental lithosphere has a higher viscosity than the mantle beneath (largely because it is colder but also because cratonic regions may be depleted in water, as discussed above), and deep continental cratonic roots have a higher viscosity than sub-oceanic mantle at the same depth (e.g., 100 km), for the same reasons. In summary, the models include high-viscosity continental blocks embedded in an otherwise constant-viscosity domain. As such, they should be regarded as a logical step towards self-consistency from 'rigid block' models, in which continents are modeled as rigid blocks in an otherwise constant-viscosity domain. However, they are not yet fully self-consistent since the continental material is inserted by the modeler rather than arising through mantle melting and differentiation, oceanic plates are not present, and rheology leading to localized deformation is not included.

*Parameters.* We assume a Boussinesq mantle with constant viscosity and other properties, except that continental material is more viscous than non-continental 'mantle' (as discussed above) by a factor of 100. Ideally a higher viscosity contrast would be preferable, but this is limited by numerical difficulties in dealing with the large viscosity jump that occurs at the edge of a continent. An 8x8x1 Cartesian domain with periodic sides boundaries is assumed, with impermeable, free-slip upper and lower boundaries. The upper boundary is isothermal whereas the lower boundary is zero flux. The (temperature-based) Rayleigh number is taken to be $10^6$, and the non-dimensional internal heating $R$ is taken to be 20 [*Tackley*, 1996a].

The buoyancy of continental material is constrained by the requirement that continents remain a 'reasonable' thickness. If the buoyancy is too high the material eventually spreads out into a layer covering the entire domain, and if the buoyancy is too low the material accumulates into a big 'blob' extending ~1000 km deep into the mantle. The appropriate chemical buoyancy is that which balances the negative thermal buoyancy due the coldness of the continental material, which physically corresponds to Jordan's isopyctic hypothesis for the continental roots [*Jordan*, 1988; *Lenardic and Moresi*, 1999]. This may not be physically necessary in more 'realistic' calculations because strong temperature-dependent viscosity would give the necessary stiffness to the continents to stop them spreading or pooling. The ratio of chemical to thermal buoyancy is expressed as the buoyancy parameter B:

$$B = \frac{\Delta\rho_{c-m}}{\rho\alpha\Delta T} \tag{1}$$

where $\Delta\rho_{c\text{-}m}$ is the density difference between 'continent' and 'mantle', $\rho$ is the mantle density, $\alpha$ is the thermal expansivity, and $\Delta T$ is the temperature difference across the mantle. In these cases $B$ is equal to 2.0.

*Numerical Method*. The equations are solved using the finite-volume multigrid code STAG3D, detailed elsewhere [*Tackley*, 1994; *Tackley*, 1996a]. Continental material is tracked using clouds of tracer particles, giving zero diffusion. This method works well provided there are at least 10s of tracers per numerical cell [*Christensen and Hofmann*, 1994; *Tackley*, 1998b; *Van Keken et al.*, 1997]. The main numerical problem, which would also apply to a field approach, is entrainment of continents by downwellings at their edges. While some entrainment may be physically realistic, the rate of entrainment is found to be resolution dependent and therefore numerically enhanced. This entrainment limits the practical length of calculations to the time for which entrainment is not large, but this time appears to be long enough for interesting dynamics to occur.

*Initial conditions*. The initial temperature field is taken from a homogeneous-composition case (i.e., with no continental material) which has reached statistically steady-state. The continents are then inserted and the subsequent development followed. Continents initially have a constant thickness of 300 km and cover 1/3 of the surface area, similar to that on Earth. Two initial configurations are used: one square supercontinent or a 4x4 array of smaller continents.

*Results*. A number of cases have been run, which will be fully explored in a future paper. Here, two cases are chosen, which are intended to be illustrative rather than definitive. Both are entirely heated from within, differing only in the initial continental configuration.

The case with 16 initial continents is shown in Plate 1 at four different times. The initial convective pattern is typical of internally-heated Boussinesq convection, being dominated by closely-spaced, time-dependent linear downwellings [*Parmentier et al.*, 1994; *Tackley*, 1996b]. When continents are inserted the continental material and also the mantle immediately under it rapidly become cold due to thermal diffusion. However, deeper underneath the continent the mantle gradually heats up because it is not being cooled so effectively by downwellings. Downwellings preferentially occur at the edge of continents. In this case, continents show no tendency to aggregate into a larger unit. Instead, they jiggle around and eventually start breaking up into smaller units due to the small-scale convection associated with internally-heated convection. Continents clearly have a large effect on the thermal structure, particularly at shallow depth. The run was continued until entrainment of continental material became significant.

The case with a single initial supercontinent is a strong contrast to this (Plate 2). Mantle warming under the supercontinent causes long-wavelength stresses which break the continent apart (which apparently happens more rapidly in internally-heated cases than with basally-heated cases [*Lowman and Jarvis*, 1999]), spreading its fragments to the 'corners' of the box. Since the box is periodic, this corresponds to a supercontinent at the opposite side of the planet. The new supercontinent then stays there for a while before splitting up again and reforming at the original position. This strong oscillation persists for several cycles (the length of the simulation) and has a periodicity of ~1 Gyr, meaning a new supercontinent aggregation every 500 Myr (based on scaling a nondimensional velocity of 500, which is about twice the rms. surface velocity and half the peak surface velocity, to 5cm/yr), perhaps similar to the timescale observed for Earth's continents [*Dalziel*, 1997]. It is clear from the rendering of temperature field that this cycle is inducing very long-wavelength heterogeneity into the mantle. There is also much small-scale structure which has the effect of breaking the continental material into small pieces, somewhat unsatisfactorily in terms of modeling Earth-like continents. It is probable that if realistic oceanic plates and lithospheric viscosities were included then these small-scale instabilities would be

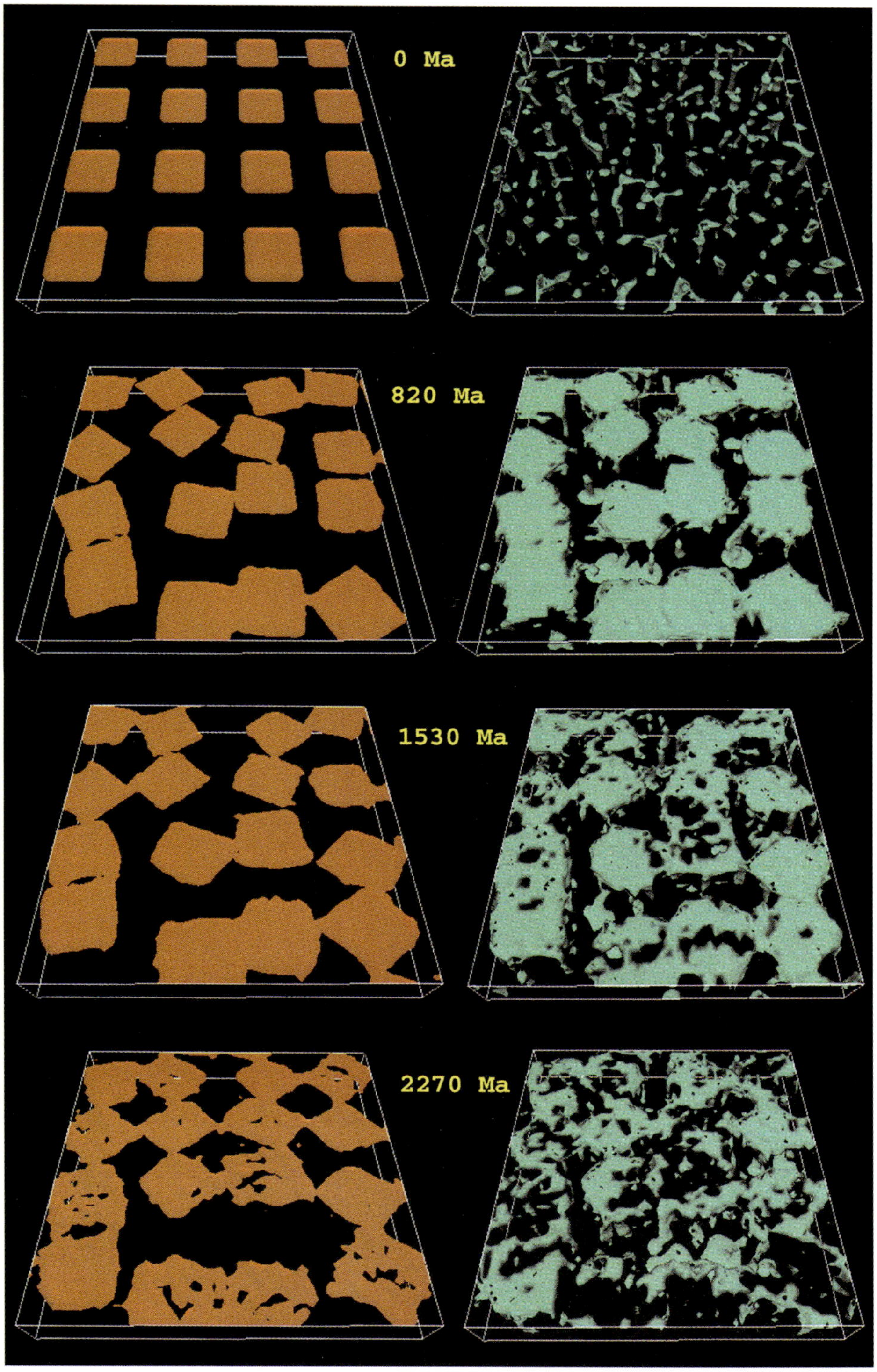


**Plate 1**. Mantle convection simulation with 16 initial continents at four times during the simulation, corresponding roughly to 0, 0.8, 1.5 and 2.3 billion years. The left plots show the continental material (isocontour $C$=0.5), and the right plots show downwellings, contouring where the temperature is 0.1 colder than the horizontal average (i.e., isocontour $T_{resid}$=-0.1)

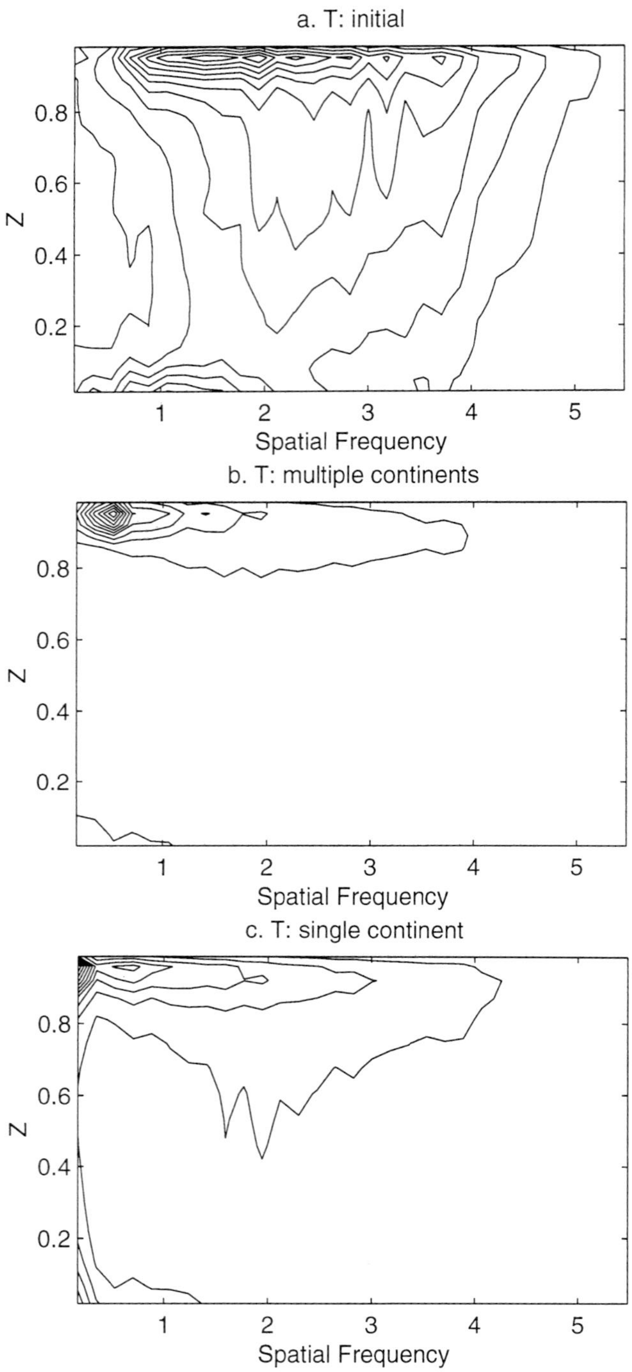


**Figure 1**. Spectral heterogeneity maps (i.e., contour plots of the spatial frequency amplitude spectrum as a function of vertical coordinate) for the continental simulations. Top: initial condition for both cases (single snapshot); Center: Time-averaged SHM for the case with multiple initial continents; Bottom: Time-averaged SHM for the case with a single initial continent. $Z$ is the vertical coordinate ranging from 0 at the bottom of the box to 1 at the top of the box. The spatial frequency is defined such at 1.0 means a horizontal wavelength of 1.0, i.e., equal to the depth of the box. Contours are scaled to the maximum amplitude of each case, in 10 equal steps.

suppressed. Scaling arguments indicate that the viscosity of continental roots must be at least 1000 times higher than regular upper mantle for their long-term survival [*Lenardic and Moresi*, 1999].

For both cases, the effect on the lengthscales of heterogeneity is profound, as evident from the spectral heterogeneity maps (SHM) illustrated in Figure 1. These are contour plots of the rms. amplitude spectrum of the temperature field as a function of vertical coordinate, $z$ [*Ratcliff et al.*, 1997; *Tackley et al.*, 1994], and are shown for the initial condition (top plot) and for each case. For each case the spectrum is time-averaged (i.e., the spectrum is calculated for 20-30 evenly-spaced frames during the calculation then averaged); individual snapshots may look more extreme. The isochemical initial condition (top frame) shows a fairly broad spectral peak centered at a spatial frequency of ~1.5, corresponding to a wavelength of 2/3 the depth of the box, roughly 2000 km. With multiple continents, the peak is sharper and shifted to roughly 3 times longer wavelength (frequency=0.5, corresponding to a wavelength of 2.0, ≈6000 km), equal to the average spacing of the continents. With a single initial continent the peak is at even longer wavelength, the longest that can be accomodated in this box, a wavelength of 8 (~24000 km). The peak amplitude is much larger with continents present, although this is not visible on the SHM.

Some features in these results are consistent with previous results in 2-D: these include the tendency of continents and mantle immediately below to become cold and for the sub-continental mantle to warm up [*Gurnis and Zhong*, 1991], for small continents to have less dynamical effect than large continents [*Zhong and Gurnis*, 1993], and the cyclicity that can be induced by supercontinent aggregation and dispersal [*Gurnis*, 1988; *Lowman and Jarvis*, 1993]. However, an important new finding is that initial condition does seem to play a role in the dynamics: if the continental material starts off as many small continents it displays no tendency to aggregate and a fairly static pattern is obtained, whereas if the material is somehow collected into a large supercontinent, strong episodicity can result and apparently persists. Basally-heated results (not presented here) display similar trends except that the supercontinent cycle dies out fairly quickly with small continental fragments settling above downwellings. Thus, it appears that supercontinent cycling is not a robust outcome of the dynamics; rather the system must be set up in a special way to induce it.

It is possible that the components of plate tectonics missing from these models, i.e., the oceanic plates, will strongly modulate the dynamics and change this conclusion. The motion of oceanic plates causes large-scale coherent motion and time-dependence which is absent in these results and may act to aggregate the continental material where it would otherwise just sit in the same place. Other factors that must be investigated in future include the effects of depth-dependent viscosity, and phase changes.

## SELF-CONSISTENT GENERATION OF PLATE BOUNDARIES

Understanding the generation of weak plate boundaries is the key to understanding plate tectonics. From field and laboratory measurements we have a good idea of the lithospheric and crustal weakening and localization mechanisms which may be important at plate boundaries, however, we do not have the firm quantitative knowledge necessary to write down the relevant equations with certainty. Thus, basic research must focus on investigating the behavior and role of various mechanisms; it is not simply a question of writing a computer code to solve already known equations.

What are the criteria for judging the success of a model in generating plate tectonics? Perhaps the most fundamental is 'plateness', i.e., the ability to generate rigid plates surrounded by weak boundary regions, with most of the deformation taking place in the boundary regions [*Weinstein and Olson*, 1992], although these boundary regions may cover as much as 15% of Earth's surface area [*Gordon and Stein*, 1992]. The plate boundaries should come in three distinct types: spreading centers, single-sided subduction zones, and transform faults. Finally, the time evolution of the plates should bear a resemblance to that of Earth's plates [*Scotese*, 1991], including, for example, sudden changes in plate motion and jumps in plate boundaries [*Bonatti et al.*, 1994].

The following sections examine the basic reasons why plate generation does not arise in mantle convection models, summarizes the mechanisms which may be important at plate boundaries, presents some parameterizations and modeling results, and discusses directions for future work.

### *The Plate Problem*

Mantle convection calculations with rheologies which are appropriate for the bulk of the mantle do not develop plate tectonics. Indeed, while it is clear that temperature-dependent viscosity is important for producing strong plates, this on its own arguably makes the system less Earth-like! Why is this? Plate 3 shows some simple calculations of mantle convection with increasing temperature-dependent viscosity contrast, taken from [*Ratcliff et al.*, 1997]. The system is Boussinesq and heated from below, with temperature-dependent viscosity being the only complexity. With constant viscosity (Plate 3a), upwellings and downwellings are symmetric and linear. With an intermediate viscosity contrast (100-1000) (Plate 3b), the system transitions to a 'sluggish lid' mode of convection, with huge cylindrical downwelling plumes and linear upwellings with plume instabilities [*Ratcliff et al.*, 1995; *Tackley*, 1993; *Trompert and Hansen*, 1996]). As viscosity contrast is increased further, the cold upper boundary layer becomes so stiff that it stops moving altogether and a rigid lid is formed [*Moresi and Solomatov*, 1995; *Ogawa et al.*, 1991; *Ratcliff et al.*, 1996]. This rigid lid is like a single plate covering the entire planet, which may be appropriate for Mars or Venus, but is not the Earth. A short-wavelength pattern of upwelling plumes and sheet-like downwellings transports heat to the base of the rigid lid, which it is conducted through. Non-Newtonian power-law rheologies display a similar trend, although the transitions occur at higher viscosity contrast [*Solomatov*, 1995]. So, temperature-dependent Newtonian or power-law rheologies which are appropriate for the interior of the mantle result in the formation of a rigid lid, not plates.

Despite this, some two-dimensional convection calculations have shown that strain-rate softening, non-Newtonian rheologies can cause weak zones (WZs) and strain rate localization above up- and down-wellings [*Christensen*, 1984; *Cserepes*, 1982], resulting in a rudimentary approximation of plates [*Weinstein*, 1996; *Weinstein and Olson*, 1992]. Why is this? In 2-D there are only convergent and divergent margins, which have concentrated buoyancy/stress sources below them to 'break' the rigid lid and focus deformation. Three-dimensional geometry (3-D) is fundamentally more challenging due to the transform (strike-slip) boundaries. There are no concentrated local buoyancy forces available to drive and localize transform boundaries; indeed, buoyancy in the mantle cannot directly drive them at all if the viscosity is purely depth-dependent [*Ribe*, 1992], as further discussed later. Thus, it is not surprising that the rheologies that produce reasonable 'plates' in 2-D do not work well in 3-D [*Christensen and Harder*, 1991; *Tackley*, 1998a; *Trompert and Hansen*, 1998; *Weinstein*, 1998]. Another point to note is that the experimentally-determined activation energies for Newtonian diffusion creep and powerlaw dislocation creep indicate that the Earth would certainly be in the rigid lid mode, if these were the only processes occuring [*Solomatov*, 1995].

Once plate boundaries do form by some mechanism, they play a dominant role in determining the form of the convection. Plate 3d shows a 'rigid lid' calculation where weak plate margins have been inserted at each end of the box, but otherwise the parameters are the same as in Plate 3c. As a result, the system behaves like an oceanic plate, moving from the left of the box to the right of the box, conductively thickening as it goes. Small-scale 'Richter-rolls' [*Richter*, 1973] are observed under the moving plate. Various researchers have investigated the influence of inserted plates on mantle convection, and much has been learned, for example [*Bunge and Richards*, 1996; *Zhong and Gurnis*, 1996]. However, these studies do not attempt to answer more fundamental questions regarding how and why plate boundaries form, what determines the sizes of plates, etc.

### *Toroidal and Poloidal Flow*

The three-dimensional mantle flow velocity field can be divided into poloidal and toroidal components. Poloidal

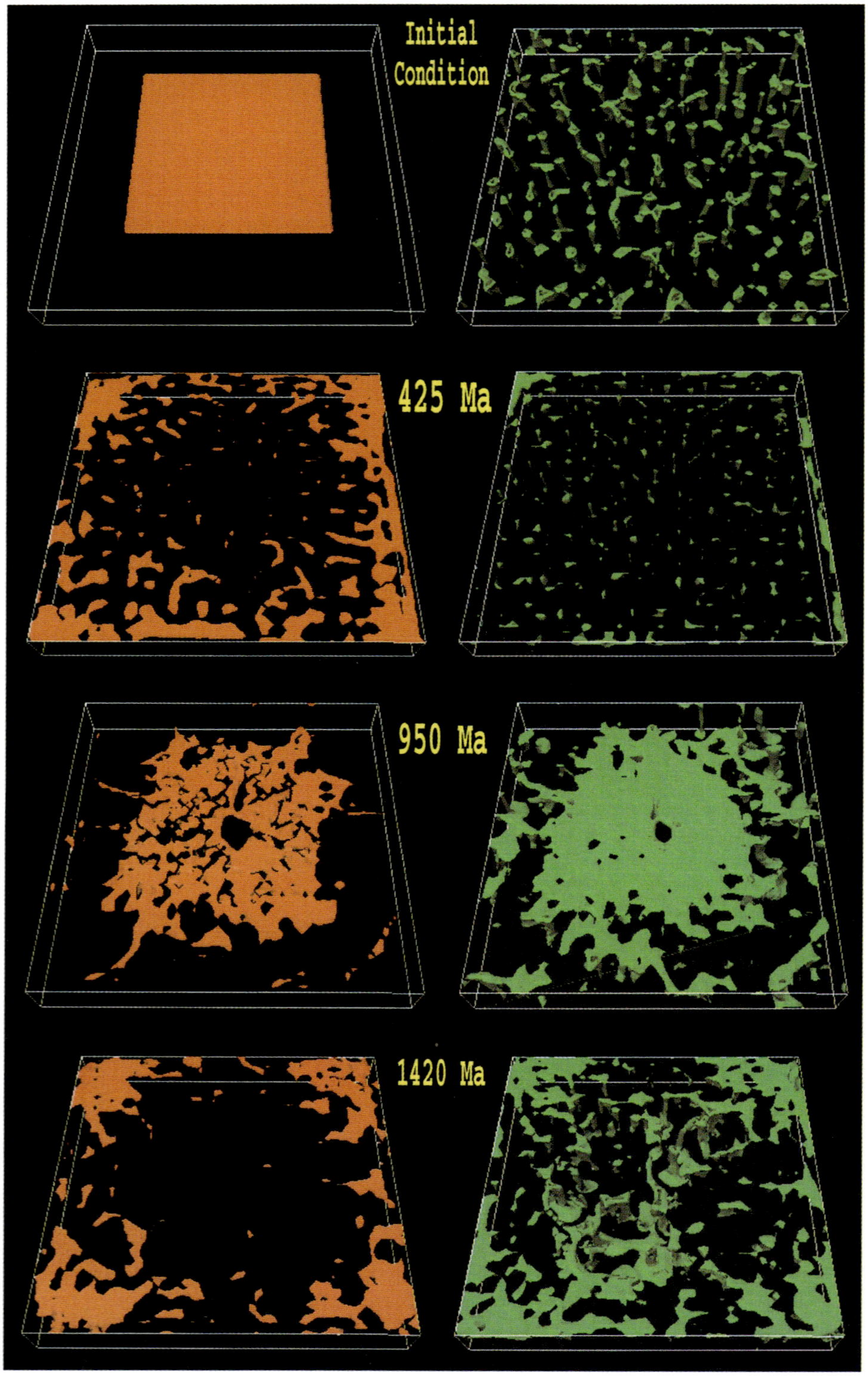


**Plate 2**. Mantle convection simulation with a single initial supercontinent, at four times during the simulation, corresponding to 0, 425, 950, and 1420 million years. The left plots show the continental material (isocontour $C$=0.9), and the right plots show downwellings, contouring where the temperature is 0.1 colder than the horizontal average (i.e., isocontour $T_{resid}$=-0.1)

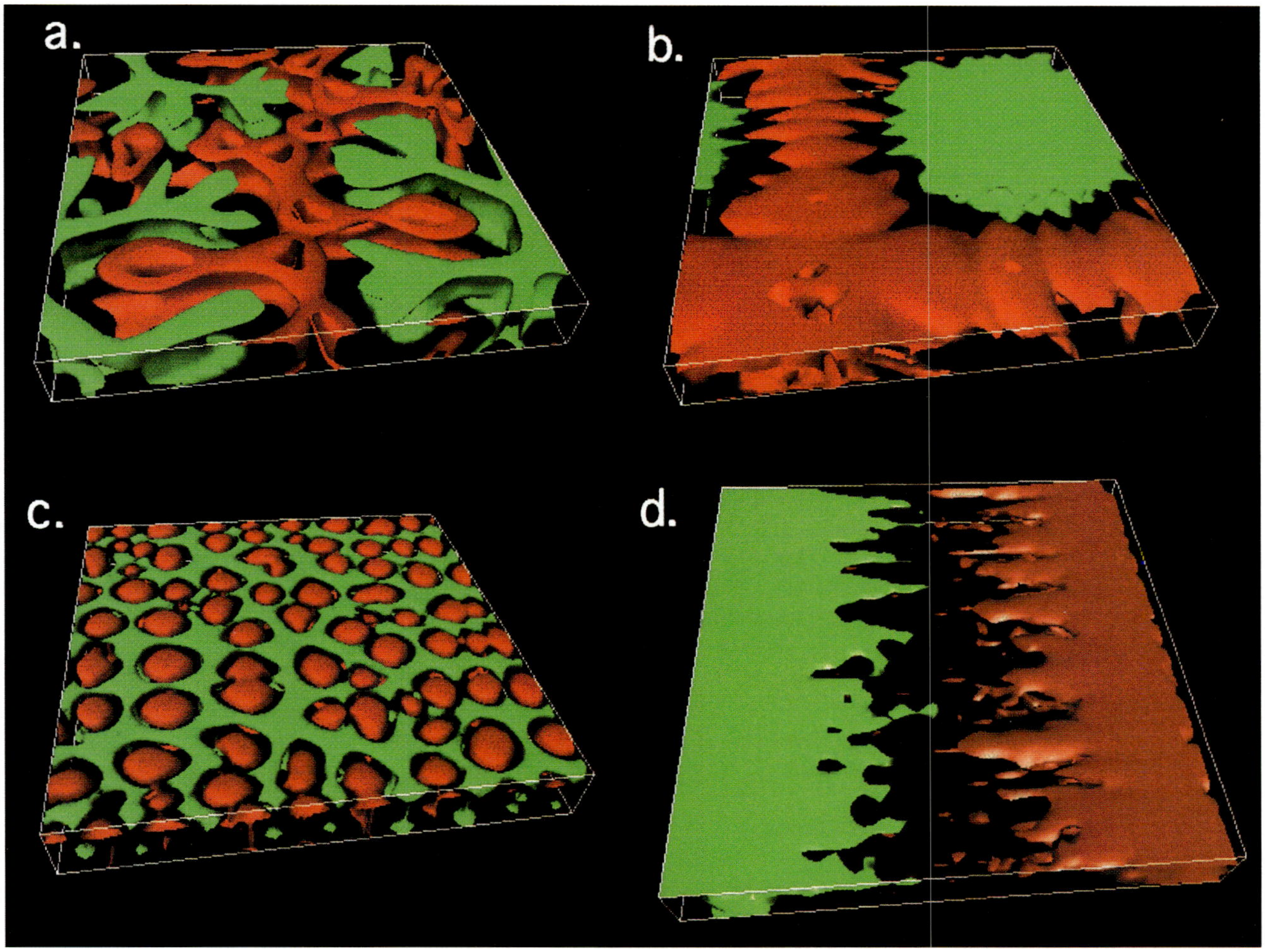


**Plate 3**. The three regimes of convection with temperature-dependent rheology, plus the effect of adding weak plate boundaries. Plotted are isocontours of residual temperature $T_{resid}$, with hot contours in red and cold contours in green. a. constant viscosity, $T_{resid}=\pm 0.1$, b. viscosity contrast of 1000, $T_{resid}=0.1$, c. viscosity contrast of $10^5$, $T_{resid}=0.05$, d. viscosity contrast of $10^5$ plus low-viscosity zones imposed at each end of the box, $T_{resid}=0.1$.

flow is associated with divergence or convergence in a horizontal plane, i.e., spreading centers and subduction zones, while toroidal flow is associated with transform (strike-slip) boundaries. Whereas poloidal flow is driven directly by buoyancy sources in the mantle, toroidal flow is coupled to buoyancy forces only by lateral variations in viscosity, or inhomogenous boundary conditions.

Curiously, the velocity field of Earth's plates has a toroidal component which is a substantial fraction of the poloidal flow [*Hager and O'Connell*, 1978]. Although the term "equipartitioning" is often used, the toroidal component is actually somewhat smaller than the poloidal component, with the amplitude ratio varying between about 0.25 and 0.5 (excluding net rotation) in the last 120 million years [*Lithgow-Bertelloni et al.*, 1993]. Where does the toroidal energy come from? One possibility is that it is generated inside the mantle by laterally varying viscosity; however, calculations have only been able to obtain a small toroidal:poloidal energy ratio this way, of order a few percent [*Balachandar et al.*, 1995; *Christensen and Harder*, 1991]. Another possibility is that it is the coupling of mantle buoyancy sources with the plates themselves which generates toroidal flow. Flow calculations with idealized plate geometries [*Gable et al.*, 1991] indicate that this is indeed the case, and show that the induced toroidal energy is maximum at the surface and decays with depth into the mantle, particularly rapidly if the lower mantle has a higher viscosity than the upper mantle. Thus, a high toroidal:poloidal ratio at the surface does not imply a high ratio deep in the mantle. These flow calculations as well as simple models [*Olson and Bercovici*, 1991] show that the toroidal:poloidal ratio is dependent on the ratio of transform boundary length to convergent/divergent boundary length.

Plate motions seem to organize themselves so as to roughly minimize the toroidal motion for a given plate geometry [*Lithgow-Bertelloni et al.*, 1993]. Thus, the significant fraction of toroidal motion is probably due to the geometries of present-day plates, rather than anything deep in the mantle. Then, of course, the question becomes: why do plates "choose" a geometry which requires a large amount of transform boundary? Toroidal motion does not transport energy, and since energy is dissipated in driving transform boundaries it is not clear why the system would arrange itself to drive them, rather than forming, for example, a system of rolls with purely poloidal motion. One idea is that the system is forced to have transform boundaries because of the sphericity: you can't fit a system of rolls onto a sphere, as you could in a box. Another idea is that continents disrupt the symmetry of the system, causing a more messy and imperfect system of plate boundaries. A more fundamental reason comes from considering dissipation of energy: it can be shown that, contrary to the simple reasoning above, transform boundaries may actually decrease the energy dissipation in the system, and therefore be beneficial [*Bercovici*, 1995a].

It should be noted that while toroidal energy is a strong feature of today's plate motions, it is not *necessary* for plate-like behavior. The most important characteristic of plates is minimal deformation in their interiors with strongly-deforming boundaries, i.e., 'plateness' as mentioned earlier. Plates with only convergent and divergent boundaries can have good plateness. The toroidal:poloidal energy reflects the ratio of strike-slip boundaries to convergent or divergent boundaries, and changes with time [*Lithgow-Bertelloni et al.*, 1993]. Thus, plateness rather than toroidal:poloidal ratio is the best indicator of plate-like behavior.

It should also be noted that, on a large scale, observed plate boundaries tend not to be purely convergent, divergent, or strike-slip, as simple models often assume, but generally contain a mixture of divergence (or convergence) and strike-slip motion. This is clear from the plate divergence and vorticity fields plotted in [*Bercovici and Wessel*, 1994]; for example the only major pure strike-slip boundary is the Pacific:North American boundary in the vicinity of the San Andreas fault. At a local scale, of course, the motion on these mixed boundaries is often accomodated by alternating segments of 'pure' motion, as at mid-ocean spreading centers, with their purely divergent ridge segments and purely strike-slip transform offsets.

### *Lithospheric Localization Mechanisms*

Plate boundaries are zones of lithospheric weakness along which concentrated deformation takes place. The key processes, which are intrinsically linked, are *shear localization* and *weakening*. A number of weakening mechanisms are observed in nature and have been proposed as being important in the formation of plate boundaries. Some of these are applicable to the brittle (seismogenic) zone, while others are applicable to the ductile deformation zone.

A traditional explanation for plate boundaries is that the lithosphere 'breaks' by brittle failure all the way through. However, few material scientists now believe that brittle faults go all the way through the lithosphere. Between the brittly-deforming upper lithosphere (perhaps 10-20 km thick) and the viscously deforming lower lithosphere, lies a zone where ductile or semi-brittle deformation is dominant. Figure 2, taken from a review paper by [*Kohlstedt et al.*, 1995], summarizes the likely strength profile in the oceanic lithosphere. In the brittle regime, lithospheric strength is determined by the frictional stress of faults, proportional to normal stress (hence depth), whereas in the viscous regime, power-law viscous (dislocation) creep prevails. The oceanic lithosphere between about 10 km and 40 km depth is in the regime of ductile or semibrittle deformation, not brittle failure. This is the strongest part of the lithosphere; thus, ductile processes determine the strength of the oceanic lithosphere. To summarize, the emerging picture of plate boundaries is one of brittle failure in the upper

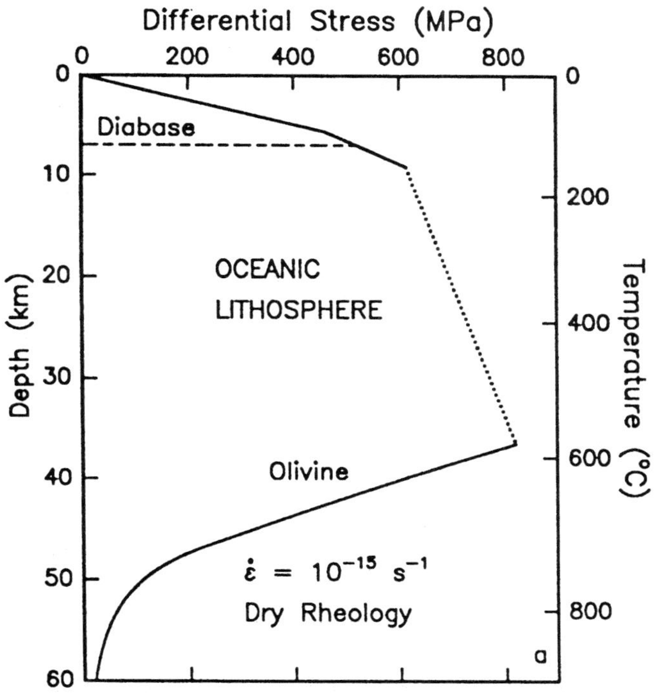


**Figure 2**. Strength envelope for 60 Ma old oceanic lithosphere (from [*Kohlstedt et al.*, 1995], Figure 9). The solid line in the top part of the lithosphere represents frictional sliding on brittle faults, the solid lines in the bottom part of the figure represent powerlaw viscous creep, and the dotted line connecting these represents ductile or semibrittle flow. The transition from brittle to ductile/semibrittle is known as the brittle-ductile transition (BDT) and the transition from that to viscous creep is known as the brittle-plastic transition (BPT).

~10-12 km, with ductile shear zones in the middle lithosphere and ductile power-law flow in the lower lithosphere and upper mantle.

While the strongest part of the lithosphere may be deforming ductily, brittle faults may play a role in seeding plate boundaries if brittle failure occurs first (because it occurs at small strains), guiding the locations for ductile shear localization [*Gilbert et al.*, 1994]. Where oceanic lithosphere bends as it goes down a subduction zone, large earthquakes nucleating in the brittle region are observed to propagate deep into the ductile region [*Kanamori*, 1970; *Kikuchi and Kanamori*, 1995]. However, in some cases, ductile shear localization is observed to precede brittle failure [*Jin et al.*, 1998]. It may be that, in general, ductile lithospheric localization is responsible for driving localization in the shallower, brittle layers. [*Batt and Braun*, 1997] showed that such a model is successful in explaining crustal observations from continental compressional orogens in terms of deep single-sided lithospheric subduction.

Localization mechanisms which may be important in brittle and ductile regimes are now discussed.

*Brittle Localization*

While this paper focuses on ductile localization, it is useful to make some points regarding elasticity and brittle deformation, for completeness and clarification. Firstly, brittle failure occurs at low strains, and secondly, deformation of an elastic material with brittle or plastic failure naturally leads to localization of shear in narrow bands [*Poliakov and Herrmann*, 1994], without any strain weakening being necessary (as it is when deforming a viscous material).

Based on laboratory experiments, the friction on faults is history-dependent, becoming weaker with increasing strain, and stronger with increasing time [*Dieterich*, 1979]. This leads to "velocity-weakening". Thus, presently active faults are much weaker than faults that have been inactive for a long time. This process may be parameterized using an evolving state variable for the memory, leading to rate and state friction laws such as the Ruina-Dieterich and Dieterich-Ruina laws [*Ruina*, 1983]. In these laws, fault strength initially recovers rapidly once slip stops, followed by a more gradual recovery as time progresses. Such recovery may have an analog in ductile weakening mechanisms discussed shortly.

Another, continuum approach to the brittle zone comes from damage theory, in which a state variable representing 'damage' (nominally associated with the opening of microcracks into voids) evolves with strain and causes a reduction in the elastic moduli [*Lyakhovsky et al.*, 1993]. This approach leads to localization in crustal brittle zone and the development of fault networks [*Lyakhovsky et al.*, 1994]. Later another type of 'damage' treatment will be applied to the ductile regime.

*Ductile Localization Mechanisms*

Ductile shear zones in the crust and upper mantle are observed in the field [*Drury et al.*, 1991; *Jin et al.*, 1998; *Pili et al.*, 1997; *Vissers et al.*, 1995], so it is possible to examine the microstructures to constrain what mechanisms were responsible for the weakening and localization. Upper mantle shear zones are commonly observed in large massifs in orogenic belts [*Drury et al.*, 1991; *Vissers et al.*, 1995], a particularly notable, 170-km long example of which is observed in west Greenland [*Sorensen*, 1983]. When formed at moderate temperatures (i.e., less than about 950°C) and high stresses, these shear zones are narrow (from cm to kilometers width) and contain mylonites, a type of rock characterized by strong grain-size reduction and (commonly) hydration [*Drury et al.*, 1991; *Pili et al.*, 1997]. Shear zones are often characterized by very small grains, hydration, and elevated temperatures. Thus, the possible mechanisms for their formation are thought to be grain-size reduction, volatile ingestion, and viscous heating. These are now examined.

*Viscous heating*. This was the first proposed method for shear localization. Viscous dissipation increases temperature, which reduces the viscosity, a positive feedback mechanism leading to higher strain rate in the weakened ration, thus more dissipation and even higher temperatures. This heating is balanced by thermal conduction of heat away from the shear zone. Two-dimensional calculations have been used to investigate this [*Fleitout and Froidevaux*, 1980; *Yuen et al.*, 1978] and show that for constant-stress boundary conditions, thermal runaway can occur and the solution diverges to infinite velocities. For more realistic constant velocity boundary conditions, however, the system reaches an equilibrium shear-zone thickness in which conduction is balanced by heat production and the viscosity is about $10^{21}$, much lower than typical lithospheric viscosities. There is evidence for this mechanism from observations [*Jin et al.*, 1998; *Obata and Karato*, 1995], and from heat flow measurements along fault zones [*Thatcher and England*, 1998].

*Grain size reduction*. The most cited mechanism for forming shear zones, this is commonly observed in the field [*Drury et al.*, 1991; *Jaroslow et al.*, 1996; *Jin et al.*, 1998; *Vissers et al.*, 1995]. Grain size reduction occurs when rock is deforming by dislocation creep, in which new, very small grains form at the boundaries between existing grains, and creep is non-Newtonian but independent of grain size. If the average grain size is reduced sufficiently, a transition may occur to diffusion creep (by diffusion of vacancies through the lattice), in which viscosity is a strong function of grain size [*Ranalli*, 1995], and thus viscosity may be reduced by several orders of magnitude, allowing localization. Competing with grain-size reduction is grain growth by annealing, which is highly temperature-dependent. These processes are well quantified through laboratory experiments [*Karato*, 1989; *Karato et al.*, 1980; *Karato et al.*, 1986], in terms of the rate of grain growth, size of recrystallized grains, and domain diagrams. However, the exact evolutionary path of the system is uncertain: dynamic recrystallization does not occur in the diffusion creep regime, so once this region is reached, subsequent evolution must involve both diffusion and dislocation creep. More research is required as to how localization occurs through this mechanism [*Jin et al.*, 1998; *Kameyama et al.*, 1997].

*Void-Volatile Weakening*. This mechanism was recently proposed by [*Bercovici*, 1998], based on field observations and ductile void growth theory [*Regenauer-Lieb*, 1999]. The basic idea is that voids produced in the deforming rock can then ingest fluid which reduces its viscosity.

It should be noted that these mechanisms are not mutually exclusive; indeed, all three are probably operating to some extent. The important question is which mechanism is dominant, the answer to which could depend on depth (perhaps void-volatile weakening is important at shallow depth, with grain-size reduction dominant at greater depth) and/or tectonic environment. Some researchers favor viscous dissipation as the dominant mechanism at all depths [*Ranalli*, 1995]. This is an area where coupled numerical modeling, field observations and laboratory experiments are necessary.

*Evolution Equations*

Each weakening mechanism can be described by a variable (e.g., average grain size, temperature, or porosity) which evolves according to some equation, the prototypical form of which is:

$$\frac{dD}{dt} = S(\sigma, \dot{e}) - H(D,T) - \kappa.\nabla^2 D \qquad (2)$$

where $S$ is the source term, $H$ is the healing (annealing) term and the final term is diffusive. '$D$' represents 'damage', which could be either grain size (or inverse grain size if higher $D$ is taken to indicate greater weakening), temperature or porosity.

To complete the parameterization, increasing 'damage' reduces the effective viscosity $\eta_{eff}$, as:

$$\eta_{eff} = \eta(D,T,p,\dot{e}) \qquad (3)$$

where $p$ is pressure and $\dot{e}$ is strain rate. These equations are coupled since viscosity determines the relationship between strain rate and stress, which both appear in the source term of (2). Note the distinction between this damage, which applies in the viscous creep regime, and the damage discussed by [*Lyakhovsky et al.*, 1993], which affects elastic moduli.

What form should these terms take? This depends on the mechanism under consideration. The following table summarizes the physical meaning of damage ($D$), possible forms (proportionalities) of source ($S$), healing ($H$), diffusive and viscosity terms for the three mechanisms, taken from [*Govers and Wortel*, 1995; *Kameyama et al.*, 1997; *Karato et al.*, 1980] for grain-size reduction and [*Bercovici*, 1998] for void-volatile weakening.

Weakening Mechanisms

| *Mech* | *D* | *S* | *H* | $\kappa$ | $\eta_{eff}$ |
|---|---|---|---|---|---|
| VH | T | $\sigma\dot{e}$ | 0 | √ | exp(E/T) |
| GSR | 1/d | $\dot{e}D(1 - Da_r)$ | $D^3$.exp(-E/T) | ~0 | $D^{-2}$ or $D^{-3}$ |
| V-V | Φ | $\sigma^2/(1 - D)$ | D.exp(-E/T)? | ~0 | (1-λD) |

$E$ is activation energy, $d$ is grain size, $\Phi$ is porosity and $a_r$ is the recrystallized grain size, which is related to stress. The $\sigma\dot{e}$ form of source term is appealing because this is the work being done on the system. A healing term proportional to $D$ leads to exponentially-decaying damage, if there is no damage production.

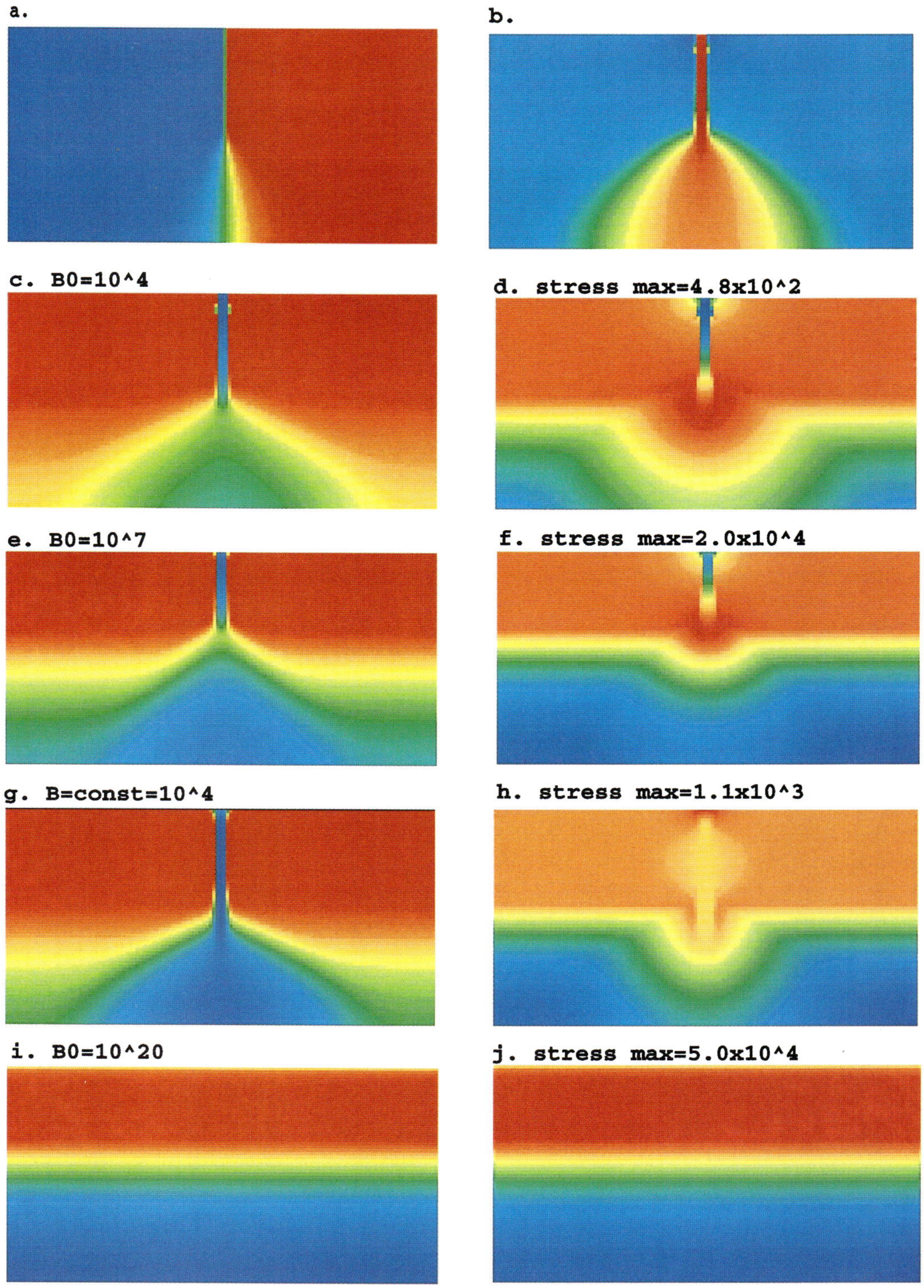


**Plate 4**. Calculations of lithospheric shear localization with a composite, SRW rheology, in a 2-D $(x,z)$ plane with motion in the y-direction, i.e., in and out of the plane. a. y-velocity, b. strain rate, c. effective viscosity, and d. stress, for the reference case with temperature-dependent healing and $B_0$=$10^4$. e. $\eta_{\text{eff}}$ and f. stress for a case with temperature-dependent healing and $B_0$=$10^7$. g. $\eta_{\text{eff}}$ and h. stress for a case with a constant healing coefficient $B$=$10^4$. i. $\eta_{\text{eff}}$ and j. stress for a case with effectively no strain-rate weakening (obtained by setting $B$=$10^{20}$).

The dependence of healing rate on temperature for GSR and VV mechanisms allows the cold upper lithosphere to retain weakness for billions of years while the warm mantle adjusts strength instantaneously. For grain-size reduction, for example, the grain growth rate has similar temperature-dependence (activation energy) as diffusion creep rate ( $\propto 1/\eta$ ) [*Karato*, 1989].

The viscosity term listed for GSR and VV mechanisms should be multiplied by the 'undamaged' viscosity to get the full viscosity law, for example, for VV weakening [*Bercovici*, 1998],

$$\eta = (1-\lambda D).\eta_u(T,p,\dot{e}), \tag{4}$$

where $\eta_u$ is the viscosity of undamaged material and $\lambda$ is the rate of change of viscosity with damage, $\leq 1$ in the analysis of [*Bercovici*, 1998].

There are several possible meanings of 'weakening': viscosity reduction, plastic yield stress reduction, or elastic modulus reduction. The above equations assume that viscous creep is the dominant mechanism. If the mechanism is instead plastic yielding, then weakening may occur in the form of reduction of the plastic yield stress, rather than reduction of viscosity. This possibility is implicitly included in the parameterization if $\eta_u$ is already an effective viscosity due to plastic yielding.

*Strain Weakening to Strain-Rate Weakening*

Clearly, the above descriptions satisfy the requirement for having memory. However, including all three mechanisms in their time-dependent form would lead to a rather complicated and unwieldy model, so for initial investigations some simplification is appropriate. One simplification is to remove the time-dependence by deriving a 'steady-state' version of the weakening, i.e., deriving the stress vs. strain rate relationship for constant strain rate. This can be obtained by setting $dD/dt$=0 and solving the coupled evolution and viscosity equations for stress as a function of strain rate and other parameters. Examples of this for thermal weakening and void-volatile weakening have been derived in [*Bercovici*, 1996; *Bercovici*, 1998], and show a very characteristic form: stress increases with strain rate up to some critical point, beyond which stress *decreases* with increasing strain rate, a behavior which we may call strain rate weakening (SRW), and has something of an analog in the brittle zone with velocity weakening on faults. To summarize, when strain weakening is combined with time-dependent healing, for the viscous dissipation or void-volatile mechanisms, the result is strain rate weakening. Curiously, SRW does not arise when this steady-state analysis is applied to a simple parameterization of grain-size reduction in the absence of any other mechanism [*L.G.J. Montesi, personal communication*, 1998]: some additional effect such as slow cooling [*Braun et al.*, 1999] or viscous dissipation [*Kameyama et al.*, 1997] must also be present. An example of SRW is now derived (most applicable to the void-volatile mechanism- for other mechanisms see the above-referenced papers), followed by a discussion of timescales and its applicability to model lithospheric localization.

To show how strain rate weakening arises, consider the simple scalar damage evolution equation:

$$\frac{dD}{dt} = a\sigma\dot{e} - bD, \tag{5}$$

where

$$\sigma = \eta_{eff}\dot{e} = (1-D)\eta_u(T,p)\dot{e} \tag{6}$$

$\eta_u$, the viscosity for 'undamaged' material, is here taken to be Newtonian. To derive the stress for steady-state deformation (constant strain-rate), the time derivative in (5) is set to zero and (6) substituted into (5):

$$0 = a\eta_u(1-D)\dot{e}^2 - bD \tag{7}$$

which leads to a solution for $D$ of:

$$D = \frac{a\eta_u\dot{e}^2}{a\eta_u\dot{e}^2 + b}. \tag{8}$$

Finally, the stress is:

$$\sigma = \eta_{eff}\dot{e} = \eta_u(1-D)\dot{e} = \frac{\eta_u\dot{e}}{1+\frac{a}{b}\eta_u\dot{e}^2} \tag{9}$$

which is similar to expressions derived by [*Bercovici*, 1998]. Note the asymptotic limits: the usual Newtonian law at low strain rates but stress reduction at high strain rates:

$$\sigma \to \begin{cases} \eta_u\dot{e} & \text{as } \dot{e}\to 0 \\ b/a\dot{e} & \text{as } \dot{e}\to\infty \end{cases} \tag{10}$$

This strain-rate weakening is compared to stress:strain-rate relationships for other well-known rheologies in Figure 3. All rheologies have been normalized such that σ=1 for $\dot{e}$=1, as follows:

$$\sigma = \eta_{eff}\dot{e} = \begin{cases} \dot{e} \\ \dot{e}^{\frac{1}{3}} \\ \min(\dot{e},1) \\ 2\dfrac{\dot{e}}{1+\dot{e}^2} \end{cases} \tag{11}$$

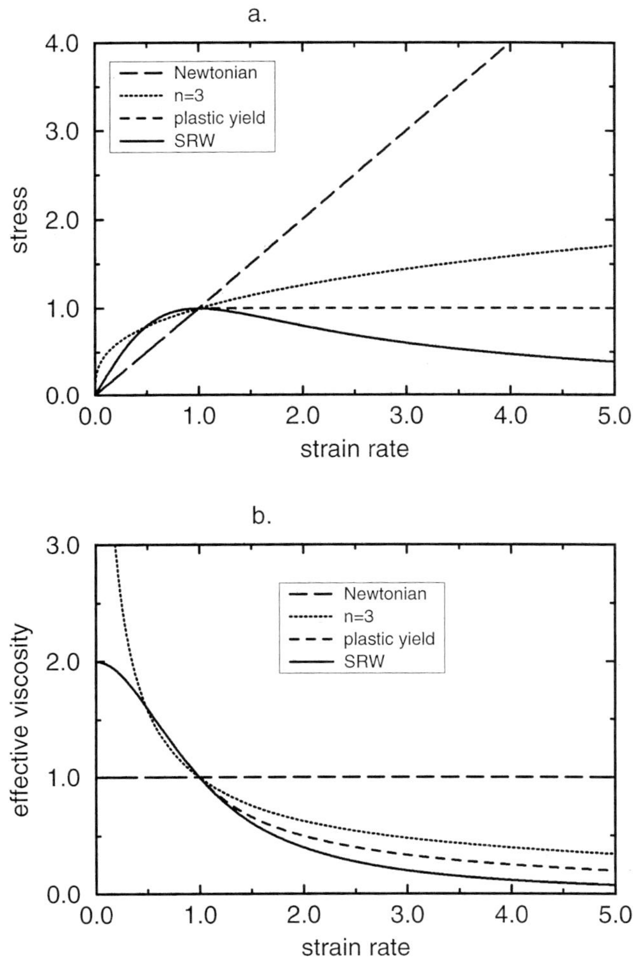


**Figure 3**. Stress vs. strain rate and $\eta_{eff}$ vs. strain rate for different rheologies, normalized so that they all pass through the point (1,1). Newtonian, powerlaw with $n$=3, Newtonian+plastic yield stress, and strain-rate-weakening (SRW).

where these apply to Newtonian, power-law ($n$=3), plastic yielding and strain-rate weakening respectively. From the lower plot of $\eta_{eff}$ vs. strain rate, it is apparent that for all except Newtonian creep, the viscosity decreases with increasing strain rate. Indeed, from this plot one is not aware of any fundamental difference between the rheologies, except that SRW has greater strain-rate softening. The fundamental difference between the rheologies is, however, clearly apparent from the stress vs. strain rate curves: for $n$=3 the stress always increases for increasing strain rate, for plastic yielding the stress remains constant for increasing strain rate – only for SRW does the stress actually decrease for increasing strain rate, and as will be shown later, this is the key to obtaining localization and weak plate boundaries.

How do the critical stress and strain rate scale with various parameters? One can derive these by differentiating the stress vs. strain-rate relationship above, finding that:

$$\dot{e}_{max} = \sqrt{\frac{b}{\eta_u a}} \quad ; \quad \sigma_{max} = \sqrt{\frac{\eta_u b}{a}} \qquad (12)$$

So, higher $\eta_u$ leads to higher $\sigma_{max}$ but lower $\dot{e}_{max}$. If $a$ and $b$ were constant, this would lead to a $\sigma_{max}$ which is maximum at the top of the lithosphere and decreases with increasing depth, the opposite trend to the usual depth-dependent yield stress used to parameterize brittle processes. In general, however, $b$ is temperature-dependent, such that it increases with increasing temperature. If $b$ has the inverse temperature-dependence as viscosity, which may be approximately the case for grain growth, then the product $\eta_u b$ is constant and $\sigma_{max}$ is independent of temperature. Thus $\sigma_{max}$ constant and $\sigma_{max}$ decreasing with depth may be regarded as reasonable limiting cases.

At this point is must be emphasized that $\sigma_{max}$ is not synonymous with yield stress, since yield stress is the instantaneous strength of the material, whereas $\sigma_{max}$ is a maximum stress arising from a dynamic balance of 'damage' production and healing.

*Timescales*. Strain-rate weakening can be regarded as the steady-state condition of the system provided it has enough time to adjust to the deformation regime. What is a realistic timescale for this? There are two timescales involved: one for weakening and one for healing. The time to reach equilibrium is the longer of these. The healing timescale may range from extremely large at the top of the lithosphere, perhaps billions of years, to very small in the mantle. Weakening requires a certain strain which may be of order 10s of percent. The weakening timescale is the time required for this total strain to occur, which depends on velocities and lengthscales. Since velocities and lengthscales change as the system evolves, this timescale is difficult to determine without full modeling. Initial conditions also play a huge role: for example, if the lithosphere were initially homogenous and strong it would take a very long time for substantial strain to occur, whereas if there is 'preexisting' weakness such as a brittle fault in the upper lithosphere, the required localized strains could occur much more rapidly. In any case, based on healing timescale alone it is unlikely that the 'steady-state' is ever reached in the top part of the lithosphere, but it may be an appropriate description for the middle and lower lithosphere.

*Validity*. Given that memory plays a large role in the evolution of plate tectonics, and that the 'steady-state' rheological condition may take an extremely long time to reach in the shallow lithosphere (these two points are essentially the same thing), under what conditions, if any,

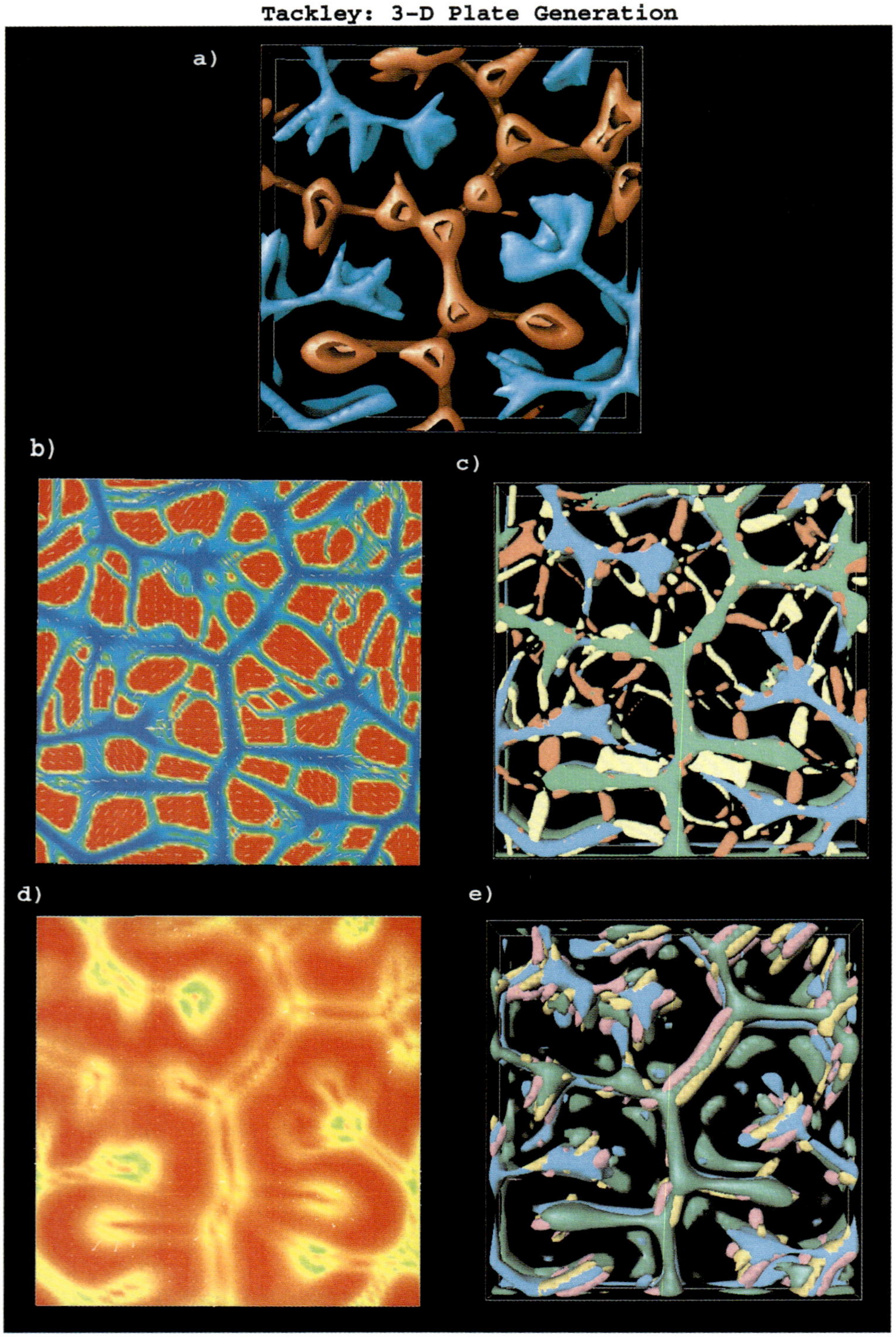


**Plate 5**. Plate formation by SRW and the failure of visco-plastic yield stress rheology in a three-dimensional instantaneous calculation with constant-thickness lithosphere, from [*Tackley*, 1998]. (a) Residual temperature isocontours 0.15 (red) and -0.15 (blue). (b) Lithospheric viscosity and velocity vectors for SRW rheology. Maximum velocity in the domain is 780. (c) Isocontours of horizontal divergence (±50, green and blue) and vertical vorticity (±25, yellow and mauve) and for SRW rheology. (d) as (b) for visco-plastic rheology. (e) As (c) for visco-plastic rheology. Horizontal divergence (±40, green and blue) and vertical vorticity (±10, yellow and mauve).

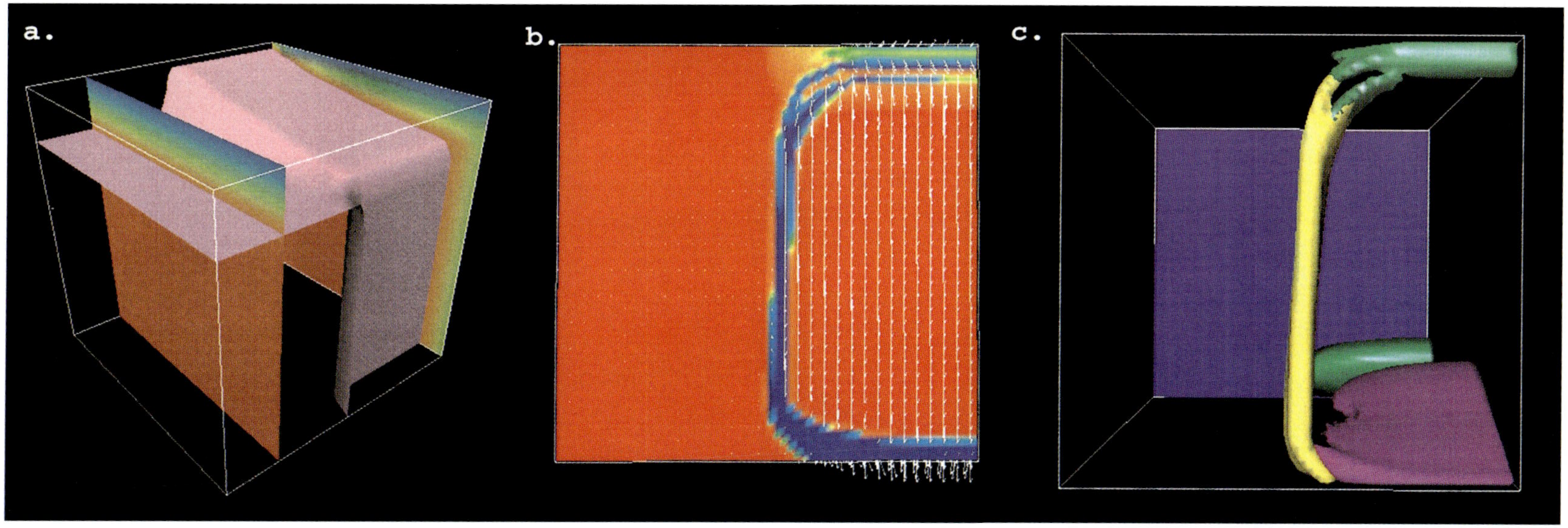


**Plate 6**. The formation of plate boundaries in a 3D instantaneous flow calculation with imposed oceanic temperature structure, and temperature-and SRW rheology. a. temperature field: isocountour $T$=0.8, and two slices, b. logarithm of viscosity, one grid level down from the surface, with variation from 1.0 to $10^3$. For rendering, $T$ was subsampled every 2 points and velocity every 4 points. c. Vorticity (yellow, -10.0), divergence (green, 10.0) and convergence (purple, 10.0).

is it valid to use a SRW rheology in numerical modeling? One possibility is for deeper regions of the lithosphere, where timescales are shorter, perhaps millions of years or less. However, the key reason for testing models with SRW is not realism but that the parameter space in which localization occurs can be rapidly identified, i.e., localization under SRW is a *prerequisite* for localization using the full time-dependent treatment of strain-weakening and healing. If strain-rate weakening does not result in localization for a given set of parameters, then the full time-dependent treatment won't either. Since these processes are difficult and costly to treat numerically, investigating the basic physics using SRW models saves a lot of computer and human time, compared to using time-dependent models. SRW should not be regarded as a realistic lithospheric rheology *per se* but rather as a necessary step on the path to realistic time-dependent rheologies.

Flow associated with spreading centers, and the bending of the slab at subduction zones are two locations where memory is probably unimportant, and it is valid to use an 'instantaneous' rheological description. Indeed, while oceanic plates in general appear to be very strong and can support loads such as several kilometer high seamounts, at subduction zones these same oceanic plates deform with an 'effective' viscosity which is similar to that of the upper mantle [e.g., *Gurnis, M., S. Zhong and J. Toth, On the competing roles of fault reactivation and brittle failure in generating plate tectonics from mantle convection, this volume*].

*2-D Results*

Investigations of plate generation using SRW rheologies were pioneered by Bercovici, who developed models in a two-dimensional horizontal sheet representing the lithosphere. (Note that although SRW rheology is often described as 'instantaneous' in the context of these models, in fact it represents the long-term steady-state solution.) Source and sinks (i.e., lines of convergence and divergence) were specified and strike-slip boundaries allowed to develop naturally. Results showed that the 'plateness' and localization was poor with powerlaw rheologies, but very good with SRW rheology [*Bercovici*, 1993]. This was later extended to spherical geometry and Earth's plate configuration [*Bercovici*, 1995b], in which a remarkably good match to Earth's present strike-slip plate boundaries was generated by specifying only the convergent/divergent motion and SRW rheology, indicating that the evolution timescales for Earth may be long enough that SRW rheology is a good approximation. Specifically considering thermal weakening, a transition to plate-like behavior was observed [*Bercovici*, 1996]. Recently, the void-volatile mechanism was introduced [*Bercovici*, 1998].

The above models assumed a sheet lithosphere, so it was not necessary to consider how the rheology changes with depth through the lithosphere and mantle. In order to apply such mechanisms to modeling a more realistic lithosphere and mantle, it is necessary to consider how various quantities change with temperature and pressure, and how the above localization mechanisms should be combined with other rheological complexities such as powerlaw dislocation creep, plastic yielding, and elasticity. In order to gain some insight as to these processes, it is convenient to consider local models of the lithosphere driven by boundary conditions, before attempting full, three-dimensional models covering the entire mantle.

Thus, in this section, we consider localization along a transform boundary by taking a two-dimensional $(x,z)$ plane, where $z$ is vertical, with motion only in the $y$-direction, i.e., in and out of the plane. The momentum equation thus becomes:

$$(\eta v_{y,x})_{,x} + (\eta v_{y,z})_{,z} = 0 \tag{13}$$

with $v_x{=}v_z{=}0$. [*Tackley*, 1998a] considered this geometry and showed how localization occurs in a prescribed thickness lithosphere underlain by an asthenosphere separated by a step viscosity jump, and a standard SRW rheology. A narrow shear zone developed through the entire lithosphere, leading to a substantial stress reduction. Here we progress to a more complex and complete rheological description, and consider how this effects the development of a shear zone through the lithosphere. The rheological description is intended to be a plausible, through not unique, way of combining various rheological behaviors.

In the following nondimensional equations, numerical values of parameters are based on the assumption of the following nondimensionalization: $\eta$ to $10^{19}$ Pa.s, length to 100 km, and strain rate to $10^{-15}$ $s^{-1}$. This leads to nondimensionalizations for other quantities: velocity to 0.315 cm/yr, time to 31.7 Myr, and stress to $10^4$ Pa ($10^{-2}$ MPa).

Firstly, a cooling half-space temperature profile is assumed, with nondimensional temperature varying from 0 (300 K) at the surface to 1 (1600 K) at depth:

$$T = erf(z) \tag{14}$$

The rheological starting point is a standard temperature-dependent powerlaw rheology,

$$\eta_{pl} = \left(\exp\left[37\left(\frac{1}{T+T_{surf}} - \frac{1}{1+T_{surf}}\right)\right]\right)^{\frac{1}{n}} \dot{e}^{\frac{1}{n}-1} \tag{15}$$

with the nondimensional activation energy of 37.0 corresponding to 400 kJ mol$^{-1}$. In the calculations presented here, n is taken to be 3 and $T_{surf}$ is 0.23 (300 K).

Next, a plastic, depth-dependent yield stress is specified, and the 'undamaged' viscosity becomes the minimum of the dislocation creep viscosity and the effective viscosity for plastic yielding:

$$\eta_u = \min\left[\eta_{pl}, \frac{\sigma_y(z)}{\dot{e}}\right] \; ; \; \sigma_y(z) = \min\left[z\sigma'_{brit}, \sigma_{duct}\right] \quad (16)$$

The depth-dependent yield stress is the minimum of the frictional resistance on faults, which is proportional to depth and represented as $\sigma'_{brit}$, and a ductile yield stress $\sigma_{duct}$ to account for ductile, semi-brittle processes in the middle lithosphere. It is uncertain what the pressure-dependence of the latter should be; the present treatment is just an estimate based on Figure 2. Numerical values are $\sigma'_{brit}$=600,000 and $\sigma_{duct}$=50,000, corresponding to dimensional values of 60 MPa/km and 500 MPa respectively.

Finally, the effective viscosity is assumed to vary linearly with damage D, i.e., equation (4) with $\lambda$=1, and a damage evolution equation with dissipative source and exponential (with time) healing is assumed:

$$\frac{dD}{dt} = \sigma : \dot{e} - B(T)D = \eta_u(1-D)\dot{e}^2 - B(T)D \quad (17)$$

where $B(T)$ is taken as either constant, or temperature-dependent with an arbitrarily chosen law picked to give reasonable variation of healing timescale ($\tau_h=1/B$) at various depths.

$$B(T) = B_0 \exp[20(T-0.5)] \quad (18)$$

A 2x1 domain is assumed, with nominal depth of 100 km. Side boundaries have a constant $v_y$ of -2 (left) and +2 (right), corresponding to a velocity difference of 1.26 cm/yr across the domain, with top and bottom boundaries stress-free. The system is initialized with some 'seed' damage along a vertical line down the center, then evolved in time according to equation (17), until it reaches a steady-state. The solution is obtained numerically using a finite-difference method on a 128x64 grid. The velocity solution for given viscosity field is obtained using a direct matrix solver. At each time instant a few subiterations on the viscosity field are necessary, due to the nonlinearities in equations (15) and (16). For numerical reasons the maximum viscosity is limited to $10^5$. So, the numerical scheme uses explicit damage evolution, not SRW rheology, although the presented results are the same because the system is evolved to a steady-state.

Plate 4 shows results for four different parameter combinations. If the healing rate $B$ is set to an extremely large value such as $10^{20}$, then the damage stays close to zero and no localization occurs (bottom case). The lithospheric stress is then 50,000, as expected because this is the ductile yield stress. As can be seen, a plastic yield stress does not cause any localization in the context of a viscous rheology. Elasto-plastic rheologies do exhibit localization, however, and this may change the picture in the upper part of the lithosphere.

For a temperature-dependent $B$ with $B_0=10^4$ (top case), the system forms a narrow, low-viscosity shear zone cutting through most of the lithosphere, with increasingly distributed deformation below this depth. The velocity field (a.) is approximately constant away from the shear zone, as indicated by the strain rate plot (b.). Effective viscosity (c.) is very low in the shear zone, particularly at shallow depth. This is because the healing rate $B$ becomes very small at shallow depth, and the stress in the asymptotic regime is proportional to $B$ (equation 14). Deeper in the shear zone, viscosity increases until the system is no longer in the 'weak' branch of the rheology. The lower lithosphere below the shear zone acts as a stress guide (d.) and thus governs the overall strength of the lithosphere. The maximum stress is 4.8x$10^2$, two orders of magnitude lower than the case with no weakening.

What effect does $B_0$ have? In parts e. and f., $B_0$ is increased by 3 orders of magnitude to $10^7$. Due to the more rapid healing, localization is more difficult and the shear zone does not extend as deep. The lower lithosphere again acts as a stress guide, and the maximum stress is 2.0x$10^4$, only a factor of 2.5 lower than the non-weakening case. So, although the picture does not look so much different, quantitatively there is a lot of difference.

What if $B$ is constant, not temperature-dependent? Parts g. and h. show the result for constant $B$ of $10^4$. As shown earlier, constant $B$ implies that the maximum stress for SRW is proportional to $\sqrt{\eta_u}$. thus decreasing with increasing depth. With constant $B$, the shear zone cuts cleanly through the lithosphere with almost constant internal viscosity, and there is no 'stress guide' through the lower lithosphere. So, in terms of producing a plate boundary, constant healing rate $B$ 'works' better than the arguably more realistic temperature-dependent healing rate.

An issue that must be considered in numerical investigations is the effect of numerical resolution. In these cases, there is no physical limit on the localization process and thus localization occurs down to the scale of the grid. With better resolution the shear zone becomes narrower and with lower internal viscosity. Convergence tests indicate that resolution does not affect the qualitative nature of the result, provided it is not very coarse, and that the maximum stress is weakly dependent on resolution, for example, the maximum stress is about 15% higher on a

64x32 grid. Weakly dependent because overall stress is not determined by the internal viscosity of the shear zone, but rather by the 'stress guide' through the lower lithosphere. It is possible to limit the localization physically by specifying a minimum viscosity [*Tackley*, 1998a], and this should probably be done.

*3-D Results*

In this section, SRW rheology is applied to three-dimensional flow calculations, with flow driven by thermal buoyancy rather than boundary conditions or imposed sources and sinks. The rheological description is simpler than that of the previous section. Two cases are presented, one using a temperature field from free convection, previously presented in [*Tackley*, 1998a], and the other using a specified temperature field designed to resemble an oceanic plate. Both are instantaneous flow calculations using a 'steady-state' SRW rheology, and thus are not intended to be realistic mantle models, but rather idealized tests designed to elucidate some basic physics.

A major conceptual difference exists between these cases and the preceding analyses in that the system is driven by fixed sources of stress (thermal buoyancy) rather than fixed sources of strain rate (velocity boundary conditions). If the 1-D and 2-D cases were driven by fixed shear stress rather than fixed velocities, solutions would diverge to infinity because stress decreases with increasing strain rate. However, in 3-D the bulk of the mantle, which is not in the SRW regime, limits the maximum velocities that can be obtained: even if the lithosphere tended towards zero strength, flow velocities would saturate and not approach infinity.

The form of SRW equations used are similar to before, but parameterized in terms of maximum stress:

$$\sigma_{ij} = 2\frac{\sigma_{\max}^2 \eta}{\sigma_{\max}^2 + \eta^2 \dot{e}^2}\dot{e}_{ij} \equiv 2\eta_{eff}\dot{e}_{ij} \quad (19)$$

where

$$\dot{e} = \sqrt{\dot{e}_{ij}\dot{e}_{ij}} \quad (20)$$

The experimental procedure is as follows: (i) A temperature field is obtained, as discussed for the individual models, (ii) The instantaneous velocity-pressure/viscosity solution for the system but with SRW rheology (and either temperature- or depth- dependence) is calculated self-consistently for the entire domain. This involves (ii.a) initializing the viscosity field to a value of 1.0 everywhere, then (ii.b) iterating between the velocity-pressure solution, calculated from the usual mantle convection equations using a finite-volume multigrid technique, both fully described elsewhere [*Tackley*, 1994; *Tackley*, 1996a], and the viscosity solution, calculated from equation (1). Roughly 30-60 iterations are necessary to achieve convergence. The subsequent time evolution of the system is not considered here.

Numerical resolution is generally 32 cells in the vertical direction, with a proportional number in the horizontal directions. Convergence tests indicate that numerical resolution does not make very much difference to the general solution provided it is high enough for strain rates to extend well into the 'weak' branch: with too low resolution the solution decays into a homogeneous rigid lid.

*Free convection temperature field.* In this case, previously reported in [*Tackley*, 1998a], a temperature field is taken from a Boussinesq, constant-viscosity, 3-D convection calculation in an 8x8x1 box with periodic sides, 100% basal heating and $Ra$=$10^5$, which has reached statistically steady-state. The lithosphere is taken to be a constant-thickness layer approximately 100 km thick with an undamaged viscosity of $10^4$, while the underlying mantle has an undamaged viscosity of 1.0.

The maximum stress $\sigma_{\max}$ is taken to be $10^5$. Solutions are not strongly dependent on the exact value, except that a homogeneous rigid lid is obtained if it is too high. Lower values lead to broader weak zones but the same basic pattern. The chosen value is just below that which would cause a rigid lid ($\sim 2(\eta\dot{e})_{\max}$ for Newtonian rheology).

The temperature field (Plate 5a) has a pattern of upwellings and downwellings characteristic of Boussinesq, constant-properties, basally-heated convection at this $Ra$. Examination of the lithospheric viscosity and velocity fields (Plate 5b) shows a pattern of strong plates separated by narrow weak zones (WZ). WZs form above the upwellings and downwellings (as expected from the stress concentration there), but more importantly, also connect the up/downwellings together. The WZs above upwellings and downwellings have a width comparable to the width of the convective feature, which is expected because the region of high stress associated with the feature has this characteristic size. These WZ are associated with strong convergence and divergence (Plate 5c), while the interconnecting zones localize down to a size which is limited by the minimum viscosity cutoff, and are associated with high vertical vorticity. Some of the plates have strong rotation which also shows up in the vorticity. While the plate boundaries appear to be well separated into divergent and strike slip in this figure, they generally include both strike-slip and convergent/divergent motion.

A visco-plastic rheology is much less effective in producing plate-like behavior (Plate 5d). WZs occur above up- and down-wellings but not in-between. Regions of high vorticity now cluster around the convergent/divergent regions, i.e., where the lid is already weakened, compatible with previous results of convection with temperature-dependent viscosity [*Balachandar et al.*, 1995; *Balachandar et al.*, 1996].

To summarize, a simple SRW rheology is very effective in producing strong plates separated by narrow WZ in a 3-D flow calculation, whereas a simple visco-plastic rheology is not. This assumption of a constant-thickness lithosphere was made for numerical reasons but constitutes a lapse of self-consistency which is corrected in the next case.

*Oceanic plate and slab.* In this experiment, a temperature field is defined based on an oceanic plate and the slab attached to it, then the instantaneous flow is calculated. Half of the box has the cooling plate, while the other half has a constant-thickness lithosphere, so that a transform margin should form between the two regions. The aim of the model is to try and obtain a velocity field which is consistent with the initial assumptions made to construct the temperature field, then advance the system forwards in time. In the present paper, only the initial solution is presented.

The domain is a 1x1x1 box. The initial nondimensional temperature field is defined for a plate traveling in the $+x$ direction as:

$$T(x,y,z)=\begin{cases}\min(T_{plate}(x,z),T_{slab}(x)) & y<0.5\\ T_{plate\_\max} & y>0.5\end{cases} \tag{21}$$

where

$$T_{plate}=erf\left(\frac{z}{2\sqrt{age}}\right) \tag{22}$$

$$T_{slab}=0.5+0.5erf\left(\frac{x-1}{2\sqrt{age_{\max}}}\right) \tag{23}$$

and

$$age=\frac{x}{v_{spread}} \tag{24}$$

$age_{\max}$ is the maximum age (at $x$=1) and $T_{\text{plate\_max}}$ is $z$-dependent temperature at this age. $v_{\text{spread}}$ is the spreading velocity, taken to be 90. This temperature field is illustrated in Plate 6a.

The 'undamaged' viscosity is taken to have an Arrhenius temperature-dependence:

$$\eta_u=\exp\left(\frac{23}{T+1}-\frac{23}{1.5}\right) \tag{25}$$

and the maximum stress in the SRW equation (24) is also temperature-dependent, with its variation taken to be mid-way between the two limited cases discussed earlier (constant or proportional to $\sqrt{\eta_u}$), i.e.,

$$\sigma_{\max}=\sigma_{\max 0}\eta_u^{\frac{1}{4}} \tag{26}$$

where $\sigma_{\max 0}=3\text{x}10^3$.

The numerical procedure is the same as that for the previous case, except that the solution is initially calculated on a 32x32x32 grid, then refined on a series of grids up to 128x128x32, with uneven z spacing to give much finer spacing in the lithosphere. $Ra=10^5$. Viscosity is cut off at 1000 and 1.0.

The result is shown in Plate 6b&c. Weak zones are formed at the spreading center, subduction zone and transform margin. Vorticity is concentrated in a line along the transform margin. Clearly, the plate margins are not perfect; for example the transform zone bends as it approaches the spreading center and subduction zone. It is thought that this behavior may be due to the large width of the weak zones; if the resolution were adjusted to allow very narrow zones (such as in the 2-D results), the transform 'fault' might be more linear.

## DISCUSSION AND CONCLUSIONS

This paper has examined the problem of incorporating plate tectonics into mantle convection models. Models including continents but no oceanic plates showed the presence of supercontinent cycles, given the right initial condition. Ductile weakening mechanisms in the mid-lithosphere are thought to be most important for generating weak plate boundaries. The most likely mechanisms are grain size reduction, void-volatile weakening and viscous dissipation. These may be described using time-dependent evolution equations, giving long-term memory. For initial investigations it is convenient to parameterize them as instantaneous strain-rate weakening. It was demonstrated how SRW leads to localization in a high-resolution local model of lithospheric shear, and in global three-dimensional models with coarser resolution.

However, this line of research is still at an early stage and there is much to be done. This section first discusses some philosophical issues which have been raised concerning the goal of self-consistent plate generation, then discusses future directions and the serious problems that must be overcome.

### *Objections to Plate Generation*

There are a number of problems and pitfalls associated with achieving the goal of self-consistent plate generation, which have led some researchers to question whether it is a worthwhile and valid goal. These are partly based on a misinterpretation of the idealizations made in the initial, very simple models (such as those presented here), which are made for the purposes of testing basic physics, rather than for realism. The assumptions of 'steady-state' rheology and an initially homogeneous lithosphere fall into this category: they are neither thought to be realistic, nor necessary for self-consistent plate generation to 'work'. Here it will be argued that there are no fundamental barriers to reaching this goal, but certainly some serious technical difficulties which may be overcome with sufficient time and effort.

*Memory.* The first possible problem is that of memory, and the lithospheric heterogeneity that arises from it. The crust and upper lithosphere (at temperatures less than about 600°C) are riddled with faults and other zones of weakness, which have been built up over billions of years of deformation. "New" plate boundaries are usually a change in style of existing plate boundaries or are formed at pre-existing zones of weakness. In contrast, mantle convection models and the initial plate generation models such as those presented here use an 'instantaneous' viscosity formulation in which a weak zone could, in principal, recover its strength immediately if deformation along it stopped. As shown earlier, this is not a fundamental problem: it is straightforward to include one or more state variables to represent lithospheric weakening mechanisms, as was done in the two-dimensional calculations. The rate of 'healing' of the weakening can be temperature-dependent, so that cold parts of the lithosphere retain a memory for billions of years whereas the mantle adjusts almost instantaneously.

Weakening and localization mechanisms are still important even in the presence of pre-existing heterogeneity, since shear zones and faults that become tectonically inactive do heal at some finite rate; indeed, the warm middle lithosphere, which is the strongest part and thus the most important for determining lithospheric strength, may 'heal' quite rapidly. Thus 'reweakening' and 'relocalization' are important if the region is reactivated at a later date. Isomorphic to the statement that pre-existing weakness controls everything is the following question and answer: Q: Why is the Earth spherical? A: Because it was spherical yesterday- it has pre-existing sphericity- it has always been spherical. While true, some essential physics is being missed by such arguments.

*Initial condition.* Associated with the issue of memory is the question of initial condition. Whereas the first, simple models assume an initially homogeneous lithosphere, it is clear that the lithosphere was always heterogeneous. As stated already, an initially homogeneous lithosphere is not at all necessary for such models, (indeed, the 2-D models presented here 'seeded' with some initial strength heterogeneity): it is simply an idealization made for the purposes of elucidating some basic physics. However, it is worth considering the likely effect of initial conditions on a simulation that is run for billions of years. Fluid dynamical systems, after sufficient time, generally adopt length scales and characteristics that are independent of initial conditions. While some initial heterogeneity is needed as a 'seed', it does not affect the general character of the final answer. If weak zones are commonplace, the system will self-organize and 'select' those which are closest to the desired breaking-points, allowing it to behave as it 'wants'. Thus, it is most unlikely that the pattern of heterogeneity 4.5 billion years ago determines present-day plate sizes and behavior.

*Small-scale details.* A common objection is that global models will not be able to reproduce the details of small-scale structure observed at real plate boundaries. It is not clear that this is true, given the rapid improvements in computing power and numerical methods, but if true, the question is: does it matter? It matters if the small-scale structure determines the large scale dynamics. If, instead, the large scale is doing "what it wants" and small-scale details are simply a stochastic reaction to it, then it does not matter.

### *Future Challenges*

Straightforward future directions include using a full time-dependent evolution of material strength and carefully investigating the role of each possible localization mechanism, by numerical models closely coupled to field observations and materials science. This section focuses on some of the most challenging and fundamental scientific and technical issues that need to be addressed.

*Small Lengthscales.* Structures that occur at plate boundaries, such as shear zones and faults, are very narrow compared to the typical resolution of a global calculation, posing both technical challenges and scientific questions. How can such narrow structures be represented? One possibility is to avoid them by adjusting the parameterization so that localization occurs to a lengthscale which is fairly large. Then it is not clear whether the model is truly representing the important physics: can a description which is appropriate for narrow shear zones be applied to regions 10s of km wide? Too-wide plate boundaries may also be 'leaky': if a shear zone is broad then material can flow in and out of it, and it will have some convergence or divergence.

*Multiple lengthscales.* Examining plate boundary regions, it is clear that lithospheric weakening must exist on different scale lengths. For example, at subduction zones the dominant deformation is along a narrow fault (or shear zone) but the slab must also be significantly weakened in order to bend around the corner. Deformation at the North America / Pacific margin is distributed over ~100 km, with a large fraction occurring on a single fault, the San Andreas fault [*Lisowski et al.*, 1991]. Thus, more than one mechanism is acting to give these multiple scale lengths.

*Elasticity*. This paper has barely touched upon the effect of elasticity which is certainly important in the crust and upper lithosphere. Localization in the elastic region, either by brittle failure or plastic flow, may play an important role in generating plate boundaries.

*Multiple timescales*. A problem with trying to incorporate the elastic region into global calculations is that elastic localization occurs at much smaller timescales than mantle convection. An extreme example is earthquakes: is it necessary to track individual earthquakes on faults? The aim must surely be to develop an implicit parameterization of these processes which allows large timesteps.

*Acknowledgments*. The author thanks Shun Karato for a thoughtful, detailed review and helpful discussions. This research also benefited from discussions with Greg Hirth and David Bercovici, and reviews by Mark Richards and Ulrich Christensen. Supported by NASA/HPCC Cooperative Agreement No. NCCS5-147 and the David and Lucile Packard Foundation.

## REFERENCES

Abbott, D.H., R. Drury, and W.D. Mooney, Continents As Lithological Icebergs - the Importance of Buoyant Lithospheric Roots, *Earth and Planetary Science Letters*, *149* (1-4), 15-27, 1997.

Balachandar, S., D.A. Yuen, and D.M. Reuteler, Localization Of Toroidal Motion and Shear Heating In 3-D High Rayleigh Number Convection With Temperature-Dependent Viscosity, *Geophysical Research Letters*, *22* (4), 477-480, 1995.

Balachandar, S., D.A. Yuen, and D.M. Reuteler, High Rayleigh Number Convection At Infinite Prandtl Number With Weakly Temperature-Dependent Viscosity, *Geophysical and Astrophysical Fluid Dynamics*, *83* (1-2), 79-117, 1996.

Batt, G.E., and J. Braun, On the thermomechanical evolution of compressional orogens, *Geophys. J. Int.*, *128*, 364-382, 1997.

Bercovici, D., A Simple-Model Of Plate Generation From Mantle Flow, *Geophysical Journal International*, *114* (3), 635-650, 1993.

Bercovici, D., On the Purpose Of Toroidal Motion In a Convecting Mantle, *Geophysical Research Letters*, *22* (23), 3107-3110, 1995a.

Bercovici, D., A Source-Sink Model Of the Generation Of Plate-Tectonics From Non-Newtonian Mantle Flow, *Journal Of Geophysical Research Solid Earth*, *100* (B2), 2013-2030, 1995b.

Bercovici, D., Plate Generation In a Simple-Model Of Lithosphere-Mantle Flow With Dynamic Self-Lubrication, *Earth and Planetary Science Letters*, *144* (1-2), 41-51, 1996.

Bercovici, D., Generation of plate tectonics from lithosphere-mantle flow and void-volatile self-lubrication, *Earth Planet. Sci. Lett.*, *154*, 139-151, 1998.

Bercovici, D., and P. Wessel, A Continuous Kinematic Model Of Plate-Tectonic Motions, *Geophysical Journal International*, *119* (2), 595-610, 1994.

Bonatti, E., M. Ligi, L. Gasperini, A. Peyve, Y. Raznitsin, and Y.J. Chen, Transform Migration and Vertical Tectonics At the Romanche Fracture-Zone, Equatorial Atlantic, *Journal Of Geophysical Research-Solid Earth*, *99* (B11), 21779-21802, 1994.

Braun, J., J. Chery, A. Poliakov, D. Mainprice, A. Vauchez, A. Tomassi, and M. Daignieres, A simple parameterization of strain localization in the ductile regime due to grain-size reduction, *J. Geophys. Res.*, *in press*, 1999.

Bunge, H.P., and M.A. Richards, The Origin Of Large-Scale Structure In Mantle Convection - Effects Of Plate Motions and Viscosity Stratification, *Geophysical Research Letters*, *23* (21), 2987-2990, 1996.

Bunge, H.P., M.A. Richards, and J.R. Baumgardner, A sensitivity study of 3-dimensional spherical mantle convection at $10^8$ Rayleigh number - Effects of depth-dependent viscosity, heating mode, and an endothermic phase change, *J. Geophys. Res.*, *102* (B6), 11991-12007, 1997.

Christensen, U., Convection With Pressure-Dependent and Temperature-Dependent Non-Newtonian Rheology, *Geophysical Journal Of the Royal Astronomical Society*, *77* (2), 343-384, 1984.

Christensen, U., and H. Harder, 3-D Convection With Variable Viscosity, *Geophysical Journal International*, *104* (1), 213-226, 1991.

Christensen, U.R., and A.W. Hofmann, Segregation of subducted oceanic crust In the convecting mantle, *J. Geophys. Res.*, *99* (B10), 19867-19884, 1994.

Cserepes, L., Numerical-Studies Of Non-Newtonian Mantle Convection, *Physics Of the Earth and Planetary Interiors*, *30* (1), 49-61, 1982.

Dalziel, I.W.D., Neoproterozoic-Paleozoic Geography and Tectonics - Review, Hypothesis, Environmental Speculation, *Geological Society Of America Bulletin*, *109* (1), 16-42, 1997.

Davies, G.F., Mantle Convection Model With a Dynamic Plate - Topography, Heat-Flow and Gravity-Anomalies, *Geophysical Journal International*, *98* (3), 461-464, 1989.

Davies, G.F., Punctuated Tectonic Evolution Of the Earth, *Earth and Planetary Science Letters*, *136* (3-4), 363-379, 1995.

Dieterich, J.H., Modeling of rock friction 1. Experimental results and constitutive equations, *J. Geophys. Res.*, *84*, 2161-2168, 1979.

Drury, M.R., R.L.M. Vissers, D. Vanderwal, and E.H.H. Strating, Shear Localization In Upper Mantle Peridotites, *Pure and Applied Geophysics*, *137* (4), 439-460, 1991.

Durrheim, R.J., and W.D. Mooney, Evolution of the Precambrian Lithosphere - Seismological and Geochemical Constraints, *Journal of Geophysical Research Solid Earth*, *99* (B8), 15359-15374, 1994.

England, P., and D. McKenzie, A Thin Viscous Sheet Model For Continental Deformation, *Geophysical Journal Of the Royal Astronomical Society*, *70* (2), 295-321, 1982.

Fleitout, L., and C. Froidevaux, Thermal and mechanical evolution of shear zones, *J. Struct. Geol.*, *2*, 159-164, 1980.

Gable, C.W., R.J. Oconnell, and B.J. Travis, Convection In 3 Dimensions With Surface Plates - Generation Of Toroidal Flow, *Journal Of Geophysical Research Solid Earth and Planets*, *96* (B5), 8391-8405, 1991.

Gilbert, L.E., C.H. Scholz, and J. Beavan, Strain Localization Along the San-Andreas Fault - Consequences For Loading

Mechanisms, *Journal Of Geophysical Research Solid Earth*, *99* (B12), 23975-23984, 1994.

Glatzmaier, G.A., and P.H. Roberts, A 3-Dimensional Self-Consistent Computer-Simulation Of a Geomagnetic-Field Reversal, *Nature*, *377* (6546), 203-209, 1995.

Gordon, R.G., and S. Stein, Global Tectonics and Space Geodesy, *Science*, *256* (5055), 333-342, 1992.

Govers, R., and M.J.R. Wortel, Extension Of Stable Continental Lithosphere and the Initiation Of Lithospheric Scale Faults, *Tectonics*, *14* (4), 1041-1055, 1995.

Gurnis, M., Large-Scale Mantle Convection and the Aggregation and Dispersal Of Supercontinents, *Nature*, *332* (6166), 695-699, 1988.

Gurnis, M., and T.H. Torsvik, Rapid Drift Of Large Continents During the Late Precambrian and Paleozoic - Paleomagnetic Constraints and Dynamic-Models, *Geology*, *22* (11), 1023-1026, 1994.

Gurnis, M., and S. Zhong, Generation Of Long Wavelength Heterogeneity In the Mantle By the Dynamic Interaction Between Plates and Convection, *Geophysical Research Letters*, *18* (4), 581-584, 1991.

Hager, B.H., and R.J. O'Connell, Subduction zone dip angles and flow driven by plate motion, *Tectonophysics*, *50*, 111-133, 1978.

Hirth, G., and D.L. Kohlstedt, Water in the Oceanic Upper-Mantle - Implications For Rheology, Melt Extraction and the Evolution of the Lithosphere, *Earth and Planetary Science Letters*, *144* (1-2), 93-108, 1996.

Houseman, G., and P. England, Crustal Thickening Versus Lateral Expulsion In the Indian-Asian Continental Collision, *Journal Of Geophysical Research-Solid Earth*, *98* (B7), 12233-12249, 1993.

Jaroslow, G.E., G. Hirth, and H.J.B. Dick, Abyssal Peridotite Mylonites - Implications For Grain-Size Sensitive Flow and Strain Localization In the Oceanic Lithosphere, *Tectonophysics*, *256* (1-4), 17-37, 1996.

Jin, D.H., S. Karato, and M. Obata, Mechanisms of Shear Localization in the Continental Lithosphere - Inference From the Deformation Microstructures of Peridotites From the Ivrea Zone, Northwestern Italy, *Journal of Structural Geology*, *20* (2-3), 195-209, 1998.

Jordan, T.H., The continental tectosphere, *Rev. Geophys. Space Phys.*, *13* (3), 1-12, 1975.

Jordan, T.H., Composition and development of the continental tectosphere, *Nature*, *274*, 544-548, 1978.

Jordan, T.H., Structure and Formation of the Continental Tectosphere, *J. Petrology*, *Special Lithospheric Issue*, 11-37, 1988.

Kameyama, M., D.A. Yuen, and H. Fujimoto, The interaction of viscous heating with grain-size dependent rheology in the formation of localized slip zones, *Geophys. Res. Lett.*, *24* (20), 2523-2526, 1997.

Kanamori, H., Seismological evidence for a lithospheric normal faulting- the Sanriku earthquake of 1933, *Phys. Earth Planet. Inter.*, *4*, 289-300, 1970.

Karato, S., Grain-Growth Kinetics In Olivine Aggregates, *Tectonophysics*, *168* (4), 255-273, 1989.

Karato, S.-I., M. Toriumi, and T. Fujii, Dynamic recrystallization of olivine single crystals during high-temperature creep, *J. Geophys. Res.*, *7*, 649-652, 1980.

Karato, S.I., M.S. Paterson, and J.D. Fitzgerald, Rheology of synthetic olivine aggregates - Influence of grain-size and water, *J. Geophys. Res.*, *91* (B8), 8151-8176, 1986.

Kikuchi, M., and H. Kanamori, The Shikotan earthquake of October 4, 1994: Lithospheric earthquake, *Geophys. Res. Lett.*, *22* (9), 1025-1028, 1995.

Kohlstedt, D.L., B. Evans, and S.J. Mackwell, Strength Of the Lithosphere - Constraints Imposed By Laboratory Experiments, *Journal Of Geophysical Research Solid Earth*, *100* (B9), 17587-17602, 1995.

Kong, X., Y. Yin, and T.M. Harrison, Evaluating the role of preexisting weaknesses and topographic distributions in the Indo-Asian collision by use of a thin-shell numerical model, *Geology*, *25* (6), 527-530, 1997.

Kuang, W.L., and J. Bloxham, An Earth-Like Numerical Dynamo Model, *Nature*, *389* (6649), 371-374, 1997.

Lenardic, A., On the Heat-Flow Variation From Archean Cratons to Proterozoic Mobile Belts, *Journal Of Geophysical Research Solid Earth*, *102* (B1), 709-721, 1997.

Lenardic, A., and W.M. Kaula, Mantle Dynamics and the Heat-Flow Into the Earths Continents, *Nature*, *378* (6558), 709-711, 1995.

Lenardic, A., and W.M. Kaula, Near-Surface Thermal/Chemical Boundary-Layer Convection At Infinite Prandtl Number - 2-Dimensional Numerical Experiments, *Geophysical Journal International*, *126* (3), 689-711, 1996.

Lenardic, A., and L.-N. Moresi, Some thoughts on the stability of cratonic lithosphere: Effects of buoyancy and viscosity, *J. Geophys. Res.*, *104* (B6), 12747-12758, 1999.

Lisowski, M., J.C. Savage, and W.H. Prescott, The Velocity-Field Along the San-Andreas Fault in Central and Southern California, *Journal of Geophysical Research Solid Earth and Planets*, *96* (B5), 8369-8389, 1991.

Lithgow-Bertelloni, C., M.A. Richards, Y. Ricard, R.J. O'Connell, and D.C. Engebretson, Toroidal-Poloidal Partitioning Of Plate Motions Since 120 Ma, *Geophysical Research Letters*, *20* (5), 375-378, 1993.

Lowman, J.P., and G.T. Jarvis, Mantle Convection Flow Reversals Due to Continental Collisions, *Geophysical Research Letters*, *20* (19), 2087-2090, 1993.

Lowman, J.P., and G.T. Jarvis, Mantle Convection Models Of Continental Collision and Breakup Incorporating Finite Thickness Plates, *Physics Of the Earth and Planetary Interiors*, *88* (1), 53-68, 1995.

Lowman, J.P., and G.T. Jarvis, Continental Collisions In Wide Aspect Ratio and High Rayleigh Number 2-Dimensional Mantle Convection Models, *Journal Of Geophysical Research Solid Earth*, *101* (B11), 25485-25497, 1996.

Lowman, J.P., and G.T. Jarvis, Effects of mantle heat source distribution on supercontinent stability, *J. Geophys. Res.*, *104* (B6), 12773-12746, 1999.

Lyakhovsky, V., Z. Benavraham, and M. Achmon, The Origin Of the Dead-Sea Rift, *Tectonophysics*, *240* (1-4), 29-43, 1994.

Lyakhovsky, V., Y. Podladchikov, and A. Poliakov, A Rheological Model Of a Fractured Solid, *Tectonophysics*, *226* (1-4), 187-198, 1993.

Mooney, W.D., G. Laske, and T.G. Masters, Crust-5.1 - a Global Crustal Model At 5-Degrees-X-5-Degrees, *Journal of Geophysical Research Solid Earth*, *103* (B1), 727-747, 1998.

Moresi, L., and V. Solomatov, Mantle Convection With a Brittle

Lithosphere - Thoughts On the Global Tectonic Styles of the Earth and Venus, *Geophysical Journal International*, *133* (3), 669-682, 1998.

Moresi, L.N., and V.S. Solomatov, Numerical Investigation Of 2d Convection With Extremely Large Viscosity Variations, *Physics Of Fluids*, *7* (9), 2154-2162, 1995.

Obata, M., and S. Karato, Ultramafic Pseudotachylite From the Balmuccia Peridotite, Ivrea Verbano Zone, Northern Italy, *Tectonophysics*, *242* (3-4), 313-328, 1995.

Ogawa, M., G. Schubert, and A. Zebib, Numerical Simulations Of 3-Dimensional Thermal-Convection In a Fluid With Strongly Temperature-Dependent Viscosity, *Journal Of Fluid Mechanics*, *233*, 299-328, 1991.

Olson, P., and D. Bercovici, On the Equipartition Of Kinetic-Energy In Plate-Tectonics, *Geophysical Research Letters*, *18* (9), 1751-1754, 1991.

Parmentier, E.M., C. Sotin, and B.J. Travis, Turbulent 3-D thermal convection in an infinite Prandtl number, volumetrically heated fluid - Implications for mantle dynamics, *Geophys. J. Int.*, *116* (2), 241-251, 1994.

Pili, E., Y. Ricard, J.M. Lardeaux, and S.M.F. Sheppard, Lithospheric Shear Zones and Mantle-Crust Connections, *Tectonophysics*, *280* (1-2), 15-29, 1997.

Poliakov, A.N.B., and H.J. Herrmann, Self-Organized Criticality Of Plastic Shear Bands In Rocks, *Geophysical Research Letters*, *21* (19), 2143-2146, 1994.

Pollack, H.N., Cratonization and Thermal Evolution Of the Mantle, *Earth and Planetary Science Letters*, *80* (1-2), 175-182, 1986.

Ranalli, G., *Rheology of the Earth*, 413 pp., Chapman and Hall, London, 1995.

Ratcliff, J.T., G. Schubert, and A. Zebib, 3-Dimensional Variable Viscosity Convection Of an Infinite Prandtl Number Boussinesq Fluid In a Spherical-Shell, *Geophysical Research Letters*, *22* (16), 2227-2230, 1995.

Ratcliff, J.T., G. Schubert, and A. Zebib, Effects Of Temperature-Dependent Viscosity On Thermal-Convection In a Spherical-Shell, *Physica D*, *97* (1-3), 242-252, 1996.

Ratcliff, J.T., P.J. Tackley, G. Schubert, and A. Zebib, Transitions in thermal convection with strongly variable viscosity, *Phys. Earth Planet. Inter.*, *102*, 201-212, 1997.

Regenauer-Lieb, K., Dilatant plasticity applied to Alpine collision: ductile void growth in the intraplate area beneath the Eifel volcanic field, *J. Geodyn. (UK)*, *27* (1), 1-21, 1999.

Ribe, N.M., The Dynamics Of Thin Shells With Variable Viscosity and the Origin Of Toroidal Flow In the Mantle, *Geophysical Journal International*, *110* (3), 537-552, 1992.

Ricard, Y., and C. Vigny, Mantle Dynamics With Induced Plate-Tectonics, *Journal Of Geophysical Research-Solid Earth and Planets*, *94* (B12), 17543-17559, 1989.

Richter, F.M., Convection and the large-scale circulation of the mantle, *J. Geophys. Res.*, *78* (35), 8735-8745, 1973.

Ruina, A., Slip Instability and State Variable Friction Laws, *Journal of Geophysical Research*, *88* (B12), 359-370, 1983.

Scotese, C.R., Jurassic and Cretaceous Plate Tectonic Reconstructions, *Palaeogeography Palaeoclimatology Palaeoecology*, *87* (1-4), 493-501, 1991.

Scotese, C.R., L.M. Gahagan, and R.L. Larson, Plate Tectonic Reconstructions of the Cretaceous and Cenozoic Ocean Basins, *Tectonophysics*, *155* (1-4), 27-48, 1988.

Solomatov, V.S., Scaling of temperature-dependent and stress-dependent viscosity convection, *Phys. Fluids*, *7* (2), 266-274, 1995.

Sorensen, K., Growth and Dynamics Of the Nordre-Stromfjord Shear Zone, *Journal Of Geophysical Research*, *88* (B4), 3419-3437, 1983.

Tackley, P.J., Effects of strongly temperature-dependent viscosity on time-dependent, 3-dimensional models of mantle convection, *Geophys. Res. Lett.*, *20* (20), 2187-2190, 1993.

Tackley, P.J., Three-dimensional models of mantle convection: Influence of phase transitions and temperature-dependent viscosity, Ph.D. Thesis, California Institute of Technology, Pasadena, 1994.

Tackley, P.J., Effects of strongly variable viscosity on three-dimensional compressible convection in planetary mantles, *J. Geophys. Res.*, *101*, 3311-3332, 1996a.

Tackley, P.J., On the ability of phase transitions and viscosity layering to induce long-wavelength heterogeneity in the mantle, *Geophys. Res. Lett.*, *23*, 1985-1988, 1996b.

Tackley, P.J., Self-consistent generation of tectonic plates in three-dimensional mantle convection, *Earth Planet. Sci Lett.*, *157*, 9-22, 1998a.

Tackley, P.J., Three-dimensional simulations of mantle convection with a thermochemical CMB boundary layer: D"?, in *The Core-Mantle Boundary Region*, edited by M. Gurnis, M.E. Wysession, E. Knittle, and B.A. Buffett, pp. 231-253, American Geophysical Union, 1998b.

Tackley, P.J., D.J. Stevenson, G.A. Glatzmaier, and G. Schubert, Effects of multiple phase transitions in a 3-dimensional spherical model of convection in Earth's mantle, *J. Geophys. Res.*, *99* (B8), 15877-15901, 1994.

Thatcher, W., and P.C. England, Ductile Shear Zones Beneath Strike-Slip Faults - Implications For the Thermomechanics Of the San-Andreas Fault Zone, *Journal Of Geophysical Research Solid Earth*, *103* (B1), 891-905, 1998.

Trompert, R., and U. Hansen, Mantle Convection Simulations With Rheologies That Generate Plate-Like Behavior, *Nature*, *395* (6703), 686-689, 1998.

Trompert, R.A., and U. Hansen, The Application Of a Finite-Volume Multigrid Method to 3-Dimensional Flow Problems In a Highly Viscous-Fluid With a Variable Viscosity, *Geophysical and Astrophysical Fluid Dynamics*, *83* (3-4), 261-291, 1996.

Trubitsyn, V.P., and V.V. Rykov, A 3-D Numerical-Model Of the Wilson Cycle, *Journal Of Geodynamics*, *20* (1), 63-75, 1995.

Turcotte, D.L., An Episodic Hypothesis For Venusian Tectonics, *Journal Of Geophysical Research Planets*, *98* (E9), 17061-17068, 1993.

Van Keken, P.E., S.D. King, H. Schmeling, U.R. Christensen, D. Neumeister, and M.P. Doin, A comparison of methods for the modeling of thermochemical convection, *Journal Of Geophysical Research Solid Earth*, *102* (B10), 22477-22495, 1997.

Vissers, R.L.M., M.R. Drury, E.H.H. Strating, C.J. Spiers, and D. Vanderwal, Mantle Shear Zones and Their Effect On Lithosphere Strength During Continental Breakup, *Tectonophysics*, *249* (3-4), 155-171, 1995.

Weinstein, S.A., Thermal-Convection In a Cylindrical Annulus With a Non-Newtonian Outer Surface, *Pure and Applied Geophysics*, *146* (3-4), 551-572, 1996.

Weinstein, S.A., The effect of convection planform on the toroidal-poloidal energy ratio, *Earth Planet. Sci. Lett.*, *155*, 87-95, 1998.

Weinstein, S.A., and P.L. Olson, Thermal-Convection With Non-Newtonian Plates, *Geophysical Journal International*, *111* (3), 515-530, 1992.

Yuen, D.A., L. Fleitout, G. Schubert, and C. Froidevaux, Shear deformation zones along major transform faults and subducting slabs, *Geophys. J. R. astr. Soc.*, *54*, 93-119, 1978.

Zhong, S.J., and M. Gurnis, Dynamic Feedback Between a Continent-Like Raft and Thermal-Convection, *Journal Of Geophysical Research-Solid Earth*, *98* (B7), 12219-12232, 1993.

Zhong, S.J., and M. Gurnis, Mantle Convection With Plates and Mobile, Faulted Plate Margins, *Science*, *267* (5199), 838-843, 1995.

Zhong, S.J., and M. Gurnis, Interaction Of Weak Faults and Non-Newtonian Rheology Produces Plate-Tectonics In a 3d Model Of Mantle Flow, *Nature*, *383* (6597), 245-247, 1996.

---

Paul J. Tackley, Department of Earth and Space Sciences, University of California, Los Angeles, 595 Charles Young Drive East, Los Angeles, CA 90095-1567

# On the Competing Roles of Fault Reactivation and Brittle Failure in Generating Plate Tectonics from Mantle Convection

Michael Gurnis

*Seismological Laboratory, California Institute of Technology, Pasadena, California*

Shijie Zhong

*Department of Earth, Atmospheric, and Planetary Sciences, Massachusetts Institute of Technology, Cambridge, Massachusetts*

John Toth[1]

*Seismological Laboratory, California Institute of Technology, Pasadena, California*

Fault reactivation plays a fundamental role in the generation of plate tectonics from mantle convection. Converging and transform plate margins are mechanically weak due to both preexisting faults and preserved shear zones within the crust and lithosphere, on the one hand, and continuous brittle failure of lithosphere, on the other hand. Transform margins are the site for nucleating new converging margins. Lithospheric earthquakes demonstrate that oceanic lithosphere within a trench can deform as fast as the upper mantle. Models demonstrate that faulted converging plate margins contribute to producing plate tectonic-like motion. Preexisting faults and a power law plastic rheology dynamically interact and together give rise to plate-like motion. Faults have an important influence over outer rise and trench topography and by comparing models with observations, interplate stress is about 10-30 MPa. Subduction can initiate on preexisting faults when interplate coupling is as high as 10-30 MPa if the oceanic plate already has slabs attached to it – as was the case for the Pacific plate when the Marianas nucleated. Modeled plates only change velocity slowly in response to the initiation of new subduction zones; the initiation of subduction in the Western Pacific at $\approx$ 45 Ma was likely the result of a change in Pacific plate motion as opposed to causing the change in the direction of plate motion. The locations of the greatest amount of present day brittle failure

[1]Now at Shell International Exploration and Production, Rijswijk, Netherlands

The History and Dynamics of Global Plate Motions
Geophysical Monograph 121

within oceanic lithosphere are not located at arbitrary positions on oceanic plates, rather brittle failure occurs in close proximity to long lived zones of preexisting weakness. Old weak structures are reused by the convecting system because it takes less energy to reactivate a preexisting structure than it does to create an entirely new plate margin from pristine, intact lithosphere.

## 1. INTRODUCTION

Plate tectonics is a kinematic description in which the outer shell of the Earth is divided into a number of plates or rigid spherical caps which move with respect to each other. Oceanic plates represent the top thermal boundary layer of mantle convection [*Davies and Richards*, 1992] and this boundary layer sinks (subducts) into the mantle at converging margins. The motion of plates (their speed and direction) is a balance between the buoyancy of cooling and thickening oceanic lithosphere and subducted slabs, on the one hand, and viscous flow, on the other [*Hager and O'Connell*, 1981]. Plate motion and the associated mantle flow appears to be a simple mode of thermal convection and has been described as a fluid dynamical process [*Turcotte and Oxburgh*, 1967]. The process, however, is not solely fluid dynamic because (a) oceanic plates are mostly rigid with little deformation within their interior and (b) an appreciable amount of plate motion occurs by strike-slip. A substantial fraction of the total dissipation associated with plate motion could occur in the bending of the oceanic lithosphere as it subducts [*Conrad and Hager*, 1999]. Why the Earth has plate tectonics is a significant unanswered question which is addressed in this paper as well as others in this volume by *Bercovici et al.* [1999] and *Tackley* [1999].

Qualitatively, one solution to the existence of plate tectonics lies in a balance between thermal convection, the thermally activated rheology of silicates at high temperatures, and brittle failure of rocks at low PT. Plate driving forces are largest at converging margins above subducting slabs so that otherwise cold and strong lithosphere fails and gives rise to weak plate margins. With this model, the lithosphere rapidly fails on geological time-scales under the action of tectonic stress. A number of studies concerning the dynamics of plate tectonics generation assume that the lithosphere fails with the application of tectonic stress [*Bercovici* 1993; 1995; *Moresi and Solomatov*, 1998; *Tackley*, 1998]. Since the rheologies used in these models do not consider the deformational history of the material, we refer to this as the *instantaneous* rheology model. We will argue that this is one fundamental component generating plate tectonics. In addition, the mechanical memory of the crust and lithosphere could also play a fundamental, but not an exclusive, role in the generation of plate tectonics. An important piece of evidence pointing toward this alternative hypothesis is that long lived zones of preexisting weakness and preexisting faults control the location of plate margins. We will argue that old weak structures are reused by the convecting system because it takes less energy to reactivate a preexisting structure than it does to create an entirely new plate margin from pristine, intact lithosphere. We will argue that it is essential that preexisting structures be incorporated into both instantaneous models of mantle flow and time-dependent, fully dynamic models simulating plate motions on time-scales less than $\sim 10^8 - 10^9$ yrs.

A long term goal of geophysics is the formulation of dynamically self consistent, time-dependent models of mantle convection in which plate tectonics naturally arises. This is an important long term objective because it not only relates to why the Earth has plate tectonics while other planets, particularly Venus, do not, but because the resolution of this problem will lead to the development of dynamic models integrating a wide variety of geological, geophysical, and geochemical observations [e.g., *Gurnis et al.*, 1998]. In this paper, we integrate results from seismology, plate tectonic reconstructions, other geophysical observations, and dynamic simulations to show that the reactivation of preexisting fault zones must play a fundamental role in controlling large-scale mantle dynamics and that use of instantaneous rheologies without consideration of deformational history is not a satisfactory solution to geodynamic problems.

## 2. DEFORMATION FIELD OF A CONVERGING MARGIN

An essential component of our hypothesis is that both preexisting structures and continuous failure of the oceanic lithosphere play fundamental roles in mantle dynamics. In order to demonstrate the significance of the continuous failure or weakening of the oceanic lithosphere we focus on the deformation field at subducting margins. Moreover, the observed character of subducting margins guides the formulation of faults in geodynamic models as described in Section 5.1. Although complex, the spatial and temporal patterns of deformation are essential constraints on how subduction operates on mantle convection time-scales.

There are two forms of seismic deformation within subduction zones: interplate and intraplate deformation. Interplate deformation (Fig. 1) is exemplified by great earthquakes, large thrust events (with moment magnitudes, $M_w$, 8.2 or greater) occurring on the interface between subducting and overriding plates; the May, 1960 Chilean earthquake is an example. Subduction zones have been classified into those which have great earthquakes (called coupled because of a substantial amount of mechanical coupling) and those which have no great earthquakes (uncoupled) [*Uyeda and Kanamori*, 1979]. Within coupled subduction zones, the overriding plate is generally in a compressive state and the slabs have moderate to shallow dip angles. The subduction zones in South America, Alaska, the Aleutians, and Kamchatka are coupled [*Kanamori*, 1986]. Coupled subduction zones occur within intermediate age lithosphere ($\approx$ 20-80 Ma) where the ratio between interplate seismic slip to the slip inferred from plate motions is unity [*Kamamori and Astiz*, 1985]. Uncoupled subduction zones are associated with back-arc opening, steeply dipping slabs and old oceanic lithosphere; the Marianas is an example. Some subduction zones are intermediate in their seismic coupling; subduction at the Tonga and N. Japan trenches are examples.

Not all seismic deformation within converging margins occurs by thrusting between subducting and overriding plates. As the subducting oceanic plate moves toward the site of interplate deformation, seismic activity increases with the occurrence of outer rise earthquakes. The outer rise is a topographic undulation ($\approx$ 200 meters in height) located a few hundred kilometers ocean ward of trenches – easily visible on the map of free-air gravity of the western Pacific (Plate 1). Outer rise earthquakes are usually tensional [*Christensen and Ruff*, 1988] and have been attributed to the bending of a strong elastic plate. When a rigid beam is bent, it will be in tension through its upper half and in compression through its lower [*Chapple and Forsyth*, 1979]. Outer rise tensional events have been interpreted as the brittle failure of the upper region of the elastic lithosphere. A bending plate which undergoes plastic failure must still be capable of supporting elastic stresses greater than $\approx$ 500 MPa within its core if trench-forebulge bathymetry is to be matched [*McAdoo et al.*, 1978; *Chapple and Forsyth*, 1979]. The normal conceptualization of the deformation field associated with subduction is that the oceanic lithosphere is strong as indicated by the presence of the outer rise (Plate 1). However, there are three reasons to question this hypothesis: the occurrence of earthquakes which completely fracture oceanic lithosphere, the space-time occurrence of outer rise earthquakes demonstrating that the outer rise is a dynamic feature, and the ability to generate trench-forebulge bathymetry dynamically with lithosphere which has no long-term elastic strength.

So called *lithospheric earthquakes* within the subducting oceanic plate have been identified [*Kikuchi and Kanamori*, 1995] and occur below and ocean ward of the seismic coupling zone (Fig. 1). Lithospheric earthquakes (large-magnitude events on faults which cut completely through the 100 km thick lithosphere) show that the oceanic lithosphere undergoes rapid, permanent deformation within the trench environment, contrary to the conventional model of a bending plate. One such event is the 1933 Sanriku ($M_w$ = 8.4) tensional event which occurred within the oceanic lithosphere ocean ward of the (intermediately coupled) Japan trench (Plate 1) [*Kanamori*, 1971]. The event ruptured down to $\approx$ 100 km, completely fracturing the oceanic lithosphere in a location where the outer rise is well developed (Plate 1). Although only 12 events between magnitude, $M_w$, 7.2 and 8.3, have been identified globally the rate of seismic moment release is substantial. The approximate location with respect to the trench, extent of rupture, and slip are illustrated schematically in Figure 1. Using the tabulation of seismic moments in *Kikuchi and Kanamori* [1995], we can estimate the volumetric seismic strain rate within the oceanic lithosphere as it slides by the surface of interplate deformation. The following estimate, which excludes the great thrust earthquakes, must be a lower limit on total strain since other deformation mechanisms could be in operation besides seismic slip (aseismic slip on faults, plastic deformation, etc.). It can be shown that

$$\dot{\epsilon} \approx \frac{\Sigma M_o}{2\mu V \Delta T} \tag{1}$$

where $\dot{\epsilon}$ is the average strain rate within the lithosphere, $\Sigma M_o$ is the cumulative seismic moment release which occurs in a volume, $V$, over a time, $\Delta T$, and $\mu$ is the shear modulus [*Bevis*, 1988]. The events occurred within a volume, $V = dLW$, where $d$ is the depth range of rupture associated with lithospheric earthquakes, $L$ is the length of trenches, and $W$ is the horizontal length scale over which the earthquakes occur. The lithospheric earthquakes occurred over a 65 yr period and if we assume that $d$=100 km, $L$=4x10$^4$ km, and $W$= 200 km, then $\dot{\epsilon} \approx$ 1x10$^{-16}$s$^{-1}$. However, as shown in Plate 1, four of these events occur in a 800 km segment along

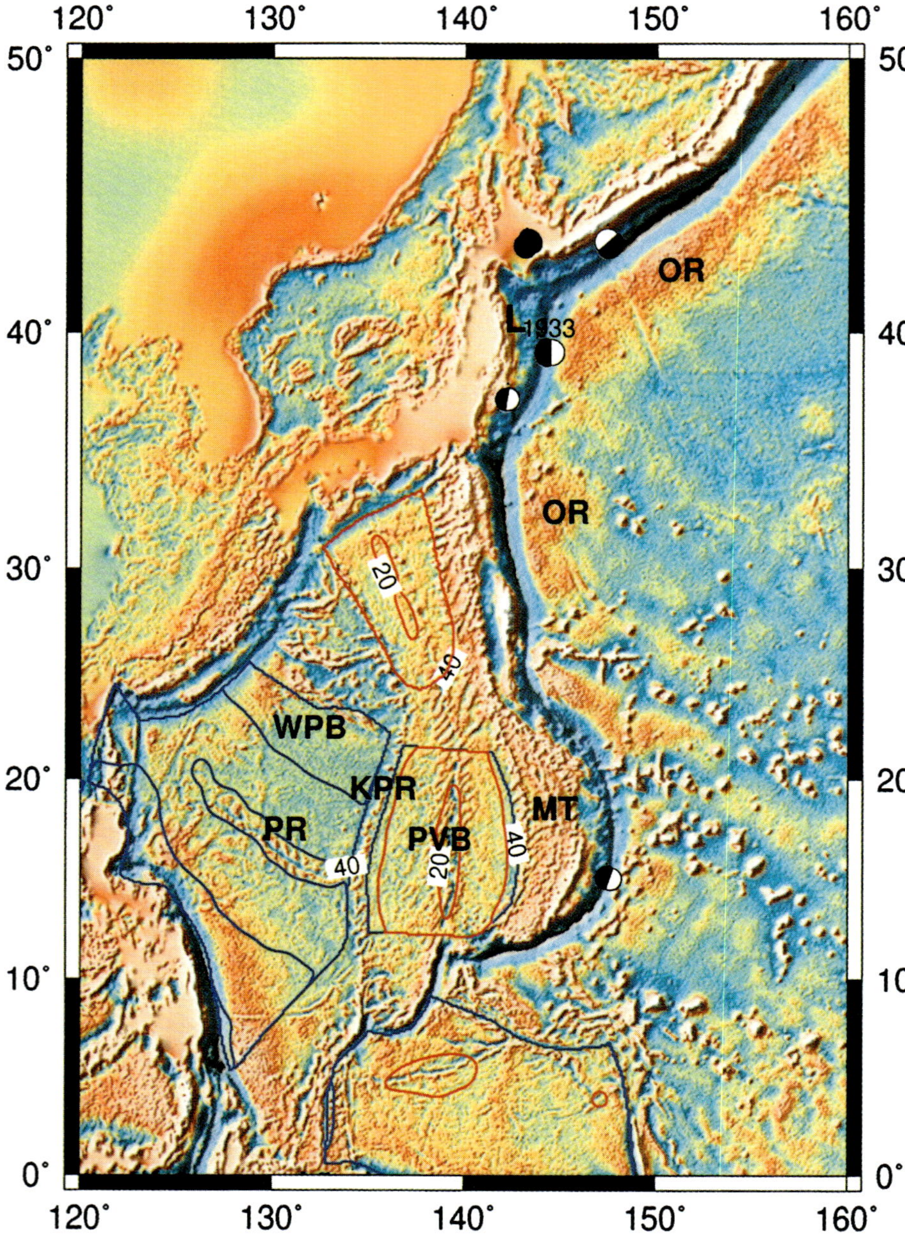


**Plate 1.** Free-air gravity (from *Sandwell and Smith* [1997]) of the Western Pacific showing features described in the text. The focal mechanisms of lithospheric earthquakes are shown in black (from references cited in *Kikuchi and Kanamori* [1995]. In the area around the Japan trench these earthquakes are labeled with an L. The 1933 Sanriku event is labeled. Contours of ages (from *Müller et al.* [1997]) for marginal basins are shown in red for 10-30 Ma crust and in blue for 40-60 Ma crust. The following features are labeled: the West Philippine Basin (WPB), Philippine Ridge (PR), Kyushu-Palau Ridge (KPR), Parece Vela Basin (PVB), the Mariana Trough (MT), and the outer rise (OR).

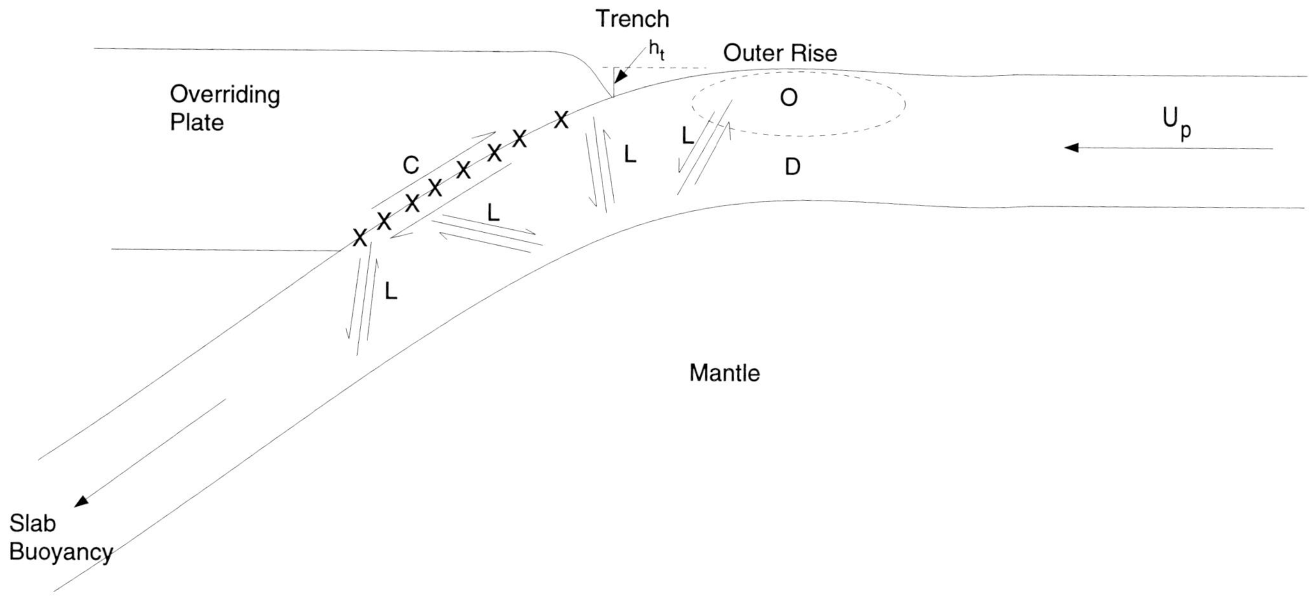


**Figure 1.** Schematic cross-section through a generic converging margins showing features described in the text. The interplate coupling zone is labeled with C, lithospheric earthquakes with L, location of the outer rise, O, area of the lower lithosphere below the outer rise, D, trench depth with $h_t$, and plate velocity with $U_p$.

the strike of the Japan trench where the strain rate is found to be $\dot{\epsilon} \approx 2\text{x}10^{-15}$ s$^{-1}$. These strain rates can be compared to those which occur in the low viscosity upper mantle which is on the order of $\approx$ 5 cm/yr $\div$ $10^3$ km = $2\text{x}10^{-15}$ s$^{-1}$. The worldwide average strain rate within the oceanic lithosphere as it traverses the trench environment is nearly as high as strain rates within the upper mantle and the lithosphere in certain locations, like the Japan trench, can deform as fast as the upper mantle.

Although it is unlikely that oceanic lithosphere can be perfectly described as a viscous material as it slides by the plane of interplate coupling, it does appear to be pervasively fractured. We can make an order of magnitude estimate on the bulk effective viscosity of the lithosphere within the trench by assuming that effective viscosity, $\eta_{eff} \sim \tau/\dot{\epsilon}$ where $\tau$ is the deviatoric stress. As we show below, the deviatoric stress within the lithosphere near the trench is approximately proportional to trench depth (e.g. $\tau \sim \rho g h_t$, where $h_t$ is trench depth, Fig. 1). $h_t$, globally, is on average about 4 km and about 2 km within the Japan trench [*Jarrard*, 1986]. Consequently, we estimate a $\eta_{eff} \sim 10^{23} - 10^{24}$ Pa s (globally averaged) and $\sim 10^{22}$ Pa s (Japan trench).

The occurrence of outer rise earthquakes in space and time show that the outer rise is a dynamic feature where stresses evolve over $\sim 10 - 100$ km scales in response to interplate interactions [*Christensen and Ruff*, 1988]. Outer rise events are primarily tensional. The less common compressional events occur in known seismic gaps or precede large interplate events [*Christensen and Ruff*, 1988]. Tensional events are consistent with the continuous pulling of the oceanic plate by slab pull, not the bending of an elastic plate. Great earthquakes are always followed by tensional outer rise earthquakes. Within uncoupled plate margins only tensional events occur. The 1952 Kamchatka ($M_w$ = 9.0) interplate event on a coupled plate margin was followed by a series of tensional outer rise events which lasted until 1981-82 when the mechanism switched to compressional [*Christensen and Ruff*, 1988]. If we assume that this 30 yr period represents the Maxwell time ($\tau_M = \eta/\mu$) of the lithosphere within the trench, then the effective viscosity of the lithosphere within the trench is $\sim 10^{20}$ Pa-s. However, this value is as small as typical upper mantle viscosities such that some of the post-seismic relaxation could be occurring within the upper mantle instead of within the lithosphere. Clearly, an explicit model of the process will be required to understand how robust this value is, but even if the Maxwell time is two orders of magnitude larger, the viscosity we obtain is still quite small.

Finally, there is no need for the oceanic lithosphere to have any elastic strength within the trench. The elastic flexure models have long been criticized because they require the lithosphere to support kilobars of stress in excess of what can be supported by lithospheric materials [*Kohlstedt et al.*, 1995]. However, if the topography is generated by viscous flow then such high stresses are not required [*DeBremaecker*, 1977; *Melosh*, 1978]. We show below that trench-forebulge topography can be dynamically maintained in models in which the lithosphere not only lacks elastic strength but has low effective viscosity within the trench environment.

Before the oceanic lithosphere enters the trench environment, it has considerable strength as indicated by distinct flexural bulges around oceanic islands. But by the time the lithosphere has passed through the trench environment, it has undergone considerable deformation. Clearly, subducting oceanic lithosphere is weak in comparison to the bulk of the lithosphere. This demonstrates that a substantial component of mantle rheology responds instantaneously on geological time-scales and this point is emphasized because this brittle failure contrasts with historical weakness described in Section 4.

## 3. THE STRENGTH OF FAULTS

If faults are substantially weaker than any other part of the plate tectonic system, they will always be used to accommodate deformation in lieu of creating new plate margins. It has long been appreciated that the large fault zones which accommodate the strain associated with plate tectonics are not only much weaker than intraplate regions but substantially weaker than predicted by extrapolation of experimental rock mechanics data.

The stress drop in large interplate earthquakes, on average 3 MPa [*Kanamori*, 1980], shows that faults are weak. Although stress drops may only reflect that released during earthquakes and not the absolute value of stress across large fault zones [*Kanamori*, 1994], estimates of absolute stress across the San Andreas transform fault in California must be 10 MPa or less because of the lack of a frictionally developed heat flow anomaly in the vicinity of the fault [*Lachenbruch and Sass*, 1988]. The weakness of large faults between plates is corroborated by the direction of maximum principle stress of small adjacent events in both strike slip [*Kanamori*, 1980] and convergent settings [*Magee and Zoback*, 1993]. The weakness of the San Andreas fault is corroborated by the complete absence of right lateral shear as inferred from stress measurement in deep boreholes [*Zoback and Healy*, 1992]. As we describe below, such estimates are consistent with 10-30 MPa absolute shear stresses on faults inferred from a comparison of dynamic models of trenches driven by the buoyancy of slabs and trench depths at ocean-ocean converging margins [*Zhong and Gurnis*, 1992; 1994].

Outside of major tectonic faults, however, lithosphere appears much stronger and capable of supporting much larger stress. The large topography in some regions especially near seamounts induce large local stresses ($\sim$ 200 MPa) [*Artyushkov*, 1973; *Jeffreys*, 1970]; since there is little observed deformation in these regions, including seismicity, the large stress implies these regions must be strong. That the lithosphere resists deformation (and remains strong) even when subject to large forces is fundamental. The properties of the lithosphere subject to deformation in trench environments and of lithosphere at seamounts must be fundamentally different. We suggest that deformation of oceanic lithosphere at trenches should not be modeled by flexure of an elastic plate.

Large faults are probably weak because the rocks making up fault zones at depth have much finer grain size than the host country rock [*Watterson*, 1975; *White*, 1976]. *White* [1976] suggested that fine grain size allowed a change in deformation from dislocation creep to one in which grain boundary sliding dominated. The exact mechanism is a subject of debate and continuing research, but reduction of grain size, especially in the presence of water, continues to be the preferred mechanism for the weakness of fault zones at depth [*Vissers et al.*, 1995; *Jaroslow et al.*, 1996; *Imber et al.*, 1997] .

## 4. FAULT LONGEVITY AND REACTIVATION

The plate tectonic record since the Mesozoic demonstrates that as oceanic plates evolve, either reuse of existing fault zones or reactivation of dormant structures is the preferred mode of deformation rather than the generation of entirely new weak margins from homogeneous and strong lithosphere. This is just a specific example of a widely recognized principle in geology that not only do major fault zones change their tectonic sense during the course of time [*White and Green*, 1986], but fault zones can remain in existence for greater than $10^9$ yrs (for example, *Sykes* [1978]). When continents breakup, new diverging plate margins (rifts) do not randomly occur within continental plates, rather they tend to follow preexisting structures, old orogenic belts in particular. The present rift system in East Africa, active since the Tertiary, has followed the margins of cratons which have been reactivated many times in the last 3 Gyr [*McConnell*, 1972]. When N. America split from

Africa during the Mesozoic, the rifting followed the Paleozoic suture zone between the two continents [*Wilson*, 1966]. Although this evolution from old suture zone to new rift is important, we focus our discussion on how fault zones evolve or have been reactivated from transform to convergence, from convergence to transform, and from divergence to strike-slip (Fig. 2).

The reuse of an active transform to create a new subduction zone can be demonstrated from the history of basins behind the Marianas trench. At the Marianas trench, the Pacific plate has been subducting under the Philippine plate for about 40 Myr. However, within the West Philippine Basin (Plate 1) there are WNW trending magnetic anomalies, the youngest being 40 Ma, and these are perpendicular to the present NS trending converging margin [*Ben-Avraham and Uyeda*, 1983]. The east-west magnetic lineations are symmetrical with respect to the Philippine Ridge an extinct ridge which ceased spreading at about 40 Ma. These magnetic anomalies abruptly terminate at the Kyushu-Palau Ridge which can be clearly seen in Plate 1. The Kyushu-Palau Ridge forms the western boundary of the Parece Vela Basin which opened between about 40 and 10 Ma. The Parece Vela Basin was the first of two back-arc basins formed by the rollback of the Mariana trench. There was extensive Eocene volcanism along the Kyushu-Palau Ridge when subduction was presumably initiated [*Stern and Bloomer*, 1992]. The arrangement of the tectonic elements, tectonic fabric, sea-floor ages, and magmatic history are all consistent with the hypothesis that a new converging margin (the Mariana trench) formed on a fracture zone (the Kyushu-Palau Ridge) and then trapped the oceanic crust behind the active trench [*Uyeda and Ben-Avraham*, 1972; *Ben-Avraham and Uyeda*, 1983; *Stern and Bloomer*, 1992].

The conversion of old fracture zones and transform faults into converging margins may be a general process which occurs repeatedly during the evolution of an ocean basin. The evolution of the Aleutian trench is similar to the Marianas: within the Bering sea there are N-S trending early Cretaceous magnetic lineations [*Cooper et al.*, 1976] which are perpendicular to the Aleutian trench, such that the trench probably formed on an E-W trending fracture zone of the Kula plate at 55 Ma [*Scholl et al.*, 1986]. Another example of an incipient subduction zone along a transform plate boundary is the Puysegur ridge and trench at the northern end of the Macquarie ridge complex [*Collot et al.*, 1995]. Oblique convergence occurs along the Puysegur ridge where subduction is propagating southwards along the strike slip boundary. Earthquake focal mechanisms show a trench-parallel sub-vertical strike-slip fault beneath the ridge axial valley, and a southeastward, shallow dipping oblique thrust [*Collot et al.*, 1995]. The development of the subduction zone is being facilitated by the development of reverse faults, which accommodate the normal convergence [*Collot et al.*, 1995].

In addition to the record on the sea-floor, *Casey and Dewey* [1984] have proposed that the evolution of transform plate boundaries into trenches explains many of the features of obducted ophiolites. Orogenic activity at continental margins is often initiated by overthrusting and obduction towards the previously stable margin. This indicates that subduction polarity was away from the continental margin, and that trench nucleation was intra-oceanic. Upon changes in poles of relative plate motion, subduction zones have a strong tendency to initiate within ocean basins by the conversion of preexisting plate boundaries to subduction zones [*Casey and Dewey*, 1984]. *Stern and Bloomer* [1992] proposed the Jurassic California arc may have formed in a similar way, subduction was initiated along a transform plate boundary.

Subduction zones probably nucleate at fracture zones because of inherent local lithospheric weakness. However, transform margins have been suggested to evolve into converging margins because of (potentially large) offsets in lithospheric age between opposite sides of the transform fault [*Stern and Bloomer*, 1992] which provides a substantial force for growing a new subduction zone. One reason *Stern and Bloomer* [1992] suggested this hypothesis is that the new Marianas subduction zone initiated at about the same time as the Pacific plate change in the direction of motion, $\approx$ 43 Ma [*Engebretson et al.*, 1985]. However, the change in Pacific plate motion appears to have happened before the nucleation of the Marianas, not after, as would be expected from the *Stern and Bloomer* [1992] model. Moreover, the descent of slabs within the upper mantle will only slowly change plate speed (and direction) [e.g., *Richards and Lithgow-Bertelloni*, 1996]. The bend in the Hawaiian-Emperor seamount chain is sharp [*Clague and Dalrymple*, 1987] implying that plate motion changed within $\sim$ 1 Myr. Later in this paper we show that plates take $\sim$ 10 Myr to fully change velocity as a new subduction zone initiates. Consequently, the force from increasing slab pull associated with the Marianas subduction zone probably could not cause Pacific plate motion to change abruptly [*Richards and Lithgow-Bertelloni*, 1996].

The converse of nucleating a converging from a transform margin is also true: preexisting fault zones evolve from thrust to strike-slip. Since the Permian, western North America was a converging margin which has progressively evolved into a transform boundary [*Atwater*, 1970]; at Cape Mendocino, California, the Cascade thrust is directly evolving into the San Andreas transform fault. Presently, the Macquarie ridge is primarily a strike-slip fault, but from at least 61 to 20 Ma it was an active spreading center on the Southeast Indian Ridge which evolved into a transform fault [*Cande et al.*, 1998]. Interestingly, as we have just described, this former spreading center (now a transform fault) is evolving into a zone of convergence with activation of thrust faults within the fault zone immediately south of the South Island of New Zealand [*Collot et al.*, 1995]. The initiation of subduction along the Puysegur ridge is occurring along a segment of a transform plate margin with little offset in age; this supports our hypothesis that subduction zones nucleate on fracture zones because of inherent weakness and not the localization of buoyancy. Clearly, for the Pacific, the largest oceanic plate, converging and strike slip boundaries have lineages related to other faults (Fig. 2).

Evidence for the reactivation of fault zones within the continental crust is substantial [*Holdsworth et al.*, 1997; *White*, 1976]. For example, at the longest time scale, the Nagssugtoqidian boundary in Greenland, made up of narrow shear zones [*Sørensen*, 1983], has had intermittent activity over 2.5 Gyr [*Watterson*, 1975] and the 500 km long Thabazimbi-Murchinson lineation through the Kaapval craton has had a history of repeated reactivation over more than 2.5 Gyr [*Good and De Wit*, 1997].

## 5. FORMULATING GEODYNAMIC MODELS WITH FAULTS AND WEAK ZONES

Our discussion demonstrates that faults and converging margins are weak, are actively reused as plate motions evolve, and become dormant only to be reused after considerable time. Weak zones and fault reactivation must have an important influence on plate and mantle dynamics. However, how faults and weak zones should be incorporated into geodynamic models is a much more difficult and unresolved issue. There is a difference of opinion between which of the following approaches is more appropriate: faults and weak zones should arise naturally from the solution of the appropriate equations [*Bercovici* 1993; 1995; *Moresi and Solomatov*, 1998; *Tackley*, 1998] or some minimum level of initial heterogeneity always needs to be incorporated into plate-mantle models (advocated here).

Models of plate-mantle dynamics are one of three types: instantaneous, transient, or long-term evolutionary. In an instantaneous model, the buoyancy forces are incorporated into a domain in which an equation of motion is solved; the results are compared to a present day quantity, such as plate motion, the geoid, or intraplate stress. Transient models couple the equation of motion with the energy equation and the results are dominated by the initial conditions; the models are compared, for example, to plate motion over million year time intervals. Transient models can be used to understand how the evolution of mantle buoyancy influences plate motion (time scales $> 10$ Myr) or they can be used to link mantle convection time-scales with the seismological character of plate boundaries (e.g. Section 2). Long term evolutionary models are only conceptualizations but one could imagine coupling the equations of motion and energy with the transport of lithology and strain history so as to generate weak zones which can then survive and be transported.

Considerable effort has been made in recent years at modeling weak zones, including incorporating faults *a priori* or letting weak zones develop on their own. Using different formulations of preexisting faults, instantaneous and transient mantle flow models have reached a level of tectonic realism allowing some important conclusions concerning the role of weak zones in mantle dynamics to be drawn. Models quantitatively demonstrate that faulted converging plate margins contribute to producing plate tectonic-like motion. Faults have an important influence over outer rise and trench topography in subduction zones and by comparing dynamic models with observed trench depths, constraints can be placed on the absolute magnitude of interplate stress. The forces and fault coupling necessary to initiate subduction at preexisting faults have been determined. Three-dimensional models which include transform and thrust faults show that weak faults and a power law (or instantaneous) rheology interact and are essential in giving rise to plate-like motion.

### 5.1. *Fault modeling methods*

Since the repeat time of great earthquakes is longer than our modern, instrumental record, it is not known if the boundary between subducting and overriding plates is a fault which changes only slowly on geologic time scales or if a different plane is used during successive earthquakes. The later possibility seems unlikely and the boundary between the two plates is almost certainly thin in proportion to the thickness of the lithosphere.

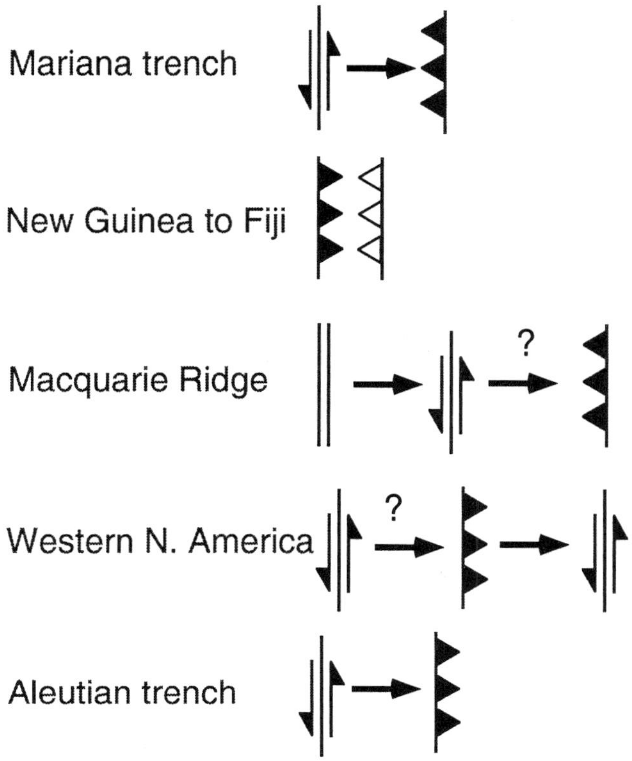


**Figure 2.** Summary of changes in the character of the active plate boundaries in the Pacific Basin.

Although at depth the interplate coupling surface may be more akin to a shear zone, from the standpoint of formulating models it is convenient to view this structure as a fault – a curvilinear surface over which there are abrupt displacements. Numerical methods have been developed to incorporate the geometric and material complexity of a converging plate boundary (Fig. 1) – a narrow fault or shear zone which extends down to $\sim 100$ km depth and curves from shallow dips near the surface to larger dips at depth – into geodynamic models. In all cases, the finite element method [e.g., *Hughes*, 1987] has been used to model faults in geodynamic contexts within viscous [*Zhong and Gurnis*, 1992; *Barr and Houseman*, 1996; *Zhong et al.*, 1998] and viscoelastic [*Melosh and Raefsky*, 1983; *Melosh and Williams*, 1989; *Toth and Gurnis*, 1998] materials.

In seismology, elastic dislocations (faults) have long been used to model the instantaneous deformation associated with earthquakes. In the first geodynamic application, relaxation following large earthquakes was determined in a viscoelastic material [*Melosh and Raefsky*, 1983]; a split node technique was used in which a single node could have a discontinuous displacement in the elastic field [*Melosh and Raefsky*, 1981] while subsequently remaining locked as the material deforms viscously. This formulation is ideal for models of the earthquake cycle but not so useful for models of longer term geodynamic processes.

Later, a slippery node technique was introduced, removing the across-fault fixed displacement, so that the fault could subsequently slip as the forces within the continuum evolved [*Melosh and Williams*, 1989]; in such a model, the geometry of the fault could change as a function of time since the fault was represented with nodes moving with the deforming Lagrangian mesh. Although this approach has been used to model converging margins on mantle convection time-scale [*Gurnis et al.*, 1996] after $\sim 10^6$ yr integration times, the mesh can become extremely contorted with voids opening at the fault tip. To overcome these problems for long integration times, *Toth and Gurnis* [1998] introduced a method to model a preexisting weak zone as a graded fault with re-meshing. The fault is modeled as a discrete boundary which is defined by a shear stiffness and a normal stiffness. The shear stiffness is effectively the amount of shear coupling between the opposing fault surfaces and controls the shear stress on the fault plane. The shear stiffness increases until the fault becomes fully locked at depth. In this way an abrupt fault tip within the upper mantle is avoided; the relative displacement across the fault gradually decreases to zero with depth as the shear stiffness on the fault increases. The discontinuous brittle deformation is gradationally replaced by continuous viscoelastic strain. In a single model it allows the elastic/brittle nature of lithospheric deformation to be integrated with viscous mantle deformation. Remeshing is essential for long ($> 10^6$ yr) integration times and an efficient, stable method has been developed for two-dimensional problems [*Toth and Gurnis*, 1998].

For integration times longer than $10^6$ years, viscous formulations have also been used. The Stokes equation appropriate for viscously dominated flows is solved in instantaneous models. Generally, faults are modeled as internal curvilinear surfaces with velocity or stress constraints placed on the fault surface. In viscous formulations where an Eulerian solution to the equations is used, this is not a problem [*Zhong and Gurnis*, 1995b]. *Zhong and Gurnis* [1992] introduced faults into a viscous model of mantle flow, assuming that normal velocities vanish on the fault, and showed that oceanic trenches could be formed with the buoyancy force of the slab. The assumption of normal velocities which vanish cannot be justified and subsequently normal velocities were assumed to be continuous while tangential velocities discontinuous [*Zhong and Gurnis*, 1994; 1996; *Zhong et al.*, 1998]. Flow on either side of a fault was coupled through frictional stress. Without the normal velocity constraint, faults nearly always have finite normal velocities. Since trenches rollback [*Chase*, 1978] (sometimes rapidly, as shown by *Bevis et al.* [1995] for the Tonga trench), converging margins are unlikely to be in steady-state, and such normal velocities are to be expected. Finite normal velocities are more naturally taken into account in transient calculations of mantle flow where the solution of the momentum and energy equations are coupled – as in any time-dependent model of mantle convection [e.g., *McKenzie et al.*, 1974]. Since faults (having finite normal velocities) are constructed with nodes, faults must be advected. In time-dependent convection models, *Zhong and Gurnis* [1995b] overcame these aspects with a hybrid Lagrangian Eulerian formulation which was shown to be convergent with migrating fault position and evolving fault geometry.

### 5.2. *Instantaneous Mantle Flow Models*

The importance of plate driving mechanisms and how mantle rheology influences plate motion, the geoid, and intraplate stress have been addressed with instantaneous models of mantle flow [*Hager and O'Connell*, 1981; *Bott*, 1983; *Vigny et al.*, 1991]. In all cases, the strategy is to incorporate the detailed geometry and magnitude of buoyancy forces and then to vary material properties in an effort to find the best fit with geophysical observations (for example). In most studies, plates have been incorporated by assuming a torque balance between plates and mantle flow. Since plates are assumed to be rigid, the existence of plates is guaranteed in torque balance models. Consequently, one of the purposes of developing fault-based methods is to overcome this substantial limitation of torque balances, as the existence of faults by themselves does not guarantee the existence of plates. In fact, we demonstrate that faults and self-weakening (or plastic) rheologies appear to dynamically interact, giving rise to plate-like motion.

A series of representative calculations showing the influence of faults on instantaneous flow are now given. For the buoyancy forces, a half-space cooling thermal boundary layer (within a region, **R**, for 3D models) and a cold subducted slab are included (Fig. 3 a,b). Outside of **R**, but excluding the slabs, the thermal structure is identical to the oldest surface age within **R**. A 100 km thick slab extends from a depth of 100 km to 670 km with a 60° dip angle. **R** is bounded by faults which dissect the upper 100 km layer, except for the side with the spreading center (Fig. 3a,b). Fault dip is shallow when the fault is above a slab, but vertical elsewhere. The 2D models can be viewed as cross sections dissecting the 3D models of a subducted slab and spreading center (Fig. 3b).

The viscosity can be a power-law rheology where the effective viscosity, $\eta_{eff}$, is given by

$$\eta_{eff} = A^{1/n}\dot{\epsilon}^{1/n-1} \tag{2}$$

where $n$ is the exponent and is unity for a Newtonian rheology; $A$ is the preexponent depending on temperature, pressure, and composition; $\dot{\epsilon}$ is the second invariant of the strain rate tensor. In the upper mantle

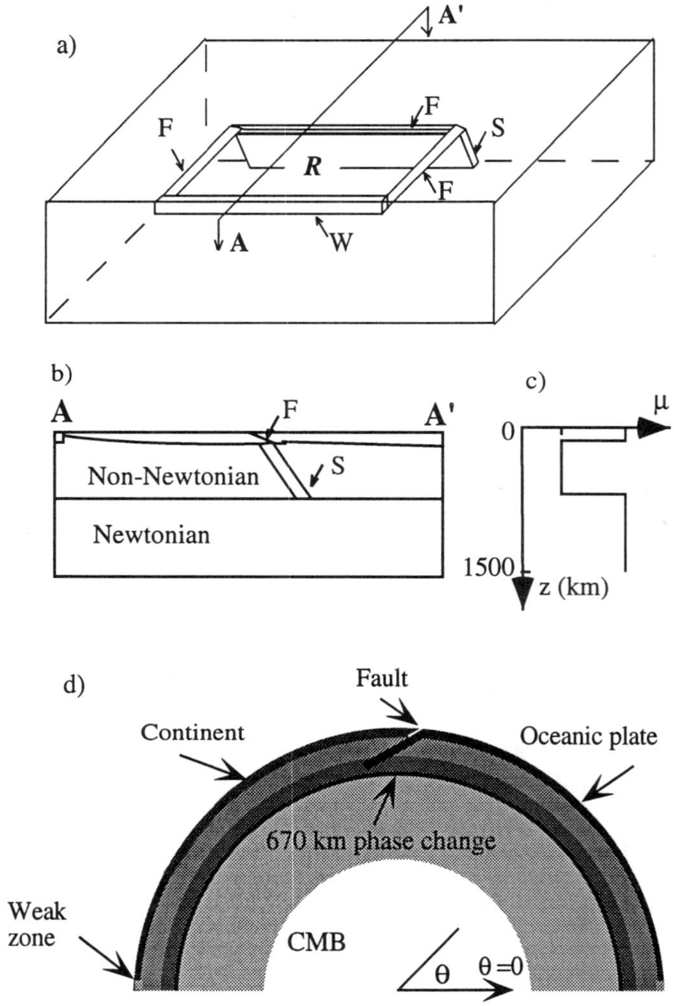


**Figure 3.** The geometry of the three-dimensional flow model with faults (a), a cross-section AA' of the model for viscosity and thermal structure (b), and the average viscosity profile (c). In (a), F, S, and W label faults, slab, and weak zone, respectively. (d) Setup for the transient mantle convection model in a 2D cylindrical geometry.

including the upper 100 km, $n = 3$, while the lower mantle viscosity is Newtonian and equal to 2x$10^{22}$ Pa s. The power law pre-exponents for slabs and upper 100 km are larger than other regions so as to implicitly account for a temperature-dependent viscosity. At the spreading center, a narrow region within the upper 100 km has a smaller pre-exponent in order to account for the higher temperature and partial melting below a ridge. These pre-exponents are chosen so that the average effective viscosities for the upper 100 km and upper mantle are about $10^{23}$ Pa s and 2x$10^{20}$ Pa s, respectively, and that the average surface velocity in $\mathbf{R}$ is close to the velocity which was used to derive the input buoyancy fields.

*5.2.1. Two dimensional models.* In order to understand the influence which faults have on instantaneous mantle flow, we begin our discussion with a two-dimensional (2D) model which has no faults. The horizontal velocity of the subducting plate is about 1.3 cm/yr, significantly larger than that of the overriding plate (Fig. 4a,b and Fig. 5a). Within a significant portion of the overriding plate ($\sim$ 500 km in width) above the slab, there is a substantial amount of in-plane strain (Fig. 4a,b and 6a); the overriding plate is in a state of a horizontal compression while the subducting plate is in horizontal extension (Fig. 4c). Looking more deeply, the shallow part of the slab is in down-dip extension (Fig. 4c) while at even greater depths (about 500 km) the slab is in down-dip compression. Dynamic topography is substantial (about 500 km in width and 2.4 km in amplitude) directly above the slab (Fig. 5b). However, this topographic depression above the slab is qualitatively and quantitatively different from oceanic trenches (Plate 1). Oceanic trenches are asymmetrical and substantially narrower in width (about 100-150 km, Plate 1) and deeper, about 4.5 km. In the top 100 km, besides a weak spreading center derived from a reduced exponent, a weak zone also develops above the slab (Fig. 4a,b). The weak zone above the slab is generated by the nonlinear viscosity which causes the otherwise strong upper 100 km layer to fail near the large slab induced buoyancy forces. In terms of the top surface horizontal velocity, this model without a fault results in a fairly good plate (termed plateness by *Zhong et al.* [1998]) in that the horizontal velocity is approximately peicewise uniform (Fig. 5a). However in terms of topography (Fig. 5b), the 'plate boundaries' are substantially more diffuse than they are in reality.

In comparison to the situation just described, a dipping fault has a substantial influence on surface topography and stress and flow within the upper part of the domain. A curved, dipping fault is included in the calculation while all other parameters are identical to the preceding case. There is no resisting shear stress on the fault. With the inclusion of a fault not only does the velocity of the subducting plate double in magnitude (Fig. 5a), but the velocity of this plate becomes substantially larger than that of the overriding plate (Fig 4e). The subducting plate velocity along the fault is

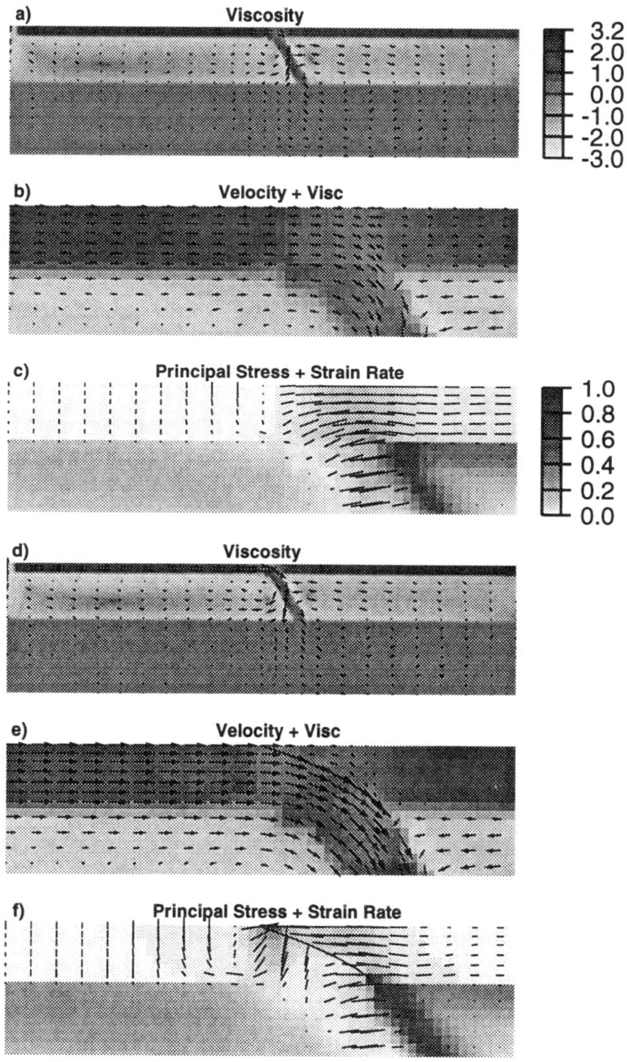


**Figure 4.** Effective viscosity and flow velocity (vectors) for the entire model domain (6,000 km by 1500 km) (a), magnified views in the vicinity of slab show effective viscosity and flow velocity (b) and second invariant of strain rate and principal compressive stress (c) for the 2D case with no fault. (d), (e), and (f) are for the case with a curved fault. In (c) and (f), the length and orientation of each line represent the magnitude and direction of principal compressive stress.

nearly parallel to the tangent of the fault (Fig. 4e), although there is still a small, finite normal component. The tangential velocity is discontinuous across the fault, and comparable with the relative motion of convergence between subducting and overriding plates (Fig. 4e and 5a). With a fault, the compressional stress is reduced in the overriding plate (Fig. 4f), compared with the case without a fault (Fig. 4c). Within the subducting plate directly beneath the fault, principal compressional stress is rotated so that it is approximately perpendicular to the fault, and in down-dip extension (Fig. 4f). Assuming an upper mantle viscosity of $4 \times 10^{20}$Pa s, we find that the average of the second invariant of the stress tensor within the oceanic lithosphere is $\approx$ 30 MPa and about double that in the forebulge. The rotation of compressional axis perpendicular to faults occurs when there is little to no resisting shear stress on the faults and is similar to the focal mechanism of small earthquakes in the vicinity of interplate faults [*Kanamori*, 1980; *Magee and Zoback*, 1993]. Moreover, the down-dip extension in the top 100 km layer is similar to seismic observations of lithospheric earthquakes (Fig. 1) and the average strain rate in the modeled region where the lithospheric earthquakes occur is $\approx 2 \times 10^{-15}$ s$^{-1}$ – somewhat larger than that inferred from lithospheric earthquakes globally but nearly that inferred for the Japan trench (Section 2). On the oceanic plate and near the fault, the topography is characterized by an outer rise ($\approx$ 0.4 km in amplitude and $\approx$ 200 km in width) and trench ($\approx$ 4 km in amplitude and $\approx$ 100 km in width) – identical to observed trench-forebulge bathymetry. On the overriding plate, topography is elevated near the fault but depressed over the slab (Fig. 5b).

For instantaneous models, these results show that plateness (defined in terms of both a piece-wise uniform horizontal velocity and the topographic sharpness of plate edges), is dramatically improved when faults with low (to nonexistent) shear stress are incorporated. These models demonstrate that trench-forebulge bathymetry can be realistically modeled with buoyancy forces with low deviatoric stresses in the plate – a substantial improvement over the ad-hoc use of displacement and moment arm boundary conditions at the edges of plates which cannot be tied to buoyancy forces. Moreover, flexural models cannot predict plate motions – a primary outcome of a geodynamic model.

*5.2.2. Three dimensional models.* Three-dimensional models are needed to examine the influence of transform faults and oblique convergence. For comparison, our first case is the 3D equivalent of our 2D case containing no faults and with slab strike parallel to the spreading center (Fig. 6a). Surface velocities in region, **R**, vary gradually in both the spreading parallel and perpendicular directions (Fig. 6b, Fig. 7a,b). The ratio of toroidal to poloidal components of the surface velocity is substantially smaller than for a perfectly rigid plate with this configuration [*Zhong et al.*, 1998]. The effective viscosity of both the diverging and converging "margins" are reduced. The effective viscosity in other surface regions varies gradually, reflecting the variations in strain rate (Fig. 6b). Topography in region, **R**, increases with age and distance from the spreading center, as expected, but over the slab there is a broad depression ($\sim$ 500 km in width) (Fig. 6c), similar to that from the 2D calculations (Fig. 5b) and completely unlike observed trench topography.

We now contrast this with a model that includes faults with zero frictional stress. The strike of the dipping fault is parallel to the spreading center but perpendicular to the strike of vertical faults (Fig. 6d). We find that surface velocity in region, **R**, bounded by faults is nearly constant (Fig. 6e, 7a,b) and parallel to the

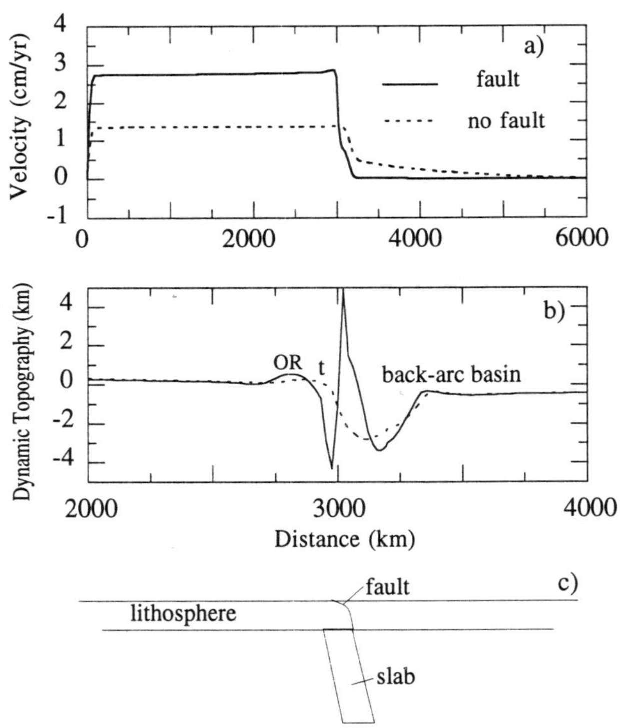


**Figure 5.** Surface velocity profiles (a) and dynamic topography (b) for the 2D instantaneous viscous flow models shown in Fig. 4. The surface expression of the fault is at 3,000 km. In b, the following are labeled for the model with no fault (solid curve): outer rise, OR, trench, t, and back arc basin. (c) A schematic showing the location of the fault and slab for the profiles in b.

strike of vertical faults. The velocity contrasts across faults are sharp (Fig. 7a,b). Unlike the case without a fault, the ratio of toroidal to poloidal surface velocities become as large as for a perfectly rigid plate with this geometry [*Zhong et al.*, 1998]. Like the 2D model, an outer rise ($\approx$ 200 km in width and 0.4 km in amplitude) and trench ($\approx$ 100 km wide and 4 km deep) topography develop near the dipping fault (Fig. 6f, 7d). Nonlinear interaction between weak faults and the stress weakening rheology results in the plate-like surface velocity. With these models, we find that weak faults and a stress weakening rheology, and weak coupling between plate and asthenosphere are all necessary in order to achieve the plate-like surface motion and that faults play an important role in producing oblique subduction.

As the frictional resistance of faults increases from 0 to 20 MPa, the surface velocity, toroidal to poloidal ratio, and plateness decrease and progressively resemble flow with no faults (Fig. 7a-c). As the magnitude of the absolute shear stress on the fault increases, trench depth decreases (Fig. 7d). In a systematic analysis of trench depth versus age of the subducting plate, fault dip angle, and Benioff zone length and dip, the best fitting interplate stress for ocean-ocean converging margins is about 10-30 MPa [*Zhong and Gurnis*, 1994], slightly larger than the stress drop of interplate earthquakes [*Kanamori*, 1980] but within the range estimated from heat flow [*Lachenbruch and Sass*, 1988].

### 5.3. *Transient mantle flow models*

Since plate geometry and motion have varied in the past [e.g., *Engebretson et al.*, 1985), time-dependent convection models are needed to address a variety of geophysical questions. One important problem is the relationship between present day mantle structure (seismic tomography) and time-dependent plate kinematics [e.g., *van der Hilst and Seno*, 1993]. Another set of questions relate to the reason subduction zones initiate where they do and how subduction initiation is related to plate evolution [e.g., *Richards and Lithgow-Bertelloni*, 1996]. A more complete understanding of the dynamics of subduction initiation is important because it allows us to determine the conditions necessary for subduction to occur (such as the strength of the driving forces) as well as those processes which inhibit subduction growth (such as resisting forces on faults). Moreover, with a better understanding of the time-evolution, predictions can be tested against the tectonic history of a former site of subduction initiation. Since the change in plate geometry must occur through variations in margin geometry, it is essential to formulate models of mantle convection that include plates with dynamically evolving margins. As mentioned in Section 5.2, in the traditional method to model plates, the torque balance approach, a fixed plate geometry must be imposed. With fault methods, these limitations can be overcome.

Time-dependent models using viscous and viscoelastic materials have been setup. For viscoelastic materials, a series of models have been formulated, first, to understand the forces necessary and fault coupling inhibiting subduction initiation and, secondly, to assess how stresses and topography vary as subduction proceeds. A preexisting, dipping fault cuts the lithosphere while the oceanic plate is pushed from the side for $\approx$ 20 Myr with force boundary conditions. In a similar way, a viscous model is formulated, integrated for $>$ 200 Myr, allowing interaction between the evolution of the slab and fault geometry and position. Viscoelastic models of plates and the mantle on $10^6$ yr time-scales allow us to bridge the gap between instantaneous models and longer term, time-dependent models of plate-mantle convection. The viscous model is entirely self-consistent in that the buoyancy forces and fault evolve

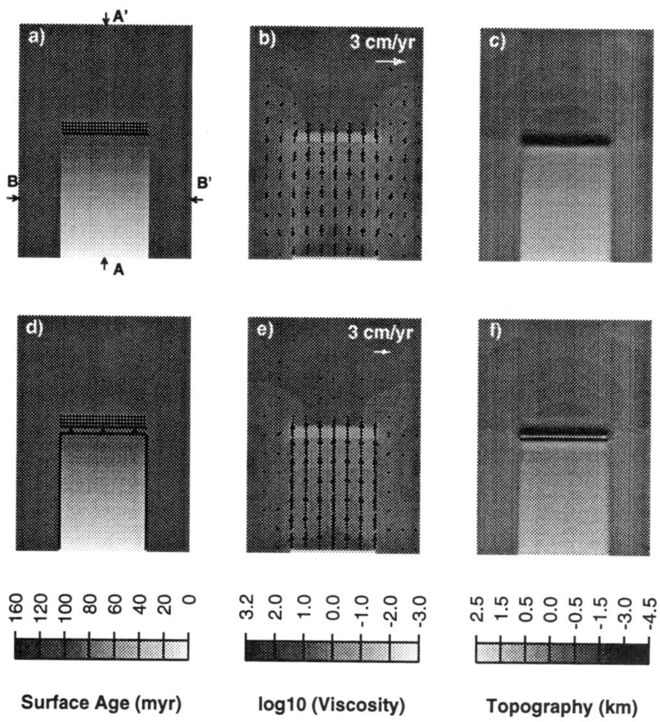


**Figure 6.** The top (a, b, and c) and bottom three panels (d, e, and f) are the mapview of input buoyancy, outputs of surface velocity and effective viscosity, and dynamic topography, for the 3D cases with no fault and with faults, respectively. In (a) and (d), the parallel lines outside **R** show the depth of slabs. In (d), the lines bounding **R** represent faults; triangles on fault lines indicate a dipping fault.

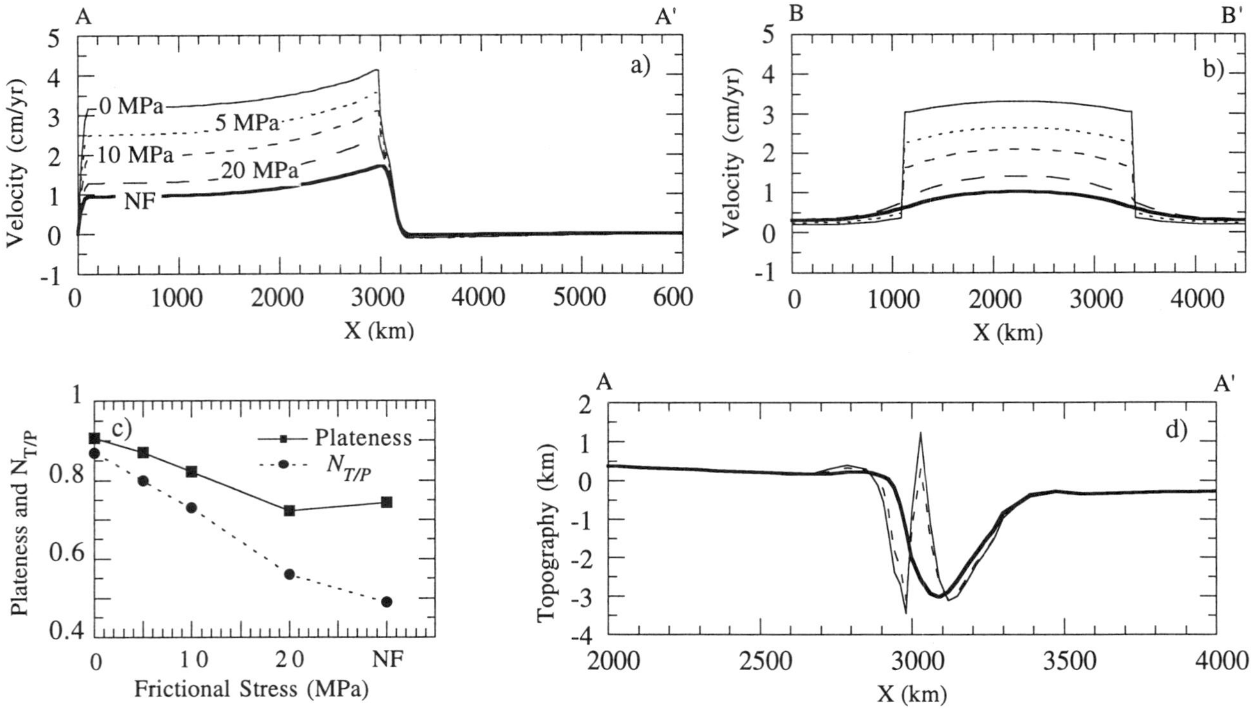


**Figure 7.** For the 3D models shown in Fig. 6, surface velocities for cross sections AA' (a) and BB' (b), and normalized ratio of toroidal to poloidal components and plateness (c) for cases with no faults (NF) and with faults of different frictional stress. In (d) dynamic topography is shown for the case with no fault as well as for a fault with either a stress of 0 MPa or 10 MPa

entirely on their own. The model in the viscoelastic material is not entirely self consistent because part of the force is imposed; however, fault geometry, shape, and position evolve self-consistently.

*5.3.1. Viscoelastic models of subduction initiation.* Within a newly nucleating subduction zone, profound changes occur at the plate boundary as slab pull comes to dominate the dynamics. Despite a fairly extensive literature, only recently have time-dependent models with realistic geometries and rheologies been computationally possible [*Toth and Gurnis*, 1998]. We highlight the major results of a detailed study by *Toth and Gurnis* [1998] in which newly initiating subduction was investigated with velocity and force boundary conditions.

The initiation of subduction in a two-dimensional viscoelastic material is studied for the first 20 Myr of plate convergence for a thermally defined oceanic lithosphere. During the first $\sim$ 10 Myr, using a ridge push appropriate for old oceanic lithosphere (4x10$^{12}$ N/m) [*McKenzie*, 1974; *Turcotte and Schubert*, 1982]), the plate is nearly everywhere in compression, except near the trench which is under extension (Fig. 8a). After 20 Myr of convergence, the slab has reached a depth of nearly 250 km and the deviatoric compression within the overriding plate, above the fault, has increased to nearly 100 MPa. Within the overriding plate, away from the trench, deviatoric stresses change from compression at 5 Myr to extension at 20 Myr. Topography, characterized by a distinctive trench and forebulge (Fig. 9d), grows with time as the buoyancy beneath the fault increases and is qualitatively the same as that which occurs in viscous models. The maximum tensional stress which occurs within the forebulge is only $\approx$ 200 MPa – substantially smaller than those which occur in elastic or elastic-plastic bending models [e.g., *Chapple and Forsyth*, 1979].

As the slab lengthens, slab pull and plate velocity increase. For example, with no resistance on the fault and a ridge push of 4x10$^{12}$ N/m, the velocity of the subducting plate is initially 0.9 cm/yr, increasing to 2.4 cm/yr by 20 Myr (Fig. 9a). With a ridge push of 6x10$^{12}$ N/m, the initial convergence velocity is 1.4 cm/yr (Fig. 9a), but by 20 Myr has increased to nearly 12 cm/yr with the slab reaching a depth of 500 km. If the ridge push is reduced to 2x10$^{12}$ N/m, the initial convergence is only 0.4 cm/yr and does not increase substantially with time (Fig. 9a). In terms of the subsurface temperature distribution, the 1200°C geotherm

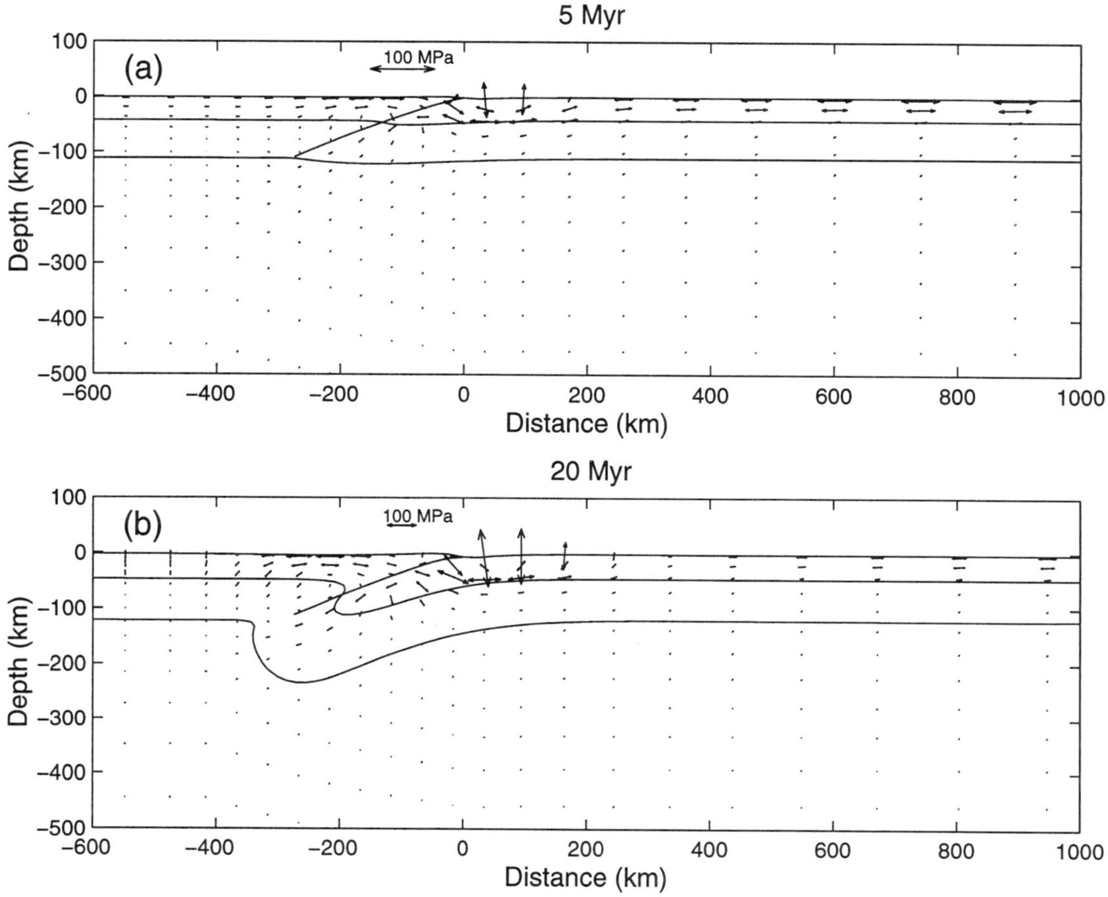


**Figure 8.** Principal deviatoric compressive stress and isotherms for the model with a constant force boundary of $4x10^{12}$ N/m . Isotherms at 600°intervals. (a) After 5 Myr little convergence has taken place and horizontal compression dominates the stress field within both the overriding and subducting plates. (b) After 20 Myr convergence rate has increased and the slab is now well developed. Down-dip compression is seen within the slab and horizontal extension is seen within the overriding plate away from the trench. Note that these views represent zoom images around the slabs and that the entire model domain was 4,000 km (width) by 1,000 km (depth).

has been perturbed downwards by less than 40 km after 20 Myr. Consequently, with this lower ridge push, the rate of convergence is so low that diffusive heating inhibits slab development, and it appears that subduction will not initiate for a significantly longer period of time. For a constant ridge push ($6x10^{12}$ N/m), when the shear stress on the fault is increased from zero to 8.8 MPa, the initial convergence velocity is reduced from 1.4 cm/yr to 0.9 cm/yr (Fig. 9b). The high shear stress prevents subduction from becoming well established as the convergence velocity after 20 Myr increased to only 1.2 cm/yr. By observing the history of plate convergence and slab lengthening, *Toth and Gurnis* [1998] were able to map out the domains of subduction initiation or failed initiation as a function of both plate driving force and interplate coupling (Fig. 10).

*5.3.2. Time-dependent mantle convection.* The time-dependent model of mantle convection which we show is in many ways similar to the suite of viscoelastic models just presented. It includes continental and oceanic plates, separated by a thrust fault with no resisting shear stress (Fig. 3d). This methodology has been used to study the influence of slab and mantle rheology [*Zhong and Gurnis*, 1995a; 1997] and thermodynamic formulation [*Ita and King*, 1998] on the history of plate motion and slab structure. An important difference is the inclusion of an endothermic phase change at 670 km depth. Initially, the fluid interior is isothermal (1600°C) with a cooling oceanic plate along the surface, zero age at $\theta = 0$. The material below the 410 km depth is assumed to be Newtonian, while the shallower regions, non-Newtonian. Like the instantaneous viscous

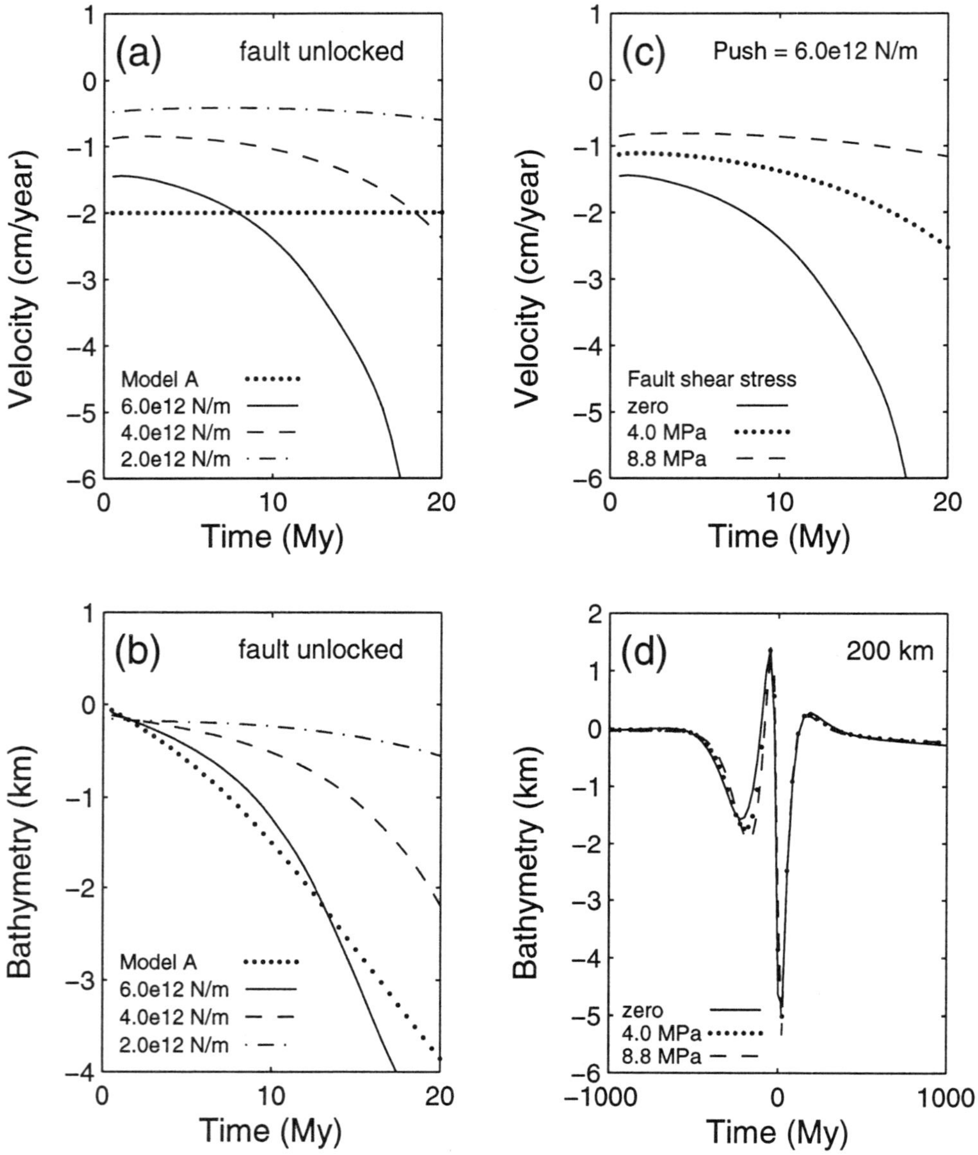


**Figure 9.** For viscoelastic models: (a) Convergence velocity and (b) bathymetry 250 km from the trench, for models with different far field forces and as well as model with a constant plate velocity. (c) Convergence velocity and (d) bathymetry after 200 km of convergence, for models with a far field force of $6\text{x}10^{12}$ N/m and different fault shear stresses.

and transient viscoelastic models, non-Newtonian viscosity is critical to the weakening of the oceanic plate as slabs slide by the thrust fault during subduction.

The negative buoyancy of the cooling oceanic plate initiates its own subduction as well as the motion of the continental plate which starts shortly after the model begins (Fig. 11a). Accompanying overriding plate motion, the faulted converging margin moves oceanward (Fig. 12b ) such that the overriding plate increases in size. When the slab encounters the phase change at 670 km (at 55 Myr) plate motion is slightly reduced (Fig. 12a). During its initial encounter with the phase change, the slab cannot penetrate below 670 km, and deforms within the transition zone (from 50 Myr to 117 Myr, Fig.11a). During this period, the subducting and overriding plates converge at relatively constant velocity, and the overriding plate increases in size at the expense of the subducting plate. By $\approx$ 200 Myr, the length of the overriding plate has increased by more than 2,000 km (Fig. 12b). The slab now sinks into the lower mantle, apparently after a sufficient mass of cold material has accumulated above the phase change (Fig. 11b) [e.g., *Tackley et al.*, 1993]. As the slab sinks into the lower mantle at $\approx$ 200 Myr the velocity of the subducting plate increases rapidly, while the velocity of the overriding plate decreases. As the slab reaches the bottom boundary (225 Myr), the velocity of subducting plate gradually decreases to a relatively constant velocity 4.0 cm/yr.

Plate-like motion is achieved for both overriding and subducting plates (Fig. 11). The overriding plate consists of continental and oceanic components, but they form a single coherent plate (Fig. 11). During the early stages of subduction, the slab is deflected above the 670 km phase change with a shallow dip angle, while the slab extends beneath the overriding plate by more than 1000 km (Fig. 11a,b). After the slab penetrates into the lower mantle, the slab has a steep dip (Fig. 11c). When the slab is deflected above the 670 km, the trench tends to migrate at a larger rate. However, when the slab penetrates into the lower mantle, the horizontal motion of the trench ceases (Fig. 12a). One of the most important conclusions of these models is the clear association between plate kinematics and the structure of subducted slabs. Moreover, slab penetration into the lower mantle is time-dependent and sometimes ephemeral.

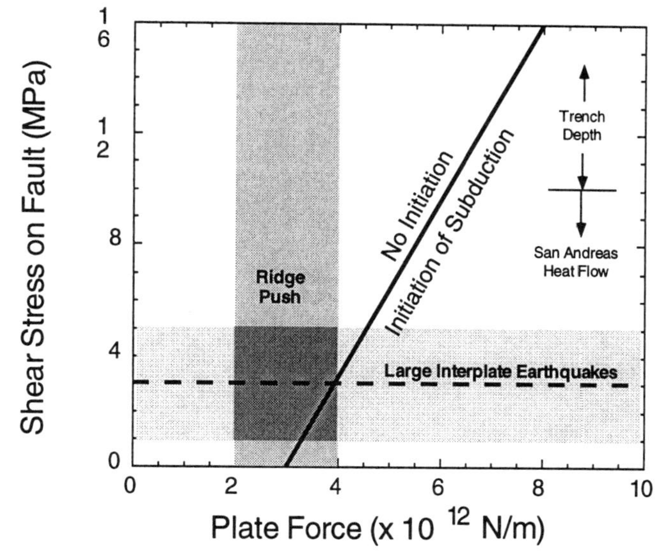


**Figure 10.** Comparison of theoretical prediction of the boundary between domains of subduction initiation and no initiation. Shear stresses on preexisting faults from three different estimates are indicated: from large interplate earthquakes, from heat flow on San Andreas fault, and from trench depths.

## 6. DISCUSSION AND CONCLUSIONS

Two important lithospheric properties motivate mantle convection models with non-Newtonian viscosity and preexisting faults. First, subducting oceanic lithosphere below trenches has a relatively small mechanical strength compared with normal oceanic lithosphere. Second, plate margins are weak not only because there is continuous brittle failure of oceanic lithosphere (instantaneous rheology), but because the lithosphere preserves weak faults and shear zones for long periods (often > 1 Gyr) of geologic time (history dependent rheology). Ultimately, these two mechanisms by which low mechanical strength is carried by the lithosphere is crucial for understanding how plate dynamics operates and how plate tectonics is formed from mantle convection.

As the oceanic lithosphere slides by the interplate coupling zone, it is normally thought of as being fairly strong as indicated by the outer rise which bears similarities to flexural bulges around oceanic islands. Even the elastic-perfectly plastic bending model assumes that the core of the oceanic lithosphere supports the outer rise elastically [*McAdoo et al.*, 1978; *Chapple and Forsyth*, 1978]. We argued that there are three reasons to question this paradigm: the occurrence of *lithospheric* earthquakes which completely fracture the oceanic lithosphere, the space-time occurrence of outer rise earthquakes which demonstrate that the outer rise is a dynamic feature, and the ability to dynamically generate trench-forebulge bathymetry with no elastic strength at all. We have poor constraints on the strain rate and effective viscosity of the trench environment, but it could have a strain rate as high as that within the upper mantle. However, it has long been known that the elastic bending model of the oceanic lithosphere, the principle argument in support of strong margins, is not unique and dynamic support of the outer rise within a weak lithosphere is possible (Fig. 5, 7).

Interplate fault zones are weak as suggested by earthquake stress drops, heat flow near faults, and the orientation of small earthquakes near interplate faults. From the geologic history of oceans and continents there are numerous examples showing how fault zones have been reactivated from transform to convergence, from convergence to transform, and from divergence to strike-slip. The time-scale between formation and reactivation can exceed a few gigayears and there are examples of reactivation on all time-scales of shorter duration.

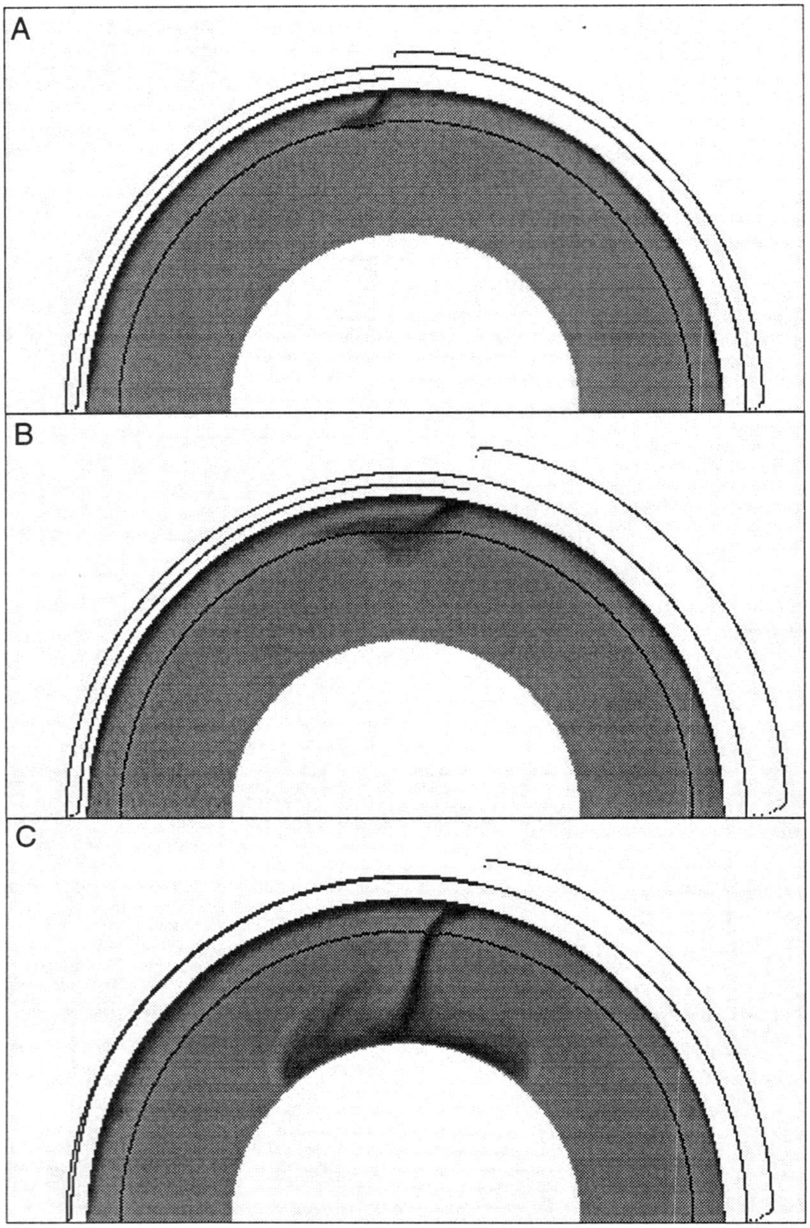


**Figure 11.** Thermal structure and surface plate velocity (solid line) resulting from the transient mantle convection model at three different times (a), (b), and (c). The dashed line indicates the zero velocity. The solid line within the fluid interior indicates the 670 km phase change.

This substantial body of geophysical and geological knowledge motivated the incorporation of preexisting faults in models of instantaneous and time-dependent mantle convection from which a number of fundamental conclusions can be drawn. With a realistic distribution of buoyancy forces within either viscous or viscoelastic materials the distinctive trench forebulge morphology easily develops. Without the ad-hoc displacement and moment arm boundary conditions on elastic plates which have long been used to curve fit bathymetry near a trench, we are able to get a nearly perfect model of the trench and forebulge. The absolute magnitudes of trench depth and plate velocity, both unobtainable with traditional flexure models, can both be predicted with the new methods with preexisting faults and realistic slab buoyancy.

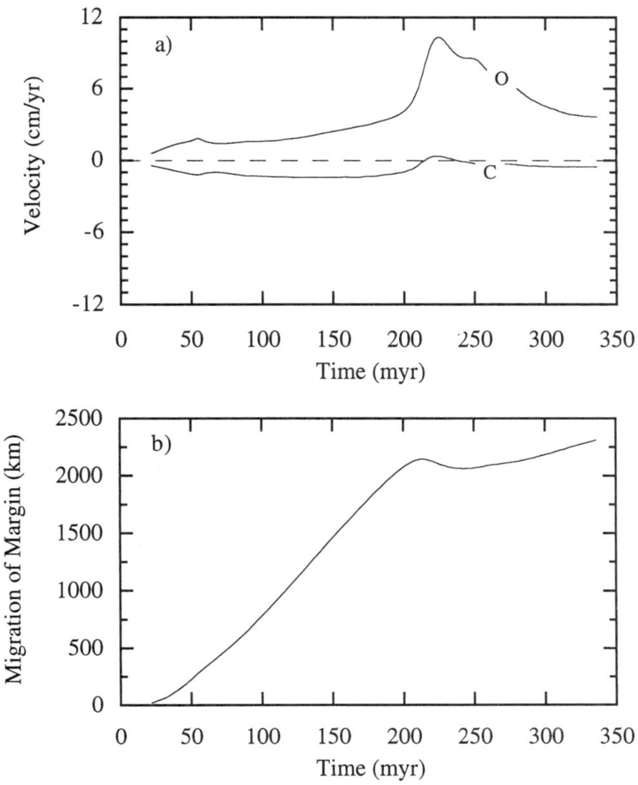


**Figure 12.** Time history of plate velocities (a) and trench migration (b) for the convection model shown in Fig. 11. Symbols O and C in (a) indicate the velocities of oceanic and continental plates, respectively. The dashed line in (a) indicates zero velocity.

A new subduction zone can initiate at a preexisting fault with reasonable ridge push and slab pull forces. For a ridge push of $4 \times 10^{12}$ N/m subduction initiates even with a fault shear stress of 3 MPa, the resistance most consistent with seismological observations. Subduction will not initiate by ridge push alone when this fault shear stress is greater than 5 MPa. However, when the Marianas and Aleutian subduction zones were forming, for example, the total force available to initiate subduction would have have been substantially larger than ridge push alone because of the force from subducted slabs attached to other parts of the Pacific plate. In this case, even with an interplate coupling estimated from dynamic models (10-30 MPa) subduction zones at preexisting faults could easily have formed (e.g. Fig. 10). Unfortunately, our estimate for the minimum forces necessary to initiate subduction is not a conservative one as it is likely that additional work will be needed to evolve a vertical fault zone into a dipping one. A challenge for the future will be the development of models to follow the evolution of a transform fault zone into a thrust fault zone. This process is apparently presently occurring at the Puysegur ridge and trench at the northern end of the Macquarie ridge complex with the development of reverse faults along the transform fault [*Collot et al.*, 1995]. It is likely that models can be tailored to this region to better understand this vital, but poorly understood, phase of the plate tectonic cycle.

With both the viscous and viscoelastic models, plate velocity only slowly increases with time. It is essential for us to determine the means by which a plate can increase in velocity by a few cm/yr in less than 1 Myr (a dramatic acceleration), as indicated by the bend in the Hawaiian-Emporer seamount chain. The modeling shows that during the initiation of subduction, the rate at which plate velocity accelerates increases with the driving force but decreases with interplate coupling. A lower bound on the time-scale at which plate velocity can dramatically accelerate has not been well determined but it is unlikely to be substantially smaller than 10 Myr if plate velocities are less than $\sim 10$ cm/yr. From the history of modeled plate convergence, it takes $> 10$ Myr for convergence to reach appreciable levels (a few cm/yr). This supports the contention that plate motion cannot quickly reorganize as new subduction zones form [*Richards and Lithgow-Bertelloni*, 1996].

We have developed 3D models with non-Newtonian viscosity and preexisting faults and are able to obtain realistic plate tectonic-like motion and realistic plate margin topography. *Bercovici* [1993] and *Tackley* [1998] have addressed this problem using a different rheology. The importance of rheology is shown by looking at how rheologies have been incorporated into dynamic plate models. A 'stick-slip' rheology, a highly nonlinear viscous rheology where $n = -1$, equation 2, has been used by *Bercovici* [1993] and *Tackley* [1998]. When $n = -1$, only a finite deviatoric stress can be supported as strain rate increases and for strain rates beyond some critical value, stress decreases. This rheology approximates the brittle failure of rocks with a viscous rheology and has been referred to as self-weakening or strain rate weakening; it is an example of what we call an instantaneous rheology. Using this rheology, *Bercovici* [1983] modeled plates in instantaneous flows with sources and sinks within a thin sheet; *Tackley* [1998] later extended the approach to 3D with more realistic internal buoyancy forces. They found that weak plate margins formed above upwellings and downwellings and within localized narrow shear bands (akin to transform faults) between upwellings and downwellings. Since the generation of weakness is proximal to downwellings we believe that this aspect of their models is akin to continuous brittle failure, which we document within the trench environment (Section 2). The sites of the greatest amount of present day brittle failure within oceanic lithosphere are not arbitrarily located, but rather occur in close proximity to long lived zones of preexisting weakness. We believe that the sole use of self-weakening rheologies is not adequate to model instantaneous plates because the weak zones in the *Bercovici* [1993] and *Tackley* [1998] models develop instantaneously within an intact medium directly in response to imposed forces. The generation of plate margins in these models from an intact lithosphere is not consistent with the geologic evolution of oceanic plate margins since the Mesozoic (Section 4).

More specifically, *Bercovici* [1995] applied the self weakening rheology to present day plate motion using a thin viscous spherical shell with sources and sinks of flow determined from observed divergence and convergence. He was able to reproduce plate motion with these constraints when a strain softening rheology was used. In these models, the weak zones form directly in response to the forces. For example, the Mariana, Macquarie Ridge, and Aleutian margins of the Pacific (as other plate margins) are weak in *Bercovici's* [1995] model; however, we have detailed the geological evolution of these regions (as well as other regions) and have shown that the locations of these margins are inherited from preexisting structures.

Furthermore, we believe that the sole incorporation of self-weakening rheologies in time dependent models of mantle convection (technically not yet achieved) in order to simulate time-dependent plate dynamics should be approached with caution. The time-dependence of mantle convection is strongly governed by the reactivation of preexisting structures. A model with only an instantaneous rheology would not be able to nucleate new plate margins in regions away from localized buoyancy forces or active plate margins. Consequently, the incorporation of history dependent rheology is essential for time-dependent models. This need for history-dependent rheology and preexisting structures motivated our development of fault methodologies. However, even these fault methods have their own limitations as they are only appropriate for transient convection calculations tailored to address specific geologic questions.

The geophysical and geological observations described here and instantaneous and transient convection models with preexisting faults suggest a complicated process for the normal evolution of plate motions – continuous brittle failure of lithosphere dynamically interacts with preexisting faults. Once a plate boundary forms, this region creates a new weak zone which later may become dormant. The lithosphere will then carry (retain) this dormant weak zone for long periods of time. The dormant structure will be the site of a new plate boundary which may be quite different from the one which it originally nucleated as. The transient models of mantle convection we have formulated do not yet approach this full treatment of the creation of mechanical weakness within the lithosphere, dormancy, and later reactivation. Without substantial modifications, it is unlikely that fault methods developed for transient models of mantle convection can be used for this purpose. A challenge for the future will be to couple the equations of motion and energy with the transport of lithology and strain history so as to generate weak zones which can survive, be transported, and be reactivated.

*Acknowledgments.* We thank Hiroo Kanamori for many discussions and unpublished information on lithospheric earthquakes. This work was supported by NSF grants EAR-9417645 and EAR-9614391. Some of the 3D models with faults were carried out on the Intel Paragon supercomputer operated by the Caltech Center for Advanced Computer Research (CACR). This represents Contribution Number 8578 of the Division of Geological and Planetary Sciences, California Institute of Technology.

## REFERENCES

Artyushkov, E. V., Stresses in the lithosphere caused by crustal thickness inhomogeneities, *J. Geophys. Res.*, *78*, 7675–7708, 1973.

Atwater, T., Implications of plate tectonics for the Cenozoic tectonic evolution of western North America, *Geol. Soc. Am., Bull.*, *81*, 3513–3536, 1970.

Barr, T. D., and G. A. Houseman, Deformation fields around a fault embedded in a non-linear ductile medium, *Geophys. J. Int.*, *125*, 473–490, 1996.

Ben-Avraham, Z., and S. Uyeda, Entrapment origin of marginal seas, in *Geodynamics of the Western Pacific-Indonesian Region*, edited by T. Hilde and S. Uyeda, pp. 91–104. Am. Geophys. Un., Washington, 1983.

Bercovici, D., A simple model of plate generation from mantle flow, *Geophys. J. Int.*, *114*, 635–650, 1993.

Bercovici, D., A source-sink model of the generation of plate tectonics from non-Newtonian mantle flow, *J. Geophys. Res.*, *100*, 2013–2030, 1995.

Bercovici, D., Y. Ricard, and M. A. Richards, The relation between mantle dynamics and plate tectonics: A primer, in *History and Dynamics of Global Plate Motions*, edited by M. A. Richards, R. Gordon and R. Van der Hilst, Geodynamic Series. Am. Geophys. Un., this volume, 1999.

Bevis, M., Seismic slip and down-dip strain rates in Wadatic-Benioff zones, *Science*, *240*, 1317–1319, 1988.

Bevis, M., et al., Geodetic observations of very rapid convergence and back–arc extension at the Tonga trench, *Nature*, *374*, 249–251, 1995.

Bott, M. H. P., Modelling the plate-driving mechanism, *J. Geol. Soc. Lond.*, *150*, 941–951, 1993.

Cande, S. C., J. Stock, C. Raymond, and D. R. Müller, New constraints on plate tectonic puzzle of the SW Pacific, *EOS, Trans. Am. Geophys. Un.*, *79*, 81–82, 1998.

Casey, J. F., and J. F. Dewey, Initiation of subduction zones along transform and accreting plate boundaries, triple junction evolution, and forearc spreading centres – implications forophiolitic geology and obduction, in *Ophiolites and Oceanic Lithosphere*, edited by I. G. Gass, S. J. Lippard and A. W. Shelton, 13, *Spec. Pub.*, pp. 269–290. Geol. Soc. Am., 1984.

Chapple, W. M., and D. W. Forsyth, Earthquakes and bending of plates at trenches, *J. Geophys. Res.*, *84*, 6729–6749, 1979.

Chase, C. G., Extension behind island arcs and motions relative to hot spots, *J. Geophys. Res.*, *83*, 5385–5387, 1978.

Christensen, D. H., and L. J. Ruff, Seismic coupling and outer rise earthquakes, *J. Geophys. Res.*, *93*, 13421–13444, 1988.

Clague, D. A., and G. B. Dalrymple, Geologic evolution of the Hawaiian–Emporer volcanic chain, *U.S. Geol. Surv. Prof. pap.*, *1350*, 5–54, 1987.

Collot, J.-Y., G. Lamerche, R. A. Wood, J. Delteil, M. Sosson, J.-F. Lebrun, and M. F. Coffin, Morphostructure of an incipient subduction zone along a transform plate boundary: Puysegur ridge and trench, *Geology*, *23*, 519–522, 1995.

Conrad, C. P., and B. H. Hager, Effects of plate bending and fault strength at subduction zones on plate dynamics, *J. Geophys. Res.*, in press, 1999.

Cooper, A. K., D. W. Scholl, and M. S. Marlow, Plate tectonic model for the evolution of the eastern Bering Sea Basin, *Geol. Soc. Am. Bull.*, *87*, 1119–1126, 1976.

Davies, G. F., and M. A. Richards, Mantle convection, *J. Geol.*, *100*, 151–206, 1992.

DeBremaecker, J.-C., Is the oceanic lithosphere elastic or viscous?, *J. Geophys. Res.*, *82*, 2001–2004, 1977.

Engebretson, D. C., A. Cox, and R. G. Gordon, Relative motions between oceanic and continental plates in the Pacific basin, *Geol. Soc. Am. Sp. Paper*, *205*, 1–59, 1985.

Good, N., and M. De Wit, The Thabazimbi-Murchison Lineameant of the Kaapvaal Craton, South Africa: 2700 Ma of episodic deformation, *J. Geol. Soc., Lond.*, *154*, 93–97, 1997.

Gurnis, M., C. Eloy, and S. Zhong, Free-surface formulation of mantle convection, part 2: Implication for subduction zone observables, *Geophys. J. Int.*, *127*, 719–727, 1996.

Gurnis, M., R. D. Müller, and L. Moresi, Dynamics of Cretaceous vertical motion of Australia and the Australian-Antarctic discordance, *Science*, *279*, 1499–1504, 1998.

Hager, B. H., and R. J. O'Connell, A simple global model of plate dynamics and mantle convection, *J. Geophys. Res.*, *86*, 4843–4867, 1981.

Holdsworth, R. E., C. A. Butler, and A. M. Roberts, The recognition of reactivation during continental deformation, *J. Geol. Soc., Lond.*, *154*, 73–78, 1997.

Hughes, T. J. R., *The Finite Element Method: Linear Static and Dynamic Finite Element Analysis*, Prentice–Hall, Inc., Englewood Cliffs, New Jersey, 1987.

Imber, J., R. E. Holdsworth, C. A. Butler, and G. E. Lloyd, Fault-zone weakening processes along the reactivated Outer Hebrides Fault Zone, *J. Geol. Soc., Lond.*, *154*, 105–109, 1997.

Ita, J., and S. D. King, The influence of thermodynamic formulation on simulations of subduction zone geometry and history, *Geophys. Res. Lett.*, *25*, 1463–1466, 1998.

Jaroslow, G. E., G. Hirth, and H. J. B. Dick, Abyssal peridotite mylonites: implications for gran-size sensitive flow and strain localization in the oceanic lithosphere, *Earth Planet. Sci. Lett.*, *256*, 17–37, 1996.

Jarrard, R. D., Relations among subduction parameters, *Rev. Geophys.*, *24*, 217–284, 1986.

Jeffreys, H., *The Earth*, Cambridge University Press, New York, 1970.

Kanamori, H., Seismological evidence for a lithospheric normal faulting-the Sanriku earthquake of 1933, *Phys. Earth Planet. Int.*, *4*, 289–300, 1971.

Kanamori, H., State of stress in the Earth's lithosphere, in *Physics of the Earth's Interior*, edited by A. Dziewonski and E. Boschi, pp. 531–554. North-Holland, Amsterdam, 1980.

Kanamori, H., Rupture process of subduction-zone earthquakes, *Ann. Rev. Earth Planet. Sci.*, *14*, 293–322, 1986.

Kanamori, H., Mechanics of earthquakes, *Ann. Rev. Earth Planet. Sci.*, *22*, 207–237, 1994.

Kanamori, H., and L. Astiz, The 1983 Akita-Oki Earthquake ($M_w$=7.8) and its implications for systematics of subduction earthquakes, *Earthq. Predict. Res.*, *3*, 305–317, 1985.

Kikuchi, K., and H. Kanamori, The Shikotan earthquake of October 4, 1994: Lithospheric earthquake, *Geophys. Res. Lett.*, *22*, 1025–1028, 1995.

Kohlstedt, D. L., B. Evans, and S. J. Mackwell, Strength of the lithosphere: Constraints imposed by laboratory experiments, *J. Geophys. Res.*, *100*, 17,587–17602, 1995.

Lachenbruch, A. H., and J. H. Sass, The stress heat flow paradox and thermal results from Cajon Pass, *Geophys. Res. Lett.*, *15*, 981–984, 1988.

Magee, M. E., and M. D. Zoback, Evidence for a weak interplate thrust fault along the northern Japan subduction zone and implications for the mechanics of thrust faulting and fluid expulsion, *Geology*, *21*, 809–812, 1993.

McAdoo, D. C., J. G. Caldwell, and D. L. Turcotte, On the elastic-perfectly plastic bending of the lithosphere under generalized loading with application to the Kuril trench, *Geophys. J. Roy. Astr. Soc.*, *54*, 11–26, 1978.

McConnell, R. B., Geological development of the rift system of eastern Africa, *Geol. Soc. Amer. Bull.*, *83*, 2549–2572, 1972.

McKenzie, D. P., The initiation of trenches: A finite amplitude instability, in *Island Arcs Deep Sea Trenches and Barck-Arc Basins*, edited by M. Talwani and W. C. Pitman, 1, *Maurice Ewing Series*, pp. 57–61. Am. Geophys. Un., Washington, 1977.

McKenzie, D. P., J. M. Roberts, and N. O. Weiss, Convection in the Earth's mantle: towards a numerical simulation, *J. Fluid Mech.*, *62*, 465–538, 1974.

Melosh, H. J., Dynamic support of the outer rise, *Geophys. Res. Lett.*, *5*, 321–324, 1978.

Melosh, H. J., and A. Raefsky, A simple and efficient method for introducing faults into finite elment computations, *Bull. Seismo. Soc. Am.*, *71*, 1391–1400, 1981.

Melosh, H. J., and A. Raefsky, Anelastic response of the Earth to a dip slip earthquake, *J. Geophys. Res.*, *88*, 515–526, 1983.

Melosh, H. J., and C. A. Williams, Mechanics of graben formation in crustal rocks: A finite element analysis, *J. Geophys. Res.*, *94*, 13,961–13,973, 1989.

Moresi, L., and V. Solomatov, Mantle convection with a brittle lithosphere: Thoughts on the global tectonic styles of the Earth and Venus, *Geophys. J. Int.*, *133*, 669–682, 1998.

Müller, R. D., U. R. Roest, J.-Y. Royer, L. M. Gahagan, and J. G. Sclater, Digital isochrons of the world's ocean floor, *J. Geophys. Res.*, *102*, 3211–3214, 1997.

Richards, M. A., and C. Lithgow-Bertelloni, Plate motion changes, the Hawaiian-Emperor bend, and the apparent success and failure of geodynamic models, *Earth Planet. Sci. Lett.*, *137*, 19–27, 1996.

Sandwell, D. T., and W. H. F. Smith, Marine gravity anomaly from ERS-1, Geosat and satellite altimetry, *J. Geophys. Res.*, *102*, 10,039–10,045, 1997.

Scholl, D. W., T. L. Vallier, and A. J. Stevenson, Terrane accretion, production, and continental growth: A perspective based on the origin and tectonic fate of the Aleutian-Bering Sea region, *Geology*, *14*, 43–47, 1986.

Sørensen, K., Growth and dynamics of the Norde Strømfjord shear zone, *J. Geophys. Res.,* , *88*, 3419–3437, 1983.

Stern, R. J., and S. H. Bloomer, Subduction zone infancy: Examples from the Eocene Izu-Bonin-Mariana and Jurassic California arcs, *Geol. Soc. Am. Bull.*, *104*, 1621–1636, 1992.

Sykes, L. R., Intraplate seismicity, reactivation of preexisting zones of weakness, alkaline magmatism, and other tectonism postdating continental fragmentation, *Rev. Geophys.*, *16*, 621–688, 1978.

Tackley, P. J., Self-consistent generation of tectonic plates in three-dimensional mantle convection, *Earth Planet. Sci. Lett.*, *157*, 9–22, 1998.

Tackley, P. J., The quest for self-consistent generation of plate tectonics in mantle convection models, in *History and Dynamics of Global Plate Motions*, edited by M. A. Richards, R. Gordon and R. Van der Hilst, Geodynamic Series. Am. Geophys. Un., this volume, 1999.

Tackley, P. J., D. J. Stevenson, G. A. Glatzmaier, and G. Schubert, Effects of an endothermic phase transition at 670 km depth in a spherical model of convection in the Earth's mantle, *Nature*, *361*, 699–704, 1993.

Toth, J., and M. Gurnis, Dynamics of subduction initiation at pre-existing fault zones, *J. of Geophys. Res.*, *103*, 18,053–18,067, 1998.

Turcotte, D. L., and E. R. Oxburgh, Finite amplitude convection cells and continental drift, *J. Fluid Mech.*, *28*, 29–42, 1967.

Turcotte, D. L., and G. Schubert, *Geodynamics*, Wiley, New York, 1982.

Uyeda, S., and Z. Ben-Avraham, Origin and development of the Philippine Sea, *Nature Phys. Sci.*, *240*, 176–178, 1972.

Uyeda, S., and H. Kanamori, Back-arc opening and the mode of subduction, *J. Geophys. Res.*, *84*, 1049–1061, 1979.

van der Hilst, R., and T. Seno, Effects of relative plate motion on the deep structure and penetration depth of slabs below the Izu-Bonin and Mariana island arcs, *Earth Planet. Sci. Lett.*, *120*, 395–407, 1993.

Vigny, C., Y. Ricard, and C. Froidevaux, The driving mechanism of plate tectonics, *Tectonophysics*, *187*, 345–360, 1991.

Vissers, R. L. M., M. R. Drury, E. H. Hoogerduijn Strating, C. J. Spiers, and D. van der Wal, Mantle shear zones and their effect on lithosphere strength during continental breakup, *Tectonophysics*, *249*, 155–171, 1995.

Watterson, J., Mechanism for the persistence of tectonic lineaments, *Nature*, *253*, 520–52, 1975.

White, S. H., The effects of strain on microstructures, fabrics, and deformation mechanisms in quartzites, *Phil. Trans. R. Soc. Lond. A*, *283*, 69–86, 1976.

White, S. H., and P. F. Green, Tectonic development of the Alpine fault zone, New Zealand: A fission track study, *Geology*, *14*, 124–127, 1986.

Wilson, J. T., Did the Atlantic close and then re-open?, *Nature*, *211*, 676–681, 1966.

Zhong, S., and M. Gurnis, Viscous flow model of a subduction zone with a faulted lithosphere: Long and short wavelength topography, gravity, and geoid, *Geophys. Res. Lett.*, *19*, 1891–1894, 1992.

Zhong, S., and M. Gurnis, Controls on trench topography from dynamic models of subducted slabs, *J. Geophys. Res.*, *99*, 15,683–15,695, 1994.

Zhong, S., and M. Gurnis, Mantle convection with plates and mobile, faulted plate margins, *Science*, *267*, 838–843, 1995a.

Zhong, S., and M. Gurnis, Towards a realistic simulation of plate margins in models of mantle convection, *Geophys. Res. Lett.*, *22*, 981–984, 1995b.

Zhong, S., and M. Gurnis, Interaction of weak faults and Non-Newtonian rheology produces tectonic plates in 3D model of mantle flow, *Nature*, *383*, 245–247, 1996.

Zhong, S., and M. Gurnis, Dynamic interaction between tectonic plates, subducting slabs, and the mantle, *Earth Interactions*, *1*(6), http://EarthInteractions.org/, 1997.

Zhong, S., M. Gurnis, and L. Moresi, The role of faults, nonlinear rheology, and viscosity structure in generating plates from instantaneous mantle flow models, *J. Geophys. Res.*, *103*, 15,255–15,268, 1998.

Zoback, M. D., and J. H. Healy, In situ stress measurements to 3.5 km depth in the Cajon Pass scientific resaerch borehole: Implications for the mechanics of crustal faulting, *J. Geophys. Res.*, *97*, 5039–5057, 1992.

M. Gurnis, Seismological Laboratory, California Institute of Technology, Pasadena, CA 91125. (e-mail: gurnis@caltech.edu)

S. Zhong, Dept. Earth, Atmospheric, and Planetary Sciences, MIT, Cambridge, MA 02139 (e-mail: szhong@rayleigh.mit.edu)

J. Toth, Shell International, Exploration and Production, Rijswijk, Netherlands (e-mail: J.TOTH@siep.shell.com)

# What the Mantle Sees: The Effects of Continents on Mantle Heat Flow

A. Lenardic[1,2], L. Guillou-Frottier[3], J.-C. Mareschal[4], C. Jaupart[5], L.-N. Moresi[6], and W. M. Kaula[7]

The effects of continents on mantle heat flow are explored through a synthesis of heat flow data analysis and theoretical modeling. The data employed are from the Canadian Shield and the Canadian Appalachians and consist of both surface heat flow measurements and measurements of crustal heat production. The combined data sets, together with gravity and seismic information, which is used to help constrain crustal structure and composition, suggest that the mantle component of surface heat flow is low, between 10 and 15 $mW/m^2$, and relatively uniform across a large portion of central to eastern Canada. Surface heat flow variations across the region covered can be well accounted for by observed variations in crustal heat production, i.e., no mantle heat flow variations need to be invoked. Laboratory experiments and numerical simulations, that explore the coupling between continents and the convecting mantle from a modeling based point of view, show how large, buoyant continents can locally constrain heat flux variations in the mantle below. Relative to what occurs in model oceanic regions, the equilibrium thermal condition imposed on the mantle by model continents is one of low and nearly uniform heat flux. Such a condition is consistent with the inferences drawn from the Canadian heat flow data. Theoretical models also imply that the local thermal constraint imposed on the mantle by continents can cause mantle heat flow to be relatively weakly correlated with tectonic age variations, lithospheric thickness variations, and local mantle flow patterns below tectonothermally stable continental regions.

## INTRODUCTION

Since the advent of plate tectonics as a theory, our understanding of the relationship between surface heat flow, plate motions, and mantle convection in oceanic regions has become well developed. Oceanic ridges are hot and stand tall. Heat flow decreases and the depth to the ocean floor increases from ridges to older oceanic basins. Both of these spatial trends result from the cooling and thickening of mantle lithosphere as it moves away from a ridge [*Sclater and Francheteau*, 1970; *Parsons and Sclater*, 1977; *Parsons and McKenzie*, 1978]. This motion and associated heat loss are an integral part of mantle convection [*Turcotte and Oxburgh*, 1967]. Thus, in oceanic regions, surface heat flow variations and shallow mantle flow patterns are linked together in, what has become, a relatively well understood way, at

[1] Department of Geology and Geophysics, University of California, Berkeley.
[2] Now at Department of Geology and Geophysics, Rice University, Houston, Texas.
[3] Bureau de Recherches Géologiques et Miniéres, Orléans, France
[4] Geotop, UQAM, Montreal
[5] Institut de Physique du Globe, Paris
[6] Australian Geodynamics Cooperative Research Centre, CSIRO Exploration and Mining
[7] Department of Earth and Space Sciences, University of California, Los Angeles

The History and Dynamics of Global Plate Motions
Geophysical Monograph 121

the most basic of levels. Although second-order problems remain to be worked out [e.g., *Stein and Stein*, 1992; *Carlson and Johnson*, 1994], unlocking the first-order connection between heat flow, topography, plate motion, and mantle convection in oceans remains one of the great, and earliest, successes of geodynamics. This accomplishment remains particularly valuable today as a cornerstone for developing a fully dynamic theory of plate motions as it directly relates aspects of the kinematic theory of plate tectonics to aspects of mantle convection, a decidedly dynamic process. It would be desirable to fill in the picture further by connecting continental heat flow to mantle convection but that task is not as easy due to the fact that subcontinental mantle motions and mantle heat flow patterns are screened from us by the long-lived scum that forms the continents themselves (continental crust and, perhaps, mantle-melt residuum).

The screening effect of continents has implications for both data based and theoretical investigations into the connection between continental heat flow and mantle convection. From a data based point of view it means that one must focus on regions were there is good coverage not only in terms of heat flow measurements but also in terms of crustal heat production measurements. Additionally, one must also have good constraints on crustal structure. All this is required if the component of continental heat flow that is coming from the mantle is to be evaluated as one must be able to constrain the crustal contribution in order to do so. From a theoretical point of view the difficulty noted above means that one must explore modeling approaches that make varied assumptions about the specifics of subcontinental mantle flow. This is required in order to see if theoretical results may be overly sensitive to the specifics of subcontinental mantle flow patterns that are not fully constrained at present by seismic tomography.

This paper addresses a specific question geared toward understanding the connection between continental heat flow and mantle convection; how do continents effect spatial variations of mantle heat flow ? We approach this question using a synthesis of heat flow data, laboratory experiments, and numerical simulations. In terms of the data analysis, we focus on eastern Canada where heat flow and crustal heat production data are now sufficient to estimate mantle heat flow. Heat flow data in this region have been multiplied by a factor of four since the compilation of *Jessop et al.* [1984]. Much of the effort has been directed toward collecting and analyzing heat flow and heat production data from the Canadian Appalachians to the Canadian Shield. We review how the new data have been used to estimate mantle heat flow. The results of these studies are then discussed in the context of laboratory experiments and numerical simulations that examine the effects of continents on mantle heat flow from a theoretical point of view.

## CONTINENTS AND MANTLE HEAT FLOW VARIATIONS: DATA

The first subsection below reviews our understanding of oceanic heat flow, discusses some often held assumptions about continental heat flow, and provides a historical review related to the use of heat flow data to determine the nature of mantle heat flow variations below continents. It ends by noting that the potential effects of mantle heat flow variations below stable continental provinces has lead to two different viewpoints: One that holds that they can significantly contribute to spatial variations of surface heat flow; another that holds that the amplitude of mantle heat flow variations below stable continental regions is small and that surface heat flow variations can be accounted for by variations in near surface crustal heat production. The second subsection discusses heat flow data from Canada and reviews how this data has been applied to discriminating between the two views just mentioned. We conclude that the Canadian data favors the view that mantle heat flow variations below stable continental regions are mild.

### *Review*

A convenient way to study mantle heat flow variations has been to use such variations to determine lithospheric thickness variations and to compare the variations so determined to lithospheric thickness variations determined using independent geophysical techniques, e.g., seismological mappings. In discussing how surface heat flow relates to lithospheric thickness we must be concrete about how the lithosphere is defined. The lithosphere may be defined by both rheological criteria and by thermal processes. The first definition relies on the strength required to support significant stresses over geological times [*Barrell*, 1914] and has led to the concept of "mechanical lithosphere". In the second definition, the lithosphere is the outer layer through which vertical heat transport occurs principally by thermal conduction [*Parsons and McKenzie*, 1978]. This introduces the "thermal lithosphere". Both definitions involve dynamical considerations and do not refer to the instantaneous, near surface temperature distribution. The lithosphere is underlain by a boundary layer where some type of mantle flow provides the heat carried by conduction above. From the standpoint of seismic tomography, it is not possible to separate the lithosphere and its basal boundary layer. In order to do so, one must use dynamical models. For the oceanic case, a model that relates first-order mantle heat flow variations to lithospheric thickness variations has been developed and before discussing continental heat flow data we first review how oceanic heat flow has been interpreted in the context of this model.

As the oceanic thermal lithosphere ages, it cools and thickens, and, as a result, mantle heat flow decreases and the depth to the ocean floor increases. Thus, the age of a plate, thermal lithospheric thickness, sea floor topography, and mantle heat flow are strongly related to one another. This result comes from a simple model that considers the thermal lithosphere to be the upper thermal boundary layer of mantle convection [*Turcotte and Oxburgh*, 1967]. Underlying assumptions are: 1) the bulk interior of the mantle is thermally well mixed so that the base of the thermal lithosphere is defined by an isotherm; 2) the top of the lithosphere is also defined by an isotherm; and 3) the oceanic lithosphere as a whole actively participates in mantle convection so that vertical heat conduction balances heat advection across its full extent (such a balance, in effect, defines an active thermal boundary layer within a convecting layer such as the mantle). This model is in accord with a large portion of oceanic data [*Parsons and Sclater*, 1977]. The dependence of mechanical lithospheric thickness on mantle heat flow in oceans is also strong as the temperature-dependence of mantle viscosity, and a near uniform crustal thickness, allow mechanical lithospheric thickening to track thermal thickening [*Watts et al.*, 1980]. Figure 1a shows a cartoon of how the age of an oceanic plate, local mantle heat flow, and the local thickness of the thermal and mechanical lithosphere are related to first-order.

The relationship between heat flow, topography, and plate age does appear to break down in the oldest sections of oceanic lithosphere. In ocean basins older than $10^8$ years, heat flow and topography tend toward a constant value. Several explanations have been offered for these observations [e.g., *Heestand and Crough*, 1981; *Phipps-Morgan and Smith*, 1992; *Doin and Fleitout*, 1996]. One of the earliest explanations relied on small-scale convection [*Parsons and McKenzie*, 1978; *Sclater et al.*, 1981]. The lower lithosphere, which is relatively weak, is assumed to thicken with time up to a point

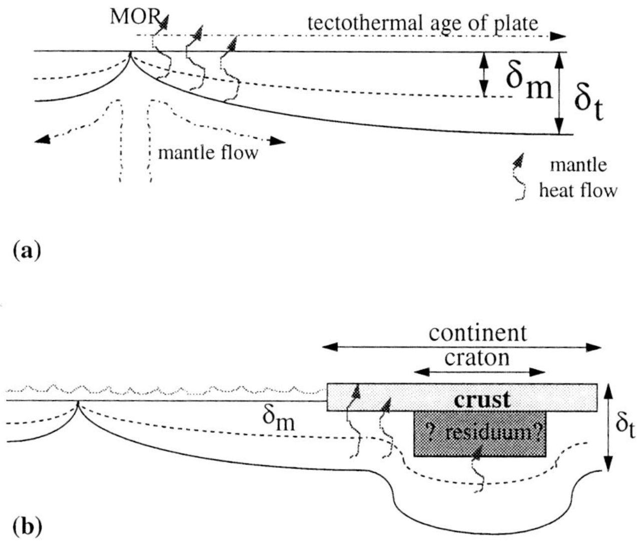


**Figure 1.** Cartoon of oceanic (a) and continental (b) lithospheric structure. Oceanic lithosphere forms at a mid-oceanic ridge, MOR, and its age increases away from the ridge. This aging is associated with decreasing mantle heat flow and increasing thermal, $\delta_t$, and mechanical, $\delta_m$, lithospheric thickness. Continental lithosphere appears to be thick below cratons relative to non-cratonic regions and this is often attributed to the presence of thick, buoyant cratonic roots composed of mantle-melt residuum. By association with inferences gained from oceanic studies it is often assumed that a strong correspondence should exist between lithospheric thickness and mantle heat flow in continents. Thus, mantle heat flow below cratons should be low relative to that below peripheral regions. This would imply that a significant component of surface heat flow variations in stable continental regions should be due to mantle heat flow variations.

at which its negative density force overcomes its viscous resisting force allowing portions to drip off into the asthenosphere thereby maintaining a constant lithospheric thickness. Theoretical considerations and laboratory experiments suggest that, for rheologic structures believed to be representative of the oceanic lithosphere, small-scale convection can set in for plate ages of 70 Ma and can maintain constant lithospheric thickness, topography, and mantle heat flow in old ocean basins [*Parsons and McKenzie*, 1978; *Davaille and Jaupart*, 1994]. It should be noted that this explanation is by no means universally accepted [e.g., *Davies*, 1988]. It is useful to review it briefly here as small-scale convection has also been invoked to explain a key observation related to continental heat flow, as discussed below.

Moving to continents, it is agreed that in tectonothermally active regions the relationship between mantle heat flow and lithospheric thickness may not be obvious as the lithosphere may be far from thermal equilibrium. However, it is thought that continental lithosphere should approach thermal equilibrium as the time since the last major magmatic and/or tectonic event increases within a region and the effects of the event on surface heat flow decay [e.g., *Vitorello and Pollack*, 1980]. Thus, for stable continental regions a relationship between tectonothermal age, thermal lithospheric thickness, and mantle heat flow has been sought [e.g., *Oxburgh*, 1981]. In terms of lithospheric thickness, continents present an additional variable: The chemical variations in the mantle. In particular, deep cratonic lithosphere may consist of chemically buoyant roots. If these roots are strong due to dehydration [*Pollack*, 1986], then mechanical lithospheric thickness variations in continents may not be dominantly controlled by temperature. Even so, it is thought that thick, strong, and chemically buoyant cratonic roots should stabilize a thick thermal lithosphere [*Jordan*, 1978] and should also shift mantle heat toward regions of thinner lithosphere [*Ballard and Pollack*, 1987; *Nyblade and Pollack*, 1993]. Thus, cratons would remain associated with thicker thermal and mechanical lithosphere and lower mantle heat flow than the stable, and generally younger, regions that surround them. These ideas are consistent with the assumption that in stable continental regions a relationship exists between mantle heat flow, tectonothermal age, and both mechanical and thermal lithospheric thickness. Figure 1b shows a cartoon of how this relationship is assumed to work.

Testing the assumptions above is not straightforward as it is impossible to directly measure mantle heat flow in continental regions. In tectonothermally stable regions surface heat flow, $q$, which can be measured, results from a combination of heat produced within the crust by radioactive elements and heat supplied to the base of the crust from the mantle below. Although the full crustal contribution can not be directly measured, near surface crustal radiogenic heat production, $A$, can. This, together with an observation, made three decades ago, allowed progress to be made in separating the components of surface heat flow in stable continental regions. The observation is that in several regions of the Earth a linear relationship between surface heat flow and surface heat production holds [*Birch et al.*, 1968]. The relationship is given by

$$q = q_r + bA \tag{1}$$

where $q_r$ is the reduced heat flow, taken to represent the heat flowing into the base of the radiogenic uppermost crust, and $b$ is a scaling parameter whose physical interpretation is less clear. It was thought to represent a length scale for the vertical distribution of heat sources in the upper crust [*Lachenburch*, 1970] but this interpretation is questionable because it does not account for lateral variations of crustal heat production [*Jaupart*, 1983]. It should be stressed that the relationship of Eq 1 was established in regions, such as the Appalachians, where $A$ is high and where the heat production of the lower crust is much lower than that of the upper crust. Recent data have shown that in many other regions the linear relationship does not hold [*Guillou et al.*, 1994].

Regions where a single relationship of the form given by Eq 1 holds are referred to as heat flow provinces [*Roy et al.*, 1968]. Figure 2a plots $q_r$ versus tectonothermal age from several such provinces. Provinces with tectonothermal ages greater than 100 million years or so are associated with tightly clustered $q_r$ values; the mean is 27 mW/m$^2$ with a standard deviation of 4 mW/m$^2$ [*Morgan*, 1984; *Pujol et al.*, 1985]. This observation lead both *Morgan and Sass* [1984] and *Pujol et al.* [1985] to conclude that surface heat flow variations in stable continental regions result primarily from variations in upper crustal heat sources, with the component of heat coming from the lower crust and mantle being nearly constant. A similar conclusion was reached somewhat earlier by *Sclater et al.* [1980; 1981] based on a more limited data set.

The earliest interpretation of the observation that reduced heat flow equilibrates to a near constant value for tectonothermal ages greater than $10^8$ years, relied on the idea that the thermal thickness of continental lithosphere also equilibrates to a near constant value [*Sclater et al.*, 1981]. That is, a constant reduced heat flow was taken to reflect a near constant mantle heat flow and this, in turn, was interpreted to reflect a near constant lithospheric thickness. In analogy with the oceans, the tendency of the lithosphere to achieve a constant thickness for ages greater than $10^8$ years was attributed to small-scale convection [*Sclater et al.*, 1981].

The inference that mantle heat flow is nearly constant below stable continental regions is by no means universally accepted. A problem raised by this interpretation is associated with observations in and around Archean cratons. There is strong seismological evidence that the thermal lithosphere below Archean cratons is very thick and that it thins considerably toward younger Proterozoic mobile belts [e.g., *Grand*, 1987] (note: Temperature varies significantly within the thermal lithosphere but not below its base and, thus, seismic tomography will provide a better estimate of thermal, as opposed to

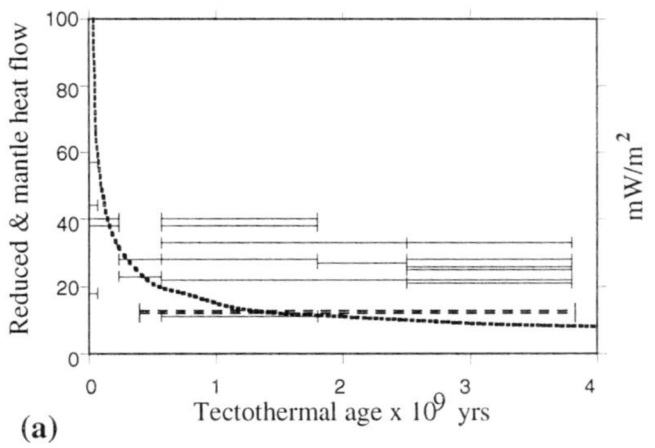


(a)

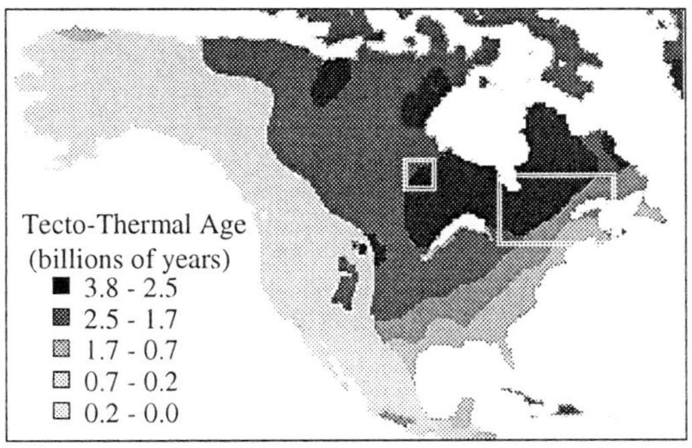


(b)

**Figure 2.** (a) Reduced heat flow in various heat flow provinces versus the tectonothermal age of a province, along with a curve of the expected mantle heat flow if the lithosphere cooled conductively with age, and data based estimates of the mantle heat flow from provinces in and around the Canadian Shield. Reduced heat flows, from *Morgan* [1984], are plotted as solid horizontal lines spanning the age range of a province. The lowest value has an error of ±73%. The average error for all other provinces with ages greater than 65 My is ±14%. Tectonothermal age is defined as the age since the last major magmatic and/or tectonic event within a region. The conductive curve assumes mantle thermal properties from *Carlson and Johnson* [1994]. The mantle heat flow estimate from several different tectonic provinces in Canada is plotted as a dashed horizontal line. The data sources on which this is based are given in the text. The area covered by the Canadian data is shown in (b) by the two double lined rectangles. The larger area covers, in order of decreasing age, the Superior, Grenville, and Appalachian tectonic provinces. The smaller area covers the Trans Hudson Orogen. Tectonothermal ages are from *Burchfiel* [1983].

mechanical, lithospheric thickness). There is also evidence, from gravity and topography modeling, that the mechanical lithosphere below cratons is thicker than in surrounding regions [e.g., *Bechtel et al.*, 1990; *Pilkington*, 1991; *Hartley et al.*, 1996; *Doucouré et al.*, 1996]. As a whole, this evidence suggests a deep structure akin to that shown in Figure 1b. This seems inconsistent with the ideas that lithospheric thickness tends toward a constant value beyond tectonothermal ages of $10^8$ years and that mantle heat flow tends toward a constant value in stable continental regions, if the relationship between lithospheric thickness and mantle heat flow is akin to that in oceans. Indeed, it has been argued that the thicker lithosphere inferred seismically below Archean cratons should result in a trend of decreasing mantle heat flow from the younger terrains that border cratons to the cratons themselves [*Ballard and Pollack*, 1987; *Nyblade and Pollack*, 1993]. Heat flow data from southern Africa [*Ballard and Pollack*, 1987] and analysis of data worldwide [*Nyblade and Pollack*, 1993] have been used in support of this hypothesis. In this interpretation, variations in deep mantle heat flow could be the primary cause of continental heat flow variations at wavelengths greater than 100 km.

### *Canadian Heat Flow Data*

The previous subsection suggests two questions: 1) is mantle heat flow nearly constant below stable continental regions or does it show significant variations between regions of differing ages (i.e., between Archean cratons and younger surrounding regions); and 2) is a near constant mantle heat flow below stable continental regions consistent with inferences that lithospheric thickness is variable across such regions. These questions relate directly to the wider one posed by this paper: How do continents affect spatial variations of mantle heat flow. To answer them we shall consider the large number of high quality heat flow data that are now available from the Canadian Shield to the Appalachians. A complete study of heat flow observations and lithospheric thickness in the Canadian Shield proper can be found in *Jaupart et al.*, [1998].

The double lined rectangles in Figure 2b show the principal areas from which extensive heat flow data has been collected over the last decade [*Mareschal et al.*, 1989; 1997; 1998; *Pinet et al.*, 1991; *Guillou et al.*, 1994; *Guillou-Frottier et al.*, 1995; 1996]. Plate 1 shows a surface heat flow map from the larger of the two regions. The data used to generate this map are documented in *Jaupart and Mareschal et al.* [1999]. All heat flow data are corrected for the effects of Pleistocene glaciation following the procedure of *Jessop* [1971]. A study of heat flow variations in a deep borehole has shown that the uncertainty on the heat flow estimate due to the effects of glaciation is small, $\approx 3 \mathrm{mW/m^2}$, and would not affect spatial heat flow trends [*Mareschal et al.*, 1999b].

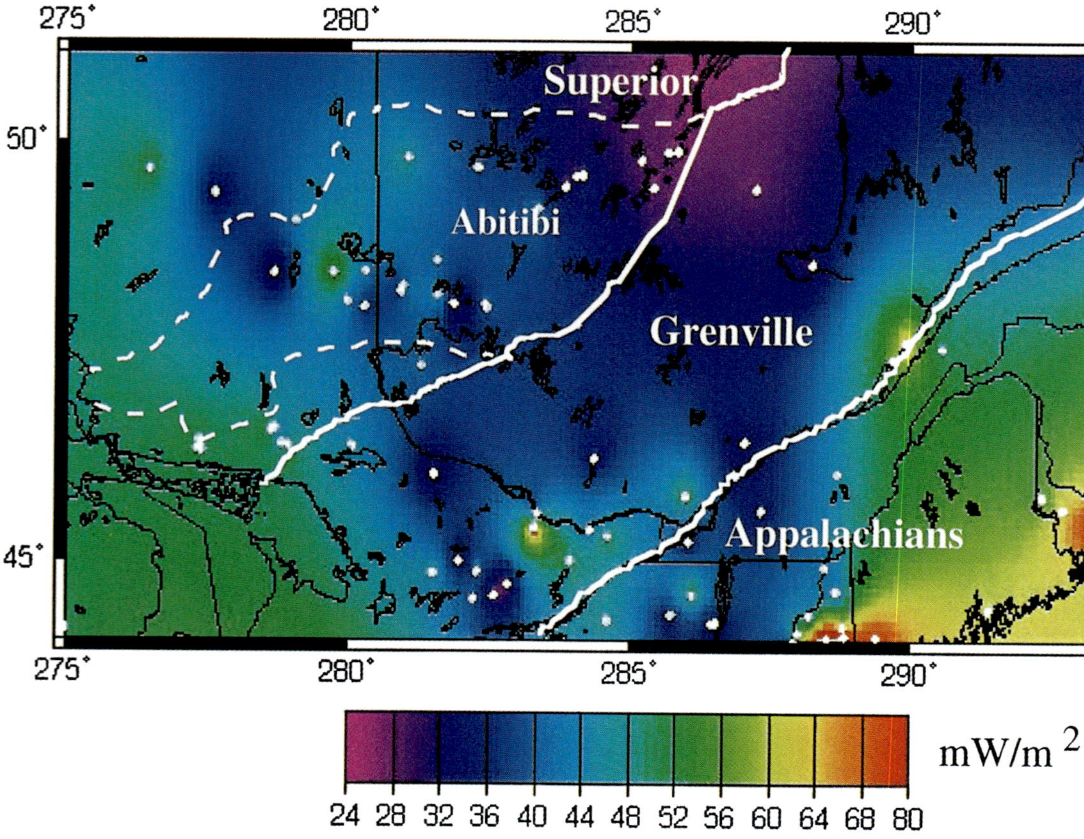


**Plate 1.** Heat flow map of eastern Canada with main tectonic provinces delineated by solid white lines. The dashed white line delineates the Abitibi greenstone belt, a subprovince of the Superior. White dots are measurement sites.

However, the differences between average heat flow values in northern and southern hemisphere shields are within this uncertainty and this should be kept in mind when comparing heat flow data from different continental shields.

The region covered in Plate 1 includes parts of the Superior and Grenville Provinces of the Canadian Shield and the Canadian Appalachians. Thus, it is associated with large age variations (tectonothermal age varies from 2.6-3.1 Ga for the Superior, to 0.9-1.1 Ga for the Grenville, to 0.4 Ga for the Appalachians). It is also associated with variations in lithospheric thickness as inferred from seismic tomography. The depth to which significant fast seismic anomalies can be detected ranges from more than 300 km below the Superior to less than 100 km below the Appalachians [*Grand*, 1987; 1994; *van der Lee and Nolet*, 1997]. Variations in the inferred elastic thickness of the lithosphere follow those of the seismically determined lithosphere. It is estimated to be roughly 120 km below the central Canadian Shield and to thin to 20-30 km towards the Canadian Appalachians [*Pilkington*, 1991]. Given these variations one would expect mantle heat flow to be lowest in the Superior and to increase as one moves off the craton *if* the physics of heat transfer below continents and oceans were the same to first-order.

To estimate the total crustal component of surface heat flow in the Abitibi belt of the Superior Province (2.7 Ga), *Pinet et al.* [1991] employed heat generation measurements made on the main rock types that make up the Superior crust. For this estimation, the crust was divided into three main rock types: greenstone, tonalitic rocks, and lower crustal granulites with 0.2, 1.1, and 0.4 $\mu W/m^3$ heat production respectively. In addition to this, *Guillou et al.* [1994] used available seismic data to constrain the thickness of the crust and gravity data as a further constraint on crustal thickness and composition. Crustal models were constructed at three distinct areas of the Abitibi belt, each separated by a few hundreds of kilometers. For each area, all model parameters (heat production within crustal layers, layer thickness', densities, and the assumed component of mantle heat) were systematically varied within the range allowed until heat flow, gravity, and seismic data were simultaneously satisfied. In this manner, *Guillou et al.* [1994] obtained an extensive set of models that fit the data. For all of them mantle heat flow was between 8-15 $mW/m^2$, with a statistically most likely value of 13 $mW/m^2$. An upper bound was obtained from measurements near the Grenville Front. There, over a wide area (80 $\times$ 80 km), five heat flow data have a constant value of 28 $\pm$1 $mW/m^2$. The exposed rock within this area, which belong to the granulite facies, are believed representative of the lower crust and have an average heat production of 0.4 $\mu W/m^3$ worldwide [*Pinet et al.*, 1991]. For the crustal thickness in this region (< 40 km), this heat production implies a mantle heat flow below the Grenville of less than 14 $mW/m^2$ [*Guillou-Frottier et al.*, 1995].

*Pilkington* [1991] has estimated that, in North America, the elastic lithosphere is thickest near the Trans Hudson Orogen (THO), an Early PaleoProterozoic suture (1.8 Ga) between the Hearne, Rae, and Superior Provinces [*Hoffman*, 1989]. This is the region covered by the smaller double rectangle in Figure 2b. Heat flow data have recently been collected from this region [*Guillou-Frottier et al.*, 1996; *Mareschal et al.*, 1997]. The data show no indication that the mantle heat flow is lower than in the East. The surface heat flow shows significant local variations with some of the highest heat flow values in the Shield observed in the Thompson Belt [*Guillou-Frottier et al.*, 1996] and the lowest in the Lynn Lake belt [*Mareschal et al.*, 1997]. This juxtaposition of high and low values points toward a crustal source and after removing the crustal heat production, the mantle heat flow beneath the THO was found to be less than 15 $mW/m^2$.

The Grenville crust (1.1 Ga) is a mixture of different terrains juxtaposed horizontally and vertically by thrust faults and tectonic displacements. This has led to the exposition at the surface of all crustal levels so that heat production in upper, middle, and lower crustal units can be determined [*Pinet et al.*, 1991; *Guillou-Frottier et al.*, 1995]. Thus, if there is no sampling bias, the tectonic fabric of the province allows the area weighted average of measured surface heat production, 0.7 $\mu W/m^3$, to be equated to the vertical average [*Pinet et al.*, 1991]. This value, along with the average surface heat flow (41 $mW/m^2$) and crustal thickness (40 km) in the Grenville, implies an average mantle heat flow of 13 $mW/m^2$ [*Guillou-Frottier et al.*, 1995]. An upper bound can also be obtained. In the Sept-Iles area, near the Canadian Appalachians, three low surface heat flow values have been measured over gabbroic and anorthositic rocks, which have no significant heat production. For an average crustal thickness in the area of 40 km and a minimum crustal heat production of 0.4 $\mu W/m^3$, the upper bound of the mantle heat flow is 16 $mW/m^2$.

Mean heat flow in the northern Appalachians (54 $mW/m^2$) is higher than in the Grenville. A heat flow heat production relationship holds only where the crust is intruded by numerous plutons. Where it can be defined, the reduced heat flow is the same in the Grenville (1.1 Ga) as in the Appalachians (0.4 Ga). This supports the view that the higher average surface heat flow in the Appalachians reflects the abundance of shallow radioactive plutons and that the mantle heat flow is the same for both regions. Indeed, the difference in the mean heat flow can be entirely accounted for by an 8 km thick upper crustal layer with 2.5 $\mu W/m^3$ heat production [*Pinet et al.*, 1991].

The data reviewed above suggest that mantle heat flow is nearly constant over a wide region of eastern North America. The estimated mantle heat flow from the three main tectonic province discussed is plotted in Figure 2a. As one can see, mantle heat flow is lower than reduced heat flow. This is to be expected as it has become clear that the lower crust contains a greater component of heat producing elements than was once thought [e.g., *Ashwal et al.*, 1987]. None the less, the inference drawn from the Canadian data supports that drawn by *Morgan* [1984; 1985] from a worldwide reduced heat flow data set. That is, heat flow variations in stable continental regions result principally from variations in near surface crustal heat production and the component of deep heat from the mantle shows, at best, mild variations. The data do not support the idea that a significant amount of mantle heat is deflected from the Superior craton to younger bordering regions.

## CONTINENTS AND MANTLE HEAT FLOW VARIATIONS: THEORY AND MODELS

This section sets out to show that the conclusions of the previous section are consistent with theoretical models. We first present simple physical arguments related to the differences that must exist between subcontinental and suboceanic mantle heat flow. We then review laboratory experiments of continent-mantle interaction that take these differences into account. The experiments show how stable and large continents can locally enforce a near constant and low heat flux condition on the convecting mantle below. Finally, we discuss results from numerical simulations that, unlike the lab experiments, also allow for continents of variable thickness. The conclusion that mantle heat flow variations should be mild below stable continental regions is shown to be robust when this added complexity is allowed for.

### *Physical Considerations*

The dynamics of a convecting system, like the mantle, can not be characterized just by the properties of the convecting material. One must also consider the material that lies above [*Sparrow et al.*, 1964; *Chapman et al.*, 1980]. The Earth is bounded from above by the hydrosphere and atmosphere, the dynamics of which maintain an effectively constant temperature at its surface, i.e., variations are insignificant at the geodynamic scale. This allows mantle convection modelers to treat the Earth's surface as an isothermal boundary where any spatial temperature variations, that could result from heat flux variations from below, are rapidly smoothed away from above. The oceanic lithosphere is, as a whole, a convective thermal boundary layer and, as such, it participates in mantle overturn, i.e., mantle convection reaches to the Earth's surface below oceans where water temperature is constant. Thus, the convecting portions of the solid Earth 'see' a constant temperature surface condition below oceans. The situation is more complicated below continents. Although continental crust, and perhaps the chemical material that may form cratonic roots, can move laterally over the mantle as continents drift, it does not participate in convective mantle overturn, i.e., it is long-lived, laterally mobile bounding material for the convecting mantle below. The viscosity of crust and root material prevents it from convectively adjusting to temperature variations on a rapid time scale relative to the mantle. Similarly, the comparable thermal diffusivity of distinct rock types also means that large temperature variations generated at the base of chemically buoyant portions of the continents, by mantle heat flow variations below, can not be rapidly diffused away. Thus, a constant temperature condition may not hold at the top of the convecting mantle in continental regions and this could lead to differences in the nature of mantle heat flow below continents and oceans.

The exact nature of the thermal coupling condition between a convecting fluid, such as the mantle, and a conducting surface layer, such as the continental crust, depends on the thermal resistances in the convecting fluid and in the surface layer (thermal resistance is directly proportional to a layers depth and inversely proportional to its thermal conductivity). The ratio of these resistances is termed a Biot number, $Bi$ [*Sparrow et al.*, 1964]. If the resistance of the convecting fluid dominates ($Bi = \infty$), then the thermal coupling condition is one of spatially constant temperature. If the resistance of the conducting layer dominates ($Bi = 0$), then the thermal coupling condition is one of spatially constant heat flux. If the ratio of resistances is comparable then the thermal coupling condition is an imperfect one. Although this does not imply constant heat flux, it does constrain lateral heat flux variations out of a convecting fluid unlike a constant temperature condition [*Sparrow et al.*, 1964]. Thus, one can expect that mantle heat flow variations below stable continental regions may be small relative to those below oceans due to the finite thermal resistance of long lived, chemically buoyant continental layers.

The simple definition of a Biot number above assumes that a convecting fluid is overlain uniformly by a conducting layer. Indeed, essentially all analytical studies on the effects of an imperfect thermal condition at the top of a convecting layer have assumed homogeneous boundary conditions [e.g., *Nield*, 1968; *Busse and Riahi*, 1980; *Riahi*, 1985; *Carriére et al.*, 1997]. For such cases relatively simple arguments can be used to assess the degree to which an imperfect condition can constrain surface heat flux variations (see, for example, the appendix of *Lenardic* [1998]). However, a fundamental characteristic of mantle convection is that the convecting fluid is overlaid by conducting bodies of finite sizes, with a constant temperature surface condition prevailing away from such bodies. That is, the surface thermal condition seen by the mantle is a heterogeneous one. Given that continents can drift relative to fixed mantle sites, it is also a time variable condition. The symmetry breaking associated with heterogeneous surface conditions generally makes analytical treatments nontractable. Thus, laboratory experiments and/or numerical simulations become particularly useful. These two approaches are used below to explore the relative degree to which an imperfect thermal condition below continents can constrain mantle heat flux variations.

*Laboratory Experiments*

In order to investigate the effect of an imperfect upper thermal condition on the dynamics of an underlying convecting fluid, laboratory experiments have been carried out in a variety of situations. Details on the experimental set-up, procedures, and physical properties of the chosen materials are described in *Guillou and Jaupart* [1995]. Figure 3 shows the two geometrical configurations that were used and with which some of the results presented below have been obtained. The "continent" is embedded in between "oceans" and the two materials of the upper plate allow temperature to be fixed on both sides of the continent (large thermal conductivity for copper), whereas no *a priori* thermal condition was imposed at the base of the continent (similar thermal conductivities for the "PVDF" material and for the fluids).

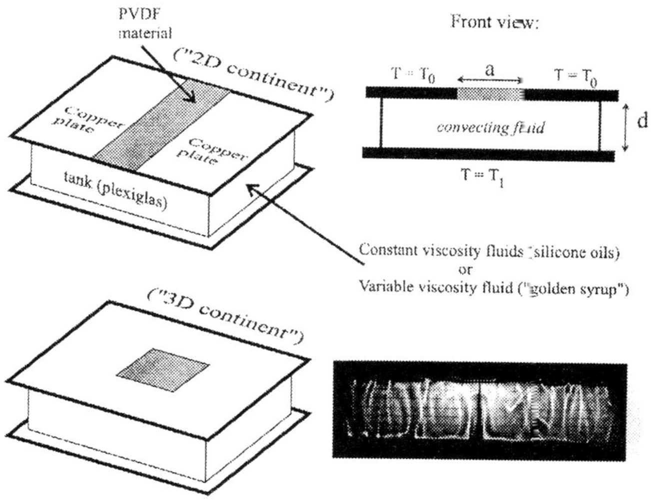


**Figure 3.** Two configurations for the experimental set-up. The tank is filled with constant or variable viscosity fluids. The upper plate is made of copper plates surrounding a material (PVDF) whose thermal conductivity is close to that of the fluids. Details on temperature measurements and heat flow measurements as well experimental procedure are described in [*Guillou and Jaupart,* 1995]. The photograph on the lower right corner shows a shadowgraph of an experiment where $a/d = 0.5$ and $Ra = 1.6 \times 10^6$. A hot central plume is going to hollow the bottom subcontinental layer (compare with numerical simulations of Figures 5 and 6). The dimensionless value for subcontinental heat flow in this experiment equals $\approx 5.0$, whereas it is decreased down to 0.9 for $a/d = 2$.

Many of the results of this, and the next, section are discussed in the context of non-dimensional parameters such as the Rayleigh number, which characterizes the intensity of convection. Although all parameters shall be defined, a full description of how the physics of mantle convection can be parameterized mathematically is not given herein as such descriptions already exist within the standard mantle convection literature [e.g., *Jarvis and Peltier*, 1982; *Busse*, 1989]. Readers who are unfamiliar with the use of non-dimensional fluid dynamical parameters and the details of mantle convection modeling in general are referred to these articles and the references therein.

Continental sizes $a/d$, Rayleigh numbers, $Ra$, and viscosity contrasts, $\mu(T_0)/\mu(T_1)$, have been varied from one experiment to the other. Prandtl numbers, $Pr$, and Biot numbers, $Bi$, have also been scaled to reproduce mantle conditions. In these experiments, the upper surface is horizontal (no continental root immersed in the convecting fluid), but geometrical effects of the mixed

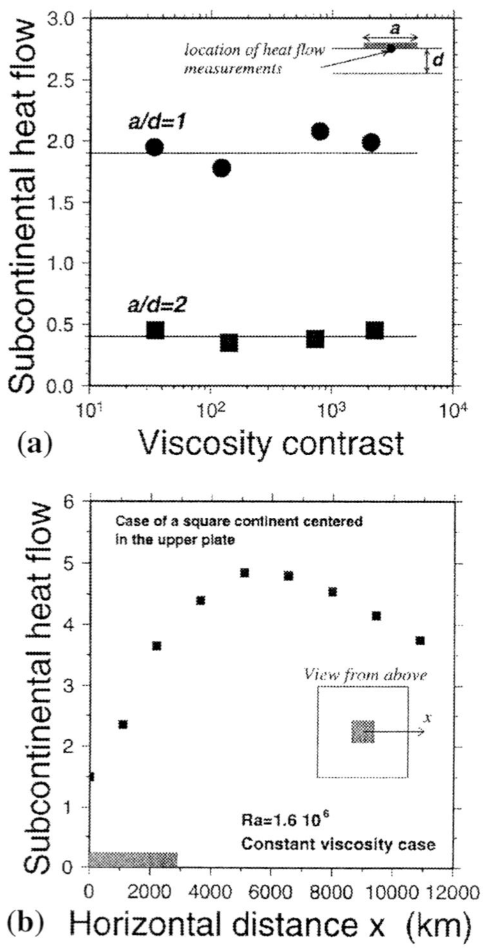


**Figure 4.** a) Dimensionless heat flow below the center of the 2-D continent (scaled with the conductive heat flow within the fluid), for two continental sizes $a/d$. A change in the viscosity contrast within the fluid does not modify the heat transfer below the continent. From the left to right of the diagram, $Ra$ is increased when it is defined with the viscosity value at mid-depth (see text). b) Spatial variation of the heat flow above the convecting fluid. Case of a square continent centered in the upper plate (3-D case), and for a similar $Ra$ value as in the Earth's mantle. A large decrease of heat flow is obtained below the continent, as in the "2D" case (see *Guillou and Jaupart,* [1995]).

upper thermal condition were investigated by varying continental sizes and tank heights. Specific effects of deep continental geometry are described in the following subsection with the use of numerical simulations.

The thermal condition below continents differs from that below oceans because temperature is not fixed by the external layer. If a conducting lid covers the whole of a convecting layer [*Sparrow et al.,* 1964; *Hewitt et al.,* 1980; *Chapman et al.,* 1980], the appropriate thermal condition is intermediate between a fixed heat flux condition and a fixed temperature condition. It can be written, in non-dimensional form, as:

$$\frac{\partial T}{\partial z} - BiT = C \qquad (2)$$

where $Bi$ is the Biot number and $C$ is a constant. As already noted, $Bi$ depends on thermal resistances of the conducting lid (the continent) and the fluid (the convecting mantle) :

$$Bi = \frac{k_c d}{k_m h} \qquad (3)$$

where $k_c$ and $k_m$ are thermal conductivities of the continent and the mantle, respectively, $d$ the thickness of the fluid layer, and $h$ is the continental thickness. One can see that for a thick and insulating continent, the above thermal condition tends to a fixed heat flux condition, i.e., $Bi$ tends to zero. On the opposite end, a thin and conducting lid would yield a fixed temperature condition.

One of the main problems with the thermal interaction between mantle and continents rests in the finite size of continents. In the case of the Earth, the conducting lids are of finite size and the above thermal condition cannot be generalized without considering the horizontal scale dependence.

In the experiments, the Biot number, which is only adequate to describe how convection and conduction are coupled, may not be specified a priori. To determine it one must determine temperature in both the lid and the convecting fluid. These temperature measurements allowed us to calculate $Bi$ values which appeared to range between 12 and 3, for a small and a large continent, respectively (see *Guillou and Jaupart,* [1995]). For a whole mantle convection regime and a buoyant continental lithosphere with a thickness of 250 km, the Biot number would be around 10 *if* the continents were not of finite size. The experimental Biot numbers showed that the finite size of continents played a major role in that the wider the continent was, the lower were the obtained $Bi$ values. In other words, a fixed heat flux condition was approached for large continental sizes. It is thus inferred that further theoretical considerations would need to incorporate a horizontal scale dependence of the Biot number.

Figure 4a shows the measured subcontinental heat flow values (scaled with the conductive heat flow value within the fluid, $k_f(T_1 - T_0)/d$) for different viscosity contrasts within the fluid. One can see that the controlling parameter is the continental size, but not the viscosity contrast or the Rayleigh number. Indeed, when $Ra$ is defined with the viscosity value at the mid-depth of the tank, $Ra$ increases with the viscosity contrast, from $10^4$ to $10^5$ in the bottom case ($a/d = 2$), and from $10^5$ to $10^6$ in the upper case ($a/d = 1$). The most important result is that the heat flow below the center of the continent is low, regardless of the value of $Ra$ and of the imposed viscosity contrast. For an $Ra$ of $10^6$, subcontinental heat flow values are a factor of 4 lower than the average value below oceans (Figure 4b). It is noteworthy that this factor depends 1) on the chosen geometry of a continent (see below) and 2) on continental width: For the largest continent and for an $Ra$ of $10^7$, the factor can reach a value of 10.

In Figure 4b, the spatial variation of the surface heat flow below the "continent" and below the "ocean" is plotted for the case of a square continent, centered in the middle of the upper plate (Figure 3). Only half of the upper surface is shown. As for previous cases a small heat flow value is measured below the center of the continent although more vigorous convection is imposed ($Ra = 1.6 \times 10^6$). However, vertical heat transfer increases towards the edges of the continent and a maximum value is reached below the ocean. Figure 4b should be compared with Figure 12 of *Guillou and Jaupart* [1995], where the two-dimensional configuration was used with the same $Ra$. The main difference is that the heat flow buffering effect was enhanced in the "2-D" case because the continental area was four times higher. For that case, spatial heat flow variations below the relatively large "continent" were very mild and a sharp increase of heat flow was obtained at the edge of the continent, whereas in Figure 4b , the increase is smooth.

Convective regimes which were observed and classified are described in *Guillou and Jaupart* [1995] as a function of Rayleigh number, i.e., as a function of the overall intensity of thermal convection. The flow patterns are increasingly complex and unstable as the Rayleigh number increases. One may point out that the low subcontinental heat flow induces a horizontal temperature gradient and thus a diverging flow from the center to the edges of the continent and this occurred without a thick "continental root". Deep plumes were then locked in beneath the continent (Figure 3). Despite the presence of these subcontinental plumes, a long wavelength convective circulation was superposed. The cell diameters, of the long wavelength circulation pattern, were comparable to the lateral extent of a continent and were shown to be regulated by continental sizes. It should be noted that within the experiments, continents were forced to remain spatially fixed and for the real Earth, were continents do not remain stationary, the observed flow patterns may be unstable. With this restriction, the experiments do suggest that the presence of continents at the upper surface of the convecting mantle can not only fix subcontinental heat flow values but can also exert control on the convective pattern within the mantle. If spatial variations of mantle heat flow are obtained near the edge of the continent, it appears that subcontinental heat flow will tend to be lower and more uniform for wider continents.

### *Numerical Simulations*

This subsection models continents in two different ways. In Figure 5 continents are modeled as accumulations of chemically buoyant material. Two buoyant components, continental crust and mantle-melt residuum, comprise model continents which reside within the upper thermal boundary layer of a chemically dense and thermally convecting mantle. This formulation allows continents to form that contain analogs to cratons underlain by thick, chemically stabilized cratonic roots and peripheral sections composed solely of crust. In Figure 6 continents are modeled as rigid, high viscosity blocks. Mechanical lithospheric thickness variations are prescribed by varying the shape of the blocks. The thicker central blocks are analogs to cratons underlain by tectospheric roots. The need for two different approaches will be made clear shortly.

Before discussing their implications, some simulation specifics need mentioning. Model equations can be found in *Lenardic and Kaula* [1996] along with full definitions of model parameters. All simulations assume: a thermal conductivity and diffusivity of 4 W/m°K and $10^{-6}$ m$^2$/s; a system temperature drop ($\Delta T$) of 2500 °C; a bottom heating thermal Rayleigh number ($Ra$), defined for bulk mantle viscosity, of $10^7$; and an internal to bottom heating $Ra$ ratio of 4.0. Side walls are reflecting; horizontal boundaries are free slip and isothermal. The simulations of Figure 5 assume reference crustal, residuum, and mantle densities of 3000, 3210, and 3300 $kg/m^3$, respectively. The temperature-dependent viscosity of mantle and residuum is set to give a factor of $10^4$ increase from the system base to the surface. Crustal viscosity is of the same functional form but one order of magnitude weaker than mantle

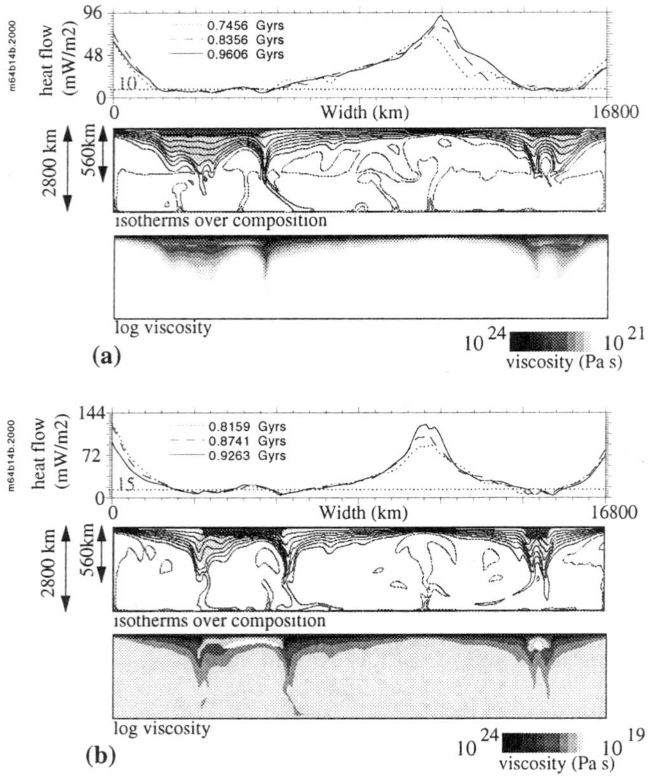


**Figure 5.** Results from 2-D simulations that include chemically light crustal and melt-residuum layers within the upper boundary layer of a thermally convecting mantle. The buoyant chemical accumulations that form within the simulations serve as model continents. The set of frames in (a) and (b) represent results from two different simulations. The three frames from each separate simulation show: Heat flux (top); isotherms over composition (middle); and viscosity (bottom). Note that the the top of middle and bottom frames are stretched vertically for visualization. For the composition fields: Dark grey is crust; light grey is residuum; white is mantle. The parameter values from simulation (a) and simulation (b) are discussed in the text.

at equivalent temperature in Figure 5a and two orders weaker in 6b. Crustal volume is 1% of the system while residuum volume is 3% in Figure 5a and 2% in 6b. Crust and residuum are not enriched in heat producing elements, i.e., surface heat flow variations are due dominantly to variations in mantle heat flow. Clearly exact heat flow values depend on assumed model parameters and dimensional constants. We explored a 'reasonable' range of these values. We also consider models to be potentially valid to the Earth only if they predict an average oceanic mantle heat flow in line with data [*Pollack et al.*, 1993]. This constrains the range of 'allowable' subcontinental mantle heat flow values within the simulations. Further modeling details can be found in *Lenardic and Kaula* [1996] and *Lenardic* [1997]. Those references present results from a suite of simulations, the heat flow implications of which will be reviewed below using the simulations of Figures 5 and 6 as representative examples.

Figure 7 shows geotherms from the simulation of Figure 5a. The 'oceanic' geotherms reflect the fact that all of the oceanic lithosphere participates in convection. Within the thermal boundary layer vertical heat flow is conductive but the thickness of the layer is controlled by advection, i.e., by the vigor of mantle convection. The 'continental' geotherms reflect the fact while the lower portion of the continental thermal lithosphere does actively participate in mantle convection, its chemically buoyant upper portion does not and, thus, heat advection is not equal to vertical heat conduction within this upper portion of the continental lithosphere. Vertical heat flow from the base of the mantle thermal boundary layer to the surface in stable continental regions does however remain conductive, i.e., the layer spanned by this full depth is an analog for the thermal lithosphere.

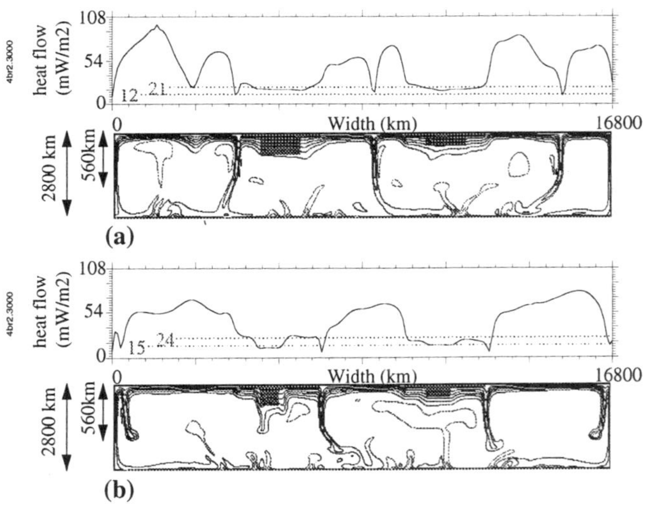


**Figure 6.** Results from simulations that assume continents to be nondeformable. High viscosity blocks of fixed dimensions define continents. The frames in (a) and (b) are from two different simulations and show: Heat flux (top); isotherms over a grey scale image highlighting the position and shape of the high viscosity continental blocks (bottom). In the grey scale images, continental blocks are dark grey while the mantle is white. For both of the simulations, the thin portions of the continental blocks are 28 km thick; the thicker portions are 140 km thick for the left block and 84 km for the right. The relative widths spanned by the thick portions of the continental blocks vary between the two simulations as shown by the bottom frames in (a) and (b).

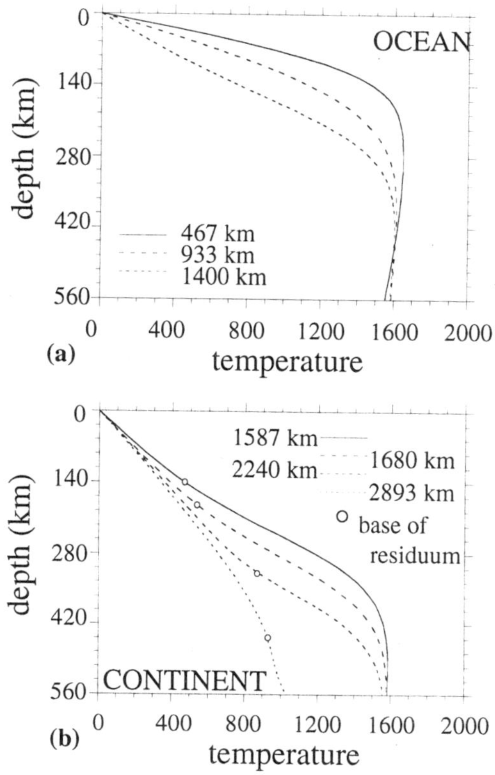


**Figure 7.** Temperature versus depth profiles from simulation (a) of Figure 5. The top and bottom sets of profiles are from model 'oceanic' and 'continental' regions, respectively. In both cases the model width at which the profiles are taken is shown in the key (the oceanic profiles are from the left most portion of the model domain in Figure 5a; the continental profiles are from the left half of the leftmost continent in Figure 5a). For the continental cases the base of the chemically buoyant portion of the continental lithosphere is also shown.

The heat flow out of the convecting mantle into the conductive continental material depends on the thickness of the mantle portion of the upper thermal boundary layer and on the temperature drop across it. This temperature drop is not constant (Figure 7b). Thus, the thickness of the mantle portion of the lithosphere and mantle heat flow are not related in a simple manner.

The simulations of Figure 5 are time-dependent and mantle convection patterns can change below buoyant continents as can lithospheric structure. If the buoyant portions of the continents are not being significantly deformed, then the time scale over which their thermal structure can change is determined by thermal diffusion. The time scale over which the thermal structure in lower portions of the continental lithosphere can change is, however, determined by a convective time scale as these portions of the lithosphere can participate in mantle convection. The slightly kinked nature of the geotherms in Figure 7b reflects the fact that convecting portions of the mantle are adjusting to the thermal conditions imposed on them by the buoyant continents from above. That is, the convective time scale is shorter than the diffusive. The effects of crust and residuum on the mantle are often considered insulating [e.g., *Anderson*, 1982] but, as Figure 7b shows, a more appropriate term is buffering. That is, if mantle heat flux were to increase locally then the temperature at the residuum-mantle interface or the crust-mantle interface would increase which would tend to decrease mantle heat flux. If the mantle portion of the lithosphere can adjust to changing thermal conditions from above more rapidly than the buoyant portions of the lithosphere can adjust to changing thermal conditions from below then this will tend to regulate mantle heat flow.

The discussion above is pertinent to evaluating the effects of drifting continents. If the thermal adjustment time in drifting, tectonically stable continents is long relative to the advective time scale of their motion, then they will be able to 'carry' their thermal structure as deep temperature variations will not be able to penetrate into their buoyant upper portions [*Lenardic and Kaula*, 1995; *Lenardic*, 1997]. Thus, although instantaneous thermal lithospheric thickness may be variable, due to the fact that continents are drifting over hotter or colder mantle, these changes may not be detectable by surface measurements.

The left continent of Figure 5b shows the beginning of a continental splitting event. The thermal plume rising below it has just started to thin the crust. As this continues, local mantle heat flow variations increase due to the fact that the crustal layer, i.e., the thermal buffer, is being advected away. In the modeling framework, this progressive continental splitting must be considered a tectonic event. The smaller continental sections that formed after the splitting event stabilized with time and showed slightly larger mantle heat flow variations than large continents, in line with laboratory results, but the variations remained mild relative to those in oceans. Thus, the tendency toward a constant mantle heat flow does not hold within the simulations if a continent is being considerably deformed but it is approached after deformation events cease, which is in line with the observations of Figure 2a.

As noted, the formulation of Figure 5 allows for the modeling of distributed continental tectonics. This potential deformability of continents might lead one to wonder if forced convection within them is tainting results. To test this, we use simulations that treat buoyant continental material as rigid (Figure 6). Continents

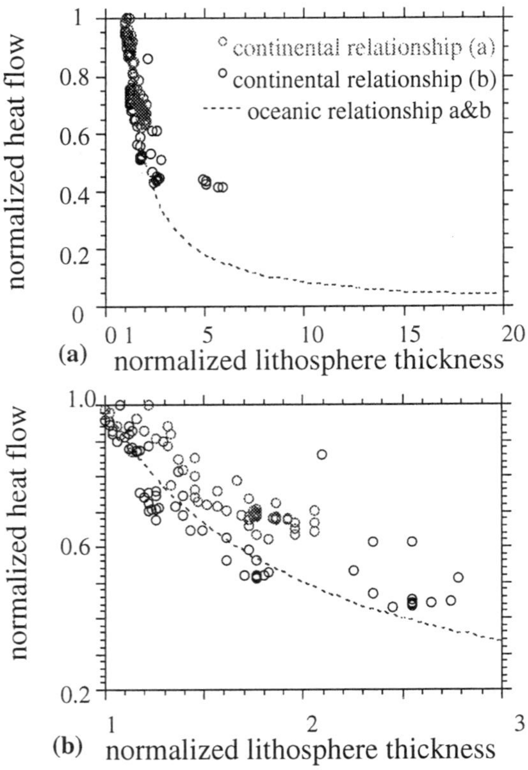


**Figure 8.** Normalized heat flow versus normalized thermal lithosphere thickness for the simulations of Figure 6. The top plot shows the full extent of the simulation values while the bottom plot shows a blow up of the thinner lithospheric thickness, higher heat flow, portion of the simulations results. The base of the thermal lithosphere is defined by extrapolating the local temperature profile in the upper portion of the mantle thermal lithosphere and determining the depth at which this profile meets the bulk internal temperature of the convecting mantle below.

are comprised of a thick central block and two thin peripheral ones. The block depths correspond to mechanical lithospheric thickness. Despite a 5 to 1 variation in the thickness of these rigid blocks in the left continent of Figure 6, the heat flow varies only by a factor of 2. That is, the relationship between mechanical lithospheric thickness and mantle heat flow is weak. The relationship between thermal lithospheric thickness and mantle heat flow is also weak relative to that in oceans (Figure 8). The flattening of the continental trend in Figure 8a reflects the fact that thermal equilibrium may not hold in regions where mantle downflows occur below thick continental sections as the time scale of downflow formation and/or motion can be less than that of conduction through the stable continental material. If dynamic mantle downflows do occur below cratons [e.g., *Polet and Anderson*, 1995; *Pari and Peltier*, 1997], then this type of behavior could further weaken the correspondence between heat flow and the seismically inferred thickness of the lithosphere.

The isotherms within the conducting continental blocks of Figure 6 illustrate why mantle heat flux variations are constrained below the continental blocks within the simulations. Within a conducting body with a constant surface temperature it is difficult to maintain isotherms that are not flat relative to those in a convective boundary layer, regardless of the thickness variations of the conducting body itself. The tendency of isotherms to flatten within the conducting continental blocks, and the fact that heat flux jumps can not occur within the system, means that heat flux at the base of the blocks, i.e., at the top of the convecting mantle, will also tend toward a relatively constant value. Thus, spatial variations of mantle heat flow below the model continents tend to be mild relative to those in oceanic regions were the entire lithosphere participates in mantle convection. The tendency of isotherms to flatten within the stable portions of continents also occurred in the models of Figure 5, again regardless of thickness variations within the chemically stabilized portions of the lithosphere. A comparison of Figures 5 and 6 shows that the more tectonically stable a continent, the greater the tendency for isotherms within it to be flat, i.e., forced convection within continents is not what is responsible for a relatively constant mantle heat flux below them.

Principal simulation results are not an artifact of 2-D geometry nor do they depend on both residuum and crust comprising continents (Figure 9). The numerical methods used for the simulation of Figure 9, as well as a discussion of the dynamics associated with this, and similar, 3-D simulations can be found in *Moresi and Lenardic* [1997]. In accord with laboratory experiments, 3-D effects can increase subcontinental heat flow values but these values remain low and nearly constant below the central part of the continent. Figure 9 also shows that although added geometric freedom allows for more complex patterns of subcontinental flow, the relationship between subcontinental mantle heat flux and thermal lithospheric thickness remains weaker than that in model oceanic regions.

## DISCUSSION AND CONCLUSIONS

The main conclusion of analysis of data from the Canadian Shield and its surroundings is that they are consistent with the idea that surface heat flow variations in stable continental regions are not due to mantle heat flow variations but to variations in crustal heat production. The data show no evidence that mantle heat is

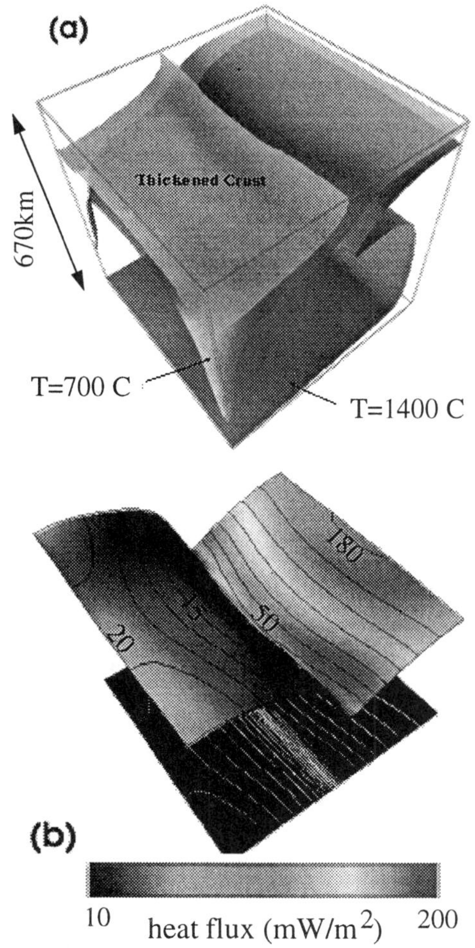


**Figure 9.** Results from an isoviscous, 3-D simulation. (a) A composition surface (light grey) showing the base of the crust along with two isothermal surfaces (darker grey) showing the thermal convection pattern in the mantle. (b) A shade plot of surface heat flux overlaid onto dynamic surface topography. In figure (b), topography contours are drawn on the topographic surface and the heat flux, at specific locations, is noted using numerical values. Notice that the heat flux in the region of continental crust varies from 10-20 mW/m$^2$ while the heat flux in the region free of continental crust varies from 10-180 mW/m$^2$. Relative crustal volume is 4%; $\Delta T = 2000$; and $Ra = 10^5$.

deflected from Archean cratons to the younger regions that border them. Rather, mantle heat flow appears to be low and spatially uniform across a broad region that spans tectonic provinces ranging in age from 0.4-2.7 Ga. This may not apply worldwide but, it should be noted, that the suggestion that significant mantle heat flow variations are associated with tectonothermal age and are the main cause of long wavelength surface heat flow variations has yet to be demonstrated with both heat flow and heat production data of the same quality and coverage as those from Canada. *Jaupart and Mareschal* [1998] have analyzed heat flow, heat production, and seismic data from South Africa and concluded that, although the average heat flow may be higher in Kaapvaal craton than in Canada, the variations in surface heat flow within the craton are due to crustal heat production. In any case, from eastern to central Canada, there is no evidence for a relationship between tectonothermal age and mantle heat flow within the uncertainty of heat flow values, i.e., a few $mW/m^2$. Additionally, evidence for lithospheric thickness variations across the region, from seismic tomography [*Grand*, 1987; 1994; *van der Lee and Nolet*, 1997] and gravity and topography modeling [*Pilkington*, 1991], suggests that the relationship between mantle heat flow and lithospheric thickness variations is also weak across a large portion of stable North America.

The main conclusion of the theory based section is that a low, relatively uniform mantle heat flow below stable continental regions is consistent with theoretical considerations of the thermal coupling between convecting portions of the solid Earth and the buoyant continents. This was demonstrated by laboratory experiments and numerical simulations that explored the interaction between continents and the thermally convecting mantle. We have obviously not explored all the complexities of mantle convection. However, the similarity of key results seen from a suite of laboratory experiments and numerical simulations, which make different simplifying assumptions, suggests that our principal conclusion is robust. Moreover, this conclusion is consistent with the available heat flow data and this leads us to believe that the equilibrium thermal condition 'seen' by the mantle below stable continental regions can approach one of nearly constant heat flux. If this is valid, then the experiments and simulations show that mantle flow patterns may not be well reflected in surface heat flow variations (e.g., in Figure 6 notice how upflows, downflows, and flow directions are well reflected in 'oceanic' heat flow variations but not in 'continental'). A second implication is that the relationship between mantle heat flow variations and lithospheric thickness variations can be weaker in continental than in oceanic regions.

Both experiments and simulations show that the degree to which a spatially near constant heat flux condition is approached depends on factors such as relative continental size and the relative size of tectospheric roots. Thus, generalizing heat flow observations from one continental region to another can be problematic. Although subcontinental heat flow variations are predicted to be mild relative to those in oceans for a range of conditions, these variations still depend on factors that may vary from one continent to another. This is why the type of heat flow and heat production data discussed above is required if any theoretical considerations that involve subcontinental heat flow are to have meaning. It is also why similar analyses from data in different continental regions are necessary to more tightly constrain the models.

The experiments and simulations discussed did not include the effect of laterally variable crustal heat production which could well account for significant mechanical lithospheric thickness variations [*Pinet et al.*, 1991]. Variable Moho temperatures are found between tectonic provinces with different near surface heat productions (see also *Mareschal et al.* [1999a]). Such Moho temperature variations would allow lithospheric thickness variations to exist across regions of constant mantle heat flow.

Alternatively, the large lithospheric thickness variations inferred, by various methods, within the Canadian Shield may indicate that the base of the long-lived, chemically buoyant portion of the lithosphere, i.e., the portion that does not participate in convection, is not necessarily an isotherm but could be defined by the condition that conductive heat flow and convective heat flow supplied from below equilibrate with one another (see the discussion in *Jaupart et al.* [1998]). In that case, one could end up with an increasing temperature at the base of the long-lived portion of Canadian lithosphere, from the east to the center. This would be akin to the situation shown by the numerical simulation of Figure 5a (see also Figure 7b). If the thickness of the chemically buoyant portion of the lithosphere has a strong effect on mechanical and seismic lithospheric thickness estimates, as seems likely, and if its base is not an isotherm, the correlation between mantle heat flow and lithospheric thickness will be weak. There is a potential complexity. If the temperature gradients discussed above become large, they could promote local upflow below the center of the continent. Clearly mantle flow patterns below continents can evolve over geologic time, as shown by flow reversals [*Lowman and Jarvis*, 1995] or changes in convective wavelength due to drift [*Zhong and Gurnis*, 1993]. Such flow changes could certainly cause mantle heat flow changes and could alter lithospheric structure. Despite these complexities, the observation remains that, at present, variations of mantle heat flow and lithospheric thickness seem weakly correlated in eastern to central Canada. Although this is unusual relative to what occurs in oceanic regions, it is not without a physically reasonable explanation, as suggested by the simulations of Figures 5 and 6, and the studies of *Pinet et al.* [1991] and *Mareschal et al.* [1998].

The discussion above suggests that the structure of deep continental lithosphere is not determined so much by mantle heat flow variations, but is, to a greater degree, set from above. The experiments discussed also suggest that not only might lithospheric structure be determined from above but that the large-scale dynamics of the mantle might also be affected. This idea is not new [e.g., *Anderson*, 1982] and several studies have explored the effects of continents on global mantle circulation patterns [e.g., *Elder*, 1967; *Gurnis*, 1988; *Lowman and Jarvis*, 1993; 1995]. We have focussed on heat flow but our conclusions re-enforce the general implications of these studies that the presence of a heterogeneous, versus a constant temperature, condition at the surface of a convecting fluid will have a strong effect on its internal dynamics. This implies that continents can not be viewed as passive features that are advected by mantle motions but must be seen as active features which influence the very nature of these motions. Thus, continents will need to be included in future efforts to unify studies of plate kinematics with models of mantle convection.

*Acknowledgments.* Support for A.L. was provided by a UC Presidents Postdoctoral Fellowship. Computing was carried out, in part, at the San Diego Supercomputer Center. Thanks to Gary Jarvis and Gene Humphries for thoughtful and helpful reviews. This paper is BRGM publication number 99016.

## REFERENCES

Anderson, D.L., Hotspots, polar wander, Mesozoic convection and the geoid, *Nature, 297,* 391-393, 1982.

Ashwal, L.D., P. Morgan, S.A. Kelley, and J.A. Percival, Heat production in an Archean crustal profile and implications for heat flow and mobilization of heat-producing elements, *Earth Planet. Sci. Lett., 85,* 439-450, 1987.

Ballard, S., and H.N. Pollack, Diversion of heat flow by Archean cratons: A model for southern Africa and Namibia, *Earth Planet. Sci. Lett., 85,* 253-264, 1987.

Barrell, J.,The strength of the Earth's crust, *J. Geology, 22,* 28-48, 1914.

Bechtel, T.D., D.W. Forsyth, V.L. Sharpton, and R.A.F. Grieve, Variations in effective elastic thickness of the North American Lithosphere, *Nature, 343,* 636-638, 1990.

Birch, F., E.R. Roy, and E.R. Decker, Heat flow and thermal history in New England and New York, in: *Studies of Appalachian Geology,* E. An-Zen, ed., pp 437-451, Wiley (Interscience), New York, 1968.

Burchfiel, B.C., The continental crust, *Scientific American,* September, 1983.

Busse, F.H., and N. Riahi, Nonlinear convection in a layer with nearly insulating boundaries, *J. Fluid Mech., 96,* 243-256, 1980.

Carlson, R.L. and Johnson, H.P., On modeling the thermal evolution of the oceanic upper mantle: an assessment of the cooling plate model, *J. Geophys. Res., 99,* 3201-3214, 1994.

Carriére, P., A. Bottaro, and P. Metzener, Wavelength selection in Rayleigh-Benard convection between horizontal boundaries of finite conductivity, *Eur. J. Mech. B/Fluids, 16,* 483-508, 1997.

Chapman, C.J., S. Childress, and M.R.E. Proctor, Long wavelength thermal convection between non-conducting boundaries, *Earth Planet. Sci. Lett., 51,* 362-369, 1980.

Davaille, A., and C. Jaupart, Onset of thermal convection in fluids with temperature-dependent viscosity: application to the oceanic mantle, *J. Geophys. Res., 99,* 19,853-19,866, 1994.

Davies, G.F., Role of the lithosphere in mantle convection, *J. Geophys. Res., 89,* 10,451-10,466, 1988.

Doin, M.P., and L. Fleitout, Thermal evolution of the oceanic lithosphere: and alternative view, *Earth Planet. Sci. Lett., 142,* 121-136, 1996.

Doucouré, C.M., M.J. de Wit, and M. F. Mushayandebvu, Effective elastic thickness of the continental lithosphere in South Africa, *J. Geophys. Res., 101,* 11,291-11,303, 1996.

Elder, J.W., Convective self-propulsion of continents, *Nature, 214,* 657-660, 1967.

Grand, S.P., Tomographic inversion for shear structure beneath the North American plate, *J. Geophys. Res., 92,* 14,065-14,090, 1987.

Grand, S.P., Mantle shear structure beneath the Americas and surrounding oceans, *J. Geophys. Res., 99,* 11,591-11,621, 1994.

Guillou, L., and C. Jaupart, On the effect of continents on mantle convection, *J. Geophys. Res., 100,* 24,217-24,238, 1995.

Guillou, L., J.-C. Mareschal, C. Jaupart, C. Gariépy, G. Bienfait, and R. Lapointe, Heat flow, gravity and structure of the Abitibi belt, Superior Province, Canada: Implications for mantle heat flow, *Earth Planet. Sci. Lett., 122,* 103-123, 1994.

Guillou-Frottier, L., J.-C. Mareschal, C. Jaupart, C. Gariepy, R. Lapointe, and G. Bienfait, Heat flow variations in the Grenville Province, Canada, *Earth Planet. Sci. Lett., 136,* 447-460, 1995.

Guillou-Frottier, L., C. Jaupart, J.-C. Mareschal, C. Gariepy, G. Bienfait, L.Z. Cheng, and R. Lapointe, High heat flow in the Trans-Hudson Orogen, central Canadian Shield, *Geophys. Res. Lett., 23,* 3027-3030, 1996.

Gurnis, M., Large-scale mantle convection and the aggregation and dispersal of supercontinents, *Nature, 332,* 695-699, 1988.

Hartley, R., A.B. Watts, and J.D. Fairhead, Isostasy of Africa, *Earth Planet. Sci. Lett., 137,* 1-18, 1996.

Heestand, R.L., and S.T. Crough, The effect of hot spots on the oceanic age-depth relation, *J. Geophys. Res., 86,* 6107-6114, 1981.

Hewitt, J.M., D.P. McKenzie, and N.O. Weiss, Large aspect ratio cells in two-dimensional thermal convection, *Earth Planet. Sci. Lett., 51,* 370-380, 1980.

Hoffmam, P., Precambrian geology and tectonic history of North America, in *The Geology of North America-An Overview,* edited by A.W. Bally and E.R. Palmer. pp. 447-512. GSA. Boulder(CO). 1989.

Jarvis, G.T., and W.R. Peltier, Mantle convection as a boundary layer phenomenon, *Geophys. J. R. Astr. Soc., 68,* 389-427, 1982.

Jaupart, C., Horizontal heat transfer due to radioactivity contrasts: causes and consequences of the linear heat flow relation, *Geophys. J. R. Astron. Soc., 75,* 411-435, 1983.

Jaupart, C., and J.-C. Mareschal, The thermal structure and thickness of continental roots, *Lithos*, in press, 1999.

Jaupart, C., J.-C. Mareschal, L. Guillou-Frottier, and A. Davaille, Heat flow and the thickness of the lithosphere in the Canadian Shield, *J. Geophys. Res., 103,* 15,269-15,286, 1998.

Jessop, A.M., T.J. Lewis, A.S. Judge, A.E. Taylor, M.J. Drury, Terrestrial heat flow in Canada, *Tectonophysics, 103,* 239-261, 1984.

Jessop, A.M., The distribution of glacial perturbation of heat flow in Canada, *Can. J. Earth Sci., 8,* 162-166, 1971.

Jordan, T.H., Composition and development of the continental tectosphere, *Nature, 274,* 544-548, 1978.

Lachenburch, A.H., Crustal temperatures and heat production implications of the linear heat flow relation, *J. Geophys. Res., 75,* 3291-3300, 1970.

Lenardic, A., On the heat flow variation from Archean cratons to Proterozoic mobile belts, *J. Geophys. Res., 102,* 709-721, 1997.

Lenardic, A., On the partitioning of mantle heat loss below oceans and continents over time and its relationship to the Archean paradox, *Geophys. J. Int., 134,* 706-720, 1998.

Lenardic, A., and W.M. Kaula, Mantle dynamics and the heat flow into the Earth's continents, *Nature, 378,* 709-711, 1995.

Lenardic, A., and W.M. Kaula, Near surface thermal/chemical boundary layer convection at infinite Prandtl number: Two-dimensional numerical experiments, *Geophys. J. Int., 126,* 689-711, 1996.

Lowman, J.P., and G.T. Jarvis, Mantle convection flow reversals due to continental collisions, *Geophys. Res. Lett., 20,* 2087-2090, 1993.

Lowman, J.P., and G.T. Jarvis, Mantle convection models of continental collision and breakup incorporating finite thickness plates, *Phys. Earth Planet. Inter., 88,* 53-68, 1995.

Mareschal, J.C., L. Guillou-Frottier, C. Gariépy, and C. Jaupart, Heat flow in the Trans-Hudson-Orogen, *1997 Lithoprobe Transect Meeting Report*, Saskatoon, 1997.

Mareschal, J.C., C. Jaupart, C. Gariépy, L.Z. Cheng, L. Guillou-Frottier, G. Bienfait, and R. Lapointe, Heat flow and deep thermal structure near the edge of the Canadian Shield, *Can. J. Earth Sci.*, in press, 1999a.

Mareschal, J.C., C. Pinet, C. Gariepy, C. Jaupart, G. Bienfait, G. Dalla Coletta, J. Jolivet, and R. Lapointe, New heat flow density and radiogenic heat production data in the Canadian Shield and the Quebec Appalachians, *Can. J. Earth Sci., 26,* 845-852, 1989.

Mareschal, J.C., F. Rolandone, and G. Bienfait, Heat flow variations in a deep borehole near Sept-Iles, Quebec, Canada: Paleoclimatic interpretation and implications for regional heat flow estimates, *Geophys. Res. Lett.,* in press, 1999b.

Morgan, P., The thermal structure and thermal evolution of the continental lithosphere, *Physics and Chemistry of the Earth, 15,* 107-193, 1984.

Morgan, P., Crustal radiogenic heat production and the selective survival of ancient continental crust, *Proc. Lunar Planet. Sci. Conf. 15th*, Part 2, *J. Geophys. Res., 90,* suppl., C561-C570, 1985.

Morgan, P., and J.H. Sass, Thermal regime of the continental lithosphere, *J. Geodyn., 1,* 143-166, 1984.

Moresi, L.-N., and A. Lenardic, Three-dimensional numerical simulations of crustal deformation and subcontinental mantle convection, *Earth Planet. Sci. Lett., 150,* 233-243, 1997.

Nield, D.A., The Rayleigh-Jeffreys problem with boundary slab of finite conductivity, *J. Fluid Mech., 32,* 393-398, 1968.

Nyblade, A.A., and H.N. Pollack, A global analysis of heat flow from Precambrian terrains: Implications for the thermal structure of Archean and Proterozoic lithosphere, *J. Geophys. Res., 98,* 12,207-12,218, 1993.

Oxburgh, E.R., Heat flow and differences in lithospheric thickness, *Philos. Trans. R. Soc. London, Ser. A, 301,* 337-346, 1981.

Pari, G., and Peltier, The free-air gravity constraint on subcontinental mantle dynamics, *J. Geophys. Res., 101,* 28,105-28,132, 1997.

Parsons, B. and D.P. McKenzie, Mantle convection and the thermal structure of the plates, *J. Geophys. Res., 83,* 4485-4496, 1978.

Parsons, B., and J.G. Sclater, An analysis of the variation of ocean floor bathymetry and heat flow with age, *J. Geophys. Res., 82,* 803-827, 1977.

Phipps Morgan, J., and W. Smith, Flattening of the seafloor depth age curve as a response to asthenospheric flow, *Nature, 359,* 524-527, 1992.

Pilkington, M., Mapping elastic lithospheric thickness variations in Canada, *Tectonophysics, 190,* 283-297, 1991.

Pinet, C., C. Jaupart, J.-C. Mareschal, C. Gariepy, G. Bienfait, and R. Lapointe, Heat flow and structure of the lithosphere in the eastern Canadian Shield, *J. Geophys. Res., 96,* 19,941-19,963, 1991.

Polet, J., and D.L. Anderson, Depth extent of cratons as inferred from tomographic studies, *Geology, 23,* 205-208, 1995.

Pollack, H.N., Cratonization and thermal evolution of the mantle, *Earth Planet. Sci. Lett., 80,* 175-182, 1986.

Pollack, H.N., S.J. Hurter, and J.R. Johnson, Heat flow from the Earth's interior: analysis of the global data set, *Rev. Geophys., 31,* 267-280, 1993.

Pujol, J., D.M. Fountain, and D.A. Anderson, Statistical analysis of the mean heat flow/reduced heat flow relationship for continents and its tectonothermal implications, *J. Geophys. Res., 90,* 11,335-11,344, 1985.

Riahi, N., Nonlinear thermal convection with finite conducting boundaries, *J. Fluid Mech., 152,* 113-123, 1985.

Roy, R.F., D.D. Blackwell, and F. Birch, and A. Lenardic, Heat generation of plutonic rocks and continental heat flow provinces, *Earth Planet. Sci. Lett., 5,* 1-12, 1968.

Sclater, J.G., and J. Francheteau, The implications of terrestrial heat flow observations on current tectonic and geochemical models of the crust and upper mantle of the Earth, *Geophys. J. Roy. astr. Soc., 20,* 509-542, 1970.

Sclater, J.G., C. Jaupart, and D. Galson, The heat flow through oceanic and continental crust and the heat loss of the Earth, *Rev. Geophys., 18,* 269-311, 1980.

Sclater, J.G., B. Parsons, and C. Jaupart, Oceans and continents: Similarities and differences in the mechanisms of heat loss, *J. Geophys. Res., 86,* 11,535-11,552, 1981.

Sparrow, E.M., R.J. Goldstein, and V.K. Jonsson, Thermal instability in a horizontal layer: effect of boundary conditions and non-linear temperature profile, *J. Fluid Mech., 18,* 513-528, 1964.

Stein, C.A., and S. Stein, A model for the global variation in oceanic depth and heat flow with lithospheric age, *Nature, 359,* 123-129, 1992.

Turcotte, D.L. and E.R. Oxburgh, Finite amplitude convection cells and continental drift, *J. Fluid. Mech., 28,* 29-42, 1967.

van der Lee, S., and G. Nolet, The upper mantle S velocity structure of North America, *J. Geophys. Res., 101,* 22,815-22,838, 1997.

Vitorello, I., and H.N. Pollack, On the variation of continental heat flow with age and the thermal evolution of continents, *J. Geophys. Res., 85,* 983-995, 1980.

Watts, A., J.H. Bodine, and M.S. Steckler, Observations of flexure and the state of stress in the oceanic lithosphere, *J. Geophys. Res., 85,* 6369-6376, 1980.

Zhong, S., and M. Gurnis, Dynamic feedback between a continent-like raft and thermal convection, *J. Geophys. Res., 98,* 12,219-12,232, 1993.

---

L. Guillou-Frottier, BRGM, Laboratoire de Géologie et de Métallogénie, 3 av. C. Guillemin, BP 6009, F-45060 Orléans Cedex 2, France; ph: (33) 2 38 64 47 91; fax: (33) 2 38 64 36 52; email: l.guillou-frottier@brgm.fr

C. Jaupart, Institut de Physique du Globe, 4 Pl. Jussieu, Paris, F-75252, France

W.M. Kaula, Department of Earth and Space Sciences, UCLA, Los Angeles, CA 90025

A. Lenardic, Department of Geology and Geophysics - MS 126, P.O. Box 1892,Rice University, Houston, TX 77251-1892; email: adrian@geophysics.rice.edu

J.C. Mareschal, Geotop, UQAM, P.O.8888, sta. Centre-Ville, Montreal, H3C3P8, Canada

L.-N. Moresi, CSIRO Exploration and Mining, 39 Fairway, Nedlands, WA 6009, Australia

# On the Determination of Self-Consistent Strain Rate Fields Within Zones of Distributed Continental Deformation

William E. Holt[1] and Bingming Shen-Tu
*Department of Geosciences, SUNY at Stony Brook, NY*

John Haines and James Jackson
*Bullard Laboratories, University of Cambridge, U.K.*

We present a method to estimate self-consistent, continuous velocity field models given geologic or seismic strain rate and space-based geodetic observations. In a least-squares inversion, geologic or seismic strain rates are matched with continuous functions, subject to the constraint that GPS or VLBI observations are also matched by the model velocity field within a defined frame of reference. We investigate the active deformation field in the western US and central and east Asia using geologic strain rates, GPS and VLBI data, and the Aegean using earthquake strain rates and SLR data. Integrated Quaternary rates of strain across the western U.S. define a Pacific-North America (PA-NA) motion that deviates counterclockwise by 6 degrees from the NUVEL-1A velocity direction, but velocity magnitudes are equal to NUVEL-1A. Inversion of recent geodetic data alone yields a PA-NA velocity direction that is 2-3 degrees counterclockwise from the NUVEL-1A PA-NA direction, but the difference between the two is not significant at the 95% confidence level. The discrepancy between total PA-NA motion obtained from the geologic data and NUVEL-1A indicates that a marginally significant amount of NE-SW directed shortening (possibly as much as 5 mm/yr both onshore and offshore) is missing in the geologic data. From the continuous velocity field solution we also determine rotation rates and seismic moment deficits within the western US. Significant moment deficits exist on the southern San Andreas Fault and in the LA-Basin-Western Transverse Range region. A joint inversion of Quaternary strain rates and GPS velocities in Asia for a self-consistent velocity field yields a solution in which south China moves at an almost uniform velocity E-SE (azimuths of 110 ± 15 degrees) relative to Eurasia at 14 ± 3 mm/yr (95 % confidence). The Sunda block moves eastward with respect to Eurasia at 16 ± 4 mm/yr with azimuth of 96 ± 20 degrees (95 % confidence). Seismic strain rates in the Aegean, from earthquakes this century, are

[1]While on leave at Lamont Doherty Earth Observatory, Palisades, NY

The History and Dynamics of Global Plate Motions
Geophysical Monograph 121

compared with total strain rates inferred from SLR data. The SLR measurements do not require strain different in orientation from that observed in earthquakes, but do require more of it. On average, the seismic strain rates are roughly two thirds of the inferred total strain rate that accommodates all relative motions in the Aegean.

## 1. INTRODUCTION

Because of Global Positioning System (GPS) technology it is now possible to directly measure relative velocities of points within zones of distributed continental deformation. These measurements, however, are only the first step toward the goal of quantifying continental kinematics. Necessary is an integrated understanding of the relationship between measured relative motions, inferred using space-based geodesy, and the strain rates and rotation rates (or velocity gradient tensor), associated with the zones of distributed deformation. Equally important is the need to quantify the relationship between the deformation field and the far-field plate motions. For example, the integrated rates of strain associated with all active structures must yield the far-field relative plate motion (e.g. the NUVEL-1A model, *DeMets et al.*, 1994, which is a 2-3 Ma average, inferred from seafloor spreading data, transform fault trends, and earthquake slip vector data). Such a complete account of the rates and style of strain that accommodates plate motion is a first-order requirement in any regionally-based seismic hazard study [*Shen-Tu et al.*, 1999].

The quantification of the velocity gradient tensor field is also an important step toward the understanding of the dynamics (or forces) affecting present-day deformation [*England and Molnar*, 1997a]. That is, the style and magnitude of strain rate is dependent on the stress field and strength (rheology) of the deforming medium. There are three primary sources affecting the stress field within the lithosphere: lateral variations in density structure [*Artyushkov*, 1973; *Fleitout and Froidevaux*, 1982; *England and McKenzie*, 1982], basal tractions [*McKenzie*, 1969; *Tovish et al.*, 1978], and boundary stresses [*Richardson*, 1992]. Lateral variations in density within the lithosphere introduce body forces that can have an important influence on the magnitude and style of deformation [*Artyushkov*, 1973; *Fleitout and Froidevaux*, 1982; *Fleitout*, 1991; *England and McKenzie*, 1982; *England and Houseman*, 1986; *Molnar and Lyon-Caen*, 1988]. *Jones et al.* [1996] have argued that much of the deformation within the Great Basin in the western United States can be attributed to these body forces associated with lateral variations in lithospheric density structure. On the other hand, recent GPS observations across the Great Basin in the western U.S. [*Thatcher et al.*, 1999] show significant velocity gradients within the western Great Basin where stresses associated with such lateral variations in density structure are supposed to be small. On this basis, *Thatcher et al.* [1999] have argued that stresses associated with plate boundary interaction (Pacific-North America motion) [e.g. *Atwater*, 1970; *Stock and Molnar*, 1988; *Bohannon and Parsons*, 1995] play a more important role in western Great Basin deformation than do lateral density variations within the lithosphere.

Most researchers agree that basal tractions, lateral density variations in the lithosphere, or boundary forces are insufficient by themselves to explain all deformation phenomenon [*Wdowinski and O'Connell*, 1991; *Flesch et al.*, 1998; *Holt et al.*, 1998; *Liu and Shen*, 1998; *Sonder and Jones*, 1999]. Instead these driving mechanisms act together to influence the style and magnitude of active deformation. Because of the wealth of GPS data available to constrain deformation gradients within zones of distributed continental deformation, a new class of dynamic problems can be constrained. It is now possible to resolve the relative role of basal tractions, lateral density variations, and boundary stresses operating within areas such as central Asia, the Aegean, and the western United States [e.g. *Flesch et al.*, 1998; *Holt et al.*, 1998]. However, inferring relative velocity vectors using the GPS technology is insufficient to solve these dynamic problems. Researchers will need to quantify the full velocity gradient tensor field since the set of dynamic solutions requires quantification of the stress tensor field, the effective rheology, and the relationship of these with the velocity gradient tensor field [e.g. *McKenzie and Jackson*, 1983; *England and Molnar*, 1997a; *Flesch et al.*, 1998]. We present here a practical approach to estimating the horizontal velocity gradient tensor field using observations from space-based geodesy, Quaternary fault slip rates, and earthquake moment tensors.

## 2. PARAMETERIZING A CONTINUOUS VELOCITY FIELD

When averaged over periods of $10^4$-$10^5$ years a crustal deformation zone involves discrete jumps in relative velocity across each fault. However, this velocity field can be approximated as continuous when horizontal distances over which relative motions are considered are a few times the thickness of the brittle elastic layer [*Jackson et al.*, 1992; *Holt and Haines*, 1993]. Because zones of distributed deformation within continental lithosphere are hundreds to even thousands of kilometers wide, we parameterize the deformation field on the surface of the earth as a continuous horizontal velocity field

$$\mathbf{u}(\hat{\mathbf{x}}) = r\mathbf{W}(\hat{\mathbf{x}}) \times \hat{\mathbf{x}} \tag{1}$$

where $\mathbf{W}(\hat{\mathbf{x}})$ is a vector rotation function, $r$ is the earth radius, and in terms of latitude $\theta$ and longitude $\phi$,

$$\hat{\mathbf{x}} = (\cos\theta\cos\phi,\ \cos\theta\sin\phi,\ \sin\theta) \tag{2}$$

is the vector of direction cosines of each point on the Earth's surface [*Haines and Holt*, 1993; *Haines et al.*, 1998]. The rotation vector function $\mathbf{W}(\hat{\mathbf{x}})$ is expanded using bi-cubic Bessel interpolation [*deBoor*, 1978] on a curvilinear grid. Unlike the polynomials used by *Haines and Holt* [1993], these functions accommodate potentially rapid spatial variations in the strain rate field [*Haines et al.*, 1998]. The horizontal rates of strain on a sphere can be written using spatial derivatives of the vector function $\mathbf{W}(\hat{\mathbf{x}})$,

$$\dot{\varepsilon}_{\phi\phi} = \frac{\hat{\Theta}}{\cos\theta} \cdot \frac{\partial \mathbf{W}}{\partial \phi} + \frac{u_r}{r}, \tag{3}$$

$$\dot{\varepsilon}_{\theta\theta} = -\hat{\Phi} \cdot \frac{\partial \mathbf{W}}{\partial \theta} + \frac{u_r}{r}, \tag{4}$$

$$\dot{\varepsilon}_{\theta\phi} = \tfrac{1}{2}\left( \hat{\Theta} \cdot \frac{\partial \mathbf{W}}{\partial \theta} - \frac{\hat{\Phi}}{\cos\theta} \cdot \frac{\partial \mathbf{W}}{\partial \phi} \right), \tag{5}$$

where $\hat{\Theta}$ and $\hat{\Phi}$ are unit vectors that point in the north and east directions, respectively. Similarly, the component of the rotation rate about a local vertical axis is

$$\dot{\omega}_r = \hat{\mathbf{x}} \cdot \tfrac{1}{2}(\nabla \cdot \mathbf{u}) = \hat{\mathbf{x}} \cdot \mathbf{W} - \tfrac{1}{2}\left( \frac{\hat{\Phi}}{\cos\theta} \cdot \frac{\partial \mathbf{W}}{\partial \phi} + \hat{\Theta} \cdot \frac{\partial \mathbf{W}}{\partial \theta} \right). \tag{6}$$

The contribution that the vertical component of velocity $u_r$ makes to the horizontal strain rates is neglected because this contribution is less than 2% for even rapid uplift rates of 10 mm/yr.

*Haines* [1982] showed that if the spatial distribution of strain rates is everywhere defined, then the full velocity gradient tensor is uniquely defined. We infer strain rates within areas (for example, see Figure 1a) using the *Kostrov* [1974] moment rate tensor summation and then match the distribution of strain rates with the bi-cubic functions to infer a continuous horizontal velocity field. From *Kostrov* [1974] the strain rate for a given volume of deforming crust is

$$\bar{\dot{\varepsilon}}_{ij} = \frac{1}{2\mu VT} \sum_{k=1}^{N} M_{ij} \tag{7}$$

where $M_{ij}$ is the moment tensor, defined by the fault orientation and slip vector [e.g. *Aki and Richards*, 1981], $\mu$ is the crustal shear modulus, $V$ is the volume containing the fault structures, and $T$ is the time interval considered. While matching horizontal strain rates with continuous functions, velocity observations from space-based geodesy can also be matched within a specified frame of reference. The fit to the distribution of strain rates is determined in a least-squares inversion. The objective function that is minimized in the inversion is

$$\begin{aligned} &\sum_{1}^{N} \left(\bar{\dot{\varepsilon}}_{ij}^{\mathrm{fit}} - \bar{\dot{\varepsilon}}_{ij}^{\mathrm{obs}}\right)\left(\bar{\dot{\varepsilon}}_{pq}^{\mathrm{fit}} - \bar{\dot{\varepsilon}}_{pq}^{\mathrm{obs}}\right)(V)_{ij,pq}^{-1} + \\ &\sum_{1}^{M} \left(u_i^{\mathrm{fit}} - u_i^{\mathrm{obs}}\right)\left(u_j^{\mathrm{fit}} - u_j^{\mathrm{obs}}\right)C_{ij}^{-1} \end{aligned} \tag{8}$$

where $ij$, and $pq$ denote tensor components of the strain rate tensor, $\bar{\dot{\varepsilon}}_{ij}^{\mathrm{obs}}$ is the average strain rate inferred for each area, using the Kostrov summation, $\bar{\dot{\varepsilon}}_{ij}^{\mathrm{fit}}$ is the average value of the fitted strain rate for the given area, defined by the bi-cubic Bessel interpolation, $V$ is the a priori variance-covariance matrix of the inferred strain rates, and $N$ is the number of grid areas. The second part of the objective function is used when GPS, VLBI or SLR velocities, $\mathbf{u}$, are incorporated as observations to be matched by the model velocity field, where $C_{ij}$ are the variance-covariance values of the observed velocity vectors, and $M$ is the number of velocity observations. In the inversion it is possible to set the spatial derivatives of the vector rotation function $\mathbf{W}$ to zero such that areas that lie on a known plate will behave rigidly. In this case the vector function for that area becomes a single rotation vector, or Euler pole. The value of this rotation vector can be solved for in the inversion, or a rotation value can be assigned. The latter is a convenient way to impose velocity boundary conditions.

### 2.1. Definition of Self-Consistent Velocity Field

By self-consistency we mean a strain rate field that is associated with a velocity field, whether that velocity field is continuous or discontinuous (i.e. associated with faulting). Such a self-consistent velocity and strain rate field satisfies St. Venant's compatibility relationships [see, for example, *Fung*, 1965]. This strain rate compatibility equation for the horizontal strain rate tensor components, which our functions satisfy everywhere, is defined in Cartesian coordinates as

$$\frac{\partial^2 \dot{\varepsilon}_{xx}}{\partial y^2} + \frac{\partial^2 \dot{\varepsilon}_{yy}}{\partial x^2} - 2\frac{\partial^2 \dot{\varepsilon}_{xy}}{\partial x \partial y} = 0 \quad (9)$$

The advantage of the constraint of this compatibility relation is that all the strain rate magnitudes for all components must be accounted for in finding a velocity field that fits observables. Often spurious observations require strain incompatible deformation fields. Thus the constraint of compatibility in our functions allows for sensible interpolation of data observations.

## 3. DEFORMATION IN THE WESTERN US INFERRED FROM QUATERNARY FAULT SLIP RATES, GPS AND VLBI VELOCITIES

We next examine the kinematics within the Pacific (PA) - North America (NA) plate boundary deformation zone using Quaternary fault slip rates, GPS and VLBI velocities. Such an analysis of a self-consistent velocity field defines the rate and style of strain within the deformation zone but it may also define deficiencies in existing data sets of geologic slip rates. That is, the integrated rates of strain inferred from known active structures must account for the full expected plate motion (e.g. the NUVEL-1A model, [*DeMets et al.*, 1994] or models of plate motion inferred from GPS observations [*Larson et al.*, 1997]).

Historically there has been some controversy about whether or not the integrated slip rates on active structures in the western US add up to the expected PA-NA motion, [see for example *Minster and Jordan*, 1987; *Argus and Gordon*, 1990]. Using Quaternary fault slip rates and space geodetic data *Humphreys and Weldon* [1994] and *Saucier and Humphreys* [1993] have argued that there are inconsistencies between far-field PA-NA plate motion predicted by plate motion models [NUVEL-1A] and motions inferred from the summation of fault slip rates across southern California. Yet broad aperture space geodetic observations [*Argus and Gordon*, 1990; *Ward*, 1990; *Bennett et al.*, 1996; *Larson et al.*, 1997] show a general agreement with the NUVEL-1A direction, suggesting that deficiencies exist in the geologic data set. *Shen et al.* [1997] and *Jackson et al.* [1997] also have argued that the narrower aperture Southern California Earthquake Center (SCEC) GPS data set shows an agreement with NUVEL-1A PA-NA motion. However, *Shen-Tu et al.* [1998] have shown that the agreement is only in magnitude, not in direction. The SCEC velocities are about 5 to 7 degrees counterclockwise of the expected PA-NA motion direction in offshore southern California [*Shen-Tu et al.*, 1998]. However, the SCEC velocity solution has recently been revised and is now more consistent with NUVEL-1A motion [*Shen-Tu et al.*, 1999].

One reason that *Humphreys and Weldon*'s [1994] result may not agree with PA-NA motion is that their method of summing slip rates across faults does not account for the complete horizontal velocity gradient tensor; they must assume rotation values for crustal blocks. In contrast our method [*Haines*, 1982; *Haines and Holt*, 1993; *Haines et al.*, 1998] allows for the determination of rotations about the vertical axis, and hence the complete horizontal velocity gradient tensor.

The area of our analysis includes the main part of the transform plate boundary zone between the Pacific and North American plates, from the San Francisco Bay segment of the San Andreas fault in the northwest to the Imperial fault located at the U.S. border in the southeast (Figure 1a). The southwestern boundary of the study area lies west of the offshore faults on the Pacific plate. These knotpoints on the far western edge of the grid in Figure 1a are made rigid (derivatives of rotation vector function **W** are set to zero) and are assumed part of the PA plate. The northeast boundary knotpoints are east of the Wasatch front fault system in Utah, east of the Colorado Plateau boundary in Arizona, and are assumed rigid and part of the NA plate. All areas between these two boundaries are allowed to strain. *Li* [1994] has shown that the velocity field inferred from the bi-cubic Bessel interpolation of strain rates is relatively insensitive to grid design as long as the grid areas sufficiently capture the spatial variations of the strain rate tensor field.

Quaternary faults are digitized from *Jennings*' [1994] "FAULT ACTIVITY MAP OF CALIFORNIA AND ADJACENT AREAS". Fault slip rates and fault parameters are collected primarily from *Peterson and Wesnousky* [1994] and published work afterwards, which is summarized in *Shen-Tu et al.* [1999]. Because quantified fault slip rates are sparse in the Basin and Range province, we use recent Very Long Baseline Interferometry (VLBI) and GPS velocities [*Dixon et al.*, 1995; *Ma and Ryan*, 1998; *Bennett et al.*, 1999; *Thatcher et al.*, 1999] to place additional constraints on deformation rates across the Basin and Range (Figures 1a and 1b).

Average horizontal strain rates are inferred for the 135 grid areas in Figure 1a using a variant of *Kostrov's* [1974] formula,

$$\bar{\dot{\varepsilon}}_{ij} = \frac{1}{2}\sum_{k=1}^{n} \frac{L_k \dot{u}_k}{A \sin \delta_k} m_{ij}^k \quad (10)$$

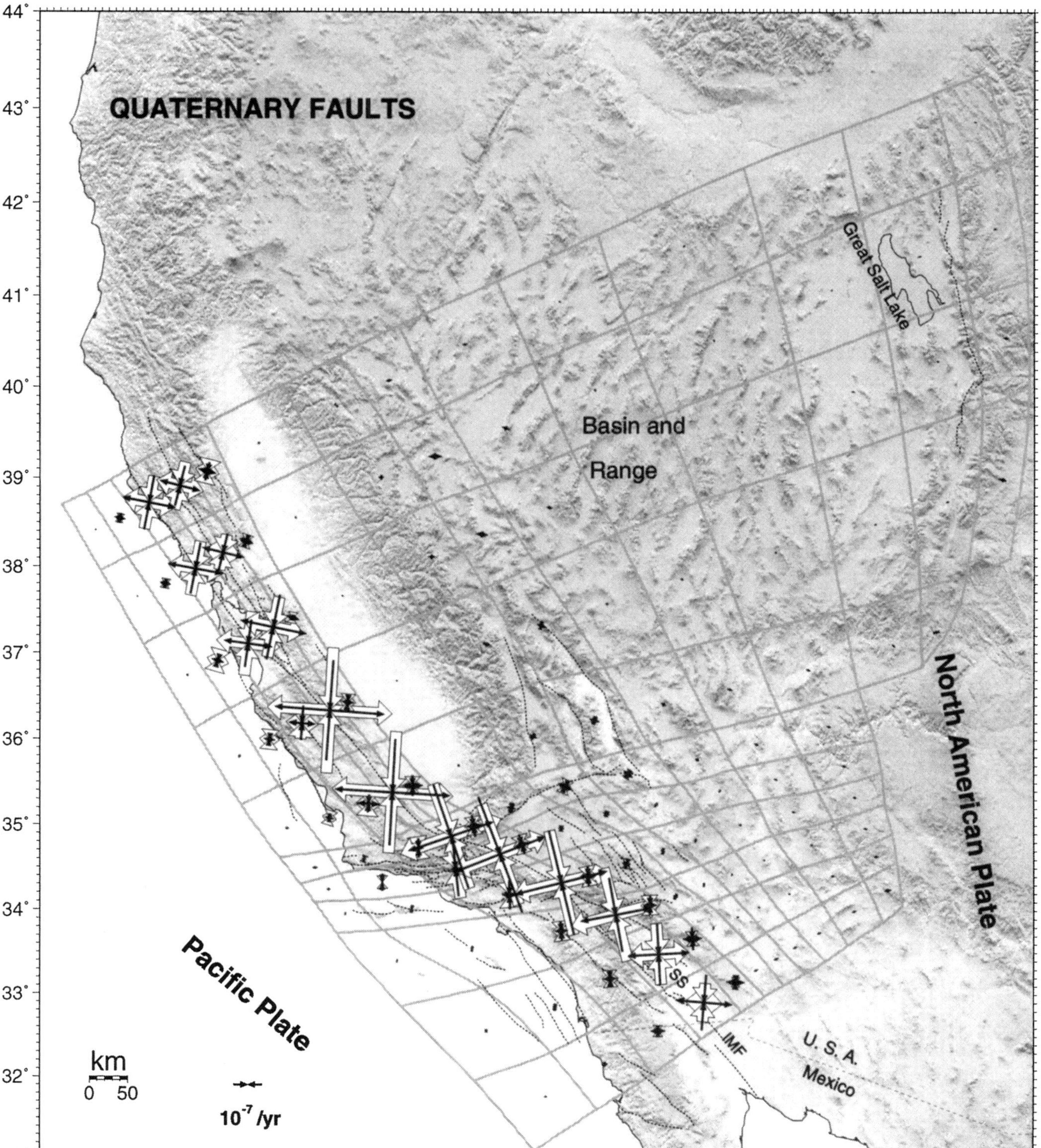


**Figure 1a.** Average principal strain rates (open arrows) calculated from slip rates on Quaternary faults (shown in dotted lines) within areas and modeled average principal strain rates (solid arrows). Model average principal strain rates are provided by the bi-cubic Bessel interpolation of the observed geological strain rates (see text). The deformation in the Great Basin is constrained by GPS and VLBI site velocities [*Dixon et al.*, 1995; *Ma and Ryan*, 1998; *Thatcher et al.*, 1999; *Bennett et al., 1999*]. Dotted lines show the Quaternary fault traces. Shaded lines represent grid boundaries.

where $n$ is the number of fault segments, each having length $L_k$, dip angle $\delta_k$, and slip rate $\dot{u}_k$, in the grid area A; $m_{ij}$ is the unit moment tensor defined by the fault orientation and unit slip vector. Note that when fault slip rates are used to infer average strain rates within areas, neither the shear modulus $\mu$ or the elastic thickness enter into the tensor summation [*Holt and Haines*, 1995; *Shen-Tu et al.*, 1995]. For more details on the Quaternary strain rate data and estimated errors, the reader is referred to *Shen-Tu et al.* [1999].

Inferred strain rates (open arrows in Figure 1a) are fitted (solid arrows) to within 1$\sigma$ standard deviation in all areas. In this solution the Pacific plate moves relative to North America at a rate of 48.1 mm/yr in a direction of N 38° W at latitude 38.7 N and 49.6 mm/yr in a direction of N 48° W at the latitude of 31.1 N (Figure 1b). The magnitude of velocity is similar to the NUVEL-1A PA-NA magnitude [*DeMets et al.*, 1994], but directions are 6 degrees counterclockwise of NUVEL-1A PA-NA velocity direction (Figure 1b), which is consistent with *Humphreys and Weldon* [1994].

Relative to North America, the Sierra Nevada block region moves about 10-13 mm/yr towards the northwest (Figure 1b). This is generally consistent with other previous geologic and geodetic estimates for Sierra Nevada block motion as well as with deformation rates accommodated across the Eastern California Shear Zone [*Sauber et al.*, 1986, 1994; *Dokka and Travis*, 1990a, 1990b; *Argus and Gordon*, 1990; 1991a; *Savage et al.*, 1990; *Feigl et al.*, 1993]. However, the relative velocity accommodated across the entire Great Basin varies from south to north along the eastern margin of the Sierra Nevada, which is inconsistent with *and Gordon's* [1991a] result of a more uniform magnitude of motion. Our velocity results along the eastern Sierra Nevada, constrained by the more recent GPS, show 10 - 13 mm/yr ± 1 mm/yr (1$\sigma$) relative to NA in direction N 42° W in the south to N 59° W in the north.

Relative convergence rate taken up across the Western Transverse Range region is about 13 mm/yr in a near N-S direction, which is in the range of the geologically and geodetically estimated convergence rates across the region [*Yeats*, 1983; *Namson and Davis*, 1988; *Donnellan et al.*, 1993; *Feigl et al.*, 1993].

### *3.1. Relative Motions Defined by GPS Data*

We next use 622 geodetic (GPS and VLBI) velocities to estimate a velocity gradient tensor field associated with geodetic velocity measurements alone. We use the bi-cubic Bessel interpolation procedure to fit velocity values and define a continuous velocity gradient tensor field solution, except that no strain rates from geologic observations are used to define the solution [see *Shen-Tu et al.*, 1999]. The geodetic velocities include 344 velocities in southern California from Southern California Earthquake Center (SCEC, velocity map version 2), 64 GPS velocities within Basin and Range province from *Thatcher et al.* [1999], 148 velocities in northern California from U.S. Geological Survey (made available on internet by Will Prescott, 1998), 36 GPS velocities in the Basin and Range from the Harvard Smithsonian solution [*Bennett et al.*, 1999], and 30 VLBI site velocities from NASA [*Ma and Ryan*, 1998]. The VLBI site velocities on the Pacific islands (MARCUS, KWAJAL26, and four VLBI sites on the Hawaii Islands, Figure 2) are also used in the model.

Model PA-NA velocities along the eastern boundary of the Pacific plate, obtained from the interpolation of GPS velocities, are 47.3 - 47.8 mm/yr in a direction of N 35° - 44° W (Figure 2). These velocity magnitudes are within 1 mm/yr of the NUVEL-1A predicted PA-NA motion [*DeMets et al.*, 1994]. Model velocity directions deviate 2-3 degrees counterclockwise from the NUVEL-1A PA-NA plate motion direction (Figure 2), but these differences are not significant at the 95% confidence limit. This result is consistent with PA-NA motion obtained by *Larson et al.* [1997] from global GPS data analysis. Therefore, version 2 of the SCEC velocity model is more consistent with the NUVEL-1A model than is version 1 [*Shen-Tu et al.*, 1998, 1999].

### *3.2. Source of Discrepancies*

Our results show that the integrated rates of geologic strain yield motions relative to North America that differ from the expected directions of PA - NA velocity given by the NUVEL-1A model. On the other hand recent geodetic data (including version 2 of the SCEC velocity model) yield a PA-NA model solution indistinguishable from NUVEL-1A (Figure 2) [*Shen-Tu et al.*, 1999]. The velocity field difference (Figure 3) between the geologic solution (Figure 1) and a velocity field solution obtained by fitting geologic strain rates with imposed NUVEL-1A velocity boundary conditions shows that a marginally significant amount of NE-SW directed shortening may be missing from the geologic data set. This inference is based on the assumption that the NUVEL-1A PA-NA motion is approximately correct, which is supported by the most recent GPS and VLBI data set (Figure 2). We infer that the strain rates most likely missing from the catalog of geologic deformation rates are in the form of blind thrusting in offshore California as well as within the coastal regions and Coastal Ranges of California. Based on this analysis we conclude that velocity field solutions obtained from the geologic data set should have the NUVEL-1A velocity

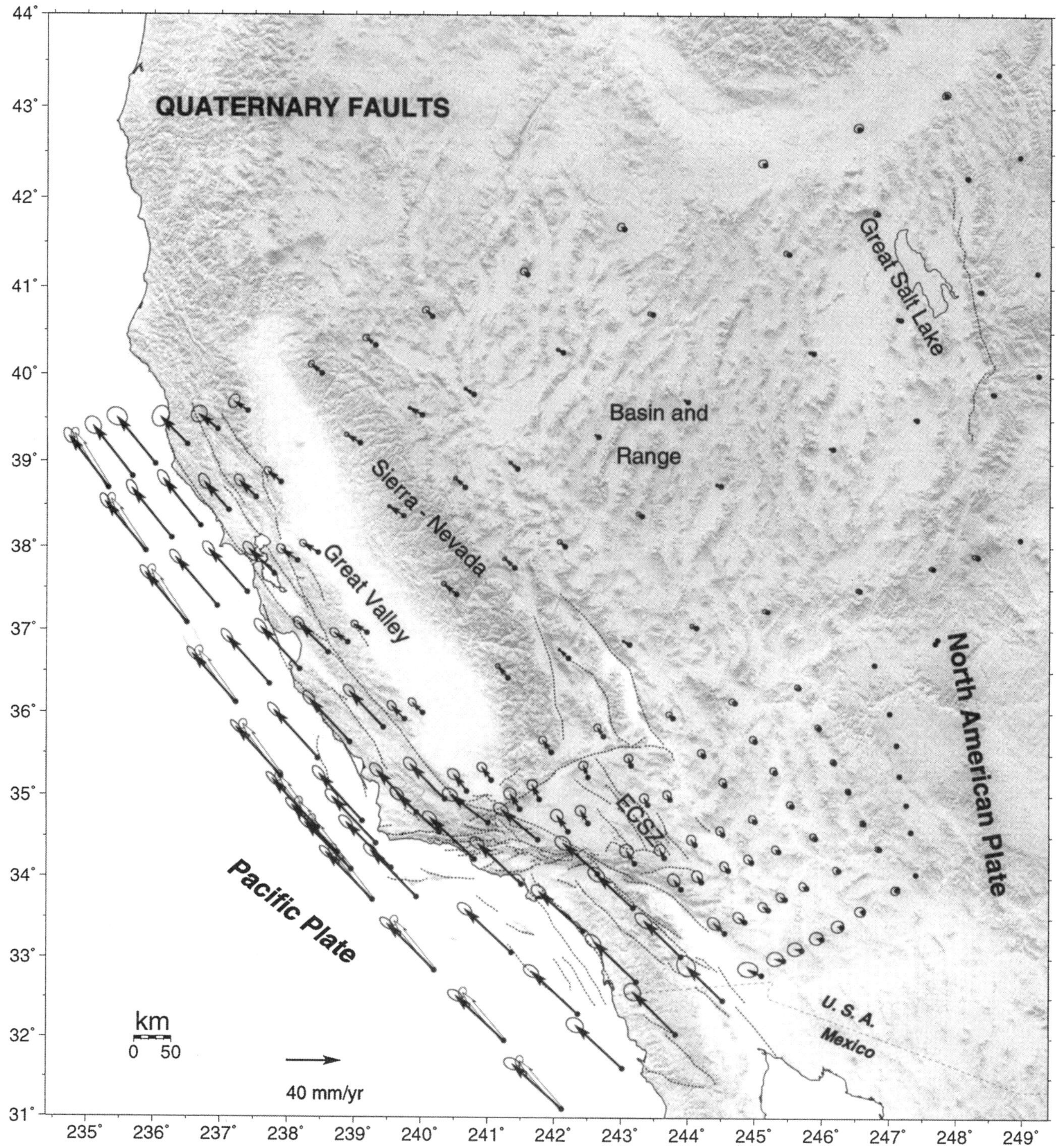


**Figure 1b.** Velocity field relative to stable North American plate obtained from bi-cubic Bessel interpolation of geologic strain rates (Figure 1a). Error ellipses are 95% confidence. Open arrows on the Pacific Plate (PA) show the PA-NA motion from the NUVEL-1A model [*DeMets et al.*, 1994] (ellipses for NUVEL-1A velocities are 95% confidence). Velocities on the PA plate are about the same magnitude as NUVEL-1A for PA-NA, but they are 6 degrees counter-clockwise in azimuth of NUVEL-1A PA-NA direction. Dotted lines represent Quaternary fault traces.

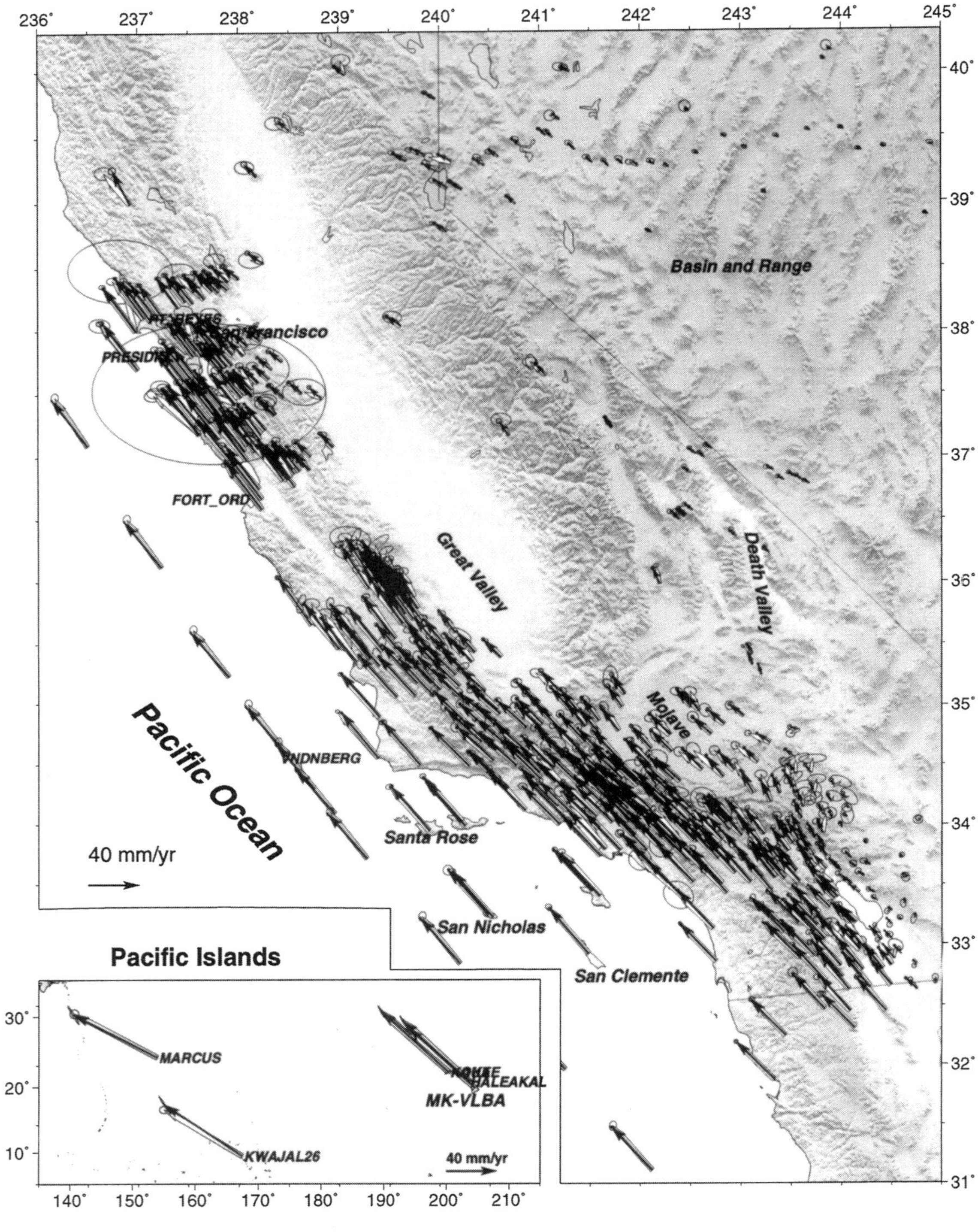


**Figure 2.** Comparison of the model velocity vectors (solid arrows) obtained from the inversion of 622 geodetic velocities with the observed velocities (open arrows) used in the model. The NUVEL-1A velocity vectors (with 95% confidence) are shown on eastern margin of Pacific plate boundary (grey arrows). The model PA-NA velocity deviates only 2-3° counterclockwise from the NUVEL-1A direction at these sites. Velocity vectors on the Pacific are shown in the sub-map at lower left. All vectors are plotted in the stable North American plate reference frame. Error ellipses are for 95% confidence limit.

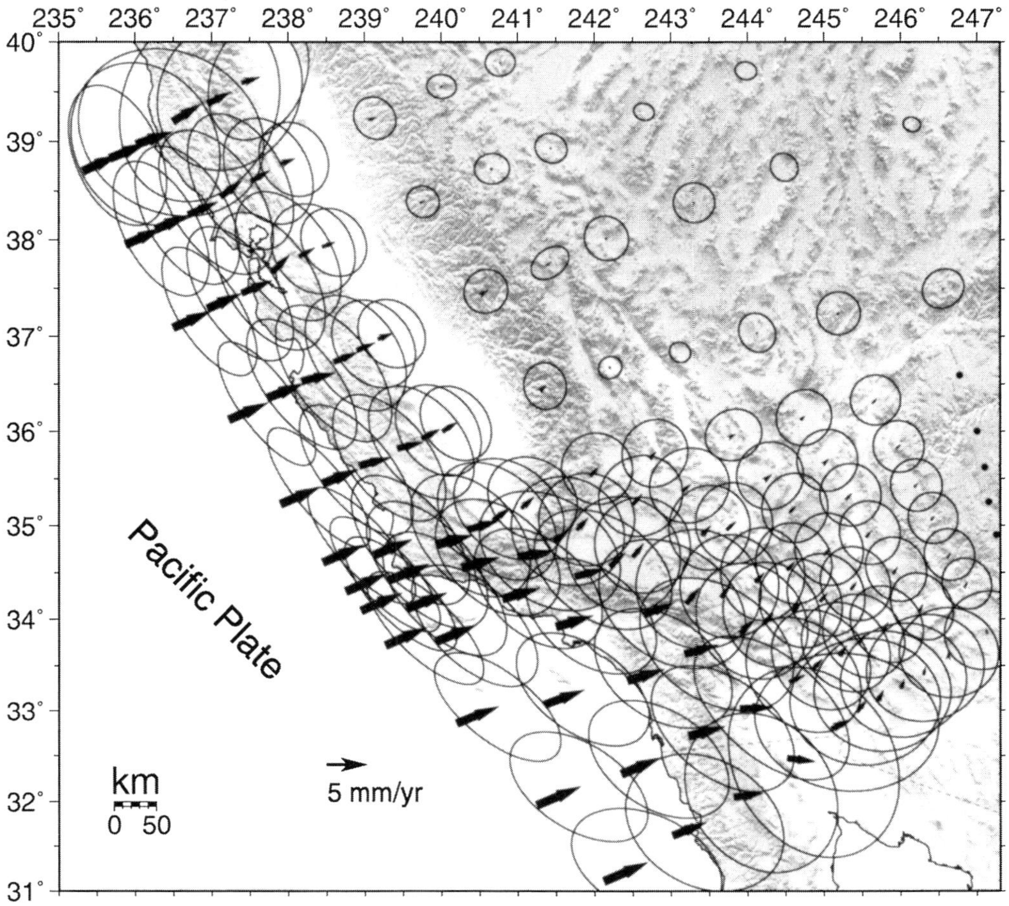


**Figure 3.** Differential velocity vectors between the velocity fields obtained from inversion of geological strain rates and the velocity field obtained from the geological strain rates with NUVEL-1A PA-NA far field motion applied as Pacific plate boundary constraints. Error ellipses are for 95% confidence limit and are associated with the velocity solution from the geologic data alone (without the NUVEL-1A constraint). Results indicate that a marginally significant amount of NE-SW directed shortening may be missing from the geologic data set. The inclusion of such shortening would explain the discrepancy (6 degrees anticlockwise) between the geologic solution and the NUVEL-1A direction.

boundary condition imposed, which accounts for shortening in offshore and coastal California (Figure 3).

### *3.3. Rotation Rates of Crustal Blocks in the Western US*

We calculate the rotation rates, $\dot{\varpi}_l$, of lines that are parallel to the strikes of major faults in each grid area, with respect to the North American plate

$$\dot{\omega}_l = \dot{\omega}_r + \dot{\varepsilon}_{\phi\phi}\cos 2\alpha + \tfrac{1}{2}\left(\dot{\varepsilon}_{\theta\theta} - \dot{\varepsilon}_{\phi\phi}\right)\sin 2\alpha \quad (11)$$

where $\alpha$ is the angle between the line direction and the $\phi$-axis, which points east, and $\dot{\varepsilon}_{\phi\phi}$, $\dot{\varepsilon}_{\theta\theta}$, and $\dot{\varpi}_r$ are the same as defined in equations (3), (4), and (6), respectively. The rotation rate for a line $\dot{\varpi}_l$ is equivalent to the rotation rate of an elongated crustal block, or blocks, with a large aspect ratio that are embedded within the crustal deformation field [*Lamb*, 1987; *Holt and Haines*, 1993]. Because the rotation rate is dependent on the orientation of the elongate block within the deformation field, rotation rates are expected to change significantly over time [*Lamb*, 1987]. Because of this, the comparison of $\dot{\varpi}_l$ with paleomagnetic rotation rates inferred from finite rotations can only remain a crude comparison until the angle-dependent behavior is taken into account. Yet even this type of comparison contains significant uncertainties because velocity gradients may change with time, and constraints on rate changes from paleomagnetic data are generally poor. The determination of $\dot{\varpi}_l$ is to be distinguished from $\dot{\varpi}_r$, which is equivalent to the rotation rate of a circular block within the velocity field [*Lamb*, 1987; *Holt and Haines*, 1993], and

**Table 1.** Instantaneous rotation rates relative to North American plate from the kinematic model in this study and rotation rates of fault blocks in southern California determined from paleomagnetic studies inferred by taking the finite rotation values divided by the age of the rocks

| | Rotation Rate (deg/m.y.) | | | | |
|---|---|---|---|---|---|
| | Model[a] | | Paleomagnetic | | |
| Region | Elongate block | Circular block($\dot{\varpi}$) | Rate | Ages of the rock | Source[b] |
| W. Transverse Ranges | 7±2 | 3.2 ±1.5 | 6 | 15 Ma | 1 |
| E. Transverse Ranges | 1.9-3.8±3 | 1.5±1 | 7 | 6 Ma | 2, 3 |
| Santa Barbara Channel | 2.3±1.5 | 2.2±1 | | | |
| Tehachapi Mountain | 2.5±2.4 | 1.3±1.5 | 3.7~5.3 | 16 Ma-23 Ma | 3 |
| Eastern Garlock Fault | 2.4-3.1±1.5 | 0.5±0.5 | | | |
| Northeastern Mojave | 1.1±2.4 | 2.4±3 | | | |
| Morro Block | 0.5±1.3 | 2.1±2 | 3 | 11 Ma | 4 |
| Santa Catalina (offshore) | 0.7±0.5 | 1.1±1 | 6 | 22 Ma | 5 |

[a]Cited errors in rotation rates stand for one standard deviation.

[b]Sources: 1, Luyendyk [1990]; 2, Carter et al. [1987]; 3, Luyendyk [1991]; 4, Khan et al. [1988]; 5, Luyendyk et al. [1988].

has been calculated for parts of southern California using geodetic results [e.g. *Feigl et al.*, 1993; *Snay et al.*, 1996; *Shen et al.*, 1996]. Unlike $\dot{\varpi}_l$ the rotation rate $\dot{\varpi}_r$ is independent of block orientation. *McKenzie and Jackson* [1983] and *Sonder et al.* [1994] have argued that the effective size of rotating blocks over time may be considerably smaller than what is reflected by currently active faults and that the rotation rate of such blocks can be approximated as that for circular blocks - $\dot{\varpi}_r$. Therefore, we also determine rotation rates for circular blocks, $\dot{\varpi}_r$, and compare them with rates for elongate blocks $\dot{\varpi}_l$ (Table 1).

Large, statistically significant rotation rates of E-W trending elongate blocks occur primarily within the Western Transverse Ranges and Eastern Transverse Ranges (Figure 4). East-west trending blocks within the Western Transverse Ranges rotate clockwise at 5 to 7 deg./Myr and rotation rates of E-W trending blocks in the Eastern Transverse Ranges are about 3 deg./Myr, clockwise. By contrast, average rotation rates of circular blocks in the Western and Eastern Transverse Ranges are 3.2 ± 1.5 deg./Myr and 1.5± 1 deg./Myr, respectively (Table 1). Rates of clockwise rotation of the eastern segment of the Garlock fault are about 1.2 - 1.8 deg./Myr, whereas they are insignificant if blocks are small and circular (Table 1). The Great Valley block, as well as blocks bounded by the Owens Valley and Death Valley have a counterclockwise rotation rate of less than 1 deg/Myr, but the rates are not significant at the 1$\sigma$ confidence level. Similar rates of rotation have been obtained by *Shen-Tu et al.* [1998] using earthquake moment tensor information and GPS velocity observations.

The clockwise rates of rotation of elongate, E-W trending blocks in the Western Transverse Ranges calculated here are consistent with those rates of rotation inferred from paleomagnetic observations (e.g. *Luyendyk*, [1990, 1991]), but rates of rotation of circular blocks are lower than that inferred from the paleomagnetic measurements (Table1). Clockwise rates of E-W trending elongate blocks in the Eastern Transverse Ranges are marginally consistent with rates inferred from 6 Myr old rocks [*Carter et al.*, 1987; *Luyendyk*, 1991) (see Table 1). Low clockwise rotation rates of about 1 deg./Myr on offshore islands in the solution presented here are not consistent with the 6 deg./Myr clockwise rotation rates inferred from paleomagnetic observations in 22 Myr old rocks [*Luyendyk et al.*, 1988]. Although not conclusive, the general agreement in sense and rate of rotation for the Western and Eastern Transverse Range regions is consistent with the possibility that velocity gradients over the past 5 - 6 Myr have been roughly constant (have not varied by more than a factor of three) within this part of the plate boundary zone. However, this hypothesis needs to be further tested, taking into account the finite rotation of elongate blocks and their rate changes through time. Velocity gradients have most

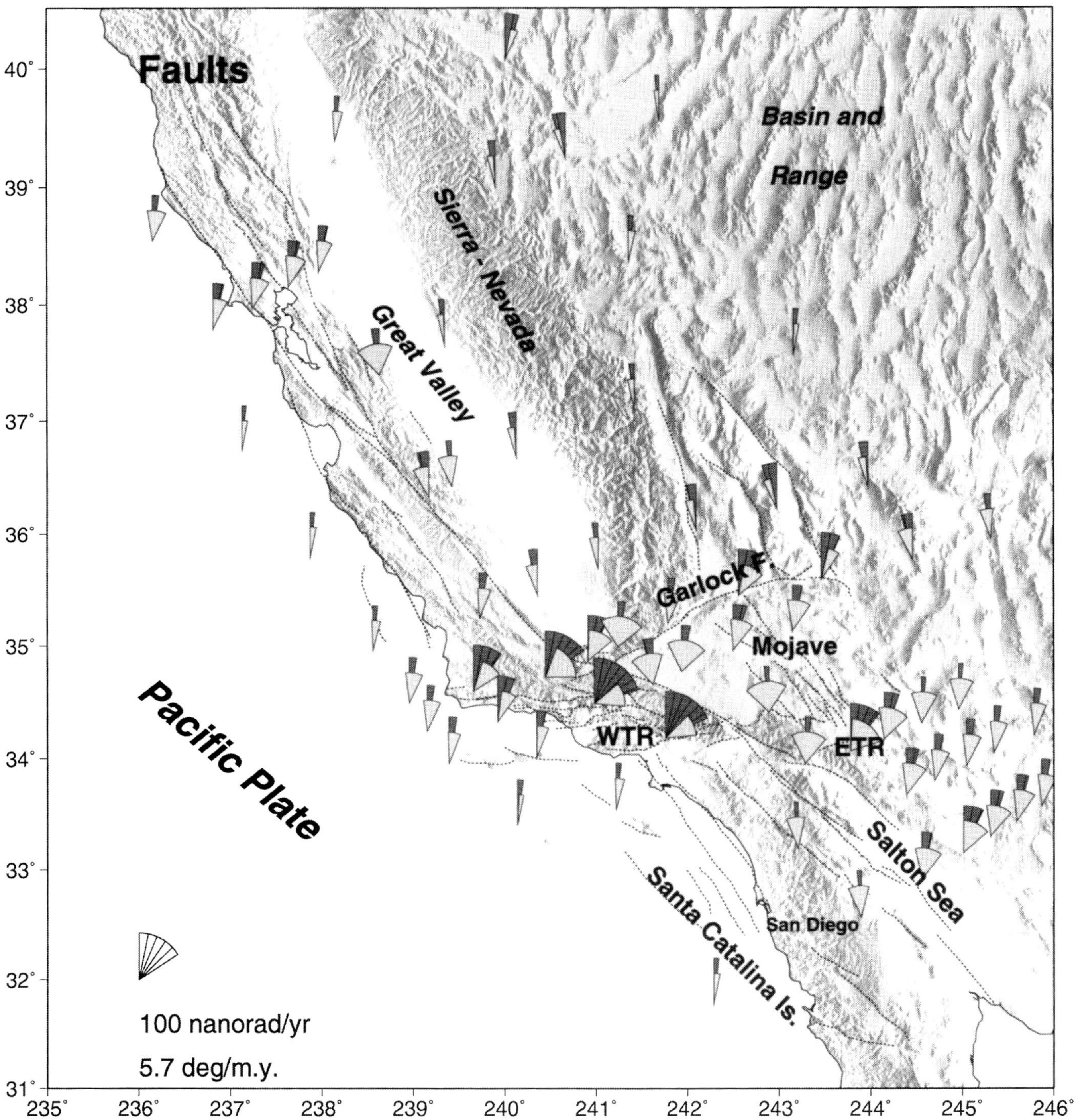


**Figure 4.** Rotation rates (represented by darker shaded wedges) of lines parallel to the major faults in each grid area calculated from the geological strain rate field solution in Figure 1 using equation (11). The light shaded wedges represent 1 sigma standard errors in the rotation rates.

probably changed in the offshore California regions since 22 Ma [e.g. *Nicholson et al.*, 1994] given the poor fit to rotation rates offshore. Equally possible is that the elongate blocks have rotated into a position in which their instantaneous rotation rate is small. It is interesting to note that nowhere are the average rotation rates of circular blocks high enough to explain rates inferred from paleomagnetic observations.

Although rotation rates are high within parts of the PA - NA deformation boundary zone, the area covered by these high rotation rates is small in comparison with the total area of the plate boundary zone. This might explain why our results are qualitatively similar to *Humphreys and Weldon* [1994].

### *3.4. Calculation of Seismic Strain Rate Deficits*

Given a self-consistent estimate of a model strain rate field in the Western US, obtained from the fitting of Quaternary strain rates and with imposed NUVEL-1A velocity boundary conditions, strain rates can be converted into an expected seismic moment rate budget [e.g. *Holt et al.*, 1995; *Tinnon et al.*, 1995; *Holt and Haines*, 1995], assuming that all strains are taken up seismically. We then use the earthquake catalog between 1850 and 1998, summarized in *Shen-Tu et al.* [1998], to determine rough moment deficit estimates by taking the difference between the model values for each area and the actual moment rates, computed from the catalog over the past 148 years.

The expected moment budget, along with the deficit, is summarized in terms of scalar moment rate in Table 2. The tensor form of the moment deficit is shown in Figure 5 [see *Shen-Tu et al.*, 1998, 1999]. Here we have taken the deficit tensor over the past 148 years for each area and expressed this in the form of a single double-couple (or faulting) mechanism. All areas, except the San Andreas Fault on the Big Bend segment (region G) and the Carrizo segment (region F), as well as the western Basin and Range province (region I), have a moment deficit (Figure 5; Table 2). There are large deficits in southern California, containing the southern San Andreas fault, San Jacinto fault, and northern segment of the Imperial fault (Figure 5). There are also significant deficits in central California, containing the San Andreas Fault, Rodgers Creek fault, Hayward fault, Calaveras fault, as well as other faults within the San Francisco Bay area. Deficits in central California can be explained, in part, by the presence of some aseismic creep along the Calaveras, Hayward, and Concord-Green Valley faults [*Savage et al.*, 1979; *Thatcher*, 1990; *Bilham*, 1996]. Deficits along the Chalome-San Juan Bautista segment of the San Andreas fault can be fully explained by aseismic creep [*Nason*, 1973; *Thatcher*, 1990; *Gladwin et al.*, 1994]. But, since the most recent large event on the southern San Andreas fault occurred in 1676 ± 35 A.D. [*Sieh and Willams*, 1990], deficits along the Southern San Andreas fault can most probably be attributed to elastic strain accumulation and events larger than $M_w = 7.0$ are required to make up the deficit (Figure 5). Notable deficits also exist within the western Transverse Ranges and Los Angeles Basin region (equivalent to $M_w =$ 6.75 to 7.25). There are small deficits in the central and northern Mojave, as well as east of Owens Valley (equivalent to $M_w$ = 6.5 to 6.75). Deficits in central and northern Mojave, as well as east and north of Owens Valley (Figure 5) are generally consistent with locations of seismic gaps in the eastern California-western Nevada seismic belt [*Ryall*, 1977; *Wallace*, 1978; 1984b; *Hill et al.*, 1985; and *dePolo et al.*, 1993].

## 4. A JOINT INVERSION OF QUATERNARY STRAIN RATES AND GPS VELOCITIES IN CENTRAL AND EAST ASIA

On a scale of the entire India-Eurasia collision zone, including eastern Asia, self-consistent strain rate solutions have previously been determined using earthquake information [*Holt et al.*, 1995] and Quaternary fault slip rates [*Peltzer and Saucier*, 1996; *England and Molnar*, 1997b]. An accurate determination of a self-consistent velocity field for Asia is important in defining the fundamental behavior of deforming continental lithosphere [*Molnar and Tapponnier*, 1975; *England and McKenzie*, 1982; *Tapponnier et al.*, 1982; *Molnar*, 1988; *England and Molnar*, 1997a]. Recent GPS and VLBI observations provide important constraints on how points are moving in a global framework within the Indo-Eurasia collision [e.g. *Freymueller et al.*, 1996; *Heki*, 1996; *Molnar and Gipson*, 1996; IGS velocities], as well as constraints on regional strain [e.g. *Abdrakhmatov et al.*, 1996; *Bilham et al.*, 1997; *King et al.*, 1997; *Calais et al.*, 1998; *Larson et al.*, 1999; *Yu et al.*, 1999]. The IGS global velocity solution of GPS vectors are available on the internet (compiled by Mike Heflin) at http://sideshow.jpl.nasa.gov/mbh/series.html. The IGS velocities referenced in this paper were obtained at this site for solution 1998.8, and are hereafter referred to as IGS. For a complete delineation of the distribution of strain rates within the India-Eurasia collision zone, it will be important to incorporate such observations from space geodesy into large-scale, self-consistent strain rate models.

**Table 2.** Minimum Moment Release Rates Associated with Horizontal Strain Rate Tensors Calculated From Earthquake Strain Rates (Observed Seismic) and the Long-Term Strain Rates

| | Region Name[a] | Area ($km^2$) | Minimum Moment Rate ($10^{24}$ dyn.cm.yr$^{-1}$)[b] | | | | |
|---|---|---|---|---|---|---|---|
| | | | Seismic | Geologic with NUVEL-1A[c] | | Geodetic | |
| | | | Observed[d] | Model | STD | Model | Deficit[e] |
| A | Central Offshore | 5.11e+04 | 0.16 (0.24) | 11.82 | 2.82 | 8.86 | 11.66 |
| B | W. Transverse Ranges | 2.96e+04 | 2.56 (4.42) | 13.52 | 2.52 | 15.05 | 10.95 |
| C | Southern Offshore | 6.36e+04 | 0.15 (0.19) | 7.34 | 2.78 | 10.98 | 7.19 |
| D | San Andreas, North | 3.74e+04 | 33.69 (35.82) | 46.67 | 3.62 | 40.09 | 12.98 |
| E | San Andreas, Parkfield | 1.28e+04 | 2.04 (2.29) | 18.88 | 1.87 | 17.80 | 16.84 |
| F | San Andreas, Carrizo | 1.27e+04 | 18.77 (18.95) | 17.66 | 1.87 | 12.03 | -1.11 |
| G | San Andreas, Big Bend | 1.32e+04 | 35.37 (38.65) | 25.41 | 2.33 | 15.31 | -9.96 |
| H | San Andreas, South | 4.11e+04 | 14.51 (15.80) | 38.52 | 3.73 | 41.66 | 24.00 |
| I | W. Basin and Range | 3.00e+05 | 45.82 (55.44) | 36.64 | 4.30 | 49.32 | -9.18 |
| J | Mojave Desert | 5.53e+04 | 2.00 (3.67) | 10.40 | 2.81 | 19.27 | 8.40 |
| K | E. Basin and Range | 3.37e+05 | 0.74 (1.81) | 18.92 | 4.41 | 21.97 | 18.18 |
| | Total | 9.55e+05 | 155.81 (177.26) | 245.77 | 10.36 | 252.36 | 89.96 |
| | Total[f] | 9.55e+05 | 133.97 (154.85) | 180.23 | 7.60 | 185.07 | 46.26 |

Abbreviations: STD, standard deviation.

[a]Region Boundaries and labels are shown in Figure 5.

[b]Minimum moment rate = $2\mu V(0.5\times|\dot{\varepsilon}_{\phi\phi}+\dot{\varepsilon}_{\theta\theta}|+\sqrt{0.25\times(\dot{\varepsilon}_{\phi\phi}-\dot{\varepsilon}_{\theta\theta})^2+\dot{\varepsilon}_{\phi\theta}^2})$, where $\mu = 3\times10^{11}$ is the shear modulus, $V$ is the volume, which is the product of the area of the region and the seismogenic depth, which is assumed to be 15 km.

[c]Model from inversion of Quaternary fault slip rates with NUVEL-1A Pacific-North America plate motion applied as a velocity boundary constraint.

[d]Values in parentheses are from scalar seismic moment summation.

[e]Moment deficit is the difference between the minimum moment rate predicted by the model obtained from geologic data with NUVEL-1A PA-NA motion applied as a velocity boundary condition and the minimum seismic moment release rate averaged over the last 148 years. The seismic moment rate is calculated based on earthquake catalogue of *Shen-Tu et al.* [1998] plus events that have occurred between September 1995 and December 1998 in the Harvard Centroid Moment Tensor catalog.

[f]A depth of 11 km is used for seismogenic thickness in calculating the minimum moment rate, instead of 15 km.

Here we determine such a model, performing a joint inversion of Quaternary strain rates and GPS and VLBI velocities.

### 4.1. Quaternary Strain Rates

Using the *Kostrov* summation outlined in section 3, strain rates are inferred within areas, which were designed in Plate 1 to separate areas of relatively distinct geologic strain. The Quaternary fault slip rates were obtained from tables in *England and Molnar* [1997b]. Minor modifications and additions to these rates were made in the following areas (see Plate 2). On the Red River Fault, right-lateral slip rates were increased from 4 ± 2 mm/yr [*Allen et al.*, 1984] to 7± 3 mm/yr [*Leloup et al.*, 1993; *Zhao*, 1995]. Fault slip rates on the Qinling fault near Ordos were increased from 5± 3 mm/yr to 7 ± 2 mm/yr [*Zhang et al.*, 1995, 1998]. Since Yunnan shows diffuse seismicity that is not limited to the Red River fault [*LeDain et al.*, 1984; *Holt et al.*, 1991], we also assign minor slip rates of 3 ± 3 mm/yr to two right-lateral faults of about 100 km length - the Qujiang Fault north of the Red River fault zone, and the Langcang Fault, south of the Red River fault zone [*LeLoup et al.*, 1995]. The right-lateral Qujiang Fault lies near the location of 1970 Tonghai $M_w$ = 7.26 event [*Zou et al.*, 1983]. South of the Red River fault, seismicity is diffuse, occurs on both right-lateral and left-lateral strike-slip structures, with events of $M_w$ > 7 frequent in this century

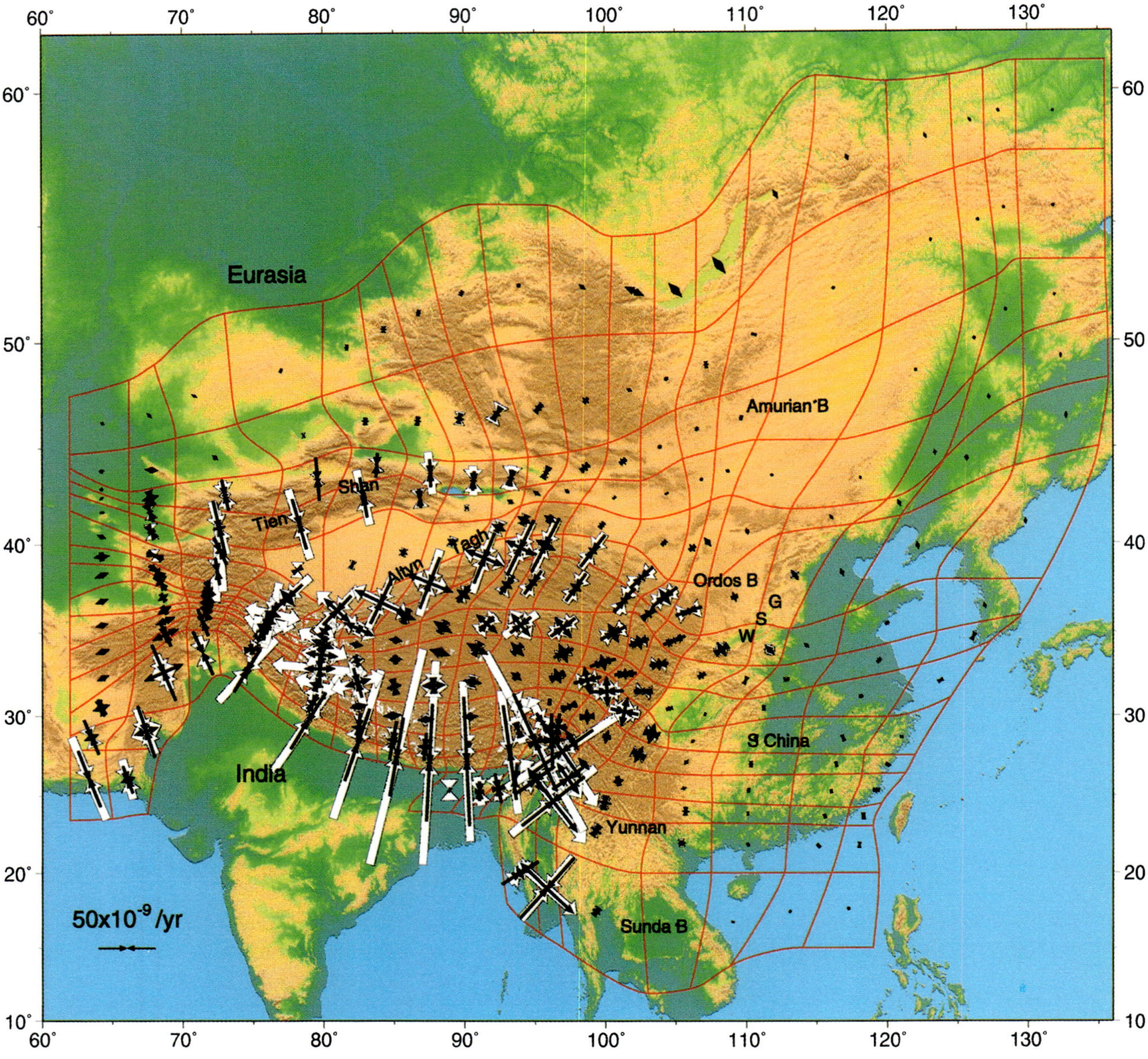


**Plate 1.** Strain rates inferred from the moment rate tensor summation of Quaternary fault data (open arrows) and principal strain rates from bi-cubic Bessel interpolation of strain rates in areas (bold arrows). Quaternary fault slip rate data are from *England and Molnar* [1997b]. See text for explanation of rates. Sunda B - Sunda Block, WSG - Weihe Shanxi graben system, Ordos B - Ordos Block, Amurian B - Amurian Block.

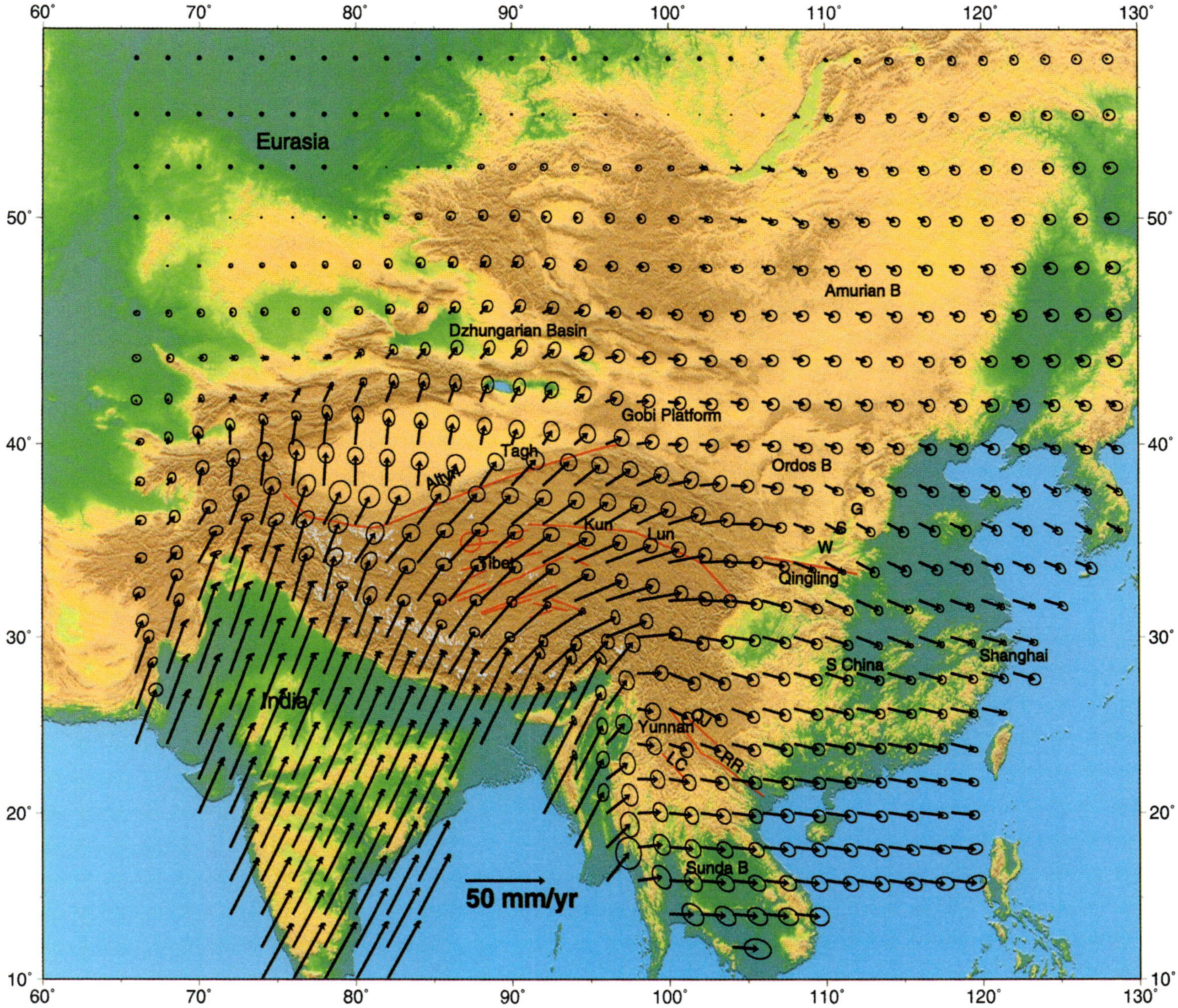


**Plate 2.** Velocity field relative to Eurasia obtained from the joint inversion of Quaternary strain rates and GPS velocities [*Abdrakhmotov et al.*, 1996; *Heki*, 1996; *Molnar and Gipson*, 1996; *King et al.*, 1997; *Calais et al.*, 1998; *Larson et al.*, 1999; *Yu et al.*, 1999; IGS]. Error ellipses are for 95% confidence level. Red lines represent active faults mentioned in the text. The lines in central Tibet are left-lateral faults from *Rothery and Drury* [1984]. Red circle shows the location of the Nov. 8, 1997 earthquake. Altyn Tagh - Altyn Tagh fault zone, Kun Lun - Kun Lun fault zone, LC - Langcang fault, QJ - Qujiang fault, RR - Red River fault zone, Qinling - Qinling fault zone, WSG - Weihe-Shanxi graben system.

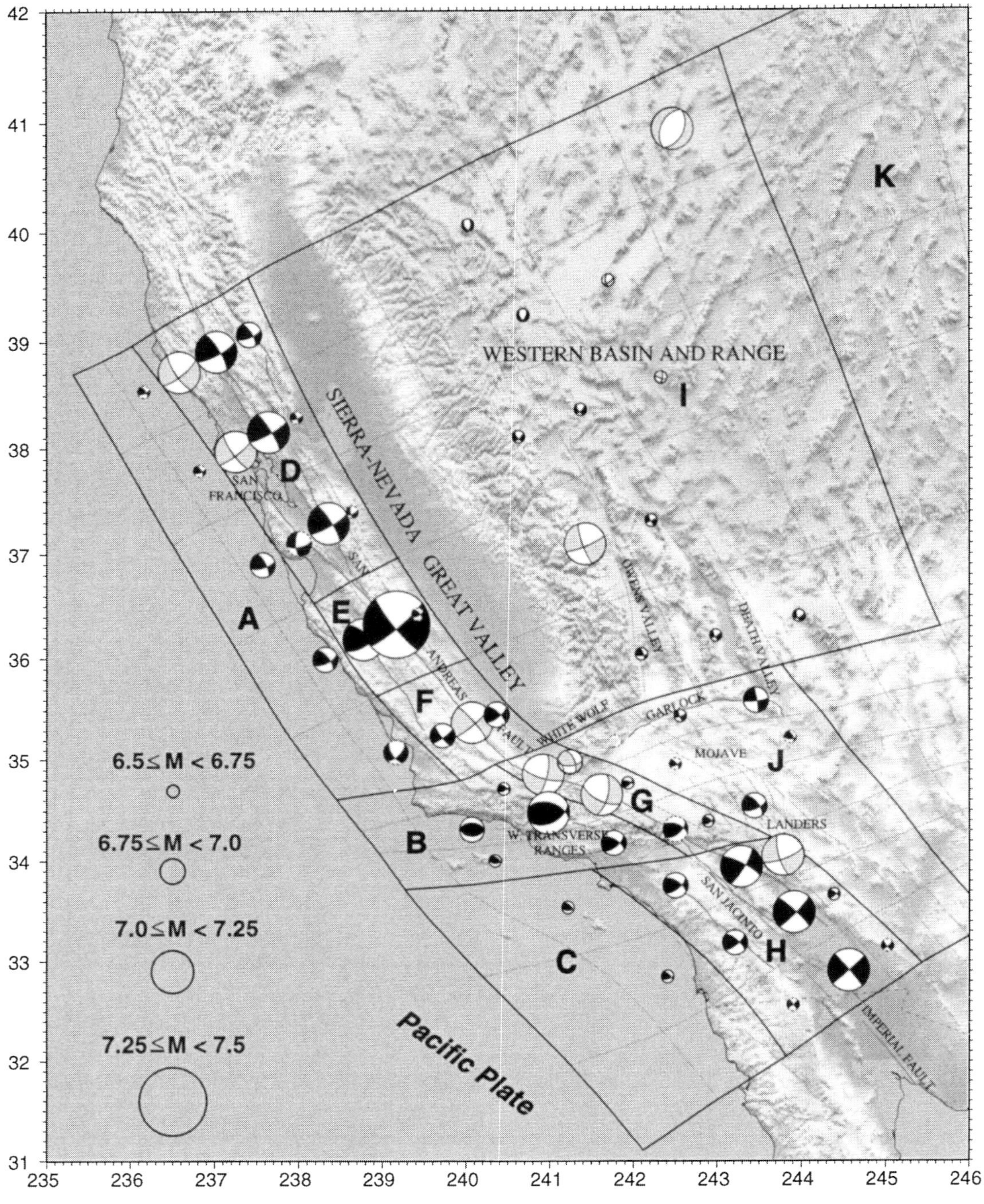


**Figure 5.** Moment deficit tensors (plotted as double-couple mechanisms) for parts of western US. The deficits are calculated by taking the difference between the model strain rate tensor in each area and the seismic strain rate tensor inferred from the seismic catalog between 1850 - 1998 [*Shen-Tu et al.* 1998], assuming a seismogenic thickness of 15 km. The model strain rates are obtained by matching the geologic strain rates with imposed NUVEL-1A velocity boundary constraints. Differential strain rate tensor elements are converted into a single double-couple mechanism by multiplying deficit rates by the time interval of the catalog - 148 years and by the region volume - area times seismogenic thickness. Black (lower hemisphere projection) focal mechanisms represent a deficit, and grey represent a surplus for the given time interval. Dashed focal mechanisms indicate that deficit is insignificant at the 2$\sigma$ confidence level.

[*LeDain et al.*, 1984; *Molnar and Deng*, 1984; *Ma*, 1986; *Holt et al.*, 1991]. The right-lateral Langcang fault, for example, lies within the area of the 1988 Langcang-Gengma ($M_s$ =7.6) series [*Chen and Wu*, 1989]. The active left-lateral structures south and west of the Red River fault [e.g. see *Lacassin et al.*, 1998] are in the data set presented by *England and Molnar* [1997b]. With the introduced rates, regional strain rates increased in Yunnan from about $1\times10^{-16}s^{-1}$ for *England and Molnar*'s [1997b] rates to about $2\times10^{-16}$ $s^{-1}$, which is still less, by about a factor of two, than the seismic strain rates in the region for this century [*Holt and Haines*, 1993]. Therefore, regional strain rates estimated for Yunnan may still be an underestimate of the total strain rates accommodated there. Indeed, GPS baseline changes across the Red River fault and surrounding areas [*Zhao*, 1995] may support higher rates of deformation in Yunnan than given by geologic slip rate estimates.

*England and Molnar* [1997a] show no active structures within east-central Tibet, south and southwest of the Kun Lun fault. Yet seismicity there shows significant, diffuse activity on east to northeast striking left-lateral strike-slip structures [*Holt et al.*, 1991; *Dziewonski et al.*, 1981, 1996; *Randall et al.*, 1995]. *Rothery and Drury* [1984] identify several of these structures, for which we assigned slip rates of 2.5± 2.5 mm/yr; the largest event in Tibet since 1951 occurred on one of these structures, west - southwest of the termination of the Kun Lun fault, on November 8, 1997 ($M_w$ = 7.5) (Plate 2). With these introduced slip rates, regional strain rates increase from $0.5\times10^{-16}$ $s^{-1}$ to $1\text{-}2\times10^{-16}$ $s^{-1}$ in east-central Tibet. Later, the influence of these introduced rates on the total solution will be shown to be small.

*4.2. GPS and VLBI Observations*

The IGS velocity for station Bangalore provides constraints for India motion in a Eurasian framework. *Larson et al.* [1999] and *Bilham et al.* [1997] have made observations in the Himalaya and at Lhasa, which provide some constraints on the amount of shortening across the central Himalaya. Observations at Lhasa [IGS; *Larson et al.*, 1999] help define the amount of eastward motion of southeastern Tibet relative to Siberia. In a Eurasian framework, this eastward motion is significant at Lhasa (23 mm/yr; IGS). *King et al.* [1997] have made measurements in southeastern Tibet, and southwest China, and these show N-S extension between stations around Chengdu and stations to the south, consistent both with earthquake mechanisms [*Holt et al.*, 1991; *Holt and Haines*, 1993] and Quaternary rates of slip [*Allen et al.*, 1991]. Several recent observations in north-central China and southern China [*Heki*, 1996; *Molnar and Gipson*, 1996; *Yu et al.*, 1999; IGS] provide new constraints on the motion of south China relative to Eurasia. GPS measurements across Lake Baikal [*Calais et al.*, 1998] provide constraints on extension rates there. Finally, *Abdrakhmatov et al.* [1996] have constrained shortening rates across the north-central Tien Shan to be of the order of 10 - 14 mm/yr, and have inferred total rates of shortening across the western Tien Shan of about 20 mm/yr.

*4.2.1. Frames of reference.*

In the modeling, the GPS data can be assigned to any frame of reference that is convenient. That is, if the data are from a global network, then the velocity observations can be placed in a frame relative to the Eurasian plate. In this case velocities are defined relative to the grid knotpoints on the northern boundary, which we assume are part of stable Eurasia. If observations are part of a smaller, local network, then the frame of reference can be solved for in the inversion procedure. That is, the relative velocities within the local network are matched, but their motion relative to Eurasia is determined in the inversion. We define the IGS velocities to lie within a fixed Eurasian plate reference frame. The IGS vectors are placed into a model Eurasian frame using the no-net-rotation pole for Eurasia (50.6° N, -112.4° E, 0.24°/Myr) [*Argus and Gordon*, 1991b]. For other networks [e.g. *Larson et al.*, 1999; *King et al.*, 1997; *Heki*, 1996; *Abdrakhmatov et al.*, 1996; *Calais et al.*, 1998; *Yu et al.*, 1999], rotation vectors are determined in the inversion that place the observations into a model Eurasian reference frame. The optimal rotations applied to the local GPS networks will be those that enable the minimum least-squares fit to both relative velocities (in Eurasian framework) and strain rates.

*4.3. The Velocity Field in Asia*

In Plate 1, knotpoints of the grid that lie south of the Himalaya, within the Indian plate, are constrained to behave rigidly and belong to one rotation vector. Knotpoints in the far north and northwestern boundary of the grid belong to rigid Eurasia. All other areas (where principal strains are shown in Plate 1) are allowed to strain.

Strain rates are concentrated along the Himalaya, along the Altyn Tagh fault, at the northern margin of Tibet, and within the Tien Shan (Plate 1). Strain rates are matched to within one standard error in all areas; shortening rates are underfitted along the central Himalaya because the constraints provided by the GPS velocities [*Larson et al.*, 1999] (Figure 6a) require a smoother variation in strain,

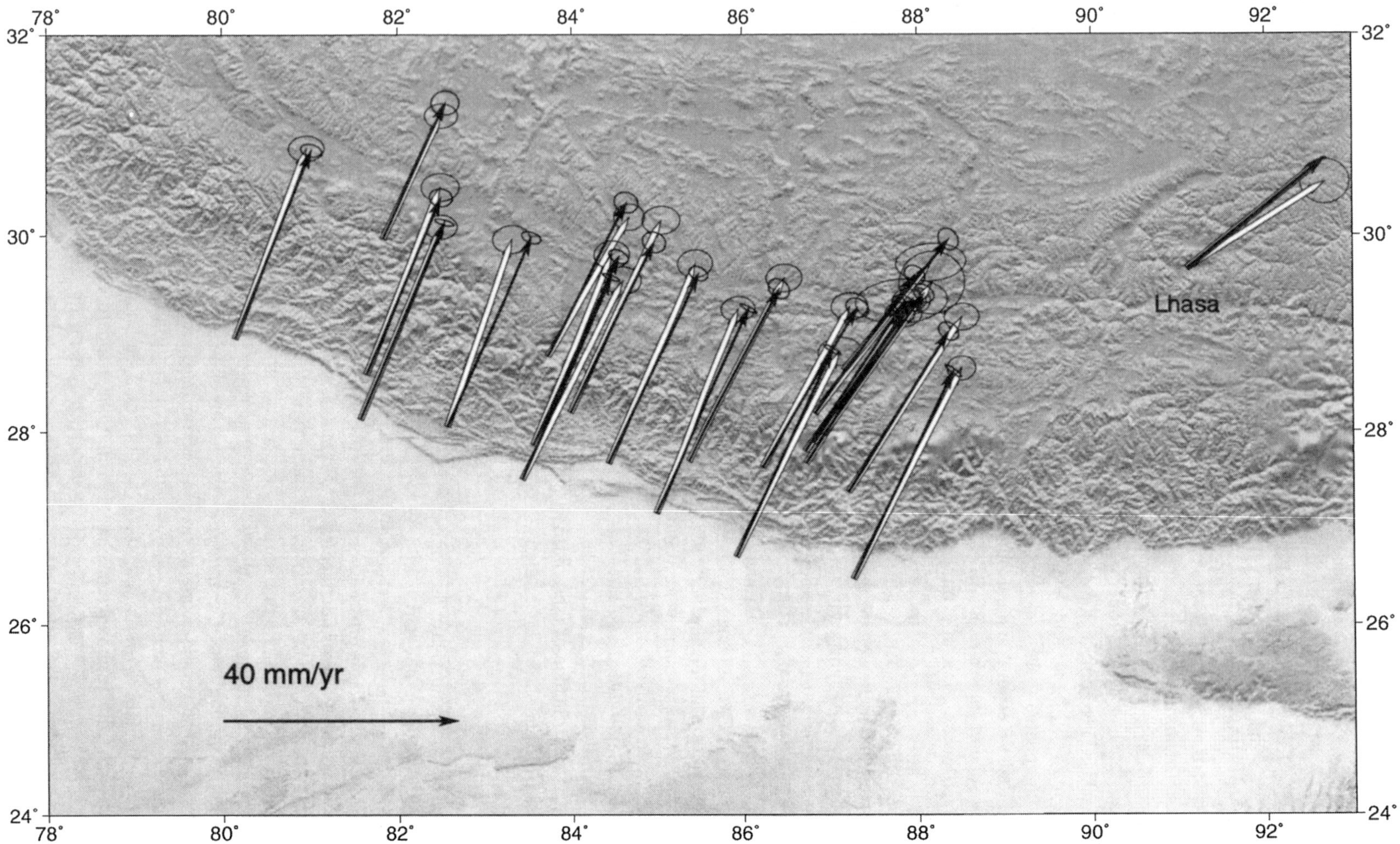


**Figure 6a.** Model velocity fit (bold velocity vectors) to GPS velocities (white vectors) in a Eurasian reference frame from solution in Plate 2. Velocities in Himalaya and southern Tibet are from *Larson et al.* [1999]; velocities at Lhasa are from *Larson et al.* [1999] and IGS. Error ellipses are for 1σ standard error. IGS velocities were obtained from web site (http://sideshow.jpl. nasa.gov/mbh/series.html, 1998.8) The frame of reference for IGS is tied to Eurasia. The frames of reference for the other observations were left free and were determined in the inversion.

due to elastic coupling [*Bilham et al.*, 1997]. Inferred slip rates of 20 ± 10 mm/yr on the Altyn Tagh fault are matched nearly in full in the solution presented in Plate 1. Below we also consider the result of fitting inferred rates of ~10 mm/yr [e.g. *England and Molnar*, 1997b] and 30 mm/yr [e.g. *Peltzer et al.*, 1989; *Avouac et al.*, 1993], respectively, on the Altyn Tagh fault.

The joint inversion of the Quaternary strain rates and GPS relative velocities yield a motion of India (IN) relative to Eurasia (EU) (Plate 2) about a pole of rotation at (30.8° N, 5.2° E, 0.38 deg./Myr), which is considerably different from the IN-EU NUVEL-1A pole of rotation (24.4° N, 17.7° E, 0.51 deg./Myr). This rotation vector fits the IGS velocity vector at Bangalore closely (39 mm/yr, N 28.5° E), and it yields total relative motions that are about 78% to 84% of the expected NUVEL-1A IN-EU model velocity magnitude along the Himalaya. The eastward component of velocity relative to Eurasia is about 3.5 - 6.5 mm/yr greater than that predicted by the NUVEL-1A IN-EU model. Surprisingly, this result is relatively insensitive to the value of left-lateral slip assumed for the Altyn Tagh fault, which is the largest easterly trending strike-slip structure in central Asia; when slip rates are assumed 10 mm/yr on the Altyn Tagh fault the rotation pole is (30.8° N, 5.22° E, 0.378 deg./Myr) and if the rate is assumed 30 mm/yr, the pole of rotation is (30.8° N, 5.18° E, 0.38 deg./Myr). The velocities within India, as well as the velocity observation at Lhasa [IGS; *Larson et al.*, 1999] (Figure 6a), place strong constraints on the component of eastward velocity of India as well as the eastward component of velocity of southeastern Tibet. Likewise GPS observations in southeast Tibet and southwest China [*King et al.*, 1997] define, in part, the low rates of strain across the Longmen Shan as well as N-S gradients of velocity east of the Eastern Himalayan Syntaxis (Figure 6b).

The predicted velocity at Shanghai is 13.6 mm/yr in a direction of 106°, which is in accord with the IGS GPS velocity at this site (Figure 6b). GPS observations at XIAN (12.3 mm/yr, azimuth 117°), TAEJ (9.4 mm/yr, azimuth 121°), and WUHN (15.2 mm/yr, azimuth 111°) (Figure 6b) are fit closely and define south China motion relative to Eurasia. New observations at continuous stations TNSM, KMNM and MZUM [*Yu et al.*, 1999] also constrain south China motion relative to Eurasia at velocities similar to those of Shanghai (Figure 6b). The rates of eastward velocity at latitudes south of 20° N (Plate 2) are not as high as those measured by the GPS GEODYSSEA (Geodynamics of South and Southeast Asia, *Wilson et al.*, 1998) network for the Sunda block [*Ambrosius et al.* 1997; *Michel et al.*, 1997; *Wilson et al.*, 1998], which show eastward components of velocity just south of 20° N of about 25 mm/yr. Eastward velocities obtained in this study for the Sunda block are 16 ± 4 mm/yr in a direction of 96° ± 20° (95% confidence).

The motion of south China is also insensitive to the assumed rates of slip on the Altyn Tagh. A combination of GPS [IGS; *Yu et al.*, 1999] and geologic [*Meyer et al.*, 1996] constraints limit the amount of eastward motion that is translated into south China.

The Amurian block [e.g. *Lesne et al.*, 1998] moves at rates of 6.5 ± 2.5mm/yr and azimuth 101° at the latitude of 45° N (Plate 2). The strain rates in Baikal and the Amurian block motion (plates 1 and 2) are constrained, in part, by the GPS observations of *Calais et al.* [1998]. The strain rates within western Mongolia are low, consistent with both field-based and modeling studies [*Baljinyam et al.* 1993; *Holt et al.*, 1995; *England and Molnar*, 1997b]. The increase in velocity within south China block is accommodated in part by extension south and east of the Ordos block (Weihe Shanxi graben system [*Zhang et al.*, 1998]), as well as by left-lateral shear on the Qinling fault system and distributed shear and extension in north central China (see Plate 2). Total N-S compression accommodated across the western Tien Shan (between 74 and 77 E) is 20 to 22 mm/yr (Plate 2), in accord with that inferred by *Abdrakhmatov et al.*, 1996] (Figure 6c).

With increased numbers of GPS observations and refined estimates of Holocene slip rates, the kinematics defined by such a self-consistent estimate of the velocity field are only going to improve. Questions remain about the amount of deformation accommodated across eastern Burma and Yunnan, the velocity of the Sunda block relative to Eurasia [e.g. *Ambrosius et al.* 1997; *Chamot-Rooke et al.*, 1997; *Michel et al.*, 1997; *Wilson et al.*, 1998], as well as the relationship of these with the forces generated by potential energy differences in the lithosphere [e.g. *England and Molnar*, 1997a] and forces generated by the relative motion between India and Eurasia [*Flesch et al.*, 1998]. Also important will be the determination of the role that the subduction zones play in affecting southeast Asia motion, which surely influences the deformation field throughout much of east Asia [*Holt et al.*, 1995].

## 5. THE VELOCITY FIELD IN THE AEGEAN DEFINED BY EARTHQUAKES AND SLR DATA.

An important parameter to determine in any seismic hazard analysis is the amount of deformation accommodated seismically and the expected style of that deformation. Because the rate of historic seismic moment release is well studied in the Aegean [*Ambraseys and Jackson*, 1990; *Jackson et al.*, 1994], it is a good area to compare both the

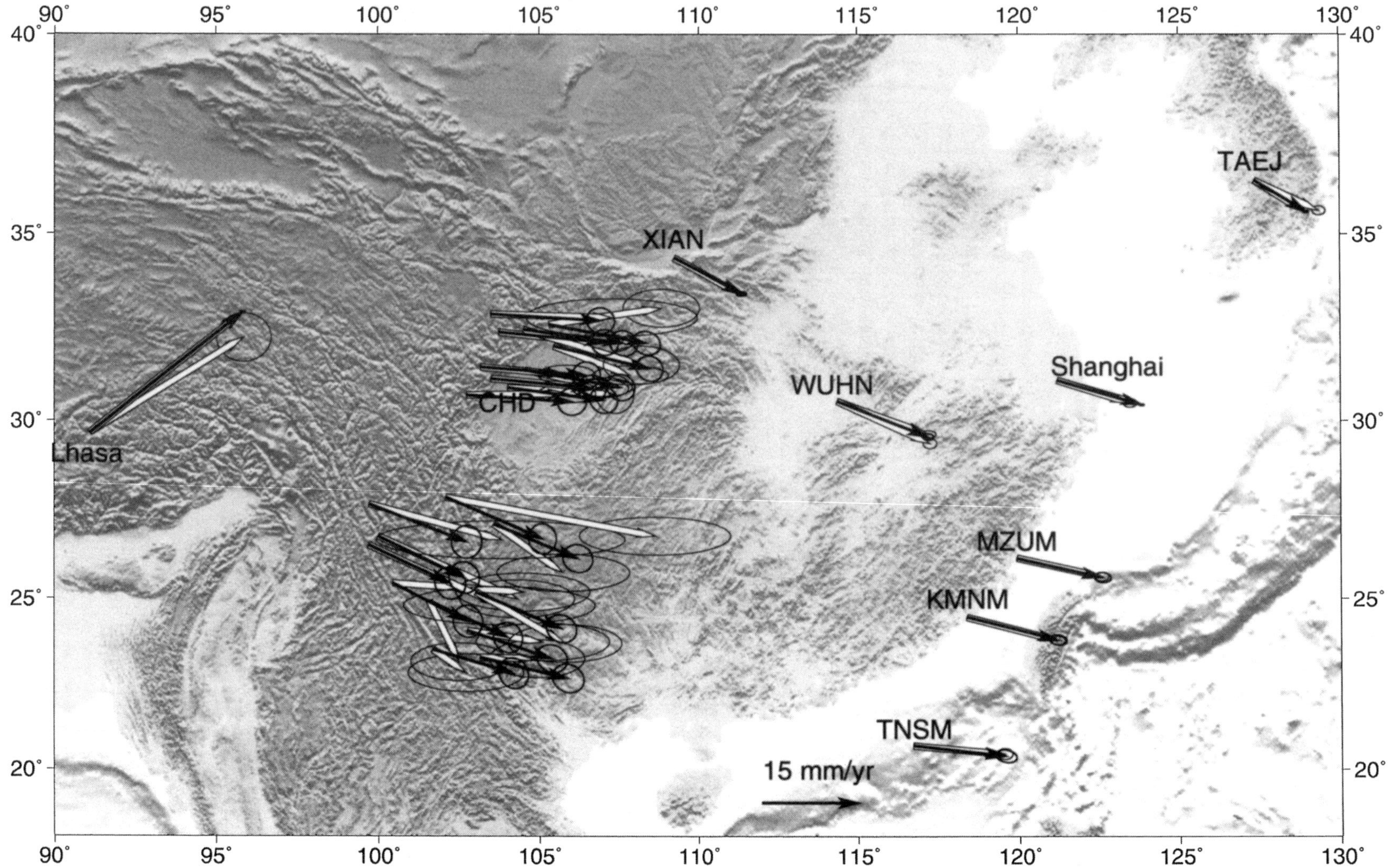


**Figure 6b.** Model velocities (bold vectors) relative to Eurasia from solution in Plates 1 and 2 plotted at GPS sites [*King et al.*, 1997; *Heki*, 1996; *Yu et al.*, 1999; IGS] in southwestern China, east China, and Korea. GPS observations are plotted as open vectors. IGS data are tied to Eurasia. Frame of reference of *King et al.* [1997], *Yu et al.*, 1999, and *Heki* [1996] were determined in the inversion. CHD is Chungdu, WUHN is Wuhan. Errors are for 1σ standard error.

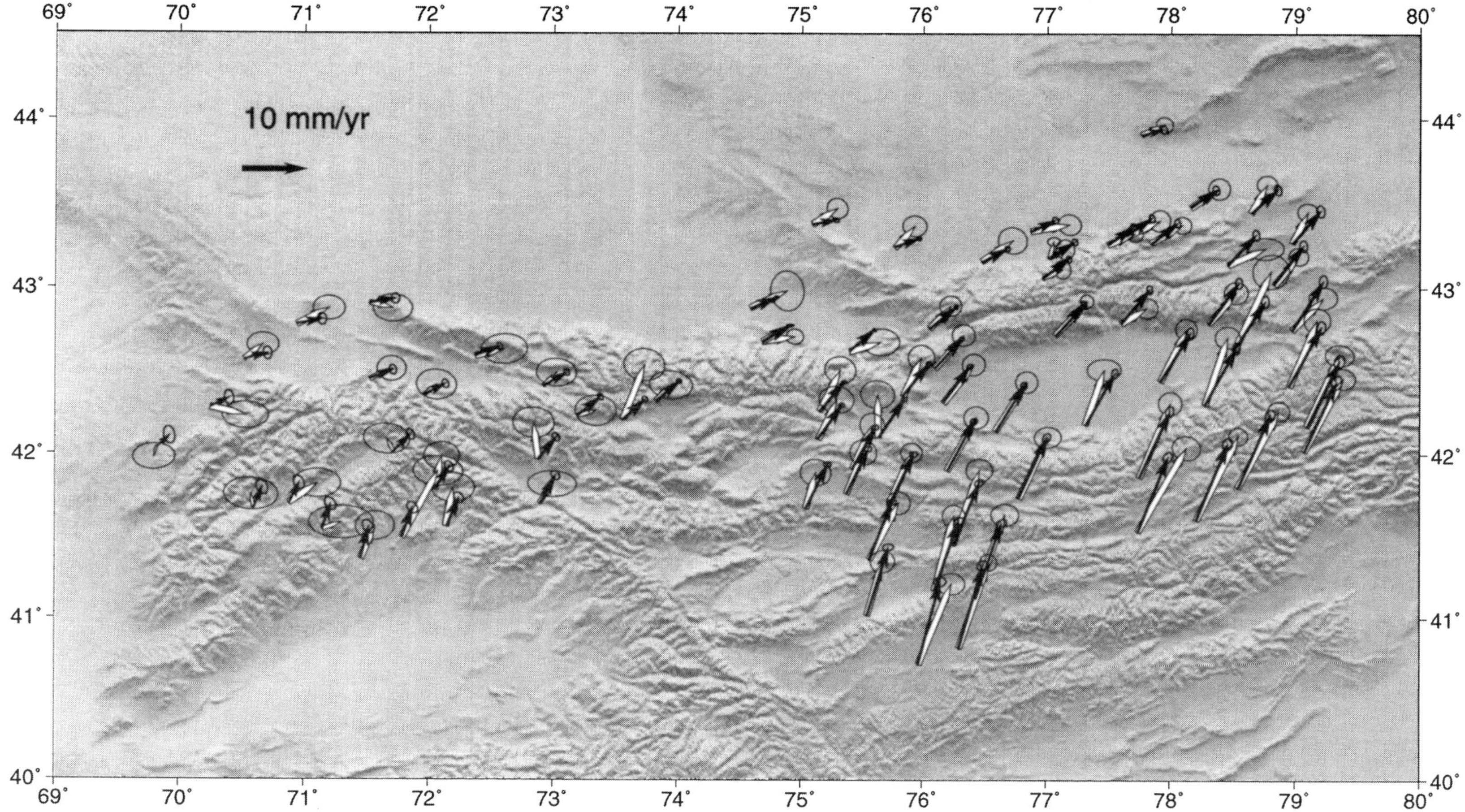


**Figure 6c.** Model velocity fit (solid velocity vectors) to GPS velocities (open vectors) in the Tien Shan [*Abdrakhmatov et al.*, 1996; IGS] in a Eurasian reference frame from solution in Plates 1 and 2. The frame of reference for *Abdrakhmatov et al.* [1996] was determined in the inversion and IGS was fixed to Eurasia. Errors are for 1σ standard error.

rates and styles of seismic strain with the total strain rate field inferred from space-based geodetic measurements [*Robbins et al.*, 1994; *Smith et al.*, 1994].

*Jackson et al.* [1992] used earthquakes with $M_s > 6.0$ in the time interval 1911 - 92 to infer a spatial distribution of seismic strain rates within the Aegean. They showed that a smooth velocity field matched to the distribution of seismic strain rates explained, to first order, the active rotations and block translations inferred within the Aegean region [e.g. *Taymez et al.*, 1991]. Later, *Jackson et al.* [1994] compared the velocity field in the Aegean with SLR velocity measurements and concluded that the SLR data did not require strain any different in style from that observed in the earthquake moment tensors; the difference lies only in the magnitude, with the earthquake strains constituting no more than 50% of the total strain inferred from the SLR measurements. Here we expand the area slightly to include more of western Turkey and we again compare with a self-consistent velocity field obtained from the SLR data [*Robbins et al.*, 1994; *Smith et al.*, 1994]. We use earthquakes over the same interval as that used by *Jackson et al.* [1992] but we make an allowance to include strains associated with smaller earthquakes [*Haines et al.*, 1998]. That is, we infer strain rates for all of the small events (less than $M_s$ = 5.95) assuming that the same frequency-moment relationship observed for a given magnitude window applies for all events below $M_s$ = 5.95.

To account for this strain rate in small earthquakes the necessary adjustment in moment to each earthquake that lies in a window between a lower moment cut-off, $M_o^a$, corresponding to a magnitude of $M_s$ = 5.95 and an upper cut-off, $M_o^b$, no larger than the largest earthquake possible in each area is

$$\overline{M}_o = M_o + \frac{M_o}{[(\frac{M_o^a}{M_o^b})^{1-B} - 1]} = \frac{M_o}{[1-(\frac{M_o^a}{M_o^b})^{1-B}]} \tag{12}$$

Here B, which is generally about 2/3, is the "b-value" in the frequency-moment relationship

$$\log N = \text{A} - \text{B}\log M_o \tag{13}$$

[*Molnar*, 1979; *Scholz and Cowie*, 1990; *Haines et al.*, 1998]. We have found that the total seismic strain rate in the Aegean, including the contribution from small earthquakes, is insensitive to the value chosen for the upper cut-off $M_o^b$. As well we have found that the adjustment for strain released in small earthquakes does not depend critically on the precise value of B. Here a value of B = 0.6 is used, which was obtained from earthquakes with $M_o > 10^{17}$ Nm in the Harvard CMT Catalog, and we have chosen $M_o^b$ to be $1.23 \times 10^{19}$ Nm (corresponding to $M_s$ = 6.85), which is only slightly larger than the largest earthquake in central Greece in the 82-year period we are considering.

An important result of this analysis is that the proportion of seismic strain released in small earthquakes of $M_s <$ 6.0 is largest in central Greece (area 9 in Figure 7b, Table 3), where there is no event in the catalog of $M_s \geq 6.9$ ($M_o > 1.23 \times 10^{19}$ Nm) and where small earthquakes account for approximately 50% of the seismic strain. Elsewhere earthquakes with $M_s < 6.0$ account for on average 15-20% of the seismic strain, because about 60-70% of the seismic strain is released in earthquakes of $M_s \geq 6.9$.

Seismic strain rates and the smooth model velocity field are shown in Figures 7a and 7b. Average strain rates were inferred [*Kostrov*, 1974] using a seismogenic thickness of 10 km and are of the order of $1 \times 10^{-15}$ s$^{-1}$ within the more rapidly straining regions of the central Aegean. To first order, the velocity field in a Eurasian framework, associated with the smooth fit to the seismic strain rates, shows a pattern similar to that inferred from space geodesy [*Reilinger et al.*, 1997], which involves a transfer of roughly west-directed vector azimuths in northwestern Turkey into southwestern relative motions in the central Aegean [e.g. *McKenzie*, 1972, 1978; *LePichon and Angelier*, 1979; *Jackson*, 1994]. Although on a more local scale there are clear differences in parts of the Aegean between the flow field associated with seismic moment release this century and the relative motions inferred from geodetic observations [*Davies et al.*, 1997], these differences are explicable in terms of the relatively short duration catalog of 82 years [*Jackson et al.*, 1994; *Haines et al.*, 1998].

Figures 7c and 7d demonstrate that the pattern of regional strain reflected in the earthquake mechanisms is a good first-order representation of the pattern of regional strain associated with the total deformation field. Here we determine a smooth velocity field for the Aegean region by inverting SLR velocities (open arrows at station-triangles in Figures 7a and 7c) [*Smith et al.*, 1994; *Robbins et al.*, 1994]. The a priori covariances of the regional strain rates are defined as anisotropic. This provides the constraint that the fitted regional strain rate tensors match the style of earthquake strain rate tensors as closely as possible, while still matching the SLR velocities [see *Haines et al.*, 1998]. Figures 7c and 7d show that the style of regional seismic strain rate tensor in the Aegean can be matched while fitting the SLR relative motions. The magnitudes of total estimated regional strain rates are roughly one and a half times the magnitudes of the seismic strain rate. Not well represented in the seismic strain rate field is the roughly

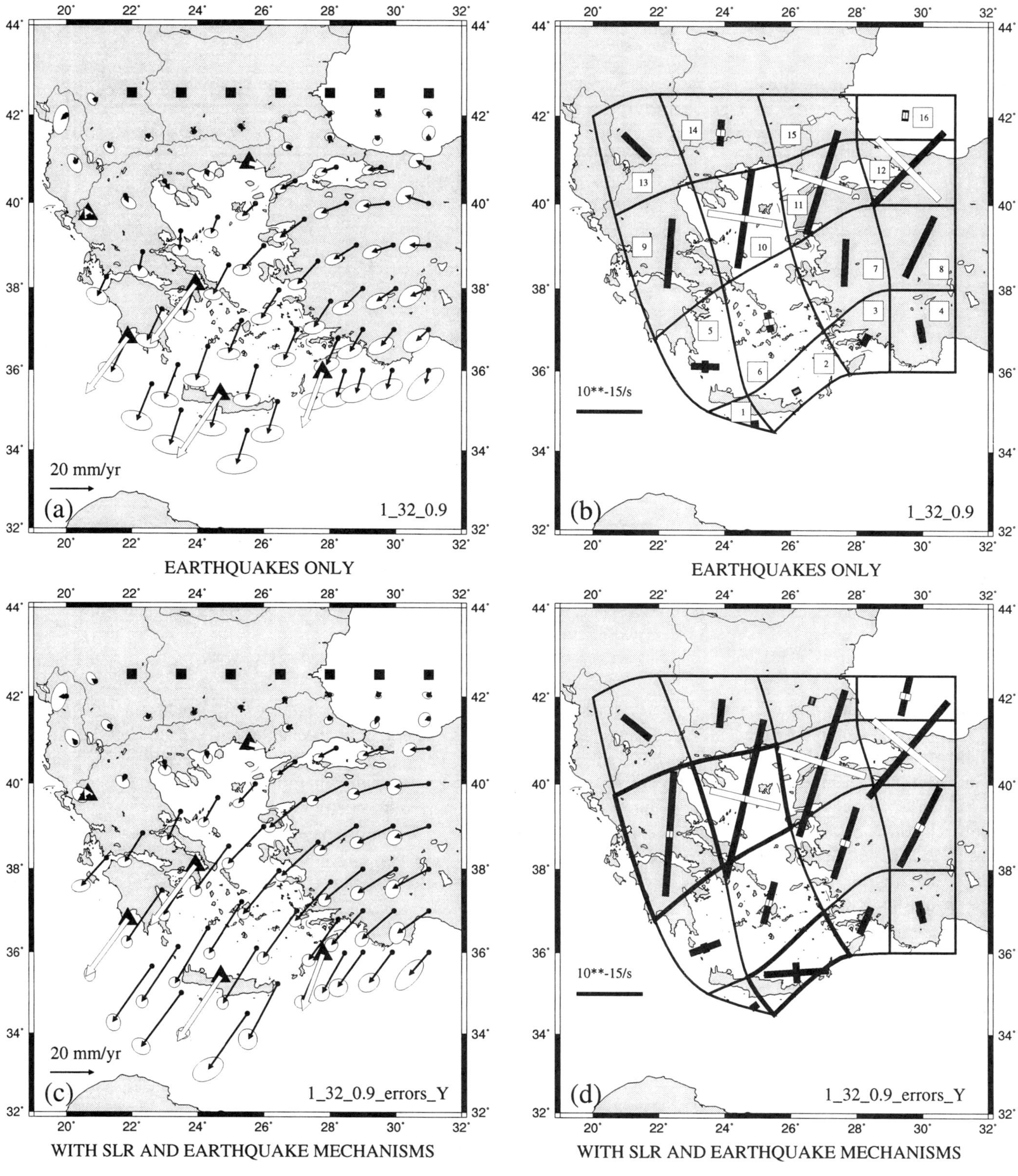


**Figure 7.** (a) Velocity field in Aegean region relative to Eurasia obtained from spline function fitting to distribution of (b) seismic strain rates (principal average strains in areas) from earthquakes of $M_s \geq 6.0$ that occurred between 1911 and 1992. The contribution from small earthquakes was also determined (see text). SLR velocities in (a) (open vectors at triangle sites) are shown for comparison. (c) Velocity field obtained from smooth fitting of SLR velocities while functions are constrained to match the style of deformation indicated by earthquake moment tensors. (d) Principal average strain rates from smooth velocity field solution, showing that distribution of SLR velocities does not require a strain rate field different in style from the earthquake strain rates, but simply more of it.

**Table 3.** Strain Rate Deficits in the Aegean in Units of $1 \times 10^{-15} s^{-1}$

| Area | Observed | Calculated (EQ) | Calculated (SLR) | Deficit | $M_s$ = 6.5/100yrs |
|---|---|---|---|---|---|
| 1 | 0.00±1.1 | 0.12±0.9 | 0.12±1.0 | | |
| 2 | 0.00±0.5 | 0.12±0.5 | 0.97±0.6 | 0.9 | 10 |
| 3 | 0.22±0.5 | 0.19±0.4 | 0.43±0.4 | 0.2 | 2 |
| 4 | 0.41±0.45 | 0.35±0.4 | 0.34±0.4 | | |
| 5 | 0.46±0.35 | 0.43±0.3 | 0.49±0.3 | | |
| 6 | 0.29±0.35 | 0.30±0.25 | 0.65±0.25 | 0.35 | 7 |
| 7 | 0.93±0.4 | 0.72±0.3 | 1.12±0.3 | 0.3 | 5 |
| 8 | 0.98±0.4 | 1.01±0.35 | 1.33±0.35 | 0.35 | 5 |
| 9 | 1.02±0.35 | 1.04±0.3 | 1.88±0.35 | 0.85 | 20 |
| 10 | 1.44±0.4 | 1.51±0.3 | 2.46±0.35 | 1.0 | 18 |
| 11 | 2.02±0.4 | 1.63±0.3 | 2.31±0.3 | 0.5 | 7 |
| 12 | 1.6±0.45 | 1.56±0.4 | 1.88±0.4 | 0.3 | 4 |
| 13 | 0.44±0.35 | 0.56±0.3 | 0.54±0.3 | | |
| 14 | 0.51±0.4 | 0.39±0.3 | 0.43±0.3 | | |
| 15 | 0.00±0.5 | 0.10±0.4 | 0.12±0.4 | | |
| 16 | 0.00±0.6 | 0.16±0.45 | 0.6±0.45 | 0.5 | 5 |

Table of observed seismic strain rate values (observed), calculated strain rate values (calculated (EQ)) from bi-cubic Bessel interpolation of distribution of seismic strain rates, calculated strain rates from interpolation of SLR velocities (calculated SLR), and the deficit. The area numbers corresponding to those in Figure 7b. Also calculated are the number of $M_s$ = 6.5 required to take up the deficit for the given area. This number of events is dependent on both the deficit and the size of the area.

arc-parallel stretching east of central Crete [e.g. *Armijo et al.*, 1992] that is required by the SLR data (Figure 7d). The deficit strain rates (total strain rates inferred from SLR in Figure 7d minus seismic strain rates in Figure 7b) are largest in areas 2, 9 and 10 (Table 3, see Figure 7b for box numbers). Although the spatial resolution on the strain rate field in 7c is not as fine as what can now be obtained from GPS in the Aegean, the velocity field shown in Figure 7c is comparable in most places (within 2 sigma) with the GPS velocities of *Reilinger et al.* [1997].

The above exercise in the Aegean suggests that the pattern of regional seismic deformation is a good approximation of the pattern of the total strain that accommodates total relative motion within the Aegean. Questions arise, though, about the missing component of deformation, not represented by seismic deformation. Accounting for small earthquakes, which is done here, does not satisfy the deficit inferred from the 82 year seismicity catalog (see Table 3). Given the randomness of the earthquake process and the potential for rare large events, one explanation could be that in this century the amount of released seismic moment is less than the long-term seismic moment rate. But there is no indication from historical sources that earthquakes larger than $M_s$ = 7.0 occur in Greece or larger than $M_s$ = 7.4 occur in western Turkey [*Ambraseys*, 1988; *Ambraseys and Jackson*, 1990] indicating that much of the deficit could be attributed to intrinsically aseismic processes. What is most probable, however, is that the deficits in some regions represents a true deficit that will be released seismically in future events (Table 3). Although higher resolution strain rate fields are now possible with dense GPS, unfortunately the duration of the instrumental catalog is probably too short to carry out anything more than regional deficit analyses, such as is done here (Table 3) [*Jackson et al.*, 1994].

## 6. CONCLUSIONS

We have demonstrated a practical approach to estimating self-consistent velocity fields, given available geologic, seismic, and geodetic observations. The solutions satisfy strain rate compatibility requirements and are important for defining the relationship between strain rates, rotation rates, and the relative motions that are directly measured using GPS. Such an analysis is fundamental for the quantification of both seismic hazards and continental dynamics.

*Acknowledgements.* We are grateful to Xavier LePichon and Nicolas Chamot-Rooke for stimulating discussions about the active tectonics of Asia, and for pointing out some published GPS rates. We thank Kristine Larson for kindly providing GPS data in the Himalaya and for providing a preprint before publication. We are grateful to Brad Hager for kindly providing the GPS observations in the Tien Shan. We are appreciative of the helpful reviews provided by Leslie Sonder and Donald Argus. W.E. Holt thanks the Lamont Doherty Earth Observatory for hosting him during sabbatical. Maps were prepared using GMT version 3.0 by Wessel and Smith. This research was funded by NSF grant EAR-9628872.

## REFERENCES

Abdrakhmatov, K.Ye., S.A. Aldazhanov, B.H. Hager, M.W. Hamburger, T.A. Herring, K.B. Kalabaev, V.I. Makarov, P. Molnar, S.V. Panasyuk,, M.T. Prilepin, R.E. Reilinger, I.S. Sadybakasov, B.J. Souter, Yu.A. Trapeznikov, V.Ye. Tsurkov,s and A.V. Zubovich, Relatively recent construction of the Tien Shan inferred from GPS measurements of present-day crustal deformation rates, *Nature, 384*, 450-453, 1996.

Aki, K., and P.G. Richards, *Quantitative Seismology: Theory and Methods*, W.H. Freeman, New York, 1980.

Allen, C.R., Z. Luo, H. Qian, X. Wen, H. Zhou, and W. Huang, Field study of a highly active fault zone: The Xianshuihe fault of southwestern China, *Geol. Soc. Am. Bull., 103*, 1178-1199, 1991.

Ambraseys, N.N., Engineering seismology, *Earthquake Eng. Struct. Dyn., 17*, 1-105, 1988.

Ambraseys, N.N., and J.A. Jackson, Seismicity and associated strain of central Greece between 1890 and 1988, *Geophys. J. Int., 101*, 663-708, 1990.

Ambrosius, B.A.C., D. Angerman, M. Becker, P. Neumaier, R. Noomen, W.J.F. Simons, Chr. Vigny, A. Walpersdorf, and P. Wilson, Final geodetic results of the GEODYSSEA project: the combination solution, *Eos Trans. AGU, 78 (46)*, Fall Meeting Supplement, F169, 1997.

Argus, D.F., and R.G. Gordon, Pacific-North American plate motion from very long baseline Interferometry compared with motion inferred from magnetic anomalies, transform faults, and earthquake slip vectors, *J. Geophys. Res., 95*, 17315-17324, 1990.

Argus, D.F., and R.G. Gordon, Current Sierra Nevada-North America motion from very long baseline interferometry: Implications for the kinematics of the western United States, *Geology, 19*, 1085-1088, 1991a.

Argus, D.F., and R.G. Gordon, No-net-rotation model of current plate velocities incorporating plate motion model NUVEL-1, *Geophys. Res. Lett., 18*, 2039-2042, 1991b.

Artyushkov, E.V., Stresses in the lithosphere caused by crustal thickness inhomogeneities, *J. Geophys. Res., 78*, 7675-7690, 1973.

Armijo, R., H. Lyon-Caen, and D. Papanastassious, East-west extension and Holocene normal fault scarps in the Hellenic arc, *Geology, 20*, 491-94, 1992.

Atwater, T., Implications of plate tectonics for the Cenozoic tectonic evolution of western North America, *Bull. Geol. Soc. Am., 81*, 3513-3536, 1970.

Avouac, J.P., and P. Tapponnier, Kinematic model of active deformation in central Asia, *Geophys. Res. Lett., 20*, 895-898, 1993.

Bennett, R.A., W. Rodi, and R.E. Reilinger, Global positioning system constraints on fault slip rates in southern California and northern Baja, Mexico, *J. Geophys. Res., 101*, 21943-21960, 1996.

Bennett, R.A., L.L. Davis, and B.P. Wernicke, Present-day pattern of Cordilleran deformation in the western United States, *Geophys. Res. Lett., 27*, 371-374, 1999.

Bilham, R., 11 mm/yr creep at Fremont, on the surface Hayward fault, and 21 m depth, *Eos Trans. AGU, 77(46)*, Fall Meet. Suppl., F516, 1996.

Bilham, R., K. Larson, J. Freymueller, and Poject Idylhim members, GPS measurements of present day convergence rates in the Nepal Himalaya, *Nature, 336*, 61-64, 1997.

Bohannon, R.G., and T. Parsons, Tectonic implications of post-30 Ma Pacific and North American relative plate motions, *Geol. Soc. Am. Bull., 107*, 937-959, 1995.

Calais, E., O. Lesne, J. Déverchère, V. San'kov, A. Lukhnev, A. Miroshnitchenko, V. Buddo, K. Levi, V. Zalutzky, and Y. Bashkuev, Crustal deformation in the Baikal rift from GPS measurements, *Geophys. Res. Lett., 25*, 4003-4006, 1998.

Carter, J., B.P. Luyendyk, and R.R. Terres, Neogene clockwise tectonic rotation of the eastern Transverse Ranges, California, *Geol. Soc. Am. Bull., 98*, 199-206, 1987.

Chamot-Rooke, N, C. Vigny, A. Walpersdorf, X. LePichon, P. Huchon, and C. Rangin, Sundaland motion detected from GEODYSSEA GPS measurements: Implications for motion at Sunda Trenches, *Eos Trans. AGU, 78 (46)*, Fall Meet. Suppl., 169, 1997.

Chen, Y., F.T. Wu, Langcang-Gengma Earthquake, a preliminary report on the November 6, 1988, Event and its aftershocks, *Eos Trans. AGU*, 70, 1527, 1540, 1989.

Davies, R., P England, B. Parsons, H. Billiris, D. Paradissis, G. Veis, Geodetic strain of Greece in the interval 1892-1992, *J. Geophys. Res., 102*, 24571-24588, 1997.

deBoor, C., *A Practical Guide to Splines*, Springer-Verlag, New York, 1978.

DeMets, C., R.G. Gordon, D.F. Argus, and S. Stein, Current plate motions, *Geophys. J. Int., 101*, 425-478, 1990.

DeMets, C., R.G. Gordon, D.F. Argus, and S. Stein, Effects of recent revisions to the geomagnetic reversal time scale on estimates of current plate motions, *Geophys. Res. Lett., 21*, 2191-2194, 1994.

DePolo, C.M., W.A. Peppin, and P.A. Johnson, Contemporary tectonics, seismicity, and potential earthquake sources in the White Mountains seismic gap, west-central Nevada and east-central California, USA, *Tectonophysics, 225*, 271-299, 1993.

Dixon, T.H., S. Robaudo, J. Lee, M.C. Reheis, Constraints on present-day Basin and Range deformation from space geodesy, *Tectonics, 14*, 755-772, 1995.

Dokka, R.K., and C.J. Travis, Late Cenozoic strike-slip faulting in the Mojave Desert, California, *Tectonics, 9*, 311-340, 1990a.

Dokka, R.K., and C.J. Travis, Role of the eastern California shear zone in accommodating Pacific-North American plate motion, *Geophys. Res. Lett., 17*, 1323-1326, 1990b.

Donnellan, A., B.H. Hager, R.W. King, and T.A. Herring, Geodetic measurement of deformation in the Ventura Basin region, southern California, *J. Geophys. Res.*, *98*, 21727-21739, 1993.

Dziewonski, A.M., G. Ekström, and M.P. Salganik, Centroid-moment tensor solutions for July-Septermber 1995, *Phys. Earth Planet. Inter.*, *97*, 3-13, 1996.

Dziewonski, A.M., T.-A. Chou, and J.H. Woodhouse, Determination of Earthquake source parameters from waveform data for studies of global and regional seismicity, *J. Geophys. Res.*, *91*, 2825-2852, 1981.

England, P., and G. Houseman, Finite strain calculations of continental deformation 2. Comparison with the India-Asia collision zone, *J. Geophys. Res.*, *91*, 3664-3676, 1986.

England, P., and D.P. McKenzie, A thin viscous sheet model for continental deformation, *Geophys. J. R. Astron. Soc.*, *70*, 295-321, 1982.

England, P.C., and P. Molnar, Active deformation of Asia: From kinematics to dynamics, *Science, 278*, 647-650, 1997a.

England, P. and P. Molnar, The field of crustal velocity in Asia calculated from Quaternary rates of slip on faults, *Geophys. J. Int.*, *130*, 551-582, 1997b.

Feigl, K.L., D.C. Agnew, Y. Bock, D. Dong, A. Donnellan, B.H. Hager, T.A. Herring, D.D. Jackson, T.H. Jordan, R.W. King, S. Larsen, K.M. Larson, M.H. Murray, Z.K. Shen, and F.H. Webb, Space geodetic measurement of crustal deformation in central and southern California, 1984-1992, *J. Geophys. Res.*, *98*, 21677-21712, 1993.

Fleitout, L., and C. Froidevaux, Tectonics and topography for a lithosphere containing density heterogeneities, *Tectonics, 1*, 21-56, 1982.

Fleitout, L., The sources of lithospheric tectonic stresses, *Phil. Trans. R. Soc. Lond. A*, *337*, 73-81, 1991.

Flesch, L.M., W.E. Holt, and A.J. Haines, The dynamics of the India-Eurasia collision zone, *Eos Trans. AGU, 79(45)*, Fall Meet. Suppl., F204, 1998.

Freymueller, J., R. Bilham, R. Burgmann, K.M. Larson, J. Paul, S. Jade, and V. Gaur, Global positioning system measurements of Indian plate motion and convergence across the Lesser Himalaya, *Geophys. Res. Lett.*, *23*, 3107-3110, 1996.

Fung, Y. C., *Foundations of Solid Mechanics.*, Prentice-Hall, Englewood Cliffs, N. J., 1965.

Gladwin, M.T., R.L. Gwyther, R.H.G. Hart, and K.S. Breckenridge, Measurements of the strain field associated with episodic creep events on the San Andreas fault at San Juan Bautista, California, *J. Geophys. Res.*, *99*, 4559-4565, 1994.

Haines, A.J., Calculating velocity fields across plate boundaries from observed shear rates, *Geophys. J. R. Astron. Soc.*, *68*, 203-209, 1982.

Haines, A.J., and W.E. Holt, A procedure for obtaining the complete horizontal motions within zones of distributed deformation from the inversion of strain rate data, *J. Geophys. Res.*, *98*, 12,057-12,082, 1993.

Haines, A.J., J.A. Jackson, W.E. Holt, and D.C. Agnew, Representing distributed deformation by continuous velocity fields, *Sci. Rept. 98/5, Inst. of Geol. and Nucl. Sci.*, Wellington, New Zealand, 1998.

Heki, K., Horizontal and vertical crustal movements from three-dimensional very long baseline interferometry kinematic reference frame: Implication for the reversal timescale revision, *J. Geophys. Res.*, *101*, 3187-3198, 1996.

Hill, D.P., R.E. Wallace, and R.S. Cockerham, Review of evidence on the potential for major earthquakes and volcanism in the Long Valley-Mono Craters-White Mountains regions of eastern California, *Earthquake Predic. Res.*, *3*, 571-594, 1985.

Holt, W.E., L.M. Flesch, B. Shen-Tu, and A.J. Haines, The kinematics and dynamics of the Western US inferred from Quaternary fault data, GPS and VLBI velocities, and topography and Geoid data, *Eos Trans. AGU, 79(45)*, Fall Meet. Suppl., F206, 1998.

Holt, W.E., and A.J. Haines, Velocity fields in deforming Asia from the inversion of earthquake-released strains, *Tectonics, 12*, 1-20, 1993.

Holt, W.E., and A.J. Haines, The kinematics of northern South Island New Zealand determined from geologic strain rates, *J. Geophys. Res.*, *100*, 17,991-18,010, 1995.

Holt, W.E., M. Li, and A.J. Haines, Earthquake strain rates and instantaneous relative motion within central and east Asia, *Geophys. J. Int.*, *122*, 569-593, 1995.

Holt, W.E., J.F. Ni, T.C. Wallace, and A.J. Haines, The active tectonics of the eastern Himalayan syntaxis and surrounding regions, *J. Geophys. Res.*, *96*, 14,595-14,632, 1991.

Houseman, G., and P. England, Finite strain calculations of continental deformation 1. method and general results for convergent zones, *J. Geophys. Res.*, *91*, 3651-3663, 1986.

Humphreys, E.D., and R.J. Weldon, Deformation across the western United States: A local estimate of Pacific-North America transform deformation, *J. Geophys. Res*, *99*, 19975-20010, 1994.

Jackson, D.D., Z. Shen, D. Potter, X.B. Ge, and L. Sung, Southern California deformation, *Science, 277*, 1621-1622, 1997.

Jackson, J.A., Active tectonics of the Aegean region, *Annu. Rev. Earth Planet. Sci.*, *22*, 239-271, 1994.

Jackson, J.A., A.J. Haines, and W.E. Holt, The horizontal velocity field in the deforming Aegean Sea region determined from the moment tensors of earthquakes, *J. Geophys. Res.*, *97*, 17,657-17,684, 1992.

Jackson, J., A.J. Haines, and W.E. Holt, A comparison of satellite laser ranging and seismicity data in the Aegean region, *Geophys. Res. Lett.*, *21*, 2849-2852, 1994.

Jackson, J., A. J. Haines, and W.E. Holt, The Accommodation of Arabia-Eurasia plate convergence in Iran, *J. Geophys. Res.*, *100*, 15, 205-15,219, 1995.

Jennings, C.W., *Fault Activity Map of California and Adjacent Areas, California,* Dept. of Conservation, Div. of Mines and Geology, Sacramento, 1994.

Jones, C.H., L.J. Sonder, and J.R. Unruh, The role of gravitational potential energy in active deformation in the southwestern United States, *Nature, 381*, 37-41, 1996.

King, R.W., F. Shen, B.C. Burchfiel, L.H. Royden, E. Wang, Z. Chen, Y. Liu, X. Zhang, J. Zhao and Y. Li, Geodetic measurement of crustal motion in southwest China, *Geology, 25*, 179-182, 1997.

Kostrov, V.V., Seismic moment, energy of earthquakes, and the seismic flow of rock, *Izv. Acad. Sci. USSR Phys. Solid Earth, Engl. Transl., 10*, 23-44, 1974.

Lacassin, R., A. Replumaz, P.H. Leloup, Hairpin river loops and slip-sense inversion on southeast Asian strike-slip faults, *Geology, 26*, 703-706, 1998.

Lamb, S.H., A model for tectonic rotations about a vertical axis, *Earth Planet. Sci. Lett., 84*, 75-86, 1987.

Larson, K.M., J.T. Freymueller, and S. Philipsen, Global plate velocities from the global positioning system, *J. Geophys. Res., 102*, 9961-9981, 1997.

Larson, K.M., R. Burgmann, R. Bilham, and J.T. Freymueller, Kinematics of the India-Eurasia collision zone from GPS measurements, *J. Geophys. Res., 104*, 1077-1093, 1999.

LeDain, A.Y., P. Tapponnier, and P. Molnar, Active faulting and Tectonics of Burma and surrounding regions, *J. Geophys. Res., 89*, 453-472, 1984.

Leloup, P.H., T.M. Harrison, F.J. Ryerson, W. Chen, L. Qi, P. Tapponnier, and R. Lacassin, Structural, petrologic and thermal evolution of a Tertiary ductile strike-slip shear zone, Diancang Shan, Yunnan, *J. Geophys. Res., 98*, 6715-6743, 1993.

Leloup, P.H., R. Lacassin, P. Tapponnier, U. Scharer, D. Zhong, X. Liu, L. Zhang, S. Ji, P.T. Trinh, The Ailao Shan-Red River shear zone (Yunnan, China), Tertiary transform boundary of Indochina, *Tectonophysics, 251*, 3-84, 1995.

LePichon, X., and J. Angelier, The Aegean Sea, *Philos. Trans. R. Soc. London Ser. A, 300*, 357-372, 1981.

Lesne, O., E. Calais, and J. Deverchere, Finite element modeling of crustal deformation in the Baikal rift zone: new insights into the active-passive rifting debate, *Tectonophysics, 289*, 327-340, 1998.

Li, M., A kinematic analysis of central and southeast Asia from the inversion of both intraplate seismic moment release and Quaternary strains, *M. Sc. thesis*, State University of New York at Stony Brook, 1994.

Liu, M., and Y. Shen, Crustal collapse, mantle upwelling, and Cenozoic extension in the North American Cordillera, *Tectonics, 17*, 311-321, 1998.

Luyendyk, B.P., Neogene-age fault slips in the continental transform zone in southern California, *Annales Tectonicae, 4*, 24-34, 1990.

Luyendyk, B. P., A model for Neogene crustal rotations, transtension, and transpression in southern California, *Geol. Soc. Am. Bull., 103*, 1528-1536, 1991.

Luyendyk, B.P., M.J. Kamerling, J.S. Hornafius, R. Haston, T. Ross, and J.N. Carter, Paleomagnetic studies in Neogene rocks from the Peninsular Ranges terrain, southern California, *Geol. Soc. Am., Abstracts with Programs, 20*, 163, 1988.

Ma, C., and J.W. Ryan, NASA Space Geodesy Program - GSFC Data Analysis - 1998, VLBI Geodetic Results 1979-1998, August 1998.

Ma, X., *Lithospheric dynamics map of China and adjacent seas: Beijing, China*, Geological Publishing House, scale 1:4,000,000 with explanatory notes,1986.

McKenzie, D.P., Speculations on the consequences and causes of plate motion, *Geophys. J. R. Astron. Soc., 18*, 1-32, 1969.

McKenzie, D., Active tectonics of the Mediterranean region, *Geophys. J. R. Astron. Soc., 30*, 109-185, 1972.

McKenzie, D., Active tectonics of the Alpine-Himalayan belt: The Aegean Sea and surrounding regions, *Geophys. J. R. Astron. Soc., 55*, 217-254, 1978.

McKenzie, D., and J. Jackson, The relationship between strain rates, crustal thinning, paleomagnetism, finite strain and movements within a deforming zone: *Earth Plan. Sci. Lett., 65*, 182-202, 1983.

Meyer, B., P. Tapponnier, Y. Gaudemer, G. Peltzer, S. Guo, and Z. Chen, Rate of left-lateral movement along the easternmost segment of the Altyn Tagh fault, east of 96° E (China), *Geophys. J. Int., 124*, 29-44, 1996.

Michel, G.W., Ch. Reigber, D. Angermann, J. Klotz, Ongoing and recent deformation in central and east Asia, *Eos Trans. AGU, 78*, Fall Meet. Suppl., 172, 1997.

Minster, J.B., and T.H. Jordan, Vector constraints on western U. S. deformation from space geodesy, neotectonics, and plate motions, *J. Geophys. Res., 92*, 4798-4804, 1987.

Molnar, P., Earthquake recurrence intervals and plate tectonics, *Bull. Seismol. Soc. Am., 69*, 115-133, 1979.

Molnar, P., Continental tectonics in the aftermath of plate tectonics, *Nature, 335*, 131-137, 1988.

Molnar, P., and P. Tapponnier, Cenozoic tectonics of Asia: Effects of a continental collision, *Science, 189*, 419-425, 1975.

Molnar, P., and Q. Deng, Faulting associated with large earthquakes and the average rate of deformation in central and eastern Asia, *J. Geophys. Res., 89*, 6203-6228, 1984.

Molnar, P., and J.M. Gipson, A bound on the rheology of continental lithosphere using very long baseline interferometry: The velocity of south China with respect to Eurasia, *J. Geophys. Res., 101*, 545-553, 1996.

Molnar, P., and H. Lyon-Caen, Some simple physical aspects of the support, structure, and evolution of mountain belts, *Spec. Pap. Geol. Soc. Am., 218*, 179-207, 1988.

Namson, J. and T. Davis, Structural transect of the western Transverse Ranges, California: implications for lithospheric kinematics and seismic risk evaluation, *Geology, 17*, 675-679, 1988.

Nason, R.D., Fault creep and earthquakes on the San Andreas fault, in *Proceedings of the conference on tectonic problems of the San Andreas fault system*, R. L. Kovach and Amos Nur (editors), Stanford University Publications in the Geological Sciences, 13, 275-285, 1973.

Nicholson C., C.C. Sorlien, T. Atwater, J.C. Crowell, and B.P. Luyendyk, Microplate capture, rotation of the western Transverse Ranges, and initiation of the San-Andreas transform as a low-angle fault system, *Geology, 22*, 491-495, 1994.

Peltzer, G., F. Saucier, Present-day kinematics of Asia derived from geologic fault rates, *J. Geophys. Res., 101*, 27943-27956, 1996.

Peltzer, G., P. Tapponnier, and R. Armijo, Magnitude of Late Quaternary left-lateral displacements along the north edge of Tibet, *Science, 246*, 1285-1289, 1989.

Peterson, M.D., and S.G. Wesnousky, Fault slip rates and earthquake histories for active faults in southern California,

*Bull. Seism. Soc. Am., 84*, 1608-1649, 1994.

Randall, G.E., C.J. Ammon, T.J. Owens, Moment tensor estimation using regional seismograms from Tibetan Plateau portable network deployment, *Geophys. Res. Lett., 22*, 1665-1668, 1995.

Reilinger R.E., S.C. McClusky, M.B. Oral, R.W. King, M.N. Toksoz, A.A. Barka, I. Kinik, O. Lenk, I. Sanli, Global positioning system measurements of present-day crustal movements in the Arabia-Africa-Eurasia plate collision zone, *J. Geophys. Res., 102*, 9983-9999, 1997.

Richardson, R.M., Ridge forces, absolute plate motions, and the intraplate stress-field, *J. Geophys. Res., 97*, 11739-11748, 1992.

Robbins, J.W., P.J. Dunn, M.H. Torrence and D.E. Smith, Deformation in the eastern Mediterranean, in *Proceedings of the First International Symposium on Deformations in Turkey*, edited by M. Sahin and P.A. Cross, 1994.

Rothery, D.A. and S.A. Drury, The neotectonics of the Tibetan Plateau, *Tectonics, 3*, 19-26, 1984.

Ryall, A.S., Earthquake hazard in the Nevada region, *Bull. Seismol. Soc. Am., 67*, 517-532, 1977.

Sauber, J., W. Thatcher, and S. Solomon, Geodetic measurement of deformation in the central Mojave Desert, California, *J. Geophys. Res., 91*, 12683-12694, 1986.

Sauber, J., W. Thatcher, S.C. Solomon, and M. Lisowski, Geodetic slip rate for the eastern California shear zone and the recurrence time of Mojave desert earthquakes, *Nature, 367*, 264-266, 1994.

Saucier, F., and E. Humphreys, Horizontal crustal deformation in southern California from joint models of geologic and very long baseline interferometry measurements, in *Contributions of Space Geodesy to Geodynamics, Geodyn. Ser., 23*, edited by D. E. Smoth and D. L. Turcotte, pp. 139-176, AGU, Washington, D. C., 1993.

Savage, J.C., M. Lisowski, and W.H. Prescott, An apparent shear zone trending north-northwest across the Mojave Desert into Owens Valley, eastern California, *Geophys. Res. Lett., 17*, 2113-2166, 1990.

Savage, J.C., W. H. Prescott, M. Lisowski, and N. King, Geodetic measurements of deformation near Holister, California, 1971-1978, *J. Geophys. Res., 84*, 7599-7615, 1979.

Scholz, C.H., and P.A. Cowie, Determination of total strain from faulting using slip measurements, *Nature, 346*, 837-839, 1990.

Shen, Z., D.D. Jackson, and B.X. Ge, Crustal deformation across and beyond the Los Angeles basin from geodetic measurements, *J. Geophys. Res., 101*, 27957-27980, 1996.

Shen, Z., D. Dong, T. Herring, K. Hudnut, D. Jackson, R. King, S. McClusky, and L. Sung, Crustal deformation measured in southern California, *Eos Trans. AGU, 87(43), 477*, 1997.

Shen-Tu, B., W.E. Holt, and A.J. Haines, Intraplate deformation in the Japanese Islands: A kinematic study of intraplate deformation at a convergent plate margin, *J. Geophys. Res., 100*, 24275-24293, 1995.

Shen-Tu, B., W.E. Holt, and A.J. Haines, Contemporary kinematics of the western United States determined from earthquake moment tensors, very long baseline interferometry, and GPS observations, *J. Geophys. Res., 103*, 18,087-18,117, 1998.

Shen-Tu, B., W.E. Holt, and A.J. Haines, Deformation kinematics in the Western United States determined from Quaternary fault slip rates and recent geodetic data, *J. Geophys. Res.*, accepted for publication, 1999.

Sieh, K.E., and P.L. Williams, Behavior of the southernmost San Andreas fault during the past 300 years, *J. Geophys. Res., 95*, 6629-6645, 1990.

Smith, D.E., R. Kolenkiewicz, J.W. Robbins, P.J. Dunn, and M.H. Torrence, Horizontal crustal motion in the central and eastern Mediterranean inferred from SLR measurements, *Geophys. Res. Lett., 21*, 1979-1982, 1994.

Snay, R.A., M.W. Cline, C.R. Philipp, D.D. Jackson, Y. Feng, and Z.-K. Shen, Crustal velocity field near the big bend of California's San Andreas fault, *J. Geophys. Res., 101*, 3173-3185, 1996.

Sonder, L.J., C.H. Jones, S.L. Salyards, and K.M. Murphy, Vertical-axis rotations in the Las Vegas Valley Shear Zone, southern Nevada: Paleomagnetic constraints on kinematics and dynamics of block rotations, *Tectonics, 13*, 769-788, 1994.

Sonder, L.J., and C.H. Jones , Western United States extension: How the West was widened, *Annu. Rev. Earth Planet. Sci, 27*, 417, 1999.

Stock, J., and P. Molnar, Uncertainties and implications of the late Cretaceous and Tertiary position of North America relative to the Farallon, Kula, and Pacific plates, *Tectonics, 6*, 1339-1384, 1988.

Tapponnier, P., G. Peltzer, A.Y. LeDain, R. Armijo, and P. Cobbold, Propagating extrusion tectonics in Asia: New insights from simple experiments with plasticine, *Geology, 10*, 611-616, 1982.

Taymez, T., J. Jackson, and D. McKenzie, Active tectonics of the north and central Aegean Sea, *Geophys. J. Int., 106*, 433-490, 1991.

Thatcher, W., Present-day crustal movements and the mechanics of cyclic deformation, in The San Andreas Fault System, California, edited by R. E. Wallace, *U.S. Geol. Surv. Prof. Paper, 1515*, 189-205, 1990.

Thatcher, W., G.R. Foulger, B.R. Julian, J. Svarc, E. Quilty, G. W. Bawden, Present Day Deformation Across the Basin and Range Province, Western United States, *Science, 283*, 1714-1718, 1999.

Tovish, A., G. Schubert, and B.P. Luyendyk, Mantle flow pressure and the angle of subduction: Non-Newtonian corner flows, *J. Geophys. Res., 83*, 5892-5898, 1978.

Tinnon, M. J., W. E. Holt, and A. J. Haines, Velocity gradients in the northern Indian Ocean inferred from earthquake moment tensors and relative plate velocities, *J. Geophys. Res., 100*, 24315-24329, 1995.

Wallace, R.E., Patterns and timing of late Quaternary faulting in the Great Basin province and relation to some regional tectonic features, *J. Geophys. Res., 89*, 5763-5769, 1984.

Wallace, R.E., Patterns of faulting and seismic gaps in the Great Basin province, *U. S. Geol. Surv. Open-File Rep., 78-943*, 1978.

Ward, S.N., Pacific-North America plate motions: New results from very long baseline interferometry, *J. Geophys. Res., 95*, 21965-21981, 1990.

Wdowinski, S., R.J. O'Connell, and P. England, A continuum model of continental deformation above subduction zones - application to the Andes and the Aegean, *J. Geophys. Res., 94*, 10331-10346, 1989.

Wilson, P., J. Rais, Ch. Reigber, E. Reinhart, B.A.C. Ambrosius, X. LePichon, M. Kasser, P. Suharto, Dato'Abdul Majid, D.A.H.O.B.H. Yaakub, R. Almeda, and C. Boonphakdee, Study provides data on active plate tectonics in southeast Asia region, *Eos Trans. AGU, 79*, 545-549, 1998.

Yeats, R.S., Large-scale Quaternary detachments in Ventura Basin, southern California, *J. Geophys. Res., 88*, 569-583, 1983.

Yu, S.-B, L.-C Kuo, R. S. Punongbayan, and E. G. Ramos, GPS observation of crustal deformation in the Taiwan-Luzon, *Geophys. Res. Lett., 26*, 923-926, 1999.

Zhang, Y.Q., J.L. Mercier, and P. Vergely, Extension in the graben system around the Ordos (China), and its contribution to the extrusion tectonics of south China with respect to Gobi-Mongolia, *Tectonophysics, 285*, 41-75, 1998.

Zhang, Y.Q., P. Vergely, J. Mercier, Active faulting in and along the Qinling Range (China) inferred from SPOT imagery analysis and extrusion tectonics of south China, *Tectonophysics, 243*, 69-95, 1995.

Zhao, S., Joint inversion of observed gravity and GPS baseline changes for the detection of the active segment at the Red River fault zone, *Geophys. Int., 122*, 70-88, 1995.

---

J. Jackson and A.J. Haines, Bullard Laboratories, University of Cambridge, U.K.

W.E. Holt and B. Shen-Tu, Department of Geosciences, State University of New York at Stony Brook, Stony Brook, NY 11794-2100

# Diffuse Oceanic Plate Boundaries: Strain Rates, Vertically Averaged Rheology, and Comparisons with Narrow Plate Boundaries and Stable Plate Interiors

Richard G. Gordon

*Department of Geology and Geophysics, Rice University, Houston, Texas*

Diffuse plate boundaries occur in both oceanic and continental lithosphere and cover ≈15% of Earth's solid surface. The fastest plate speeds accommodated across diffuse oceanic plate boundaries are ≈15 mm/yr. The smallest strain rates averaged across narrow plate boundaries are at least $10^2$ times larger than the largest strain rates across diffuse oceanic plate boundaries and at least $10^4$ times larger than those across stable plate interiors. The effective viscosity ($\eta_{eff}$) of the lithosphere is estimated from the ratio of vertically averaged shear stresses to strain rates for three tectonic settings: (i) oceanic transform fault zones, for which $\eta_{eff} = 3\times10^{16}$ to $5\times10^{19}$ Pa s, comparable to estimates for the asthenosphere, (ii) diffuse oceanic plate boundaries, for which $\eta_{eff} = 1\times10^{23}$ to $6\times10^{23}$ Pa s, ≈10 times larger than for diffuse continental plate boundaries, and (iii) stable plate interiors, for which $\eta_{eff} = 1\times10^{24}$ to $2\times10^{27}$ Pa s.

The rheology of oceanic lithosphere over times longer than earthquake cycles is modeled as a plastic layer overlying a layer that deforms by creeping flow [*Martinod and Davy*, 1992]. Oceanic lithosphere deforms when the yield strength of the upper lithosphere is exceeded. The vertically averaged rheology of deforming oceanic lithosphere can be approximated by a power-law fluid for which $\dot{\varepsilon}_s \propto (\tau_s)^n$ where $\dot{\varepsilon}_s$ is the rate of shear strain and $\tau_s$ is the shear stress. If the ratio of the yield strength of the upper lithosphere to the force required to deform the lower lithosphere at a strain rate of $10^{-16}\,s^{-1}$ is varied from $10^{-2}$ to $10^2$, the calculated value of $n$ varies from ≈3 to ≈300.

The map-view aspect ratio of a deforming zone in a thin sheet of power-law fluid is proportional to $n^{-1/2}$ [*England et al.*, 1985]. A profile of displacement versus distance inferred from a seismic profile across the Central Indian Basin (India-Capricorn diffuse oceanic plate boundary), where the lithosphere is about 60-Myr old, indicates that the lithosphere is being deformed in response to a rigid moving boundary near the north end of the basin. If so, the vertically averaged rheology of the lithosphere in the Central Indian Basin is described

The History and Dynamics of Global Plate Motions
Geophysical Monograph 121

by a power-law fluid with $n \approx 8$, which suggests that the yield strength of the upper lithosphere is a few times larger than the force required to deform the lower lithosphere. Owing to a lack of data, other diffuse oceanic plate boundaries can at best be interpreted qualitatively. Whereas the total aspect ratios (total along-strike length / total across-strike width) for diffuse oceanic plate boundaries in 60- to 90-Myr-old lithosphere are 1.25–1.8, they are 0.73–0.82 in young lithosphere adjacent to an active mid-ocean ridge, The latter aspect ratios suggest smaller values of $n$, indicating a relatively weaker brittle upper lithosphere relative to the force required to deform the lower lithosphere, similar to results for diffuse continental plate boundaries, for which $n \approx 3$ [*England and Molnar*, 1991]. In contrast, the East Africa Rift has a large aspect ratio, 2.3, which suggests its lithosphere behaves more like 60- to 90-Myr-old oceanic lithosphere than like other diffuse continental plate boundaries.

## INTRODUCTION

In many papers that examine the tectonics of the continents, it is common to contrast the rigidity of oceanic lithosphere with the nonrigidity of continental lithosphere. Unlike these other papers, here I emphasize the nonrigidity of oceanic lithosphere, which in places exhibits distributed deformation over horizontal scales of hundreds to thousands of kilometers with strain rates that overlap, but tend to be lower than, those in diffuse continental plate boundaries. Moreover, in the spirit of the Chapman conference that spawned the present volume, I attempt to go beyond kinematics to examine some of the implications of the geometry, style, and rate of deformation of these diffuse oceanic plate boundaries for the rheology and dynamics of oceanic lithosphere.

Tectonicists are familiar with thinking of the lithosphere as being rigid or elastic, except in narrow zones, which are the plate boundaries, and which define the edges of the plates. On the other hand, many geodynamicists (e.g., *Bercovici et al.* [2000], *Tackley* [2000], *Gurnis et al.* [2000]) model the lithosphere as a viscous fluid. I hope that the present analysis, which in part views deforming oceanic lithosphere as a viscous fluid, and which follows successful efforts in treating deforming continental lithosphere as a viscous fluid (e.g., *England and McKenzie* [1982], *Sonder and England* [1986], *England and Molnar* [1991]), can help to bridge the gulf between plates as viewed by tectonicists and as viewed by geodynamicists.

Here I address several fundamental questions about the occurrence, rheology, and dynamics of diffuse oceanic plate boundaries: Why do they occur where they do? What are the strain rates in diffuse plate boundaries and how do these compare with strain rates in stable plate interiors and narrow plate boundaries? Is deformation distributed symmetrically across the strike of the diffuse oceanic plate boundaries? How strong is the lithosphere in diffuse plate boundaries? How does its vertically averaged effective viscosity compare with that of stable plate interiors and narrow plate boundaries? What is the rheology of diffuse oceanic plate boundaries? What controls the widths of diffuse oceanic plate boundaries?

## GEOMETRY OF DIFFUSE PLATE BOUNDARIES

### *Convergent-Divergent Pivots*

Figure 1 is a world map showing diffuse plate boundaries superposed on traditional depictions of narrow plate boundaries. In particular note the boundaries between (i) the North American and South American plates, (ii) the Nubian and Somalian plates, (iii) the Indian and Capricorn plates, and (iv) the Capricorn and Australian plates. (Three of these cases are entirely within oceanic lithosphere, but only the convergent portion of the Nubia-Somalia boundary is in oceanic lithosphere.) In each case the diffuse plate boundary has two parts, a zone of convergence and a zone of divergence. Between these two zones lies the pole of rotation of the plates bounding the diffuse plate boundary. Here I propose the term *convergent-divergent pivot* for a pole of rotation that lies between paired convergent and divergent zones, as it does for these 4 cases. (A fifth case is provided by the pole of rotation between the North American and Eurasian plates, but there the divergent zone is mainly narrow.)

### *Composite and Component Plates*

Plates are evidently more strongly coupled across diffuse plate boundaries, especially diffuse oceanic plate boundaries, than they are across narrow plate boundaries. This strong coupling suggests that in some sense the plates discussed above and their intervening diffuse plate boundaries are a coherent, if nonrigid, mechanical entity with edges defined by narrow plate boundaries. It is confusing to refer to the rigid plates and the partly nonrigid mechanical entities by the same name.

*Royer and Gordon* [1997] thus proposed the use of the terms "component plate" and "composite plate" to express these different concepts. In their terminology, the Indian, Capricorn, and Australian plates, each of which is rigid or nearly rigid, as far as is known, are component plates. In contrast, the Indo-Australian plate is a composite plate comprising these three component plates, the diffuse plate boundaries that separate the components, and a hypothesized diffuse plate boundary west of the Macquarie Ridge that is hypothesized to absorb some of the motion between the Pacific plate and the Australian component of the Indo-Australian plate [*DeMets et al.*, 1988]. Composite plates are delimited by traditional narrow plate boundaries. For example, because the Pacific plate is separated from the Indo-Australian composite plate by a narrow boundary (i.e., along the Macquarie Ridge), it is not considered to be part of the Indo-Australian composite plate. Component plates are at least partly delimited by the edge of a diffuse plate boundary, but may be partly delimited by narrow boundaries. Other composite plates include the American composite plate, with North American and South American component plates, and an African composite plate with Nubian and Somalian component plates [*Gordon*, 1998; *Chu and Gordon*, 1999].

### *Locations of Diffuse Oceanic Plate Boundaries*

Why are diffuse oceanic plate boundaries located where they are? Early work assumed that their locations were due to the existence of pre-existing zones of weakness [e.g., *Sykes*, 1978; *Stein*, 1979; *Bergman and Solomon*, 1980]. For example, a boundary in the Indian Ocean was assumed largely to follow the Ninetyeast Ridge or the fracture zone adjacent to it [*Stein and Okal*, 1978; *Minster and Jordan*, 1978; *Stein and Gordon*, 1984; *Wiens et al.*, 1985]. Data now available, however, indicate that deformation is not limited to these hypothesized zones of weakness, although discrete weak elements, such as the fracture zone adjacent to the Ninetyeast Ridge, are part of the broad zone of deformation [*Petroy and Wiens*, 1989; *Gordon et al.*, 1990; 1998; *DeMets et al.*, 1994; *Royer and Gordon*, 1997] (Figure 1).

Most models for stresses in the Indo-Australian composite plate indicate very high levels of deviatoric NW-SE compression in the Indo-Australian zone of horizontal contraction and some models indicate deviatoric NW-SE tension near the zones of Indo-Australian stretching (Figure 2) [e.g., *Cloetingh and Wortel*, 1985; *Coblentz et al.*, 1998]. This coincidence of high stresses in these models with the locations of the deformation zones suggests that the diffuse plate boundaries occur where vertically averaged deviatoric stresses are higher than elsewhere, presumably above a critical level above which strain rates are much higher than in stable intraplate environments. In detail, however, old weaknesses accommodate much of the deformation. In the Central Indian basin, for example, north-south shortening appears to be accommodated on three different sets of faults [*Bull and Scrutton*, 1992; *Chamot-Rooke et al.*, 1993]. The first two sets are "high-angle" thrust faults, one set north-dipping and the other set south-dipping, both with dips from 40° to 50°. These two sets of faults are thought originally to have been normal faults, which were formed during seafloor spreading and later reactivated as thrust faults [*Weissel et al.*, 1980; *Bull and Scrutton*, 1990]. The third set of faults were low-angle thrust faults, with dips of 25° to 35°, which cluster where deformation is a maximum, and are believed to be a new set of faults that were initiated and have grown in response to the north-south compressive stresses [*Bull*, 1990; *Chamot-Rooke et al.*, 1993].

Unlike the Central Indian Basin, in the sparsely surveyed Wharton basin thrust-faulting earthquakes strike obliquely to existing seafloor fabric and therefore presumably occur on a new set of faults formed in response to NW-SE compression [cf., figure 2 of *Royer and Gordon*, 1997].

### *Relative Plate Speeds*

The maximum velocity between the component plates bounding diffuse oceanic plate boundaries range in magnitude from 3 to 15 mm/yr [*Gordon*, 1998]. These rates are smaller than typical rates of motion across narrow plate boundaries. For example, well-determined rates of seafloor spreading range from 12 mm/yr to 160 mm/yr with a median rate of 40 mm/yr, and convergence rates between the stable interiors of plates meeting at deep sea trenches range from a low of 20 mm/yr along the southern Chile trench, where the Antarctic plate underthrusts the South American plate, to 110 mm/yr along the Australia-Pacific plate boundary with a median rate of 70 mm/yr [*DeMets et al.*, 1990; *Gordon*, 1995]. Thus the fastest motion accommodated across a diffuse plate boundary is only slightly faster than the slowest rate of seafloor spreading and slower than the convergence rate across any trench (Figure 3).

## STRAIN RATES

### *The Importance of Investigating Strain Rates*

Investigations of the driving forces of plates have traditionally relied on two sets of observations: the angular velocities of the plates relative to the hotspots and intraplate stress orientations [*Forsyth and Uyeda*, 1975; *Zoback*, 1992; *Richardson*, 1992; *Lithgow-Bertelloni and Richards*, 1995]. Neither of these data sets is sensitive to the rheology of the lithosphere. In contrast, strain rates must be sensitive to the rheology and may be sensitive to the level of deviatoric stresses.

### *Stable Plate Interiors*

The estimation of strain rates in stable plate interiors is problematic for two reasons: First, the strain rates are so low that they are beneath the threshold of detection by direct measurements [*Argus and Gordon*, 1996; *Dixon et al.*, 1996; *Gordon*, 1998]. Second, the straining that does occur is concentrated in zones of weakness, leaving unresolved the even lower strain rates outside these zones of weakness. Thus, the following discussion can only be a rough guide to the actual rates of strain. The rare occurrence of earthquakes in stable plate interiors indicates that deformation rates are not zero. A lower bound on the strain rates in the stable plate interior of the central and eastern United States, excluding sites of large historical

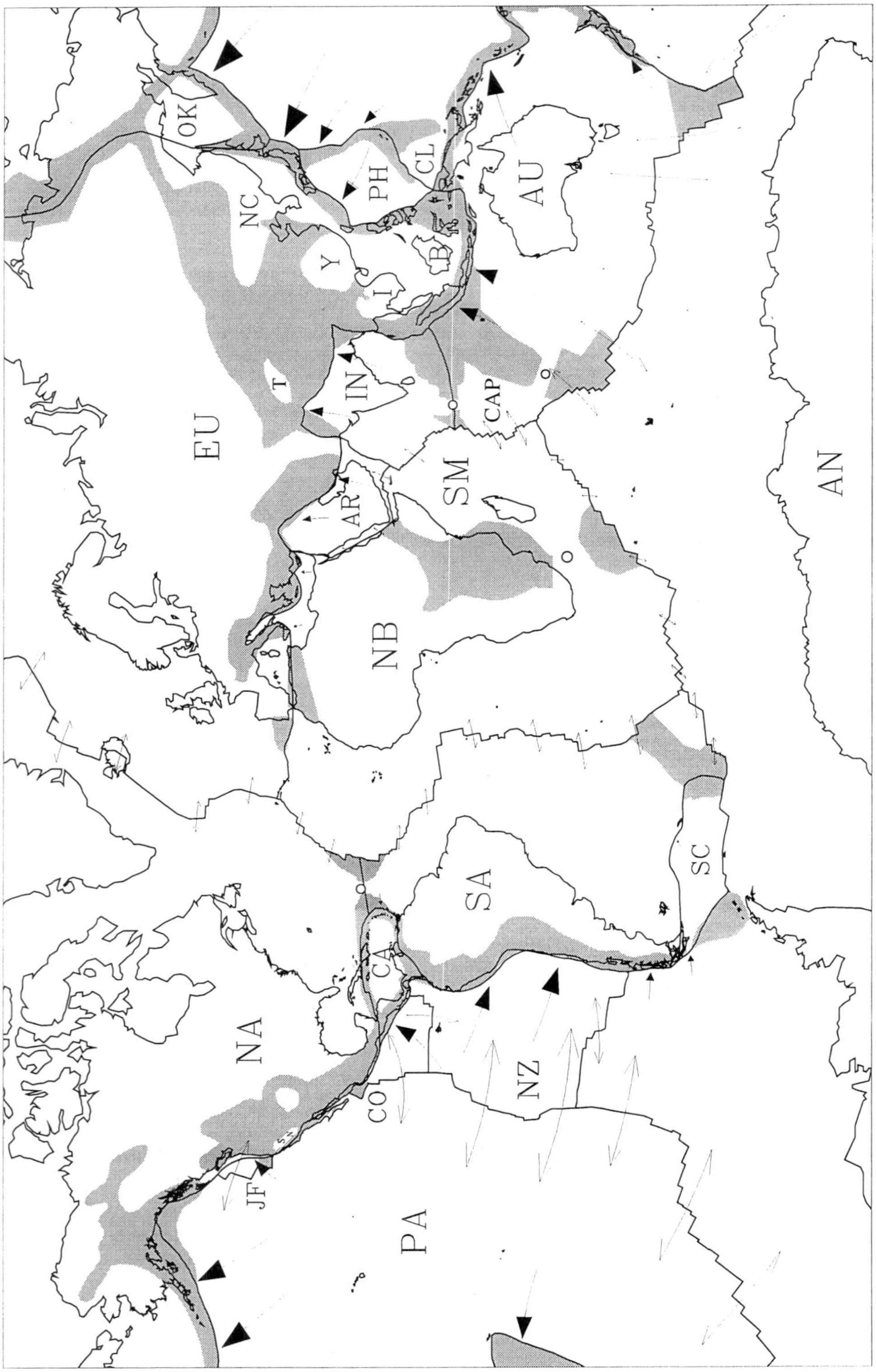
OK
NC
PH
CL
AU
Y
I
B
EU
T
IN
CAP
SM
AR
NB
AN
SA
SC
CA
NA
CO
NZ
JF
PA

earthquakes, comes from seismic moment release and is $3\times10^{-20}$ to $3\times10^{-19}\,s^{-1}$, which may have additional uncertainty of several orders of magnitude [*Anderson*, 1986]. The average strain rate of stable oceanic plate interiors can also be estimated from seismic moment release. *Wiens and Stein* [1983] present a global catalog of 25 oceanic "intraplate" earthquakes that occurred during the 15 year interval starting in 1964 and ending in 1978. *Wiens and Stein* [1983] calculate a corresponding rate of seismic strain release of $\sim10^{-19}\,s^{-1}$, but eight of these 25 earthquakes are now interpreted to have occurred in diffuse oceanic plate boundaries. Two of these eight occurred in the North America-South America diffuse plate boundary and the other six occurred in the India-Capricorn-Australia diffuse plate boundaries. Eliminating these eight earthquakes leaves a seismic strain rate of $5\times10^{-20}\,s^{-1}$. I take this to be a lower bound, albeit highly uncertain, on the strain rate of the stable interiors of oceanic plates.

*Gordon* [1998] discusses two upper bounds on the strain rate of stable plate interiors based on VLBI geodesy [*Argus and Gordon*, 1996] and on plate circuit closures, which give an upper bound of $\approx1\times10^{-17}\,s^{-1}$. The upper bound and lower bounds thus differ by a factor of $\sim10^{2}-10^{3}$, which may have important implications for plate reconstructions. For example, an average strain rate of $10^{-17}\,s^{-1}$ across a global plate motion circuit spanning a complete great circle path over the 65 Myr of Cenozoic time gives a displacement of 1000 km, several times larger than the other uncertainties in such reconstructions, and more than the inferred displacement between the Hawaiian and Atlantic hotspots [*Molnar and Stock*, 1987; *Raymond et al.*, 2000]. If the strain rate was instead $10^{-20}\,s^{-1}$, the corresponding displacement would be merely 1 km, much smaller than the uncertainties in reconstructed points in global plate motion circuits. Evidently, the actual rate of strain of stable plate interiors is critical to the interpretation of results from global plate reconstructions [*Gordon*, 1998].

### *Narrow Plate Boundaries (Oceanic Transform Faults)*

The type of narrow plate boundary for which it is easiest to quantify the strain rate is oceanic transform faults. (Other narrow plate boundaries deserve careful analysis but are not considered here.) Velocities across transform faults mainly range from 12 to 160 mm/yr and across-strike widths of oceanic transform fault zones, i.e., the zone of active faulting, range from ≈500 m to ≈2 km. Their width at depth is unknown. Combining these parameters indicates that strain rates averaged across transform fault zones range from $2\times10^{-13}\,s^{-1}$ to $3\times10^{-11}\,s^{-1}$. Thus, whereas the smallest strain rates in narrow plate boundaries are 4 orders of magnitude larger than the upper bound obtained for strain rate in stable plate interiors, the largest strain rates in narrow plate boundaries are nearly 9 orders of magnitude larger than the lower bound on strain rates in stable plate interiors. This enormous variation in regionally averaged strain rates (Figure 4) is one of many challenges to generating plate tectonics from models of mantle convection [cf., *Bercovici*, 1993; *Bercovici et al.*, 2000; *Gurnis et al.*, 2000; *Tackley*, 2000].

---

**Figure 1.** Map showing idealized narrow plate boundaries, velocities between plates, and regions of deforming lithosphere, which can be regarded as diffuse or wide plate boundaries. Plate velocities are shown by arrows. The length of the arrows shows what the displacement would be if the plates were to maintain their present relative angular velocity for 25 million years. The plate separation rate across mid-ocean ridges is shown by symmetrical diverging arrows with unclosed arrowheads at both ends. The plate convergence rate is shown by asymmetrical arrows with one solid arrowhead, which are shown on the underthrust plate where convergence is asymmetric and the polarity is known. Each convergence arrow points toward the overthrust plate. The outlines of deforming regions are approximate and the existence of some deforming zones is speculative. Separate small plates or blocks are labeled in southeast Asia, but their uncertainly located and possibly nonexistent boundaries are not shown in their entirety. Deformation has been inferred from seismicity, topography, other evidence of faulting, and nonclosure of plate motion circuits. Most deep sea trench boundaries are treated as if they have a narrow element (i.e., the thrust fault on top of the subducting plate) and a diffuse component in which distributed deformation occurs in the leading edge of the overriding plate. These deforming regions, which constitute diffuse plate boundaries, cover about 15% of Earth's surface. Future observations may demonstrate that deforming lithosphere covers an area larger or smaller than shown here. Idealized plate boundaries (solid curves) are after *Argus and Gordon* [1991] with the addition of the Scotia plate and several other minor changes. Plate abbreviations: B, Borneo; AN, Antarctica; AR, Arabia; AU, Australia; CA, Caribbean; CAP, Capricorn; CL, Caroline; CO, Cocos; EU, Eurasia; I, Indo-China; IN, India; JF, Juan de Fuca; NA, North America; NB, Nubia; NC, North China; NZ, Nazca; OK, Okhotsk; PA, Pacific; SA, South America; SC, Scotia Sea; SM, Somalia; Y, Yangtze. Four small open circles show the location of the poles of rotation for North America-South America, Nubia-Somalia, India-Capricorn, and Capricorn-Australia relative motion. These four poles of rotation are convergent-divergent pivots (see text), which lie in the middle of the diffuse plate boundaries that separate their respective plate pairs. Modified with permission from *Gordon and Stein* [1992], *Gordon* [1995], and *Gordon* [1998]. Copyright 1992 American Association for the Advancement of Science. Copyright 1995 American Geophysical Union. Copyright 1998 Annual Reviews.

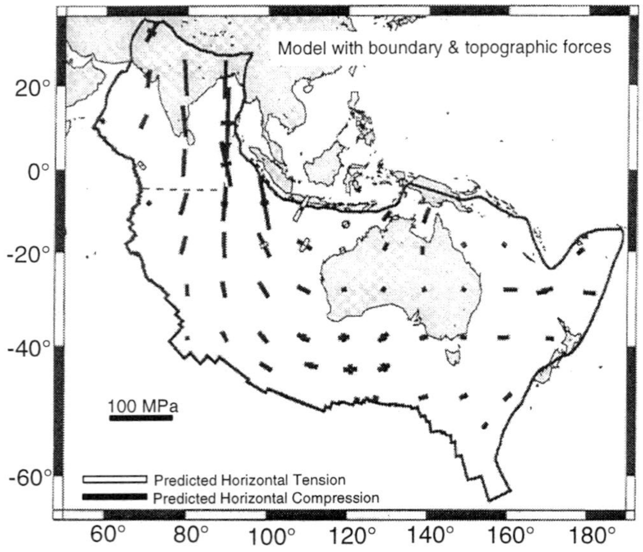


**Figure 2.** Predicted stresses in the Indo-Australian composite plate from one of several models analyzed by *Coblentz et al.* [1998]. Solid and open bars indicate nonlithostatic compression and tension, respectively. Boundary forces were applied along ridges and along collisional and subduction boundaries. Topographic forces were included. Basal drag was used to balance the torque acting on the plate. Maximum compressive stresses, which occur in the Central Indian Basin, approach 110 MPa assuming a nominal plate thickness of 100 km. Modest ridge-parallel tension is predicted near part of the Central Indian Ridge. The model matches many broad details of the observed pattern of intraplate stress, but also mismatches many details. Modified from *Coblentz et al.* [1998]. Copyright 1998 American Geophysical Union.

### *Diffuse Oceanic Plate Boundaries*

Strain rates across diffuse oceanic plate boundaries are probably largest where the largest relative plate speeds are accommodated across them. For example, across the eastern part of the boundary between the Indian and Capricorn plates, a speed of about 10 mm/yr is accommodated across a zone about 1000 km wide, which indicates an average strain rate in this region of $\approx 3\times10^{-16}$ $s^{-1}$. Calculations across other diffuse plate boundaries in regions far from the relevant pole of rotation indicates strain rates of $3\times10^{-17}$ to $3\times10^{-16}$ $s^{-1}$. Thus the largest strain rates across diffuse oceanic plate boundaries are about 30 times larger than the upper bound on strain rates of stable plate interiors, and about 700 times smaller than the smallest strain rates across oceanic transform fault zones.

### *Discussion*

The largest strain rates across diffuse oceanic plate boundaries are between those for stable plate interiors and those for narrow plate boundaries. The largest strain rates in diffuse continental plate boundaries [*England and Molnar*, 1997; *Davies et al.*, 1997] are perhaps a factor of 10 or so bigger than the largest strain rates in diffuse oceanic plate boundaries, but still nearly two orders of magnitude smaller than strain rates averaged across oceanic transform fault zones (Figure 4). Although there is a clear gap between strain rates obtained for diffuse oceanic plate boundaries and for narrow plate boundaries, the range of strain rates in diffuse oceanic plate boundaries must overlap those of stable plate interiors. This is the case because strain rates across diffuse oceanic plate boundaries are vanishingly small near the poles of rotation, which tend to lie in the diffuse plate boundary itself.

## EFFECTIVE VISCOSITIES

### *Rheology of Deforming Lithosphere*

Because plate interiors and diffuse plate boundaries undergo permanent deformation, as manifested for example by intraplate earthquakes, herein I treat the lithosphere as inelastic. Given the existence of permanent deformation, there are two main alternative approaches. First, one can treat the lithosphere of a single deforming zone as composed of many fault-bounded elastic blocks between which small velocities are accommodated. Second, one can consider the lithosphere to be a continuously deforming medium [*Molnar and Tapponnier*, 1975; *England and McKenzie*, 1982]. This second alternative seems a preferable approximation for diffuse oceanic plate boundaries where the faults and fault-bounded blocks can number in the hundreds or more and the spacing between the faults is generally small compared with the thickness of the lithosphere.

Whereas the upper part of the lithosphere deforms by brittle faulting, the lower part of the lithosphere deforms by high-temperature creep. The deformation of the upper lithosphere, when averaged over time intervals longer than earthquake cycles, behaves in effect as a plastic rheology, i.e., it fails at a constant yield stress [*Martinod and Davy*, 1992; *Molnar et al.*,

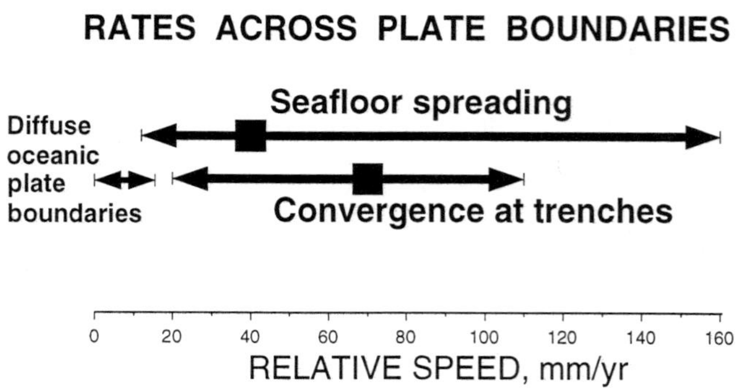


**Figure 3.** Comparison of the range of maximum speeds accommodated across seven diffuse plate boundaries (Capricorn-Australia, North America-South America, Nubia-Somalia, Nubia-Eurasia, India-Capricorn, Sierra Nevada-North America, and India-Australia) with the range of spreading rates and convergence rates across narrow plate boundaries. Redrawn from *Gordon* [1998].

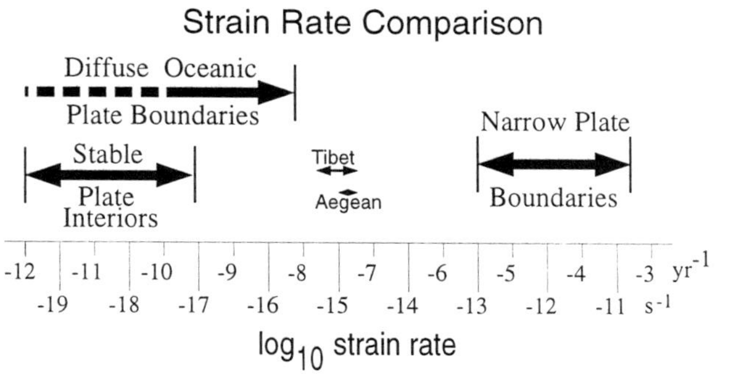


**Figure 4.** Comparison of strain rates in several tectonic settings discussed in the text. The range of strain rates labeled "Narrow plate boundaries" is inferred from relative plate speeds across, and widths of, oceanic transform fault zones offsetting midocean ridges. The range of strain rates labeled "Stable plate interiors" are upper and lower limits based on seismic moment release and space geodesy. The upper limit of strain rates shown for "Diffuse plate boundaries" comes from observed widths and maximum displacement rates across diffuse oceanic plate boundaries. Strain rates for two continental diffuse plate boundaries, the Tibetan Plateau [*England and Molnar*, 1997] and the Aegean [*Davies et al.*, 1997], are shown for comparison.

1993]. The lower lithosphere, if it deforms by movement of dislocations in crystals as indicated by laboratory experiments [e.g., *Weertman*, 1970; *Brace and Kohlstedt*, 1980; *Kirby*, 1983], behaves as a power-law fluid, for which $\dot{\varepsilon}_s \propto (\tau_s)^n$ where $n \approx 3$, $\dot{\varepsilon}_s$ is the rate of shear strain, and $\tau_s$ is the shear stress. These two layers operate in parallel to determine the overall resistance of the lithosphere to deformation (Figure 5).

In many prior papers the vertically averaged rheology of the entire continental lithosphere has been approximated by a thin sheet of viscous fluid [*Bird and Piper*, 1980; *England and McKenzie*, 1982; *Vilotte et al.*, 1982]. In one approach, the lithosphere is treated as a power-law fluid obeying the following constitutive equation

$$\overline{\tau}_{ij} = B\dot{E}^{(\frac{1}{n}-1)}\dot{\varepsilon}_{ij} \qquad (1)$$

where $\overline{\tau}_{ij}$ is the $ij^{th}$ component of vertically averaged deviatoric stress, $\dot{\varepsilon}_{ij}$ is the $ij^{th}$ component of the strain rate, assumed independent of depth, and $B$ is a constant [*England and McKenzie*, 1982]. $\dot{E}$ is the second invariant of the strain-rate tensor and is equal to $(\dot{\varepsilon}_{ij}\dot{\varepsilon}_{ij})^{1/2}$ with the convention of summation over repeated indices. Equation (1) expresses the rheology of a power-law fluid, with $n$ being the power-law exponent, and the expression $\frac{1}{2}B\dot{E}^{(1/n\ -1)}$ giving the effective viscosity, which decreases with increasing magnitude of the strain rate, a characteristic usually referred to as shear thinning.

Equation (1) can represent a wide range of flow laws from that of a Newtonian fluid when $n=1$ to a perfectly plastic material as $n \rightarrow \infty$. *Sonder and England* [1986] showed that this equation, with suitable adjustments to the values of $n$ and $B$, well approximates the vertical average of the rheology of continental lithosphere failing by slip on faults in its upper crust and by a combination of creep and brittle failure in the lower lithosphere, according to the strain rate and temperature there. The exponent $n$ represents no individual deformation mechanism, but instead represents the vertical averaging of deformation involving slip on faults with deformation by high-temperature creep [*Sonder and England*, 1986]. The success of this approximation does not reflect any fundamental physical process, but is merely a consequence of the ease of fitting a straight line through a plot of the logarithm of one quantity versus the logarithm of another [*England and Molnar*, 1991]. Here I adopt a similar approach, not for deforming continental lithosphere, but for deforming oceanic lithosphere.

## *A Qualitative Rheologic Model for Oceanic Lithosphere*

The simple rheologic model for oceanic lithosphere used here is that of an upper plastic layer overlying a layer of power-law fluid with $n=3$ (Figure 5). At the depth in the lithosphere at which the transition occurs from brittle deformation to creeping flow, the stress required to deform by either mechanism is equal (Figure 6). The relative contributions of the two layers to the resistance to deformation of the lithosphere depends on the yield stress as a function of depth of the upper layer and on the stress as a function of depth required to deform the lower layer at a given strain rate (Figure 6).

If the power-law approximation for the vertically averaged rheology of the entire lithosphere is valid, a log-log plot of strain rate versus force per unit length should approximate straight lines, as appears to be the case in Figure 7, which shows results for this simple model for strain rates from $10^{-17}\,s^{-1}$ to $10^{-15}\,s^{-1}$, which spans the strain rates of interest in diffuse oceanic plate boundaries. The slopes vary systematically

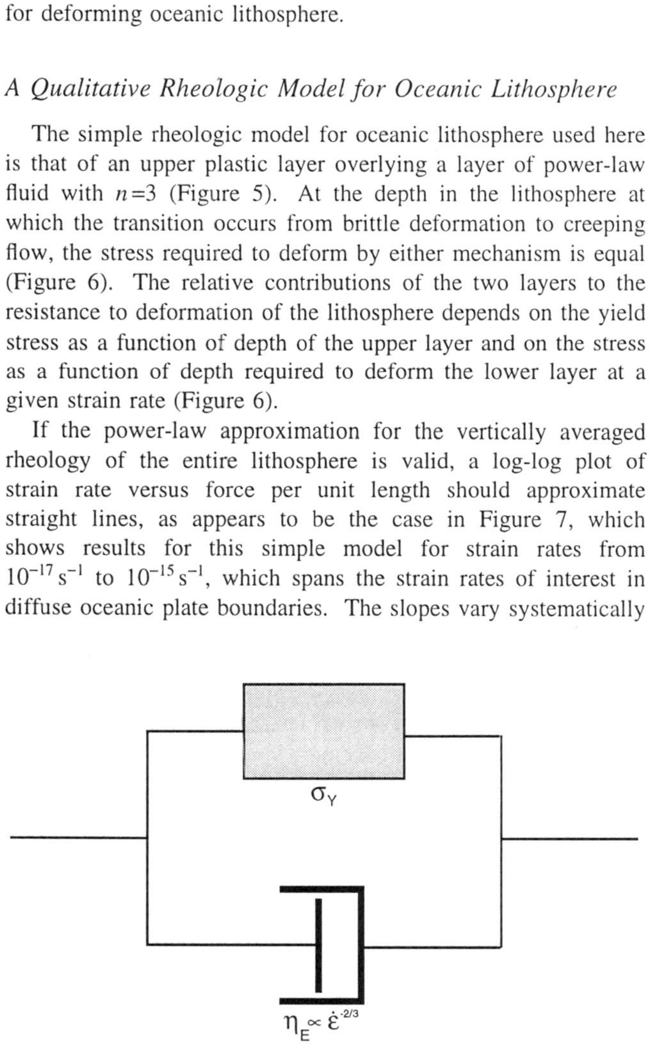


**Figure 5.** Simple conceptual model for the rheology of the lithosphere, that of a plastic medium characterized by a yield stress, $\sigma_Y$, acting in parallel with a dash-pot in a power-law fluid with an exponent of about 3 [*Martinod and Davy*, 1992; *Martinod and Molnar*, 1995]. In oceanic lithosphere, the plastic medium represents the upper lithosphere characterized by brittle or semi-brittle failure [e.g., *Kohlstedt et al.*, 1995] whereas the dash-pot represents the lower lithosphere, which deforms by creeping flow.

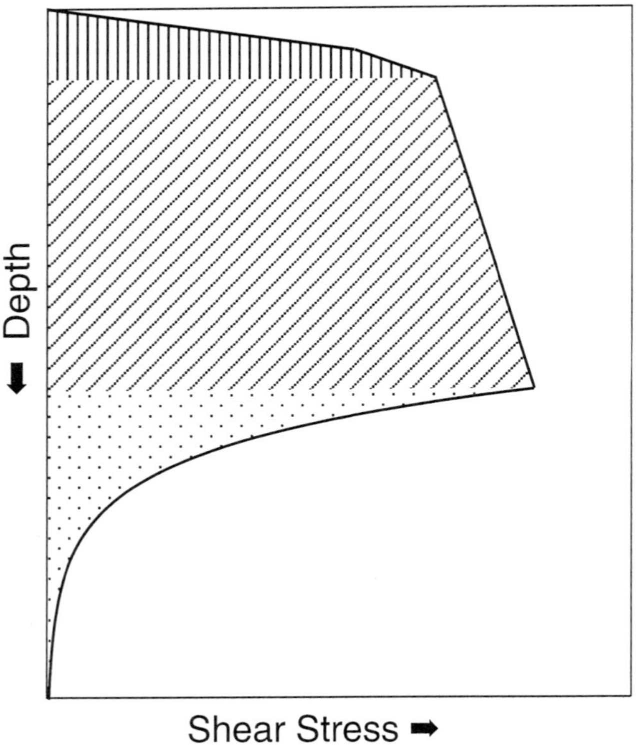


**Figure 6.** Qualitative strength envelope for oceanic lithosphere, redrawn from quantitative strength envelope of *Kohlstedt et al.* [1995]. Uppermost oceanic lithosphere has brittle rheology (vertical lines). Below that is a thick layer characterized by semi-brittle failure (diagonal lines) [*Kohlstedt et al.*, 1995]. Lowest portion of the lithosphere deforms by creeping flow (stippled). The brittle and semi-brittle layers together constitute the upper lithosphere effectively characterized by perfectly plastic deformation over time intervals longer than the earthquake cycle [*Martinod and Davy*, 1992; *Martinod and Molnar*, 1995].

with the ratio of the yield strength of the upper plastic layer to the force required to deform the lower layer of power-law fluid at a reference strain rate of $10^{-16}$ $s^{-1}$ (Figure 7).

Figure 8 displays the results for the same models but over a range of strain rates varying by a factor of $10^{10}$, which is the same range of strain rates as those observed in the lithosphere (Figure 4). Figure 8 shows that the power-law approximation cannot be extrapolated indefinitely to higher or lower strain rates and deforming forces. When the deforming force is less than the strength of the upper lithosphere, deformation essentially ceases, indicating that plate interiors are essentially rigid when less than the critical force is applied.

*Stable Plate Interiors*

If one has an estimate of both strain rate and deviatoric stress, the effective viscosity can be estimated from the ratio of the two. Strain rates were discussed above. To estimate the shear stresses, I assume that the force per unit length in stable plate interiors is between the ridge push force, $\approx 3\times10^{12}$ N m$^{-1}$ [*Richardson*, 1992], and the force that the Tibetan Plateau exerts on adjacent lithosphere, $8\pm2\times10^{12}$ N m$^{-1}$ [*England and Houseman*, 1989; *Molnar et al.*, 1993]. Assuming a nominal 100-km thickness of the lithosphere, this gives an vertically averaged shear stress of 15–50 MPa. Combining the upper value for shear stress with the lower end of values for seismic strain rate gives an upper bound on effective viscosity of $1.7\times10^{27}$ Pa s. Similarly combining the lower end of the range of shear stresses with the upper bound on strain rate gives a lower bound on average effective viscosity of about $1.5\times10^{24}$ Pa s. If the lithosphere of stable plate interiors is thicker or thinner than 100 km, the average effective viscosity is proportionally lower or higher. The relaxation time of the lithosphere has been estimated to be tens of millions of years or longer [*Beaumont*, 1978; 1981; *Watts et al.*, 1982], which indicates that the effective viscosity of the lithosphere in a vertically averaged sense is at least $10^{25}$ Pa s [*Ranalli*, 1995], which is consistent with the estimate above.

*Oceanic Transform Fault*

Assuming values for vertically averaged shear stress across an oceanic transform fault zone of 1 to 10 MPa, the effective viscosity is $3\times10^{16}$ to $5\times10^{19}$ Pa s, about 8 orders of magnitude smaller than that for stable plate interiors. If the transform fault zone is wider at depth, the strain rates would be smaller, and the estimated effective viscosity would be larger. On the other hand, if the transform fault zone is narrower at depth, the strain rates would be larger, and the estimated effective viscosity would be even smaller. In any event, the high strain rates across transform fault zones imply that the effective viscosity, at least for the upper lithosphere, is small.

*Diffuse Oceanic Plate Boundaries*

The best example of diffuse oceanic plate boundaries comes from the Indo-Australian composite plate. Models for intraplate stresses vary widely in the estimated level of stress that drives this deformation. On the one hand, *Cloetingh and Wortel* [1985] found average stresses above the lithostatic stress of 300 to 500 MPa. On the other hand, *Coblentz et al.* [1998] present some models in which the stresses above the lithostatic stress are as low as a few tens of MPa. Here I focus on independent means of calibration for these stresses.

*Martinod and Molnar* [1995] use the presence of lithospheric folds in the Central Indian Basin, where sediments fill the topographic depressions, to show that the force per unit length deforming the upper lithosphere must be at least $4.8(\pm1.3)\times10^{12}$ N m$^{-1}$, giving a lower bound of $3.5\times10^{12}$ N m$^{-1}$ for this force. They also show that a force per unit length of $1.1(\pm0.3)\times10^{13}$ N m$^{-1}$ would be required to form folds in regions free of sediment. They note that the folds die away to the south in the Central Indian Basin, as do the sediments. If this is not a coincidence, this implies that the force per unit length deforming the upper lithosphere must be less than $1.1(\pm0.3)\times10^{13}$ N m$^{-1}$

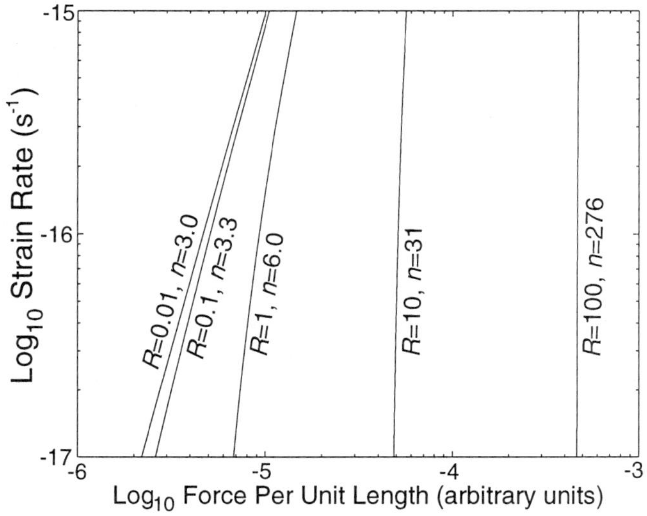


**Figure 7.** Strain rate versus force per unit length for simple rheological model for the lithosphere shown in Figure 6. "$R$" is the ratio of the yield strength of the plastic layer (upper lithosphere) to the force required to deform the fluid layer (lower lithosphere) at a reference strain of $10^{-16}\,\mathrm{s}^{-1}$. The fluid layer is assumed to obey a power-law rheology, i.e., $\dot{\varepsilon}_s \propto (\tau_s)^n$ where $n$=3, $\dot{\varepsilon}_s$ is the rate of shear strain, and $\tau_s$ is shear stress. The slope of each curve, which corresponds to the appropriate value of $n$ for the power-law rheology of vertically averaged oceanic lithosphere, is listed. The slopes vary systematically with $R$. If $R$ is 0.1 or 0.01, the overall behavior resembles that of a power-law fluid with n ≈ 3. If $R$ is 1, the vertically averaged rheology is approximated by a power-law fluid with n ≈ 6. If $R$ is 10 or 100, the vertically averaged rheology is approximated by highly non-Newtonian, power-law fluid.

in the Central Indian Basin, which gives an upper bound of $1.4\times10^{13}\,\mathrm{N\,m^{-1}}$.

Here I interpret these bounds differently for two reasons. First, as discussed in a section below, the strain rate (and therefore the vertical average of stress) diminishes from north to south in the Central Indian Basin. Thus, folds may be absent to the south merely because the applied force per unit length decreases to the south. Second, unlike *Martinod and Molnar* [1995], I am convinced by the evidence that lithospheric folding occurs in the Wharton Basin [*Petroy and Wiens*, 1989; *Stein et al.*, 1989; *Royer and Gordon*, 1997; J.-Y. Royer and R. G. Gordon, manuscript in preparation], where sediments are thin. Thus, at least in the Wharton Basin, the yield strength of the upper lithosphere is at least $1.1(\pm0.3)\times10^{13}\,\mathrm{N\,m^{-1}}$. I know of no reason why the lithosphere of the Central Indian Basin should be significantly weaker than that of the Wharton Basin. I thus take $8\times10^{12}\,\mathrm{N\,m^{-1}}$ to be a lower bound on the strength of the lithosphere in the Central Indian Basin. This lower bound is not significantly greater than the force per unit length currently exerted by the Tibetan Plateau on the Indo-Australian composite plate of $8\pm2\times10^{12}\,\mathrm{N\,m^{-1}}$ [*England and Houseman*, 1989; *Molnar et al.*, 1993]. Thus, the different interpretation I make is not inconsistent with the hypothesis of *Molnar et al.* [1993] and *Martinod and Molnar* [1995] that the rise of the Tibetan Plateau at about 8 Ma caused the episode of faulting and lithospheric folding in the Central Indian Basin that continues today.

To estimate an approximate upper bound on the strength of oceanic lithosphere, I use published strength envelopes for 60-Myr-old oceanic lithosphere based on laboratory experiments on dry olivine [*Kohlstedt et al.*, 1995]. Integration of their strength curve for oceanic lithosphere gives strength for the upper lithosphere of $\approx2.3\times10^{13}\,\mathrm{N\,m^{-1}}$, about twice as large as *Martinod and Molnar*'s [1995] upper bound.

If the lithosphere is deforming at a high enough strain rate, the force required to deform the lower lithosphere by creeping flow cannot be neglected relative to the force required to deform the upper lithosphere. For the rheological profile for oceanic lithosphere shown by *Kohlstedt et al.* [1995] (cf., Figure 7), simple calculations show that the force per unit length required to deform the lower lithosphere, expressed as a fraction of the strength of the upper lithosphere, is $\approx1.4\,h_0/z_0$ times the strength of the upper lithosphere, where $h_0$ is the change in depth over which the effective viscosity of the lower lithosphere decreases by 1/e and $z_0$ is the thickness of the upper lithosphere. *Martinod and Molnar* [1995] find that $h_0/z_0$ is 0.16 (±0.09). Adding the upper limit of the force to deform the lower lithosphere to the upper bound strength of the upper lithosphere gives an upper bound force per unit length of $3.4\times10^{13}\,\mathrm{N\,m^{-1}}$ required to deform the entire lithosphere. Adding the lower limit of the force to deform the lower lithosphere to the lower

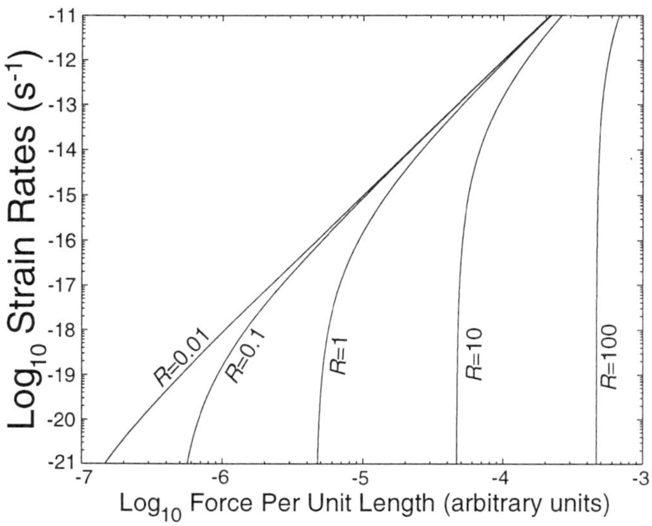


**Figure 8.** Strain rate versus force per unit length for simple rheological model for the lithosphere shown in Figure 6. "$R$" is the ratio of the assumed yield strength of the perfectly plastic layer to the force required to deform the power-law fluid with $n$=3 at a reference strain of $10^{-15}\,\mathrm{s}^{-1}$. This figure is the same as Figure 8, except that here 10 orders of magnitude of strain rate are shown versus 3 orders of magnitude of strain rate in Figure 8.

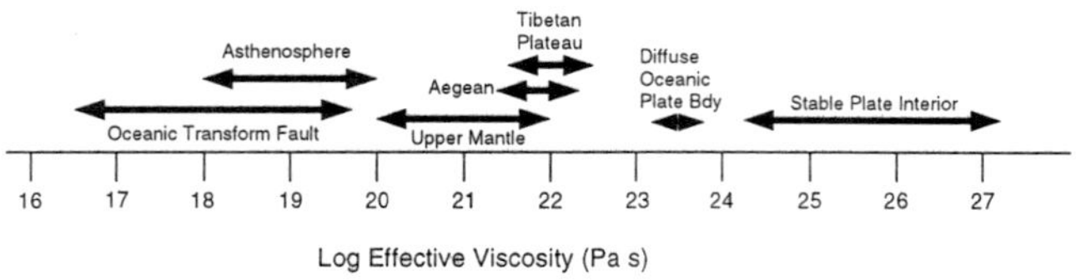


**Figure 9.** Effective viscosities of the vertically averaged lithosphere in various tectonic settings compared with limits on the effective viscosity of the asthenosphere and the upper mantle beneath the asthenosphere. The effective viscosities for oceanic transform fault zones, diffuse oceanic plate boundaries, and North America stable plate interior are inferred from the limits on strain rates (Figure 4) and estimates of vertically averaged deviatoric stresses. The estimates for the diffuse oceanic plate boundaries are applicable to the regions with the largest strain rates. Effective viscosity of the Tibetan plateau was estimated by *England and Molnar* [1997] and that for the Aegean by *Davies et al.* [1997].

bound given above for the strength of the upper lithosphere gives a lower bound force per unit length of $8.8\times10^{12}$ N m$^{-1}$.

Assuming a nominal 100-km thickness of the lithosphere gives vertically averaged shear stresses of 44 MPa to 170 MPa. At a strain rate of $3\times10^{-16}$ s$^{-1}$, the effective viscosity is $1.5\times10^{23}$ Pa s for 44 MPa and $5.7\times10^{23}$ Pa s for 170 MPa.

*Discussion*

Figure 9 compares the vertically averaged effective viscosity of the lithosphere in various tectonic settings. The inferred effective viscosity for the diffuse oceanic plate boundary in the Central Indian Basin is about 10 times larger than that for the diffuse continental plate boundaries in Tibet [*England and Molnar*, 1997] and in the Aegean [*Davies et al.*, 1997]. This result is a consequence of assuming deviatoric stresses in the Central Indian Basin that are about the same as, or slightly larger than, those that drive deformation in Tibet and in the Aegean while the strain rates in the Central Indian Basin are about a factor of 10 smaller than in the continental settings. As a consequence of shear thinning, both continental and diffuse oceanic plate boundaries have effective viscosities smaller than those of stable plate interiors.

Some results presented in Figure 9 may have implications for efforts to generate plate tectonics in mantle convection models. For example, the effective viscosity of oceanic transform fault zones is at least 5 orders of magnitude less than that for stable plate interiors and indistinguishable from that estimated for the asthenosphere. This is inconsistent with the assumption made in some models for plate generation from mantle flow that the viscosity of the asthenosphere is negligible in comparison with that in narrow plate boundaries [*Bercovici*, 1993; 1995].

Moreover, the effective viscosity of transform fault zones is at least 2 orders of magnitude less than that of diffuse continental plate boundaries and at least 3 orders of magnitude less than that of diffuse oceanic plate boundaries (Figure 5). *Bercovici* [1993] has argued that it is not possible to generate narrow plate boundaries with a power-law rheology like that used here for diffuse oceanic plate boundaries. To be able to generate narrow plate boundaries from mantle flow successfully, *Bercovici* [1993] has instead hypothesized that the lithosphere has a self-lubricating rheology. In a shear-thinning rheology, although effective viscosity decreases with increasing strain rate, the stress needed to deform the fluid increases monotonically with increasing strain rate. In contrast, a self-lubricating rheology is strain-rate weakening—above a critical value of strain rate, stresses decreases with increasing strain rate.

One self-lubricating rheology is described by the following constitutive law

$$\tau \propto \frac{\dot{\varepsilon}}{1+\dot{\varepsilon}^2/\gamma^2} \qquad (2)$$

where $\dot{\varepsilon}$ is strain rate and $\gamma$ is the strain rate at which the deviatoric stress $\tau$ is greatest [*Whitehead and Gans*, 1974; *Bercovici*, 1993; *Tackley*, 1998; 2000; *Bercovici et al.*, 2000]. In this self-lubricating rheology, the fluid acts like a highly viscous power-law fluid at strain rates less than $\gamma$, but the stresses decrease with increasing strain rate for strain rates larger than $\gamma$ [*Bercovici*, 1993; also see figure 12 of *Bercovici et al.*, 2000]. The observed gap in strain rates between the largest strain rate in a diffuse oceanic plate boundary ($3\times10^{-16}$ s$^{-1}$) and the smallest strain rate in a narrow plate boundary ($2\times10^{-13}$ s$^{-1}$) (Figure 4) may define this transition in lithospheric behavior between a power-law-like fluid at lower strain rates and a strain-rate-weakening fluid at higher strain rates. If so, $\gamma$ must lie between $3\times10^{-16}$ s$^{-1}$ and $2\times10^{-13}$ s$^{-1}$, but perhaps closer to the lower value, because once the critical strain rate is exceeded, self-lubrication should lead to strain-rate weakening followed by strain localization and potentially additionally increased strain rates.

## ACROSS-STRIKE WIDTHS OF DIFFUSE OCEANIC PLATE BOUNDARIES

*Exponent and Aspect Ratios of Deforming Zones*

The across-strike width of a deforming zone in a thin sheet of power-law fluid depends on the exponent, $n$, of the power law [*England and McKenzie*, 1982]. *England et al.* [1985] showed that deformation associated with a boundary of length $L$ should diminish exponentially with distance from that boundary over a characteristic distance that is proportional to $L$ and inversely proportional to $n^{1/2}$. *England et al.* [1985] show for a converging or diverging boundary that the across-strike width, $w$, at which the strain rate falls to 1/e of its maximum is given by

$$w = 4L/(\pi n^{1/2}) \qquad (3)$$

where $L$ is the along-strike length of a deforming zone accommodating convergence or divergence in lithosphere with a vertically averaged rheology given by equation (1). For a boundary accommodating simple shearing, $w$ is a factor of 4 smaller [*England et al.*, 1985]. Equation (3) implies that if $n\approx3$, as expected if the force required to deform the lower lithosphere was much larger than the yield strength of the upper lithosphere, the 1/e width of the deforming zone should be about three-fourths of its length. In contrast, if $n=100$, in which case the yield strength of the upper lithosphere is much larger than the force required to deform the lower lithosphere, then the 1/e width would only be about one-eighth the length. Thus the map-view aspect ratios may contain information on the relative sizes of the resistance to deformation of the upper and lower lithosphere in diffuse oceanic plate boundaries.

Equation (3) is subject to the caveat that buoyancy forces caused by the deformation are negligible. The main source of buoyancy forces caused by the deformation in continental lithosphere is the resulting variation in the thickness of continental crust, which is of course absent in oceanic lithosphere. Thus it may be reasonable to neglect buoyancy forces due to the deformation when examining the widths and lengths of oceanic deforming zones, as is done below, especially since the strains may be no more than about 5%.

*England and Molnar* [1991] examined the lengths and widths of continental deforming zones with a goal of estimating $n$ for continental lithosphere. Because of the strong buoyancy forces in continental zones of divergence or convergence they restricted their analysis to zones of transcurrent deformation, concentrating on the Pacific-North America plate boundary in California, the Pacific-Australia boundary in the South Island of New Zealand, and in Oregon and Washington in the U.S. Pacific northwest. They found approximate values for $n$ of 3, 3, and 4.5, respectively, indicating that viscous deformation by high-temperature creep is much more important in determining the resistance to deformation of the lithosphere in those regions than is friction on faults. Their results are perhaps analogous to the curves labeled $R=0.01$ or $R=0.1$ in Figure 7. Here I attempt a similar analysis for diffuse oceanic plate boundaries.

### *Symmetric versus Asymmetric Deformation*

In the analysis of a continental deforming zone, the deformation is usually assumed to be asymmetric or one-sided with a moving rigid boundary along one edge of the deforming zone or diffuse plate boundary. The resulting solutions for deformation in a thin sheet of power-law fluid indicate that the deformation rate is largest immediately adjacent to the moving rigid boundary and that the rate decays exponentially away from the boundary (Figure 10b). In the case of Pacific-North America interaction in California, the oceanic Pacific plate may act as the rigid moving boundary while continental lithosphere in California and adjacent Arizona deform [*England and Molnar*, 1991]. In Oregon and Washington, the oceanic Juan de Fuca plate acts as the rigid moving boundary while continental lithosphere in Oregon and Washington deform [*England and Wells*, 1991].

For diffuse oceanic plate boundaries, the appropriate boundary conditions are unclear. It is also unclear whether the deformation is one-sided, two-sided, or in between. Figure 10 illustrates two idealized possibilities. Figure 10a shows a hypothetical deformation profile without rigid boundaries. For this profile it is assumed that deformation rate reaches a maximum near the center of the zone and smoothly decays with distance away from the center of the zone. Figure 10b shows the one-sided deformation rate and displacement rate corresponding to a moving rigid boundary along one edge of a deforming zone, as assumed by *England et al.* [1985].

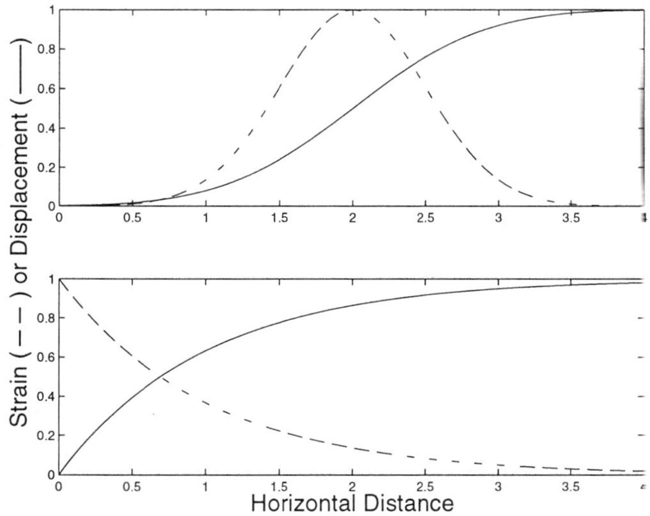


**Figure 10.** Two hypothetical deformation profiles across a diffuse oceanic plate boundary. Dashed curves show strain (or strain rate) whereas solid curves show cumulative displacement (or velocity). (a) Hypothetical symmetric deformation profile with largest strain rate in its center. (b) Deformation profile expected for the response of a thin sheet of power-law fluid to a rigid moving boundary on its left end.

Nearly all diffuse oceanic plate boundaries lack useful data for estimating the spatial variation in deformation, deformation rate, displacement, or displacement rate. The single exception is the convergent portion of the India-Capricorn boundary [*Chamot-Rooke et al.*, 1993; *Van Orman et al.*, 1995].

### *India-Capricorn Convergence Zone*

The converging portion of the India-Capricorn diffuse oceanic plate boundary is blanketed with sediments of the Bengal Fan, which provides a medium in which the deformation can be imaged. Two long north-south seismic reflection profiles traverse all the deforming zone in the Central Indian basin [*Chamot-Rooke et al.*, 1993; *Van Orman et al.*, 1995]. The longer of these profiles is 2100-km long, extends from 14°S to the coast of Sri Lanka, strikes 3° CW of north, and was obtained along 81.5°E during the 1991 Phèdre cruise on the French research vessel *Marion Dufresne* [*Chamot-Rooke et al.*, 1993]. Deformation was observed over a zone 900 km in N-S extent, from just north of the equator to 8°S, an interval which contains 134 faults or folds identified in the sediments and assumed to connect with crustal faults. The average spacing between faults is about 7 km. The horizontal offset on the faults could not be resolved from the seismic profile. Instead

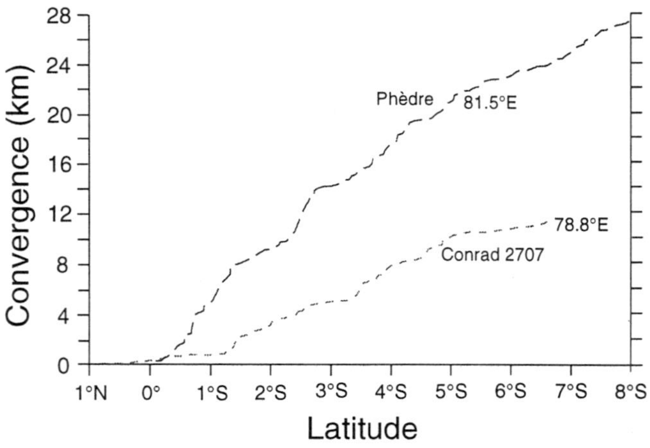


**Figure 11.** Cumulative north-south convergence on two north-south profiles through the India-Capricorn convergent plate boundary: Phèdre, medium-dashed curve [*Chamot-Rooke et al.*, 1993]; *Conrad* 2707, short-dashed curve [*Van Orman et al.*, 1995]. Neither profile becomes horizontal at the southern end, which suggests that not all the deformation has been successfully determined from the seismic profiles. *Chamot-Rooke et al.* [1993] noted that the southern limit of the boundary could not be strictly defined because the sediment thickness is almost negligible south of 8°S. They suggest that the contribution from faults farther south does not exceed a few kilometers. Redrawn from *Van Orman et al.* [1995].

the throw or vertical offset was estimated for each fault, with the largest throws being about 500 m, except for a single throw of about 1100 m. Throws smaller than about 50 m could not be resolved. The total throw was combined with an assumed range of the mean dip of the crustal faults [*Bull and Scrutton*, 1990; 1992] to estimate a total convergence accommodated by shortening across the thrust faults of 27.4 ± 5 km [*Chamot-Rooke et al.*, 1993].

Deformation is not symmetric with respect to the center of the deformation zone [*Chamot-Rooke et al.*, 1993]. Much of the deformation is concentrated in the northern part of the profile just south of the equator. A sharp southern boundary cannot be defined because the sediment thickness is almost negligible south of 8°S; *Chamot-Rooke et al.* [1993] conclude that no more than a few kilometers of convergence is accommodated south of 8°S. Thus, from south to north there is a gradual increase in deformation with a sharp drop in deformation near the equator. The displacement versus distance for this profile (Figure 11) more closely resembles the one-sided profile of Figure 10b than it does the symmetric profile of Figure 10a.

*Van Orman et al.* [1995] present the other north-south profile that completely crosses the deformation zone. Their single-channel seismic profile, from R.V. *Robert D. Conrad* cruise 2707, is along 78.8°E, 300 km west of the Phèdre profile. They found that the zone of north-south shortening to be 823 km in N-S extent along this profile with a total of 127 thrust faults between 0.8°N and 6.6°S. The average spacing between faults is 6.5 km. As was true for the Phèdre profile, the horizontal offset on the faults could not be resolved from the seismic profile. Instead the throw or vertical offset was estimated for each fault, with a range from 9 to 623 m with a mean throw of 73.3 m. It was not possible to resolve throws smaller than about 10 m. The total observed throw was combined with an assumed dip of the crustal faults of 40° [*Bull and Scrutton*, 1990; 1992] to estimate a total convergence accommodated by shortening across the thrust faults of 11 ± 2 km. In contrast to the Phèdre profile, there is no obvious asymmetry, with the plot of displacement versus distance (Figure 11) being perhaps as much like Figure 10a as like Figure 10b. As *Van Orman et al.* [1995] emphasize, the convergence accommodated across the *Conrad* 2707 profile is much smaller than across the Phèdre profile, but the north-south width of the deforming zone is nearly the same. They also analyze two other *Conrad* profiles at 79.4°E and 81.0°E, which only partly cross the zone of deformation. They show that there is a systematic progression from west to east of increasing throw on the faults from 74 m to 177 m. They further show that the average mean fault spacing is an approximately constant 7 km. It appears that much of the shortening in the Central Indian Basin is accommodated along a select group of re-activated pre-existing normal faults bounding the abyssal hills [*Weissel et al.*, 1980; *Bull and Scrutton*, 1992; *Van Orman et al.*, 1995]. These results indicate that the across-strike (i.e., north-south) width of the deforming zone is not strongly affected by the amount of convergence and are consistent with my assumption that the width of the deforming zone is a function of the along-strike length of the deforming zone and of $n$.

The along-strike length, $L$, of the India-Capricorn convergence zone is about 1000 km, from 77°E to 86°E. The western limit is given by the location of the Indo-Capricorn pole of rotation [*Royer et al.*, 1997]. Deformation obviously continues east of 86°E, but the style of deformation changes from thrust-faulting west of 86°E to strike-slip faulting east of 86°E. Because the region that deforms by thrust faulting conforms to the assumptions made for a thin sheet of power-law fluid, but the region east of 86° does not, I consider the east region to be a separate zone that mainly deforms in horizontal plane strain.

For an exponential decay of deformation with distance, 63% of the displacement occurs where the deformation drops to $1/e$ of its maximum value. Figure 11 indicates that this distance is about 450 km, as measured from the northern limit of deformation, along the Phèdre profile. Substituting these values into equation (3) and solving for $n$ gives $n \approx 8$, larger than those found for continental deforming zones, but small enough that both the strength of the upper lithosphere and the effective viscosity of the lower lithosphere are contributing to the dynamics of the deformation.

The north-south *Conrad* 2707 profile at 78.8°E also spans the entire deforming zone, but the profile may differ from that expected in response to a rigid moving boundary (i.e., Figure 10b). Nonetheless its width ought to be related to $n$, and is probably proportional to $n^{-1/2}$, although the constants of proportionality may be different from those in equation (3). I note

Table 1: Parameters for Map-View Aspect Ratios of Diffuse Oceanic Plate Boundaries

| Plate-Pair | Width, $W$ (km) | Width, $w$ (km) | Length, $L$ (km) | $L/W$ | $n$ |
|---|---|---|---|---|---|
| *60- to 90-Myr old Oceanic Lithosphere* | | | | | |
| India-Capricorn (CIB) | 800[1] | 450 | 1000 | 1.25 | 8 |
| India-Capricorn (Chagos) | 220 | 110 | 280 | 1.27 | 11 |
| Capricorn-Australia | 800[2] | 400 | 1200 | 1.50 | 15 |
| N. America-S. America | 500[4] | 250 | 900 | 1.8 | 21 |
| *Young (Near-Ridge) Oceanic Lithosphere* | | | | | |
| Nubia-Somalia | 1700[3] | 850 | 1400 | 0.82 | 4.4 |
| Capricorn-Australia | 1600[2] | 800 | 1200 | 0.75 | 3.6 |
| N. America-S. America | 1100[4] | 550 | 800 | 0.73 | 3.4 |
| *Continental Lithosphere* | | | | | |
| Nubia-Somalia | 2200[6] | 1100 | 4200 | 1.9 | 24 |

$W$ is the best available estimate of the total across-strike width of a diffuse plate boundary. $w$ is an estimate (or guess) of the across-strike distance in which the deformation rate falls from its maximum to 1/e of its maximum. Only the India-Capricorn boundary in the Central Indian Basin (CIB) has data that allow $w$ to be estimated. In the other cases $w$ is merely guessed to be half of $W$. $L$ is measured along the strike of the boundary. $n$, the estimated exponent of the effective power law, is calculated from the formula $n=(4L/\pi w)^2$ [*England et al.*, 1985]. References: (1) *Chamot-Rooke et al.* [1993]; *Van Orman et al.* [1995]. (2) *Royer and Gordon* [1997]. (3) *Chu and Gordon* [1999]. (4) *Argus* [1990]. (5) *Royer et al.* [1997]. (6) Estimated from figure 2 of *Chu and Gordon* [1999].

that 63% of the displacement occurs in a zone about 350 km wide (between offset 220 km and 570 km in Figure 11), which includes the part of the profile with the highest deformation rate. If one were to use this value, instead of 450 km, $n \approx 13$, illustrating the high sensitivity of $n$ to the estimated width of the deforming zone.

*Comparison with Map-View Aspect Ratios of Other Diffuse Oceanic Plate Boundaries*

The other diffuse plate boundaries lack data for estimating the distribution of displacement or strain across them. It is possible, however, to estimate their overall along-strike length and across-strike width (Table 1). First, the India-Capricorn convergence zone, discussed above, has a length of 1000 km and a total (not 1/e) width of about 800 km, giving an aspect ratio of 1.25. Three other deformation zones occur in oceanic lithosphere not immediately adjacent to a ridge. The first is the convergent zone between the Capricorn and Australian plates. From figure 2 of *Royer and Gordon* [1997], I estimate the width of the zone to be 800 km, and its length to be 1200 km, giving an aspect ratio of 1.5. The second is the zone of convergence between the North American and South American plates. I take the length of this zone to be the distance from the pole of rotation to the Lesser Antilles trench, about 900 km (figure 11 of *Gordon* [1998]). The across-strike width of the deforming zone is uncertain; my guess for this distance is 500 km based on a speculative interpretation of the satellite-derived gravity field in the deformation zone. This gives an aspect ratio of about 1.8. The fourth is part of the India-Capricorn zone of divergence near Chagos Bank, where the lithosphere deforms by normal faulting [*Stein*, 1978; *Wiens*, 1986]. The along-strike length, about 280 km, of this part of the deforming zone is small, however, as is the across-strike width, about 220 km. The aspect ratio of 1.3 is uncertain. Thus this group of aspect ratios range from about 1.25 to 1.8—in every case the along-strike length appears to exceed the across-strike width, albeit with large uncertainty. In the absence of additional observations, the 1/e width of each of these zones is unknown, although it must be less than the full width of the zones. For the purpose of making comparisons, I estimate $n$ assuming that the 1/e width of each of these zones is equal to half the full width I have estimated, which indicates that $n$ ranges from 10 to 21 (Table 1).

Another set of four diffuse plate boundaries are terminated by mid-ocean ridges and are thus, in part, in very young oceanic lithosphere. The best constrained across-strike width of a diffuse oceanic plate boundary in young lithosphere is that of the India-Capricorn boundary along the Central Indian Ridge. Reconstructions of magnetic anomaly and fracture zone crossings show that the boundary is about 600 km in north-south extent [*Royer et al.*, 1997]. This plate boundary is problematic, however, because the limited available evidence, mainly earthquake mechanisms, suggests that this region is undergoing horizontal plane strain, so I exclude it. The next best constrained near-ridge diffuse oceanic plate boundary is the Capricorn-Australia divergent boundary. I take its along-strike length to be the distance from the pole of rotation to the Southeast Indian Ridge, about 1200 km (figure 2 of *Royer and Gordon* [1997]). I take its across-strike width to be about 1600 km, as indicated by the distribution of normal faulting earthquakes accommodating extension (perpendicular, not parallel, to the spreading direction) [figure 2 of *Royer and Gordon*, 1997]. This gives an aspect ratio of 0.75. The third zone is the divergent portion of the North America-South America boundary, which lies along the Mid-Atlantic Ridge. Its along-strike length is 900 km, based on its distance from the North America-South America pole of rotation [figure 11 of *Gordon*, 1998]. *Argus* [1990] has interpreted the azimuths of transform faults along the Mid-Atlantic ridge as indicating that the north-south extent of the boundary is very large, perhaps about 1100 km. This distance is speculative because the uncertainty in the estimated azimuths of the transform faults is poorly known. Using this value gives an aspect ratio of 0.82. The fourth zone is the diffuse convergent plate boundary between Nubia and Somalia along the Southwest Indian Ridge. I take the along-strike length of the deforming zone to be the distance from the pole of rotation to the Southwest Indian Ridge, about 1400 km (figure 2 of *Chu and Gordon* [1999]). From the large separation between their best-fitting location for a hypothetical narrow boundary and the location of a thrust-faulting earthquake with a magnitude ($M_w$) of 6.8, *Chu and Gordon* [1999] speculate that the width is about 1700 km. This gives an aspect ratio of 0.82. Thus the aspect ratios for the three zones range from 0.75 to 0.82, all smaller than those for the deforming zones in older oceanic lithosphere, which ranged from 1.25 to 1.8. This suggests that the ratio of the strength of the upper lithosphere to the force required to deform the lower lithosphere is smaller for very young lithosphere than it is for older oceanic lithosphere. I again estimate $n$ assuming that the 1/e width of each of these zones is equal to half the full width I have estimated, which indicates that $n$ ranges from 3.6 to 4.4. These small values of $n$ suggest that the deformation of very young oceanic lithosphere is more strongly controlled by the effective viscosity of the lower lithosphere than by the strength of the upper lithosphere, which is presumably thin in young oceanic lithosphere.

The East Africa Rift obviously does not fit into either category considered above because it occurs in a continent. I estimate the along-strike length of the boundary to be 4200 km (from the Nubia-Somalia pole of rotation to Afar). The across-strike width is variable; using figure 2 of *Chu and Gordon* [1999] as a guide, I estimate its width at its maximum extent to be about 2200 km. These combine to give an aspect ratio of 1.9, which is slightly higher than that estimated for any of the oceanic regions. If the 1/e width of this zone were equal to half the full width I have estimated, then $n \approx 24$. If the true 1/e width of the zones is similar to this half-width, the results would suggest that the deformation occurring across the East Africa Rift is controlled more, perhaps much more, by the strength of the brittlely deforming upper lithosphere than by the effective viscosity of the lower lithosphere.

This result differs considerably from the results obtained by *England and Molnar* [1991] for three other continental deformation zones. It suggests that the vertically averaged rheology of the continental lithosphere across and near the East Africa Rift is very different from that in coastal California (and Arizona), Oregon and Washington, and New Zealand where the deformation of the lithosphere appears to be controlled by the effective viscosity of the lower lithosphere. Some other evidence comes from the depth of earthquakes. As far as I know, in the regions considered by *England and Molnar* [1991], earthquakes are limited to a depth of 12 to 15 km, which gives a rough guide to the depth extent of the brittle upper lithosphere. In contrast, earthquakes occur to a depth of 25–30 km in the diffuse continental plate boundary between Nubia and Somalia [*Shudofsky*, 1985]. These earthquakes do not occur in the classic rift structures of the eastern rift in Ethiopia or Kenya, but the orientation of the normal faulting mechanisms are consistent with extension in about the same direction as is presumed to occur in the rifts. In part from published surface heat flow measurements from the region, *Shudofsky et al.* [1987] propose that these earthquakes occur in a strong brittlely deforming lower crust, suggesting that the brittle upper lithosphere is thicker and stronger in the East Africa Rift than in the regions of continental deformation considered by *England and Molnar* [1991]. Given the relatively cold temperatures inferred for the lower crust from the heat-flow measurements and from the occurrence of earthquakes in the lower crust, it is surprising that no earthquakes occur in the upper mantle. Their absence implies that temperatures in the lithospheric mantle are higher than inferred from extrapolation of the heat flow data [*Shudofsky et al.*, 1987]. The higher inferred temperature is consistent with (but by no means requires) the existence of an unusually thin lower lithosphere beneath the East Africa Rift. This combination might explain the high value of $n$ inferred here.

## CONCLUSIONS

The smallest strain rates across narrow boundaries are more than two orders of magnitude greater than the largest strain rates averaged across diffuse plate boundaries and 4 orders of magnitude larger than the upper bound on strain rate averaged across stable plate interiors. The upper range of strain rates averaged across diffuse plate boundaries is a factor of 20 larger than the upper bound on strain rate averaged across a stable plate interior. The lower limit on strain rates in diffuse plate boundaries, however, must overlap that for stable plate interiors. The large gap in strain rates between narrow ($\geq 10^{-13}$ s$^{-1}$) and diffuse oceanic plate boundaries ($\leq 2\times10^{-16}$ s$^{-1}$) is consistent with the self-lubricating rheology considered by *Bercovici* [1993] to obtain platelike behavior in models of mantle flow. If so, this gap in strain rate bounds the critical strain rate, $\gamma$, at which the maximum stress occurs.

The effective viscosity of the lithosphere was estimated from the ratio of vertically averaged shear stresses to strain rates for three tectonic settings: (i) oceanic transform fault zones, which have effective viscosities of $3\times10^{16}$ to $5\times10^{19}$ Pa s, comparable to estimates for the asthenosphere, (ii) diffuse oceanic plate boundaries, which have effective viscosities of $1\times10^{23}$ to $6\times10^{23}$ Pa s, about a factor of 10 higher than the effective viscosities for diffuse continental plate boundaries, and (iii) stable plate interiors, which have effective viscosities of $1\times10^{24}$ to $2\times10^{27}$ Pa s. The variations in these estimates are dominated by the variation in strain rates over many orders of magnitude.

Map-view aspect ratios of the deforming zones of diffuse oceanic plate boundaries divide into two groups. Sixty- to 90-Myr-old oceanic lithosphere has aspect ratios of 1.25 to 1.8 (taken here as the ratio of the total along-strike length to the total across-strike width of the deforming zone). In contrast, the aspect ratios for diffuse oceanic plate boundaries in young lithosphere near mid-ocean ridges range from 0.75 to 0.82. The aspect ratio for the East Africa Rift (Nubia-Somalia diffuse plate boundary) is 1.9. These aspect ratios may be related to $n$, the exponent in a power-law rheology for the vertically averaged lithosphere (cf. equation (3)). Data for a rigorous estimate exist only for the India-Capricorn zone of convergence, which indicate $n \approx 8$. If one makes the arbitrary assumption that the 1/e widths are half the total width, $n$ is greatest (24) for the East Africa Rift, next highest (8 to 21) for 60- to 90-Myr-old oceanic lithosphere, and the least (3.6 to 4.4) for young oceanic lithosphere. The latter is similar to estimates ($\approx 3$) for typical deforming regions of continental lithosphere. These variations in $n$ indicate variations in the ratio of the strength of the brittle upper lithosphere relative to the force required to deform the lower lithosphere, which deforms by creeping flow. In particular, the results indicate that this ratio is higher for 60- to 90-Myr-old deforming oceanic lithosphere than it is in deforming oceanic lithosphere near mid-ocean ridges.

*Acknowledgments.* This work was supported by NSF grants OCE-9819354 and EAR-9903763 and completed while I was a Visiting Miller Research Professor at the University of California at Berkeley. I thank Roland Bürgmann, Mark Jellinek, and Mark Richards for helpful reviews.

## REFERENCES

Anderson, J. G., Seismic strain rates in the central and eastern United States, *Seismol. Soc. Amer. Bull., 76,* 273–290, 1986.

Argus, D. F., Current plate motions and crustal deformation, unpublished Ph.D. dissertation, Northwestern University, Evanston, 163 pp., 1990.

Argus, D. F., and R. G. Gordon, No-net-rotation model of current plate velocities incorporating plate motion model NUVEL-1, *Geophys. Res. Lett., 18,* 2038–2042, 1991.

Argus, D. F., and R. G. Gordon, Tests of the rigid-plate hypothesis and bounds on intraplate deformation using geodetic data from very long baseline interferometry, *J. Geophys. Res., 101,* 13,555–13,572, 1996.

Beaumont, C., The evolution of sedimentary basins on a viscoelastic lithosphere: theory and examples, *Geophys. J. R. Astron. Soc., 55,* 471–497, 1978.

Beaumont, C., Foreland basins, *Geophys. J. R. Astron. Soc., 65,* 291–329, 1981.

Bercovici, D., A simple model of plate generation from mantle flow, *Geophys. J. Int., 114,* 635–650, 1993.

Bercovici, D., A source-sink model of the generation of plate tectonics from non-Newtonian mantle flow, *J. Geophys. Res., 100,* 2013–2030, 1995.

Bercovici, D., Y. Ricard, and M. Richards, The relation between mantle dynamics and plate tectonics: A primer, in *History and Dynamics of Global Plate Motions, AGU Monograph Series* edited by M. Richards, R. G. Gordon and R. D. van der Hilst, American Geophysical Union, Washington, 2000.

Bergman, E. A., and S. C. Solomon, Oceanic intraplate earthquakes: Implications for local and regional intraplate stress, *J. Geophys. Res., 85,* 5389–5410, 1980.

Bird, P., and K. Piper, Plane-stress finite-element models of the tectonic flow in southern California, *Phys. Earth Planet. Inter., 21,* 158–175, 1980.

Brace, W. F., and D. L. Kohlstedt, Limits on lithospheric stress imposed by laboratory experiments, *J. Geophys. Res., 85,* 6248–6252, 1980.

Bull, J. M., Structural style of intra-plate deformation, Central Indian Ocean Basin: evidence for the role of fracture zones, *Tectonophysics, 184,* 213–228, 1990.

Bull, J. M., and R. A. Scrutton, Fault reactivation in the central Indian Ocean and the rheology of oceanic lithosphere, *Nature, 344,* 855–858, 1990.

Bull, J. M., and R. A. Scrutton, Seismic reflection images of intraplate deformation, Central Indian Ocean, and their tectonic significance, *J. Geol. Soc. London, 149,* 955–966, 1992.

Chamot-Rooke, N., F. Jestin, B. de Voogd, and Phèdre Working Group, Intraplate shortening in the central Indian Ocean determined from a 2100-km-long north-south deep seismic reflection profile, *Geology, 21,* 1043–1046, 1993.

Chu, D., and R. G. Gordon, Evidence for motion between Nubia and Somalia along the Southwest Indian Ridge, *Nature, 398,* 64–67, 1999.

Cloetingh, S., and R. Wortel, Regional stress field of the Indian plate, *Geophys. Res. Lett., 12,* 77–80, 1985.

Coblentz, D. D., S. Zhou, R. R. Hillis, R. M. Richardson, and M. Sandiford, Topography, boundary forces, and the Indo-Australian intraplate stress field, *J. Geophys. Res., 103,* 919–931, 1998.

Davies, R., P. England, B. Parsons, H. Billiris, D. Paradissis, and G. Veis, Geodetic strain of Greece in the interval 1892-1992, *J. Geophys. Res., 102,* 24,571–24,588, 1997.

DeMets, C., R. G. Gordon, and D. F. Argus, Intraplate deformation and closure of the Australia-Antarctica-Africa plate circuit, *J. Geophys. Res., 93,* 11,877–11,897, 1988.

DeMets, C., R. G. Gordon, D. F. Argus, and S. Stein, Current plate motions, *Geophys. J. Int., 101,* 425–478, 1990.

DeMets, C., R. G. Gordon, and P. Vogt, Location of the Africa-Australia-India triple junction and motion between the Australian and Indian plates: Results from an aeromagnetic investigation of the Central Indian and Carlsberg ridges, *Geophys. J. Int., 119,* 893–930, 1994.

Dixon, T. H., A. Mao, and S. Stein, How rigid is the stable interior of the North American plate?, *Geophys. Res. Lett., 23,* 3035–3038, 1996.

England, P., and G. Houseman, Extension during continental convergence, with application to the Tibetan Plateau, *J. Geophys. Res., 94,* 17,561–17,579, 1989.

England, P. C., and D. P. McKenzie, A thin viscous sheet model for continental deformation, *Geophys. J. R. Astron. Soc., 70,* 295–321, 1982.

England, P. C., and P. Molnar, Inferences of deviatoric stress in actively deforming belts from simple physical models, *Philos. Trans. R. Soc. London, Ser. A.,, 337,* 151–164, 1991.

England, P., and P. Molnar, Active deformation of Asia: From kinematics to dynamics, *Science, 278,* 647–650, 1997.

England, P., and R. Wells, Neogene rotations and quasicontinuous deformation of the Pacific Northwest continental margin, *Geology, 19,* 978–981, 1991.

England, P. C., G. A. Houseman, and L. J. Sonder, Length scales for continental deformation in convergent, divergent, and strike-slip environments: analytical and approximate solutions for a thin viscous sheet model, *J. Geophys. Res., 90,* 3351–3357, 1985.

Forsyth, D. W., and S. Uyeda, On the relative importance of the driving forces of plate motion, *Geophys. J. R. Astron. Soc., 43,* 163–200, 1975.

Gordon, R. G., Present plate motions and plate boundaries, in *Global Earth Physics: A Handbook of Physical Constants, AGU Reference Shelf 1* edited by T. J. Ahrens, pp. 66–87, AGU, Washington, D. C., 1995.

Gordon, R. G., The plate tectonic approximation: Plate nonrigidity, diffuse plate boundaries, and global plate reconstructions, *Ann. Rev. Earth Planet. Sci., 26,* 615–642, 1998.

Gordon, R. G., and S. Stein, Global tectonics and space geodesy, *Science, 256,* 333–342, 1992.

Gordon, R. G., C. DeMets, and D. F. Argus, Kinematic constraints on distributed lithospheric deformation in the equatorial Indian Ocean from present motion between the Australian and Indian plates, *Tectonics, 9,* 409–422, 1990.

Gordon, R. G., C. DeMets, and J.-Y. Royer, Evidence for long-term diffuse deformation of the lithosphere of the equatorial Indian Ocean, *Nature, 395,* 370–374, 1998.

Gurnis, M., S. Zhong, and J. Toth, On the competing roles of fault reactivation and brittle failure in generating plate tectonics from mantle convection, in *History and Dynamics of Global Plate Motions, AGU Monograph Series* edited by M. Richards, R. G. Gordon and R. D. van der Hilst, American Geophysical Union, Washington, 2000.

Kirby, S. H., Rheology of the lithosphere, *Rev. Geophys. Space Phys., 21,* 1458–1487, 1983.

Kohlstedt, D. L., B. Evans, and S. J. Mackwell, Strength of the lithosphere: Constraints imposed by laboratory experiments, *J. Geophys. Res., 100,* 17,587–17,602, 1995.

Lithgow-Bertelloni, C., and M. A. Richards, Cenozoic plate driving forces, *Geophys. Res. Lett., 22,* 1317–1320, 1995.

Martinod, J., and P. Davy, Periodic instabilities during compression or extension of the lithosphere, 1, Deformation modes from an analytical perturbation method, *J. Geophys. Res., 97,* 1999–2014, 1992.

Martinod, J., and P. Molnar, Lithospheric folding in the Indian Ocean and the rheology of the oceanic plate, *Bull. Soc. géol. France, 166,* 813–821, 1995.

Minster, J. B., and T. H. Jordan, Present-day plate motions, *J. Geophys. Res., 83,* 5331–5354, 1978.

Molnar, P., and J. Stock, Relative motions of hotspots in the Pacific, Atlantic and Indian Oceans since late Cretaceous time, *Nature, 327, No 6123,* 587–591, 1987.

Molnar, P., and P. Tapponnier, Cenozoic tectonics of Asia: Effects of a continental collision, *Science, 189,* 419–426, 1975.

Molnar, P., P. England, and J. Martinod, Mantle dynamics, uplift of the Tibetan Plateau, and the Indian Monsoon, *Rev. Geophys., 31,* 357–396, 1993.

Petroy, D. E., and D. A. Wiens, Historical seismicity and implications for diffuse plate convergence in the northeast Indian Ocean, *J. Geophys. Res., 94,* 12,301–12,319, 1989.

Ranalli, G., *Rheology of the Earth,* 413 pp., Chapman and Hall, London, 1995.

Raymond, C. A., J. M. Stock, and S. C. Cande, Fast Paleogene motion of the Pacific hotspots from revised global plate circuit constraints, in *History and Dynamics of Global Plate Motions, AGU Monograph Series* edited by M. Richards, R. G. Gordon and R. D. van der Hilst, American Geophysical Union, Washington, 2000.

Richardson, R. M., Ridge forces, absolute plate motions, and the intraplate stress field, *J. Geophys. Res., 97,* 11,739–11749, 1992.

Royer, J.-Y., and R. G. Gordon, The motion and boundary between the Capricorn and Australian plates, *Science, 277,* 1268–1274, 1997.

Royer, J.-Y., R. G. Gordon, C. DeMets, and P. R. Vogt, New limits on the motion between India and Australia since chron 5 (11 Ma) and implications for lithospheric deformation in the equatorial Indian Ocean, *Geophys. J. Int., 129,* 41–74, 1997.

Shudofsky, G. N., Source mechanisms and focal depths of east African earthquakes using Rayleigh-wave inversion and body-wave modelling, *Geophys. J. R. Astron. Soc., 83,* 563–614, 1985.

Shudofsky, G., S. Cloetingh, S. Stein, and R. Wortel, Unusually deep earthquakes in East Africa: constraints on the thermo-mechanical structure of a continental rift system, *Geophys. Res. Lett., 14,* 741–744, 1987.

Sonder, L. J., and P. C. England, Vertical averages of rheology of the continental lithosphere: Relation to thin sheet parameters, *Earth Planet. Sci. Lett., 77,* 81–90, 1986.

Stein, C. A., S. Cloetingh, and R. Wortel, Seasat-derived gravity constraints on stress and deformation in the northeastern Indian Ocean, *Geophys. Res. Lett., 16,* 823–826, 1989.

Stein, S., An earthquake swarm on the Chagos-Laccadive Ridge and its tectonic implications, *Geophys. J. R. Astron. Soc., 55,* 577–588, 1978.

Stein, S., Intraplate seismicity on bathymetric features: The 1968 Emperor Trough earthquake, *J. Geophys. Res., 84,* 4763–4768, 1979.

Stein, S., and R. G. Gordon, Statistical tests of additional plate boundaries from plate motion inversions, *Earth Planet. Sci. Lett., 69,* 401–412, 1984.

Stein, S., and E. A. Okal, Seismicity and tectonics of the Ninetyeast Ridge area: Evidence for internal deformation of the Indian plate, *J. Geophys. Res., 83,* 2233–2245, 1978.

Sykes, L. R., Intraplate seismicity, reactivation of pre-existing zones of weakness, alkaline magmatism, and other tectonism postdating continental fragmentation, *Rev. Geophys. Space Phys., 16,* 621–688, 1978.

Tackley, P., Self-consistent generation of tectonic plates in three-dimensional mantle convection, *Earth Planet. Sci. Lett., 157,* 9–22, 1998.

Tackley, P., The quest for self-consistent generation of plate tectonics in mantle convection models, in *History and Dynamics of Global Plate Motions, AGU Monograph Series* edited by M. Richards, R. G. Gordon and R. D. van der Hilst, American Geophysical Union, Washington, 2000.

Van Orman, J., J. R. Cochran, J. K. Weissel, and F. Jestin, Distribution of shortening between the Indian and Australian plates in the central Indian Ocean, *Earth Planet. Sci. Lett., 133,* 35–46, 1995.

Vilotte, J. P., M. Daignieres, and R. Madariaga, Numerical modeling of intraplate deformation: Simple mechanical models of continental collision., *J. Geophys. Res., 87,* 10,709–10,728, 1982.

Watts, A. B., G. D. Karner, and M. S. Steckler, Lithospheric flexure and the evolution of sedimentary basins, *Phil. Trans. R. Soc. Lond., A305,* 249–281, 1982.

Weertman, J., Creep laws for the mantle of the Earth, *Rev. Geophys. Space Phys., 8,* 145–168, 1970.

Weissel, J. K., R. N. Anderson, and C. A. Geller, Deformation of the Indo-Australian plate, *Nature, 287,* 284–291, 1980.

Whitehead, J. A., and R. F. Gans, A new, theoretically tractable earthquake model, *Geophys. J. R. Astron. Soc., 39,* 11–28, 1974.

Wiens, D. A., Historical seismicity near Chagos: A complex deformation zone in the equatorial Indian Ocean, *Earth Planet. Sci. Lett., 76,* 350–360, 1986.

Wiens, D. A., and S. Stein, Age dependence of oceanic intraplate seismicity and implications for lithospheric evolution, *J. Geophys. Res., 88,* 6455–6468, 1983.

Wiens, D. A., C. DeMets, R. G. Gordon, S. Stein, D. Argus, J. F. Engeln, P. Lundgren, D. Quible, C. Stein, S. Weinstein, and D. F. Woods, A diffuse plate boundary model for Indian Ocean tectonics, *Geophys. Res. Lett., 12,* 429–432, 1985.

Zoback, M. L., First- and second-order patterns of stress in the lithosphere: The World Stress Map Project, *J. Geophys. Res., 97,* 11,703–11,728, 1992.

R. G. Gordon, Department of Geology and Geophysics—MS 126, Rice University, P.O. Box 1892, Houston, TX 77251-1892. (email: rgg@rice.edu)

# Mesozoic/Cenozoic Tectonic Events Around Australia

R. Dietmar Müller[1], Carmen Gaina[1], Anahita Tikku[2], Dona Mihut[3], Steven C. Cande[4], and Joann M. Stock[5]

We use an absolute and relative plate motion model for the plates around Australia to identify major plate tectonic events, evaluate their causes, and investigate their effects on anomalous intraplate subsidence or uplift and on the history of oceanic crustal accretion. An event at ~136 Ma is marked by the onset of sea floor spreading between Greater India and Australia. At about this time long-lived subduction east of Australia ceased, probably due to subduction of the Phoenix-Pacific spreading ridge, changing this plate boundary to a transform margin. Between 130 and 80 Ma, Australia and East Antarctica moved eastward in the Atlantic-Indian mantle hotspot reference frame. This can be plausibly linked to ridge push from the NW-SE oriented spreading center NW of Australia and to the inferred geometry and continued subduction of the Phoenix plate beneath the West Antarctic margin. A drastic change in spreading direction between the Indian and Australian plates from NE-SW to N-S occurred at about 99 Ma, possibly caused by a change in absolute motion of the Pacific Plate. Chron 27 (~61 Ma) marks the onset of relative motion between East and West Antarctica, and a change in the relative motion between Australia and Antarctica. It may be linked to the subduction of a segment of the Neo-Tethyan Ridge. Both events caused anomalous subsidence on the Northwest Shelf of Australia. The almost stationary position of Australia w.r.t. the mantle from ~80 Ma to ~40 Ma may reflect the progressive subduction of the Pacific-Phoenix ridge to the east of New Zealand preceding 80 Ma, resulting in a diminished trench suction force east of Australia. Preliminary reconstructions to close the Pacific-Australian plate circuit based on recently collected geophysical data indicate that a tectonic event at 43 Ma may mark the onset of renewed subduction east of Australia. At the same time spreading in the Wharton Basin between India and Australia ceased, and tectonic reactivation is recorded in the Bass Strait. Excess late Tertiary subsidence on the northwest shelf of >500 m matches the anomalous depth of the

[1] School of Geosciences, Division of Geology and Geophysics, University of Sydney, Australia
[2] Laboratory for Satellite Altimetry National Oceanic and Atmospheric Administration, MD, USA.
[3] PGS Tensor A/P, Singapore
[4] Scripps Institution of Oceanography, University of California San Diego, La Jolla, CA
[5] California Institute of Technology, Pasadena, CA

The History and Dynamics of Global Plate Motions
Geophysical Monograph 121

Argo abyssal plain ocean floor. This anomalous subsidence may express large-scale intraplate deformation in the Indian Ocean. Asymmetries in oceanic crustal accretion around Australia are caused mainly by hotspot-ridge and coldspot-ridge interaction.

## 1. INTRODUCTION

The Australian and Indian plates are unique in that their absolute plate motion rates have been much higher than those of any other plates which include a large continent. Changes in plate boundary configurations and plate driving forces have clearly caused considerable fluctuations in paleo-intraplate stresses, as all major plate tectonic events are accompanied by intraplate tectonic reactivation. However, cause and effect are not well understood.

Recently, our models of the tectonic history of Australia and its neighboring plates have improved considerably for a number of reasons. Satellite altimetry has revolutionized our knowledge of the tectonic structure of the ocean floor where ship data are sparse. Additional magnetic and seismic data have recently been collected off Antarctica [*Ishihara et al.*, 1996; *Mizukoshi et al.*, 1986; *Saki et al.*, 1987; *Tsumuraya et al.*, 1985], around the Macquarie Triple Junction and in the Tasman Sea [*Cande et al.*, 1998] in the Tasman Sea and off West Australia [*Symonds et al.*, 1996]. These data are being used to construct more detailed plate tectonic models. We have combined some recently published work [*Cande et al.*, in press; *Cande et al.*, 1998; *Gaina et al.*, 1998a; *Gaina et al.*, 1999; *Gaina et al.*, 1998b; *Mihut and Müller*, 1998; *Müller et al.*, 1998a; *Müller et al.*, 1998b; *Müller et al.*, 1997; *Tikku*, 1998; *Tikku and Cande*, 1999] with other work in progress to present digital grids of the age of the ocean floor (Figure 1), spreading rates (Figure 2), asymmetries in crustal accretion (Figure 3) and plate tectonic reconstructions to identify major plate tectonic events, evaluate their cause, and investigate their effect on anomalous intraplate subsidence or uplift and on the history of oceanic crustal accretion.

Of particular interest is the relationship between anomalous intraplate subsidence or uplift and either mantle convection or lithospheric processes. The importance of intraplate stresses for the tectonic history and of sedimentary basins and fault reactivation has been recognised only recently. One of the main paradigms of plate tectonics was thought to be that there is a limited number of large tectonic plates that behave rigidly through long geological time intervals. Instead recent analyses of diffuse plate boundaries [*Gordon*, 1998; *Gordon et al.*, 1998] intra-plate earthquakes [*Crone et al.*, 1997; *Denham et al.*, 1979] and borehole breakouts [*Hillis*, 1991; *Hillis et al.*, 1997] on the Australian plate and elsewhere have shown clearly that forces applied at plate boundaries propagate into large plates, and produce substantial intra-plate deformation. A spectacular example of oceanic intraplate deformation, resulting in an oceanic fold and thrust belt, can be found in the central Indian Ocean [*Royer and Gordon*, 1997]. Many tentative links between plate tectonic events and tectonic reactivation on the Australian plate have been proposed [*Cloetingh et al.*, 1992; *Etheridge et al.*, 1991]. *Etheridge et al.* [1991] pointed out that in many Australian sedimentary basins nearly all oil and gas fields are associated with tectonic reactivation, highlighting the need to understand changing plate driving forces through time and their impact on intraplate stresses from an exploration point of view.

## 2. TIMESCALE

For the work presented here, we used the *Gradstein et al.* [1994] and *Cande and Kent* [1995] time scales for Mesozoic and Cenozoic magnetic anomaly identifications, and the AGSO timescale [*Young and Laurie*, 1996] for the stratigraphy and tectonic subsidence analysis of well data from the northwest shelf of Australia. The AGSO timescale was chosen for this purpose, because the regional biostratigraphy is tied to it. Table 1 summarizes the differences between the former two timescales and the AGSO timescale for the times of some major tectonic events. The difference between the timescales is largest for chron M0, and negligible in the Cenozoic.

## 3. HISTORY OF OCEANIC CRUSTAL ACCRETION AROUND AUSTRALIA

We have used the isochrons by *Gaina et al.* [1998a] and *Gaina et al.* [1999] in the Tasman and Coral seas, respectively, from *Mihut* [1997] west of Australia, and from *Müller et al.* [1997] to construct grids of the age of the ocean floor, half-spreading rates, and asymmetries in crustal accretion (Figures 1-3). The isochrons are based on magnetic anomaly identifications and fracture zone picks (Figures 4a-d), which were used jointly to derive finite rotations describing relative plate motions around Australia (Table 2). We plot reconstructed magnetic anomaly and fracture zone picks, as well as selected small circles computed from stage rotation poles for each isochron time, keeping one plate fixed. Then best-fit continuous isochrons are constructed, connected by transforms, in the framework of one fixed plate (see also *Müller et al.*

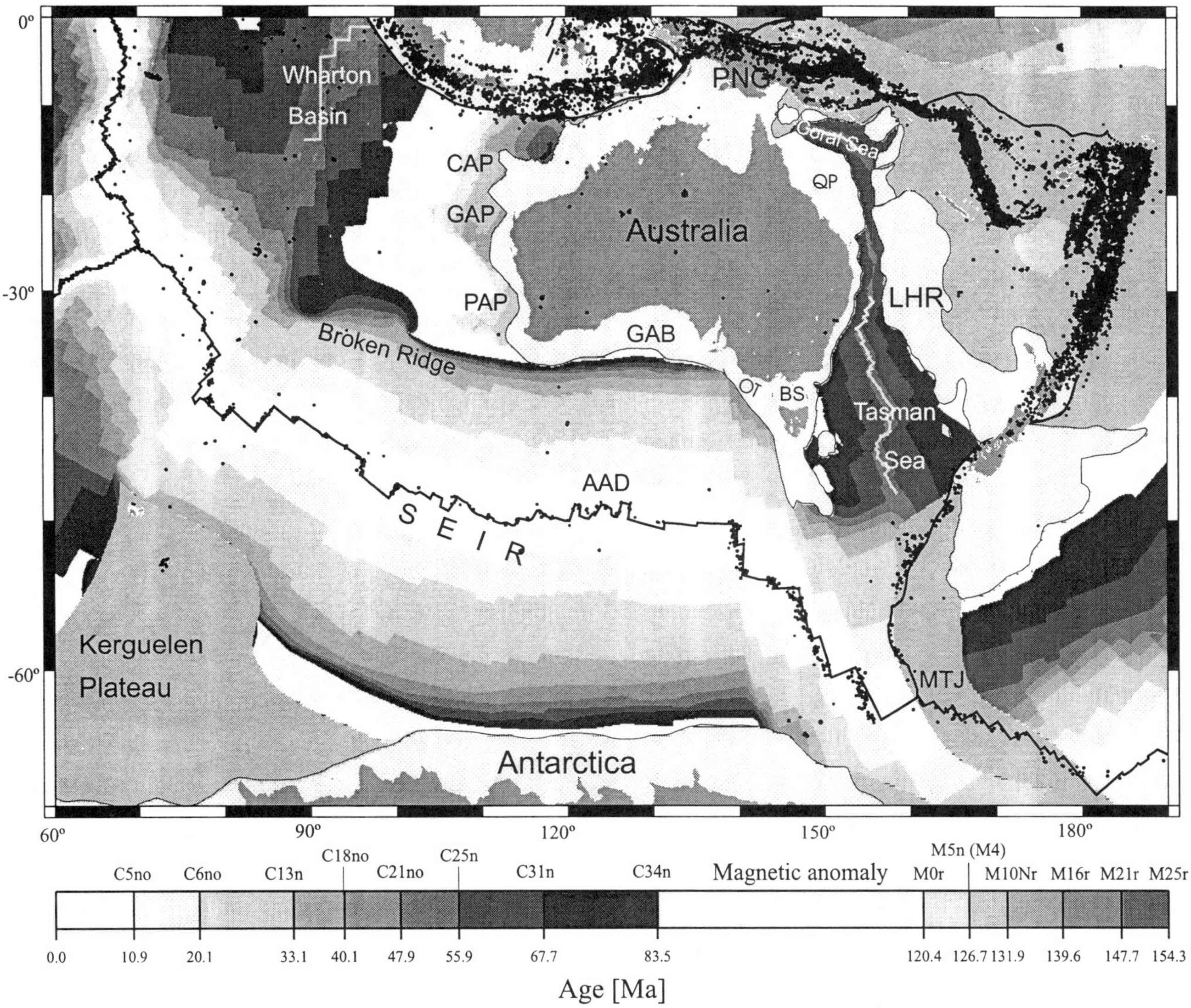


**Figure 1**. Age of the ocean crust, earthquake epicenters and plate boundaries around Australia. Continental crust below sea level is shown in light gray; oceanic crust of unknown age is medium gray. CAP, Cuvier Abyssal Plain; GAP, Gascoyne Abyssal Plain; PAP, Perth Abyssal Plain; GAB, Great Australian Bight; OT, Otway Basin; BS, Bass Basin; QP, Queensland Plateau; LHR, Lord Howe Rise; MTJ, Macquarie Triple Junction. Magnetic anomalies correspond to the young end of periods with normal polarity (labelled "n") for chrons younger than the Cretaceous Normal Superchron (from 120.4 to 83.5 Ma) or reversed polarity (labelled "r") for M-sequence anomalies, with the exception of anomaly M4, which actually corresponds to the young end of anomaly M5n. An "o" label indicates that the old end, rather than the young end of a period of normal polarity was chosen for magnetic anomaly identification.

[1991]). A complete set of isochrons for all conjugate plate pairs was derived by rotation of every isochron to their present day position.

Grids for crustal age, half-spreading rates and relative crustal accretion (expressing spreading asymmetry) were constructed by rotating adjacent pairs of isochrons on a given plate into a coordinate system in which the stage pole of motion between the two isochrons is the geographic north pole. For each pair of adjacent isochrons we determine the percentage of crustal accretion by dividing the angular distance between two isochrons by the half-stage pole angle assuming spreading symmetry. We create intermediate isochrons between adjacent isochron pairs by interpolating crustal ages linearly along small

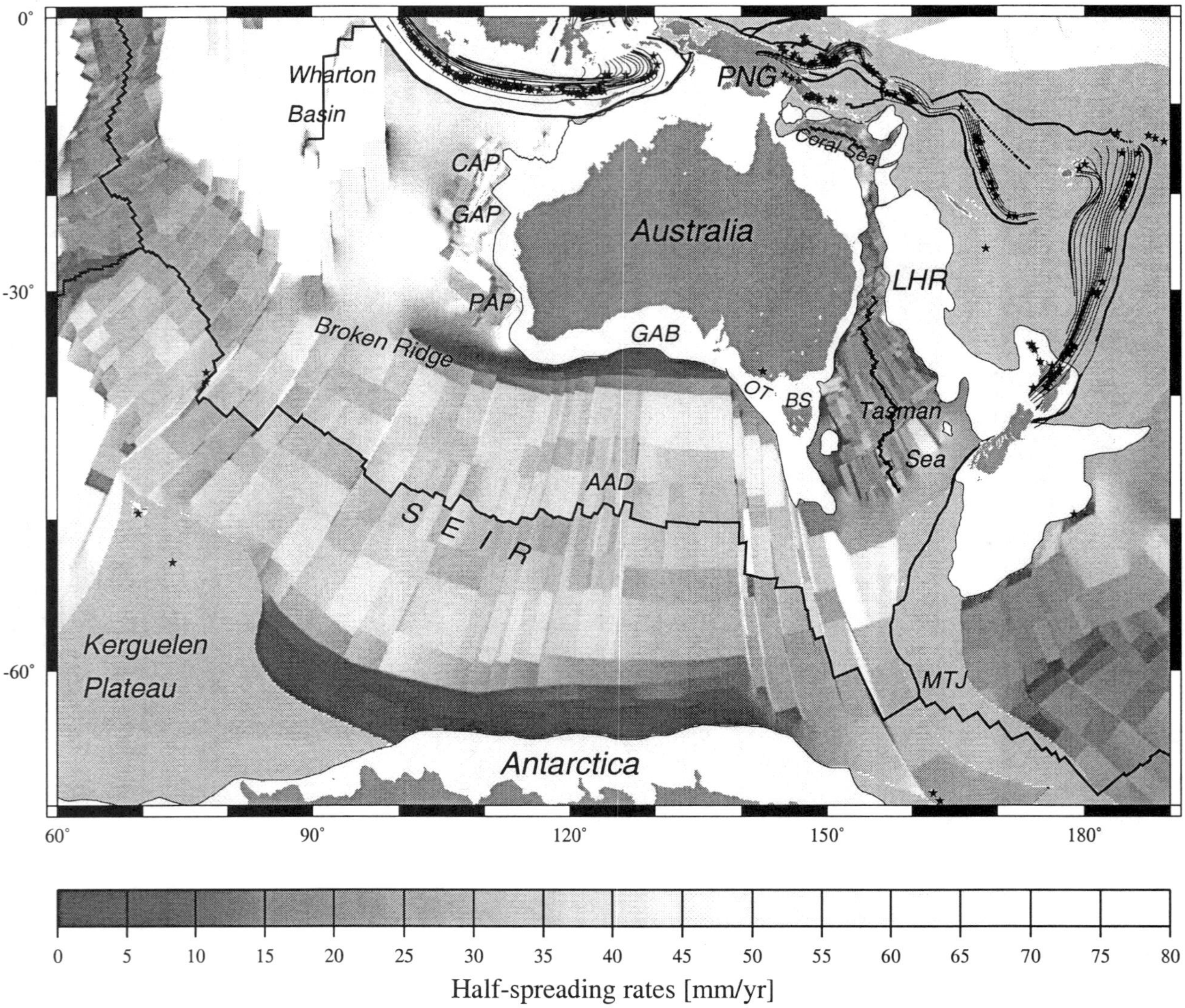


**Figure 2.** Half-spreading rates, contours of subducted slabs [*Gudmundsson and Sambridge*, 1998], active volcanoes (stars) and plate boundaries around Australia.

circles. Half-spreading rates on a particular ridge flank are based on the distance between two adjacent isochrons divided by their age difference. We rotate the complete set of isochron data for each stage back into the present-day geographic reference frame. To interpolate the data onto regular grids, we assume that the isochrons are continuous, which is implemented by densely interpolating between observation points along each isochron and gridding the resulting values onto a regular grid at a resolution of 0.1 degrees using continuous curvature splines in tension [*Smith*, 1990] with a tension factor of 0.7. The resulting grids are shown on Figures 1, 2 and 3, respectively, together with earthquake epicenters (Figure 1), contours of subducted slabs [*Gudmundsson and Sambridge*, 1998] and locations of volcanoes (Figure 2), as well as major hotspots (Figure 3).

The grids provide insights in crustal accretion processes which cannot be obtained by investigating active ridges alone. The largest asymmetries in crustal accretion are observed west of the Perth abyssal plain west of Australia. Here two major ridge propagation events during the Cretaceous Normal Superchon (KNS) accreted large segments of the Indian Plate to the Australian Plate [*Mihut*, 1997; *Müller et al.*, 1998a]. Virtually all crust created during the KNS in the spreading corridor west of the Perth Abyssal Plain between India and Australia was left on the Australian Plate. The most likely cause for these large asymmetries are consecutive ridge propagation events

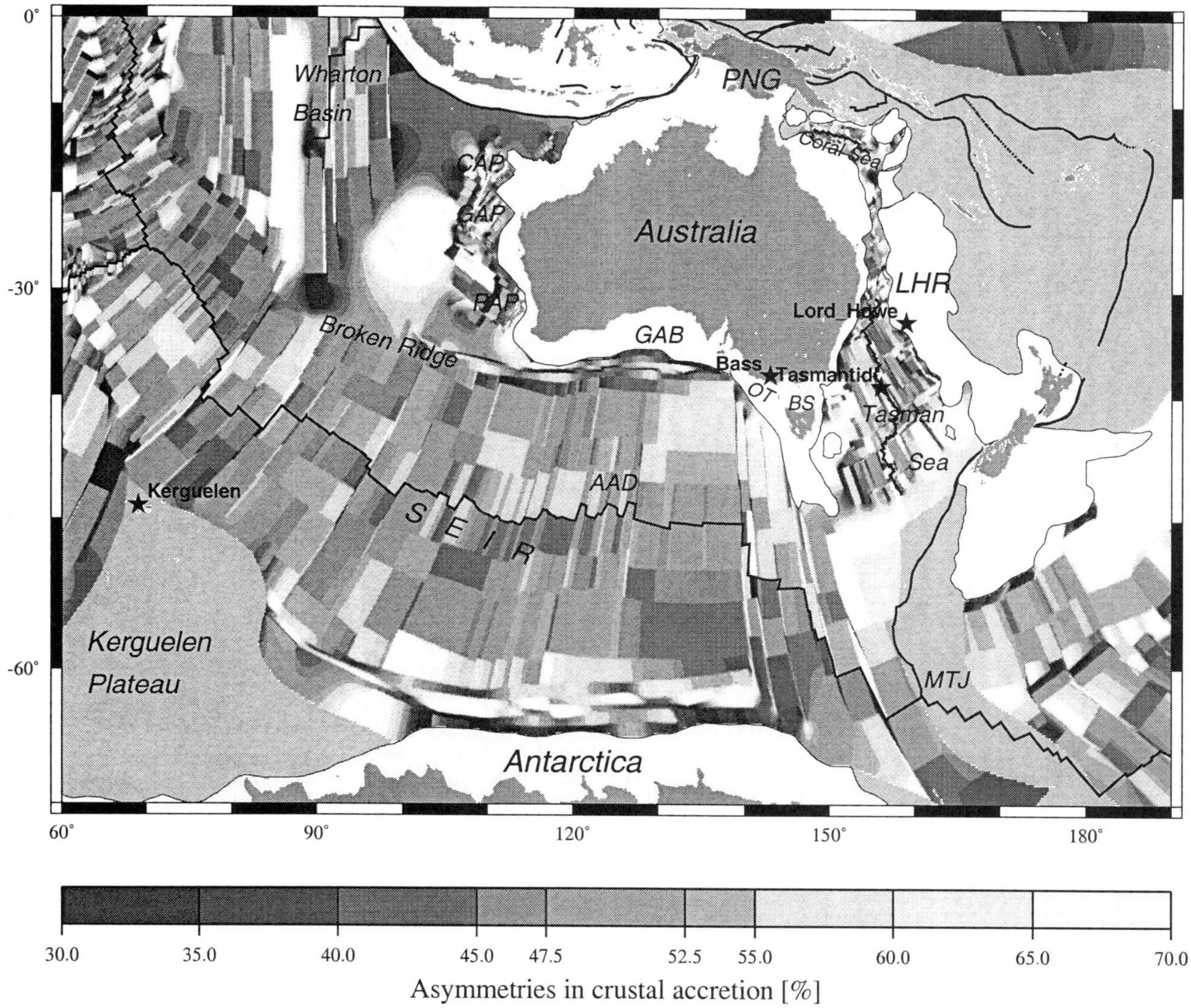


**Figure 3.** Asymmetries in oceanic crustal accretion and major hotspots around Australia. Asymmetry is expressed as the percentage of crustal area on one plate relative to the total area of crust formed for a given time range and length along the ridge axis between conjugate plates. Symmetric spreading results in values of 50% on both plates, whereas extreme asymmetry results in 100% on one plate and 0% on the conjugate plate.

towards the Kerguelen hotspot, which would have been located between the Elan Bank and India at 99 Ma (Figure 5a), assuming that it is now situated at Kerguelen Island (Figure 3). *Müller et al.* [1998b] found that most observed long-term asymmetries in spreading are likely caused by ridge-hotspot interaction, rather than by the migration of the spreading ridge over a fixed or slowly moving mantle [*Stein et al.*, 1977]. *Stein et al.'s* [1977] model suggests that trailing flanks of mid-ocean ridges, i.e. ridge flanks on plates that move more slowly relative to the mantle than their counterparts, should always show excess accretion. Their argument is based on minimizing energy dissipation. However, the extreme asymmetries in crustal accretion observed in the Perth Abyssal Plain are not accompanied by asymmetries of similar magnitude further north in the Gascoyne and Cuvier abyssal plains, which would be expected if they were caused by absolute ridge migration.

Along the central Indian Ridge, pronounced asymmetries in spreading are recorded in the westernmost segment on Figure 3, i.e. the segment that has recorded relative motion between India and Antarctica, west of Kerguelen. This segment shows consistent excess accretion on the trailing Antarctic ridge flank, in accord with *Stein et al.'s* [1977] model. This ridge also moved at

**Table 1.** Ages for key times discussed in this paper from *Gradstein et al.* [1994], *Cande and Kent* [1995], and *Young and Laurie* [1996]. Our magnetic anomaly identifications and models, and the ages of well data from the Queensland and Marion plateaus are based on the former two timescales, whereas the ages of well data from the northwest shelf of Australia were determined using the biostratigraphy from *Young and Laurie* [1996].

| Magnetic anomaly | Ages from Gradstein et al. [1994][1], Cande and Kent [1995][2] | Ages from Young and Laurie [*Young and Laurie*, 1996] |
|---|---|---|
| M27 | 156.0[1] | 153.0 |
| M14 | 135.9[1] | 133.9 |
| M0 | 120.4[1] | 114.9 |
| | 99.0[1] | 96.4 |
| C34 | 83.5[1] | 83.0 |
| C27 | 61.2[2] | 61.1 |
| C5 | 9.9[2] | 10.0 |
| C3 | 5.0[2] | 5.0 |

high rates of 30-80 mm/y relative to the mantle since chron 34 (83 Ma), higher than any other spreading ridge. *Müller et al.* [1998b] interpreted this as suggestive of a minimum "absolute" ridge migration speed for *Stein et al's* [1977] mechanism to have an effect on the asymmetry of spreading.

In contrast, the spreading corridor between Kerguelen and the western Broken Ridge, just to the east, shows the opposite behavior, i.e. a consistent crustal deficit on the Antarctic plate. This is most likely due to ridge-plume interaction, whereby the Kerguelen hotspot may cause ridge jumps to the south by pressure release melting of material flowing toward the ridge, causing excess dike intrusion on the ridge flank proximal to the hotspot [*Müller et al.*, 1998b]. Together with the large spreading asymmetries found in the Perth Abyssal Plain, these observations document a history of interaction between the Kerguelen hotspot and adjacent spreading ridges which can be traced back to about 110 Ma. This is the time at which the first major ridge propagation event occured in the Perth Abyssal Plain, accompanied by an increase in roughness of the oceanic basement (Figure 4a). A causal connection between this ridge propagation event and the arrival of the Kerguelen plume head is plausible, as the oldest basalts recovered from the Kerguelen Plateau have been dated as about 110 m.y. [*Pringle et al.*, 1997; *Storey*, 1995].

Seafloor further east, between the eastern Broken Ridge and the Australian Antarctic Discordant Zone (AAD), shows no consistent long-term asymmetries, presumably because it is not influenced by a plume, and neither has the ridge migrated fast enough to produce the effect suggested by *Stein et al.* [1977]. The spreading segments adjoining the AAD show little asymmetry in accretion, until some time after chron 6 (19 Ma) [*Marks et al.*, 1999]. After chron 5 (10.9 Ma), we observe excess accretion on the Australian flank in virtually all spreading corridors which are part of the AAD (Figure 3).

*Gurnis et al.* [1998] developed a dynamic model that accounts for the existence of the AAD as a consequence of the spreading ridge drawing up cold mantle material from an old N-S striking subducted slab that has partially stagnated over at the 660 km endothermic phase transition. In their model, the ridge is currently not centered on the volume of drawn up slab material, but offset to the north. This is because the ridge has continually migrated north while pulling up old slab material. This model prediction is confirmed by *Marks et al.* [1999], who showed that the residual depth anomaly is also offset to the south of the ridge. In this context, considering that the AAD has become more pronounced through time if *Gurnis et al.'s* [1998] model is correct, the observed spreading asymmetries, starting some time after 19 Ma, may be interpreted as ridge jumps towards the center of the dynamic topography low while it increased in amplitude. However, the process driving ridge jumps towards a "cold spot" is unclear.

Further east, between the AAD and the George V Fracture Zone, we find the opposite pattern, i.e. excess accretion on the "leading" Australian Plate before 11 Ma, but symmetric spreading thereafter. Firstly, this indicates that ridge migration over the mantle cannot account for these asymmetries, as it would result in excess accretion on

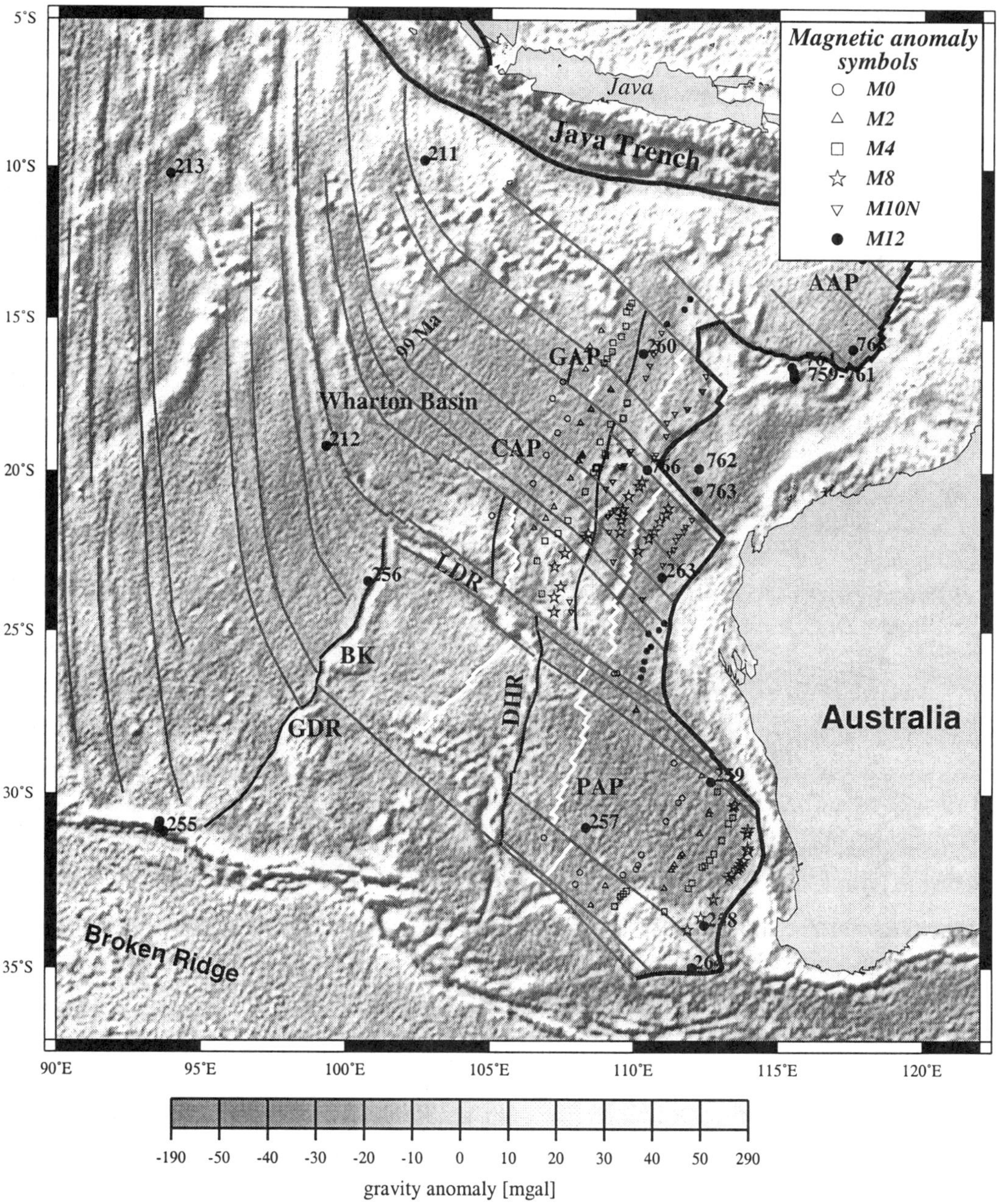


**Figure 4a.** Magnetic anomaly identifications from ship crossings and fracture zone picks overlain on marine gravity anomalies from satellite altimetry [*Sandwell and Smith*, 1997] west of Australia. Extinct ridges are shown as white staircase patterns, and pseudofaults are bold black lines. Well locations are shown as bold black dots; DSDP and ODP sites are numbered. Abbreviations are: AAP, Argo abyssal plain; GAP, Gascoyne abyssal plain; CAP, Cuvier abyssal plain; PAP, Perth abyssal plain; GDK, Gulden Draak Knoll; BK, Batavia Knoll; DHR, Dirk Hartog Ridge; LDR, Lost Dutchman Ridge.

the trailing Antarctic Plate. Secondly, this observation suggests that the V-shaped patterns visible in gridded gravity anomalies east of the AAD (Figure 4b) are not caused by ridge propagators, as suggested by *Phipps-Morgan and Sandwell* [1994]. They interpreted the V-shaped limbs north of the ridge as the wakes of extinct rifts and the conjugate traces on the Antarctic Plate as pseudofaults, consistent with the present-day transform offsets of the mid-ocean ridge. However, this scenario would result in excess accretion on the Antarctic Plate. In

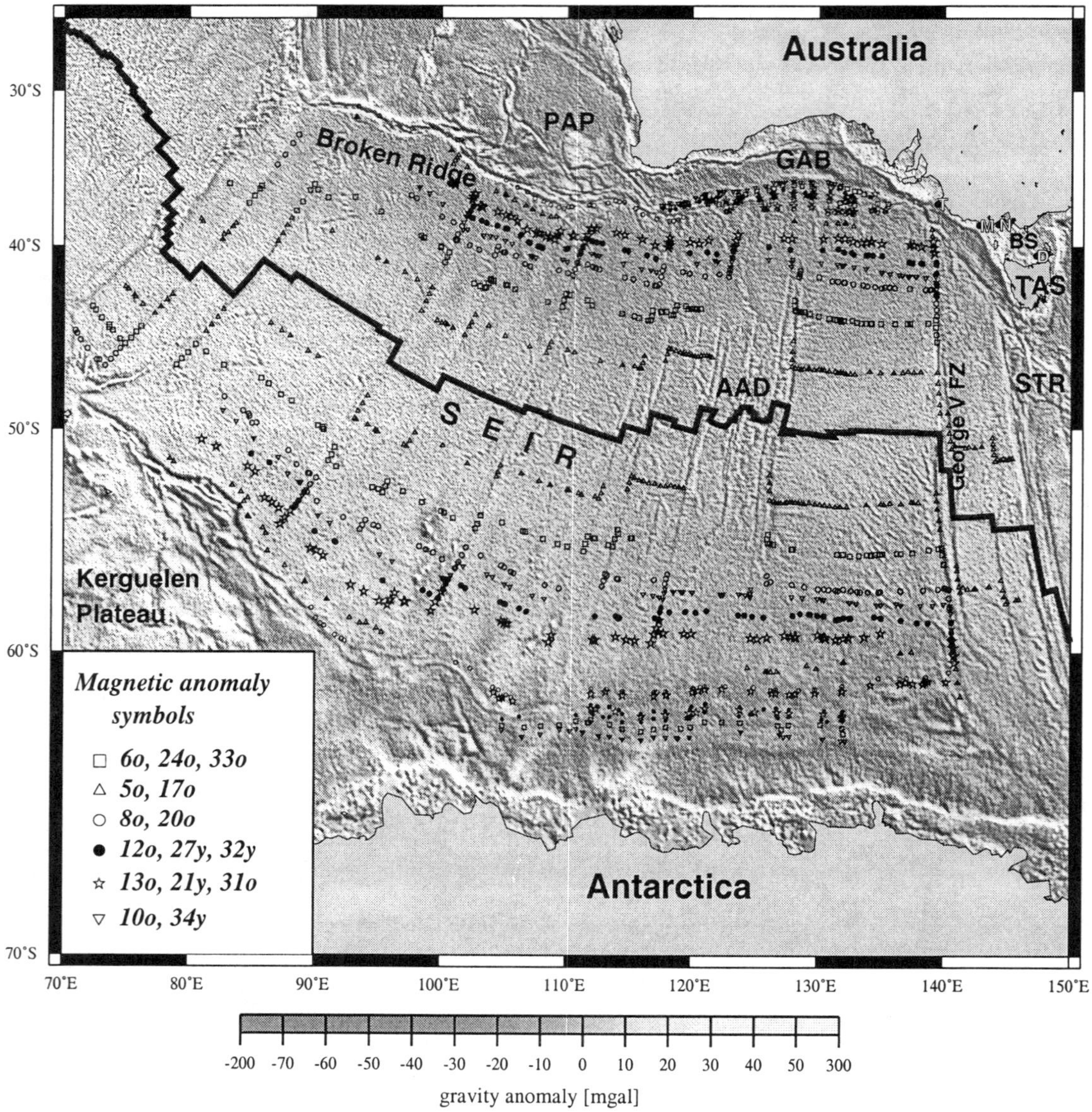


**Figure 4b**. Magnetic anomaly identifications from ship crossings and fracture zone picks overlain on marine gravity anomalies from satellite altimetry [*Sandwell and Smith*, 1997] south of Australia. Annotations are: SEIR, South East Indian Ridge; AAD, Australian Antarctic Discordance; GAB, Great Australian Bight; BS, Bass Strait; T, Troast; M, Minerva; N, Nerita 1; D, Durroo; TAS, Tasmania; STR, South Tasman Rise.

contrast, we find the opposite before 11 Ma, and spreading symmetry after 11 Ma. Therefore it is more likely that the V-shaped traces observed in the gravity field are due to the migration of the ridge over small-offset volcanic spreading centers, as described by *Schouten et al.* [1987]. However, the eastward component of ridge migration east of the AAD is not large enough to account for the steep angle of the observed V-shaped traces. The V-shapes could be created by a combination of eastward ridge migration and westward flow of the asthenosphere. The required westward flow of Pacific asthenosphere may rather be an expression of an evolving spreading ridge over a shallow wedge of Indian mantle material, dipping to the west (see Figure 9 of *Gaina et al.*, this vol.). This wedge may have been created by lateral viscous drag of Indian asthenosphere over Pacific mantle to the east by the Late Cretaceous eastward motion of Australia and Antarctica [*Gurnis et al.*, this vol.]. The early Southeast Indian Ridge in this area would have first sampled Indian mantle from a thin westward dipping wedge, which was progressively replaced by Pacific mantle from underneath this wedge. This would explain both the apparent westward migration

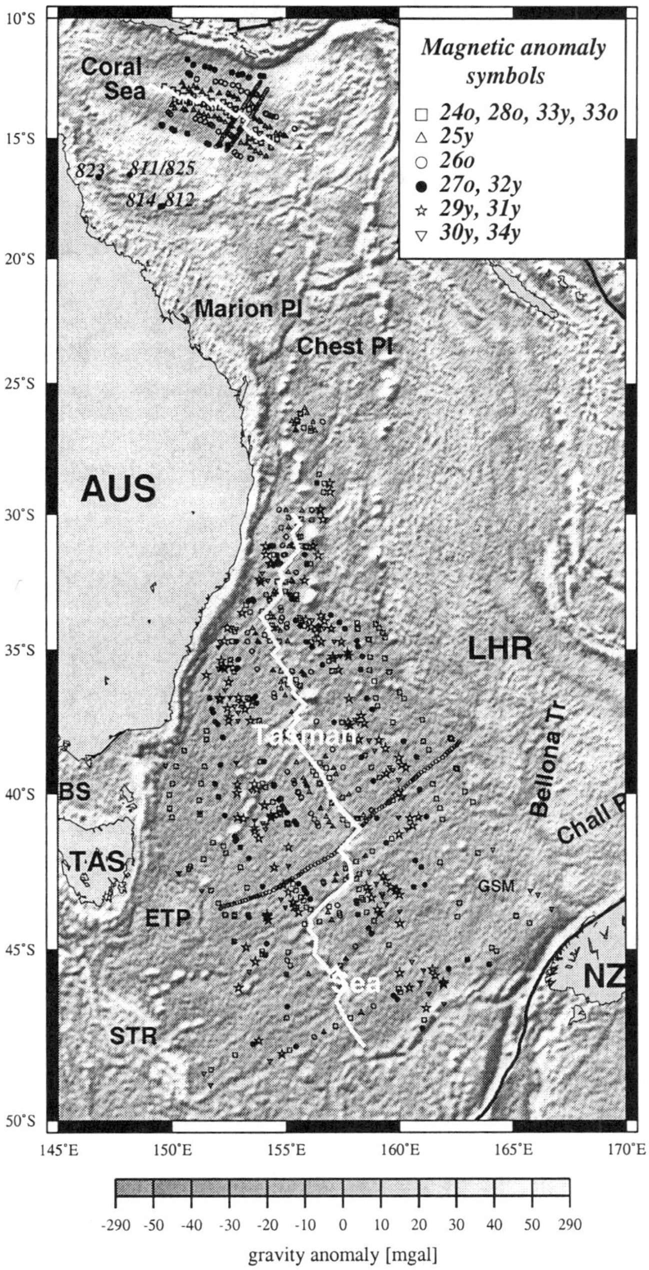


**Figure 4c.** Magnetic anomaly identifications from ship crossings and fracture zone picks overlain on marine gravity anomalies from satellite altimetry [*Sandwell and Smith*, 1997] east of Australia. Abbreviations are: Chest Pl, Chesterfield Plateau; LHR, Lord Howe Rise; Chall Pl, Challenger Plateau; ETP, East Tasman Plateau; NZ, New Zealand; GSM, Gilbert Seamount.

of Pacific mantle, and, together with an eastward component of ridge migration in the Tertiary [*Gurnis et al.*, 1998] the observed V-shapes in the gravity field, if small-scale convection under the ridge resulted in a series of small volcanic centers. Isochrons east of the George V Fracture Zone (Figure 4b) are currently not well constrained, and the asymmetries shown on Figure 3 have to be interpreted with caution. The prominent asymmetries in accretion visible in the Tasman Sea (Figures 4c, 4d) are discussed in a separate paper [*Gaina et al.*, this vol.].

## 4. RELATIVE AND ABSOLUTE PLATE MOTIONS

Relative and absolute plate motions for the plates around Australia since the Early Cretaceous are based on the rotations summarized in Table 2. The oldest magnetic anomaly found in the Perth, Gascoyne and Cuvier abyssal plains west of Australia that can be interpreted with confidence is M10N (~132 Ma) ( (Figure 4). *Mihut* [1997] interpreted anomalies as old as M14 east of the Wallaby Plateau in the Cuvier abyssal Plain. However, the relationship between these oldest magnetic lineations and seaward dipping reflectors based on new multichannel deep seismic data collected by the Australian Geological Survey Organization (AGSO) (*P. Symonds, pers. com.*) is still under investigation. Figures 5 and 6 show the positions of the plates in an Atlantic-Indian hotspot frame starting at 132 Ma, rather than a paleomagnetic reference frame. *Van Fossen and Kent* [1992] showed that a 12° latitudinal difference between the two reference frames my have existed in the Atlantic from 122 to 88 Ma, implying that the actual latitude of Australia might have been 12° farther south at 130 Ma compared to the position shown in Figures 5 and 6.

It is interesting to note that the breakup age between Greater India and Australia roughly corresponds to the time when subduction stopped east of Australia at around 130 Ma, probably due to subduction of the Phoenix-Pacific spreading ridge [*Bradshaw*, 1991] (see *Gurnis et al.* [this vol.] for a discussion of data constraining the subduction history east of Australia). The subsequent opening of the nascent central Indian Ocean was partly accomplished by the joint eastward motion of Australia and Antarctica in an Atlantic ocean mantle reference frame and partly by the northwestward motion of India (Figures 5a, 5b, and 6). However, these observations are at odds with *Duncan and Clague's* [1985] model for westward Pacific absolute plate motion from 150 to 100 Ma, as it would require convergence between the Australian and Pacific plates from 130 to 100 Ma, if all hotspot have been fixed relative to each other. This discrepancy highlights the possibility of large-scale differential mantle motion as suggested by *Tarduno and Gee* [1995] and *DiVenere and Kent* [1999], rather than true polar wander, to account for *Van Fossen and Kent's* [1992] observations and for the lack of evidence for Australian-Pacific convergence from 130-100 Ma.

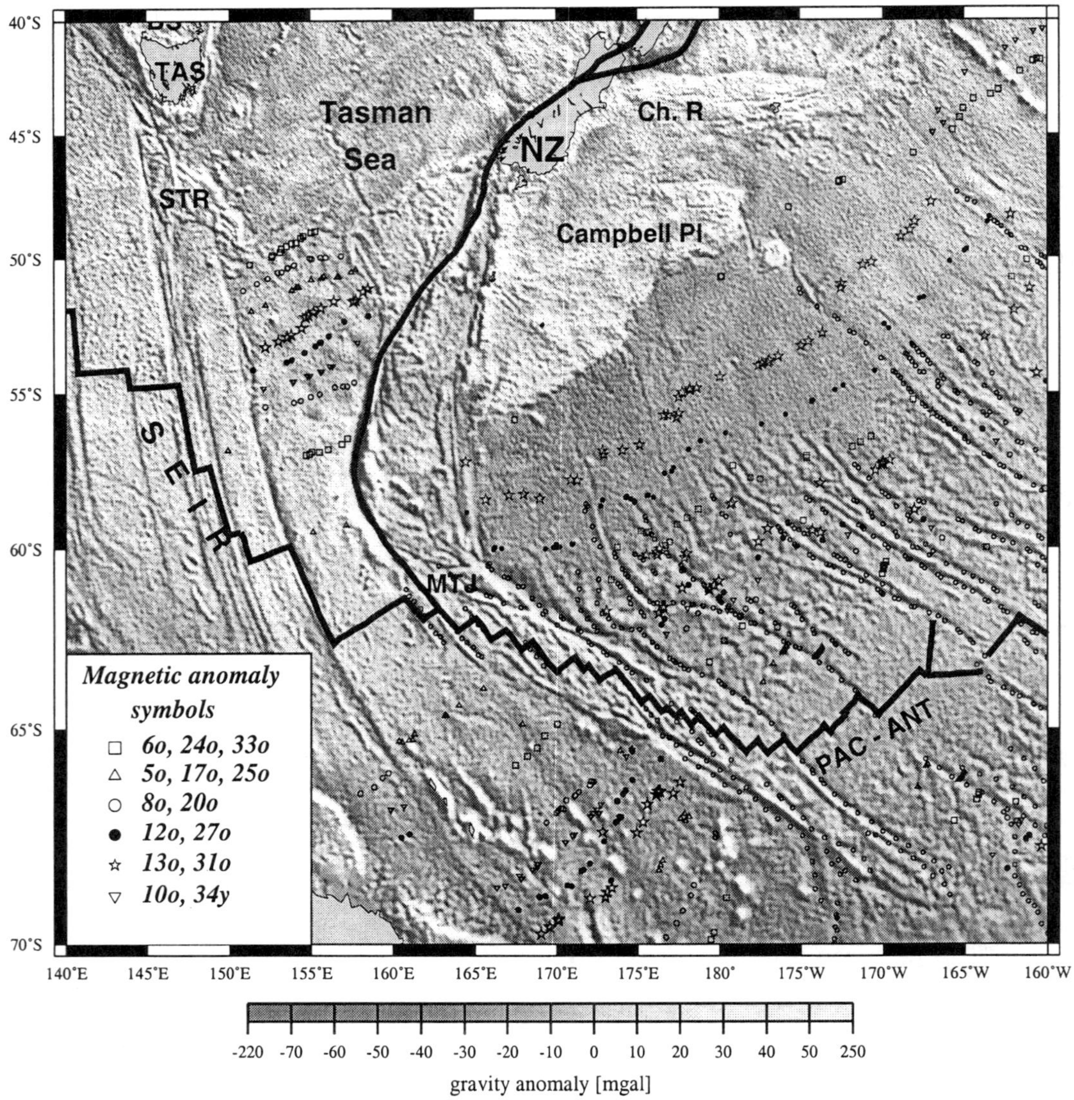


**Figure 4d.** Magnetic anomaly identifications from ship crossings and fracture zone picks overlain on marine gravity anomalies from satellite altimetry [*Sandwell and Smith*, 1997] in the southwest pacific. Abreviations are: MTJ, Macquarie Triple Junction; Ch. R, Chatham Rise; PAC-ANT, Pacific-Antarctic Ridge.

The pattern of spreading in the Early Cretaceous west of Australia persisted until about 99 Ma in the Cretaceous Magnetic Quiet Zone, when a drastic clockwise change in spreading direction from NW-SE to N-S occurred, resulting in the roughly north-south oriented spreading in the Wharton Basin (Figure 1). This event is expressed by a major bend in many fracture zones and some unusual linear troughs, ridges and knolls visible in *Sandwell and Smith's* [1997] gravity anomaly grid southeast of the Wharton Basin (Figure 4a). Preceding the 99 Ma event, a number of ridge propagators accreted large portions of the Indian Plate onto the Australian Plate, propagating northward (see also *Mihut and Müller* [1998] and *Mihut* [1997] for a more detailed discussion of these events). The timing of the bend in the fracture zones that documents this major change in spreading direction can only be constrained indirectly. *Powell et al.* [1988] proposed an age of 96 Ma for the event, based on extrapolation of spreading rates and directions. A ridge north of Batavia Knoll, 150 km SE of the major bend in fracture zones was sampled at Deep Sea Drilling Project Site 256, and microfossils yielded a minimum age of 102 Ma [*Luyendyk and Davies*, 1974]. The Albian-Cenomanian boundary is placed at 100 Ma in the time scale used by *Luyendyk and Davies* [1974], whereas this boundary is dated as 98.9 Ma by *Gradstein et al.* [1994], reducing *Luyendyk & Davies'*

**Table 2.** Finite Rotations for Different Plates around Australia

| *Age,* Ma | *Latitude* +°N | *Longitude* +°E | *Angle,* deg |
|---|---|---|---|
| *North Lord Howe Rise Relative to Australia (90-84 Ma)** | | | |
| 84.0 | 3.51 | – 43.6 | 12.27 |
| *North Lord Howe Rise Relative to Middle Lord Howe Rise(84-64 Ma)* | | | |
| 84.0 | 8.10 | -35.62 | -2.72 |
| *Middle Lord Howe Rise Relative to Australia* | | | |
| 53.3 | 14.19 | 130.41 | -0.723 |
| 55.8 | 15.93 | 133.47 | -2.112 |
| 57.9 | 16.93 | 136.23 | -3.792 |
| 61.2 | 4.65 | 131.51 | -4.432 |
| 62.5 | 4.71 | 132.68 | -5.168 |
| 64.0 | 0.19 | 130.37 | -5.461 |
| 65.6 | 3.99 | 131.80 | -6.735 |
| 67.7 | 9.04 | 134.46 | -8.83 |
| 71.1 | 14.72 | 139.04 | -13.08 |
| 73.6 | 9.53 | 137.20 | -12.94 |
| 79.0 | 0.37 | 133.82 | -13.00 |
| 83.0 | 1.57 | 133.42 | -13.04 |
| 84.0 | 4.50 | – 42.26 | 14.96 |
| 86.0 | 4.06 | – 42.35 | 15.51 |
| 90.0 | 3.27 | – 42.59 | 16.61 |
| *Challenger Plateau Relative to Middle Lord Howe Rise (90-77 Ma)* | | | |
| 86.0 | 60.83 | 63.50 | -1.00 |
| 90.0 | 72.27 | 63.13 | -1.39 |
| *North Dampier Ridge Relative to Australia (90-71.4 Ma) ** | | | |
| 71.4 | 11.84 | – 44.20 | 11.20 |
| *Middle 1 Dampier Ridge Relative to Australia (90-72.4 Ma)* | | | |
| 72.4 | 13.05 | – 41.52 | 12.49 |
| *Middle 2 Dampier Ridge Relative to Australia (90-74.2 Ma)* | | | |
| 74.2 | 7.25 | – 44.15 | 11.46 |
| *South Dampier Ridge Relative to Australia( 90-79 Ma)* | | | |
| 79.0 | 5.28 | – 43.88 | 11.87 |
| *Gilbert Seamount Complex Relative to East South Tasman Rise (90-79 Ma)* | | | |
| 79.0 | 0.52 | 131.47 | -13.10 |
| 90.0 | 7.5 | -50.4 | 14.87 |

**Table 2** (continued)

| *Age,* Ma | *Latitude* +°N | *Longitude* +°E | *Angle,* deg |
|---|---|---|---|
| *Chesterfield Plateau Relative to Australia (90-71.6 Ma)* | | | |
| 71.6 | 7.57 | 112.18 | – 5.03 |
| *Chesterfield Plateau Relative to North Lord Howe Rise (71.6-64 Ma)* | | | |
| 71.6 | 22.48 | -24.74 | -8.83 |
| *East Tasman Plateau Relative to Australia (90-84 Ma)* | | | |
| 84.0 | 60.20 | -38.13 | -2.70 |
| *East South Tasman Rise Relative to Antarctica (90-60 Ma)* | | | |
| 60.0 | 10.00 | 36.42 | -25.50 |
| *West South Tasman Rise Relative to Antarctica(90-40.1Ma)* | | | |
| 40.1 | 26.19 | 22.65 | -30.63 |
| *Tasmania Relative to Australia (120-95 Ma)* | | | |
| 95.0 | 90.0 | 0 | 0 |
| 120.0 | 28.43 | 177.10 | -1.15 |
| *Small Plate Between Middle Lord Howe Rise and Challenger Plateau Relative to Challenger Plateau (90-86 Ma)* | | | |
| 90.0 | 56.51 | 64.76 | 0.59 |
| *Small Plate Between Middle Lord Howe Rise and Challenger Plateau Relative to middle Lord Howe Rise (86-77 Ma)* | | | |
| 86.0 | 65.08 | 108.98 | -0.83 |
| *Louisiade Plateau Relative to Australia (61.3-53.3 Ma)* | | | |
| 53.35 | -7.746 | 144.566 | -2.085 |
| 55.89 | 7.155 | 124.346 | -1.895 |
| 57.91 | 1.144 | 132.092 | -4.630 |
| 61.28 | -3.001 | 137.843 | -10.650 |
| *Finite rotations for different tectonic elements in the Tasman Sea and Coral Sea are from Gaina et al. [1998] and Gaina et al. [1999]* | | | |
| *North India Relative to Australia (131-99)* * | | | |
| 99.0 | 6.89 | 0.33 | -44.68 |
| 120.4 | 5.26 | -176.95 | 53.96 |
| 124.0 | 6.56 | -176.61 | 55.28 |
| 126.7 | 7.79 | -176.29 | 56.60 |
| 129.0 | 8.64 | -176.14 | 57.82 |
| 130.9 | 9.43 | -176.06 | 59.20 |

* *Mihut, [1997]*

**Table 2** (continued)

| *Age,* Ma | *Latitude* +°N | *Longitude* +°E | *Angle,* deg |
|---|---|---|---|
| *Pacific relative to Marie Bird Land (West Antarctica) (86-0 Ma)** | | | |
| 0.78 | 64.25 | -79.06 | 0.68 |
| 2.58 | 67.03 | -73.72 | 2.42 |
| 5.89 | 67.91 | -77.93 | 5.42 |
| 8.86 | 69.68 | -77.60 | 7.95 |
| 12.29 | 71.75 | -73.77 | 10.92 |
| 17.47 | 73.68 | -69.85 | 15.17 |
| 24.06 | 74.72 | -67.28 | 19.55 |
| 28.28 | 74.55 | -67.38 | 22.95 |
| *Pacific relative to Marie Bird Land (West Antarctica) (86-0 Ma)** | | | |
| 33.54 | 74.38 | -54.74 | 27.34 |
| 42.54 | 74.90 | -51.31 | 34.54 |
| 47.91 | 74.52 | -50.19 | 37.64 |
| 53.35 | 73.62 | -52.50 | 40.03 |
| 61.1 | 71.38 | -55.57 | 44.90 |
| 67.7 | 69.33 | -53.44 | 51.50 |
| 86.0 | 65.77 | -47.44 | 67.10 |
| ** Cande et al. [1995]* | | | |
| *Pacific relative to the mantle (150-86 Ma) ** | | | |
| 86.0 | 63.50 | -50.30 | 58.99 |
| 100.0 | 44.80 | -77.0 | 63.45 |
| 150.0 | 54.80 | 91.50 | 81.36 |
| ** Duncan and Clague [1985]* | | | |
| *East Antarctica Relative to West Antarctica (61.2-33.5) ** | | | |
| 33.54 | -18.15 | -17.85 | 0.696 |
| 43.08 | -18.15 | -17.85 | 1.7 |
| ** Cande et al. (in prep)* | | | |
| *East Antarctica Relative to Australia (135-0)* | | | |
| * 10.95 | 11.74 | 38.16 | 5.80 |
| ** 43.80 | 15.07 | 31.78 | 24.55 |
| 46.30 | 14.00 | 33.34 | 24.70 |
| 52.70 | 10.39 | 35.59 | 25.15 |
| 60.9 | 9.95 | 36.52 | 25.55 |
| *** 79.1 | 5.12 | 39.80 | 25.57 |

**Table 2** (continued)

| *Age,* Ma | *Latitude* +°N | *Longitude* +°E | *Angle,* deg |
|---|---|---|---|
| 83.0 | 2.05 | 40.79 | 27.12 |
| **** 95.0 | 1.4 | 40.5 | 27.35 |
| 135.0 | 0.55 | 39.77 | 28.13 |

* *Cande et al. [in prep]*
** *Tikku [in prep]*
*** *Tikku and Cande [1999]*
**** *this study*

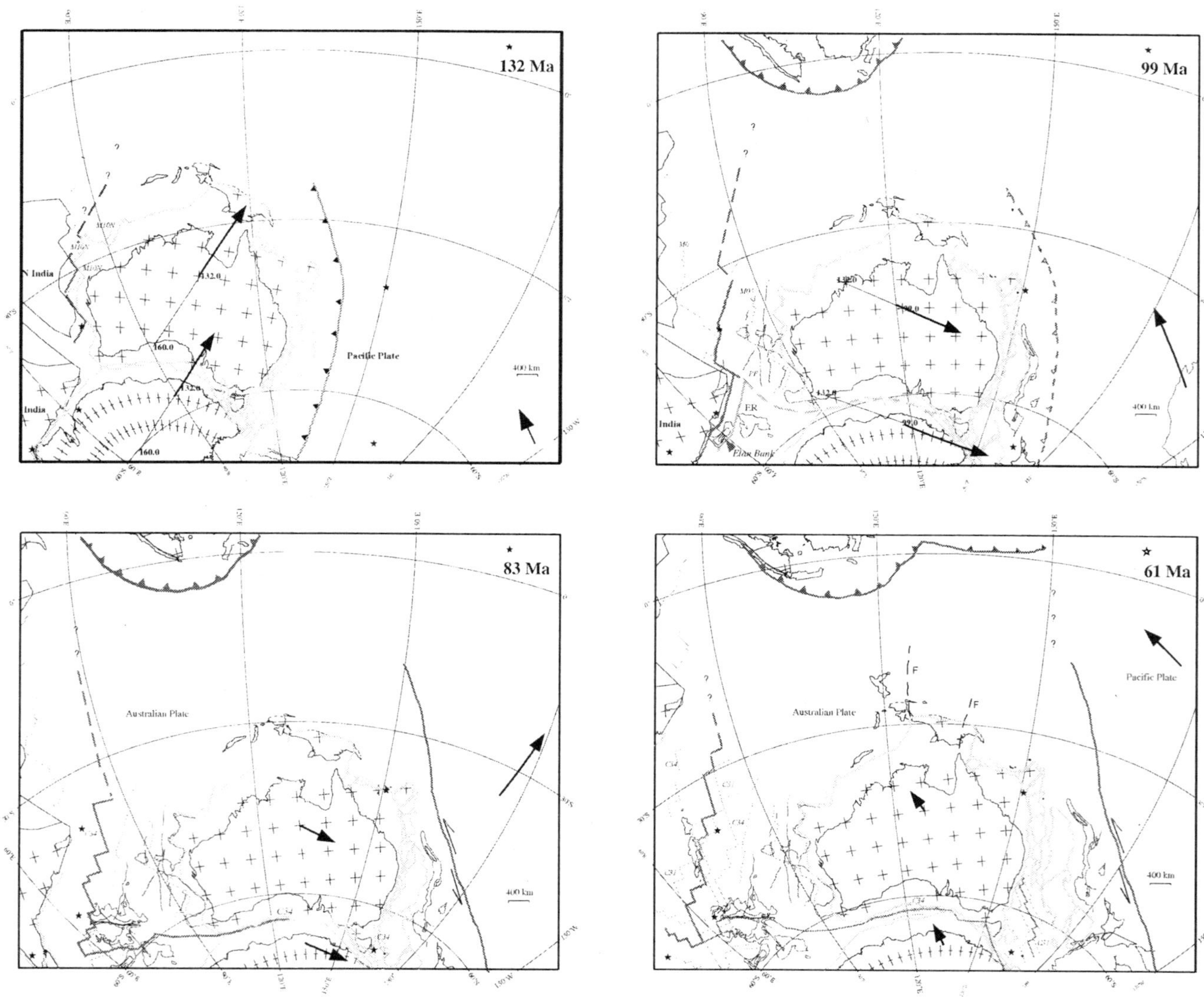


**Figure 5a.** Plate reconstructions in the Atlantic-Indian hotspot frame from *Müller et al.* [1993]. Pacific Plate absolute motion at 132 and 99 Ma is shown in the Pacific hotspot reference frame, as the closure of the plate circuit between the Pacific Plate and Atlantic-Indian hotspots is impossible to due the absence of Pacific passive margins at these times. Black stars represent hotspot locations. Arrows indicate the absolute plate motion of Australian, Antarctic and Pacific plates. Abbreviations are: PF, pseudofault; ER, extinct ridge, F, fault; BR, Broken Ridge; WB, Wharton Basin; NR, Ninetyeast Ridge; BA, Banda Arc; OP, Ontong-Java Plateau; DR, Dampier Ridge; CP, Campbell Plateau; CR, Chatham Rise; MTJ, Macquarie Triple Junction; TK, Tonga-Kermadec.

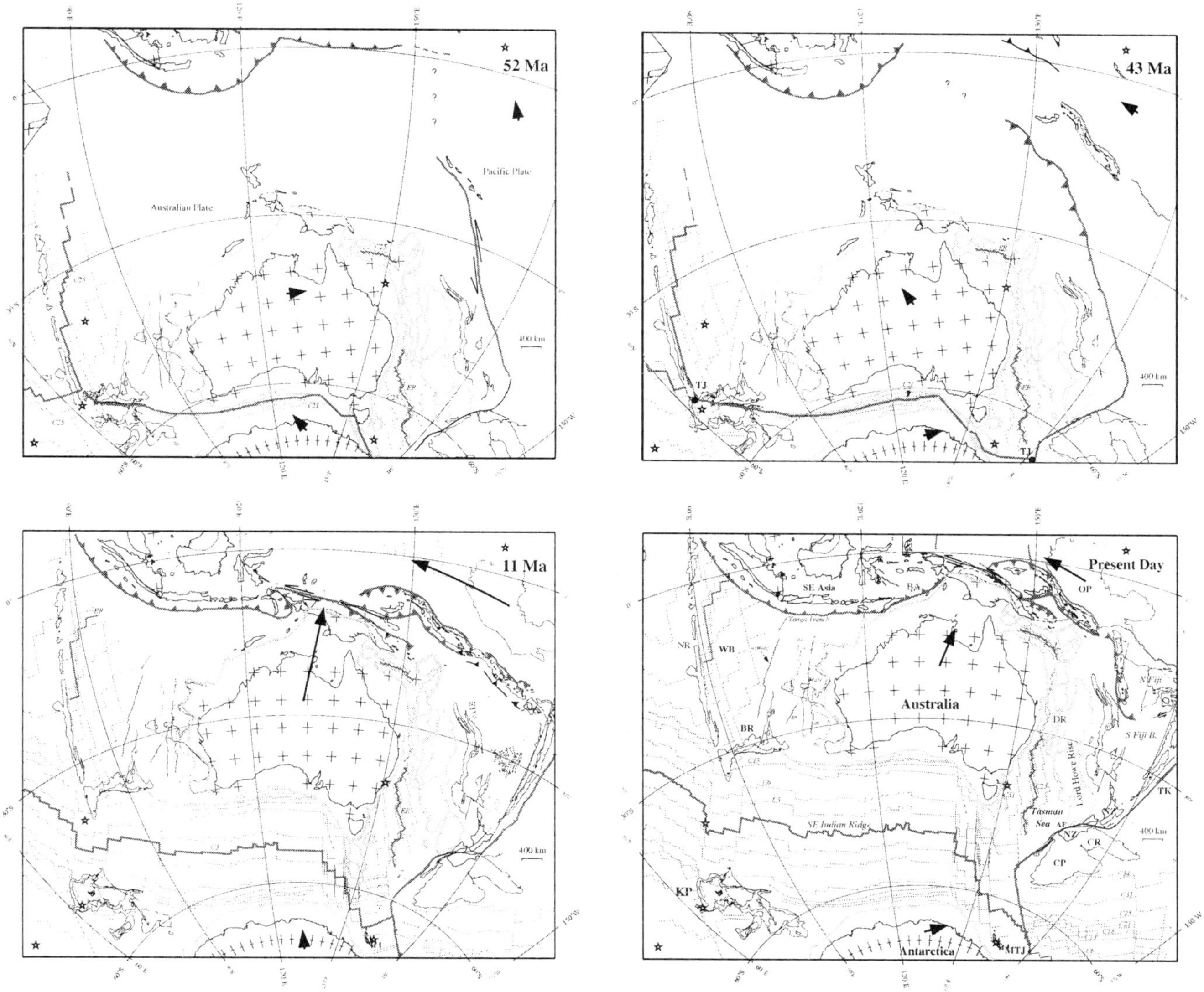


**Figure 5b.** Plate reconstructions in the Atlantic-Indian hotspot frame from *Müller et al.* [1993]. Abbreviations are: BR, Broken Ridge; WB, Wharton Basin; NR, Ninetyeast Ridge; BA, Banda Arc; OP, Ontong-Java Plateau; DR, Dampier Ridge; CP, Campbell Plateau; CR, Chatham Rise; MTJ, Macquarie Triple Junction; TK, Tonga-Kermadec.

[1974] basement age estimate at Site 256 to a minimum of about 101 Ma. Our modeled half-spreading rate between Greater India and Australia after chron M0 is 36 mm/yr. This spreading rate is not merely an extrapolation of M-sequence spreading rates, but based on a number of arguments outlined in *Müller et al.* [1998a] to create a self-consistent set of reconstructions between Australia, Antarctica, India, Africa and Madagascar. Combining the DSDP minimum age of 101 Ma of the ridge north of Batavia Knoll with this spreading rate yields a minimum age of 97 Ma for the fracture zone bend, located 150 km NW of the dated ridge.

A major unconformity on the southern margin of Australia , interpreted as the breakup unconformity by *Veevers* [1984], is well exposed and straddles the Albian-Cenomanian boundary (98.9 Ma, [*Gradstein et al.*, 1994]) between the volcanogenic Otway Group and the quartzose Sherbrook Group, dated by means of the *Phimopollenites pannosus* palynological zone [*Veevers*, 1984]. The difference between this age and the minimum age of 97 Ma derived above for the plate reorganisation between Greater India and Australia is within the combined errors of these age estimates. Therefore, we argue that it is likely that the two events occurred contemporaneously, and assign the Albian-Cenomanian boundary age of ~99 Ma to the major fracture zone bend south of the Wharton Basin. Tectonic lineations expressed in gravity anomalies south of the Great Australian Bight (Figure 4b) indicate that the

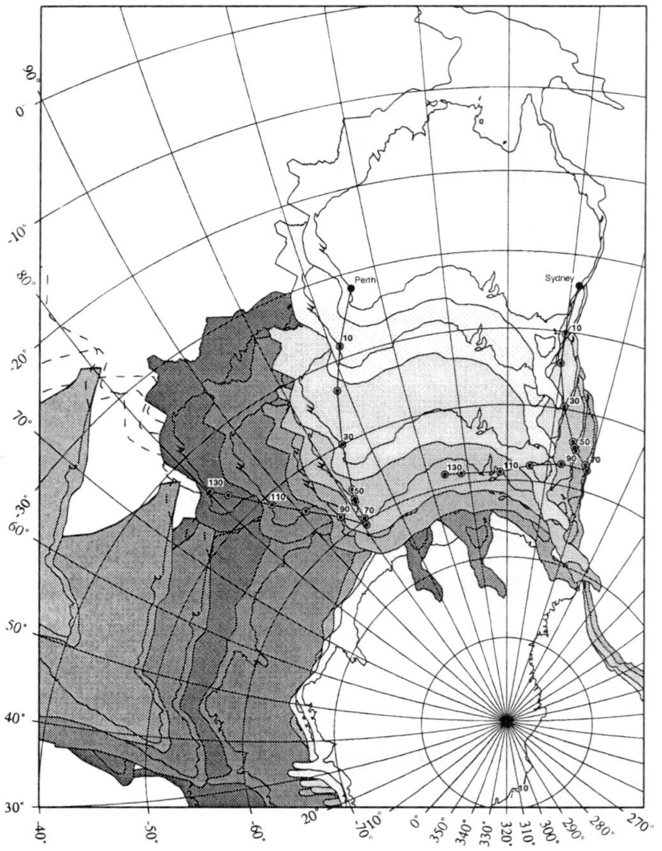


**Figure 6.** Australian, Indian and East Antarctic plate motions in the Atlantic-Indian hotspot frame from *Müller et al.* [1993]. Absolute plate motion is shown in 20 m.y. intervals starting at 130 Ma by tracking the positions of both the coastlines and the boundaries between continental and oceanic crust through time. The plates are gray-shaded using the darkest gray for the 130 Ma plate positions, and incrementally lighter grays for younger reconstruction times. In addition, we show the positions of Perth and Sydney through time in 10 m.y. intervals.

relative motion between the Australian and East Antarctic plates was oriented NW-SE initially, and changed to N-S roughly at chron 27 (61.2 Ma) [*Tikku*, 1998].

The main unresolved question regarding the opening between India, Australia and Antarctica is whether sea floor spreading between India and Antarctica commenced contemporaneously with the formation of the Gascoyne and Cuvier abyssal plains, as depicted in *Veevers* [1984]. If it did, then Greater India and Australia would have constituted a two-plate system, and we should find Mesozoic M-sequence magnetic anomalies in the Enderby Basin off Antarctica west of Kerguelen and off the east coast of India. However, the Enderby Basin is extremely poorly mapped, with no M-sequence anomalies identified to date, while low-amplitude magnetic anomalies off eastern India have recently been interpreted as reflecting the magnetic quiet zone [*Gopala Rao et al.*, 1997].

*Powell et al.* [1988] suggested that Greater India may have separated from Australia/Antarctica in the early Cretaceous around a stage rotation pole located not far west off the southern tip of India. If this scenario were correct, then small circles about this stage pole should follow the fracture zones in the Perth, Cuvier and Gascoyne abyssal plains west of Australia. However, *Müller et al.* [1998a] found a large discrepancy between small circle and fracture zone azimuths, implying that this model does not agree with the tectonic fabric of the ocean floor formed during this time west of Australia (Figure 4a). Further, the counter-clockwise rotation of India required by this scenario would have given rise to transpression between India and Madagascar, and for this there is no evidence.

Instead, the fracture zones west of Australia indicate that Greater India separated from Australia about a stage rotation pole at roughly 17°S and 10°E in the reference frame of the Australian Plate [*Müller et al.*, 1998a]. If Greater India represented one single plate, the associated northward motion of India would have resulted in about 1000 km of left-lateral strike-slip between India and Madagascar from 136 Ma to 99 Ma. This may seem conceivable, except that India's pre-breakup position with respect to Madagascar is fairly well constrained by conjugate fracture zones visible in the marine gravity anomaly grid. This pre-breakup position, preceding formation of the Mascarene basin between the Seychelles and Madagascar, requires India to be located slightly further south relative to Madagascar than in its fit reconstruction [*Müller et al.*, 1998a]. Consequently, the left-lateral strike-slip between India and Madagascar required to model sea floor spreading between Greater India and Australia as a two-plate system would have to be succeeded by an even greater amount of right-lateral strike-slip before breakup between India and Madagascar. This model also results in compression between India and Arabia from 136 to 120 Ma, for which there is no evidence. Consequently, we reject this model as unlikely.

The most plausible, internally consistent model for the breakup of eastern Gondwanaland involves a large transform fault between southern and northern Greater India, which would have formed in the Early Valanginian, when sea floor spreading started west of Australia, while rifting between India and Antarctica remained slow until breakup in the Aptian at about 120 Ma (Figure 5a). The transform fault would have separated northern Greater India (NGI) from southern Greater India (SGI) until motion between the two plates stopped at about 120 Ma. The triple junction between NGI, SGI and Antarctica would have been located southwest of the Naturaliste

Plateau. This model implies that the ocean floor west of the Naturaliste Plateau, southwest of the southernmost isochrons in the Perth Abyssal Plain (Figures 1, 4a) would reflect post-120 Ma Greater India-Antarctica spreading.

Breakup between India and Antarctica after rather than before 120 Ma is also supported by seismic data and the biostratigraphy of sediments from the Mac. Robertsen shelf, northwest of Prydz Bay, East Antarctica [*Truswell et al.*, 1999], which is conjugate to the northeastern Indian margin. They found evidence for half-graben formation on the Antarctic margin until at least until the early Aptian, which started at 121 Ma [*Gradstein et al.*, 1994]. Truswell et al. [1999] dated the onset of marine conditions in this area to Aptian-Paleocene, i.e. the age of the oldest offlapping, marine sediments on the outer shelf, in accordance with our tectonic model.

Chron 27 (61.2 Ma) also marks the onset of faster seafloor spreading east of the Balleny Fracture Zone [*Cande et al.*, in press], compared with the remainder of the Southeast Indian Ocean, indicating the formation of a triple junction which accommodated the incipient motion between East and West Antarctica. The spreading direction between Australia and Antarctica changed from NW-SW to N-S, accompanied by the initiation of rifting between Broken Ridge and the Kerguelen Plateau as well as a magmatic pulse creating conjugate volcanic ridges on both plates [*Tikku*, 1998]. Another tectonic event, the joining of the Bellinghausen plate to West Antarctica, was recorded at this time in the Pacific Ocean [*Cande et al.*, 1998]. The 61 Ma event also affected the Tasman Sea basin resulting in a counterclockwise change in the spreading direction, which initiated a period of transform motion along the central and northern parts of the eastern Australian margins [*Gaina et al.*, 1998b]. *Klepeis* [1999] showed that at about 60 Ma dextral transpression was initiated in Fiordland, New Zealand, as a result of a change in the plate boundary geometry. The Coral Sea started opening off northeast Australia at the same time [*Gaina et al.*, 1999] (Figure 5a). Chron 24 young (52 Ma, Figure 5b) marks the cessation of seafloor spreading along the entire eastern and northeastern Australian margin [*Gaina*, 1998; *Gaina et al.*, 1998a] and the onset of a compressional event in the Papua-Rennell-New Caledonia region [*Kroenke*, 1984].

The almost stationary position of Australia w.r.t. the mantle from ~80 Ma to ~40 Ma (Figures 5a, 5b) may reflect the progressive subduction of the Pacific-Phoenix ridge to the east of New Zealand preceding 80 Ma, resulting in a diminished trench suction force east of Australia. However, we stress that the timing of the apparent bend in the absolute plate motion path of Australia (Figure 6) is not well constrained. It essentially reflects the bend between the New England seamount chain and the Corner seamounts at roughly 80 Ma in the central North Atlantic [*Müller et al.*, 1993], since the Walvis Ridge as the only other constraint for pre-80 Ma hotspot track models in the Atlantic-Indian oceans is too broad to be useful for constraining rapid changes in absolute plate motion directions.

In the late Tertiary Australia has moved north rapidly at rates between 40 and 60 mm/yr. It is important to realize that the absolute motion of Australia relative to the mantle is quite different from the relative motion of Australia away from Antarctica, as expressed in magnetic lineations and fracture zones along the Southeast Indian Ridge. This is because the absolute motion of both Antarctica and Australia since about 40 Ma has had an eastward component (Figure 6), while Australia has moved northward relative to Antarctica. Therefore, fracture zones of the Southeast Indian Ridge reflect only relative, but not the absolute plate motion of Australia.

## 5. ANOMALOUS SUBSIDENCE AND UPLIFT

Passive margins are expected to exhibit thermal, exponential subsidence after rifting ceases [*Jarvis and McKenzie*, 1980; *McKenzie*, 1978]. Instead, tectonic subsidence curves on Australia's continental shelves show many large, unexpected deviations from thermal subsidence, many of which occurred at times of major plate tectonic reorganizations. We investigate regional anomalous subsidence and uplift postdating the onset of sea floor spreading and examine how these tectonic events may be related to plate tectonic or mantle processes. The tectonic subsidence analysis is based on 108 industry wells on the northwest shelf of Australia, 4 wells on Australia's southern margin, and 5 Ocean Drilling Program (ODP) Sites on the northeast shelf of Australia. The magnitude of subsidence or uplift in a region depends on the net contributions from sediment and water loading/unloading and vertical tectonic movements. We have isolated the tectonic component of subsidence/uplift by removing the isostatic component caused by sediment and water loading, using a standard backstripping methodoloy [*Sclater and Christie*, 1980], based on the age, thickness and lithology of each stratigraphic unit, the present depth below sea-level of each unit, a porosity-depth relationship for each lithology, and an estimated range for the palaeo water depth at the time of deposition of each unit.

To identify at which times anomalous subsidence or uplift was most widespread on the Northwest shelf, we created histograms which show the number of wells which exhibit either anomalous subsidence or uplift (Figure 7) for 10 my time periods since 130 Ma. The subsidence histogram shows three distinct peaks for the periods 100-90 Ma, 70-60 Ma and 10 Ma to present, also clearly visible

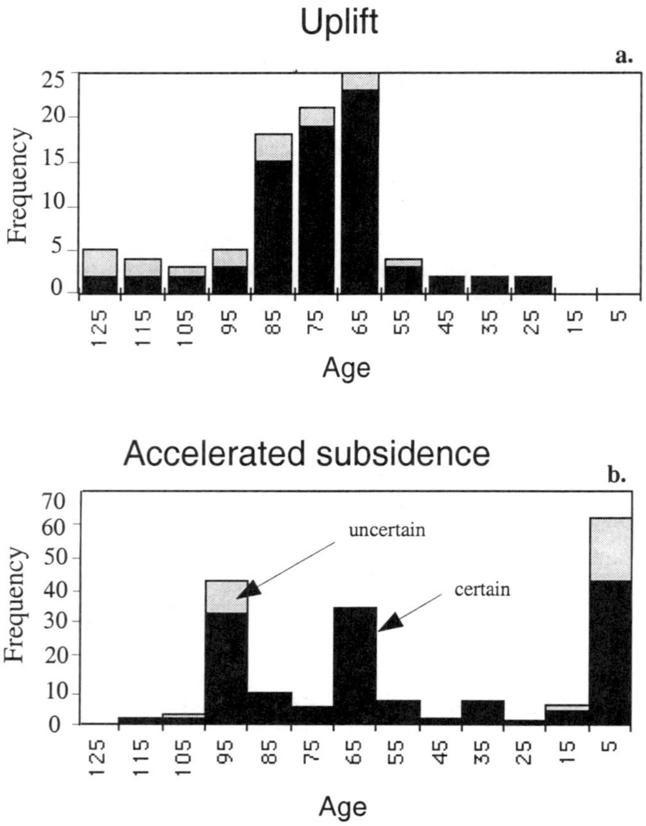


**Figure 7**. Histogram showing frequency of accelerated subsidence or uplift events on the North West Shelf. Note tectonic subsidence events at 100-90 Ma, 70-60 Ma and 10 Ma - present and uplift events at 70-60 Ma.

in many of the subsidence curves shown in Figure 8, whereas the uplift histogram shows widespread anomalous uplift from 90 to 60 Ma (Figure 7).

The anomalously high subsidence (or uplift) for the three time intervals from 100-90 Ma, 70-60 Ma and 10 Ma to present were then block averaged using a median filter and gridded using a natural spline with a grid interval of 0.1 degrees. The smoothed spatial distribution of tectonic subsidence and uplift as shown in the contour plots in Figure 9 gives a qualitative, regional "low-pass filtered" image of tectonic reactivation. The results are meaningful despite the shortcomings of this simple method, as (1) it is based on a large number of wells, and (2) the anomalous vertical motions observed (up to 700 m within a few million years, Figure 9) are up to an order of magnitude larger than the expected thermal subsidence for the same time period. We also show the tectonic subsidence for four wells with a relatively high-resolution stratigraphy covering the period from 130 Ma to present from the Otway and Bass basins, southeastern margin of Australia (Figure 10).

Available ODP data from the Queensland and Marion Plateau off northeast Australia only cover the last 20 million years. *Müller et al.* [1999] integrated geophysical logs, biostratigraphic and lithological information and seismic reflection data to correlate the pattern of Miocene hardgrounds to 3rd order eustatic sea-level lowstands at ODP Sties 812 and 814 to better estimate ages where biostratigraphic information is poor. They backstripped the sediment sequence to compute tectonic subsidence histories for five sites on the Queensland and Marion plateaus and in the Queensland Trough (Figure 11).

## 6. TECTONIC EVENTS

### *6.1. 99 MA*

A tectonic event between 100 and 90 Ma (Figure 9a) resulted in tectonic subsidence up to 550 m on the Northwest shelf of Australia in the Malita Graben-Sahul Trough, extending south to the Londonderry High, and moderate anomalous tectonic subsidence (not exceeding 200 m) in the Browse Basin. At the same time, inversion in the Bass Basin north of Tasmania occurred [*Hill et al.*, 1995]. These events are clearly related to the plate reorganisation between India and Australia that we date at about 99 Ma. Most of the tectonic subsidence recorded at this time has been permanent. This indicates that the change in intra-plate stresses triggered by the 99 Ma event resulted in renewed local lithospheric thinning. In contrast, this time marks the cessation of rifting in many wells along the southern margin of Australia (Troas-1, Durroon-1, Figure 10, see also *Hegarty et al.* [1988]), and likely corresponds to the breakup-unconformity in the Otway Basin [*Veevers*, 1984]. These observations are consistent with an approach of the Neo-Tethyan ridge north of India towards the trench as a cause for this event. As the subducted plate becomes smaller and younger in age, the ridge push force decreases, and vanishes when the ridge is subducted.

### *6.2. 61 MA*

The time period from 70-60 Ma is characterised by accelerated tectonic subsidence of the Sahul and Ashmore platforms and the Londonderry High, while the intervening grabens (Malita, Vulcan) show either slight uplift or no anomalous subsidence (Figure 9b). The magnitudes of subsidence anomalies at this time are smaller than at 100-90 Ma, ranging from up to 100 m of uplift to 100-200 m of

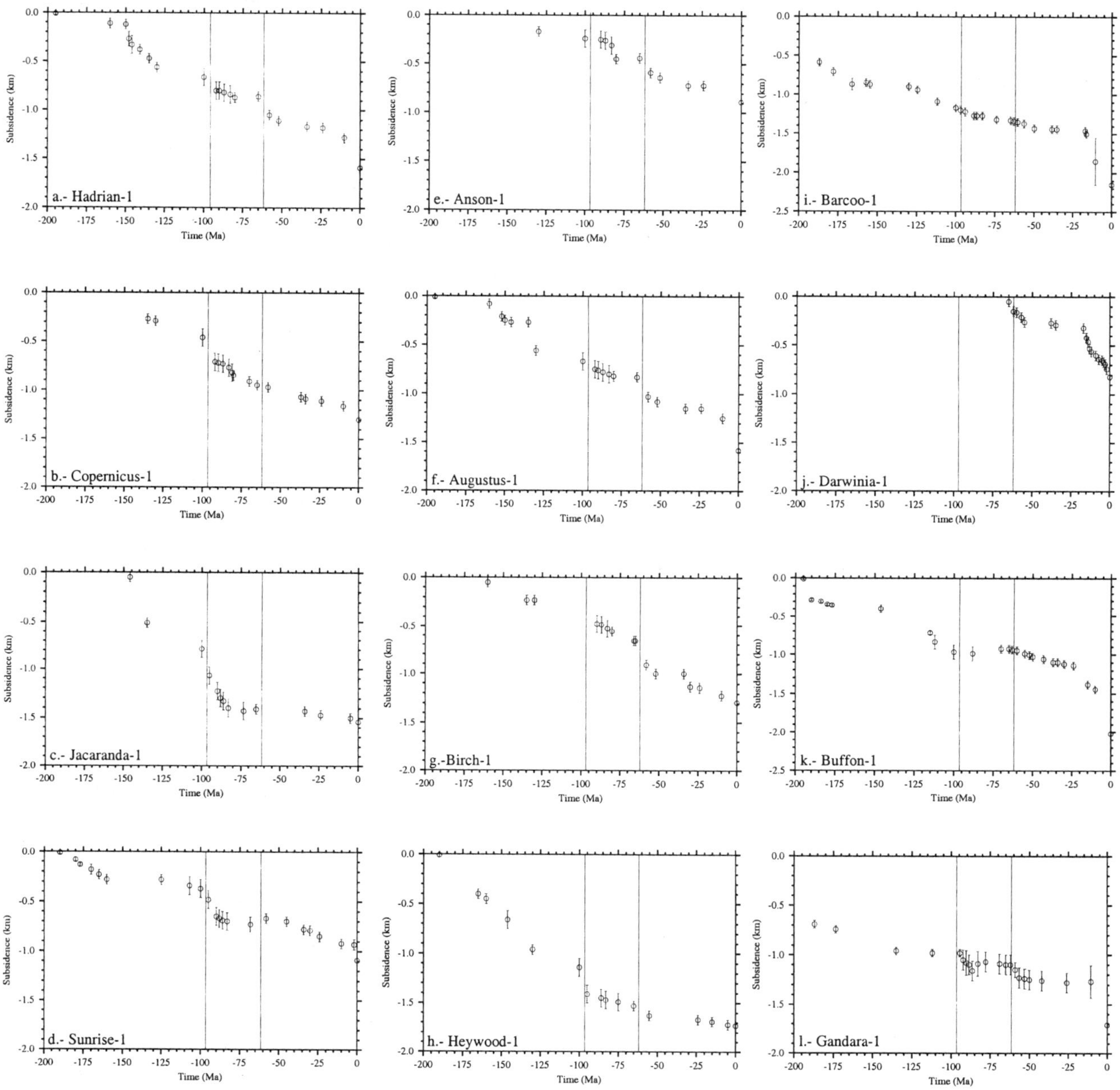


**Figure 8**. Selected backstripped wells from the northwest shelf of Australia. Well locations are shown on Figures 4b and 9. Water depth uncertainties are shown by vertical error bars. Note deviations from exponential subsidence at about 99 Ma and 61 Ma (thin black vertical lines), and 10 Ma – present.

subsidence. The amplitude and spatial distribution of anomalous subsidence and uplift between 70 and 60 Ma may reflect elastic plate bending caused by extensional in-plane stresses, in that platforms exhibit subsidence, whereas rifted grabens show uplift or no subsidence anomalies. This event may be related to a number of plate tectonic events at (61.2 Ma) described above. Both the extensional elastic plate deformation on the northwest shelf, as well as changes in plate kinematics suggest a decrease in southward oriented ridge push force north of Australia, pointing to the subduction of some segments of the Neo-Tethyan spreading ridge north of Australia. Both the 99 and 61 Ma events may have originated from the stepwise subduction of the Neo-Tethyan Ridge, first north

A)

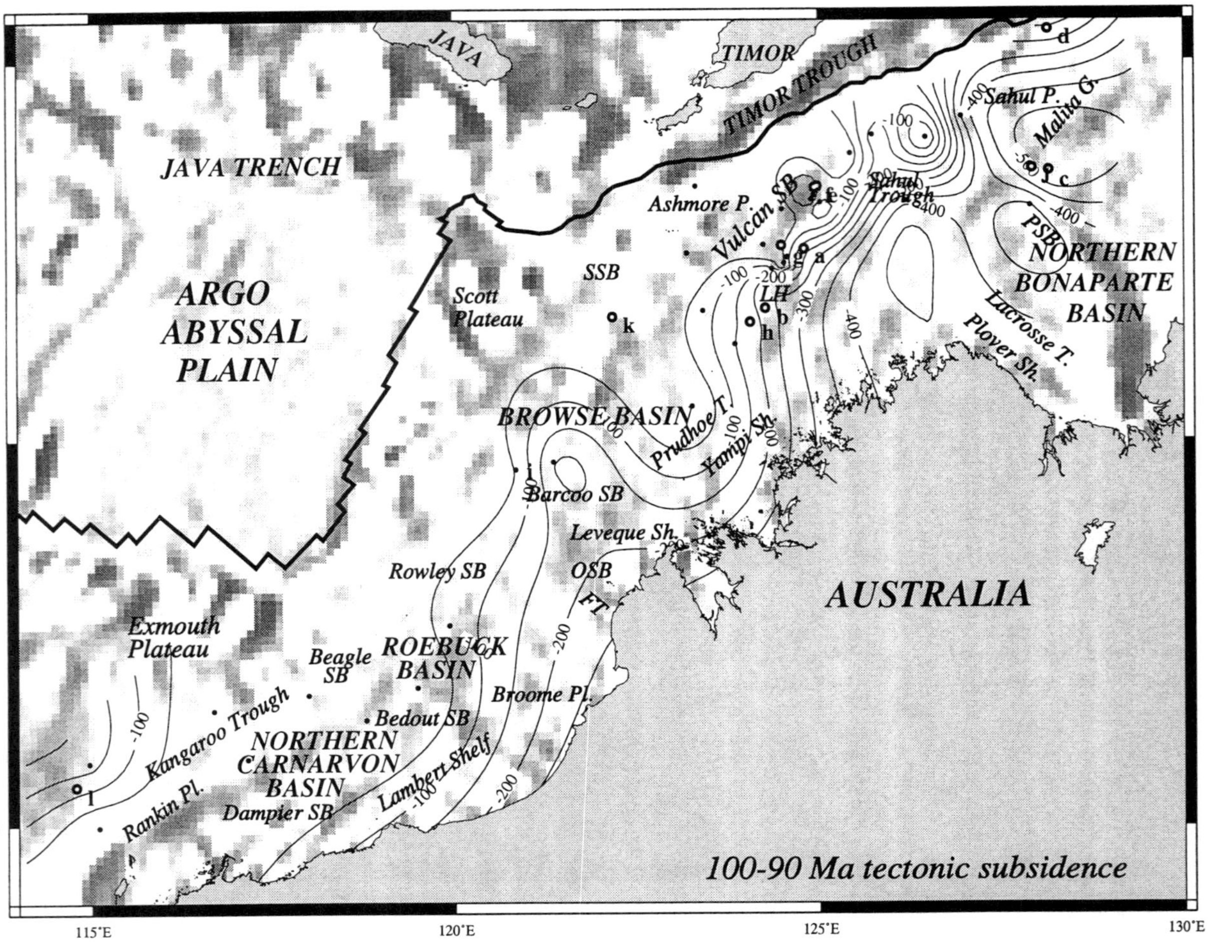


**Figure 9.** Contoured tectonic subsidence on the North West Shelf at 100-90 Ma (a), 70-60 Ma (b) and 10 Ma – present (c). Positive values indicate uplift, whereas negative values represent subsidence. Black dots represent well locations and letters denote wells shown on Figure 8. Abbreviations are: SB, Sub Basin; T, Terrace; Sh, Shelf; G, Graben; P, Platform; FT, Fitzroy Trough; PSB, SSB, OSB, Petrel Seringapatam and Oobagooma sub-basins; (subdivision of North West Shelf after *Hocking et al.* [1994]. See text for discussion.

of India at 99 Ma and then north of Australia at 61 Ma. Alternatively, the 99 Ma event could be related to a major change in absolute motion of the Pacific Plate documented for about the same time [*Duncan and Clague*, 1985].

### *6.3. 43 MA*

This time marks the well-known bend of the Hawaiian-Emperor Chain [*Clague et al.*, 1989]. *Norton* [1995] proposed that the bend had no causal connection to relative plate motions as there is no obvious bend in the Farallon-Pacific fracture zones and a lack of tectonic events along the margins of the northern Pacific Ocean. It has also been suggested that the collision between the Indian plate and Eurasia, dated at about the same time, caused the sudden change in the Pacific plate motion [*Patriat and Achache*, 1984], as spreading in the Wharton Basin between India and Australia ceased at chron 20 (~43 Ma) [*Liu et al.*, 1983]. *Lithgow-Bertelloni and Richards* [1998] tested this hypothesis by modelling global plate velocities and concluded that it is unlikely that the two phenomena were related. Their results confirmed the hypothesis also proposed by *Gordon et al.* [1978] that the 43 Ma event was caused by the sudden change of the Pacific-Australian margin from transform to subduction along preexisting transform faults.

Preliminary reconstructions that include relative motion between East and West Antarctica [*Cande et al.*, 1998] based on recently collected geophysical data north of the Ross Sea and south of the Tasman Sea indicate that a

B)

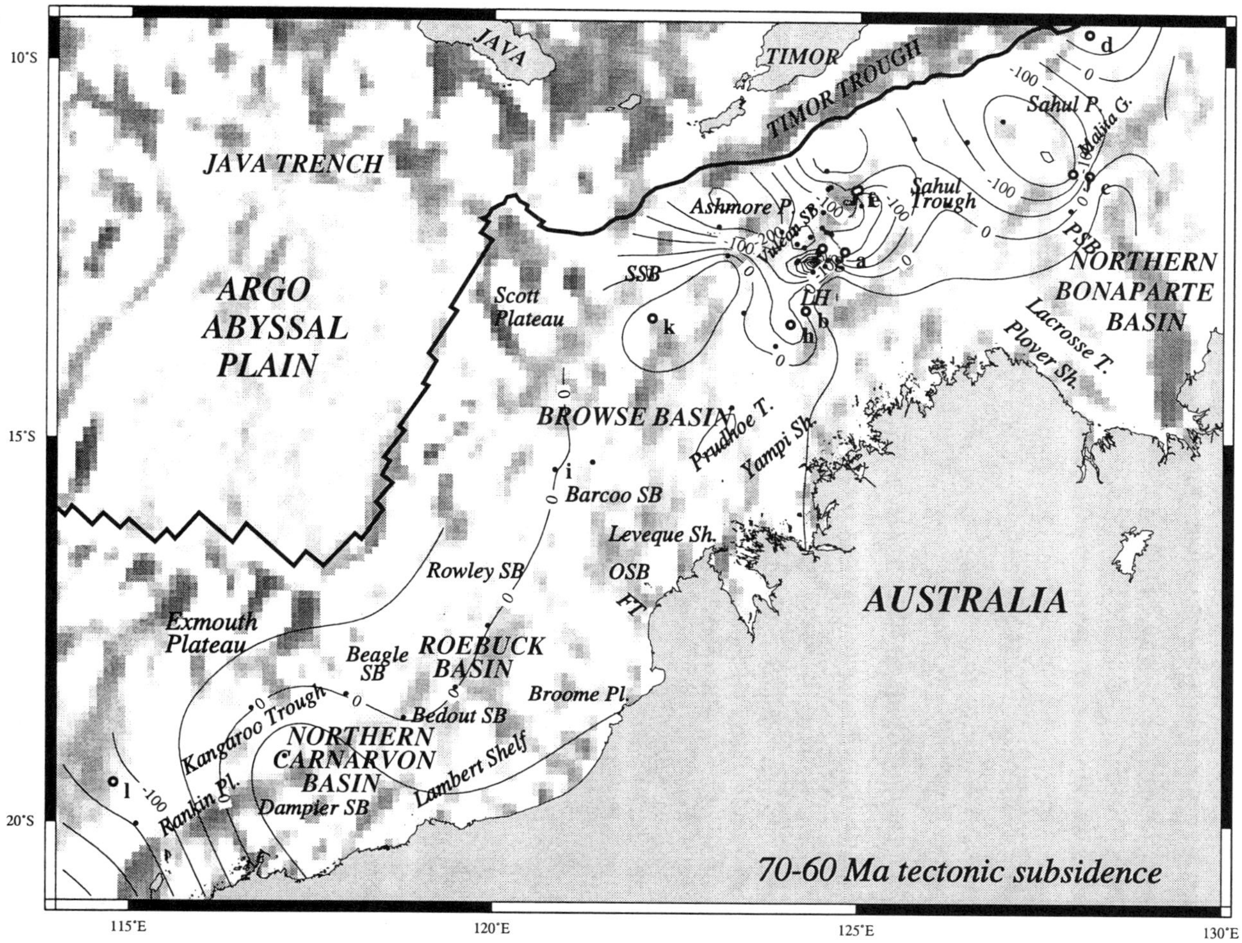


**Figure 9b.** (Continued.)

tectonic event at 43 Ma may indeed mark the onset of renewed subduction east of Australia [*Gaina*, 1998] (Figure 5b). There is abundant evidence that compression affected the eastern and northeastern Australian plate close to 43 Ma. Overthrusting of New Caledonia that has been dated Middle-Late Eocene is the result of a compressional event [*Collot et al.*, 1987]. Middle Eocene foredeep sediments and olistostromes record the collision of the Norfolk-New Caledonia Ridge continental ribbon with the forearc region of the intra-oceanic Loyalty-d'Entrecasteaux arc [*Aitchison et al.*, 1995]. In addition, *Kroenke et al.* [1993] suggested that the age of hiatuses described for rock samples from DSDP Site 289 and ODP Site 803 (both on the Ontong Java Plateau) are close to 43 Ma, indicating a tectonic event in the Southwest Pacific. The tectonic event that occurred at 43 Ma also influenced the northern Australian margin. *Pigram and Symonds* [1991] constructed basement subsidence curves from three wells located offshore eastern Papuan Basin, which show a sudden increase in subsidence at 43 Ma, most evident for the northernmost well. Similarly, two wells in the Bass Strait exhibit accelerated subsidence at this time (Figure 10).

In summary, contrasting *Norton* [1995], there is a wealth of evidence for a tectonic event at the southwestern margin of the Pacific Plate at 43 Ma. However, even if subduction between the Pacific and Australian plates was re-initiated at around 43 Ma, this does not prove that it was the ultimate cause of the Hawaiian-Emperor bend. Exactly how to reconcile relative and absolute plate motions, and large-scale differential motion in the mantle, is the subject of ongoing work, in particular involving the reconstruction of the history of relative motion between East and West Antarctica.

C)

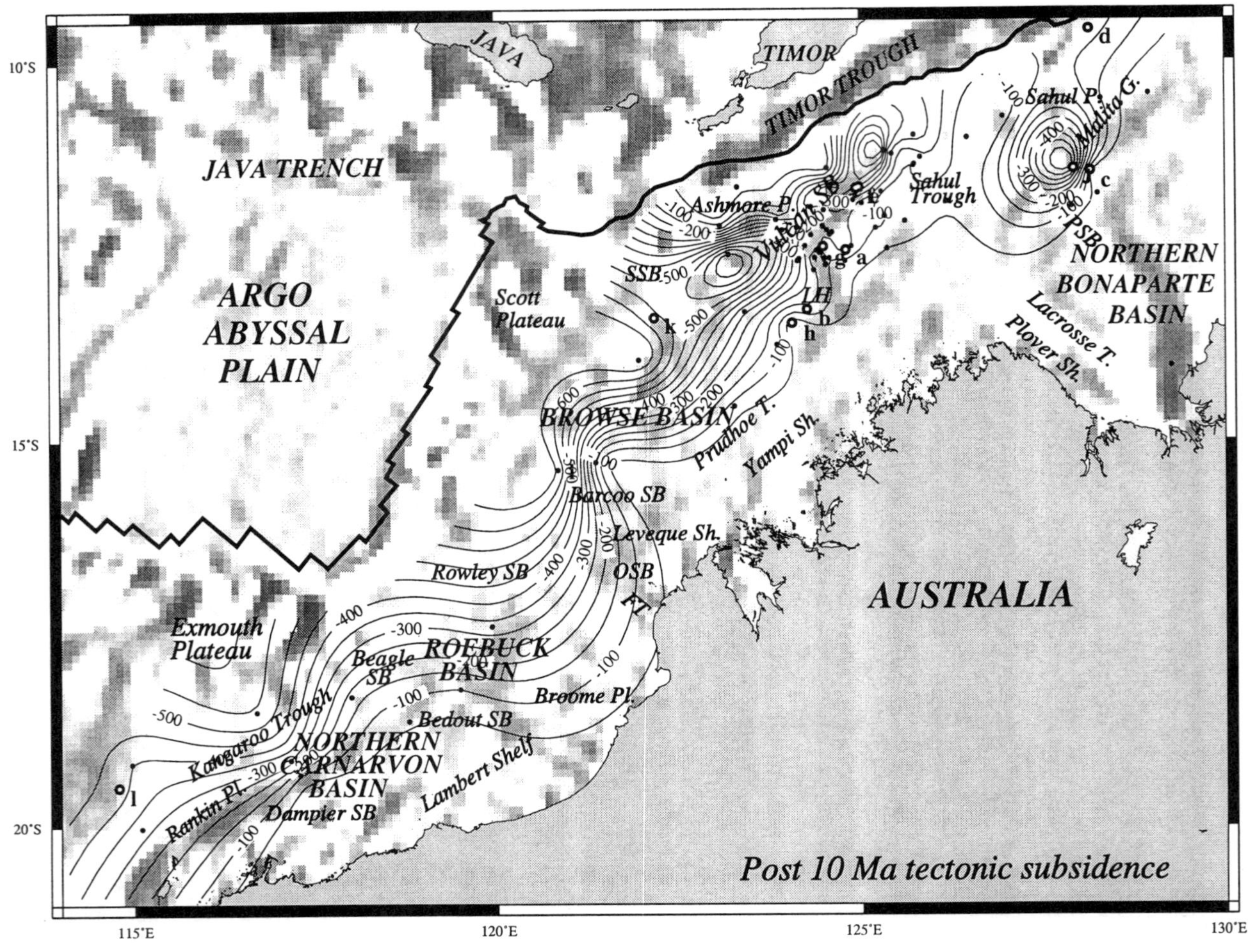


**Figure 9c.** (Continued.)

*6.4. 25 MA TO PRESENT*

At 23 Ma subduction along the Melanesian trench ceased after the Ontong Java Plateau docked with the Melanesian Arc [*Kroenke*, 1984]. From this time until 10 Ma subduction was accommodated along the Tonga-Kermadec Trench (Figure 5b). At 12 Ma, after a period of tectonic quiescence along the Melanesian arc, the Vitiaz-New Hebrides-Fiji-Lau-Tonga arc split and led to extension behind the Vitiaz Trench. Subduction along New Guinea ended and subduction and volcanism started along the Solomon Islands [*Yan and Kroenke*, 1993] (Figure 5b). In the last 10 Ma the Vitiaz Trench became inactive, and subduction was transferred to the New Hebrides Trench, which began a clockwise rotation (Figure 5b).

Subduction along the Solomon Trough and other changes in plate boundaries north of Australia favoured an extensional regime northeast of Papuan Peninsula and created the Woodlark Basin starting at 5 Ma [*Taylor et al.*,

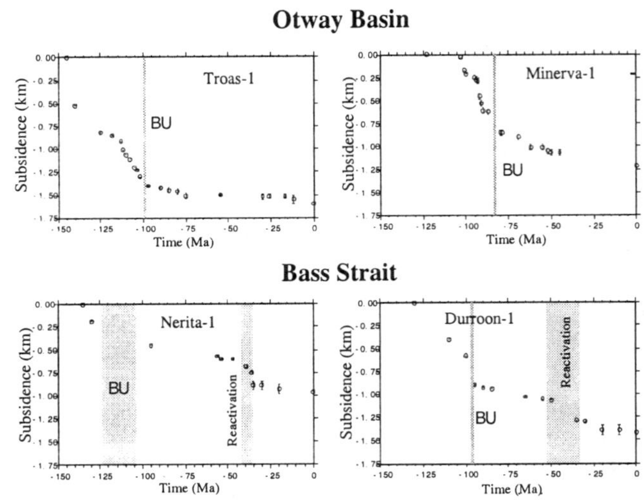


**Figure 10.** Water loaded tectonic subsidence for four wells in the Otway and Bass basins. BU: Breakup unconformity. Note Paleogene accelerated subsidence in Bass Basin.

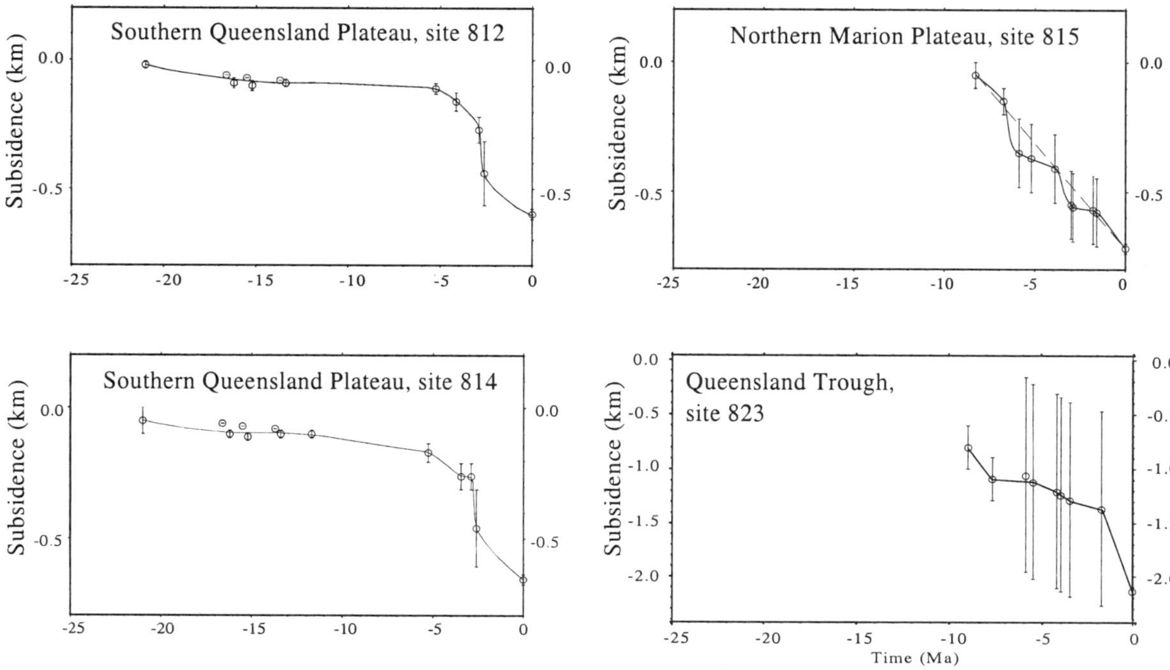


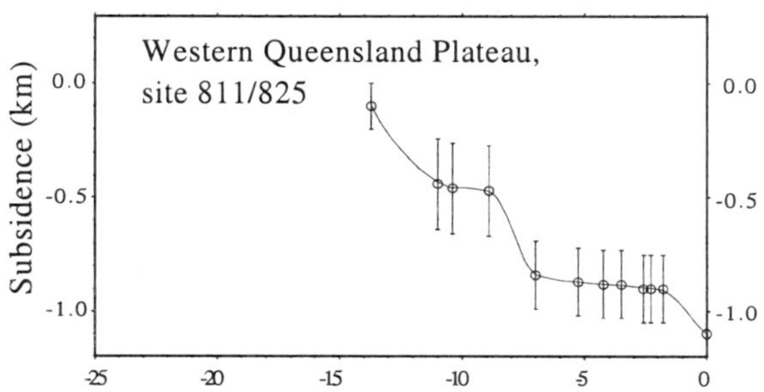


**Figure 11**. Water loaded tectonic subsidence for ODP Leg 133 sites 812, 814, 811/825, 815 and 823. Note subsidence events after 9 Ma.

1995]. Trench roll-back determined the opening of the Lau Basin (6-5 Ma) [*Scholl*, 1995] between the Lau-Colville and Tonga-Kermadec volcanic arcs. Collision along the Banda Orogen northwest of Australia created a foreland basin including the Timor Trough and the outer North West Shelf post-5 Ma [*Lorenzo et al.*, 1998] (Figure 5 b).

Widespread tectonic reactivation occurred on Australia's northwest shelf in the Late Miocene-Pliocene. Regional tectonic subsidence (Figure 9c) shows widespread accelerated post 10 Ma subsidence of the outer northwest shelf, stretching from the Vulcan Graben to the Northern Carnarvon Basin. Wells in the Roebuck Basin and the northern Carnarvon Basin show accelerated subsidence commencing as early as 20 Ma (e.g. Barcool, Darwinial, Figure 8). The tectonic subsidence we show represents a minimum estimate as sea level has generally fallen in the Miocene/Pliocene [*Haq et al.*, 1987]. Including a falling eustatic sea level would increase the backstripped tectonic subsidence.

Anomalous subsidence and/or uplift has been linked to wrench and inversion tectonics and foreland basin formation [*Etheridge et al.*, 1991; *Lorenzo et al.*, 1998; *Shuster et al.*, 1998]. Miocene to recent anomalous subsidence in the Cartier Trough and the Malita Graben has likely been caused by reactivation of rift structures by sinistral shear, as a consequence of oblique collision north of Australia [*Shuster et al.*, 1998].

Recent collision along the Banda Orogen has created a foreland basin including the Timor Trough and the outer North West Shelf [*Lorenzo et al.*, 1998]. Sea floor faulting due to foreland basin formation occurs within a 300 km radius south of the axis of the Timor Trough and modelled values of the elastic plate thickness suggest that plate flexure due to thrust loading should not affect the North West Shelf south of latitude ~14-15°S. Therefore it is unexpected to find up to 500 m of Late Miocene-Pliocene subsidence in the western Roebuck Basin, the western Browse Basin, and the Vulcan Graben (Figure 9c), up to 1000 km south of the active plate boundary. Regional

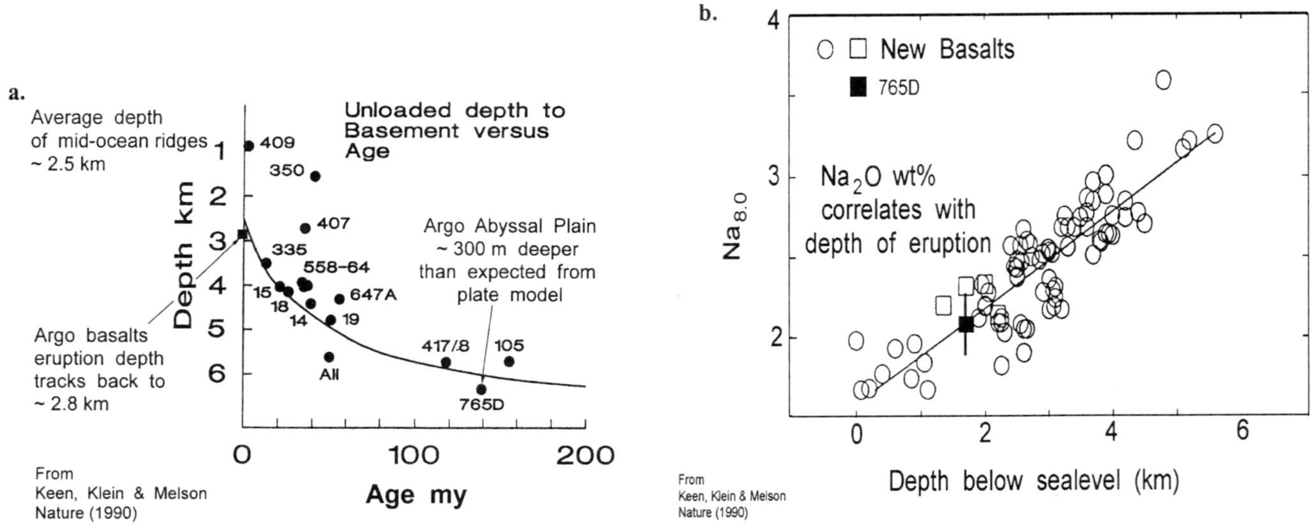


**Figure 12**. The Argo abyssal plain is also anomalously deep by about 300, assuming that it formed at an average ridge elevation of 2500 m, subsided according to the plate model, and has an average thickness of 7-7.1 km (a). However, the normalized oceanic crustal $Na_2O$ content as well as other major element oxides of the Argo basalts at ODP site 765d in the Argo Abyssal Plain, aged about 140 Ma, indicate that it formed at much shallower depths at 1.8±0.5 km, at least 200 m shallower than average ocean crust (b) [*Keen et al.*, 1990]. This means that the crust at ODP site 765d is at least 500 m too deep if it has subsided according to the plate model.

accelerated subsidence, which started at least at 20 Ma, cannot be accounted for by simple foreland basin modelling, as the wavelength of the elastic deformation caused by thrust loading along the Banda Arc does not exceed a few 100 km [*Lorenzo et al.*, 1998].

Accelerated subsidence from 25-20 Ma to the present has also been observed on the western Indian margin, where the rate of subsidence increased both on the continental shelf and seaward of the shelf edge, resulting in up to one km of tectonic subsidence in 20 million years [*Agraval and Rogers*, 1988]. *Royer and Chang* [1991] suggested that the break-up of the Indo-Australian plate into two plates began between chrons 6 and 5 (20-10 Ma). *Royer and Gordon* [1997] showed that at least since 11 Ma a third plate, named the Capricorn Plate, has formed by further fragmentation of the Australian Plate. This process has created an intra-plate deformation zone with an "ocean fold and thrust belt" extending from the central Indian Ocean to the Northwest shelf [*Royer and Gordon*, 1997]. More recently, *Gordon et al.* [1998] established a minimum age for the relative motion between the Indian and Capricorn plates at about 18 Ma.

The large area affected by Miocene-Pliocene accelerated subsidence on the North West Shelf, together with contemporaneous accelerated subsidence off western India, suggests that both are related to the complex breakup of the Indo-Australian Plate into the Indian, Australian and Capricorn plates, rather than foreland and pull-apart basin formation alone. In this context, it is interesting to note that reverse faults have been mapped in the ocean crust of the Argo Abyssal Plain [*Kritski et al.*, 1999], east of the area considered to be part of the diffuse intraplate deformation zone by *Royer and Gordon* [1997]. The Argo abyssal plain (Figure 4a) is also anomomalously deep by about 300 m, assuming that it formed at an average ridge elevation of 2500 m, subsided according to the plate model [*Keen et al.*, 1990], and has an average crustal thickness of 7-7.1 km based on seismic refraction data [*Kritski et al.*, 1999] (Figure 12a). However, the normalized oceanic crustal $Na_2O$ content as well as other major element oxides of the Argo basalts at ODP site 765d [*Keen et al.*, 1990] in the Argo Abyssal Plain, aged about 140 Ma, indicate that it formed at much shallower depths of 1.8±0.5 km, at least 200 m shallower than average ocean crust (Figure 12b). It follows that the Argo abyssal plain is at least 500 m deeper than expected, assuming that its original ridge depth was at least as shallow as 2.3 km.

Assessed in isolation, this observation may indicate that the Argo Abyssal Plain has subsided according to the thermal boundary layer model, as ocean crust 140 m.y. old would show an excess depth of ~600 m if it subsided as a thermal boundary layer compared with the plate model [*Parsons and Sclater*, 1977]. However, the observed excess subsidence in the Argo Abyssal Plain is of the same order as the tectonic subsidence that occurred within the last 10 million years on the outer northwest shelf of

Australia (Figure 9c). This coincidence may indicate that both subsidence anomalies are due to the same cause. In this case the cause would be unrelated to the thermal boundary layer model, as all excess subsidence would have occurred recently, since about 20 Ma. The wavelength of this deformation points to viscous, rather than elastic processes, as an elastic plate thickness typical for old ocean crust of 30-40 km [*Bodine et al.*, 1981] would be unable to produce the observed deformation as a result of plate boundary forces. Elucidating this regional anomalous subsidence and Miocene reactivation on the North West Shelf and the Indian Plate will require modelling the effects of the breakup of a large plate with a laterally varying yield-stress envelope, as well as viscous flow in the mantle.

The northeast shelf of Australia, i.e. the Queensland Plateau and Trough and the Marion Plateau (Figure 11) also exhibits a phase of accelerated tectonic subsidence in the late Miocene/Pliocene. Five ODP sites show "foreland-basin type" subsidence rates of up to 145 m/my., even though this area is located about 800 km away from the Papuan foreland basin to the north, and about 1000 km away from the thrust load which caused the Papuan Basin to form (Figure 11). Post 9 Ma subsidence amounts to 1300±200 m in the Queensland Trough, 650±200 m on the western margin of the Queensland Plateau, post 5 Ma subsidence of 500±30 m on its southern margin, and 660±50 m on the northern margin of the Marion Plateau [*Müller et al.*, 1999].

A conventional foreland-basin model, in which a foreland basin forms in front of an advancing thrust sheet, cannot explain the observed tectonic subsidence, even if we include compressional in-plane forces and variations in equivalent elastic plate thickness [*Müller et al.*, 1999]. A published model [*Karner et al.*, 1993] for simulating the flexural response to in-plane forces of a margin with an overall structure similar to that given by the Queensland Plateau and Trough also fails to account for the observed large amplitudes of passive margin subsidence [*Müller et al.*, 1999].

It is more likely that the observed post 9 Ma tectonic subsidence of the Queensland and Marion plateaus and the Queensland Trough is largely caused by dynamic surface topography due to Australia's northeastern margin overriding a slab burial ground, modulated by flexural deformation resulting from collision north of Australia. A high-resolution upper mantle shear wave tomography model by van der Hilst et al. [1998] displays a NNW-SSE trending band of anomalously high velocities in the upper mantle at depths of 300 km in Queensland, landward of the southern Queensland and Marion plateaus. An upper mantle shear wave tomography model by Zhang and Tanimoto [1993] confirms that this band of high velocities extends down to 500 km and may stretch from the Queensland Plateau to Indonesia. Taking into account the absolute plate motion history of Australia and the subduction history north of Papua New Guinea, a band of observed positive p and s-wave anomalies in the upper mantle stretching from the Queensland Plateau to Indonesia may reflect subducted slab material originating from Oligocene-Eocene subduction north of the Caroline back-arc basin. The resulting basement subsidence has probably been modulated by flexural deformation resulting from collision north of Australia, even though this effect cannot be quantified in the absence of detailed knowledge of the rifting history of the margin.

## 7. CONCLUSIONS

We have combined a relative and absolute plate kinematic model with grids of oceanic crustal age, spreading rates and asymmetries as well as tectonic subsidence analyses of the northwest and northeast shelves of Australia to identify major plate tectonic events, evaluate their causes, and investigate their effects on anomalous intraplate subsidence or uplift and on the history of oceanic crustal accretion. From this analysis it appears that the main events that have caused plate reorganizations are spreading ridge subduction, plate collisions, and subduction initiation. Further it is clear that Australia has suffered from large fluctuations in intra-plate stresses due to plate boundary reorganizations, but also due to viscous mantle processes, causing transient vertical tectonic motions. The overriding of subducted slabs may have not only been responsible for Cretaceous flooding of eastern Australia, its subsequent re-ermergence and the formation of the Australian-Antarctic discordant Zone [*Gurnis et al.*, 1998], but is also expressed in asymmetries in accretion at the AAD, and may have caused the drowning of carbonate platforms on the Queensland and Marion plateaus northeast of Australia. Viscous coupling between the lithosphere and the mantle may also be involved in the observed anomalous depth of the Argo Abyssal Plain and the outer northwest shelf of Australia. Plume-lithosphere interaction has caused large asymmetries in crustal accretion north of Kerguelen and in the Tasman Sea. Combining plate kinematics with mantle convection models and geological data in the future should provide us with better constraints for understanding the driving forces of absolute plate motions, and their consequences in terms of the history of uplift, subsidence and deformation of the crust.

*Acknowledgments.* We thank John Veevers, Richard Gordon, and an anonymous reviewer for constructive comments which improved the quality of the manuscript. This work was supported by the Australian Research Council, the Australian Geological Survey Organization and the University of Sydney.

## REFERENCES

Agraval, A., and J.J.W. Rogers, Structure and tectonic evolution of the western continental margin of India: Evidence from subsidence studies for a 20-0 Ma plate reorganization in the Indian Ocean, in *Basement Tectonics 8: Characterization and comparison of ancient and Mesozoic continental margins*, edited by M.J. Bartholomew, D.W. Hyndman, D.W. Mogk, and R. Mason, pp. 583-590, Kluwer Academic Publishers, Butte, Montana, 1988.

Aitchison, J.C., G.L. Clarke, S. Meffre, and D. Cluzel, Eocene arc-continent collision in New Caledonia and implications for regional SW Pacific tectonic evolution, *Geology*, *23*, 161-164, 1995.

Bodine, J.H., M.S. Steckler, and A.B. Watts, Observations of flexure and rheology of the oceanic lithosphere, *J. Geophys. Res.*, *86*, 3695-3707, 1981.

Bradshaw, J.D., Cretaceous dispersion of Gondwana: continental and oceanic spreading in the south-west Pacific-Antarctic sector, in *Geological evolution of Antarctica*, edited by M.R. Thomson, J.A. Crome, and J.W. Thomson, pp. 581-585, Cambridge University Press, 1991.

Cande, S., J. Stock, R.D. Müller, and T. Ishihara, Cenozoic Motion between East and West Antarctica, *Nature*, in press.

Cande, S., J. Stock, C. Raymond, and R.D. Müller, New constraints on an old plate tectonic puzzle of the Southwest Pacific, *EOS*, *79*, 81-82, 1998.

Cande, S.C., and D.V. Kent, Revised calibration of the geomagnetic polarity timescale for the Late Cretaceous and Cenozoic, *J. Geophys. Res.*, *100* (B4), 6093-6095, 1995.

Clague, D.A., G.B. Dalrymple, T.L. Wright, F.W. Klein, R.Y. Koyanagi, R.W. Decker, and D.M. Thomas, The Hawaiian-Emperor Chain, in *The Eastern Pacific Ocean and Hawaii*, edited by E.L. Winterer, D.M. Hussong, and R.W. Decker, pp. 187-217, Geological Society of America, Boulder, 1989.

Cloetingh, S., C. Stein, P. Reemst, F. Gradstein, P. Williamson, N. Exon, U. von Rad, and i. , Continental margin stratigraphy, deformation, and intraplate stresses for the Indo-Australian region, in *Proceedings of the Ocean Drilling Program, Scientific Results*, edited by F.M. Gradstein, J.N. Ludden, and e. al., pp. 671-713, 1992.

Collot, J.Y., A. Malahoff, J. Recy, G. Latham, and F. Missegue, Overthrust emplacement of New Caledonia ophiolite: geophysical evidence, *Tectonics*, *6*, 215-232, 1987.

Crone, A.J., M.N. Machette, and J.R. Bowman, Episodic nature of earthquake activity in stable continental regions revealed by palaeoseismic studies of Australian and North American Quaternary faults, *Aus. J. Earth Sci.*, *44*, 203-214, 1997.

Denham, D., L.G. Alexander, and G. Worotnicki, Stress in the Australian crust: Evidence from Earthquakes and in situ stress measurements, *BMR J. Aus. Geol. and Geophys.*, *4*, 289-295, 1979.

DiVenere, V., and D.V. Kent, Are the Pacific and Indo-Atlantic hotspots fixed? Testing the plate circuit through Antarctica, *Earth Planet. Sci. Let.*, *170*, 105-117, 1999.

Duncan, R.A., and D.A. Clague, Pacific Plate Motion Recorded by Linear Volcanic Chains, in *The Pacific Ocean*, edited by Nairn, Stehli, and Uyeda, pp. 89-121, 1985.

Etheridge, M., H. McQueen, and K. Lambeck, The role of intraplate stress in Tertiary (and Mesozoic) deformation of the Australian continent and its margins: A key factor in petroleum trap formation, *Expl. Geophys.*, *22*, 123-128, 1991.

Gaina, C., Tectonic history of microplates east of Australia, Ph.D. thesis, The University of Sydney, Sydney, 1998.

Gaina, C., R.D. Müller, and S. Cande, Absolute plate motions, mantle flow and volcanism at the boundary between the Pacific and Indian ocean mantle domains since 90 Ma, *AGU Geophysical Monograph*, this vol.

Gaina, C., R.D. Müller, J.-Y. Royer, J. Stock, J. Hardebeck, and P. Symonds, The tectonic history of the Tasman Sea: a puzzle with 13 pieces, *J. Geophys. Res.*, *103*, 12413-12433, 1998a.

Gaina, C., R.D. Müller, J.-Y. Royer, and P. Symonds, The tectonic evolution of the Louisiade Triple Junction, *J. Geophys. Res.*, *104*, 12973-12939, 1999.

Gaina, C., W. Roest, R.D. Müller, and P.A. Symonds, The opening of the Tasman Sea: A gravity anomaly grid animation, *Earth Interactions*, *2-004*, 1-23, 1998b.

Gopala Rao, D., K.S. Krishna, and D. Sar, Crustal evolution and sedimentation history of the Bay of Bengal since the Cretaceous, *J. Geophys. Res.*, *102*, 17747-11768, 1997.

Gordon, R.G., The plate tectonic approximation: Plate nonrigidity, diffuse plate boundaries, and global plate reconstructions, *Annu. Rev. Earth Planet. Sci.*, *26*, 615-642, 1998.

Gordon, R.G., A. Cox, and C.E. Harter, Absolute motion of an individual plate estimated from its ridge and trench boundaries, *Nature*, *274*, 752-755, 1978.

Gordon, R.G., C. Demets, and J.Y. Royer, Evidence For Long-Term Diffuse Deformation of the Lithosphere of the Equatorial Indian Ocean, *Nature*, *395* (6700), 370-374, 1998.

Gradstein, F.M., F.P. Agterberg, J.G. Ogg, J. Hardenbol, P. van Veen, J. Thierry, and Z. Huang, A Mesozoic time scale, *J. Geophys. Res.*, *99* (B12), 24,051-24,074, 1994.

Gudmundsson, O., and M. Sambridge, A regionalized upper mantle (RUM) seismic model, *J. Geophys. Res.*, *103*, 7121-7136, 1998.

Gurnis, M., L. Moresi, and R.D. Müller, Models of mantle convection incorporating plate tectonics: The Australian region since the Cretaceous, *AGU Geophysical Monograph*, *in press*, this vol.

Gurnis, M., R.D. Müller, and L. Moresi, Cretaceous vertical motion of Australia and the Australian-Antarctic discordance, *Science*, *279* (5356), 1499-1504, 1998.

Haq, B.U., J. Hardenbol, and P.R. Vail, Chronology of fluctuating sea levels since the Triassic, *Science*, *235*, 1156-1167, 1987.

Hegarty, K.A., J.K. Weissel, and J.C. Mutter, Subsidence history of Australia's southern margin: constraints on basin models, *Am. Assoc. Pet. Geol. Bull.*, *74*, 615-633, 1988.

Hill, K.C., K.A. Hill, G.T. Cooper, A.J. O'Sullivan, P.B. O'Sullivan, and M.J. Richardson, Inversion around Bass Basin, SE Australia, in *Special Publication*, edited by J.G. Buchanan, and P.G. Buchanan, pp. 525-537, Geological Society, 1995.

Hillis, R., 1991, Australia-Banda Arc collision and in situ stress in the Vulcan Sub-basin (Timor Sea) as revealed by borehole breakout data, *Expl. Geophys.*, *22*, 189-194, 1991.

Hillis, R.R., S.D. Mildren, C.J. Pigram, and D.R. Willoughby, Rotation of horizontal stresses in the Australian North West contienntal shelf due to collision of the Indo-Australian and Eurasian plates, *Tectonics*, *16*, 323-335, 1997.

Hocking, R.M., A.J. Mory, and I.R. Williams, An Atlas of Neoproterozoic and Phanerozoic basins of Western Australia, in *The sedimentary Basins of West Australia*, edited by P.G. Purcell, and R.R. Purcell, pp. 21-43, Perth, 1994.

Ishihara, T., M. Tanahashi, M. Sato, and Y. Okuda, Preliminary report of geophysical and geological surveys of the West Wilkes Land Margin, *Proc. NIPR Symp. Antarct. Geosci.*, *9*, 91-108, 1996.

Jarvis, G.T., and D.P. McKenzie, Sedimentary basin formation with finite extension rates, *Earth Planet. Sci. Lett.*, *48*, 42-52, 1980.

Karner, G.D., N.W. Driscoll, and J.K. Weissel, Response of the

lithosphere to in-plate-force variations, *Earth Planet. Sci. Lett.*, *114*, 397-416, 1993.
Keen, M., E. Klein, and Melson, Ocean-ridge basalt compositions correlated with paleobathymetry, *Nature*, 1990.
Klepeis, K., The effects of extensional and transpressional plate boundary processed on the evolution of the lower crust: the Cretaceous to Tertiary history of Fiordland, New Zealand, in *Penrose Conference*, American Geophysical Union, New Zealand, 1999.
Kritski, A., R.D. Müller, G. Christesen, and C. Collins, Velocity structure of the Argo abyssal plain, Northwest shelf of Australia, in *European Geophysical Society, XXIV General Assembly*, European Geophysical Society, The Hague, 1999.
Kroenke, L.W., Cenozoic tectonic development of the Southwest Pacific, in *Tech. Bull.-U.N. Econ. .Soc .Comm. Asia Pac.,Comm. Co-ord. Jt. Prospect Miner. Resour. South Pac. Offshore Areas*, 1984.
Kroenke, L.W., J.M. Resig, and R.M. Leckie, Hiatus and tephrochronology of the Ontong Java Plateau: correlation with regional tectono-volcanic events, in *Proceedings of the Ocean Drilling Program, Scientific Results*, edited by W.H. Berger, L.W. Kroenke, and L.A. Mayer, pp. 423-444, Ocean Drilling Program, Texas A&M University, College Station, TX, 1993.
Lithgow-Bertelloni, C., and M. Richards, The dynamics of Cenozoic and Mesozoic Plate motions, *Rev. Geophys.*, *36*, 27-38, 1998.
Liu, C.S., J.R. Curray, and J.M. McDonald, New constraints on the tectonic evolution of the Eastern Indian Ocean, *Earth Planet. Sci. Lett.*, *65*, 331-342, 1983.
Lorenzo, J.M., O'Brian, J. Stewart, and K. Tandon, Inelastic yielding and forebulge shape across a modern foreland basin: North West Shelf of Australia, Timor Sea., *Geophys. Res. Let.*, *in press*, 1998.
Luyendyk, B.P., and T.A. Davies, Results of DSDP Leg 26 and the geologic history of the southern Indian Ocean, in *Init. Repts. of DSDP 26*, edited by T.A.Davies et al., 1974.
Marks, K.M., J.M. Stock, and K.J. Quinn, Evolution of the Australian-Antarctic discordance since Miocene time, *J. Geophys. Res.*, *104*, 4697-4981, 1999.
McKenzie, D., Some remarks on the development of sedimentary basins, *Earth Planet. Sci. Lett.*, *40*, 25-32, 1978.
Mihut, D., Breakup and Mesozoic seafloor spreading between the Australian and Indian plates, Ph.D. thesis, The University of Sydney, Sydney, 1997.
Mihut, D., and R.D. Müller, Volcanic margin formation and Mesozoic rift propagators in the Cuvier Abyssal Plain off Western Australia, *J. Geophys. Res.*, *103* (B11), 1998.
Mizukoshi, I., H. Sunouchi, T. Saki, S. Sato, and M. Tanahashi, Preliminary report of geological and geophysical surveys off Amery Ice Shelf, East Antarctica, *Mem. Natl. Inst. Polar Res.*, *43*, 48-61, 1986.
Müller, R.D., V. Lim, and A. Isern, Late Tertiary tectonic subsidence on the Queensland Plateau: response to dynamic topography?, *Mar. Geology*, *in press*, 1999.
Müller, R.D., D. Mihut, and S. Baldwin, A new kinematic model for the formation and evolution of the Northwest and West Australian margin, in *The sedimentary basins of western Australia 2*, edited by P.P.G.a. R.R, pp. 55-72, Petroleum Exploration Society of Australia, Perth, 1998a.
Müller, R.D., W.R. Roest, and J.-Y. Royer, Asymmetic seafloor spreading expresses ridge-plume interactions, *Nature*, *396*, 455-459, 1998b.
Müller, R.D., W.R. Roest, J.-Y. Royer, L.M. Gahagan, and J.G. Sclater, Digital isochrons of the world's ocean floor, *J. Geophys. Res.*, *102*, 3211-3214, 1997.
Müller, R.D., J.-Y. Royer, and L.A. Lawver, Revised plate motions relative to the hotspots from combined Atlantic and Indian Ocean hotspot tracks, *Geology*, *16*, 275-278, 1993.
Müller, R.D., D.T. Sandwell, B.E. Tucholke, J.G. Sclater, and P.R. Shaw, Depth to basement and geoid expression of the Kane Fracture Zone: a comparison, *Mar. Geophys. Res.*, *13* 105-129, 1991.
Norton, I.O., Plate motions in the North Pacific: the 43 Ma nonevent, *Tectonics*, *14* (5), 1080-1094, 1995.
Parsons, B., and J.G. Sclater, An analysis of the variation of ocean floor bathymetry and heat flow with age, *J. Geophys. Res.*, *82* (5), 803-827, 1977.
Patriat, P., and J. Achache, India-Eurasia collision chronology has implications for crustal shortening and driving mechanisms of plates, *Nature*, *311*, 615-621, 1984.
Phipps-Morgan, J.P., and D.T. Sandwell, Systematics of ridge propagation south of 30°S, *Earth Planet. Sci. Lett.*, *121*, 245-258, 1994.
Pigram, C.J., and P.A. Symonds, A review of the timing of the major tectonic events in the New Guinea Orogen, *J. Southeast Asian Earth Sci.*, *6*, 307-318, 1991.
Powell, C.M., S.R. Roots, and J.J. Veevers, Pre-breakup continental extension in East Gondwanaland and the early opening of the eastern Indian Ocean, *Tectonophys.*, *155*, 261-283, 1988.
Pringle, M., M.F. Coffin, and M. Storey, Estimated Melt Production of the Kerguelen Hot Spot, in *EOS Trans. Amer. Geophys. Union*, pp. F728, 1997.
Royer, J.-Y., and T. Chang, Evidence for relative motions between the Indian and Australian plates during the last 20 m.y. from plate tectonic reconstructions: implications for the deformation of the Indo-Australian plate, *J. Geophys. Res.*, *96* (B7), 11779-11802, 1991.
Royer, J.-Y., and R. Gordon, The Motion and Boundary between the Capricorn and Australian Plates, *Science*, *277* (29), 1268-1274, 1997.
Saki, T., Y. Tamura, S. Tokuhashi, T. Kodato, I. Mizokushi, and H. Amano, Preliminary report of geological and geophysical surveys off Queen Maud Land, East Antarctica, *Proc. NIPR Symp. Antarct. Geosci.*, *1*, 23-40, 1987.
Sandwell, D.T., and W.H.F. Smith, Marine gravity anomaly from Geosat and ERS-1 satellite altimetry, *J. Geophys. Res.*, *102*, 10039-10054, 1997.
Scholl, D.W., Initiation of new arc-backarc systems of crustal generation in response to ridge-trench collision: example of the Lau-Tonga region by activation of rapid trench rollback, *Eos, Transactions, AGU*, *76*, 614, 1995.
Schouten, H., H.J.B. Dick, and K.D. Klitgord, Migration of mid-ocean-ridge volcanic segments, *Nature*, *326*, 835-839, 1987.
Sclater, J.G., and P.A.F. Christie, Continental Streching - an explanation of the post mid-Cretaceous subsidence of the Central Graben of the North Sea, *J. Geophys. Res.*, *85*, 3740-3750, 1980.
Shuster, M.W., S. Eaton, L.L. Wakefield, and H.J. Kloosterman, Neogene tectonics, greater Timor Sea, offshore Australia: Implications for trap risk, *The APPEA Journal*, *38* (1), 351-379, 1998.
Smith, W.H.F.a.P.W., Gridding with continuous curvature splines in tension, , *55*, 293–305, 1990.
Stein, S., H.J. Melosh, and J.B. Minster, Ridge migration and asymmetric sea-floor spreading, *Earth Planet. Sci. Lett*, *36*, 51-62, 1977.
Storey, B.C., The role of mantle plumes in continental breakup: Case histories from Gondwanaland, *Nature*, *377* (6547), 301-308, 1995.
Symonds, P.A., S. Planke, J.B. Colwell, and A.J. Crawford, Volcanic evolution of the western Australian continental margin, in *Australian Geol. Convention*, pp. 19-23, Canberra, 1996.
Tarduno, J.A., and J. Gee, Large-scale motion between Pacific and Atlantic hotspots, *Nature*, *378* (6556), 477-480, 1995.

Taylor, B., A. Goodliffe, F. Martinez, and R. Hey, Continental rifting and initial sea-floor spreading in the Woodlark basin, *Nature, 374* (6522), 534-537, 1995.

Tikku, A.A., Late Cretaceous to Early Tertiary tectonics of the Southeast Indian Ocean, Ph.D. thesis, University of California, San Diego, 1998.

Tikku, A.A., and S.C. Cande, The oldest magnetic anomalies in the Australian-Antarctic Basin: Are they isochrons?, *J. Geophys. Res., 104*, 661-677, 1999.

Truswell, E.M., M.E. Dettmann, and P.E. O'Brian, Mesozoic palynofloras from the Mac. Robertson shelf, East Antarctica: geological and phytogeographic implications, *Ant. Sci., 11* (2), 239-255, 1999.

Tsumuraya, Y., M. Tanahashi, T. Saki, T. Machihara, and N. Asakura, Preliminary report of the marine geophysical and geological surveys off Wilkes Land, Antarctica in 1983-1984, *Mem. Natl. Inst. Polar Res., 37*, 48-62, 1985.

van der Hilst, R.D., B.L.N. Kennett, and T. Shibutani, Upper mantle structure beneath Australia from portable array deployments, in *Structure and evolution of the Australian continent*, edited by J. Brown, J. Dooley, B. Goleby, R. van der Hilst, and C. Klootwijk, pp. 39-57, American Geophysical Union, 1998.

Van Fossen, M.C., and D.V. Kent, Paleomagnetism of 122 Ma plutons in New England and the Mid-Cretaceous paleomagnetic field in North America: True polar wander or large-scale differential mantle motion?, *J. Geophys. Res., 97* (B13), 19651-19661, 1992.

Veevers, J.J., *Phanerozoic Earth History of Australia*, 418 pp., Oxford University Press, New York, 1984.

Yan, C.Y., and L.W. Kroenke, A plate tectonic reconstruction of the Southwest Pacific, 0 100 Ma, in *Proceedings of the Ocean Drilling Program, Scientific Results*, edited by W.H. Berger, L.W. Kroenke, and L.A. Mayer, pp. 697, College Station, TX, 1993.

Young, G.C., and J.R. Laurie, *An Australian Phanerozoic timescale*, 279 pp., Oxford University Press, Melbourne, 1996.

Zhang, Y.S., and T. Tanimoto, High-resolution global upper mantle structure and plate tectonics, *J. Geophys. Res., 98* (B6), 9793-9823, 1993.

R.D. Müller, C. Gaina, D. Mihut , Division of Geology and Geophysics, School of Geosciences, University of Sydney, Edgeworth David Building, F05, The University of Sydney, New South Wales 2006, Sydney, Australia. (dietmar@es.usyd.edu.au; carmen@es.usyd.edu.au; dona@es.usyd.edu.au)

A. Tikku, Laboratory for Satellite Altimetry National Oceanic and Atmospheric Administration, SSMC3 E/OC2, 1315 East-West Highway, Silver Spring, MD 20910.

S. C. Cande, Scripps Institution of Oceanography, La Jolla,, CA 92093-0215, USA

J.Stock, California Institute of Technology, 252-21, Pasadena, CA 91125.

# Absolute Plate Motion, Mantle Flow, and Volcanism at the Boundary Between the Pacific and Indian Ocean Mantle Domains Since 90 Ma

Carmen Gaina and R. Dietmar Müller

*Division of Geology and Geophysics, School of Geosciences, University of Sydney, Sydney, New South Wales, Australia.*

Steven C. Cande

*Scripps Institution of Oceanography, La Jolla, CA 92093-0215.*

The Tasman Sea and the southeastern Indian Ocean are situated at the interface between the Indian and Pacific oceans mantle domains. Even though the present location of the boundary between these two reservoirs at the Australian Antarctic Discordance (AAD) is well constrained by geochemical data, its dynamic behaviour is less well understood. Here, we investigate the relationship between volcanism in and south of the Tasman Sea, absolute plate motion, spreading asymmetries and microcontinent formation. We test three hypotheses: since 90 Ma, the upper mantle east of the AAD and west of the Macquarie Triple Junction was attached to 1) the Indian-Atlantic hotspot reference frame, 2) the Pacific hotspot reference frame 3) to the Pacific hotspots from 90 to 40 Ma, and subsequently to the Atlantic-Indian hotspots, due to partial separation of the two mantle domains by subducting slabs. We find that hostspot tracks in this area are best modeled using a Pacific hotspot reference frame, confirming hypothesis (2). If a hotspot is now located close to the Balleny Islands, it could not have produced the Tasmanian seamount chain irrespective of which reference fame is chosen, contrary to previous suggestions. Seafloor spreading asymmetries and microcontinent formation in the Tasman Sea are best accounted for by plume-ridge interactions, rather than the absolute motion of the spreading ridge relative to the mantle. The short-lived nature of most hotspot volcanism in this area, as well as its geochemistry, points to a paleo-superswell in the upper mantle as its source, rather than deep seated plumes, with the exception of the Tasmantid seamounts.

## 1. INTRODUCTION

The location of the Tasman Sea at the boundary between the Pacific and Indian oceans offers a unique opportunity to study the effect of active margin processes on the dynamic behavior of the interface between two large mantle domains. A fundamental question is: where exactly was the boundary between these two mantle domains located through time?

*Gurnis et al.* [1998] suggested that during the last 130 million years Australia has overridden a subducted slab, which is now located under the Australian Antarctic Discordance Zone (AAD). The location of this slab marks

The History and Dynamics of Global Plate Motions
Geophysical Monograph 121

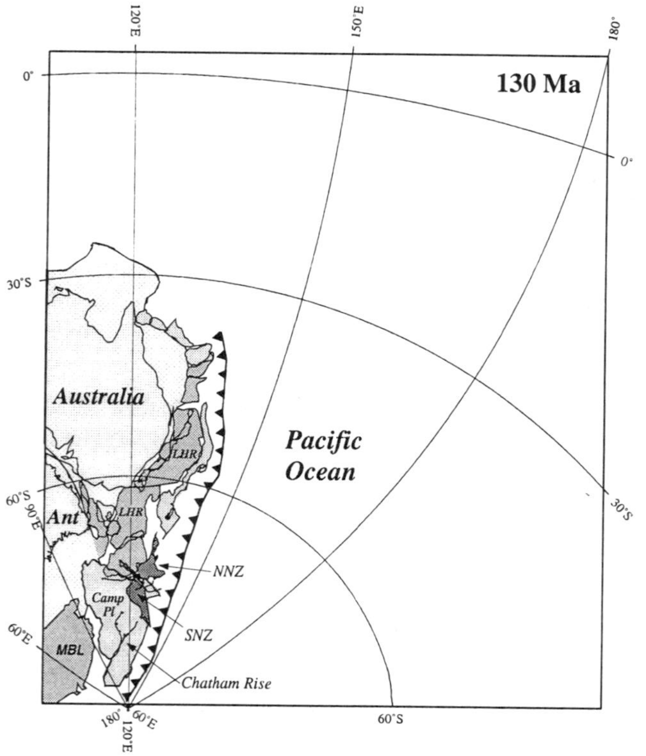


Figure 1. Reconstruction of eastern Australia at 130 Ma (general stereographic projection with center at 120°E and 30°S). The Lord Howe Rise (LHR), the two islands of the New Zealand (NZ), the Campbell Plateau (Camp. Pl), and the Chatham Rise were part of the eastern Gondwana together with the East and West Antarctica (abbreviated as Ant and MBL). The Phoenix plate was subducted under eastern Gondwana (filled triangles).

the long-lasting Paleozoic boundary between Indian and Pacific mantle, while the Phoenix plate was subducted east of Australia (Figure 1). Considering that this boundary is now located at the AAD south of Australia (Figure 2), it follows that the mantle underlying the area between the AAD and the Pacific Ocean should represent Pacific type mantle. This is supported by the sharp transition of Indian to Pacific isotope and trace element signatures at the AAD [*Klein et al.*, 1988; *Pyle et al.*, 1992]. However, it does not necessarily follow that the mantle between the AAD and the Pacific Ocean has also moved dynamically in concert with the Pacific mantle, which may have moved relatively fast (3 cm/yr) with respect to the spin axis and to the hotspots in the Atlantic ocean [*Tarduno and Gee*, 1995] in the Late Cretaceous and early Tertiary.

As subduction has been active east of the Tasman Sea, probably for the last 40-50 m.y. [*Kroenke*, 1984; *Eissen et al.*, 1998], and as the bulk of the ancient subducted slab under the AAD is now in the lower mantle, one might expect that since 50 – 40 Ma subduction east of the Australian plate has formed a partial barrier for large-scale dynamic flow between the two mantle domains, at least in the upper mantle. *West et al.*, [1997] and *Christie et al.*, [1998] suggested that the source of the AAD is actively moving westwards, driven by mantle flow from the Pacific Ocean. In contrast, *Gurnis et al.* [this. vol] have proposed that the apparent westward migration of the isotope anomaly at the AAD is a consequence of the eastward component of Australia's absolute plate motion over a stationary mantle source, as well as of the shallow dip of the subducted slab, which has been drawn out to the east by Late Cretaceous eastward motion of Australia and Antarctica, as shown in Figures 1 and 2.

In the following, three hypotheses will be tested. The mantle underneath the Tasman Sea has moved in concert with: 1) Pacific mantle since 90 Ma, 2) Atlantic-Indian ocean mantle since 90 Ma, and 3) Pacific mantle initially, but became attached to Atlantic-Indian mantle after subduction started along the Tonga and Kermadec trenches.

This test will be based on: a) models for the motion of the Pacific plate [*Müller et al.*, 1991] and the plates around the Atlantic-Indian oceans [*Müller et al.*, 1993] with

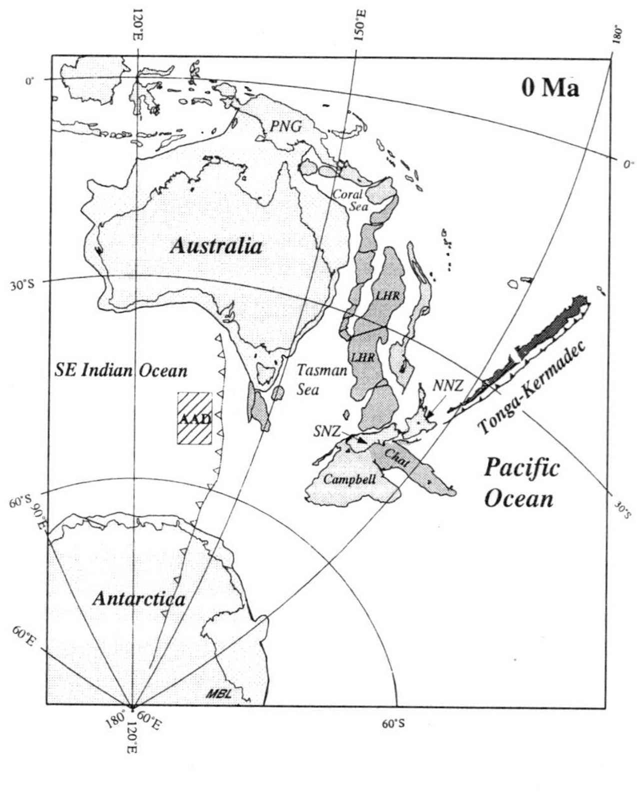


Figure 2. Present day location of 130 Ma subduction boundary (open triangles) and active subduction at the Tonga-Kermadec trench (general stereographic projection with center at 120°E and 30°S). AAD: Australian-Antarctic discordance. See Figure 1 for abbreviations.

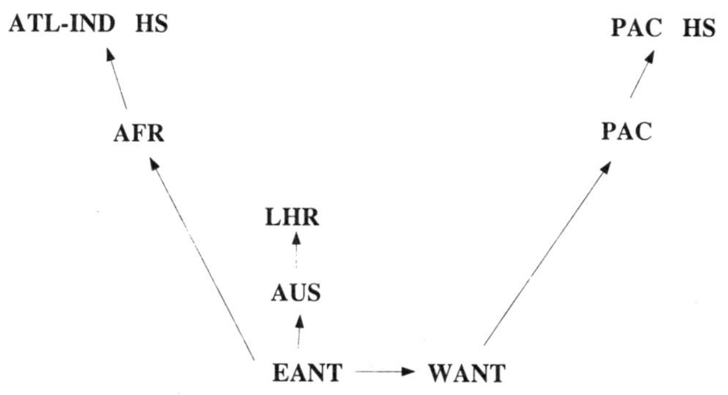


**Figure 3**. Plate circuit used to test hypotheses on the history of lateral upper mantle flow beneath the Tasman Sea. ATL-IND HS- Atlantic Indian Hotspot reference frame, PAC HS- Pacific Hotspot reference frame, EANT, WANT-East and West Antarctica, AUS-Australia, PAC-Pacific, LHR-Lord Howe Rise.

respect to their underlying mantle; b) closure of the Pacific, West Antarctica, East Antarctica, Australia, Lord Howe Rise plate circuit (see Figure 3) based on rotations derived by *Cande et al.* [1995], *Gaina et al.* [1998], and a rotation to reconstruct the relative motion between East and West Antarctica based on new geophysical data north and south of the Macquarie Triple Junction [*Cande et al., in press*]; and c) seamount chains in the Tasman Sea as mapped from gravity anomalies [*Sandwell and Smith*, 1997] and dated by *McDougall et al.* [1981], *McDougall and Duncan* [1988], *Duncan and McDougall* [1989], and *Jenkins et al.* [1992]. An absolute plate motion model constructed by *Müller et al.* [1991] has been chosen for Pacific plate motions relative to the sub-Pacific mantle, because it provides a better fit for the geometry of seamount chains in the Pacific Ocean compared to other recently published models (e.g. *Wessel and Kroenke*, [1997], and *Koppers* [1998], see Figure 4). This model predicts relative motion between the Atlantic-Indian hotspots and Pacific hotspots.

## 2. VOLCANISM IN THE TASMAN SEA

The Tasman Sea has been affected by volcanism both during its opening as well as after seafloor spreading became extinct. There are three distinct linear seamount chains that have likely been created by hotspot related volcanism. The oldest seamounts dated in the Tasman Sea are Zeehan-Heemskirk, close to the Lord Howe Rise flank (Figure 5a), with K-Ar ages of 71.3-59.1 Ma and the Janszoon Seamount dated 50-60 Ma [*Jenkins et al.*, 1992]. These seamounts together with other undated guyots form two linear volcanic chains (Tasmanian seamounts) that cross the Tasman Sea (Figure 5a). The Lord Howe Rise segment trends E-W, and the Australian segment has a NE-SW direction visible in the linear gravity anomalies (Figure 5a). Further south lies Soela Seamount formed as a result of Paleogene hotspot volcanism on the East Tasman Plateau [*Lanyon et al.,* 1993; *Crawford et al.*, 1997]. In the northern Tasman Sea a N-S oriented volcanic chain, the Tasmantid seamounts, displays ages ranging from 7 to 24 Ma [*McDougall and Duncan,* 1988]. Another southward-younging seamount chain is present farther north, on the Lord Howe Rise. Lord Howe Island at the southern end of this chain is an eroded basaltic shield volcano that was built 6.9 –6.4 Ma ago [*McDougall et al.*, 1981]. It has been suggested that the three volcanic chains in the Tasman Sea are hotspot traces that describe the absolute motion of the Australian plate. Several studies related the Tasmanian seamounts to the Balleny hotspot [*Jenkins et al.*, 1992; *Lanyon et al.*, 1993; *Sutherland*, 1994], which may now lie close to Antarctica (Figure 5b).

*Duncan and McDougall* [*1989*] have been the first to suggest a relationship between volcanism on the Balleny Islands and south Tasman Sea volcanism. *Jenkins et al.* [1992] extrapolated this idea and tried to tie the origin of older volcanoes from the Tasmanian seamount chain to the Balleny hotspot. *Lanyon et al.* [1993] postulated that initial Tasman Sea opening might have been synchronous with inception of the Balleny plume. They suggested that the location of this hotspot during the eastern Gondwana break-up and the resemblance of the geochemical signature of the Tasmanian seamounts and the Soela seamount with the Balleny province demonstrate that the Balleny hotspot activity is responsible for the Tasmanian seamount chain. *Sutherland* [1994] proposed two alternative hotspot traces that might explain the geometry of Tasmanian volcanic chain, one that accounts for an age progression from an initial Balleny hotspot and another one with a bilateral formation of seamounts away from the spreading ridge during the Tasman Sea opening and unilateral after spreading ceased.

The Tasmantid volcanic chain is also considered a hotspot trace. *Duncan and McDougall* [*1989*] suggested that the southward decrease in age along with recently reported seismicity [*Denham*, 1985] south of the younger seamount are strong evidence for a hotspot origin. They also postulated that the Lord Howe seamount chain is likely a hotspot trail. *Müller et al.* [1993] showed that the Tasmantid seamount chain geometry can be modeled reasonably well by assuming an absolute Australian plate motion in an Atlantic-Indian hotspot reference frame. However, their model fails to predict the geometry of the Tasmantid seamounts older than 16 Ma. Recently, *Sutherland* [1998] modeled a hotspot track that averages the absolute plate motions in an Atlantic-Indian hotspot and in a Pacific hotspot reference frames. He suggested

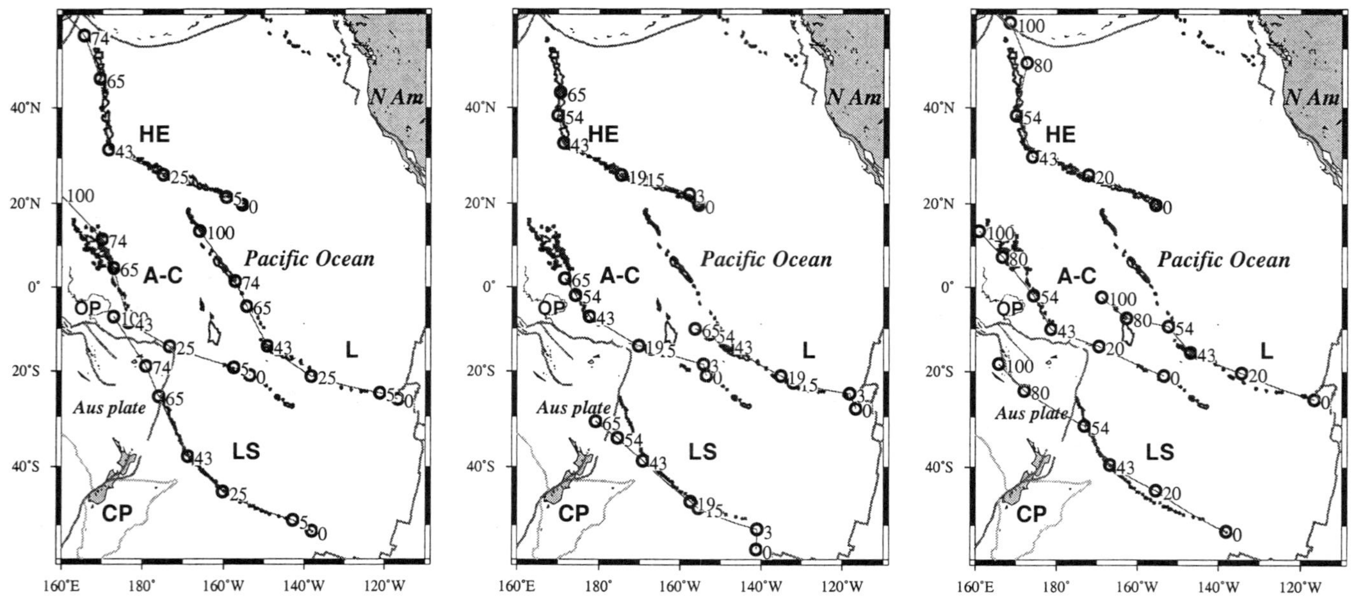


*a) Mueller et al. [1991]* *a) Wessel and Kroenke [1997]* *c) Koppers [1998]*

**Figure 4**. Three models for hotspot tracks on the Pacific plate: a) *Müller et al.* [1990], b) *Wessel and Kroenke* [1997], and c) *Koppers* [1998]. The model by *Müller et al.* [1991] fits the geometry of the Hawaiian-Emperor chain (HE), of the Louisville chain (LS), of the Austral-Cook Islands (A-C) and the Line Islands (L). Hotspot tracks based on *Wessel and Kroenke*'s [1997] (based on the age-independent hotspotting technique) and *Kopper*'s [1998] models fail to fit the geometry of the older segment of the Louisville chain, the younger part of the Austral-Cook Islands (A-C) and the Line Islands (L). OP is Ontong-Java Plateau, CP is Campbell Plateau.

that this is the most suitable solution to match observed geometry of the Tasmantid seamount.

## 3. THE RELATIONSHIP BETWEEN VOLCANISM AND HOTSPOTS

The Atlantic-Indian hotspot reference frame cannot account for an oblique seamount chain in the Tasman Sea (as the Tasmanian seamounts), since the Australian plate was nearly stationary relative to the Atlantic-Indian hotspots between 80-43 Ma and then moved north/northeastward [*Müller et al*, 1993; *Gurnis et al.*, 1998]. Therefore, two other hypotheses were tested, i.e. a Pacific hotspot reference frame and a mixed Pacific-Indian-Atlantic reference frame, in which the Tasman Sea mantle moved with Pacific mantle initially, but became attached to Indian ocean mantle after onset of subduction. Preliminary reconstructions of the SW Pacific area, including relative motion between East and West Antarctica, suggest that the onset of subduction east of the Australian plate occurred around 43 Ma [*Gaina*, 1998]. Therefore, the present model will test the hypothesis that at around 40 Ma the upper mantle beneath the Tasman Sea ceased to be attached to the Pacific mantle and gradually became attached to the Indian mantle domain. This time also marks the onset of fast spreading between Australia and East Antarctica.

The result of these models is presented in Figures 5a and 5b, showing the hotspot trail with respect to Pacific mantle as well as the mixed absolute reference frame. The present day location of a possible hotspot in the two scenarios is on the Antarctic plate. In the Pacific hotspot absolute frame, the location of a present day hotspot would be close to the small Antarctic active volcanic province (the Pleiades, *Kyle*, [1990], Figure 6). Both models predict the positions and ages of the two Tasmanian seamounts that were dated (the Zeehan-Heemskirk, 59-71 Ma and the Janszoon, 50-60 Ma) quite well.

Another model was constructed to fit the NE-SW oriented seamount chain present on the Australian flank of the Tasman Sea, including the Janszoon seamount (50-60 Ma) and the Soela seamount (36 Ma). This model relies on the assumption that the oblique seamount chain formed after seafloor spreading ceased in the Tasman Sea and the underlying mantle belonged to the Pacific mantle for the first 10 m.y. (50-40 Ma) and subsequently to the Atlantic-Indian mantle (Figures 5a and 5b). In this case, the present day hotspot location is predicted to be NW of the Balleny Islands.

The models created to explain the origin of the Tasmanian seamounts show that the Pacific and Indian mantle domains were not moving substantially relative to each other for the last 40 Ma (Figures 5a and 5b). This is in agreement with *Sager and Pringle* [1987] who found little difference in the paleolatitudes of the central Pacific derived from either hotspot tracks or paleomagnetic data after ~40 Ma. This observation implies that the relatively fast recent westward motion of Pacific mantle east of the AAD postulated by *Pyle et al.* [1992], *West et al.* [1997], and *Christie et al.*, [1998] cannot be accounted for by relative motion of these two large mantle domains.

The Tasmantid seamounts, whose ages range from 7 to 24 Ma, are modeled assuming either an Atlantic-Indian or a Pacific hotspot reference frame for the last 40 m.y. A model based on Pacific hotspot tracks shows an improved match with the geometry of the Tasmantid seamount chain compared with *Müller et al.*'s [1993] model (Figure 7). The Atlantic-Indian hotpot reference frame model is able to predict the younger seamount positions (7 and 14 Ma), but its path for older times (16 to 24 Ma) diverges from the observed seamount geometry (Figure 7).

It is not well known how a subducted slab behaves at the boundary between the upper and lower mantle. Some studies show that slabs penetrate the lower mantle [*Creager and Jordan*, 1986]; others indicate that slabs are deflected by that interface [*Van der Hilst et al.*, 1991; *Fukao et al.*, 1992]. *Van der Hilst* [1995] and [Gudmundsson, 1998 #9272] imaged subducted lithosphere in the mantle beneath the Tonga trench and concluded that in the north slabs are deflected in the transition zone, but in the south the slabs penetrate the lower mantle (Figure 8, see also Figure 2 of Müller et al., this vol.).

A layered mantle may exhibit different convection patterns in each layer. A study by *Olson et al.* [1990] showed that the pattern of convection in the upper mantle is compatible with surface plate motions, whereas the lower mantle convection conforms more closely to that in a purely viscous fluid. *Olson et al.* [1990] presented a model of the lower mantle convection pattern under the Pacific plate with a large convection cell that includes the area covered by the eastern Australian continent. This model is consistent with the idea that eastern Australia has overridden Pacific mantle after 130 Ma [*Gurnis et al.*, 1998]. However, the spatial pattern of distinct upper mantle reservoirs appears to be more complex than suggested by this simple model. *Hergt and Hawkesworth* [1994] showed that the youngest oceanic floor in the Lau Basin has Atlantic-Indian geochemical signature, whereas older Lau Basin seafloor has Pacific mantle affinities. Other authors [*Crawford et al.*, 1995] report a similar phenomenon in the Central New Hebrides Island Arc, with an isotopic shift in the last 2-3 Ma from Pacific type lavas to Atlantic-Indian type astenosphere. Reviewing the geochemical results from basalts collected from several western Pacific marginal basins over the past 20 years, *Hickey-Vargas et al.*, [1995] concluded that these basins shared a mantle source related to source of Indian Ocean ridge basalts. They suggested that Atlantic-Indian type astenosphere flowed to the north, northeast, northwest and southeast of Australia in the last 50 Ma. In another recent study *Turner and Hawkesworth* [1998] proposed that according to geochemical tracers in the Lau Basin, the Atlantic-Indian mantle has migrated eastward at rates of 45-65 mm/yr in the last 6 Ma.

Three DSDP sites, situated in the SW Tasman Sea, revealed two different MORB isotope signatures. Ocean floor basalts drilled from Site 280A and 282 (Figure 8) show Indian Ocean mantle signature, and are dated 64 Ma and 60 Ma, respectively, whereas samples recovered from DSDP hole 283 (Figure 8) are reported to present Pacific mantle type [*Pyle et al.*, 1995]. These observations raise the question of whether the two upper mantle domains migrated relative to each other or whether the presence of the two types of basalts close to each other is rather the effect of absolute plate motions. *Zhang et al.* [1999] suggested that the Atlantic-Indian type mantle represents a long-term feature underlying the eastern part of Gondwana, including eastern Australia, and argue that therefore one would expect to find Atlantic-Indian MORB-type basalts east of the AAD. They consider that the hypothesis of eastward migration of the Atlantic-Indian mantle (as *Pyle et. al.*, [1995] suggested) is unrealistic, since it would have required a migration rate of more that 10 cm/yr. *Zhang et al.* [1999] also show that Cenozoic basalts (55-14 Ma) in eastern Australia (including Tasmania) have mantle isotope signatures of Pacific mantle. Therefore, the basalts sampled at DSDP Site 280A and 282 [*Pyle et al.*, 1995] are rather an exception considering that all other Tertiary basalt samples in this region have Pacific MORB-type signatures.

*Gurnis et al.* [1998] tested the hypothesis that a subducted slab is responsible for the AAD, using a three-dimensional model of mantle convection combined with a plate tectonic model. In their model the initially sharp boundary between the two mantle domains (Atlantic-Indian and Pacific) became diffuse as a result of the fast Late Cretaceous eastward motion of Australia and Antarctica. The resulting shear traction created a thin Indian mantle wedge overlying Pacific mantle east of the AAD which is shown at 60 Ma on Figure 9. After the onset of seafloor spreading between Australia and Antarctica ocean crust east of the AAD was created by partial melting from this Indian mantle reservoir, forming basalts with a geochemistry more similar to Indian Ocean than to Pacific mantle. While this thin mantle-wedge reservoir was slowly being exhausted by continuing seafloor spreading, Pacific mantle from underneath this wedge gradually replaced the wedge above, until Pacific

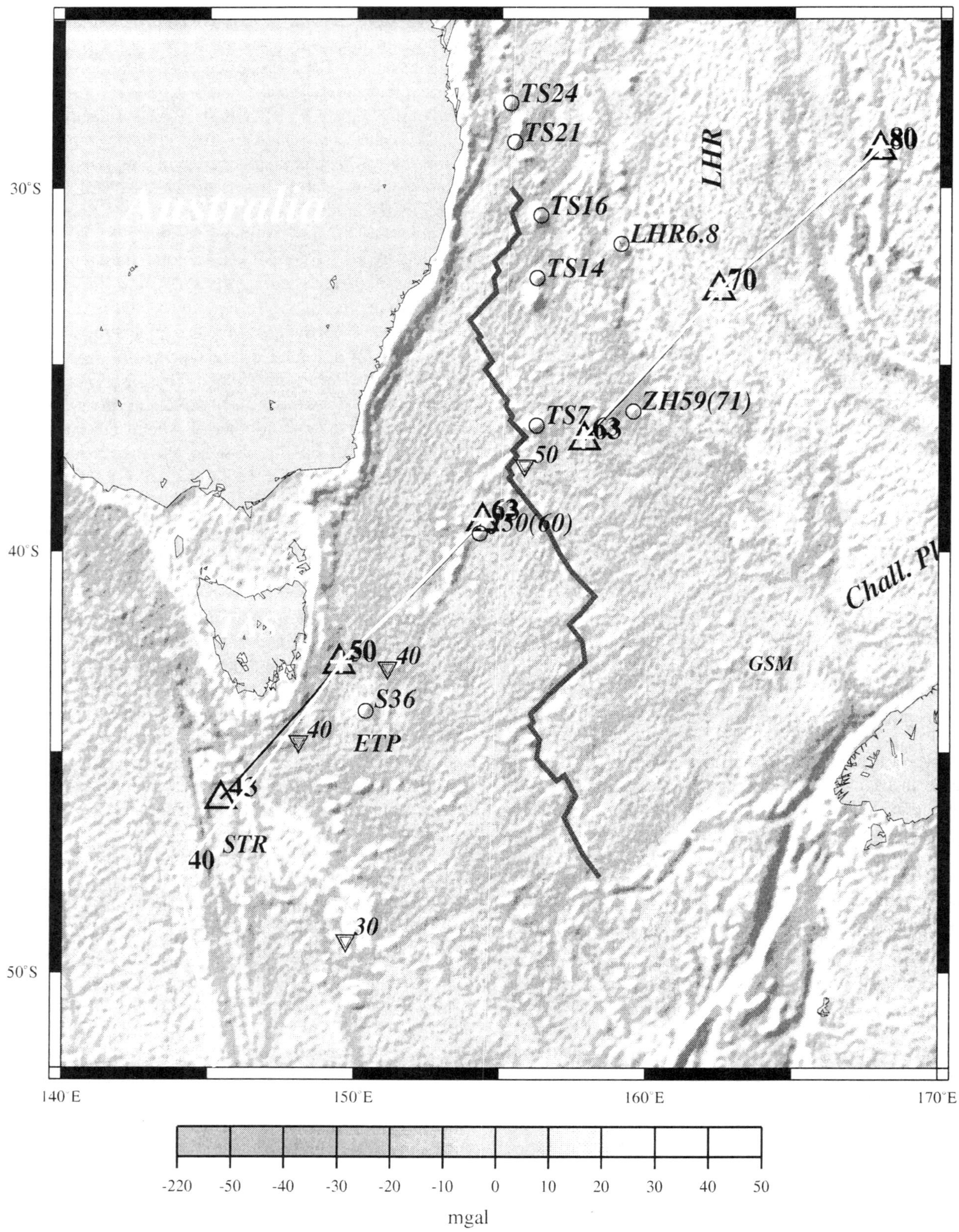
TS24
TS21
LHR
80
30°S
Australia
TS16
LHR6.8
TS14
70
ZH59(71)
TS7
63
50
63
S50(60)
40°S
Chall. Pl
50
40
GSM
S36
40
ETP
43
STR
40
30
50°S
140°E
150°E
160°E
170°E
-220
-50
-40
-30
-20
-10
0
10
20
30
40
50
mgal

mantle only is sampled to create ocean crust east of the AAD. As the Indian mantle wedge is thinnest underneath eastern Australia and the South Tasman Rise (Figure 9), we expect to see a complete replacement of Indian by Pacific mantle south of the South Tasman Rise first, and last at the AAD. This results in an apparent westward migration of Pacific mantle east of the AAD, and may account both for a V-shaped boundary between Indian and Pacific mantle tracers east of the AAD and for basalts with Indian mantle signatures at DSDP Sites 280A and 282.

Based on stage poles derived from the new set of rotations, we have computed the direction of relative motion between the Pacific and Atlantic-Indian hotspots (Figure 8). The direction of motion of Pacific hotspots relative to the Atlantic-Indian hotspots shows several notable changes, at about 60 Ma, 40 Ma and 20 Ma. From 80 to 60 Ma, the Pacific mantle migrated westward, and around 60 Ma changed its direction toward S, SW. At 40 Ma another change in direction of Pacific mantle occurred, most notable in the Australian-Antarctic Basin, accompanied by a major slowdown in relative motion of the two mantle domains.

This model, together with our analysis of hotspot tracks in the Tasman Sea, suggests that the mantle beneath and south of the Tasman Sea has likely moved in concert with the Pacific mantle in the Tertiary and that subduction along the Tonga-Kermadec trenches has not yet acted as a substantial obstacle to the coupling between the Pacific mantle and southeast Indian Ocean mantle east of the AAD. The model by *Gurnis et al.* [1999] outlined above also alleviates the need for active westward flow of Pacific asthenosphere towards the AAD, for which there is no obvious driving force.

It has been suggested that the Balleny hotspot is responsible for the Tasmanian seamount chain [*Jenkins et al.*, 1992; *Lanyon et al.*, 1993; *Sutherland*, 1994]. Here this hypothesis is tested following the same three assumptions that were previously described. The track obtained by assuming a Pacific hotspot absolute framework (Figure 10) crosses the southeastern Tasman basin and does not match the volcanic seamount chain further north. In a fixed Atlantic-Indian hotspot framework, the geometry of this seamount chain is not matched either, but the 40 Ma position of the modeled track may explain volcanism southeast of the South Tasman Rise (Figure 10). A third solution is based on the assumption that the underlying Tasman Sea mantle would have been part of the Pacific plate from 90 to 40 Ma, and then became attached to the Indian-Atlantic mantle (40 to 0 Ma) (Figure 10). The time of the hotspot ridge crossing has been computed using revised Tasman Sea reconstructions [*Gaina et al.* 1998]. None of the three scenarios is able to predict the Tasmanian seamount chain geometry. Thus, there is no evidence that the Balleny hotspot was responsible for the formation of these seamounts, if it is located at the Balleny Islands today. Indeed, the suspicious alignment of the Balleny Islands with the large-offset Balleny Fracture Zone and the complete lack of any seamount chain away from the Balleney Fracture Zone with a geometry that could be related to a "Balleny Hotspot" using absolute plate motion models suggests that there may be a "Balleny hotspot" at the surface, by no deep seated mantle plume connected to it.

## 4. DEEP SEATED MANTLE PLUMES OR SUPERSWELL ?

Even though the seamount chains in the Tasman Sea can be accounted for by using models for the motion of plates over either fixed or slowly moving hotspots as discussed above, it is clear that these hotspots were not long-lived.

**Figure 5a**. Hotspot track models superimposed on gravity anomalies with details of the Tasmanian seamount track. Dated seamounts in the Tasman Sea are shown by open circles, the attached numbers represent their ages (the Tasmanian seamounts: the Zeehan-Heemkirsk – ZH 59(71) Ma, the Janszoon – J50(60) Ma; the Soela seamount (S) on the East Tasman Plateau is 36 my old; the N-S trending Tasmantid seamounts (TS) range from 7 to 24 Ma and one seamount of the Lord Howe Rise parallel seamount chain (LHR) has been dated as 6.8 Ma. The Tasmanian volcanic chain has been modeled in two ways. The track that connects black open triangles indicates hotspot traces for the Pacific hotspot reference frame. This model matches the two dated Tasmanian seamounts well. The present day hotspot position required by this model is located under Antarctica, close to the Pleiades volcanic province (Figure 6), and the ridge would have crossed this hotspot at about 43 Ma. If the hotspot that produced the Tasmanian chain were fixed to the Indian-Atlantic mantle in the last 40 Ma, then the present day position would be SE of the Balleny Islands (white crosses). The NE-SW trending volcanic chain visible on the gravity anomaly in the Tasman Sea on the Australian flank is matched by a hotspot model that assumes the hotspot was fixed to the Pacific mantle between 50 and 40 Ma and subsequently to the Indian-Atlantic mantle (open inverted triangles). This model also predicts the age and position of the Soela seamount on the East Tasman Plateau. Abbreviations are: Chall-Challenger Plateau; ETP - East Tasman Plateau; GSM - Gilbert Seamount Complex; LHR - Lord Howe Rise; STR - South Tasman Rise; TAS - Tasmania; MTJ-Macquarie Triple Junction, NZ – New Zealand.

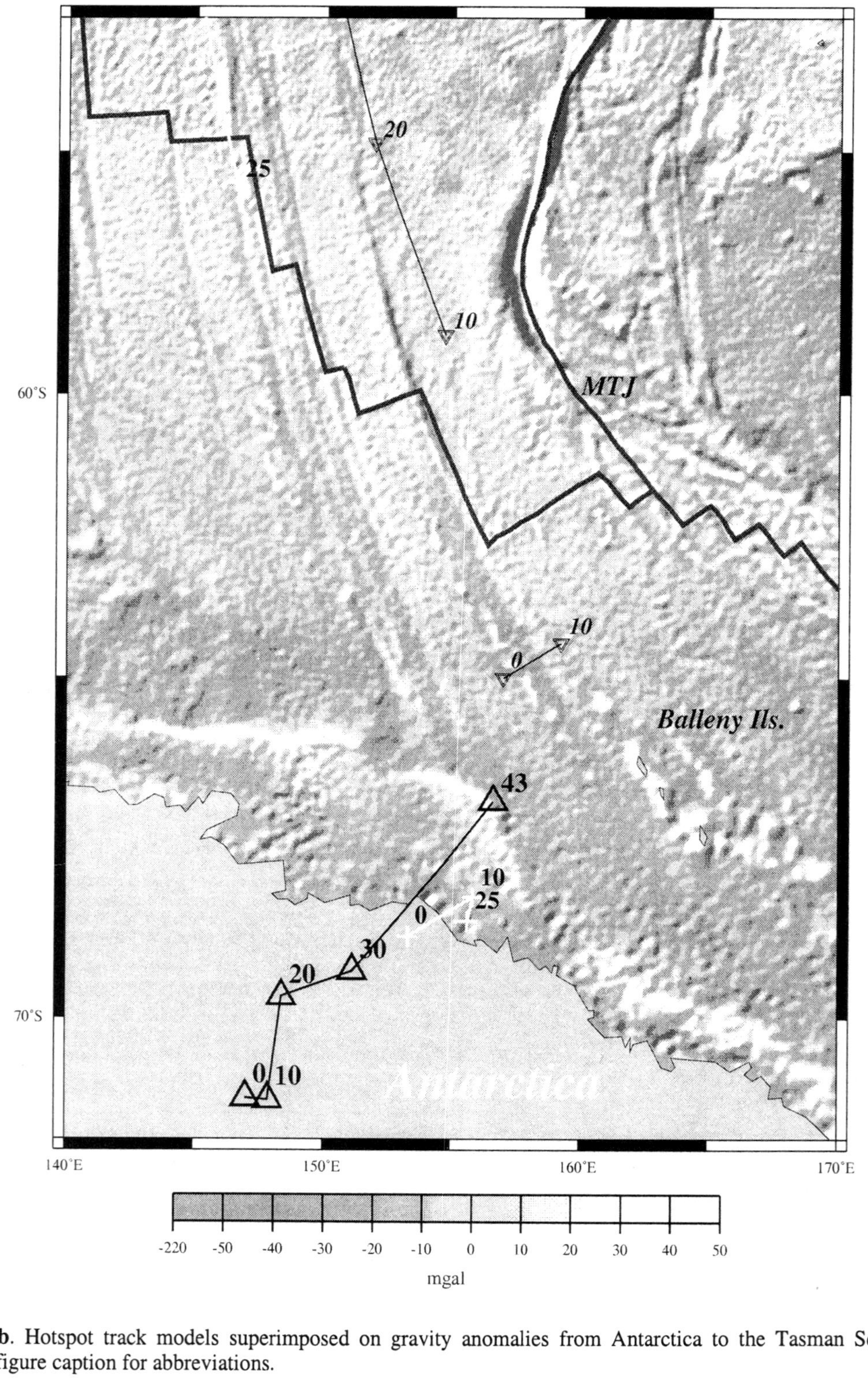


**Figure 5b**. Hotspot track models superimposed on gravity anomalies from Antarctica to the Tasman Sea. See previous figure caption for abbreviations.

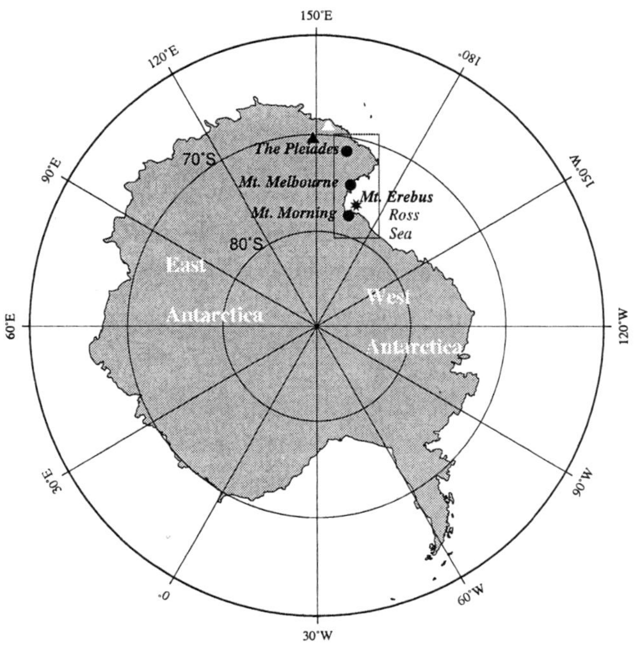


**Figure 6**. Holocene and present volcanism west of the Ross Sea. The Pleiades volcanic province may be the recent manifestation of a hotspot (black triangle) that produced the Tasmanian volcanic chain (white triangle represents present day location of a hotspot in an Atlantic-Indian reference frame, filled circles are dormant volcanoes, the star symbolises an active volcano).

There is no single seamount chain in or south of the Tasman Sea that can be unequivocally traced to a present day hotspot by a continuous seamount chain. Therefore, it appears that the classical mantle plume hypothesis based on long-lived hotspots as proposed by *Morgan* [1971] may not apply to hotspot volcanism in this area.

An alternative explanation to long-lived hotspots for the origin of volcanism in the Tasman Sea is a superswell. *McNutt* [1998] defined a superswell as a dynamic feature caused by hot, buoyant material in the upper few hundred kilometers beneath the plate. Therefore, the superswell is an upper mantle phenomenon, whereas long-lived hotspots are believed to originate from the lower mantle. Superswells create volcanic chains that may show age-progression, but usually only for a short period of time, and volcanism is more scattered than that resulting from a deep-seated hotspot. The existence of a superswell under the Pacific plate has been postulated to explain volcanism of the Darwin Rise and French Polynesia, which may form by rising diapirs in the upper mantle, creating short tracks of seamounts in the direction of absolute plate motion [*McNutt*, 1998].

Similar characteristics of volcanism have been observed in the Tasman Sea, New Zealand and Antarctica. The Tasmanian volcanic chain displays a linear geometry on the Australian plate and presumably shows age-progression (only one seamount –Janszoon – has been dated), but an exact present day location for a hotspot responsible for this volcanism is not obvious (see previous discussion). Furthermore, Tertiary volcanic activity has been described close to the South Tasman Rise, on the New Zealand microcontinent and on Antarctica [*Lanyon et al.*, 1993; *Crawford et al.*, 1997]. Volcanism in the vicinity of the Balleny Islands and in the Tasman Sea cannot be produced by a single hotspot as the models here demonstrate (see Figures 5 and 10).

*Sutherland* [1994] proposed a multiple hotspot model for this region, which incorporates both primary lower mantle hotspots as well as secondary upper mantle upwellings. The SW Pacific superswell may have originally extended to the western boundary of the Pacific mantle, marked by a long-lived subduction zone, whose trace is now located at the AAD [*Gurnis et al.*, 1998] (Figure 2). It is interesting to note the majority of seamounts and onshore basaltic volcanism in the Tasman Sea area (except the Tasmantid seamounts), New Zealand, Antarctica and Balleny Islands have an isotopic signature similar to seamounts in the Pacific Ocean, i.e. HIMU (high $\mu$) to EMII (enriched mantle II) [*Staudigel et al.*, 1991; *Lanyon et al.*, 1993, *Hart et al.*, 1997)]. This suggests a genetic relationship of scattered volcanism through time from the eastern Southeast Indian Ocean to French Polynesia. Both may reflect the remainder of a massive upwelling event from a deeper thermal boundary (e.g. the core-mantle boundary). In the area between the Tasman Sea, New Zealand, Marie Byrd Land East Antarctica such an event may have created a "fossil plume" in the upper mantle, as suggested by *Hart et al.* [1997]. They proposed that the plume emplacement event did not result in volcanism at its time of emplacement, but is now sampled by small-scale convection processes in the upper mantle, resulting in short-lived, scattered volcanism. However, it is not clear how a large plume could be emplaced without volcanism. Alternatively, the "fossil plume" material in the upper mantle may be related to an actual flood basalt event, such as the one that created the Jurassic Ferrar magmatic province, which stretches from New Zealand to Tasmania and East Antarctica *[Mortimer et al.*, 1995; *Encarnación et al.*, 1996). Even though these flood basalts show clear signs of the involvement of either continental crust or sub-continental mantle in their genesis [*Hergt et al.*, 1991; *Molzahn et al.*, 1996], their short emplacement time and volume suggests that they were related to a mantle thermal event, rather than subduction processes [*Encarnación et al.*, 1996; *Duncan et al.*, 1997].

In contrast, the Tasmantid isotope and trace element data trend towards EMI (enriched mantle type I), regarded as Dupal anomaly [*Eggins et al.*, 1991]. *Castillo* [1988]

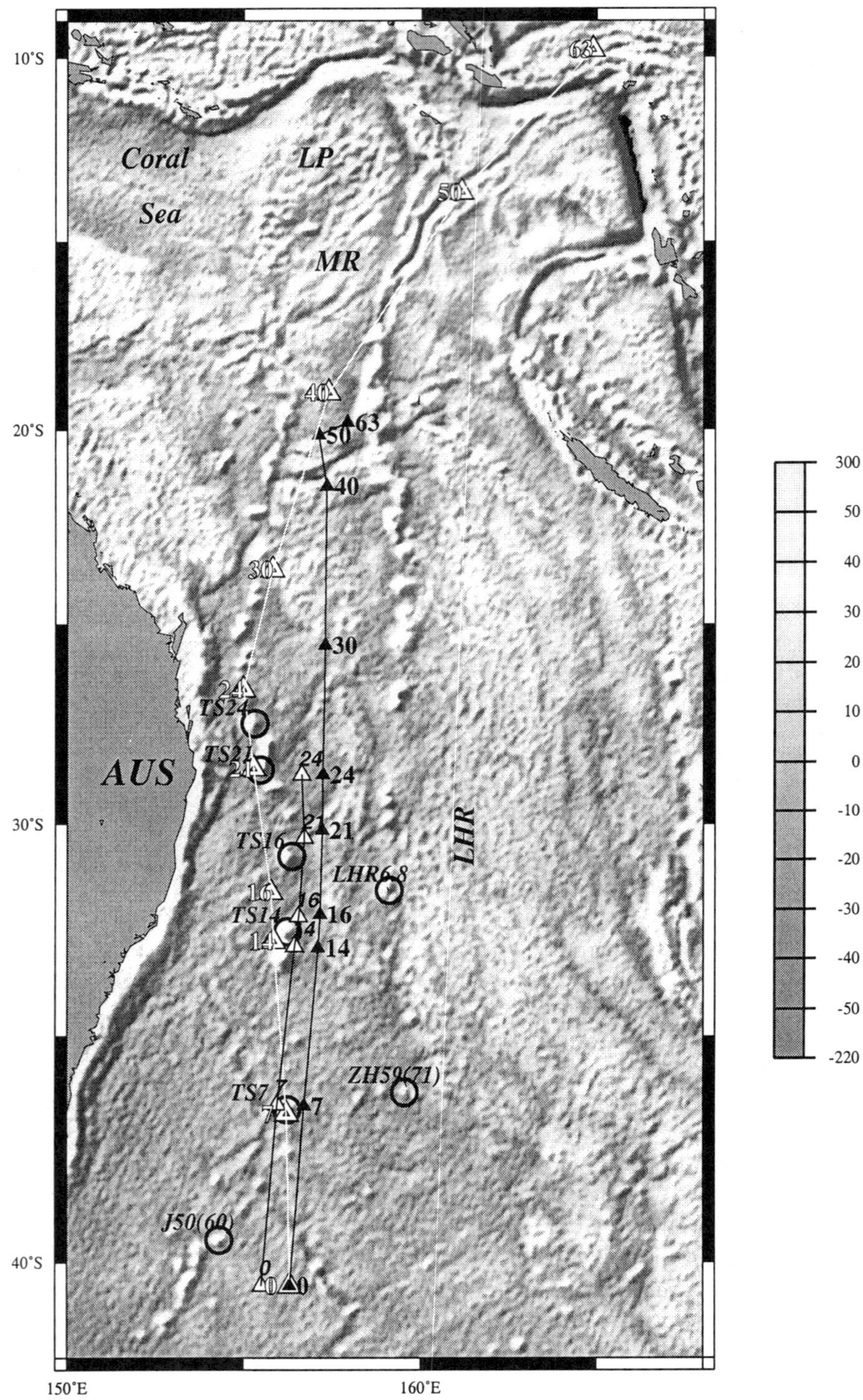


**Figure 7**. Modeled hotspot tracks shown on gridded gravity anomalies from *Sandwell and Smith* [1997]. The Tasmantid seamount chain geometry was modeled for two different absolute reference frames: the traces generated by a hotspot fixed to the Atlantic-Indian mantle are indicated by black triangles, whereas a hotspot fixed to the Pacific mantle would create a trace shown as open white triangles. The hotspot track modeled by *Müller et al.*, [1993] is shown by filled white triangles. Abbreviations are as in Figure 5a.

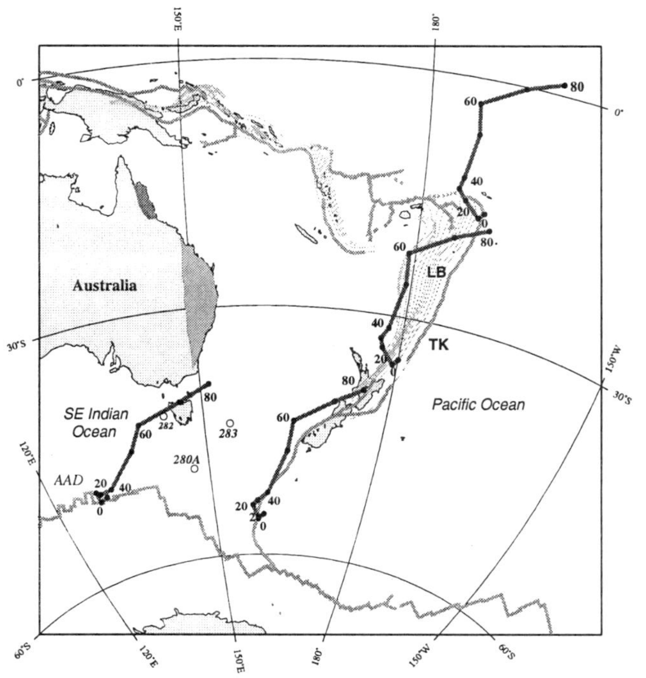


**Figure 8.** Pacific mantle motion relative to a fixed Atlantic-Indian mantle for eight stages from 80 Ma to present. Present plate boundaries are drawn as thick gray lines, contours of subducted slabs [*Gudmundsson and Sambridge*, 1998] in thin light gray lines, location of DSDP holes are filled circles. The gray filled region in eastern Australia and Tasmania [*Zhang et al.*, 1999] represents the area where Cenozoic volcanism with Pacific MORB mantle signature has been recorded. The light gray filled area in northeast Australia shows the location of Atlantic-Indian mantle-type volcanic rocks [*Zhang et al.*, 1999].

proposed a correlation between two Dupal anomaly maxima and two large-scale regions of low seismic velocity in the lower mantle. One of the two Dupal maxima corresponds to the central Pacific and the low seismic velocity domain extends to eastern-northeastern Australia. This observation, together with our model based on a Pacific hotspot frame that best predicts the age and geometry of the Tasmantid chain, suggests that the hotspot responsible for these seamounts originated from the Pacific lower mantle.

## 5. CAUSES OF ASYMMETRIC SEAFLOOR SPREADING IN THE TASMAN SEA AND CORAL SEA

In the framework of volcanism and seafloor spreading at the boundary between the Indian and Pacific mantle domains, it is also of interest that the Tasman Sea shows marked asymmetries in spreading. Asymmetries in accretion may be partly governed by the speed of ridge migration relative to the upper mantle. Based on a fluid mechanical model, *Stein et al.* [1977], suggested that the trailing plate should accrete faster than the leading plate (the plate that moves faster relative to the mantle than their conjugates). *Müller et al.* [1998] tested this idea based on a global data set and plate model. They found no convincing correlation between ridge migration, rates and spreading asymmetries, with the exception of the western Southeast Indian Ridge between India and Antarctica, which migrated away from Antarctica at rates averaging approximately 50 mm/yr. The Tasman Sea is an interesting case to further test this idea, as its spreading ridge migrated eastward, away from a nearly stationary

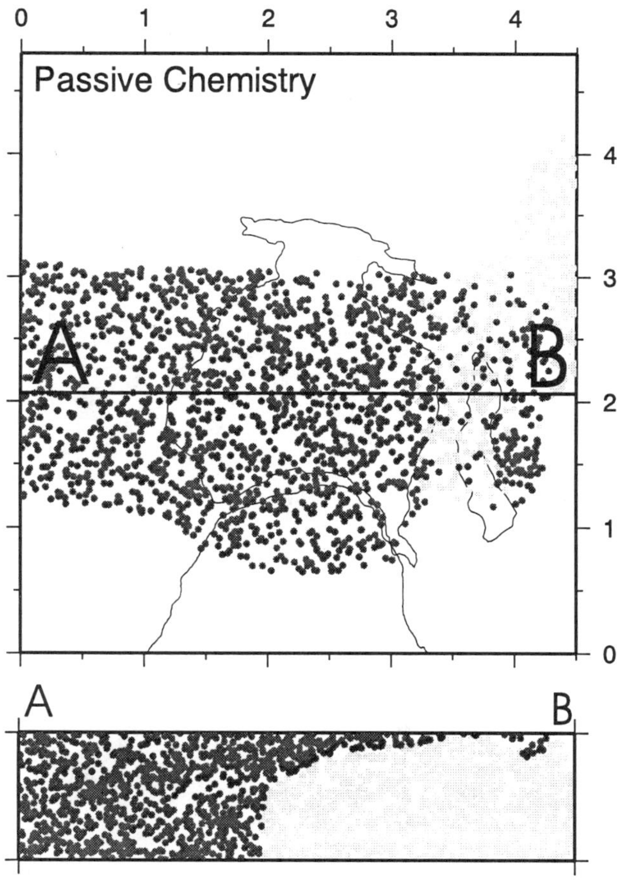


**Figure 9.** The regional distribution of Pacific (dark gray) and Atlantic-Indian (light gray) mantle tracers at 60 Ma and a vertical profile showing the depth distribution of the same tracers along a profile (A-B), based on the geodynamic model by *Gurnis et al.* [this vol.]. Note a thin wedge of Atlantic-Indian mantle overlying Pacific mantle east of the AAD. In their model, this wedge is created by lateral viscous drag of Indian asthenosphere over Pacific mantle to the east by the Late Cretaceous eastward motion of Australia and Antarctica.

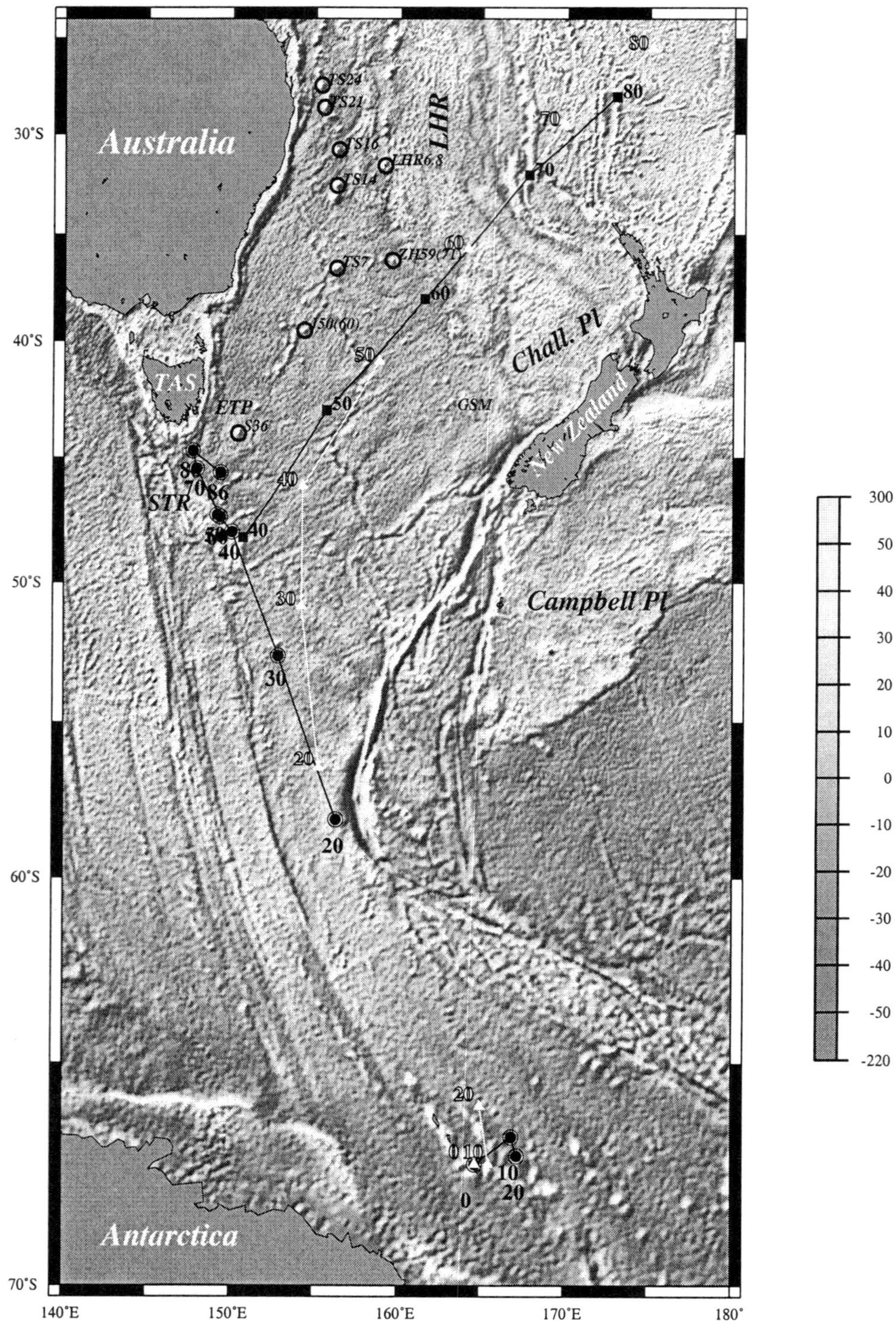


**Figure 10.** The Balleny hotspot tracks were modeled for different absolute reference frames, superimposed on the gravity anomaly grid. Black circles outline hotspot positions based on an Atlantic - Indian hotspots reference frame; the white track shows the Balleny hotspot traces in the Pacific hotspot reference frame and the black squares display the hotspot trace, if the hotspot would have been fixed to the Pacific mantle between 80 and 40 Ma and to the Atlantic - Indian mantle between 40 and 0 Ma. None of the modeled hotspot tracks predict the Tasmanian seamount volcanic chain. The present day position of the Balleny hotspot is NE of Antarctica. Abbreviations are as in Figure 5.

Australian plate, while the sub-Pacific mantle flowed to the southwest, towards the ridge.

We have investigated the causes of these asymmetries by first creating a digital age grid that interpolates isochrons, which describe the tectonic evolution of the oceanic basins east and northeast of Australia (Tasman Sea, Coral Sea, Cato Trough and the Louisiade Trough, Figure 11). Isochrons were constrained by reconstructed magnetic anomaly and fracture zone crossings, aided by small circles computed from stage rotation poles for each isochron time, and by the geometry of continental oceanic boundaries (*Gaina et al.*, 1998).

A grid of crustal accretion rates was constructed based on the method outlined by *Müller et al.* [1998] at a grid resolution of 0.05 degrees. The crustal accretion grid is illustrated in Figure 12 with dark grays indicating deficits in accretion and light grays showing excess accretion.

The data show a clear zonation of asymmetries in both the Tasman Sea and Coral Sea. Long term seafloor spreading asymmetries in the Tasman Sea are found in three areas: the northern region with excess accretion west of the ridge, the central region, which displays the same characteristic for older times, but more symmetric spreading for younger times, and the southern Tasman Sea, which shows excess crustal accretion on the eastern ridge flank.

The northernmost part of the Tasman Sea, the Cato Trough and the Louisiade Trough have a very complicated isochron geometry due to major changes in seafloor spreading direction and ridge jumps [*Gaina et al.*, 1999]; therefore the pattern of spreading asymmetries is extremely complicated. The western part of the Coral Sea is characterised by excess crustal accretion on the northeastern flank (Louisiade plate) and the central area is characterised by a deficit in accretion on the Louisiade plate (Figure 12).

The closely spaced isochrons in the Tasman Sea and the Coral Sea allow a detailed analysis of asymmetries for short geological time scales and small spatial scales. For this purpose the Tasman Sea and Coral Sea oceanic crust have been divided into several conjugate polygons that correspond to three major spreading corridors (Figure 12), consistent with the observed trends of seafloor asymmetry. Spreading asymmetries by stage and cumulative were plotted for each area (Figures 13 and 14). Any areas on conjugate plate pairs equal in time range and length along the ridge show percentages of crustal accretion which should sum up to 100%. Due to errors introduced by selecting conjugate polygon outlines, the percentage of plate accretion for conjugate plate pairs differs by 0.9-2.2% in the Tasman Sea and by 0.2-2.4% in the Coral Sea for individual stages.

The most striking seafloor spreading asymmetry is found in the southern Tasman Sea (Figure 13), with excess crust on the eastern flank (the Lord Howe Rise plate) between chron 33y (73 Ma) and chron 32y (71.2 Ma), and a reversal in asymmetry for the following stage (67.8-71.2 Ma). The central Tasman Sea shows more symmetric spreading for the same time period. Other small-scale crustal asymmetries occur between chron 31 (67.8 Ma) and chron 27 (61.2 Ma) in all polygons considered. Another change in spreading asymmetry characterises the stage 26o – 25y (56.7 –55.8 Ma) with excess accretion on the Lord Howe Rise flank for the southern and central Tasman Sea and a deficit in accretion for the same flank in the northern Tasman Sea. Overall, the cumulative spreading asymmetry is less than 10% in the Tasman Sea (Table 1).

Asymmetry in the Coral Sea is seen mostly in the western and central basin (Figure 12). Excess crust on the western Louisaide plate is observed for the time interval 55.8 –52.0 Ma (chron 25y – 24y), whereas the central Louisiade flank exhibits excess accretion only shortly before seafloor spreading ceased (53.3 – 52.0 Ma) (Figure 14). The cumulative spreading asymmetry in the Coral Sea is less than 5% (Table 2).

The rates of mid-ocean ridge migration relative to the mantle, parallel to the spreading direction, have been calculated, using the Pacific Ocean absolute plate motion model from *Müller et al.* [1991] (Figures 15 and 16), as our previous analysis has shown that this area was most likely always underlain by mantle attached to the Pacific hotspots. The largest spreading asymmetries on the Lord Howe Rise plate in the southern Tasman Sea (between chron 33y –32y, Figure 13) corresponds to an increase in ridge migration rate (Figure 15). An abrupt increase in ridge migration rate at chron 31y (67.7 Ma) resulted in excess accretion on the eastern flank in the northern Tasman Sea (see Figures 13 and 15). In the central and southern Tasman Sea the ridge migration rate increased abruptly three times from chron 34y (83 Ma) to chron 32y (71.1 Ma), then decreased smoothly until chron 24o (53.3 Ma) when another sharp increase in the rate migration took place (Figure 15). In the northern Tasman Sea the increase in ridge migration rates occurred between chrons 32y and 31y (71.1 – 67.7 Ma).

The exceedingly high ridge migration rate relative to the mantle (up to 70 mm/yr, Figure 14) is the result of the convergence of the Tasman Sea ridge and Pacific mantle. As a result, we would expect excess accretion on the trailing Australian plate for the entire Tasman Sea. This expectation is matched by the central Tasman Sea, but not by the northern and southern Tasman Sea. Alternatively, hotspot-ridge interaction may play a role for causing the observed asymmetries in crustal accretion. This appears to

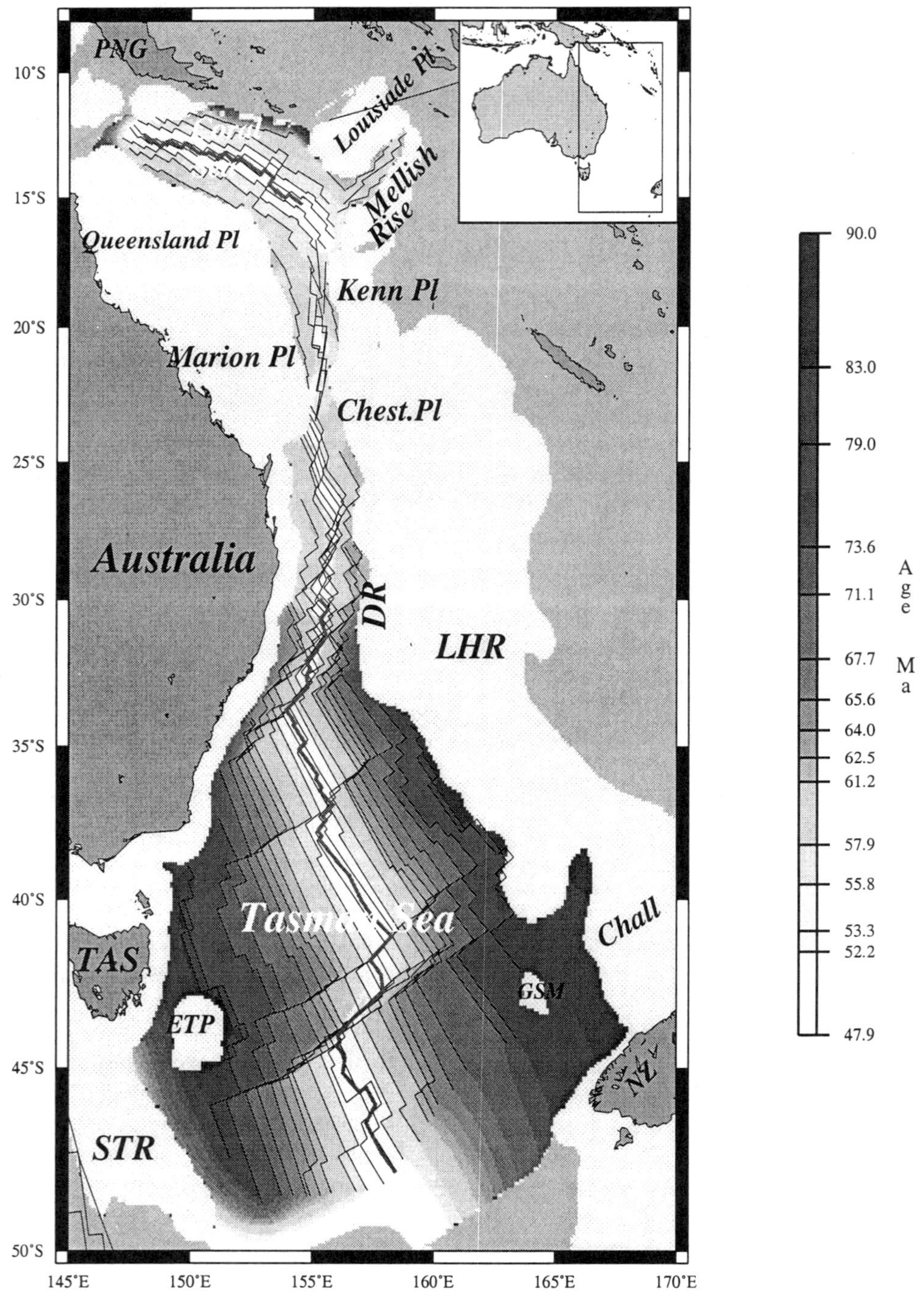


**Figure 11.** Age grid offshore eastern Australia based on *Gaina et al.*'s [1998] plate model. The isochrons were gridded using a grid-node interval of 0.05 degrees. In the southernmost Tasman Sea this grid is not constrained by isochrons. Black arrows show the direction of astenosphere flow of plume material towards mid-ocean ridge during the opening of the Tasman Sea. Abbreviations are: Chest Pl.-Chesterfield Plateau; Chall - Challenger Plateau; DR - Dampier Ridge; ETP - East Tasman Plateau; GSM - Gilbert Seamount Complex; LHR - Lord Howe Rise; STR - South Tasman Rise; TAS – Tasmania.

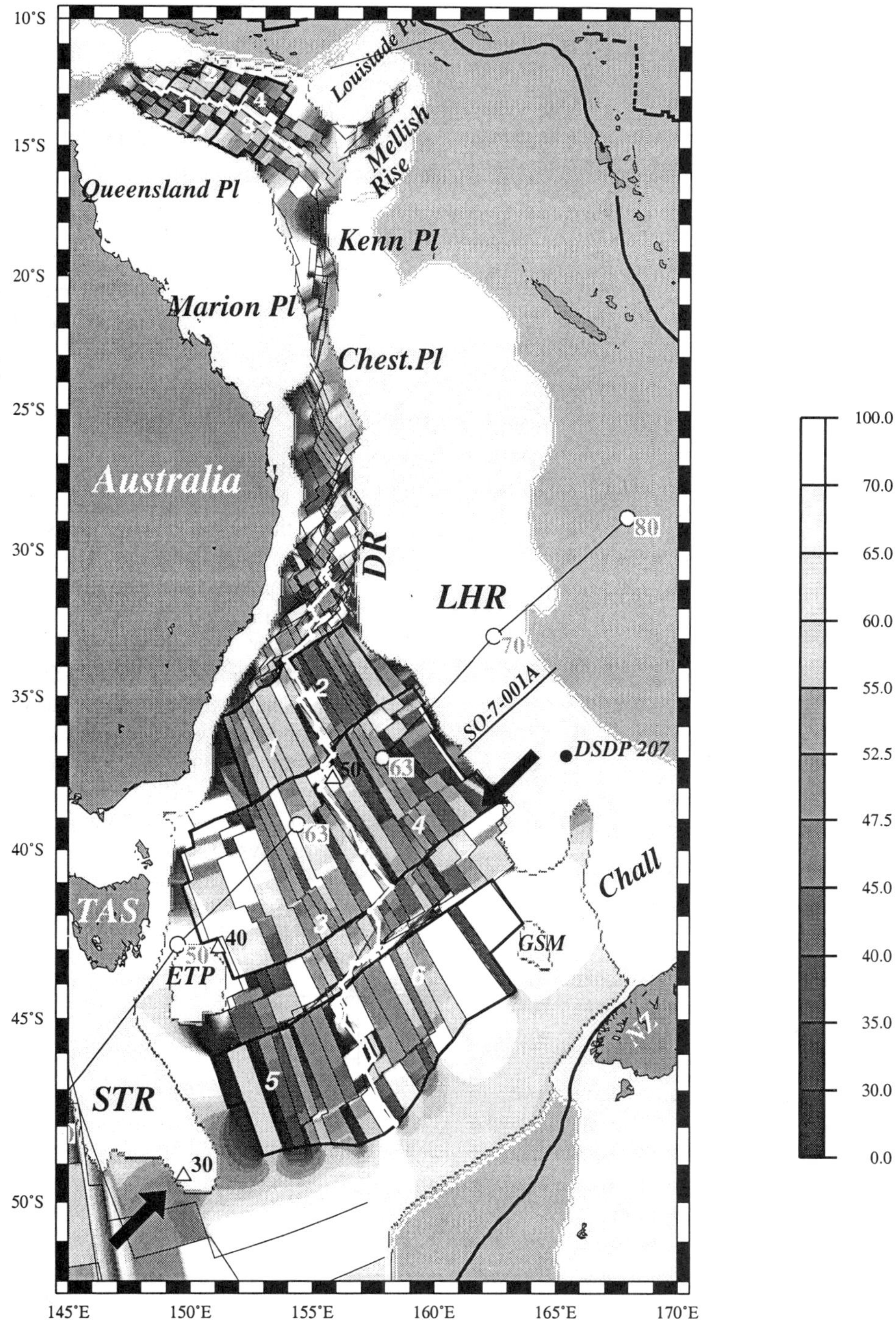


**Figure 12.** Gridded percentages of crustal accretion. White areas outline excess ocean crust, and dark gray areas deficits in crustal accretion, compared to symmetric spreading. Continental margins are light grey, and areas of insufficient data coverage are dark grey. Dark greys indicate deficit accretion and light greys show excess accretion. Hotspot tracks models for a Pacific hotpot absolute frame is shown by white circles; hotspot traces in a mixed absolute reference frame (i.e. 50-40 Ma Pacific hotspots reference frame, 40-0 Atlantic - Indian hotspot reference frame) are shown by white triangles. Areas analysed for seafloor spreading asymmetries are outlined by dark grey polygons and numbered from 1 to 6 in the Tasman Sea and from 1 to 4 in the Coral Sea. Extinct spreading ridges are white. Isochons are shown as thin black thin lines. The grid of the southernmost Tasman Sea is not constrained by isochrons. On the central Lord Howe Rise (LHR) is shown the location of DSDP 207 ocean drilling site and the SO-7-001A seismic line from Sonne survey.

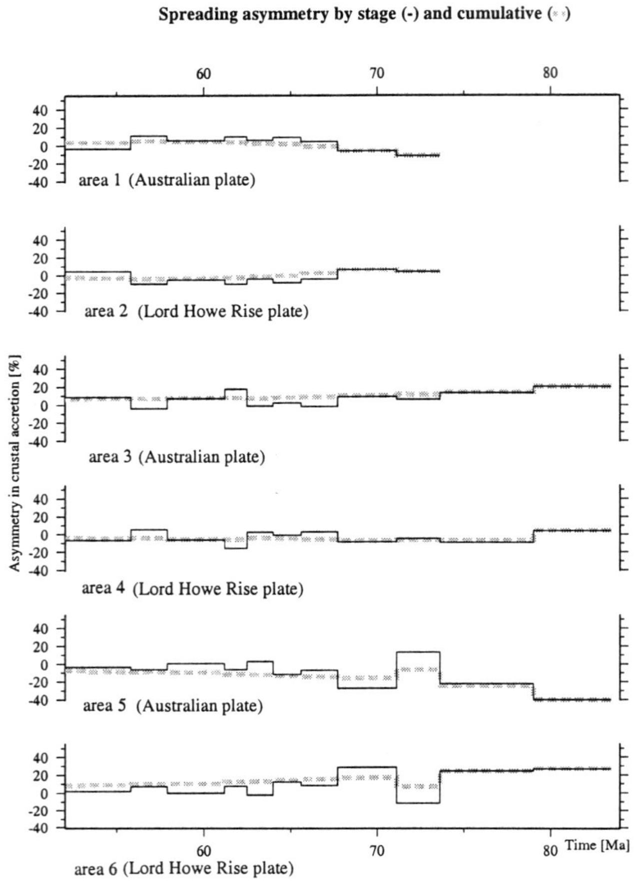


**Figure 13.** Spreading asymmetry by stage and cumulative for areas in the Tasman Sea shown in Figure 12. Small scale asymmetries are large in the southern Tasman Sea between chron 33y and 31y (73-67 Ma), but the overall cumulative asymmetry is less than 10% in the Tasman Sea.

be the case for most long-term spreading asymmetries in the ocean basins [*Müller et al.*, 1998]. We showed previously that the hotspot model developed for the Tasmanian volcanic chain requires a hotspot close to the central Lord Howe Rise during the first 20 m.y. of seafloor spreading in the Tasman Sea. The presence of the East Tasman Plateau (Figure 12), a continental fragment surrounded by ocean floor, which formed after a ridge jump toward the east in the central Tasman Sea, may have also required a thermal anomaly to initiate the ridge jump which isolated the microcontinent. The Lord Howe Rise experienced several phases of volcanism. Rhyolites and tuffs recovered from the DSDP Site 207 (Figure 12) are dated approx. 94 Ma [*McDougall and van der Lingen*, 1974] and may represent the younger phase of volcanism. *Willcox et al.* [1980] interpreted seismic data acquired from the central Lord Howe Rise and identified Late Cretaceous-early Tertiary intrusions. These data support a scenario in which a hotspot may have been present in the vicinity of the central Lord Howe Rise during Late Cretaceous-early Tertiary and influenced seafloor spreading in the northern and central Tasman Sea. This hotspot is now most likely located underneath the Pleiades volcanic province in northern Victoria Land in East Antarctica (Fig. 6), and caused the formation of the Tasmanian seamount chain, as outlined earlier (Fig. 5). However, as the rates of ridge migration in a Pacific mantle reference frame were relatively high, the mechanism suggested by *Stein et al.* [1977] may have also contributed to spreading asymmetry in this area.

The southern Tasman Sea exhibits a pattern opposite to that of the remainder of the basin, with large excess accretion on the eastern flank. The excess oceanic crust was accreted between chron 33y and 31y (73 – 67.7 Ma). In this case the ridge axis in the southern Tasman Sea would have propagated toward the west and created excess oceanic crust on the Lord Howe Rise flank. Our model describes the ridge propagation event that detached the Gilbert Seamount Complex from the Australian plate at about 77 Ma [*Gaina et al.*, 1998]. Similar events or more subtle ridge jumps would have led to excess accretion on the southern Lord Howe Rise plate, due to the presence of a thermal anomaly situated close to the South Tasman Rise. *Weaver et al.* [1994] speculated that the mantle plume known to exist beneath the Cenozoic alkaline basalt field in Marie Byrd Land may have existed in Cretaceous

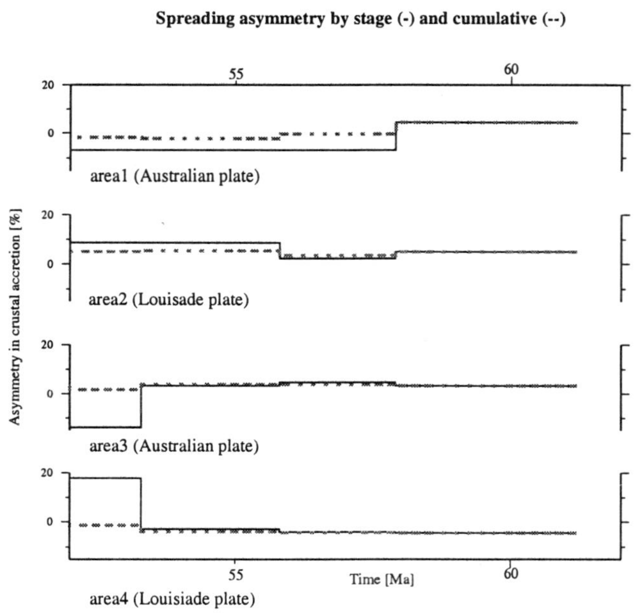


**Figure 14.** Spreading asymmetry by stage versus time for conjugate areas in the Coral Sea (see Figure 12). The cumulative asymmetry is less than 5%.

**Table 1.** Spreading Asymmetry by Stage and Cumulative in the Tasman Sea

| Stage | Stage asymmetry | Cumulative asymmetry | Stage asymmetry | Cumulative asymmetry |
|---|---|---|---|---|
| | Area 1 | | Area 2 | |
| 73.6 – 71.1 | -11.4 | -11.4 | 4.4 | 4.4 |
| 71.1 – 67.7 | -5.8 | -5.8 | 6.7 | 6.6 |
| 67.7 – 65.6 | 4.6 | -1.2 | -4.4 | 2.5 |
| 65.6 – 64.0 | 9.4 | 1.6 | -8.2 | -0.1 |
| 64.0 – 62.5 | 5.9 | 2.6 | -4.4 | -1.0 |
| 62.5 – 61.2 | 10.0 | 3.8 | -9.9 | -2.4 |
| 61.2 –57.9 | 5.8 | 4.4 | -5.2 | -3.2 |
| 57.9 – 55.8 | 11.0 | 5.3 | -9.6 | -4.0 |
| 55.8 – 53.3 | -3.5 | 4.1 | 4.7 | -3.1 |
| 53.3 – 52.0 | -3.5 | 4.0 | 4.7 | -2.9 |
| | Area 3 | | Area 4 | |
| 83.5 – 79.0 | 20.3 | 20.3 | 3.9 | 3.8 |
| 79.0 – 73.6 | 13.0 | 14.0 | -9.2 | -7.3 |
| 73.6 –71.1 | 6.0 | 11.4 | -5.0 | -6.6 |
| 71.1 – 67.7 | 9.2 | 10.8 | -8.6 | -7.2 |
| 67.7 – 65.6 | -1.8 | 8.9 | 2.5 | -5.8 |
| 65.6 – 64.0 | 2.2 | 8.2 | -1.1 | -5.3 |
| 64.0 – 62.5 | -10 | 7.4 | 2.3 | -4.6 |
| 62.5 – 61.2 | 17.7 | 8.1 | -15.8 | -5.5 |
| 61.2 –57.9 | 6.9 | 7.9 | -6.4 | -5.6 |
| 57.9 – 55.8 | -4.0 | 6.9 | 5.2 | -4.7 |
| 55.8 – 53.3 | 8.7 | 7.0 | -6.8 | -4.9 |
| 53.3 – 52.0 | 8.7 | 6.7 | -6.8 | -4.5 |
| | Area 5 | | Area 6 | |
| 83.5 – 79.0 | -40.2 | -40.2 | 26.8 | 26.8 |
| 79.0 – 73.6 | -22.4 | -24.9 | 24.3 | 24.4 |
| 73.6 –71.1 | 13.2 | -6.2 | -11.3 | 7.2 |
| 71.1 – 67.7 | -27.5 | -16.2 | 28.9 | 17.0 |
| 67.7 – 65.6 | -6.9 | -14.1 | 8.2 | 15.2 |
| 65.6 – 64.0 | -11.8 | -13.8 | 12.5 | 14.8 |
| 64.0 – 62.5 | 3.0 | -12.2 | -2.2 | 13.8 |
| 62.5 – 61.2 | -6.3 | -11.6 | 7.7 | 12.5 |
| 61.2 –57.9 | 0.5 | -9.5 | 0.0 | 10.3 |
| 57.9 – 55.8 | -6.4 | -9.1 | 7.3 | 9.9 |
| 55.8 – 53.3 | -3.7 | -8.5 | 1.8 | 9.1 |
| 53.3 – 52.0 | -3.7 | -7.4 | 1.8 | 8.3 |

**Table 2.** Spreading Asymmetry by Stage and Cumulative in the Coral Sea

| Stage | Stage asymmetry | Cumulative asymmetry | Stage asymmetry | Cumulative asymmetry |
|---|---|---|---|---|
| | Area 1 | | Area 2 | |
| 61.2 –57.9 | 4.7 | 4.7 | -5.0 | -5.0 |
| 57.9 – 55.8 | -6.9 | -1.4 | 2.3 | 3.5 |
| 55.8 – 53.3 | -6.9 | -3.2 | 8.6 | 5.3 |
| 53.3 – 52.0 | -6.9 | -2.6 | 8.6 | 5.0 |
| | Area 3 | | Area 4 | |
| 61.2 –57.9 | 3.2 | 3.2 | -4.4 | -4.4 |
| 57.9 – 55.8 | 4.6 | 3.8 | -4.1 | -4.3 |
| 55.8 – 53.3 | 3.3 | 3.7 | -3.0 | -3.9 |
| 53.3 – 52.0 | -13.9 | 1.6 | 17.8 | -1.4 |

time and controlled the position of break-up between New Zealand and Marie Byrd Land. *Stern and ten Brink* [1989] estimated that a thermal event affected the Ross Embayment at 70 Ma and initiated a thermal uplift. Also, *Sutherland* [1994] suggested that a thermal anomaly may have been located between the southern ends of the Challenger Plateau and South Tasman Rise, now situated under Mt. Erebus. The proximity of this Antarctic thermal anomaly during the early seafloor spreading in the Tasman Sea may have accounted for the formation of the Gilbert Seamount Complex microplate by a westward ridge jump and subsequent excess accretion on the southern Lord Howe Rise plate, by means of continued ridge jumps towards the plume. Our plate model shows that if this thermal anomaly still exists, and is fixed to Pacific hotspots, then it would be located in the vicinity of Lake Vostok, East Antarctica. This may provide an intriguing cause for the origin of this large sub-glacial lake [Siegert, 1998 #9425].

Even though the observed spreading asymmetries east of the East Tasman Plateau may be accounted for by *Stein et al.*'s [1977] mechanism, their model does not predict ridge jumps onto continental lithosphere creating a microcontinent. The ridge jump that detached the East Tasman Plateau from the Lord Howe Rise and the following asymmetry in seafloor spreading is therefore more likely a consequence of a hotspot close to the Lord Howe Rise. A similar scenario might have taken place in the case of the Jan Mayen microcontinent in the Norwegian-Greenland Sea, where the ridge jumped towards the Iceland hotspot, detached the Jan Mayen microplate from Greenland, leading to excess crust formation on the eastern flank [*Talwani and Eldholm*, 1977]. Later, the ridge crossed the hotspot and excess crust accretion switched from the Jan Mayen to Greenland [*Müller et al*, 1998]. The Seychelles in the Indian Ocean were isolated from India when the Reunion hotspot was located under the western margin of India. Both spreading

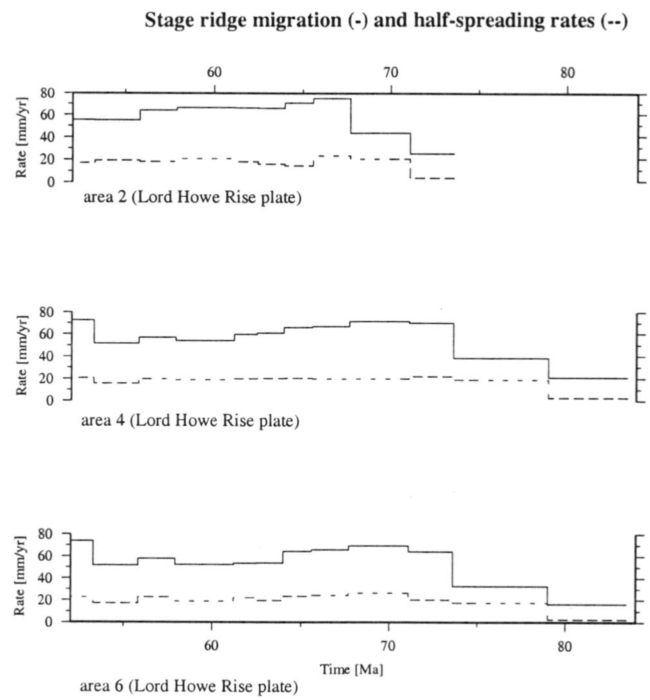


**Figure 15**. Ridge migration and half spreading rates for the leading plate (the Lord Howe Rise) for three spreading corridors in the Tasman Sea (Pacific hotspot reference frame). The ridge normal component of the migration of mid-ocean ridge relative to the mantle versus time is computed for points on the extinct ridge in each area analysed separately, as outlined on Figure 10.

ridges subsequently crossed the mantle plume, resulting in a change in the sense of asymmetries in accretion [*Müller et al.*, 1998]. In all cases ridge-hotspot interaction caused the formation of a microplate followed by a prolonged period of asymmetric spreading.

In the Coral Sea ridge, migration rate diagrams display constantly high values, around 60 –70 mm/y, in the Pacific hotspot reference frame (Figure 16), as the Coral Sea ridge migrated towards the sub-pacific mantle, which moved to the southeast. In this case *Stein et al.*'s [1977] model would predict large asymmetries, but the observed values are low (see Table2). *Sutherland* [1994] and *Yan and Kroenke* [1993] suggested that the opening of the Coral Sea might have been influenced by same hotspot that generated the Tasmantid volcanic chain, with a hotspot that was situated northeast of the Louisiade Plateau at 65 Ma (pre-breakup of the Coral Sea). More recently, *Sutherland* [1998] published a revised hotspot model for explaining the volcanism in the northern Tasman Sea and eastern Australian margin. He proposed that a system of hotspots might have been active in the proximity of the northeastern Australia by the time of continental breakup. The model presented in this paper, based on the most up to date set of rotations, including the relative motion between the East and West Antarctica [*Cande et al.*, in press], shows that, irrespective which hotspot frame is chosen, the position of the hotspot at 65 Ma is far away from the Coral Sea (Figure 7). Therefore, the deficit in crustal accretion on the leading plate in the central Coral Sea (Figure 11) can not be explained by the presence of a known hotspot. In this case, taking into account that the asymmetry is relatively small in scale, it might have been rather caused by long-lived asymmetries in small-scale convection under the ridge [*Rouzo et al.*, 1995] or variations in melt supply [*Sempéré et al.*, 1995].

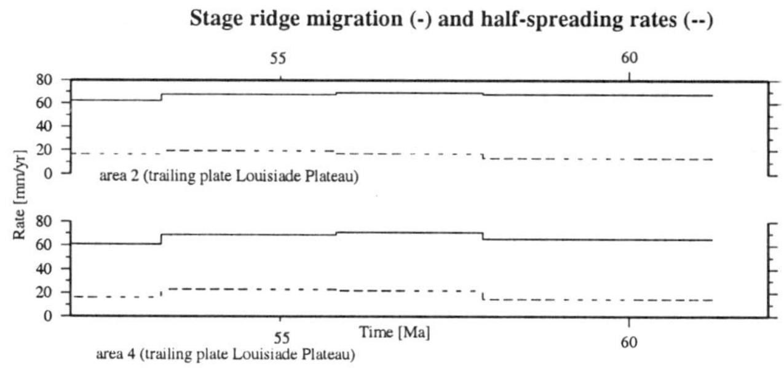


**Figure 16**. Ridge migration and half spreading rates for two selected regions in the Coral Sea (Pacific absolute reference frame).

## 6. CONCLUSIONS

Our study aimed to investigate the relationship between volcanism in and south of the Tasman Sea, absolute plate motions, and the dynamics of the two mantle domains: the Atlantic-Indian and Pacific. For the late Tertiary, the evidence for different mantle domains flow rests on modeling the path of the Tasmantid seamount chain, which appears to originate from a lower mantle source [*Eggins et al.*, 1991]. This confirms that lower mantle flow in this area has not been substantially affected by subduction east of Australia, as the slabs are partially deflected at the upper/low mantle interface [*Van der Hilst*, 1995]. The Tasmantid seamount chain geometry is best matched by a hotspot in a Pacific reference frame, but the hotspot path modeled in an Atlantic-Indian reference frame does not differ substantially from the Pacific one for the last 16 m.y. This reflects that the relative motion between the two systems of hotspots was relatively small during this time.

Apart from the Tasmantid seamounts, the origin of volcanism in the Tasman Sea and surrounding regions (New Zealand, Antarctica, Balleny Islands) likely resides in the upper mantle [*Lanyon et al.*, 1993]. Three hypotheses have been tested in order to investigate the type of mantle that generated the Tasman Sea volcanism. A mantle cell that was attached to the Atlantic-Indian mantle from 90 Ma to present cannot explain the Tasmanian seamount chain geometry. In the case of a mixed Pacific-Indian-Atlantic reference frame, the model fails to predict the older Tasmanian seamounts, but the position and age of the Janszoon and Soela seamounts are well matched assuming a present day hotspot location southwest of the Macquarie triple junction. No recent volcanism has been described for this area, but recent seismic activity has been reported close to that location [*Wiens and Wysession*, 1998; *Nettles et al.*, 1999]. If a hotspot is now located near the Balleny Islands, it could not have created the Tasmanian seamounts, irrespective of which hotspot reference frame is chosen.

Most of volcanism in the Tasman Sea and south of the Tasman Sea (New Zealand, Antarctica) matches superswell type upper mantle sources, i.e. the hotspot tracks are short-lived and seamounts have a similar geochemistry to other Pacific seamounts. Our results help to differentiate between two hypotheses: (1) The Australian-Antarctic Discordance (AAD) exists because Pacific asthenosphere has recently flown westward into an area that was previously underlain by Atlantic-Indian mantle [*West et al.*, 1997; *Christie et al.*, 1998; *Zhang et al*, 1999], and (2) The source of the AAD is a subducting slab that marks the long-lived boundary between two mantle domains. Only model 2 is consistent with all observations, and in particular accounts for the spatial distribution of isotopic basalt signatures east of the AAD.

The border between the Pacific and Atlantic-Indian mantle types remained almost stationary through time, and the apparent Pacific mantle migration to the west is due to absolute plate motion over a subducted slab with varying inclination through time in the upper mantle. Part of the Indian mantle reservoir originally overlying a westward dipping slab east of Australia was drawn out laterally by viscous drag caused by Late Cretaceous eastward motion of Australia and Antarctica, thus decreasing the initial dip of the slab in the late Cretaceous. However, this model does not account for young ocean floor in the Lau Basin, which shows Atlantic-Indian geochemical signatures.

The origin and trends of seamounts chains in the Tasman Sea can best be accounted for by hotspots in a Pacific hotspot frame. We conclude that the mantle beneath the Tasman Sea is not only Pacific mantle in a geochemical sense, but also moved dynamically in concert with sub-Pacific mantle since the opening of the Tasman Sea.

A Pacific hotspot model is also able to explain the observed asymmetry in crustal accretion in the Tasman Sea. If the Pleiades volcanic province located in Antarctica is the result of a hotspot, then the same hotspot would have been active at 80 Ma east of the Lord Howe Rise in a Pacific mantle reference frame, and resulted in microcontinent formation and excess accretion on the Australian Plate by means of ridge jumps towards the hotspot. The southern Tasman Sea displays excess accretion on the Lord Howe Rise plate that may be explained by the presence of a thermal anomaly on the Antarctic plate which was located SW of the Tasman Sea in the Late Cretaceous.

*Acknowledgments.* The results presented here are part of the first author's Ph.D. project, sponsored by the Australian Geological Survey Organisation, with support from the Division of Geology and Geophysics, School of Geosciences, the University of Sydney. This work was supported by the University of Sydney (CG, RDM) and Scripps Institution of Ocenography (SC). We thank Lin Sutherland, Anthony Crawford, and Richard Gordon for many constructive comments which improved the quality of the manuscript.

## REFERENCES

Cande, S.C., C.A. Raymond, J. Stock, and W.F. Haxby, Geophysics of the Pitman fracture zone and Pacific Antarctic plate motion during the Cenozoic, *Science, 270*, 947-953, 1995.

Cande, S.C., Seafloor spreading constraints on Cenozoic motion between East and West Antarctica, Abstract volume, Penrose Conference, 25-31 March, New Zealand, 1999.

Cande, S., J. Stock, R.D. Müller, and T. Ishihara, Cenozoic Motion between East and West Antarctica, *Nature*, in press.

Castillo P., The Dupal anomaly as a trace of the upwelling lower mantle, *Nature, 336*, 667-670, 1988.

Christie D. M., B. P. West, D.G. Pyle, and B.B. Hanan, Chaotic topography, mantle flow and mantle migration in the Australian-Antarctic discordance, *Nature, 394*, 1998.

Crawford, A.J., Lanyon, R., and Elmes, M., Geochemistry and significance of basalic rocks dredged from the South Tasman Rise and adjacent seamounts, *Aus. J. Earth Sci.*, 44, 539-541, 1997.

Crawford, A.J., Briqueu, L., Laporte, C., and Hasenaka, T., Coexistence of Indian and Pacific Oceanic upper mantle reservoirs beneath the central New Hebrides Island Arc, Active margins and marginal basins of the Western Pacific, *Geophysical Monograph 88*, 199-217, 1995.

Creager, K.C., and T.H. Jordan, Slab penetration into the lower mantle beneath the Mariana and other island arcs of the Northwest Pacific, *J. Geophys. Res., 91*, 3573-3589, 1986.

Denham, D., The Tasman Sea earthquake of 25 November 1083, and stress in the Australian plate, *Tectonophys., 111*, 329-338, 1985.

Duncan, R. A., Hotspots in the southern oceans-An absolute frame of reference for motion of the Gondwana continents, *Tectonophys.*, 74, 29-42, 1981.

Duncan, R., and I. McDougall, Volcanic time-space relationships, in *Intraplate Volcanism in Eastern Australia and New Zealand*, edited by R. Johnson, pp. 43-54, Cambridge Univ. Press, New York, 1989.

Duncan, R.A., P.R. Hooper, J. Rehacek, J.S. Marsh, and A.R. Duncan, The Timing and Duration of the Karoo Igneous Event, Southern Gondwana, *J. Geophys. Res., 102*, 18127-18138, 1997.

Encarnación, J., T.H. Fleming, D.H. Elliot, and H.V. Eales, Synchronous Emplacement of Ferrar and Karoo Dolerites and the Early Breakup of Gondwana, *Geology, 24*, 535-538, 1996.

Eggins, S.M., D.H. Green, and T.J. Fallon, The Tasmantid seamounts: Shallow melting and contamination of an EMI mantle plume, *Earth Planet. Sci. Lett., 107*, 448-462, 1991.

Eissen J. P., A.J. Crawford, J. Cotton, S.Meffre, H. Bellon, M. Delaune, Geochemistry and tectonic significance of basalts in the Poya Terrane, New Caledonia, *Tectonophys, 284*, 203-219, 1998.

Fukao, Y., M. Obayashi, H. Inoue, and M. Nenbai, Subducting slabs stagnant in the mantle transition zone, *J. Geophys. Res., 97*, 4809-4822, 1992.

Gaina, C., R.D. Müller, J.-Y. Royer, J. Stock, J. Hardebeck, and P. Symonds, The tectonic evolution of the Tasman Sea: A tectonic puzzle with thirteen pieces, *J. Geophys. Res., 103,* B6, p12413-12433 , 1998.

Gaina, C., Tectonic history of microplates east of Australia, Ph.D. thesis, The University of Sydney, Sydney, 1998.

Gaina, C., R.D. Müller, J.-Y. Royer, and P. Symonds, The tectonic evolution of the Louisiade Triple Junction, *J. Geophys. Res., 104,* B6, p12927-12939 , 1999.

Gudmundsson, O., and M. Sambridge, A regionalized upper mantle (RUM) seismic model, *J. Geophys. Res., 103*, 7121-7136, 1998.

Gurnis, M., R.D. Müller and L. Moresi, Cretaceous Vertical Motion of Australia and the Australian-Antarctic Discordance, *Science, 279*, 1499-1504, 1998.

Gurnis, M., L. Moresi, and R.D. Müller, Models of mantle convection incorporating plate tectonics: The Australian region since the Cretaceous, *this vol.*

Hart, S.R., Blusztajn, J., LeMasurier, W.E., Rex, D.C., Hobbs Coast Cenozoic volcanism; implications for the West Antarctic rift system, *Chemical Geology, 139*, 223-248, 1997.

Hergt, J.M., D.W. Peate, and C.J. Hawkesworth, The petrogenesis of Mesozoic Gondwana low-Ti flood bsaslts, *Earth Planet. Sci. Lett., 105*, 134-148, 1991.

Hergt, J.M., and Hawkesworth, C.J., Pb-, Sr-, and Nd-Isotopic evolution of the Lau Baisn: Implications for mantle dynamics during backarc opening, Proceedings of theOcean Drilling Programs, Scientific results, 135, 505-517, 1994.

Hickey-Vargas, R., Hergt, J.M., and Spadea, P., The Indian Ocean-type isotopic signature in Western Pacific marginal basins: Origin and significance, Active margins and amrginal basins of the Western Pacific, *AGU Geophysical Monograph, 88*, 175-197, 1995.

Jenkins, C.J., G.H. Packham, T.C. Hubble, P.G. Quilty, and C.J. Adams, Trace of the Balleny hotspot in the Tasman Sea and Southern Ocean from 70 Ma to present, *First Australian Marine Geoscience Meeting Abstracts*, 9-11 July, Aus. Nat. Univ., Canberra, 1992.

Klein, E.M., C.H. Langmuir, A. Zindler, H. Staudigel, and B. Hamelin, Isotope evidence of a mantle convection boundary at the Australian-Antarctic Discordance, *Nature, 333*, 623-629, 1988.

Koppers, A.A.P., $^{40}Ar/^{39}Ar$ geochronology and isotope geochemistry of the West Pacific seamount province: implications for Pacific plate motion and chemical geodynamics, PhD., Vrije Universiteit Amsterdam, the Netherlands, 263 p, 1998.

Kroenke, L.W., Cenozoic tectonic development of the Southwest Pacific, *Tech. Bull.-U.N. Econ., Soc. Comm. Asia Pac., Comm. Co-ord. Jt. Prospect Miner. Resour. South Pac. Offshore Areas, 6*, 1984.

Kyle, P.R., The Pleiades, in Volcanoes of the Antarctic plate and southern oceans, W.E. LeMesurier and J.W. Thomson (Eds.), *Ant. Res. Series, 48*, 60-64, Am. Gephys. Un., Washington D.C., 1990.

Lanyon, R., Varne, R., Crawford, A., Tasmanian Tertiary basalts, the Balleny plume, and the opening of the Tasman Sea (southwest Pacific Ocean), *Geology, 21*, 555-558, 1993.

McDougall, I. and G.J. Van der Lingen, Age of the rhyolites of the Lord Howe Rise and evolution of the southwest Pacific Ocean, *Eath Planet. Sci. Lett., 21*, 117-126, 1974.

McDougall, I., B.J.J. Embelton, and D.B. Stone, Origin and evolution of the Lord Howe Island, soutwest Pacific ocean, *J. of the Geol. Soc.of Aust., 28*, 155-176, 1981.

McDougall, I. and R.A. Duncan, Age progressive volcanism in the Tasmantid seamounts, *Earth Planet. Sci. Lett., 89*, 207-220, 1988.

McNutt, M.K., Superswells, *Rev. of Geophys., 36,* 211-245, 1998.

Morgan, W.J., Convection plumes in the lower mantle, *Nature, 230*, 42-43, 1971.

Mortimer, N., D. Parkinson, J.I. Raine, C.J. Adams, I.J. Graham, P.J. Oliver, and K. Palmer, Ferrar Magmatic Province Rocks Discovered in New-Zealand - Implications For Mesozoic Gondwana, *Geology, 23*, 185-188, 1995.

Müller, R.D., J.-Y. Royer, and L.A. Lawver, A review of absolute plate motion models from Jurassic to present day, *Technical Report, 112,* Univ. of Texas Inst. of Geophysics, Austin, Texas, 1991.

Müller, R.D., J.-Y. Royer, L.A. Lawver, Revised plate motions relative to the hotspots from combined Atlantic and Indian Ocean hotspot tracks, *Geology, 21*, 275-278, 1993.

Müller, R.D., W. Roest and J.-Y. Royer, Asymmetric seafloor spreading expresses ridge-plume interactions, *Nature, 396*, 455-459, 1998.

Nettles, M., T.C. Wallace, and S.L. Beck, The March 25, 1998 Antarctic plate earthquake, *Geophys. Res. Let., 26*, 2097-2100, 1999.

Olson, P., P.G. Silver, and W.C. Carlson, The large-scale structure of convection in the Earth's mantle, *Nature, 344*, 209-215, 1990.

Pyle, D.G., D.M. Christie, and J.J. Mahoney, Resolving an isotopic boundary within the Australian-Antarctic Discordance, *Earth Planet. Sci. Lett., 112*, 161-178, 1992.

Pyle, D.G., D.M. Christie, J.J. Mahoney, and R.A Duncan, Geochemistry and geochronology of ancient southeast Indian and southwest Pacific seafloor, *J. Geophys. Res, 100,* 22,261-22,282, 1995.

Rouzo, S., M. Rabinowicz, and A. Briais, Segmentation of mid-ocean ridges with an axial valley induced by small-scale mantle convection, *Nature, 74*, 795-798, 1995.

Sager, W.W. and Pringle, M.S., 1987, Palaeomagnetic constraints on the origin and evolution of the Musicians and South Hawaiian seamounts, central Pacific Ocean, *in* Keating, B.H. et al., eds., Seamount islands and atolls, *AGU Geophysical Monograph 43*, 133-162.

Sandwell, D.T., and W.H.F. Smith, Marine gravity anomaly from ERS-1, Geosat and satellite altimetry, *J. Geophys. Res., 102*, 10039-10045, 1997.

Sémperé, J.-C., Blondel, P., Briais, A., Fujiwara, T., Geli, L., Isezaki, N., Pariso, J.E., Parson, L., Patriat, P., and Rommevaux, C., The Mid-Atlantic ridge between 29°N and 31°30'N in the last 10 Ma, *Earth Planet. Sci. Lett.*, 130, 45-55, 1995.

Siegert, M.J., and J.K. Ridley, An Analysis of the Ice-Sheet Surface and Subsurface Topography Above the Vostok Station Subglacial Lake, Central East Antarctica, *J. Geophys. Res., 103*, 10195-10207, 1998.

Staudigel, H., K.-H. Park, M. Pringle, J.L. Rubenstone, W.H.F. Smith, and A. Zindler, The longevity of the South Pacific isotopic and thermal anomaly, *Earth Planet. Sci. Lett., 102*, 24-44, 1991.

Stein, S., H.J. Melosh and J.B. Minster, Ridge migration and asymmetric sea-floor spreading, *Earth Planet. Sci. Lett., 36*, 51-62, 1977.

Stern, T.A., and U.S ten Brink, Flexural uplift of the Transantarctic mountains, *J. Geophys. Res., 94*, 10315-10330, 1989.

Storey, B.C., and P.R. Kyle, An Active Mantle Mechanism For Gondwana Breakup, *Trans. Geol. Soc. South Africa, 100*, 283-290, 1997.

Sutherland, F.L., Tasman Sea evolution and hotspot trails, *Evolution of the Tasman Sea, Proceedings of the Tasman Sea Conference*, eds.G.J. van der Linden, K.M. Swanson and R.J. Muir, 35-51, 1994.

Sutherland, F.L., Origin of north Queensland Cenozoic volcanism: Relationship to long lava flow basaltic fields, Australia, *J. Gephys. Res., 103*, 27347-27358, 1998.

Tarduno, J.A., and J. Gee, Large-scale motion between Pacific and Atlantic hotspots, *Nature, 378*, 477-480, 1995.

Talwani M., and O. Eldholm, Evolution of the Norwegian-Greenland Sea, *Geol. Soc. Am. Bull., 88*, 969-999, 1977.

Turner, S., and Hawkesworth, C., Using geochemistry to map mantle flow beneath the Lau Basin, *Geology*, 26, 1019-1022, 1998.

Van der Hilst, R., Complex morphology of subducted lithosphere in the mantle beneath the Tonga trench, *Nature, 374*, 154-157, 1995.

Van der Hilst, R., R. Engdahl, W. Spakman, and G. Nolet, Tomographic imaging of subducted lithosphere below northwest Pacific island arcs, *Nature, 353*, 37-43, 1991.

Weaver, S.D., B.C. Storey, R.J. Pankhurst, S.B. Mukasa, V.J. DiVenere, and J.D Bradsaw, Antarctica-Nw Zealand rifting and Marie Byrd Land lithospheric magmatism linked to ridge subduction and mantle plume activity, *Geology, 22*, 811-814, 1994.

Wessel P. and L. Kroenke, A geometric technique for relocating hotspots and refining absolute plate motions, *Nature, 387*, 365-369, 1997.

Wessel, P., and W.H.F. Smith, Free software helps map and

display data, *Eos Trans. AGU*, *72*, 441, 445-446, 1991.

West, B.P., S.D. Wilcock, J.-C. Sempere and L Geli, Three dimensional structure of astenospheric flow beneath the Southeast Indian Ridge, *J. Geophys Res., 102*, 7783-7802, 1997.

Wiens, D.A. and M.E. Wysession, Recent oceanic intraplate earthquake in Balleny Sea was largest ever detected, *Eos, 79*, no. 30, 1998.

Willcox, J.B., P.A. Symonds, K. Hinz, and D. Bennett, Lord Howe Rise, Tasman Sea-preliminary geophysical results and petroleum prospects, *BMR J. Aust. Geol. Geophys.*, *5*, 225-236, 1980.

Yan, C.Y., and L.W. Kroenke, A plate reconstruction of the southwest Pacific, 0-100 Ma, *Proceeding of the ODP Program, Scientific Results*, 130, 697-709, 1993.

Zhang, M., S.Y. O'Reilly, D. Chen, Location of Pacific and Indian mid-ocean ridge-type mantle in two time slices: Evidences from Pb, Sr and Nd isotopes for Cenozoic Australian basalts, *Geology, 27*, 39-42, 1999.

C. Gaina and R.D. Müller, Division of Geology and Geophysics, School of Geosciences, University of Sydney, Edgeworth David Building, F05, The University of Sydney, New South Wales 2006, Sydney, Australia. (carmen@es.usyd.edu.au; dietmar@es.usyd.edu.au)

Steven C. Cande, Scripps Institution of Oceanography, La Jolla, CA 92093-0215, USA (cande@gauss.ucsd.edu).

# Models of Mantle Convection Incorporating Plate Tectonics: The Australian Region since the Cretaceous

Michael Gurnis

*Seismological Laboratory, California Institute of Technology*

Louis Moresi

*Australian Geodynamics Cooperative Research Centre, CSIRO Exploration and Mining*

R. Dietmar Müller

*Department of Geology and Geophysics, The University of Sydney*

We propose that the anomalous Cretaceous vertical motion of Australia and distinctive geochemistry and geophysics of the Australian-Antarctic Discordance (AAD) were caused by a subducted slab which migrated beneath the continent during the Cretaceous, stalled within the mantle transition zone, and is presently being drawn up by the Southeast Indian Ridge. During the Early Cretaceous the eastern interior of the Australian continent rapidly subsided, but must have later uplifted on a regional scale. Beneath the AAD the mantle is cooler than normal, as indicated by a variety of observations. Seismic tomography shows an oblong, slab-like structure orientated N-S in the transition zone and lower mantle, consistent with an old subducted slab. Using a three-dimensional model of mantle convection with imposed plate tectonics, we show that both of these well documented features are related. The models start with slabs dipping toward the restored eastern Australian margin. As Australia moves east in a hot spot reference frame from 130-90 Ma, a broad dynamic topography depression of decreasing amplitude migrates west across the continent causing the continent to subside and then uplift. Most of the slab descends into the deeper mantle, but the models show part of the cooler mantle becomes trapped within the transition zone. From 40 Ma to the present, wisps of this cool mantle are drawn up by the northwardly migrating ridge between Australia and Antarctica. This causes a circular dynamic topography depression and thinner crust to develop at the present position of the AAD. The AAD is unique

The History and Dynamics of Global Plate Motions
Geophysical Monograph 121

within the ocean basins because it is the only place where a modern ridge has migrated over the position of long term Mesozoic subduction. Our study demonstrates the predictive power of mantle convection models when they incorporate plate tectonics.

## 1. INTRODUCTION

Mantle convection is the means by which heat is removed from the deep interior of the planet. Plate tectonics defines the dominant mode of mantle convection and subduction is responsible for creating most density heterogeneity within the mantle as imaged by seismic tomography. The oceanic lithosphere is part of mantle convection and the plates drive themselves through a combination of the buoyancy force associated with cooling oceanic lithosphere (often referred to as ridge push) and subducting oceanic lithosphere (slab pull). There is now little serious debate concerning the dynamics controlling first order features of instantaneous plate kinematics. It is now time to move on to equally important but unsolved problems in mantle dynamics, including why we have plate tectonics (as opposed to some other tectonic mode which would remove planetary heat), what controls the time-dependence of mantle convection, and the connection between mantle dynamics and the wide range of observations which convection models have not traditionally been compared with. In this paper, we describe a regional study pertaining to the latter class of problems: what is the time-dependent nature of mantle convection as constrained by geological and geochemical observations?

From plate kinematics, we know that mantle convection must be time-dependent. At the largest-scale, all continents aggregated during the Late Paleozoic into a single supercontinent, Pangea, and then dispersed during the Mesozoic [*Van der Voo*, 1993]. There was probably at least one earlier supercontinent, referred to as Rodinia [*Dalziel*, 1992]. On a somewhat smaller scale, $\approx 43$ Ma ago, the Pacific plate apparently changed direction from a northerly to a westerly direction [*Engebretson et al*, 1985]. During the Cenozoic, there has been extensive trench rollback of oceanic trenches [*Chase*, 1978] and ridge jumps [*Mammerickx and Sandwell*, 1986]. Since plate tectonics is a clear expression of mantle convection, these time-dependent aspects of plate kinematics must represent time-dependent mantle convection – there must be causes and consequences of these surface phenomena in the Earth's mantle. In fact, it is hard to imagine that such time-dependent convection would not be associated with the vertical motion of mantle buoyancy forces. Beyond the kinematic evidence, it is generally believed that we have limited direct observational constraints on time-dependent mantle convection.

In this paper, we have focused on the Australian plate because there are large-scale features which cannot be explained by application of plate tectonics, but which may be associated with the radial motion of buoyancy within the mantle. The two perplexing features which we focus on here include the complete subaerial exposure of the Australian continent (i.e. a regional sea-level minimum) during the Cretaceous global sea level maximum and the existence of a cold spot, often interpreted as a convective down welling, beneath the Australian-Antarctic spreading center. Without considering the ultimate cause, *Veevers* [1982] had earlier suggested that these two features were related. Here, we will specifically argue that the two features are related to the overriding of an ancient, long-lived subduction zone by the Australian plate. With the use of numerical models of mantle convection, we will argue that anomalous sea-level fluctuations, sea-floor depths, isotopic variations, and lower mantle seismic variations, can be understood within the context of time-dependent mantle convection. We show that a range of reasonable models can easily account for the overall features of Australian vertical motion and a localized topographic depression ranging from 200 to 800 m in the present position of the AAD. The implication of this regionally focussed study is that disparate data types can be understood within the context of time-dependent mantle convection when convection models are explicitly tailored to observed paleogeography.

This paper is organized in the following manner. We first describe the stratigraphic, marine geophysical and geochemical, and seismic tomography observations which characterize this region. These observations dictate the kinds of predictions we will attempt to make from our forward models and consequently how such models should be formulated. Since the forward models will be initial and boundary value problems, the results are dependent on the paleogeography of the Australian region. We describe the model formulation including some of the technical challenges and limitations of our approach. We systematically step through a single computation in terms of physical processes and associated predictions. However, we show that the predictions are dependent on both intrinsic unknowns of the mantle (like mantle viscosity) and uncertainties in paleogeography. An abbreviated dscription of our worked appeared in *Gurnis et al.* [1998].

## 2. AUSTRALIAN VERTICAL MOTION

Considerable insight can be gained into the time-dependence of mantle convection, especially the temporal evolution of slab structure, by looking at the vertical motion of continents, either the motion of one continent with respect to others or the tilting within a single continent [*Gurnis*, 1992]. The vertical motion of the Earth's surface associated with the rearrangement of mantle buoyancy forces have rates generally less than 100 m/Myr. Consequently, we must rely on observations from sedimentary geology to constrain such motions [*Gurnis*, 1992]. Unfortunately, stratigraphic architecture depends on a number of variables, including sediment supply, climate, eustasy (or global sea–level), and local or regional vertical motions. Consequently, the best proxies for continental vertical motion come from relative sea–level (RSL) change, often observed as anomalous marine inundation of one continent with respect to the mean inundation of all continents [*Hallam*, 1992]. One of the best examples of such motion is the anomalous subsidence of the western United States during the Cretaceous [*Bond*, 1976; *Cross and Pilger*, 1978; *Mitrovica et al.*, 1989] which has been quantitatively linked first to a shallowing dip angle of the Farallon plate subducting beneath North America and the subsequent uplift associated with the demise of subduction [*Mitrovica et al.*, 1989; *Burgess et al.*, 1997; *Lithgow-Bertelloni and Gurnis*, 1997] .

Australia shows substantial oddities in the pattern of marine inundation which have been linked to the vertical motion of that continent. During the Cretaceous, the marine inundation of Australia and global (e.g. eustatic) sea-level curves are out of phase (Fig. 1A). Maximum flooding of Australia occurred in the Late Aptian to Early Albian, 120-110 Ma [*Paine*, 1990; *Struckmeyer and Brown*, 1990], when large fractions of the craton, especially in the east, and margins experienced marine inundation (Fig. 1C). During the Late Cretaceous, Australia became progressively exposed, with a flooding minimum in the Campanian, 80-70 Ma, when eustatic sea-level was inferred to be near a maximum [*Veevers*, 1984]. Relative sea-level curves of Australian basins are also discrepant; inundation curves for the western margin of Australia resemble eustatic sea level curves, while inundation inferred from the Eromanga and Surat Basins in the east correlate poorly with eustatic variations [*Stuckmeyer and Brown*, 1990]. The Eromanga and Surat Basins (dashed line, Fig. 1C) are part of an extensive intracratonic basin system containing up to 3.5 km of Cretaceous and Jurassic sediments and occupy a combined area of $1.5 \times 10^6$ km$^2$ [*Exon and Senior*, 1976].

These apparently incongruous relative sea-level curves support the hypothesis that Australia experienced several episodes of craton-wide vertical motion as well as regional vertical movement during the Cretaceous [*Veevers*, 1984]. Estimates of Australian vertical motion indicate the occurrence of both a large-scale vertical translation of the entire continent and a regional subsidence event [*Veevers*, 1984; *Bond*, 1976; *Russell and Gurnis*, 1994]. Firstly, during the Early Cretaceous, the vertical motion appears to have been confined to a smaller portion of the east in the area defined by the Eromanga and Surat Basins. Secondly, near the end of the Cretaceous, Australia, was approximately 250 m higher than it is today (Fig. 1B). Some time during the Cenozoic, Australia appears to have subsided as a single unit, perhaps with very little tilting [*Russell and Gurnis*, 1994], as the continent moved northward after it separated from Antarctica.

Superimposed on this continent wide motion is the subsidence and subsequent uplift of the Eromanga Basin, specifically, and the eastern margin, generally. The Eromanga Basin, slowly subsiding during the Jurassic, experienced a rapid pulse of subsidence at 100 Ma, amounting to 500 m of tectonic subsidence in 10 Myr [*Gallagher and Lambeck*, 1989]. This rapid subsidence is well displayed through the detailed geohistory analysis of *Gallagher and Lambeck* [1989] for the Walkandi-1 borehole within the western portion of the Eromanga Basin (Fig. 2). Many of the sediments from the Early Cretaceous are marine and associated with a general marine incursion. During the end of the Early Cretaceous, marine sediments alternate with non-marine (volcano-clastics derived from the eastern margin), followed by a depositional hiatus when global sea level was highest [*Gallagher and Lambeck*, 1989]. The thermal history of sediments inferred from apatite fission tracks suggest that the Cretaceous section was probably deposited, but has since been eroded in both the Surat Basin and the basins directly on the eastern margin [*Gallagher et al.*, 1994]. This is consistent with the tilting up toward the east of marine strata of areas which were at or below sea level during the Cretaceous [*Gallagher et al.*, 1994; *Russell and Gurnis*, 1994].

## 3. PRESENT DAY STATE OF THE AUSTRALIAN REGION

The present day physical and chemical state of a region complements the constraints on mantle convection provided by relative sea–level change. Residual topography, the geoid, and seismic structure have been used to constrain mantle convection in a number of previous studies and we present an overview of such observations

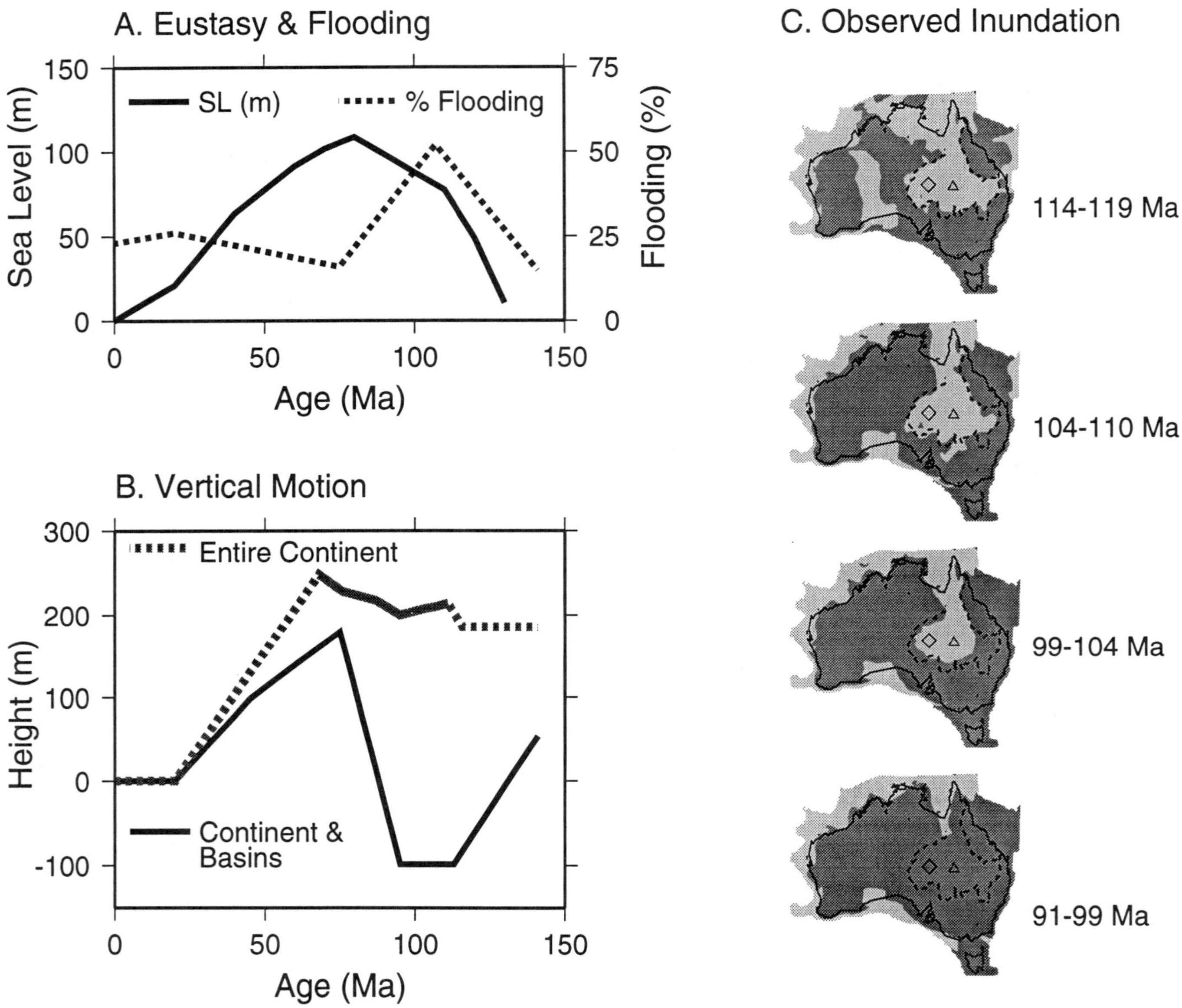


**Figure 1.** A. Eustatic, or global, sea level (solid line) inferred from passive margins in the Atlantic ocean [*Watts and Steckler*, 1979] and the percent area of Australia above 200 m below present sea-level flooded [*Veevers*, 1984]. B. Estimated vertical motion of Australia from hypsometric analysis. The motion of the entire continent, but excluding the eastern interior basins, is shown as the grey shaded line [*Russell and Gurnis*, 1994]. The vertical motion including the eastern basins is shown as the solid line (from *Veevers* [1984]). C. Paleogeographic reconstructions of Australia during the Cretaceous [*BMR Paleogeographic Group*, 1990]; light shading denotes marine inundation while medium shading denotes those parts of the continent that were subaerially exposed. Dashed line is the outline of the Eromanga and Surat basins and the Walkandi-1 [*Gallagher and Lambeck*, 1989] and Wareena-1 [*Gallagher*, 1990] boreholes are denoted with a diamond and triangle, respectively.

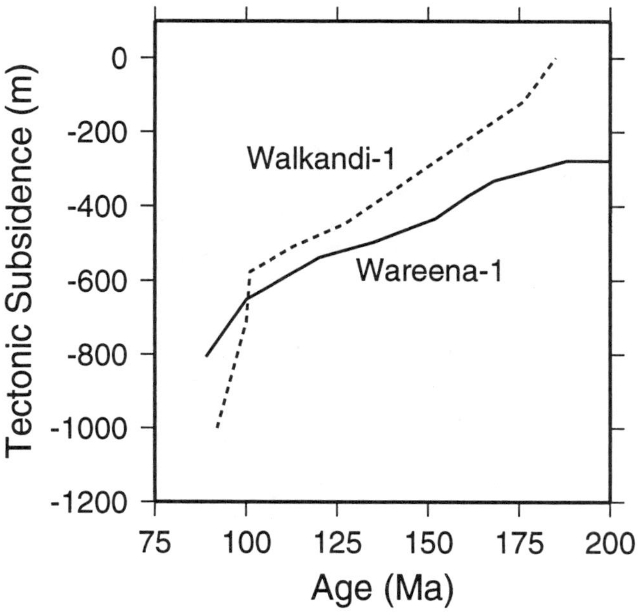


**Figure 2.** Tectonic subsidence for two boreholes within the Eromanga basin, the Walkandi-1 [*Gallagher and Lambeck*, 1989] and Wareena-1 [*Gallagher*, 1990]. Their positions are shown on Figure 1.

for the southern ocean here. Geographical variations in the isotopic composition of oceanic basalts is yet another powerful set of constraints which hitherto has not yet been quantitatively compared to physical models.

### 3.1. *Australian-Antarctica Discordance*

The Southeast Indian Ocean formed with oceanic spreading between Australia and Antarctica primarily since 45 Ma, although initial seafloor spreading dates to about 90 Ma [*Cande and Mutter*, 1982]. The Australian-Antarctic Discordance (AAD) is an unusual segment of the Southeast Indian Ridge (SEIR) between 120° and 128° E, south of the Great Australian Bight (GAB). It is characterized by extremely rugged ridge flank morphology, axial depths 800 m deeper than normal, and a series of closely spaced, N-S trending fracture zones (Fig. 3A; *Weissel and Hayes* [1974]). The prevailing view since the work of *Weissel and Hayes* [1974] is that the AAD represents either a cold-spot or a cold down welling within the mantle but that this concept has often been difficult to grasp since ridges are thought of as regions of upwelling mantle.

The clearest means by which to view the AAD is through residual topography – the topography which results when the normal subsidence of the oceanic lithosphere is removed from observed bathymetry. The oceanic lithosphere is the simple conductive thermal boundary layer of mantle convection [e.g., *Davies and Richards*, 1992], and bathymetry increases as $age^{1/2}$ where age is that since formation at the mid-oceanic ridge. We believe that the preponderance of evidence concerning the oceanic lithosphere is most consistent with it behaving as a cooling, semi-infinite half-space. However, because bathymetry tends to flatten for ages greater than 80 Ma, the half-space model is often not accepted and a so-called plate model is preferred (see discussion in *Davies* [1988]). Fortunately, the question of which model to use for the oceanic lithosphere is moot in this region because essentially all of the oceanic lithosphere between Australia and Antarctica has formed since 45 Ma (Fig. 3B). Following *Pribac* [1991] bathymetry is corrected for conductive cooling using

$$h_r = h_o - (2600 + 220a^{1/2}) \quad (1)$$

where $h_r$ is residual depth (in meters), $h_o$ is observed bathymetry, and $a$ is age of the oceanic lithosphere (in Myr). In addition, we correct for sediment loading using the procedure in *Schroeder* [1984]. When corrected for normal subsidence and sediment loading, the resulting residual depth is characterized by a prominent linear north-south trending depression from the GAB through to the AAD [*Cochran and Talwani*, 1977] and down to nearly the Antarctic margin with depths exceeding 1000 meters (Fig. 3C). Previously, it does not seem to have been widely recognized that the depth anomaly is nearly "V" shaped and symmetrical with respect to the SEIR. In comparison with global residual oceanic topography, the AAD is the most prominent, excessively deep, bathymetric anomaly not associated with present day subduction [*Pribac*, 1991]. The only other region with equally large residual depth is the East Philippine Sea which is controlled by both a thinner oceanic crust and dynamic topography from present day subduction [*Gurnis*, 1993]. The character of the SEIR depth anomaly has the appearance of a property (such as thinner crust) fixed with respect to the plates as they move slowly eastwardly with respect to a fixed point – much as a fixed hot spot gives rise to thicker crust which is symmetrical with respect to a spreading center.

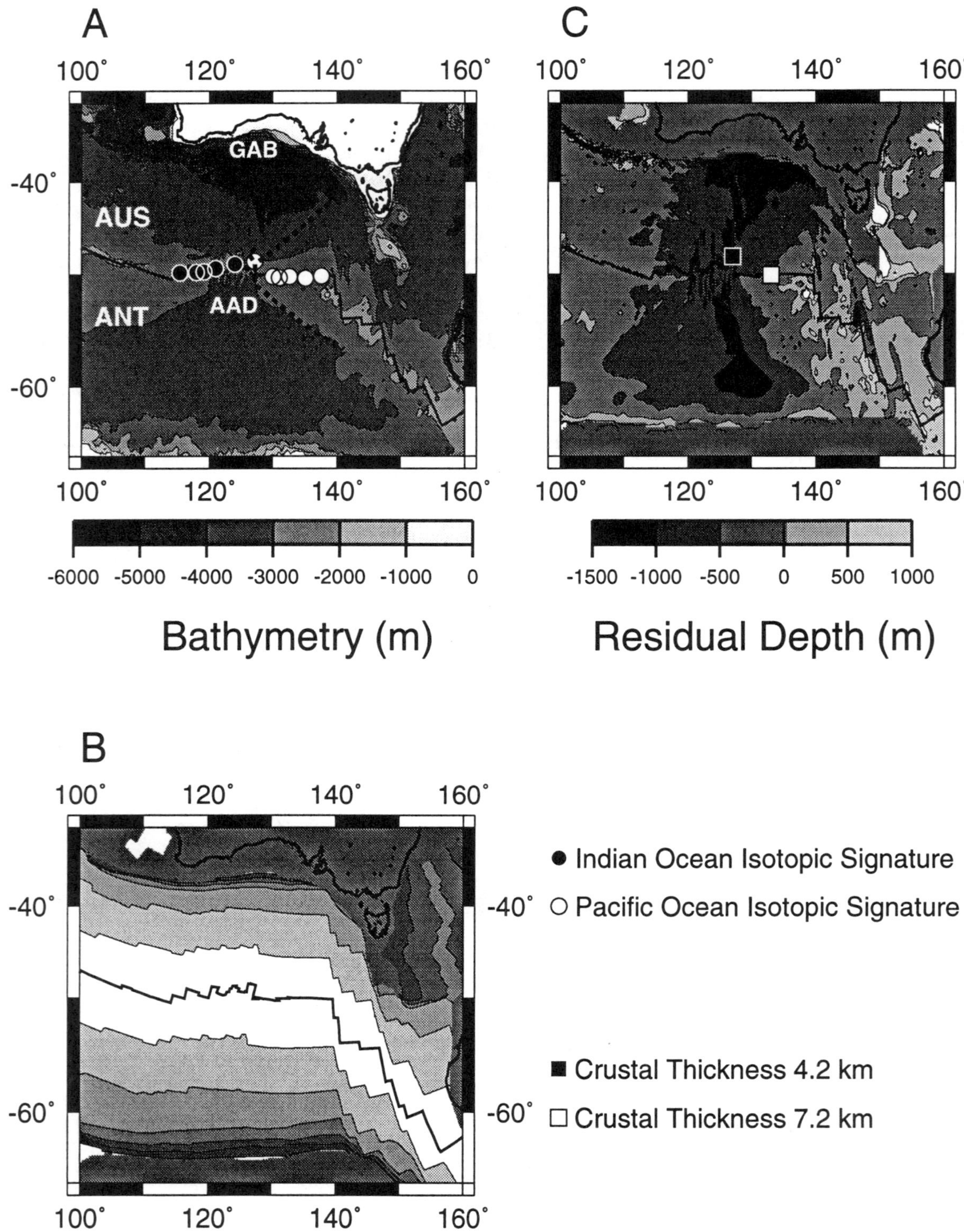


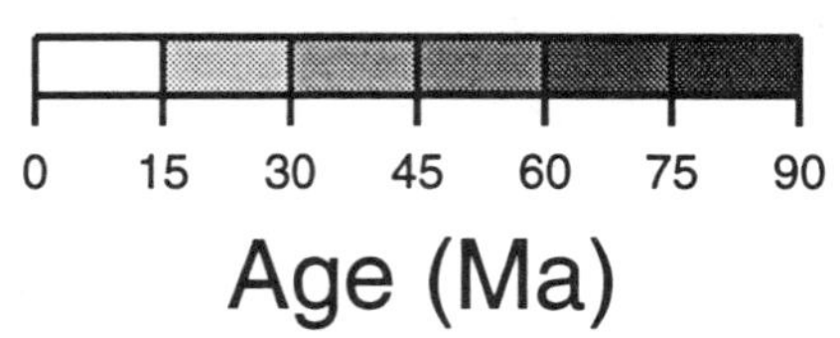


**Figure 3.** Observed characteristics of the Southeast Indian Ocean. A. Observed bathymetry (ETOPO5) and isotopic signature of the mid-ocean ridge basalts [*Klein et al.*, 1988]. B. Age of the ocean floor from *Müller et al.* [1997]. C. Residual topography (see text for details). Crustal thickness values from [*Tolstoy et al.*, 1995].

There are different ways to distinguish between thinner crust and dynamic topography, including petrological models as a proxy for crustal thickness, direct measurement of thickness from seismology, and the gravity field. Major element systematics indicate that basalts making up the AAD are consistent with lower temperature mantle rising adiabatically under the ridge and crossing the solidus at lower pressure and experiencing a lower degree of melting [*Klein and Langmuir*, 1987]. As expected from the petrological model, the crust within the AAD has been measured from wide angle seismic refraction to be thinner than an adjacent part of the SEIR [*Tolstoy et al.*, 1995], the AAD having a thickness of 4.2 km compared to the SEIR with a thickness of 7.2 km (Fig. 3C). The thermal anomaly responsible for the thinner crust can be estimated. Thermodynamic calculations and petrologic models show that oceanic crustal thickness increases as 0.06 km for each 1°C increase in mantle temperature [*Langmuir et al.*, 1992; *Asimov*, 1997]. If we assume a difference of thickness of 3 km then a temperature difference of 50°C is implied. However, if we assume a difference between the AAD thickness and an average 6 km thickness of oceanic crust [*McClain and Atallah*, 1983] then only a 30°C anomaly is implied.

Assuming isostasy, it appears that less than half of the 1000 meter depth anomaly (Fig. 3C) can be accounted for by thinner crust. Using the observed difference in crustal thickness [*Tolstoy et al.*, 1995] and a simple isostatic balance ($\rho_c = 2900$ kg m$^{-3}$ and $\rho_m = 3250$ kg m$^{-3}$) a bathymetric difference of 470 m is predicted, significantly smaller than the observed $\sim$ 1000 m. Using a more sophisticated isostatic balance and observed major element chemistry, *Lecroart et al.*, 1997] predict a 400 m difference between observed and isostatic topography. Consequently, there may be comparable contributions to topography from both thinner oceanic crust and viscous stresses, e.g., dynamic topography, which arise from a cold and heavy mantle anomaly. This is the most reasonable interpretation of excess bathymetry. We suspect from the symmetry of the residual depth, which extends to at least 45 Ma crust both north and south of the ridge, that the cold spot has been in existence for at least (and probably longer than) 45 Myr. Therefore, the portion of the mantle which is cold would be expected it to extend into the upper mantle a least a few hundred kilometers. If this truely reflects the structure deeper in the mantle, then it should contribute toward dynamic topography.

The existence of cold mantle beneath the AAD is corroborated by the along ridge distribution of isotopic variations. Along the SEIR, Sr, Pb, and Nd isotopic systematics of basalts reveal that there are two distinct isotopic provinces: one to the west of the AAD characteristic of the Indian ocean and one to the east characteristic of the Pacific (Fig. 3A). These isotope systematics indicate long term ($10^8$ to $10^9$ yr enrichment and depletion of mantle reservoirs. The two provinces together are nearly globally encompassing [*Hart*, 1984]. The isotope boundary is sharp but does become gradational on a scale of approximately 40 km [*Pyle et al.*, 1992]. *Klein et al.* [1988] hypothesize that such a sharp isotopic boundary could be formed by down welling of material flowing east-west, parallel to the trend of the SEIR, toward the AAD. Relative down welling of the mantle is clearly consistent with the balance of the evidence which points toward cold mantle beneath the AAD. Further, *Klein et al.* [1988] hypothesize that the large-scale Indian Ocean isotope province [*Hart*, 1984] which comes to an abrupt termination at the AAD, could well be a remnant of the former Gondwanaland supercontinent with the isotope systematics consistent with sediment subduction, among other processes, beneath continental lithosphere [*Klein et al.*, 1988; *Pyle et al.*, 1992; *Rehkamper and Hofmann*, 1997].

### *3.2. Seismic Structure of the Mantle in the Australian Region*

The shallow (< 100 km depth) seismic structure of the Australian region is not particularly unusual (Fig. 4) – cratons are seismically fast and ridges and back-arc basins slow. Unfortunately, there are not large seismic sources in Antarctica which would provide crossing rays of the SEIR, so we rely upon the global Rayleigh wave inversion (Fig. 4) of *van Heijst* [1997]. Earlier, *Forsyth et al.* [1987] used a small number of relatively shallow and small events on the SEIR and inverted for 1-D structure from Rayleigh waves received in Australia for those paths which crossed the region with anomalously deep bathymetry and compared these to an inversion for paths which did not cross the anomalous region. They conclude that the mantle beneath the segment with deep bathymetry was characterized by unusually fast shear wave velocity in the depth range 20–40 km. This is not inconsistent with the global inversion at 50 km depth where we see that the AAD segment is faster than the other ridge segments (Fig. 4).

A distinctive fast anomaly oriented N-S is evident directly beneath the AAD and becomes coherent at all depths > 500 km and extends down to at least 900 km, the depth limit on resolution from the global model [*van Heijst*, 1997]. Inspection of the entire global inversion

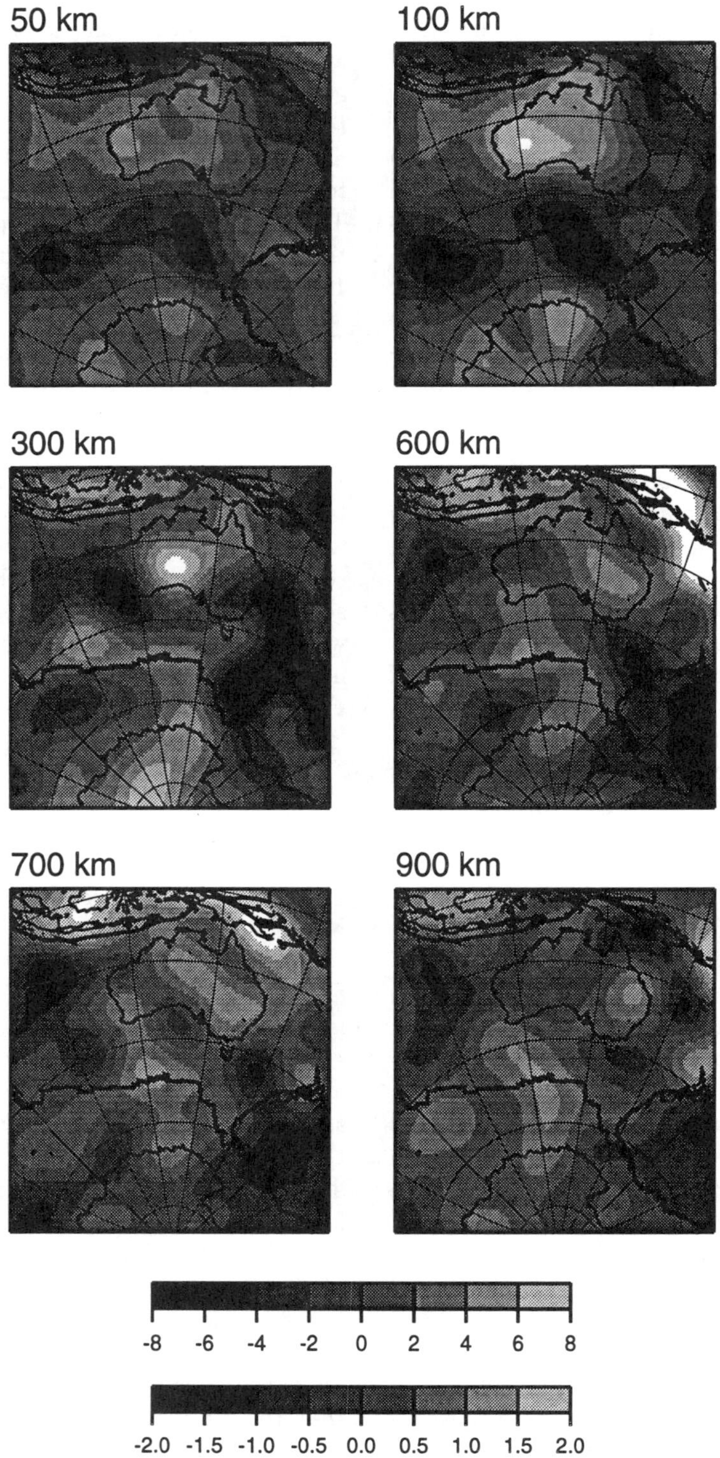


**Figure 4.** Shear velocity in percent from the global Rayleigh wave inversion of *van der Heijst* [1997]. The 8 % scale is for 50 km and 100 km depths while the 2% scale is for all deeper levels.

of the uppermost lower mantle shows that all substantial fast shear velocity anomalies are associated with Mesozoic to present subduction, as has been shown with body wave tomography [*Grand et al.*, 1997]. Unfortunately, the global P- and S- wave models [*Grand et al.*, 1997] have poor resolution in our study region through the upper mantle, but the resolution is better at the core mantle boundary and both P- and S- waves show a distinct seismic anomaly beneath the AAD. Most fast anomalies at the CMB are associated with Mesozoic subduction [*Grand et al.*, 1997].

In summary, there is some indication of fast seismic velocity within the shallow upper mantle beneath the AAD, consistent with the lower temperatures inferred from the anomalous bathymetry, thinner crust, and petrological models. However, the most coherent anomaly beneath the AAD is an oblong, slab-like structure orientated N-S in the transition zone and lower mantle. Considering the association of all other lower mantle fast seismic anomalies with old subduction, we suggest that the sub-AAD anomaly could also be an old subducted slab.

## 4. PLATE TECTONIC EVOLUTION OF THE AUSTRALIAN PLATE

Throughout the Paleozoic and for much of the Mesozoic, the eastern Australian margin constituted a long-lived active margin with a subduction zone located east of the Lord Howe Rise/Norfolk Ridge (Fig. 5A) [*Symonds et al.*, 1996]. Widespread Triassic and middle Jurassic volcanogenic sediments in a foreland basin along the eastern Gondwanaland margin were derived from a magmatic arc along a west-dipping subduction zone [*Jones and Veevers*, 1983; *Symonds et al.*, 1996]. The tectonic setting of the margin remained convergent until the Late Jurassic/Early Cretaceous, despite some local rifting events. The time of cessation of subduction has been subject of some controversy. The Whitsunday-Proserpine-Grahams Creek volcanic belt extending for about 900 km along the northeast Australian coast, are volcanic and intrusive intermediate to silicic, high-K calc-alkaline rocks, dated as 132-95 Ma [*Ewart et al.*, 1992]. They had been interpreted by *Veevers* [1984] as andesitic, subduction-related volcanism. However, more recently, *Ewart et al.* [1992] and *Bryan et al.* [1997] have shown that their calc-alkaline signature is inherited from a crustal source, and that volcanism occurred in a low-relief extensional environment, with no evidence for a high-standing Andean-type magmatic arc. Therefore, it is now thought that subduction of the Phoenix Plate ceased off eastern Australia at about 130 Ma. Further south, in New Zealand, oblique subduction of the Phoenix Plate continued at least for another 20 Myr. Here, the earliest extensional tectonism is dated as 105-100 Ma [*Laird*, 1994], but is possibly as old as 110 Ma [*Tulloch and Kimbrough*, 1989]. *Bradshaw* [1991] suggested that the cessation of subduction was caused by subduction of the Phoenix-Pacific spreading ridge. As the slab that is now underlain the AAD corresponds to subduction off eastern Australia, we assume in our model that subduction ceased at 130 Ma.

The nature of the post-130 Ma plate boundary east of the Lord Howe Rise is not well known, but closure of the Pacific-Antarctic-Australian plate circuit back to about 86 Ma, including a preliminary model for relative motion between East and West Antarctica from *S. C. Cande* [pers. comm.], suggests that it was a strike-slip margin after 86 Ma [*Gaina*, 1998]. The margin maintained its strike-slip character until about 43 Ma, when convergence between the Pacific Plate and the Australian Plate may have been reinitiated [*Gaina*, 1998].

## 5. SET UP OF MANTLE CONVECTION MODELS WITH IMPOSED PLATE TECTONICS

### 5.1. *Overview of formulation*

Mesozoic paleogeographic reconstructions showing subduction near the restored eastern Australian margin and the detection of a slab-like structure within the lower mantle beneath the AAD, the approximate former position of subduction in a fixed hotspot reference frame (Fig. 5B), leads us to advance the hypothesis that a subducted slab moved through the mantle over the last 130 Myr, resulting in the anomalous Cretaceous vertical motion of Australia and a cold zone centered at the AAD. We would like to test this slab hypothesis by making comparisons between the observed Cretaceous inundation of Australia and present day geochemical and geophysical observations within the Southeast Indian Ocean, on the one hand, and dynamic models, on the other. A 'mantle convection' problem has to be formulated so that models start with a subducted slab initially at a continental margin with evolving plates on the surface.

The problem we set up is similar to a traditional mantle convection model [e.g., *McKenzie et al.*, 1974] in that the conservation equations are solved to determine velocity and stress throughout the domain with time-dependence simulated by interleaving the solution of the momentum and energy equations. However, by attempting to construct a model which is intended to be compared with observations, tectonic plates (following

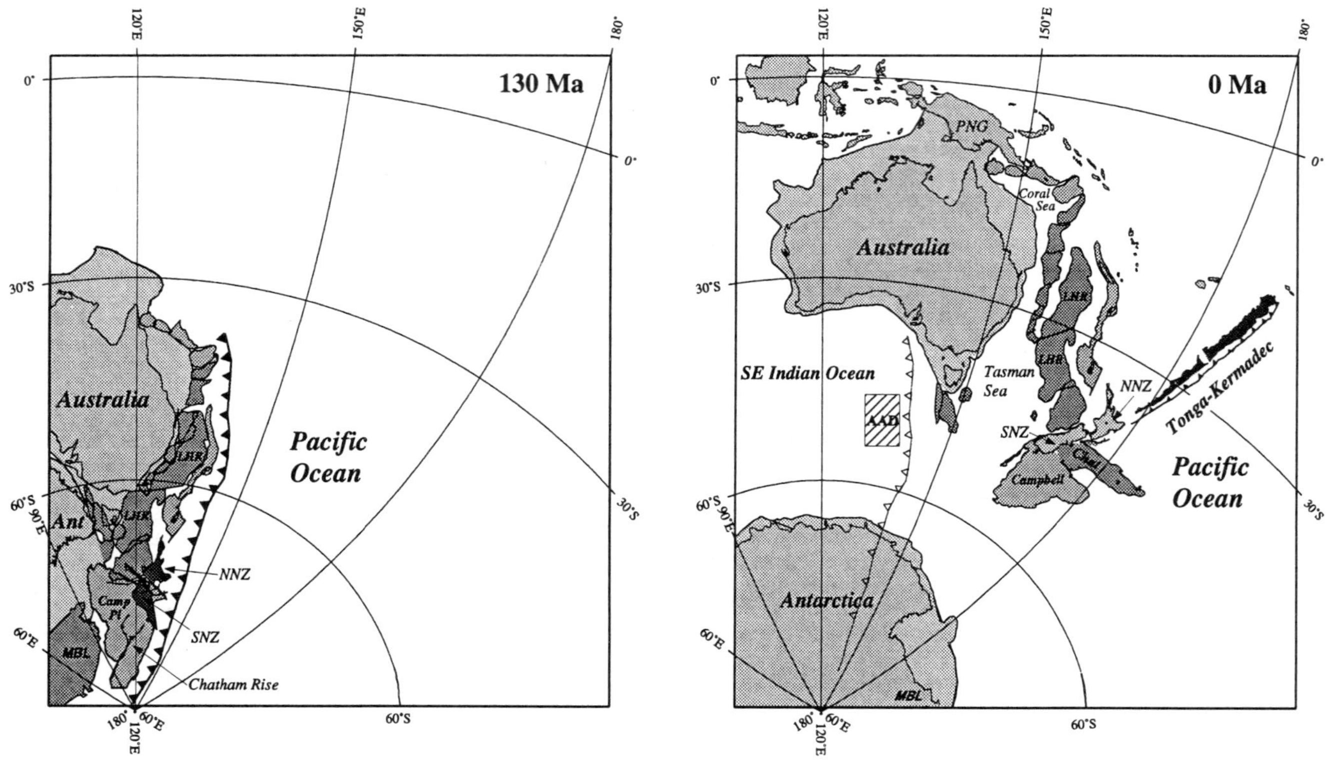


**Figure 5.** a. Reconstruction of Gondwanaland at 130 Ma based on the absolute plate motion model relative to hotspots in the Atlantic-Indian oceans by *Müller et al* [1993], and relative plate motions from *Gaina* [1998] and *Müller et al* [1998]. Before this time, a westward dipping subduction zone was located east of the Lord Howe Rise (LHR), which became extinct at about 130 Ma. The projection is general sterographic, with the projection center at 120°E, 30°S. b. Present-day plate configuration in the southwest Pacific Ocean, with the 130 Ma location of the subduction zone superimposed, now situated just east of the AAD.

past inferred motions) must be incorporated. A self-consistent dynamic formulation could have plate kinematics and evolving plate geometry as predictions. Although this has been attempted with evolving faults in two-dimensions [e.g., *Zhong and Gurnis*, 1995], it is computationally difficult in three dimensions and a good match with observed plate kinematics can not be guaranteed. A fully dynamic model capable of predicting absolute plate motions would need to include all the important slabs and plate boundaries together with poorly constrained time-dependent, plate boundary constitutive relations amplifying the errors in the region of interest. So as to avoid the difficulty of matching observed velocities with a fully dynamic model, time-dependent plate kinematics can be incorporated as boundary conditions. Models with velocity boundary conditions and evolving buoyancy have been formulated so that imposed plates neither speed up nor slow down the flow from what would normally be generated with a free-slip top surface [e.g., *Davies*, 1986]. Unfortunately, this methodology is potentially problematical because the kinematic boundary conditions amount to an external force which the Earth does not have. Since lithosphere freely slides by the hydrosphere, the Earth has a free-slip top surface and the top surface does no work on the system.

The imposed plate velocity method has been assessed by *Han and Gurnis* [1999]. Fully dynamic models of plate evolution were created using a fault methodology and a free-slip upper surface. Since both evolving plate geometry and plate velocity are outputs of the fully dynamic model, velocity from the dynamic model can be imposed on a kinematic model. Comparing often complex slab structure resulting from dynamic and kinematic models, *Han and Gurnis* [1999] found that internal structure was nearly identical for integration times up to nearly 100 Myr. The velocity boundary condition, however, often did an order of magnitude more work on the system than did the slab. This result should be viewed with respect to dynamic models where slabs always do most of the work on the fluid interior. In kinematic models, this additional work was dissipated within the lithosphere [*Han and Gurnis*, 1999]. Although the criticism made of the kinematic approach is valid (i.e. velocity boundary conditions do substantial work on the system), the BC does not incorrectly alter the evolution of internal buoyancy compared to a model having plates generated self-consistently. Clearly, the imposed velocity approach is ideal for the present study investigating the descent of a slab through the mantle when the evolution of plate motions are known.

Topography, present day and time-variable, is a primary observable. Crustal and lithospheric thickness variations and dynamic topography are all important sources of changing topography. Isostatic contributions are well known – dynamic topography generated by viscous flow in the mantle less so. From the reference frame of one plate, contributions to topography work differently. Crustal thickness variations are fixed with respect to plates, although a plate itself can undergo internal deformation (e.g. non plate-like characteristics). Topography generated by conductive cooling of oceanic lithosphere is also fixed with respect to plates, and is almost entirely determined by the age of the lithosphere.

Dynamic topography is not fixed with respect to plates, as plates move with respect to changing dynamic topography [*Gurnis*, 1990]. Like most fluid dynamic problems, our mantle convection models employ an Eulerian solution of the equations on a fixed mesh. Stratigraphic observations (like subsidence inferred at a single borehole) are fixed to plates moving with respect to the Eulerian mesh and in this case a Lagrangian reference frame is more appropriate. The topographic 'history' of points fixed with respect to plates can be found by following individual parcels of material within the flow as they are carried over a changing dynamic topography field by the background flow and boundary conditions. Using this procedure, we can immediately extract tectonic subsidence curves. For the pattern of marine flooding, a Lagrangian coordinate system attached to a continent is needed and a sediment model can then act on this grid of points. Such sediment models can be complex, involving not only sedimentation, but erosion and sediment transport [*Burgess et al.*, 1997].

We have adopted a simplified approach to tracking topography due to lithospheric thickening. Since our interpretation of the geologic evolution of the eastern Australian margin has no active subduction from 130 to 50 Ma, the buoyancy created from diffusion across the top boundary will not be inserted into the mantle. This means that the diffusive thermal boundary layer of the mantle can be removed and this is accomplished by fixing the temperature at the top of the box to the average mantle temperature. However, we will make the upper 100 km of the mantle have a high effective viscosity because although it is not needed to enforce the plate motions, the vertical distribution of viscosity strongly modulates the topographic response from the underlying flow [*Richards and Hager*, 1984]. In making comparisons between computed model topography and observations, we subtracted the normal thermal subsidence of the oceanic lithosphere from the observed bathymetry (Section 3.1, Fig. 3).

Another influence on topography is crustal thickness variations. Within our modeled continents, only sedimentation changes crustal thickness. For the oceans, a synthetic lithosphere is created from the modeled margin between the Australian and Antarctic plates. Oceanic crust is created with a thickness depending on mantle temperature below the ridge using the relation between mantle temperature and crustal thickness from *Langmuir et al.* [1992] and *Asimov* [1997] described above. Crustal thickness is computed by sampling the temperature within a 200 km region beneath the ridge. A crustal thickness and chemistry were assigned to points on the Australian and Antarctic plates and subsequently moved according to the plate model. When we make comparisons with the residual bathymetry for the Southeast Indian Ocean described above, we will sum the dynamic topography and the isostatic topography resulting from crustal thickness variations. In both cases the topography will be comparable since the cooling oceanic lithosphere has been eliminated.

### 5.2. Numerical solution of mantle convection

Equations for momentum, mass, and energy conservation are solved to obtain the evolution of the system. The infinite Prandtl number approximation, which is appropriate for extremely viscous fluids, produces an equation of motion which has no inertial terms (usually called the Stokes' equation),

$$\nabla \cdot \boldsymbol{\tau} - \nabla p = \left[ \alpha(T - T_0) - \sum_k \Gamma_k \frac{\Delta\rho_k}{\rho_0} \right] g\rho_0 \hat{\mathbf{z}} \quad (2)$$

in which $p$ is the pressure, $g$ is gravitational acceleration, $\rho_0$ is the density at a reference temperature $T_0$, $T$ is temperature, $\Gamma_k$ and $\Delta\rho_k$ are the phase function (a marker which indicates the equilibrium phase at a given location) and density jump for the $k$th phase change, and $\hat{\mathbf{z}}$ is a unit vector in the vertical direction. Across the $k$th phase change, the phase function, $\Gamma_k$, increases by 1. The deviatoric stress tensor is given by

$$\tau_{i,j} = \frac{\eta}{2}(\partial u_i/\partial x_j + \partial u_j/\partial x_i) \tag{3}$$

where $\eta$ is the dynamic viscosity, $u_i$ is a component of velocity, and $x_i$ is a component of the Cartesian coordinate system. For an incompressible fluid, the continuity equation becomes

$$\nabla \cdot \mathbf{u} = 0. \tag{4}$$

The energy equation is

$$\frac{\partial T}{\partial t} = \kappa \nabla^2 T - \mathbf{u} \cdot \nabla T + Q \tag{5}$$

where $t$ is time, $\kappa$ is thermal diffusivity, and $Q$ is the volumetric heat production which is assumed to be zero for the present study.

The equations are non-dimensionalized such that with the boundary and initial conditions, the problem is determined by the following non-dimensional parameters.

$$Ra = \frac{g\alpha\rho_0 \Delta T D^3}{\kappa \eta_{um}} \tag{6}$$

$$R_k = \frac{\Delta\rho_k g D^3}{\kappa \eta_{um}} \tag{7}$$

$$\gamma_k' = \frac{\gamma_k \Delta T}{\rho_0 D g} \tag{8}$$

where $Ra$ is the thermal Rayleigh number, $R_k'$ and $\gamma_k'$ are the phase change Rayleigh number and normalized Clapeyron slope for the $k$th phase change, $\eta_{um}$ is the viscosity of the upper mantle, and $D$ is the depth of the box.

The dynamic viscosity, $\eta$, in (3) can be temperature-dependent. We use the following non-dimensionalized form of the Arrhenius relation [e.g., *Ranalli*, 1995]

$$\eta = \eta_m e^{\frac{E'}{T+T_1'}} \tag{9}$$

where $E'$ and $T_1'$ are non-dimensional free parameters and $\eta_m$ is the nominal viscosity of the lithosphere, upper mantle, and lower mantle (Table 2). $E'$ can be approximately related to the activation energy, $E_a$.

The equation of motion (2) is elliptic and thus has properties which make special solution methods necessary for three dimensional problems. In particular, stresses are propagated instantaneously through the fluid such that at each time step the response to a local change in buoyancy is felt everywhere. The distributed response of the fluid to stresses makes explicit solution methods impracticably slow - it takes far too long for information to propagate back and forth across the numerical grid to reach equilibrium. Instead, the system of equations are solved implicitly by formulating a discrete representation on a grid and solving the resulting matrix equation. A number of discretization procedures are possible but the finite element method is used because it is robust in the face of discontinuous layering and strong variations in material properties.

The most robust way to solve the finite element matrix representation of (2) and (4) is a direct matrix solver which uses techniques such as Crout elimination [*Hughes*, 1987] optimized to take advantage of the sparse nature of the finite element matrices. The principal difficulty with this strategy is that with large scale problems (such as realistic 3D problems with many grid points), direct solution methods require a computational effort proportional to $N^3$ where $N$ is the number of unknowns. As $N$ becomes large, the solution time rapidly becomes unmanageable. Iterative methods, in which a series of approximate solutions converge on an accurate answer, are often more efficient for large problems but are usually less robust than direct solution methods. Conjugate gradient schemes, for instance, can solve a problem with $N$ unknowns in a time proportional to $N \log N$. The most efficient scheme is multigrid which uses a series of nested grids and can propagate information rapidly over all physical scales in the problem; its solution time scales, ideally, as $N$ [*Press et al.*, 1992; *Golub and Van Loan*, 1989]. General iterative methods can be temperamental, particularly when strong variations in material properties occur between discretization points. The code we use, CITCOM, uses a multigrid solver which has been optimized in 2D and 3D for Stokes' flow problems in which the viscosity varies strongly across the grid [*Moresi and Solomatov*, 1995; *Moresi and Gurnis*, 1996].

The energy equation (5) is parabolic and requires different numerical treatment from the equation of motion. The diffusive term and the volumetric heat production term can be integrated in time from the initial condition without significant difficulty. However the advective transport term can pose greater difficulties because the advection of quantities on the mesh is not just a question of moving values from one node to the next. Inevitably, values from the middle of an element will be transported to a node during a timestep. This constitutes an interpolation, although it is not usually written out explicitly, and repeated interpolation does not preserve information. The result is an artificial diffusion

Table 1. Summary of Controlling Parameters

| Parameter | Symbol | Value |
|---|---|---|
| Depth of Box | $D$ | 2000 km |
| Acceleration of gravity | $g$ | 10 m s$^{-2}$ |
| Coefficient of thermal expansion | $\alpha$ | 2x10$^{-5}$ K$^{-1}$ |
| Temperature scale | $\Delta T$ | 1400 K |
| Ambient density | $\rho_0$ | 3300 kg m$^{-3}$ |
| Thermal diffusivity | $\kappa$ | 10$^{-6}$ m$^2$ s$^{-1}$ |
| Dynamic viscosity | $\eta_{um}$ | 10$^{21}$ Pa-s |

Table 2. Viscosity Model

| Depth Range (km) | Non-Dimensional Viscosity |
|---|---|
| 0 – 100 | 100 |
| 100 – 670 | 1 |
| 670 – 2000 | 100 |

which tends to smooth the temperature field. There are a number of schemes designed to reduce this effect including Streamline-Upwind, Petrov-Galerkin [*Brooks*, 1981] which we use.

Dynamic topography was computed on the fixed mesh by assuming a balance between the vertical stress at the uppermost plane of nodes and the restoring force due to supposed deflections of the surface. We use the Consistent Boundary Flux (CBF) method [*Zhong et al.*, 1993] which has been extensively benchmarked with both radial and lateral variations in viscosity [*Moresi et al.*, 1996]. Solutions to the equations of motion do not change when a constant is added to the pressure. This in turn means that the dynamic topography has no absolute reference level so the output of the simulations always has the horizontal average of the topography removed. An alternative approach would be to assume a frame of reference associated with the sides of the box. There is only a small difference between these two approaches in the predicted subsidence and uplift associated with a falling slab. To ensure that no spurious topography signal is introduced by the imposed velocity boundary conditions, we compute surface stresses using the buoyancy field at any given time with a no-slip upper surface boundary condition.

### *5.3. Application of the method to the Australian region*

The procedures described above can be used for many different convection problems, but to test the slab hypothesis we need to specifically tailor our regional Cartesian model. Plate kinematics, including plate margins, are transformed from a spherical surface to a Cartesian plane using an Azimuthal Lambert equal area projection. We used the continental outlines, isochrons and referenced rotations for relative plate motions from *Müller et al.* [1997] and absolute plate rotations based on hotspot tracks from *Müller et al.* [1993]. From 130 Ma to the present, the plate geometry is imposed as velocity boundary conditions which are updated every 10 Myr. All of the models are based on a box depth of 2000 km (Table 1); this depth is less than that of the whole mantle, but has a volume greater than the mantle under an equivalent area of Earth's surface. We generally assumed a layered viscosity model (Table 2) which is consistent with the observed long-wavelength geoid associated with subduction zones [*King*, 1995]. Reflecting boundary conditions were used on all four side walls; on the base, we have used free-slip and isothermal conditions.

The initial condition of the subduction zone consists of a rectangular slab with a dip, $\theta_s$, an age, $\tau_s$, and a trench at a constant distance, $X_s$, from the left boundary of the box (Table 3). Having rotated the Lord Howe Rise back to its pre-rift position (Fig. 5a) our model trench was located parallel to the continental margin (Fig. 6). Clearly, this represents a simplification of the much more extensive Gondwanaland-Pacific margin, where convergence may have extended further toward New Guinea and certainly encompassed West Antarctica and South America. The subducted slab had a dip, $\theta_s$, through the upper mantle, but below 670 km the slabs were always vertical. This is consistent with active subduction shown by both seismic tomography [*van der Hilst et al.*, 1997] and fully dynamic convection models with plates [*Zhong and Gurnis*, 1995].

A unique observation of this region is the abrupt change in isotope systematics along the SEIR. All hypotheses explaining the feature essentially have two large-scale geochemical reservoirs within the mantle meeting at the AAD. With our model formulation, we can test the hypothesis that an initially sharp boundary between two large-scale geochemical provinces, delineated by subduction before Gondwanaland breakup, caused the present boundary observed at the AAD. The two provinces contain passive tracers, i.e. neutrally buoyant so they do not influence the flow, inserted through the mantle with one group to the west (left) of the slab (Gondwanaland tracers) and the other to the right (east) of the slab (Pacific tracers).

The changing dynamic topography of this region evident from epeirogenic motion of Australia (Fig. 1B) is constrained by the pattern of marine inundation (Fig. 1C). We model this inundation directly by using a simple model in which sediment accumulates at a constant rate for those points below sea-level. Because the model we formulate is regional, eustatic sea level

Table 3. Summary of Mantle Convection Models

| | | | | | 410 km | | 670 km | | |
|---|---|---|---|---|---|---|---|---|---|
| Case | Notes | $\theta_s$ | $\tau_s$(Ma) | $X_s$ | $\gamma'$ | $R$ | $\gamma'$ | $R$ | Grid |
| 1 | A34 | 30° | 100 | 2.75 | 0.043 | 1.78 | -0.089 | 2.85 | 129x65x33 |
| 2 | A32 | 30° | 100 | 2.75 | 0.043 | 1.78 | -0.089 | 2.85 | 65x65x33 |
| 3 | A26 | 30° | 100 | 2.75 | 0.043 | 1.78 | -0.089 | 2.85 | 33x65x33 |
| 4¶ | A31 | 30° | 100 | 2.75 | 0.043 | 1.78 | -0.089 | 2.85 | 129x65x33 |
| 5 | A24 | 30° | 100 | 2.75 | 0.000 | 1.78 | -0.089 | 2.85 | 33x65x33 |
| 6 | A25 | 30° | 100 | 2.75 | 0.085 | 1.78 | -0.089 | 2.85 | 33x65x33 |
| 7 | A37 | 30° | 100 | 2.75 | 0.000 | 1.78 | 0.000 | 2.85 | 65x65x33 |
| 8 | A35 | 90° | 100 | 2.75 | 0.043 | 1.78 | -0.089 | 2.85 | 65x65x33 |
| 9 | A36 | 30° | 100 | 3.25 | 0.043 | 1.78 | -0.089 | 2.85 | 65x65x33 |
| 10 | A27 | 30° | 50 | 2.75 | 0.043 | 1.78 | -0.089 | 2.85 | 33x65x33 |
| 11 | A28 | 30° | 200 | 2.75 | 0.043 | 1.78 | -0.089 | 2.85 | 33x65x33 |
| 12† | A38 | 30° | 100 | 2.75 | 0.043 | 1.78 | -0.089 | 2.85 | 65x65x33 |
| 13‡ | A32 | 30° | 100 | 2.75 | 0.043 | 1.78 | -0.089 | 2.85 | 65x65x33 |

¶Temperature-dependent viscosity; $E'$=34, $T_1'$=0.195
†No slab in lower mantle
‡Slab width in lower mantle 150 km, half the value used in Cases 1-11

must be imposed. We use the global sea-level curve of *Watts and Steckler* [1979] which has been inferred from the stratigraphy of passive margins in the Atlantic, althought it is similar to other eustatic curves [*Hallam*, 1992]. Despite a general opinion to the contrary, in a dynamic planet spatial variations in sea level are much more strongly influenced by dynamic topography than the geoid [*Gurnis*, 1990; *Lithgow-Bertelloni and Gurnis*, 1997]. Consequently, we assume that at any given time the sea surface is 'flat' and tracks the eustatic change. For each point fixed to Australia sediment accumulates at a constant rate, $S$, between the solid and sea surfaces, a difference which is referred to as accommodation space. The sediment and water loads are isostatically compensated. We use sedimentation rates of 0 – 20 m/Myr which are consistent with the range observed for the north American platform during the Phanerozoic [*Sloss*, 1988]. The initial solid surface at 130 Ma is modeled as either a constant height or has the shape of the present day topography but offset with a constant freeboard. We add dynamic topography at 130 Ma and subsequently the surface height changes as sediment accumulates and dynamic topography changes.

## 6. DYNAMIC MODELS: CRETACEOUS TO THE PRESENT

We explore a variety of cases varying the uncertainties associated with the initial configuration of subduction ($\theta_s$, $\tau_s$, $X_s$) as well as uncertainties intrinsic to the mantle (viscosity model, $\gamma_k'$, $R_k$). In particular, we find that the results are sensitive to viscosity stratification through the transition zone and the strength of the phase transition at 670 km depth.

### 6.1. *Plate kinematics and topography*

We first consider a nominal model, Case 1, of a slab in which the trench is on average 1400 km east of the present coastline of Queensland (Fig. 6) and with an age of 100 Ma and a dip of 30°. Case 1 has a Rayleigh number of 7.4 x $10^6$ (based on upper mantle and transition zone viscosity of $10^{21}$ Pa-s, and other parameters summarized in Table 1), no temperature-dependent viscosity and a factor 100 jump in viscosity at 670 km (i.e., viscosity model V2, Table 2). Using the values in Table 1, the dimensional Clapeyron slopes and density jump become 2.0 MPa/K and 5% at 410 km and -4.2 MPa/K and 8% at 670 km. In the marginal area off the coast there is a calculated prominent linear topographic depression (Fig. 6 & 7A) which can be associated with an oceanic trench. The topography of this trough is significantly wider and shallower than oceanic trenches owing to the neglect of faulted plate margins which are required to reproduce the details of observed trenches; however, in the backarc region, there is little quantitative difference in dynamic topography between cases with and without faults [*Zhong and Gurnis*, 1996]. From 130 Ma to the present, the slab descends under its own weight, is sheared by the shifting plates, and later partially drawn upward under a ridge.

We describe how the plate motions change over the course of the model; the plate motions are imposed as

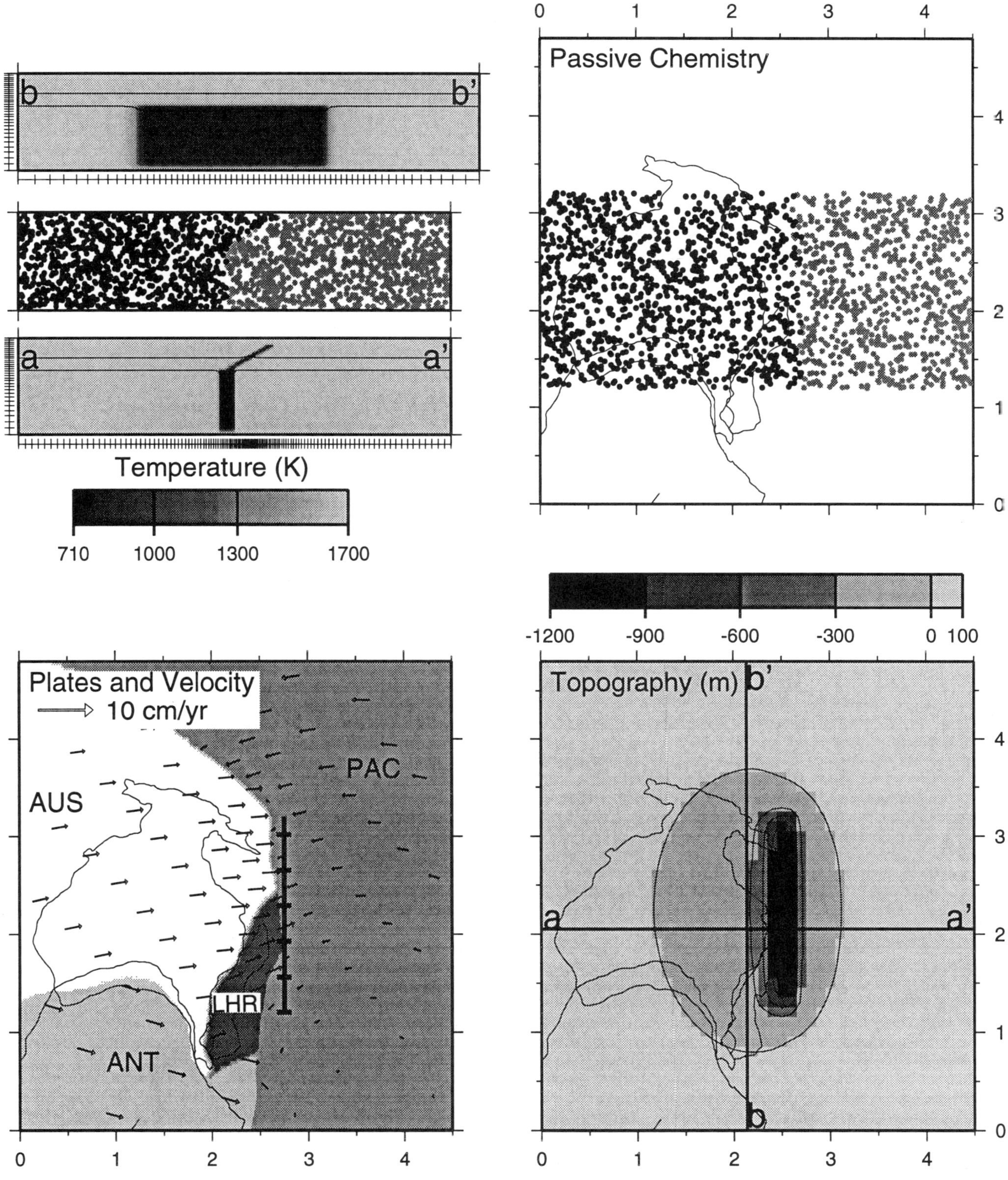


**Figure 6.** Initial condition for Case 1 with plate stencils (ANT for Antarctica, AUS for Australia, LHR Lord Howe Rise, and PAC for Pacific) and imposed velocity vectors shown on the lower left. The position of the trench is shown by the vertical line with the crossing tick marks such that the slab extends from y=1.2 to 3.2. Major tick marks on the box edges have a 2000 km spacing. The domain is Cartesian with horizontal dimensions 4.5 x 4.8 and a depth of 1. In the upper left, the initial configuration of the slab is shown by two intersecting cross-sections (a-a' and b-b'); the center of the slab delineates the boundary between "Gondwanaland" tracers shown as black dots and "Pacific" tracers shown as grey dots. In the upper right, the tracers are plotted for those depths less than $10^{-5}$.

boundary conditions, while the topography is predicted by the model. First, there is a general motion toward the Pacific (toward the right in the Cartesian model) by the three plates which are part of Gondwanaland (AUS, LHR, and ANT) with velocities exceeding 5 cm/yr with respect to the hot spot reference (Fig. 7A - 7C). Second, following a period of relative quiescence (during which time Australia rifts from Antarctica but moves only slowly northward), Australia starts moving away from Antarctica at about 45 Ma with nearly constant velocities between 7 and 8 cm/yr. While the plates change their positions, dynamic topography continuously evolves as the former slab changes its structure and depth distribution (Fig. 8A-D).

Inspection of the changing topographic signature as the imposed plate kinematics evolve reveals two fundamental features; these features are displayed by nearly all cases we explore and do not depend strongly on a range of plausible parameters, as described below. During the first phase of plate motion, Australia migrates over the topographic depression generated by the slab. From 120 to 110 Ma, the eastern interior of Australia is over the former position of subduction which is still a down welling (Fig. 8A) with up to $\sim$ 350 meters of overlaying dynamic topography (Fig. 7B). There is a continued slow migration of the topographic depression westward over the continent while the amplitude of the topography decays; this means the continent, mostly in the east, uplifts by about 200 m from 110 to 60 Ma. The subsidence and uplift of the continent is shown well by the history of a point fixed to Australia. In Figure 9 (lower right) we show the changing dynamic topography, or tectonic subsidence, for the position of the Walkandi-1 borehole. From 130 to 120 Ma this point subsides by $\sim$ 80 m but then from 100 to 90 Ma this point uplifts by $\sim$ 100 m.

During the second phase, while Australia starts in a nearly stationary position over a mild, $\sim -100$ m, nearly continental-scale topographic depression (Fig. 7D), as the SEIR starts to spread faster, a strongly concentrated, approximately circular depression, forms at the ridge near the Great Australian Bight (Fig. 7E). The amplitude of this concentrated depression increases from 45 Ma onward and the increase in its magnitude accelerates over the last 20 Myr (Fig. 7F).

During the first phase of westward motion, the slab, which initially has a 30°dip (Fig. 6), partially stalls above the phase transition at 670 km depth (Fig. 8A). From 60 Ma (Fig. 8B) to the present cold subducted material becomes drawn upward beneath the ridge (Fig. 8 C-D). At the present, the position of the ridge (R in Fig. 8D) is directly above the area where mantle, a few tens of a degree cooler than normal, is being drawn upward. The intense circular topographic depression (Fig. 7F) is caused by a linear source of negative buoyancy orthogonal to the upwelling flow.

### 6.2. *Integration with sedimentation*

Dynamic topography changes for points fixed to plates because dynamic topography changes in time with respect to the box edges (comparable to a hot spot reference frame) and because the plates change their position. From this dynamic topography, we predict the marine inundation of Australia following the methods outlined in Section 5. The stratigraphic predictions are given as paleogeographic maps, inundation curves, borehole subsidence curves, and isopach maps (Fig. 9). The inundation of the continent is displayed as negative accommodation space.

For Case 1 with a sedimentation rate, $S$, of 5 m/Myr, we find that the eastern portion of Australia is flooded at 120 Ma with a nearly uniform tilt downwards toward the east (Fig. 9). This marine inundation is directly related to the rapid eastward motion of Australia over the slab (Fig. 7A-B). From 120 to 100 Ma the inundation becomes broad and migrates toward the west with decreasing accommodation spaces. By 90 Ma, however, only the Gulf of Carpentaria and smaller portions of the west (mostly the northern and southern margins in the west) are covered by a thin layer of water such that by 80 Ma the continent has been nearly completely exposed, despite the increasing height of global sea level which peaks at $\sim$ 70 Ma (Fig. 1A). From 80 Ma to the present, the model continent remains completely exposed.

Given a dynamic topography prediction, the pattern of inundation is influenced by both the higher frequency component of topography due to crustal thickness variations and to sedimentation. As an end member case, we assumed the initial topography of Australia to be at a uniform 150 m height with respect to the ocean floor and from this the inundation was computed from simply adding the dynamic topography and eustatic sea-level change as a function of time (Fig. 10B). We see that a 'wave' of flooding spreads westward over Australia from 120 to 80 Ma. The inundation in this case is unlike that which was observed (Fig. 1C) – the observed flooding being confined much more to the east in

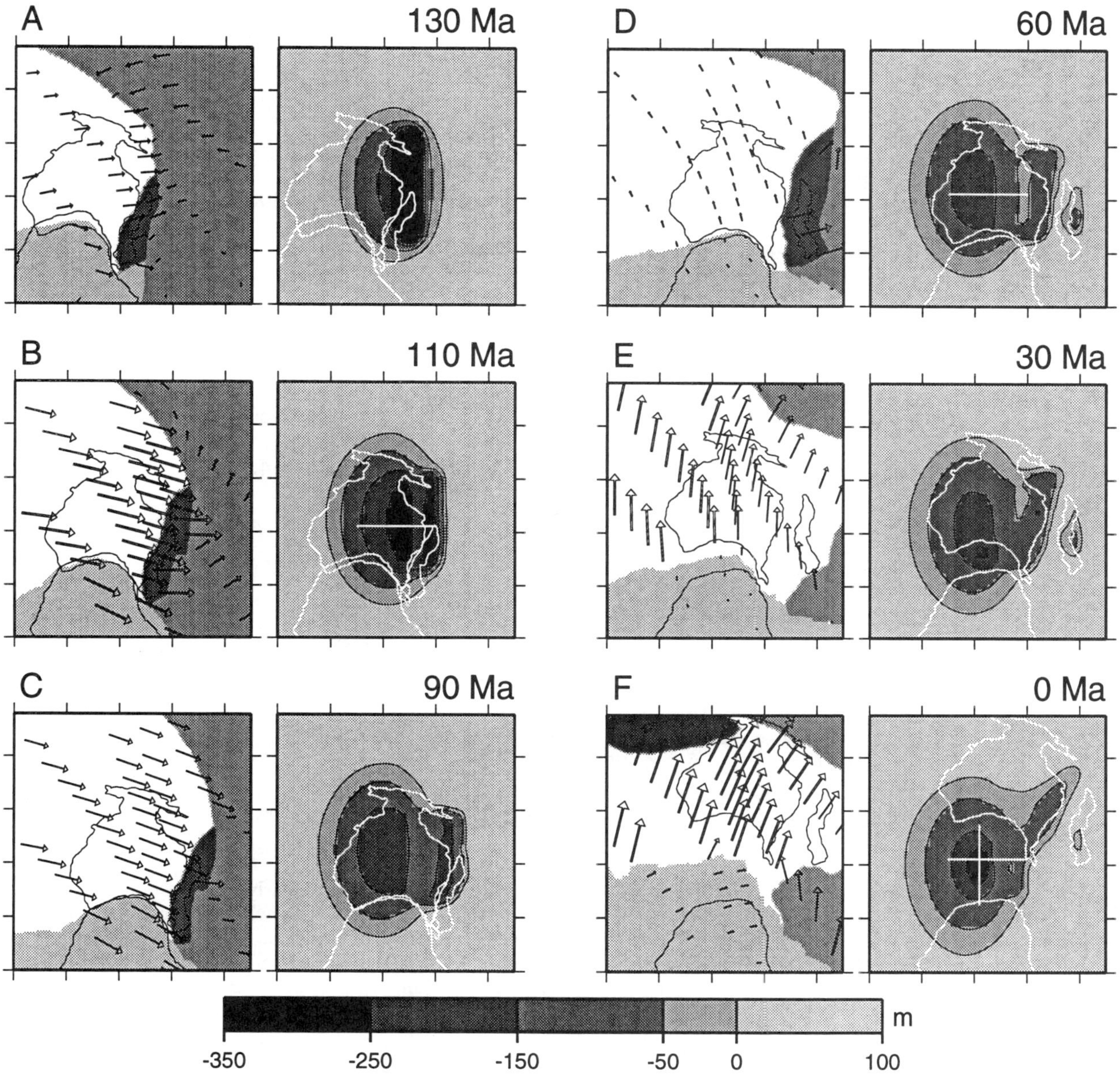


**Figure 7.** Plate stencils and imposed plate velocity (left) and predicted dynamic topography at six times for Case 1. The white lines on some of the topography plots denote the temperature cross sections shown in Fig. 8. At 130 Ma, the grey scale for topography is saturated; see Fig. 6 for the complete topography scale.

those areas which are presently topographically lowest. By increasing sedimentation rate from 5 to 10 m/Myr (compare Fig. 10A with 10C), marine inundation becomes more confined to the east with earlier exposure of the continent during the sea level rise. In all cases, the continent becomes exposed because with sedimentation and an increasing sea level, there is uplift on a large-scale as dynamic topography decays.

### *6.3. Present day geophysics and geochemistry*

With this dynamic model (Case 1) we predict the present day structure of oceanic lithosphere between Australia and Antarctica. Tracers and mantle temperatures are sampled beneath the ridge and moved with the two plates from 50 Ma to the present at 10 Myr time intervals. On the modeled ridge an abrupt

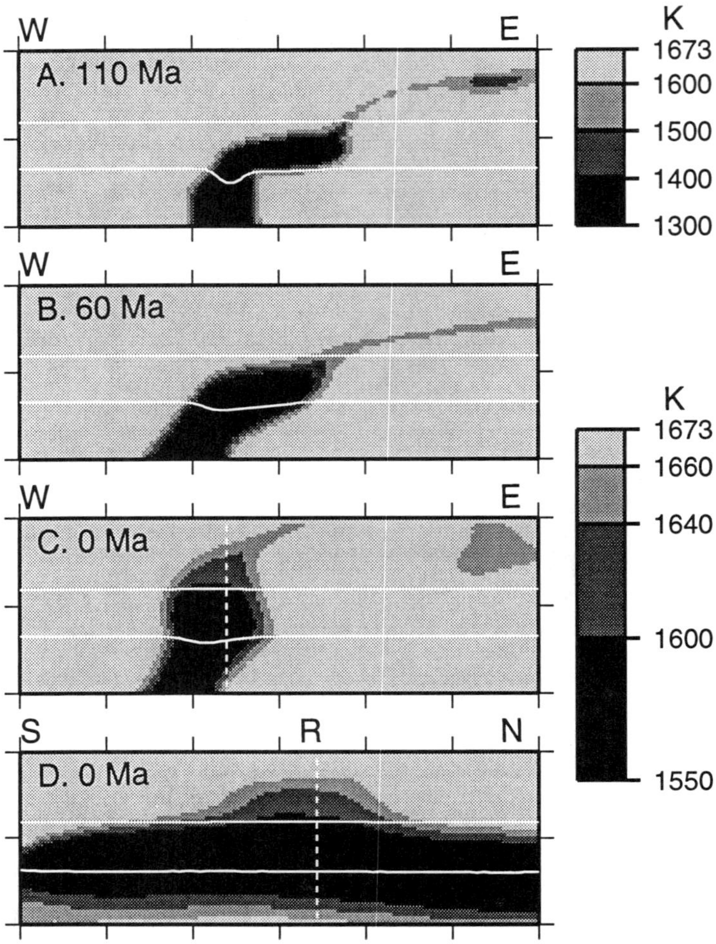


**Figure 8.** Vertical cross sections of temperature for standard model. A-C are nearly east-west profiles. D and C are two orthogonal sections with respective lines of intersection shown by white vertical dashed lines. Locations of cross sections are shown as white lines in Fig. 7. The upper scale is only for A (representing a large temperature span) while the lower scale is for B-D. The tick marks have a 500 km spacing. The horizontal white lines mark the positions of the 410 km and 670 km phase transitions.

change in chemistry is clearly evident as delineated by the sampled passive tracers (Fig. 11A) in close proximity to the depression in dynamic topography between 120°and 130°E. Moreover, the boundary between these two provinces has migrated to the west with an apparent velocity of 3-4 cm/yr, being partly a consequence of the ~ 2 cm/yr eastward component of Australian/Antarctic plate motion in the fixed hot-spot reference (Fig. 7F).

Two regions of thinner crust are evident (Fig. 11B) reflecting the sampling of cooler mantle. The two anomalies are oriented N-S but merge at ~ 45 Ma lithosphere. The anomaly to the east is due to the fragment of the slab which was sheared beneath the plates earlier in the computation. On Figure 8A-C, notice the cool anomaly in the upper right. The more pronounced western anomaly is a 300 km wide region of thinned oceanic crust which develops at the ridge such that the center of thinned crust is essentially aligned with the distinct discontinuity in passive chemistry (Fig. 11B). The thinned crust does not extend uniformly north-south from the ridge, rather, like the passive chemistry the thinned crust shows an apparent migration to the west. The thinning of the crust accelerates at about 20 Ma from thicknesses only a few hundred meters less than normal 10 Myr after rifting to over one kilometer presently. The thinned crust is caused by the drawing upward of a fraction of the old cold slab which had partially stagnated in the transition zone beneath the ridge (Fig. 8). The onset of significant thinning corresponds to the time it takes the slab to be drawn up from the transition zone (Fig. 8B-D) starting at 45 Ma when rifting between Australia and Antarctica accelerated (Fig. 7D-7E). Finally, both dynamic stress

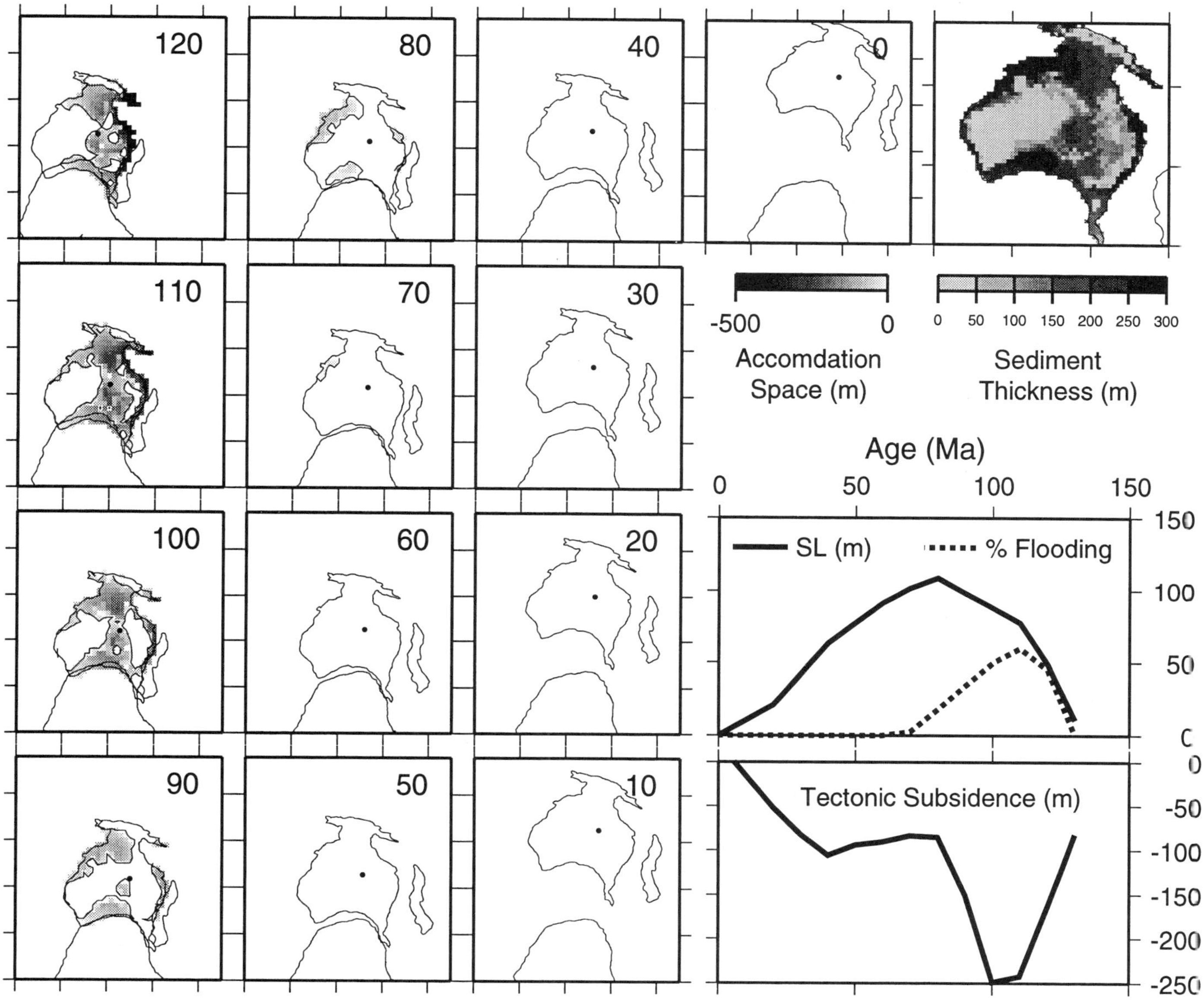


**Figure 9.** Maps (120 - 0 Ma) showing predicted marine inundation of Australia for Case 1 plotted as accommodation space with sedimentation rate of 5 m/Myr. Areas with gray scale spanning from light to dark are those areas covered by shallow seas; white regions denote those parts of the continent predicted to be exposed. In the upper right is predicted sediment thickness and in the lower right are the imposed eustatic sea level and predicted fractional inundation of the entire continent. Also shown is the tectonic subsidence for the Walkandi-1 borehole with the location shown as black dots in the maps, 120-0 Ma.

and thinner crust give rise to a distinct topographic depression, nearly 500 m in amplitude, on the ridge in the vicinity of 125°E (Fig. 11C).

### *6.4. Influence of mantle rheology and phase transitions*

The predictions of the vertical motion of Australia and the character of the sea-floor between Australia and Antarctica are sensitive to assumed mantle characteristics. We vary the temperature-dependence and radial variation of viscosity, and strength of phase transitions at 410 km and 670 km depth.

Silicate rheology is strongly thermally activated [e.g., *Ranalli*, 1995], so that the use of a constant lateral viscosity may appear to be a limitation of the nominal model. However, it is unlikely that slabs are as strong as predicted by a simple thermally activated rheology.

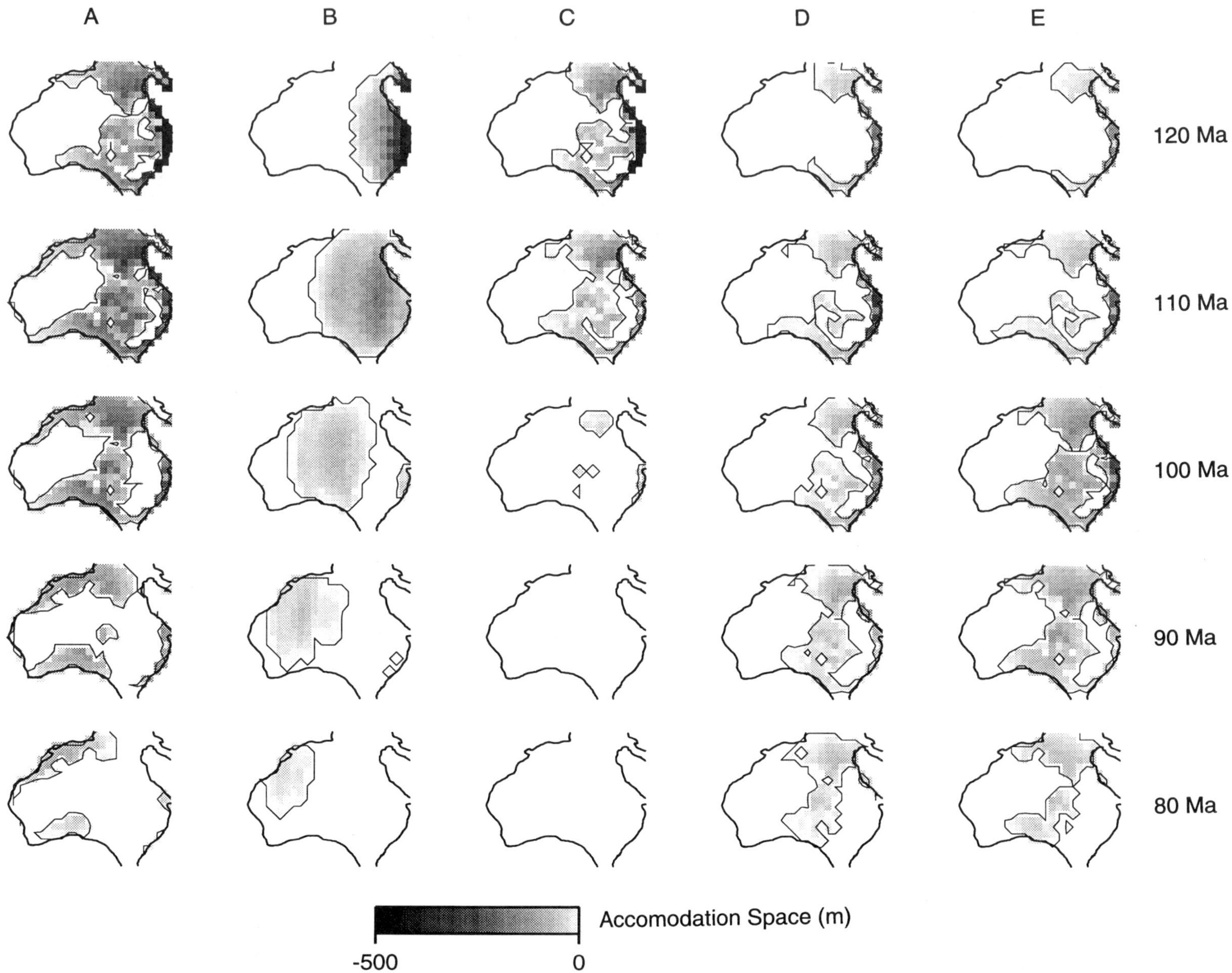


**Figure 10.** Predicted pattern of marine inundation for Australia from 120 Ma to 80 Ma in terms of accommodation space. A. Case 1 $S = 5$ m/Myr. B. Case 1, $S = 5$ m/Myr but with an initially flat surface. C. Case 1, $S = 10$ m/Myr. D. Case 8 with $\theta_s = 90°$ and $S = 5$ m/Myr. E. Case 9 with $X_s = 3.25$ and $S = 5$ m/Myr.

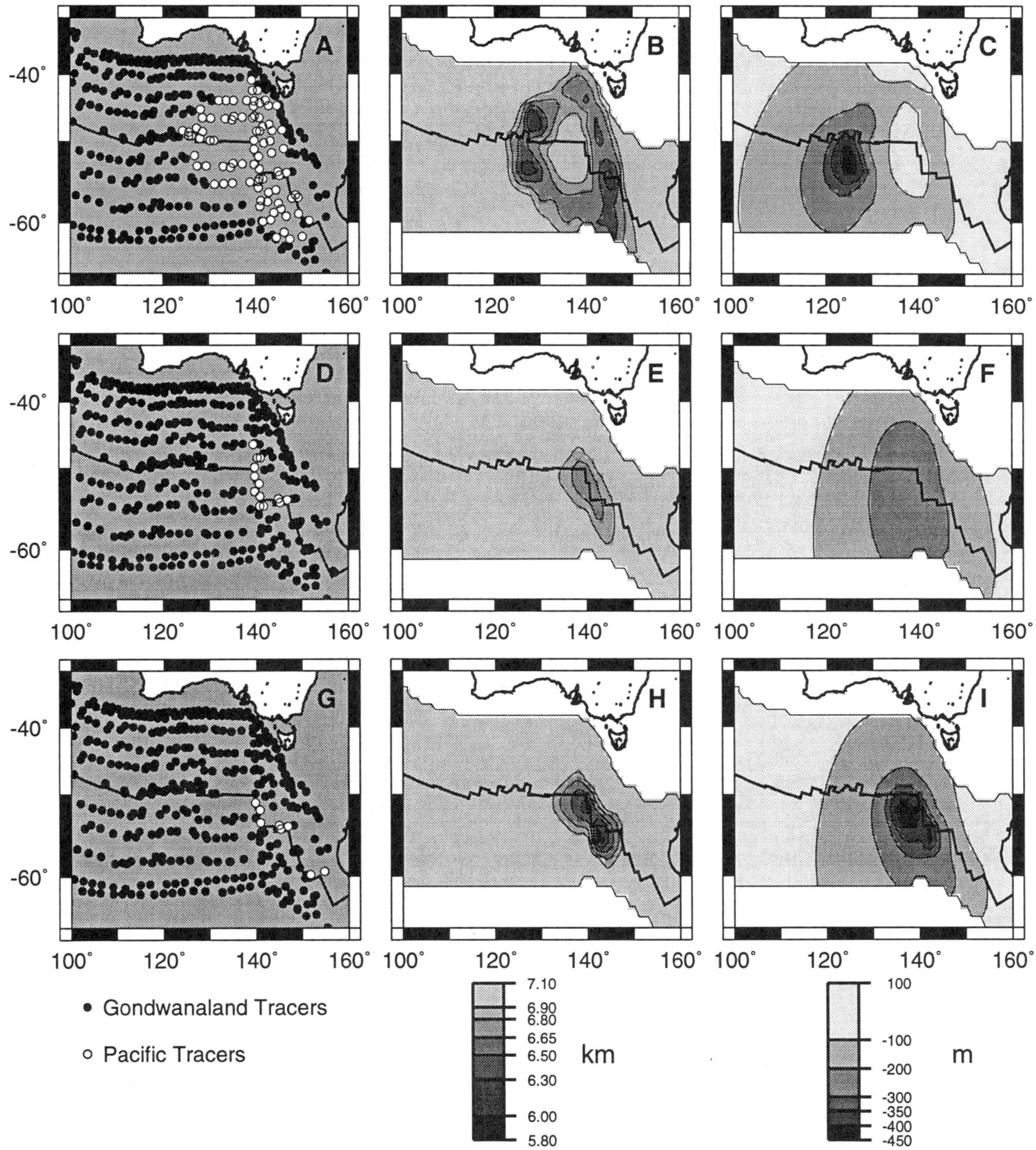


**Figure 11.** A-C. Predictions from Case 1. A. Tracers sampled at the ridge and rotated to present positions. B. Inferred crustal thickness based on mantle temperature. C. Total topography on the top surface driven by both viscous flow and crustal thickness. D-F. Same quantities for a identical model except for an initial vertically dipping slab, Case 8. G-I Same quantities except for a 30° dipping slab at $X_s$=3.25 for Case 9.

Theoretically, non-linear viscosity is expected to limit the variation induced by temperature-dependent viscosity [*Christensen*, 1984]. Moreover, strong slabs tend to give long wavelength geoid lows over slabs, not geoid highs as observed [*Moresi and Gurnis*, 1996]. With a model which incorporated a temperature-dependent viscosity, Case 4, we find that the results become less realistic. With $E' = 34$ and $T_1' = 0.195$ in equation (9) the dimensional $E_a$ becomes 400 kJ/mol. Within a given layer, the maximum lateral variations in viscosity (e.g., viscosity of slab to background viscosity) are limited to 100. Slab descent diminishes to only a few mm/yr in the upper mantle and stagnates within the lower mantle because the slab base cannot deform and bend at the bottom boundary – not unlike the stagnation of the top thermal boundary layer in thermal convection with temperature-dependent viscosity [*Christensen*, 1984]. The magnitude of dynamic topography increased by about 20% compared to the nominal case for all times, but more importantly the decay of dynamic topography from 120 Ma to 60 Ma and its subsequent reappearance when Australia and Antarctic separated is not nearly as pronounced with strong as compared to weak slabs. Dynamic topography is significantly less time-dependent as strong slabs resist deformation.

The influence of the phase change at 410 km is minor. When the 410 km phase change is eliminated (Case 5), dynamic topography decreases by 20% from the nominal model and when the Clapeyron slope is doubled to 4.0 MPa/K (Case 6), dynamic topography increases by 25%. However, after 130 Myr of plate evolution, dynamic topography is only slightly reduced in cases with stronger 410 km phase changes, reflecting the propensity of the phase change to draw the slab into the transition zone and then subsequently to resist the drawing upward of cold fluid by the diverging plates. However, in no case are any of the major conclusions drawn from the nominal model changed with moderately larger or smaller Clapeyron slopes at 410 km depth.

The influence of the phase change and viscosity jump at 670 km are substantial. Holding the viscosity contrast between the upper and lower mantle constant and eliminating both phase transitions (Case 7), we find that it is no longer possible to trap a significant fraction of the slab above 670 km depth for significant time periods compared to a case with both phase changes present. The trapping of the slab by positive buoyancy of the phase change at 670 km depth, especially when the slab is sheared out by the eastward motion of Gondwanaland from 130 Ma to 90 Ma (e.g. Fig. 8A, Fig. 7A-7C) plays a fundamental role in the appearance of a distinct topographic depression and thin crust along an isolated part of the SEIR. The elimination of the phase change at 670 km has only a minor influence on the initial flooding of Australia from 120 to 80 Ma and its subsequent exposure at $\sim 70$ Ma, but strongly influences predictions made for the Southeast Indian Ocean. We are able to predict a mild undulation in topography, but are unable to reproduce the thinner crust near 130°E.

### 6.5. *Influence of initial conditions*

The sensitivity of model outcomes which results from uncertainty in Jurassic paleogeography is explored by varying the initial value of the following: position of trench, initial dip angle of the slab, amount of slab within the lower mantle, and age of subducting lithosphere. Changing the characteristics of the slab has a direct influence on the inundation of Australia during the Cretaceous. Increasing slab dip from 30° to vertical, while maintaining the position of trench (Case 8), or moving the trench an additional 1,000 km from the Australian margin while keeping slab dip fixed (Case 9), delays the time of inundation, as shown in Fig. 10D and 10E, respectively. In both cases maximum flooding occurs at 100–90 Ma and flooding persists past 80 Ma (Fig. 10D-E). The prolonged flooding in the eastern interior is inconsistent with observed inundation (Fig. 1C). In addition, these two paleogeographic models have a substantial influence on the predicted position of the topographic and geochemical anomaly along the present day ridge. With the vertical dipping slab, Case 8, little of the upper mantle slab becomes trapped within the transition zone, despite the strong shearing of the upper mantle by the fast eastward motion from 120-100 Ma (Fig. 7C). Consequently, there is only an insignificant amount of cool mantle remaining within the upper mantle and transition zone when Australia moves away from Antarctica at 45 Ma such that the predicted crustal thickness and depth anomalies are insignificant (Fig. 11E-F). Moreover, the anomaly is shifted approximately 2,000 km to the east in comparison to the case with the dipping slab. The model in which the initial trench is about $\sim 2,400$ km from Queensland in the Jurassic also predicts a geophysical anomaly along the SEIR which is substantially displaced to the east (Fig. 11H-I).

The model with an initial vertically dipping slab shows little apparent westward migration of the geochemical anomaly (Fig. 11D) in comparison to the model with an initial 30°dipping slab (Fig. 11A). This helps to explain why the apparent migration in Case 1 (Fig. 11A) is larger than the overall 2 cm/yr eastward motion of the Australian and Antarctic plates. Part of the migration is the result of initial conditions – as the tracers are extracted from deeper depths as time progresses, the boundary between Gondwanna and Pacific tracers shifts to the west (Fig. 6).

We find that decreasing slab age from 100 to 50 Ma has little influence because with the reduced negative buoyancy of the slab the rate of its descent decreases (Case 10). Although the slab has less negative buoyancy, the source is closer to the surface and the net change on dynamic topography is small. During the final 20 Myrs of the model, when cold material is drawn up by the ridge, fluid then present within the transition zone has a smaller temperature difference but it is drawn up more effectively. Consequently, the age of the slab has little influence on our conclusions. Finally, the initial amount of slab in the lower mantle is uncertain, but has a substantial influence on our predictions. With only an upper mantle slab (i.e. without slab extension into the lower mantle, Case 12), we find a strong influence on the characteristics of the present day ridge. None of the slab is pulled into the lower mantle so that a much larger fraction of the fluid remains to both decrease the temperature beneath the ridge and dynamically pull the topography downward. Crustal thickness and topography variations substantially increase with smaller amounts of lower mantle buoyancy (Case 12, 13).

## 7. DISCUSSION AND CONCLUSIONS

We have carried out a series of dynamic models with imposed plate tectonics and evaluated the hypothesis that both the anomalous marine inundation of Australia during the Cretaceous and the existence of a cold zone beneath the AAD are related to an ancient subducted slab. From our extensive modeling, we conclude that Australia first over rode this slab in the Cretaceous, but later the slab stagnated in the transition zone before being drawn upward over the last 20 Myr by the SEIR. Predictions from dynamic models match observed quantities.

For a range of reasonable values, we are able to match observed inundation of Australia (Fig. 1A) with simple dynamic models of convection with imposed plate tectonics (Fig. 9). The uplift of Australia is a natural consequence of dynamic topography which decreases with the demise of subduction on the Pacific margin of Australia during the Cretaceous. A similar conceptual model has previously been proposed by *Gallagher et al.* [1994] to explain thermal and sedimentary histories of eastern Australian basins. With both the stratigraphic constraints and the predicted character of the AAD, we reject two alternative paleogeographic reconstructions – one with a vertically dipping slab and the other with the trench ≈2,400 km from the Cretaceous margin. Our preferred model has the slab within about 1,500 km of the restored coast of Queensland in the Jurassic and dipping at a moderate angle.

The most significant limitation of the present models is the inability to explain the permanent subsidence of the Eromanga and Surat basins. Qualitatively, we judge from previous work [*Gallagher et al.*, 1994; *Russell and Gurnis*, 1994] that about half of the Cretaceous subsidence of these basins was recovered during the Cenozoic while half persists to the present. By itself, dynamic topography is incapable of explaining permanent subsidence of sedimentary basins since it is reversible. But it is the reversible nature of dynamic topography which allows us to so easily explain the uplift and exposure of the eastern half of the continent, even as global sea level rises. The linking of the formation of an intracratonic basin (Eromanga) to subduction dynamics may create new opportunities to solve more general problems of enigmatic intracratonic basin subsidence [*Sloss*, 1990].

The largest scale, Cenozoic subsidence of Australia is not specifically modeled but we believe it is an important clue in deciphering the geodynamics of the Australian plate. This subsidence may be related either to an unusually warm sub Gondwanaland mantle or to the motion of Australia toward the subduction zones presently north of Australia. The Gondwanaland supercontinent was high standing over much of the Phanerozoic [*Veevers*, 1995] and Africa and East Antarctica today have excessively high hypsometries compared to all other continents [*Cogley*, 1984]. It is possible that the Gondwanaland mantle was a few 10's of degrees hotter than average mantle and the motion of Australia off the resulting dynamic topography high caused it to subside during the Cenozoic [*Gurnis*, 1994]. Alternatively, subduction driven global flow models show that Australia could have subsided by about 200 meters since 20 Ma as it moved toward the extensive system of Indonesian and New Guinea subduction [*Lithgow-Bertelloni and Gurnis*, 1997].

While we are hypothesizing that Australia overrode a subducted slab during the Cretaceous, we note that Australia did not experience a period of volcanism and Laramide–type compression which characterized North America when it over rode the putative flat laying Farallon slab directly beneath the lithosphere [*Bird*, 1984].

Rather, the Jurassic and Early Cretaceous of Australia were typified by a low-relief extensional environment, with no evidence for a high-standing Andean-type magmatic arc. However, further to the east within the interior of N. America, the continent may have overridden a more inclined slab, resulting in a long wavelength downward tilt and subsequent rebound [*Mitrovica et al.*, 1989] as we are now hypothesizing for Australia. The slab beneath Australia was probably much deeper than the slab beneath N. America – in fact best results are obtained when the slab is on average a few hundred kilometers beneath the lithosphere. Moreover, when Australia moved eastward in the direction of the converging margin during the Cretaceous, the margin was probably $\sim 1,000$ km from the continent and consequently we would not expect the slab to be immediately beneath the continental lithosphere.

The same dynamic models are also able to match well the present day geochemistry and geophysics of the Southeast Indian ridge, including the prominent topography on the ridge (Fig. 11C), thin crust at $\sim 130°$ E (Fig. 11B), the onset of anomalous tectonics over the last tens of millions of years (Fig. 11B), the sharp jump of an otherwise long wavelength isotopic signature along the ridge and the apparent westward migration of the isotopic anomaly at a few cm/yr (Fig. 11B). We imply that the intensity of the dynamic stress (dynamic topography) increased over the last tens of million years, commensurate with the onset of thin oceanic crust, as the cold mantle was advected closer to the surface. This is consistent with the increasing roughness of seafloor topography and development of fracture zone offsets in the AAD over the Neogene [*Weissel and Hays*, 1974]. However, *Marks et al.* [1999] have recently shown that residual topography apparently has not increased since 20 Ma while the fracture zone offsets have; our models are probably inconsistent with this observation.

The primary control on the development of anomalous ridge structure by an old slab is the presence of an endothermic phase change at 670 km depth along with a jump in effective viscosity between transition zone and lower mantle which tends to trap a flat laying slab before the cool mantle is partially drawn upward by the new spreading center beneath the opening Australian and Antarctic plates. Results from seismic tomography and dynamic models have shown that trapping of slabs within the transition zone maybe a common and fundamental attribute of subduction dynamics. This is best displayed beneath the Izu Bonin Arc south of Japan where the slab is flat lying and trapped in the transition zone over a horizontal distance of about 700 km [*van der Hilst and Seno*, 1993], much like the structures displayed by the dynamic models (Fig. 8A). In both the mantle and dynamic models, these flat laying structures are best displayed in converging margins which are migrating oceanward in an absolute reference frame [*van der Hilst*, 1995; *Zhong and Gurnis*, 1995, 1997] probably identical to the extensional tectonics of the Australian eastern margin during the Cretaceous [*Williams and Korsch*, 1996].

A substantial body of observational evidence and modeling support our contention that the AAD formed by sampling mantle subjected to long term Mesozoic subduction. This poses the question of whether there are any other regions where a ridge has migrated over a former position of Mesozoic subduction. In Fig. 12 we show the location of Jurassic to Cretaceous subduction fixed with respect to hot spots and present day plate margins. The AAD is clearly unique in being the location where an old subduction zone and a present day ridge are perpendicular to one another. Only ridges associated with back-arc spreading (e.g., on the Scotia and Carribean plates), overlay positions of Mesozoic subduction (Fig. 12). An AAD-like anomaly has been suggested along a western segment of the Pacific-Antarctic ridge at about 185°W [*Marks and Stock*, 1994]. This segment has an axial valley along the ridge, somewhat similar to that along the AAD, and there is a small slope to the depth versus age$^{1/2}$ relation suggesting lower temperatures within the mantle [*Marks and Stock*, 1994]. Although this ridge segment does not cross inferred Mesozoic subduction it does meander close to the early Cretaceous position (Figure 12) and overlaps the strong free-air gravity and geoid low of the Ross Sea which may be associated with the overall position of Mesozoic subduction [*Chase and Sprowl*, 1983].

The model we are proposing for the formation of the AAD is distinctly different from other models particularly recent ones based on isotope geochemistry [*West et al.*, 1997] in which the source of the AAD is actively moving toward the west and driven by mantle flow from the Pacific to Indian Oceans. During the opening of the SEIR the trapped portion of the slab within the transition zone moves little horizontally and the apparent migration is partly due to the motion of the plates toward the east in the hot spot frame and partly due to the initial shallow dip of the subducted slab. The AAD is somewhat similar to the track of a hot spot except here we are dealing with the track of a cold spot. However, the cold region is elongated NS, perpendicular to the ridge and is a 'spot' because the linear cold anomaly

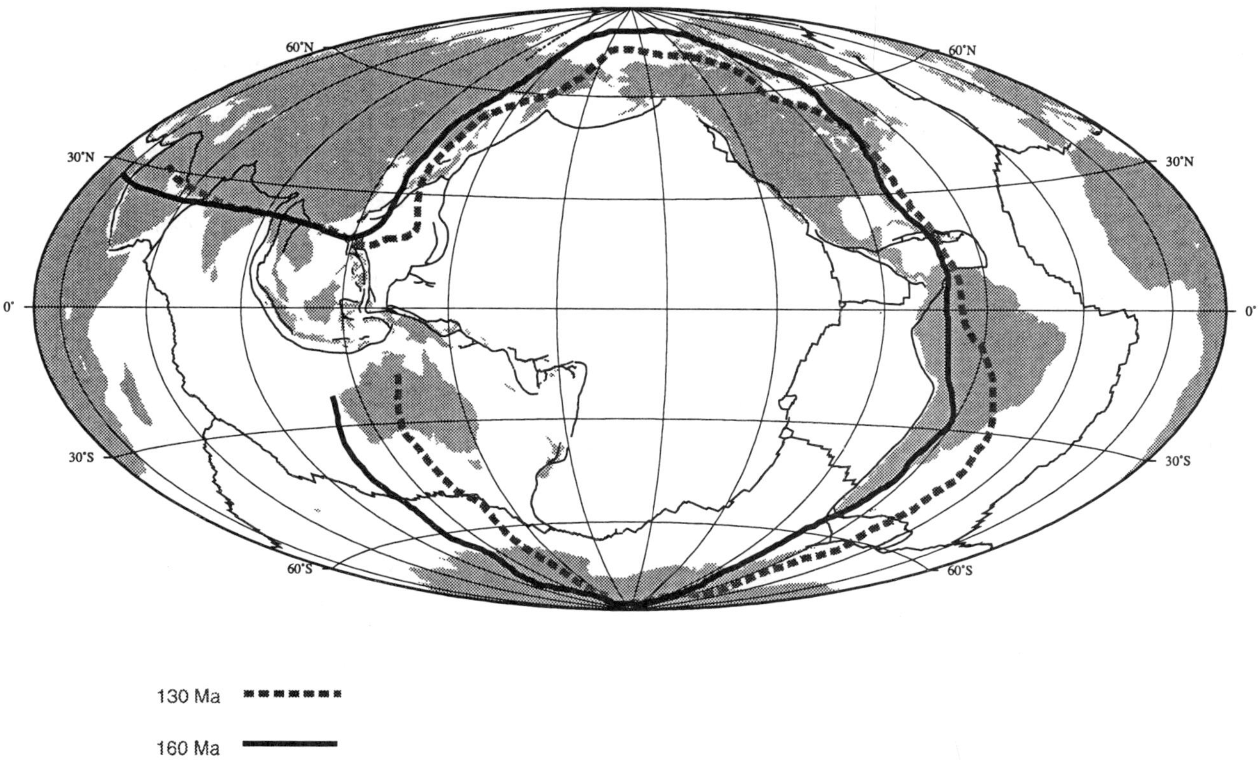


**Figure 12.** Present day plate bounaries (thin solid line) and present day continental outlines shown along with the inferred position of Early Cretaceous (130 Ma, thick grey dotted line) and Late Jurassic (160 Ma, thick black line) subduction. For 130 Ma the location of trenches is based on *Müller et al.'s* [1993] absolute plate motion model. The 160 Ma trench locations are based on a preliminary reconstruction derived by assuming that the various present-day hotspots were associated with flood basalts and dike swarms in the Jurassic. In particular this model places the Vema hotspot at the Karoo igenous province, Fernando da Cunha near the central Atlantic dike swarm, and the Bouvet hotspot near the Wedell Sea [*L. M. Gahagan, pers. comm.*].

and the diverging flow (ridge) are orthogonal. This geometry is substantially different from the interaction of a plume with a ridge. In order to avoid any confusion, it is probably best not to refer to this structure as a cold spot – a 'cold zone' maybe a better term.

Individually, the correspondence between a predicted and observed quantity is not perfect – the importance of our result is the ability to predict a wide suite of disparate observations. These include the subsidence and uplift of Australia during the Cretaceous, the westward migration of an isotopic boundary in the ocean basin, a distinct crustal thickness, isotopic, and topographic structure at the approximate position of the present AAD, and the disruption of the oceanic crust since the start of the Neogene. Mantle or paleogeographic parameters have not been adjusted in order to improve the fit with observed quantities. However, by finding the best match between the wide range of observations and model predictions fundamental constraints can be placed on intrinsic mantle properties (like rheology) and the Mesozoic paleogeography of the Gondwanaland-Pacific margin. Given the strong association between essentially all substantial high seismic velocity mantle anomalies with Mesozoic subduction [*Grand et al.*, 1997], the approach outlined here connecting paleogeography with mantle convection models has wide applicability and can be used to address many problems in continental geology, isotope geochemistry, plate tectonics, and mantle dynamics.

*Acknowledgments.* We thank H. van Heijst for access to his tomographic models and A. Lenardic, C. Beaumont, and R. Pysklywec for helpful comments on this manuscript. This represents Contribution Number 8577 of the Division of Geological and Planetary Sciences, California Institute of Technology. Some of the work reported here was conducted as part of the Australian Geodynamics Cooperative Research Centre and is published with the consent of the Director, AGCRC.

## REFERENCES


Asimow, P. D., A thermodynamic model of adiabatic melting of the mantle, Ph.D. thesis, California Institute of Technology, Pasadena, 1997.

Bird, P., Laramide crustal thickening event in the Rocky Mountain foreland and great plains, *Tectonics*, *3*, 741–758, 1984.

BMR Paleogeographic Group, Cretaceous Paleogeographic Maps, Paleogeographic Series 14, Record 1990/30, Bureau of Mineral Resources, Canberra, 1990.

Bond, G., Evidence for continental subsidence in North America during the late Cretaceous global submergence, *Geology*, *4*, 557–560, 1976.

Bond, G., Speculations on real sea-level changes and vertical motions of continents at selected times in the Cretaceous and Tertiary periods, *Geology*, *6*, 247–250, 1978.

Bradshaw, J., Cretaceous dispersion of Gondwana: continental and oceanic spreading in the south-west Pacific-Antarctic sector, in *Geological evolution of Antarctica*, edited by M. R. Thomson, J. A. Crome and J. W. Thomson, pp. 581–585. Cambridge Univ. Pr., 1991.

Brooks, A. N., A Petrov-Galerkin Finite Element Formulation for Convection Dominated Flows, Ph.D. thesis, California Institute of Technology, Pasadena, CA, 1981.

Bryan, S. E., Early Cretaceous volcano-sedimentary successions along the eastern Australian continental margin: Implications for the break-up of eastern Gondwana, *Earth Planet. Sci. Lett.*, *153*, 85–102, 1997.

Burgess, P. M., M. Gurnis, and L. Moresi, Formation of sequences in the cratonic interior of North America by interaction between mantle, eustatic and stratigraphic processes, *Geol. Soc. Am. Bull.*, *108*, 1515–1535, 1997.

Cande, S. C., and J. C. Mutter, A revised identification of the oldest sea-floor spreading anomalies between Australia and Antarctica, *Earth Planet. Sc. Lett.*, *58*, 151–160, 1982.

Chase, C. G., Extension behind island arcs and motions relative to hot spots, *J. Geophys. Res.*, *83*, 5385–5387, 1978.

Chase, C. G., and D. R. Sprowl, The modern geoid and ancient plate boundaries, *Earth Planet. Sci. Lett.*, *62*, 314–320, 1983.

Christensen, U., Convection with pressure and temperature dependent non Newtonian rheology, *Geophys. J.*, *77*, 343–384, 1984.

Cochran, J. R., and M. Talwani, Free-air gravity anomalies in the world's oceans and their relationship to residual elevation, *Geophysical Journal*, *50*, 495–552, 1977.

Cogley, J. G., Deglaciated hypsometry of Antarctica, *Earth Planet. Sci. Lett.*, *67*, 284–296, 1984.

Cross, T. A., and R. H. Pilger, Tectonic controls of Late Cretaceous sedimentation, western interior, USA, *Nature*, *274*, 653–657, 1978.

Dalziel, I. W. D., Antarctica: A tale of two supercontinents?, *Ann. Rev. Earth Planet. Sci.*, *20*, 501–526, 1992.

Davies, G. F., Mantle convection undersimulated plates: effects of heating modes and ridge and trench migration, and implications for the core–mantle boundary, bathymetry, the geoid and Benioff zones, *Geophys. J. R. astron. Soc.*, *84*, 153–183, 1986.

Davies, G. F., Ocean Bathymetry and mantle convection: 2. Small-scale flow, *J. Geophys. Res.*, *93*, 10481–10488, 1988.

Davies, G. F., and M. A. Richards, Mantle convection, *J. Geol.*, *100*, 151–206, 1992.

Engebretson, D. C., A. Cox, and R. G. Gordon, Relative motions between oceanic and continental plates in the Pacific basin, *Geol. Soc. Am. Sp. Paper*, *205*, 1–59, 1985.

Ewart, A., R. W. Schön, and B. W. Chappell, The Cretaceous volcanic-plutonic province of the central Queensland (Australia) coast - rift related 'calc-alkaline' province, *Trans. Roy. Soc. Edinburgh Earth Sci.*, *83*, 327–345, 1992.

Exon, N. F., and B. R. Senior, The Cretaceous of the Eromanga and Surat Basins, *BMR Journal of Australian Geology and Geophysics*, *1*, 33–55, 1976.

Forsyth, D. W., R. L. Ehrenbard, and S. Chapin, Anomalous upper mantle beneath the Australian–Antarctic discordance, *Earth Planet. Sci. Lett.*, *84*, 471–478, 1987.

Gaina, C., Tectonic history of microplates east of Australia, Ph.D. thesis, The University of Sydney, Sydney, 1998.

Gaina, C., R. D. Müller, J.-Y. Royer, J. Stock, J. Hardebeck, and P. Symonds, The tectonic history of the Tasman Sea: a puzzle with 13 pieces, *J. Geophys. Res.*, *103*, 12,413–12,433, 1998.

Gallagher, K., Permian to Cretaceous subsidence history along the Eromanga-Brisbane Geoscience Transect, in *The Eromanga-Brisbane Geoscience Transect*, edited by D. M. Finlayson, no. 232 in Bulletin, pp. 133–151. Bur. Min. Res., Canberra, 1990.

Gallagher, K., and K. Lambeck, Subsidence, sedimentation and sea-level changes in the Eromanga Basin, Australia, *Basin Research*, *2*, 115–131, 1989.

Gallagher, K., T. A. Dumitru, and A. J. W. Gleadow, Constraints on the vertical motion of eastern Australia during the Mesozoic, *Basin Resaerch*, *6*, 77–94, 1994.

Golub, G. H., and C. F. Van Loan, *Matrix Computations*, Johns Hopkins University Press, Baltimore, 1989.

Grand, S. P., R. D. van der Hilst, and S. Widiyantoro, Global seismic tomography: A snapshot of convection in the Earth, *GSA Today*, *7*(4), 1–7, 1997.

Gurnis, M., Bounds on global dynamic topography from Phanerozoic flooding of continental platforms, *Nature*, *344*, 754–756, 1990.

Gurnis, M., Long-term controls on eustatic and epeirogenic motions by mantle convection, *GSA Today*, *2*, 141–157, 1992.

Gurnis, M., Depressed continental hypsometry behind oceanic trenches: A clue to subduction controls on sea level change, *Geology*, *21*, 29–32, 1993.

Gurnis, M., Epeirogenic motion of Australia in the Mesozoic and Cainozoic: A mantle dynamics perspective, in *Geoscience Australia – 1994 and Beyond*, Abstracts Number 37, pp. 154–156. Geological Society of Australia, abstract, 1994.

Gurnis, M., R. D. Müller, and L. Moresi, Dynamics of Cretaceous vertical motion of Australia and the Australian-Antarctic discordance, *Science*, *279*, 1499–1504, 1998.

Hallam, A., *Phanerozoic Sea-Level Changes*, Columbia University Press, New York, 1992.

Han, L., and M. Gurnis, How valid are dynamic models of subduction and convection when plate motions are prescribed?, *Phys. Earth Planet. Int.*, *110*, 235–246, 1999.

Hart, S. R., A large-scale isotope anomaly in the Southern Hemisphere mantle, *Nature*, *309*, 753–757, 1984.

Hughes, T. J. R., *The Finite Element Method: Linear Static and Dynamic Finite Element Analysis*, Prentice–Hall, Inc., Englewood Cliffs, New Jersey, 1987.

Jones, J. G., and J. J. Veevers, Mesozoic origins and antecedents of Australia's Eastern Highlands, *J. Geol. Soc. Aust.*, *30*, 305–322, 1983.

King, S. D., Models of mantle viscosity, in *Mineral Physics and Crystallography: A Hand Book of Physical Constants*, AGU Reference Shelf 2, pp. 227–236. Am. Geophys. Un., 1995.

Klein, E. M., and C. H. Langmuir, Global correlations of ocean ridge basalt chemistry with axial depth and crustal thickness, *J. Geophys. Res.*, *92*, 8089–8115, 1987.

Klein, E. M., C. H. Langmuir, A. Zindler, H. Staudigel, and B. Hamelin, Isotope evidence of a mantle convection boundary at the Australian-Antarctic Discordance, *Nature*, *333*, 623–629, 1988.

Laird, M., Geological aspects of the opening of the Tasman Sea, in *Evolution of the Tasman Basin*, edited by G. Van der Lingen, K. Swanson and R. Muir. A. A. Balkema, Rotterdam, 1994.

Langmuir, C. H., E. M. Klein, and T. Plank, Petrological systematics of mid-ocean ridge basalts: Constraints on melt generation beneath ocean ridges, in *Mantle Flow and Melt Generation*, edited by J. Phipps Morgan, D. K. Blackman and J. M. Sinton, 71, *Geophysical Monograph*, pp. 183–280. American Geophysical Union, Washington, D.C., 1992.

Lecroart, P., A. Cazenave, Y. Ricard, C. Thoraval, and D. G. Pyle, Along-axis dynamic topography constrained by major-element chemistry, *Earth Planet. Sci. Lett.*, *149*, 49–56, 1997.

Lithgow-Bertelloni, C., and M. Gurnis, Cenozoic subsidence and uplift of continents from time-varying dynamic topography, *Geology*, *25*, 735–738, 1997.

Mammerickx, J., and D. Sandwell, Rifting of old oceanic lithosphere, *J. Geophys. Res.*, *91*, 1975–1988, 1986.

Marks, K. M., and J. M. Stock, Variations in ridge morphology and depth-age relationships on the Pacific-Antarctic ridge, *J. Geophys. Res.*, *99*, 531–541, 1994.

Marks, K. M., J. M. Stock, and K. J. Quinn, Evolution of the Australian-Antarctic discordance since Miocene time, *J. Geophys. Res.*, *104*, 4967–4981, 1999.

McClain, J. S., and C. A. Atallah, Thickening of the oceanic crust with age, *Geology*, *14*, 574–576, 1983.

McKenzie, D. P., J. M. Roberts, and N. O. Weiss, Convection in the Earth's mantle: towards a numerical simulation, *J. Fluid Mech.*, *62*, 465–538, 1974.

Mitrovica, J. X., C. Beaumont, and G. T. Jarvis, Tilting of continental interiors by the dynamical effects of subduction, *Tectonics*, *8*, 1079–1094, 1989.

Moresi, L., and M. Gurnis, Constraints on the lateral strength of slabs from three-dimensional dynamic flow models, *Earth Planet. Sci. Lett.*, *138*, 15–28, 1996.

Moresi, L., S. Zhong, and M. Gurnis, The accuracy of finite element solutions of Stokes' flow with strongly varying viscosity, *Phys. Earth Planet. Int.*, *97*, 83–94, 1996.

Moresi, L. N., and V. S. Solomatov, Numerical investigation of 2D convection with extremely large viscosity variations, *Phys. Fluid*, *7*, 2154–2162, 1995.

Müller, R., D. Mihut, and S. Baldwin, A new kinematic model for the evolution of the West and Northwest Australian margin, in *West Australian Basins Conference*, edited by P. Purcell and R. R. Purcell, Perth. Petroleum Exploration Society of Australia, 1998.

Müller, R. D., J.-Y. Royer, and L. A. Lawver, Revised plate motions relative to the hotspots from combined Atlantic and Indian hotspot tracks, *Geology*, *21*, 275–278, 1993.

Müller, R. D., U. R. Roest, J.-Y. Royer, L. M. Gahagan, and J. G. Sclater, Digital isochrons of the world's ocean floor, *J. Geophys. Res.*, *102*, 3211–3214, 1997.

Paine, A. G. L. (ed.), *Australia: Evolution of a Continent*, Bureau of Mineral Resources, Australia, Canberra, 1990.

Press, W. H., B. P. Flannery, S. A. Teukolsky, and W. T. Vetterling, *Numerical Recipes in C*, Cambridge University Press, New York, 2nd edn., 1992.

Pribac, F., Superswells Due to Mantle Convection, Ph.D. thesis, Australian National University, Canberra, 1991.

Pyle, D. G., M. Christie, D, and J. J. Mahoney, Resolving an isotopic boundary within the Australian-Antarctic discordance, *Earth Planet. Sci. Lett.*, *112*, 161–178, 1992.

Ranalli, G., *Rheology of the Earth*, Chapman and Hall, New York, 2nd edn., 1995.

Rehkämper, M., and A. W. Hofmann, Recycled ocean crust and sediment in Indian Ocean MORB, *Earth Planet. Sci. Lett.*, *147*, 93–106, 1997.

Richards, M. A., and B. H. Hager, Geoid anomalies in a dynamic Earth, *J. Geophys. Res.*, *89*, 5987–6002, 1984.

Russell, M., and M. Gurnis, The planform of epeirogeny: Vertical motions of Australia during the Cretaceous, *Basin Research*, *6*, 63–76, 1994.

Schroeder, W., The empirical age-depth relation and depth anomalies in the Pacific Ocean basin, *J. Geophys. Res.*, *89*, 9873–9884, 1984.

Sloss, L. L., Tectonic evolution of the craton in Phanerozoic time, in *Sedimentary Cover - North American Craton: U. S.*, edited by L. L. Sloss, The Geology of North America D-2, pp. 25–51. Geological Society of America, Boulder, 1988.

Sloss, L. L., Epilog, in *Interior Cratonic Basins*, edited by M. W. Leighton, D. R. Kolata, D. F. Oltz and J. J. Eidel, Memoir 51, pp. 799–805. The American Association of Petroleum Geologists, Tulsa, 1990.

Struckmeyer, H. I. M., and P. J. Brown, Australian Sealevel Curves Part 1: Australian Inundation Curves, Record 1990/11, Bureau of Mineral Resources, Canberra, 1990.

Symonds, P., J. B. Colwell, H. Struckmeyer, J. Willcox, and P. Hill, Mesozoic rift basin development off eastern Australia, in Mesozoic Geology of the Eastern Australia Plate, in *Mesozoic Geology of the Eastern Australian Plate Conference*, Extended Abstracts No. 43, pp. 528–542, Sydney. Geological Society of Australia, 1996.

Tolstoy, M., A. J. Harding, J. A. Orcutt, and J. Phipps Morgan, Crustal thickness at the Australian Antarctic discordance and neighboring South East Indian ridge, in *EOS*

*Fall Meeting Supplement*, 76, p. 570. 1995.

Tulloch, A., and D. L. Kimbrough, The Paparoa metamorphic core complex, New Zealand: Cretaceous extension associated with fragmentation of the Pacific margin of Gondwana, *Tectonics*, *8*, 1217–1234, 1989.

van der Hilst, R., Complex morphology of subducted lithosphere in the mantle beneath the Tonga trench, *Nature*, *374*, 154–157, 1995.

van der Hilst, R., and T. Seno, Effects of relative plate motion on the deep structure and penetration depth of slabs below the Izu-Bonin and Mariana island arcs, *Earth Planet. Sci. Lett.*, *120*, 395–407, 1993.

van der Hilst, R. D., S. Widiyantoro, and E. R. Engdahl, Evidence for deep mantle circulation from global tomography, *Nature*, *386*, 578–584, 1997.

Van der Voo, R., *Paleomagnetism of the Atlandtic, Tethys and Iapetus Oceans*, Cambridge University Press, New York, 1993.

van Heijst, H. J., New constraints on the seismic structure of the Earth from surface wave overtone phase velocity measurements, Ph.D. thesis, University of Oxford, 1997.

Veevers, J. J., Australian-Antarctic depression from the mid-ocean ridge to adjacent continents, *Nature*, *295*, 315–317, 1982.

Veevers, J. J. (ed.), *Phanerozoic Earth History of Australia*, Clarendon Press, Oxford, 1984.

Veevers, J. J., Emergent, long–lived Gondwanaland vs. submergent, short–lived Laurasia: Supercontinental and Pan–African heat imparts long–term buoyancy by mafic underplating, *Geology*, *23*, 1131–1134, 1995.

Watts, A. B., and M. Steckler, Subsidence and eustasy at the continental margin of eastern North America, in *Deep Drilling Results in the Atlantic Ocean: Continental Margins and Paleoenvironments*, edited by M. Talwani and W. Hays, Maurice Ewing Series 3, pp. 218–234. American Geophysical Union, Washington, D.C., 1979.

Weissel, J. K., and D. E. Hayes, The Australian-Antarctic Discordance: New Results and Implications, *J. Geophys. Res.*, *79*, 2579–2587, 1974.

West, B. P., S. D. Wilcock, J.-C. Sempére, and L. Géli, Three-dimensional structure of asthenospheric flow beneath the Southeast Indian Ridge, *J. Geophys. Res.*, , *102*, 7783–7802, 1997.

Williams, N., and R. Korsch, The Mesozoic geology of the eastern Australian plate, in *Mesozoic Geology of the Eastern Australian Plate Conference*, Extended Abstracts No. 43, pp. 564–568, Sydney. Geological Society of Australia, 1996.

Zhong, S., and M. Gurnis, Mantle convection with plates and mobile, faulted plate margins, *Science*, *267*, 838–843, 1995.

Zhong, S., and M. Gurnis, Interaction of weak faults and Non-Newtonian rheology produces tectonic plates in 3D model of mantle flow, *Nature*, *383*, 245–247, 1996.

Zhong, S., and M. Gurnis, Dynamic interaction between tectonic plates, subducting slabs, and the mantle, *Earth Interactions*, *1*(6), http://EarthInteractions.org/, 1997.

Zhong, S., M. Gurnis, and G. Hulbert, Accurate determination of surface normal stress in viscous flow from a consistent boundary flux method, *Phys. Earth and Planet. Int.*, *78*, 1–8, 1993.

---

M. Gurnis, Seismological Laboratory, California Institute of Technology, Pasadena, CA 91125. (email: gurnis@caltech.edu)

L. Moresi, Australian Geodynamics Cooperative Research Centre, CSIRO Exploration and Mining, Nedlands, WA 6009 Australia. (e-mail: louis@ned.dem.csiro.au)

R. Dietmar Müller, Department of Geology and Geophysics, The University of Sydney, Sydney, NSW 2006, Australia. (e-mail: dietmar@es.su.oz.au)

# Mapping the lithosphere and asthenosphere with surface waves: Lateral structure and anisotropy

Göran Ekström

*Department of Earth and Planetary Sciences, Harvard University, Cambridge, Massachusetts*

A review of the basic theory and practice of mapping lateral variations in surface wave phase velocities is presented, with some examples from the work of *Ekström et al.* [1997]. The importance of crustal corrections is demonstrated by comparison of observed phase velocities with those calculated from the recent crustal model of *Mooney et al.* [1998]. Phase velocity maps including terms of azimuthal anisotropy are presented, and an experiment suggests that in areas of poor path coverage, a bias is introduced when azimuthal effects are neglected in the inversion. The inversion procedure for obtaining a model of three-dimensional velocity heterogeneity in the mantle from surface wave dispersion data is discussed in the context of the new mantle model S20A [*Ekström and Dziewonski*, 1998]. This model exhibits significant lateral variations in radial anisotropy in the top 200–300 km of the mantle, resulting from the incompatibility of Rayleigh and Love wave dispersion data. This result suggests an important new direction for tomographic imaging of the upper mantle.

## 1. INTRODUCTION

The plates of plate tectonics are the rigid portions of the Earth's crust and upper mantle which participate in global mantle convection in the role of an upper boundary layer. A fundamental observation of plate tectonics is that the Earth's surface is divided into oceanic and continental lithosphere, and that these two types of lithosphere participate in the mantle's convection circulation in different ways: oceanic lithosphere is created at spreading centers and consumed through subduction, while the bulk of continental lithosphere appears to remain at the Earth's surface through time. How this differentiation came about, and how it has been maintained for four billion years, is one of the central questions of geology. Many of the controlling processes, such as partial melting and mixing, take place in the top part of the upper mantle, and a satisfying answer will involve a detailed characterization of the state and composition of the lithospheric plates and the underlying asthenosphere, as well as a thorough theoretical understanding of the physics of mantle flow.

The structure and thickness of a plate can be characterized in many ways, considering its thermal, compositional and elastic properties. Since most of the Earth is inaccessible to direct examination, much of our knowledge about the properties of the plates comes from remote sensing, in particular the imaging of seismic velocities by tomography. The crust and the shallowest part of the mantle have been characterized successfully by seismic reflection and refraction studies, but at greater depths these techniques have limited utility. Most of the seismic energy from surface explosions is refracted to great distances and the interpretation of the mantle refracted phases that are recorded tends to be difficult and nonunique. Near subduction zones, it is possible to

The History and Dynamics of Global Plate Motions
Geophysical Monograph 121

use both shallow and deep earthquakes as illuminating natural sources, and to obtain the crossing ray-coverage required to determine the 3-D structure, but this approach is limited to only a few regions of the world.

The principal seismological constraints on the large-scale elastic structure of the top few hundred kilometers of the mantle therefore come from Love and Rayleigh surface waves. The propagation speed of a Love or Rayleigh wave at a certain period, traveling along the surface of the Earth, depends on radial averages of the local elastic Earth structure, rather than on the seismic velocities at a specific depth. Slow propagation of Rayleigh waves at 60 s period might, for example, be due to slow $P$ wave velocities in the crust and shallow mantle, or to slow $S$ wave velocities in the asthenosphere, reflecting the relative sensitivity of the surface wave speed to the intrinsic velocities at different depths (see, for example, Chapter 9 of *Dahlen and Tromp* [1998]). It is only by measuring propagation speeds of surface waves over a range of periods, and thereby observing the dispersion of the wave, that the radial variations of intrinsic seismic velocities can be determined.

Figure 1 shows a typical teleseismic seismogram from a distant ($\Delta$=85°) shallow focus earthquake. The surface waves are generally the largest amplitude phase in teleseismic seismograms, especially at long periods. Figure 1 illustrates the dispersive properties of Rayleigh waves: when the seismogram is filtered in different frequency bands, the arrival times of the Rayleigh wavepackets change. The arrival times of these wavegroups provide information which can be used to infer the average radial velocity structure along the path traversed by the surface wave. Typical variations in surface wave speeds at a given period are relatively small; around 3% at 45 s and 1% at 175 s period. For seismograms such as the one in Figure 1, variations in travel times are therefore of the order of 75 s at 45 s period and 30 s at 175 s period. It is clear that great care is required to make precise and useful observations of surface wave dispersion.

In this article I first provide, in Section 2, a brief overview of earlier work on the measurement of surface wave dispersion and its interpretation in terms of heterogeneous Earth structure. The focus here is on global studies; a large literature also exists dealing with short-period surface waves and crustal structure. In Section 3, I present the theory needed for interpreting observations of surface wave travel times in terms of lateral phase velocity variations. Section 4 summarizes some of the results obtained in a recent study [*Ekström et al.*, 1997] of

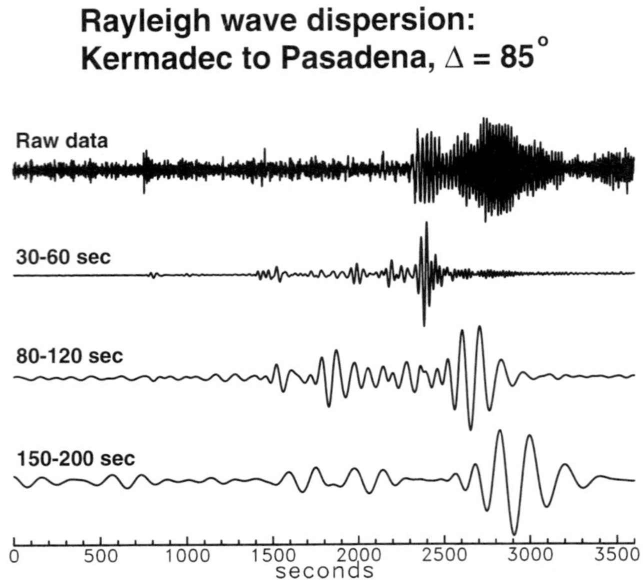


**Figure 1.** A teleseismic seismogram for an earthquake in the Kermadec Islands recorded at Pasadena, CA. The top trace shows the raw vertical component seismogram from the long-period (LH) channel. The three seismograms below show the same data filtered in different frequency bands: 30–60 s, 80–120 s, and 150–200 s. The dispersion of the Rayleigh wave is evident, with the energy arriving substantially later for the longest period component of the wavepacket.

global intermediate period surface waves, and presents results of an extension of that study to consideration of azimuthal anisotropy. Section 5 describes how surface wave dispersion data can be used in an inversion for three-dimensional (3-D) mantle structure, and Section 6 contains an examination of the recently published 3-D earth model S20A [*Ekström and Dziewonski*, 1998], with some implications for the anisotropic elastic structure of the Pacific plate and its underlying asthenosphere.

## 2. HISTORICAL OVERVIEW

Regional variations in the propagation speeds of teleseismically surface waves have been noted since the early part of the century [*Tams*, 1921a, b]. In the 1960s, the magnitudes of these global variations for waves in the period range 10–150 s could be summarized and discussed in terms of gross lateral differences in the layering and structure of the crust and upper mantle [e.g., *Oliver*, 1962; *Dorman*, 1969]. Variations at longer periods were also observed and interpreted in terms of lateral variations of upper mantle elastic and anelastic structure [e.g., *Toksöz and Anderson*, 1966; *Kanamori*,

1970; *Dziewonski*, 1971; *Wu*, 1972]. With the start of digital data collection from new global seismograph networks [*Agnew et al.*, 1976; *Peterson et al.*, 1976] and the development of new techniques for analyzing surface wave dispersion through waveform inversion [e.g., *Dziewonski and Steim*, 1982], efforts began to map the global variations in long-period phase velocities and the corresponding three-dimensional (3-D) structure of the Earth using tomographic techniques [e.g., *Masters et al.*, 1982; *Nakanishi and Anderson*, 1982, 1983, 1984; *Woodhouse and Dziewonski*, 1984; *Tanimoto and Anderson*, 1985; *Nataf et al.*, 1986; *Tanimoto*, 1986; *Wong*, 1989; *Montagner and Tanimoto*, 1990, 1991]. These studies have made use of a variety of seismic data to image the Earth's interior; surface waves have, however, provided the most valuable constraints on upper mantle structure.

Although good recordings of intermediate-period surface waves are much more common than good very-long-period recordings, the shortest-period surface waves included in the global tomographic studies referenced above is 75 s, and most of the studies consider only surface waves with periods greater than 150 s. The explanation for the emphasis on longer-period observations is twofold. First, while shorter-period surface waves are sensitive to velocity variations in the shallow mantle, they also depend critically on the structure and thickness of the crust, which are not very well known on the global scale. By limiting surface wave observations to periods which have less sensitivity to the crust and applying simple a priori crustal corrections to the data, these studies have sought to avoid contamination by crustal effects. Second, it is more difficult to measure the dispersion of surface waves at intermediate periods than at long periods. In particular, establishment of the total propagation phase of a teleseismic surface wave at a particular period, such that there is no remaining ambiguity in the integral number of cycles, is a nontrivial task.

There are, in addition, several complicating aspects of surface wave propagation in a laterally heterogeneous Earth which are more significant for shorter-period waves, but which are not very well understood or predicted, such as focusing, refraction of the apparent ray path, and scattering [e.g., *Evernden*, 1954; *Capon*, 1970; *Sobel and von Seggern*, 1978; *Lay and Kanamori*, 1985; *Woodhouse and Wong*, 1986; *Snieder and Nolet*, 1987; *Laske*, 1995; *Larson et al.*, 1998]. These complications usually do not prevent us from making measurements of dispersion, but do suggest that the accurate interpretation of these measurements in terms of Earth structure will depend on whether the chosen theoretical model for surface wave propagation in a heterogeneous Earth is adequate.

Recently, interest in incorporating shorter-period surface waves in global studies has grown [*Trampert and Woodhouse*, 1995, 1996; *Zhang and Lay*, 1996; *Laske and Masters*, 1996; *Ekström and Dziewonski*, 1995, 1998], primarily due to the high resolving power, both radially and laterally, that these waves can afford in the shallowest mantle.

## 3. THEORY

The propagation of fundamental mode Rayleigh and Love waves can be well explained using ray theory on a sphere [e.g., *Tromp and Dahlen*, 1992, 1993]. The surface wave seismogram $u(\omega)$ is written

$$u(\omega) = A(\omega)\exp[i\Phi(\omega)], \tag{1}$$

where $A(\omega)$ and $\Phi(\omega)$ are the amplitude and phase, respectively, of the wave as functions of the angular frequency $\omega$. For a given source–receiver geometry, the phase $\Phi$ is the sum of four terms,

$$\Phi = \Phi_S + \Phi_R + \Phi_C + \Phi_P, \tag{2}$$

where $\Phi_S$ is the source phase calculated from the source mechanism and geometrical ray takeoff azimuth, $\Phi_R$ is the receiver phase, $\Phi_C$ is the static phase contribution from each ray focus, and $\Phi_P$ is the propagation phase

$$\Phi_P(\omega) = \int \frac{\omega}{c(\omega)} ds, \tag{3}$$

where $c$ is the local phase velocity and the integration follows the ray path. The amplitude $A$ can be expressed as

$$A = A_S A_R A_\Delta A_Q, \tag{4}$$

where $A_S$ is the magnitude of the excitation at the source, $A_R$ is the receiver amplitude, $A_\Delta$ is the geometrical spreading factor, and $A_Q$ is the decay factor due to attenuation along the ray path. When the location and focal mechanism of the earthquake are known, a theoretical reference seismogram $u^0(\omega)$ based on a spherical Earth model can be calculated and written as

$$u^0(\omega) = A^0(\omega)\exp[i\Phi^0(\omega)]. \tag{5}$$

The propagation phase for the reference surface wave is

$$\Phi_P^0(\omega) = \frac{\omega R\Delta}{c^0} = \frac{\omega X}{c^0}, \tag{6}$$

where $c^0$ is the spherical Earth phase velocity, $\Delta$ is the angular epicentral distance, $R$ is the radius of the Earth, and $X$ is the propagation path length measured along the great circle. The observed surface wave $u(\omega)$ can be expressed as a perturbation with respect to the reference seismogram

$$u(\omega) = [A^0(\omega) + \delta A(\omega)] \exp i[\Phi^0(\omega) + \delta\Phi(\omega)]. \quad (7)$$

Assuming no other contributions, the phase anomaly $\delta\Phi$ can be attributed to a perturbation in the propagation phase,

$$\Phi_P = \Phi_P^0 + \delta\Phi = \frac{\omega X}{c^0 + \bar{\delta c}}, \quad (8)$$

where $\bar{\delta c}$ is the apparent average phase velocity perturbation, calculated for the distance $X$ along the great circle. Note that $\delta\Phi$ can equally well be thought of as a travel time residual $\delta T$ since

$$\delta\Phi = \omega\, \delta T. \quad (9)$$

Observations of phase anomalies are interpreted as having accumulated along the path between a source at $(\theta_S, \phi_S)$ and a receiver at $(\theta_R, \phi_R)$

$$\delta\Phi = -\frac{\omega}{c^0} \int_{(\theta_S,\phi_S)}^{(\theta_R,\phi_R)} \frac{\delta c(\theta,\phi)}{c^0} ds, \quad (10)$$

where the approximation $1/(1 + \delta c/c^0) \approx 1 - \delta c/c^0$ has been introduced to make travel time residual observations depend linearly on velocity perturbations. This approximation is customary and acceptable since $\delta c$ is small; it can be avoided by considering slowness, rather than velocity, variations.

The velocity perturbation $\delta c(\theta, \phi)$ on the right hand side of equation (10) can be parameterized in many different ways in order to formulate an inverse problem of finite dimensions. The oldest and simplest method is the so-called pure-path method. The observed phase anomaly is modeled as having accumulated as a result of phase velocity variations in $M$ different tectonic or geographic regions,

$$\delta\Phi = -\frac{\omega}{c^0} \sum_{i=1}^{M} X^i \frac{\delta c^i}{c^0}, \quad (11)$$

where $X^i$ is the path length and $\delta c^i$ is the phase velocity variation in the $i$th tectonic region. This equation can be generalized to represent a uniform division of the surface of the Earth into, for example, 5° × 5° patches, but has traditionally been used to invert for variations in a small number of tectonic environments.

An alternative to the pure-path approach is a smooth, global parameterization in terms of spherical harmonics. The predicted propagation phase anomaly between a source and a receiver is then written as in equation (10), and the phase velocity perturbation is expressed in complex spherical harmonics

$$\frac{\delta c(\theta,\phi)}{c^0} = \sum_{l=0}^{L} \sum_{m=-l}^{m=l} C_{lm} Y_{lm}(\theta,\phi), \quad (12)$$

where the $Y_{lm}$ are the fully normalized surface spherical harmonics [*Edmonds*, 1960] and $L$ is the maximum degree of the expansion.

In order to estimate the parameters defining the geographical phase velocity variations from a set of observed phase anomalies, a misfit function $\chi^2$ can be defined,

$$\chi^2 = \sum_{j=1}^{N} \frac{1}{\sigma_j^2} \left( \delta\Phi_j - \frac{\omega}{c^0} \sum_{i=1}^{M} X_j^i \frac{\delta c^i}{c^0} \right)^2, \quad (13)$$

where $j$ is the index of the observation, $N$ is the total number of observations, and $\sigma_j$ is the uncertainty associated with each observation. The optimal perturbations in phase velocity in each region, $\delta c^i$, are obtained by minimization of $\chi^2$. The spherical harmonic coefficients $C_{lm}$ can be obtained in an analogous fashion.

When the number of observations is large in comparison with the number of parameters to be estimated, the inverse problem is often overdetermined, and no regularization of the solution is necessary. When regularization is necessary, this is easily introduced by minimizing the sum of $\chi^2$ and some additional function which reflects the size or roughness of the phase velocity model. For example, if we define the global rms gradient of the phase velocity map as

$$\mathcal{R}_{\text{rms}} = \left[ \frac{1}{4\pi} \int_S (\nabla \frac{\delta c}{c^0}) \cdot (\nabla \frac{\delta c}{c^0})\, d\Omega \right]^{1/2}, \quad (14)$$

we can choose to minimize the quantity

$$\left(\chi^2 + \gamma \mathcal{R}_{\text{rms}}^2\right), \quad (15)$$

where $\gamma$ is an arbitrary parameter which defines the trade-off between fitting the observations and obtaining a smooth model.

## 4. GLOBAL MAPS OF PHASE VELOCITY

The main obstacle to obtaining high quality global maps of phase velocities has been the difficulty of measuring the propagation phase at shorter periods. Phase velocity variations are larger at short periods, which means that phase delays often will be several cycles at teleseismic distances; it is not unusual for phase delays to be 10 cycles at 35 s period. Great care is therefore needed to accurately measure the total phase at these periods.

Recently *Ekström et al.* [1997] collected a large number of global dispersion measurements of Love and Rayleigh waves in the period range 35–150 s using a technique they called iterative frequency band expansion. This method, which is similar to one developed by *Trampert and Woodhouse* [1995], was designed specifically to address the difficulties of making reliable measurements at shorter periods (T < 100 s). The number of measurements collected in the study ranged from 15,473 to 37,739 depending on the period.

*Ekström et al.* [1997] used their dispersion measurements to determine phase velocity maps represented in a spherical harmonic expansion up to degree 40. The maps explain 80–96% of the variance in Love wave phase anomalies and 70–90% of the Rayleigh wave anomalies. The smallest variance reduction is obtained for the longest periods. The measurement technique and the resulting maps are described in the original paper, and here I focus on two aspects of the maps: (1) The extent to which the geographical variations of phase velocity depend on, and are correlated with, the thickness and structure of the crust, and (2) the effects of azimuthal anisotropy. Both of these effects are potentially of great importance in determining 3-D velocity structure from surface wave dispersion data.

### 4.1. The Effect of the Crust

In tomographic studies of the Earth's mantle, surface wave dispersion measurements and phase velocity maps are commonly used to constrain upper mantle velocity variations. For this use, it is necessary to account for crustal effects in the observations before inferring from them the deeper mantle structure. This correction is usually made by subtracting from the observations the predicted effects of an a priori crustal structure, often one which corresponds to the average oceanic or continental crust. The crustal signal is strong for shorter period waves, and decreases at longer periods. Since the rms variations in observed phase velocities also decrease with increasing period, however, it is not obvious that the relative importance of the crustal signal decreases with increasing period.

In order to investigate the representation of the crust in the phase velocity maps more quantitatively, we compare dispersion predicted from the global crustal model

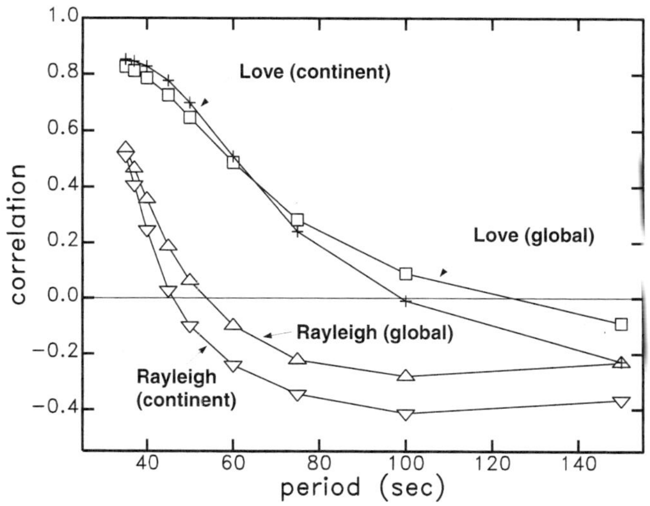


**Figure 2.** Correlation between the observed phase velocity maps of *Ekström et al.* [1997] and the corresponding maps predicted from the crustal model of *Mooney et al.* [1998]. At long periods the correlation for continental areas is worse than for the global average because of the (on average) faster mantle material that underlies the continental crust.

CRUST-5.1 of *Mooney et al.* [1998] with the observations. The dispersion predicted from CRUST-5.1 is calculated by substituting the CRUST-5.1 values for $v_P$, $v_S$, and $\rho$ into PREM, adjusting the depth of the Moho and accounting for topography and bathymetry. We then evaluate the surface wave dispersion in each modified local Earth structure. The 5° by 5° global dispersion map which results is expanded in spherical harmonics up to degree 40.

The correlation between observed phase velocities and those due to the crust (Figure 2) does follow the expected trend of a greater effect at short periods and for Love waves. The 0.83 correlation coefficient for Love waves at 35 s shows that while Love waves, even at this period, have some sensitivity to mantle structure, the observed variations are dominated by the crust. The correlation of the observations with the crustal model decreases with increasing period, and for periods longer than 50 s for Rayleigh waves and 100 s for Love waves, the correlation is negative. When we calculate the correlation for continental regions only, it is seen to deteriorate faster with increasing period, so that for Rayleigh waves the correlation becomes negative at around 45 s, reflecting a lack of correlation between the thickness of the continental crust and the seismic velocities of the underlying mantle.

The good correlation at short periods between predicted and observed phase velocities indicates that the CRUST-5.1 model is quite accurate. However, it is also useful to consider areas where the predicted phase velocities do not agree well with those observed. One such area is Tibet. Figure 3 shows the observed and predicted Love wave phase velocities for Eurasia. While in most areas the agreement between the two maps is good, phase velocities across the southeastern part of the Tibetan plateau are noticeably slower in the observed than in the predicted map. The model CRUST-5.1 does include an appropriately deep (70 km) Moho in this area, so the most likely source for the discrepancy is the average $S$ velocity in the crust, which in CRUST-5.1 is approximately 3.7 km $s^{-1}$. There are several indications that the middle crust in southern Tibet contains substantial fractions of melt [*Kind et al.*, 1996; *Chen et al.*, 1996], which would very significantly lower the $S$ velocity. A second area for which we find a discrepancy between observations and predictions is the Caspian depression, where predicted velocities are slower than those observed. Here, the difference may be related to higher $S$ velocities than those inferred to exist in the very thick column of sediments. In some areas, the differences are likely to reflect lateral heterogeneity in the mantle, since these waves also have some sensitivity to the structure at depth.

Though the specific causes of the discrepancies remain unresolved at this time, the comparison of the observations of *Ekström et al.* [1997] with the model of *Mooney et al.* [1998] indicates that surface waves can provide both a test of, and useful constraints on, the global structure of the Earth's crust.

### *4.2. Effects of Azimuthal Anisotropy*

It is well established that azimuthal anisotropy significantly affects the propagation of intermediate period Rayleigh waves [e.g., *Forsyth*, 1975; *Nishimura and Forsyth*, 1988], in particular waves traversing oceanic lithosphere. Two issues which remain unclear are the magnitude of the anisotropy, and the bias introduced by ignoring the effects of anisotropy in the calculation of isotropic phase velocities.

The dispersion data set used by *Ekström et al.* [1997] can be used to address both of these issues. For surface wave propagation across a weakly anisotropic Earth we write, following *Smith and Dahlen* [1973],

$$\delta\Phi = -\frac{\omega}{c^0}\int_{(\theta_S,\phi_S)}^{(\theta_R,\phi_R)}\left[\frac{\delta c(\theta,\phi)}{c^0} + A(\theta,\phi)\cos 2\zeta(\theta,\phi) + B(\theta,\phi)\sin 2\zeta(\theta,\phi) + C(\theta,\phi)\cos 4\zeta(\theta,\phi) + D(\theta,\phi)\sin 4\zeta(\theta,\phi)\right]\,ds, \quad (16)$$

where $\zeta(\theta,\phi)$ is the local ray angle. We expand the quantities $\delta c/c^0$, $A$, $B$, $C$, and $D$ in complex spherical harmonics,

$$\frac{\delta c(\theta,\phi)}{c^0} = \sum_{l=0}^{L}\sum_{m=-l}^{m=l} C_{lm}Y_{lm}(\theta,\phi); \quad (17)$$

$$A(\theta,\phi) = \sum_{l=0}^{L}\sum_{m=-l}^{m=l} C_{lm}^{A}Y_{lm}(\theta,\phi), \quad (18)$$

and so on. The expansion in this case is to degree $L = 20$, and we invert the data using similar smoothing to that employed by *Ekström et al.* [1997].

Figure 4 shows the results of this global inversion for the Pacific ocean, where azimuthal anisotropy is observed to be the greatest. We obtain the largest anisotropy near the East Pacific Rise, where the fast direction agrees with the current direction of spreading. The maximum anisotropy is $\pm$ 0.9% at 35 s period and $\pm$ 0.8% at 100 s period, similar to the values obtained by *Forsyth* [1975]. Interestingly, we find that at shorter periods, the fast direction correlates better with the direction of fossil spreading than with current plate motions, while at longer periods the reverse is true. Another striking feature is the sudden decrease in the anisotropic signal west of 165°W.

To address the issue of tradeoff between isotropic and anisotropic structure, we invert the data set using the anisotropic and the isotropic parameterizations, both expanded to $L = 20$, and compare the isotropic part of the resulting maps. Figure 5 shows the result of this exercise for Rayleigh waves at 60 s period. The difference map (top panel) suggests that the tradeoff is most significant in the southern hemisphere, even though the anisotropic signal is equally large in the northern Pacific. The likely explanation is that paths across the southern Pacific tend to have a dominant direction, while the northern Pacific is sampled more homogeneously. We conclude that neglecting azimuthal anisotropy in the parameterization has an effect on the isotropic models, but that in regions of uniform coverage, the effect is not very significant.

To conclude this section, it should be pointed out that the determination of azimuthal anisotropy is quite difficult and the results of the mapping tend to be nonunique. Often the improvement in fit to the dispersion data after adding the complexity of laterally varying anisotropic structure is quite small, and can equally well be accomplished with the addition of smaller scale isotropic variations.

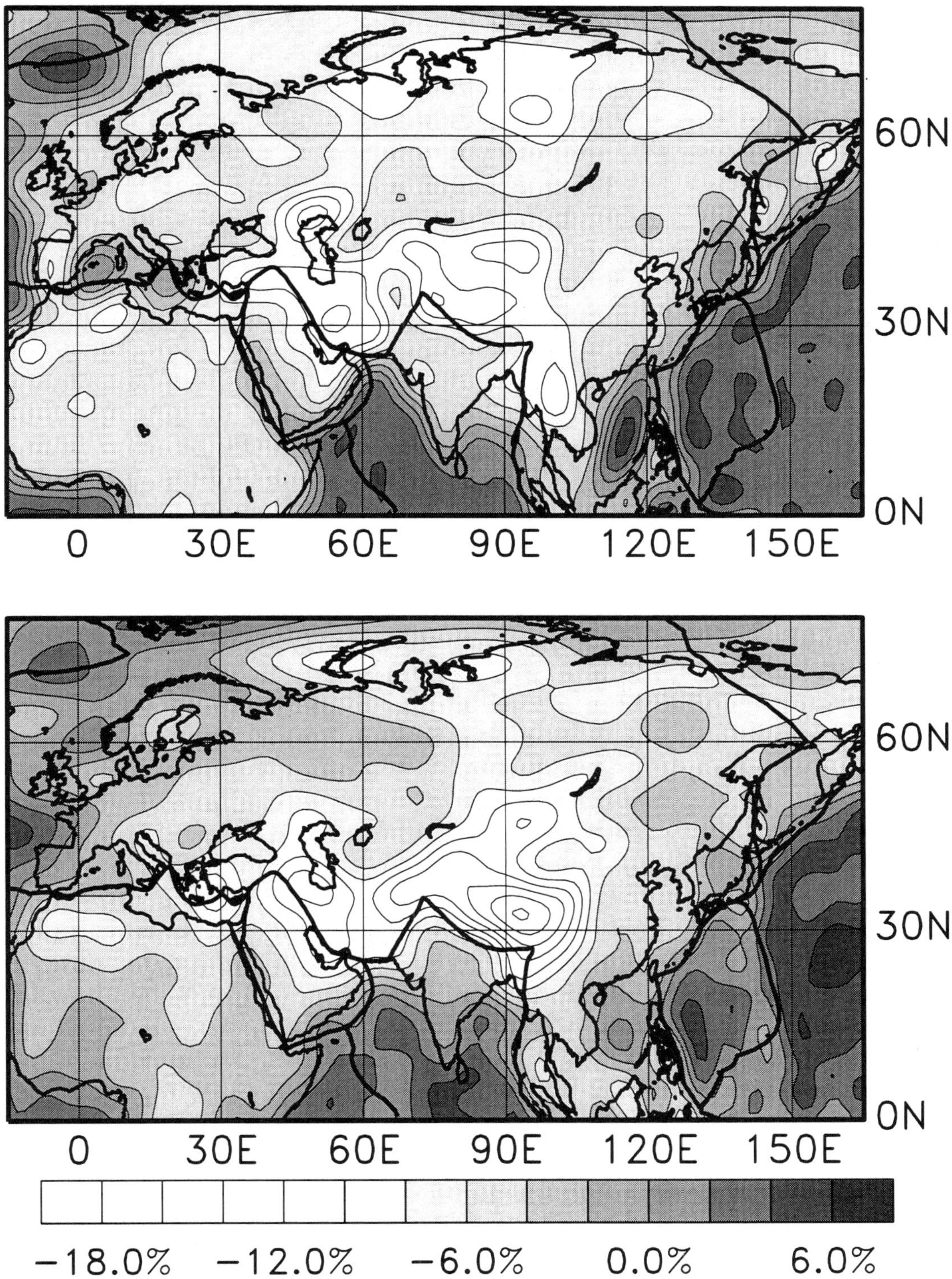


**Figure 3.** Regional map showing Love wave phase velocities at 35 s period. The bottom map shows the observed variations [*Ekström et al.*, 1997] and the top shows the predictions from *Mooney et al.*'s [1998] crustal model.

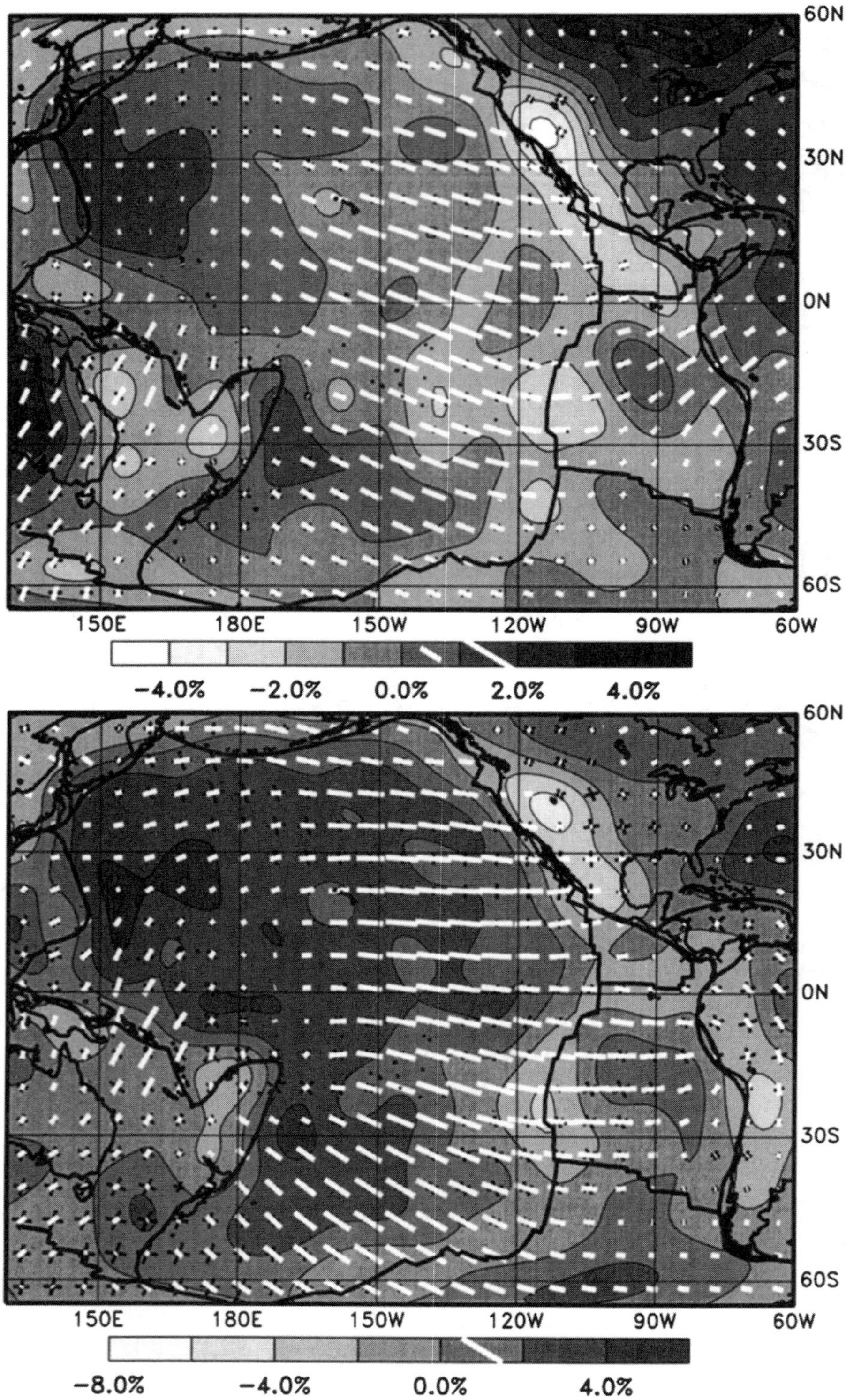


**Figure 4.** Maps showing the isotropic and anisotropic variations of phase velocity in the Pacific Basin. The top map shows Rayleigh waves at 100 s period and the lower map shows Rayleigh waves at 35 s period. The lines show fast directions of propagation: thick white lines correspond to the $2\zeta$ variations and the thin lines to the $4\zeta$ variations. The $4\zeta$ variations are significantly smaller, as predicted by theory.

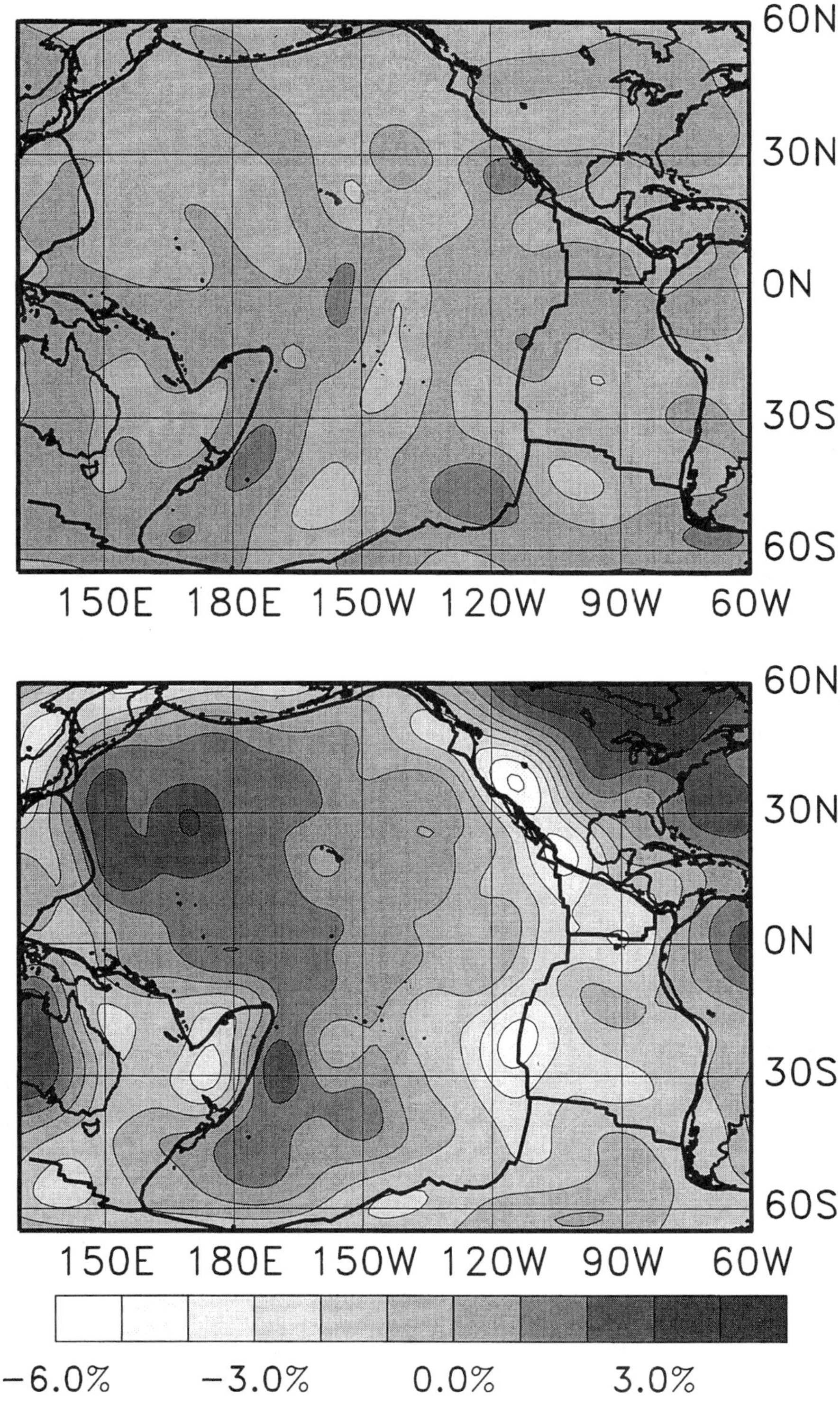


**Figure 5.** Bottom map shows the isotropic Rayleigh wave phase velocity at 60 s period obtained in an inversion which included azimuthal anisotropy. The top map shows the difference between the bottom map and the equivalent map obtained using an isotropic parameterization. Differences are generally less than ±1%.

## 5. UPPER MANTLE STRUCTURE

The surface wave phase velocity perturbation at a location $(\theta, \phi)$ depends on perturbations in the radial seismic velocity and density structure,

$$\frac{\delta c(\theta,\phi)}{c^0} = \int_0^R \left[ K_S(r) \frac{\delta v_S(r,\theta,\phi)}{v_S^0(r)} + K_P(r) \frac{\delta v_P(r,\theta,\phi)}{v_P^0(r)} + K_\rho(r) \frac{\delta \rho(r,\theta,\phi)}{\rho^0(r)} \right] dr \tag{19}$$

where $K_S$, $K_P$, and $K_\rho$ are the differential kernels for $v_S$, $v_P$ and $\rho$, respectively, and $v_S^0$, $v_P^0$, $\rho^0(r)$ are the values in the reference model. In writing this equation we have made the assumption that the kernel functions of the starting model are adequate and independent of position. This assumption is very common in seismic tomography, but should probably be reexamined for intermediate period surface waves, for which variations in crustal structure can cause large variability in the surface wave eigenfunction. In the case of anisotropy, a more complicated but analogous expression applies.

The basic relation (equation 10) presented in the previous section for determining the two-dimensional variations of phase velocity from observations of phase anomalies can be extended to three dimensions. For a phase anomaly $\delta\Phi$ we write

$$\delta\Phi = -\frac{\omega}{c^0} \int_{(\theta_S,\phi_S)}^{(\theta_R,\phi_R)} \left[ \int_0^R \left\{ K_S(r) \frac{\delta v_S(r,\theta,\phi)}{v_S^0(r)} + K_P(r) \frac{\delta v_P(r,\theta,\phi)}{v_P^0(r)}(r) + K_\rho(r) \frac{\delta \rho(r,\theta,\phi)}{\rho^0(r)} \right\} dr \right] ds. \tag{20}$$

Equations 19 and 20 suggest two different strategies for obtaining the 3-D velocity structure from surface wave phase anomalies. A two-step approach can be adopted, in which phase velocity maps are determined in the first step and a mapping to radial structure at each geographical point $(\theta, \phi)$ is done in a second step. Alternatively, a direct inversion of phase anomalies for 3-D structure can be performed. The former approach has the advantage that the inverse problem to be solved is much smaller. In fact, the problem reduces to one of solving a separate inverse problem for each geographical location or for each spherical harmonic coefficient. Also, this approach has the advantage that the close relationship between the data and the model allow for a more straightforward error analysis. The latter approach, though computationally more demanding, is in general preferable, since it avoids the difficulty of accounting for the regularization in step one of the inversion when inverting for the radial structure. This is the method used in the analysis described in the next section.

## 6. A 3-D, ANISOTROPIC MODEL OF THE MANTLE

In a recent study, *Ekström and Dziewonski* [1998] presented a new 3-D model of the Earth called S20A. This model was based on a large number of different types of observations, but the intermediate period dispersion measurements of *Ekström et al.* [1997] discussed in Section 4 provided the primary constraints on the heterogeneity of the top 400 km of the mantle.

The main finding of *Ekström and Dziewonski* [1998] was that large lateral variations in radial anisotropy are needed to reconcile the Love and Rayleigh wave dispersion measurements. In particular, the asthenosphere beneath the central Pacific was identified as a region of very large radial anisotropy ($v_{SV}$ much slower than $v_{SH}$). In this section, I describe the parameterization and inversion procedure used in the determination of the 3-D model S20A, present a cross section through the center of the Pacific, and discuss the seismological constraints on the average upper mantle structure of continents and oceans in the context of the aging of the Pacific lithosphere and asthenosphere.

### *6.1. Parameterization and Inversion of Model S20A*

The S20A model parameterization describes 3-D *S*-wave velocity perturbations with respect to PREM [*Dziewonski and Anderson*, 1981]. Similarly to earlier studies [*Dziewonski et al.*, 1996], the model was expanded using basis functions

$$\delta v_S(r,\theta,\phi)/v_0(r) = \sum_{k=0}^{K}\sum_{\ell=0}^{L}\sum_{m=0}^{\ell} f_k(r) \times p_\ell^m(\cos\theta)(\,_kA_\ell^m \cos m\phi + \,_kB_\ell^m \sin m\phi), \tag{21}$$

where $v_0(r)$ is the PREM reference velocity, $f_k(r)$ are normalized Chebyshev polynomials, $p_\ell^m(\cos\theta)$ the associated Legendre polynomials, $\theta$ is colatitude, and $\phi$ is longitude. The upper and lower mantle are described by separate sets of coefficients, and the maximum degree in the radial expansion is $K = 7$ for the upper mantle and $K = 5$ for the lower mantle. In the horizontal direction,

Table 1. Data Used in Deriving the Anisotropic Model S20A

| Data Type | Observations | Sensitivity | Variance Reduction |
|---|---|---|---|
| Travel Times | | | |
| *SS–S* | 5,124 | *SV* | 72% |
| *ScS–S* | 3,543 | *SV* | 61% |
| *S–SKS* | 3,567 | *SV* | 55% |
| *SKKS–SKS* | 2,232 | *SV* | 32% |
| *S* | 26,462 | *SV* | 37% |
| *SS* | 11,417 | *SV* | 39% |
| *ScS* | 4,422 | *SV* | 49% |
| *ScSScS* | 1,230 | *SV* | 40% |
| Long-Period Waveforms | | | |
| *Body waves*, $T > 45$ s (Z, L) | 13,997 | *SV* | 28% |
| *Mantle waves*, $T > 85$ s (Z, L) | 5,523 | *SV* | 55% |
| *Mantle waves*, $T > 85$ s (T) | 2,665 | *SH* | 59% |
| *Mantle waves*, $T > 135$ s (Z, L) | 4,422 | *SV* | 40% |
| *Mantle waves*, $T > 135$ s (T) | 1,710 | *SH* | 47% |
| *Mantle waves*, $T > 200$ s (Z, L) | 4,978 | *SV* | 14% |
| *Mantle waves*, $T > 200$ s (T) | 1,853 | *SH* | 23% |
| Surface Wave Dispersion | | | |
| *Rayleigh (R1)*, 35–150 s | 27,948–37,240 | *SV* | 68–90% |
| *Love (G1)*, 35–150 s | 15,189–23,947 | *SH* | 76–96% |
| *Rayleigh (R1, R2)*, 150–300 s | 18,295–10,256 | *SV* | 30–71% |
| *Love (G1, G2)*, 150–300 s | 9,143–3,830 | *SH* | 45–73% |

the maximum spherical harmonic degree is $L = 20$, for both the upper and lower mantle.

Table 1 lists the four types of data used in the model inversion: (1) Absolute and differential travel times measured from digital seismograms using a cross correlation technique [*Dziewonski et al.*, 1996]. (2) Complete long-period waveforms in several frequency bands. These data are similar to the data used in the early Harvard tomographic models of the upper mantle [*Woodhouse and Dziewonski*, 1984], but the data set has been much expanded. (3) Surface wave dispersion measurements [*Ekström et al.*, 1997] in the period range 35–150 s. (4) Additional long-period (150–300 s) Love and Rayleigh wave dispersion measurements. The datasets of greatest importance for resolving the anisotropy of the upper mantle are the measurements of Love and Rayleigh wave dispersion (3 and 4).

*Ekström and Dziewonski* [1996] found that Love and Rayleigh wave dispersion data could not simultaneously be fit using only isotropic variations in $v_S$, suggesting that variations in radial anisotropy may be significant. Five elastic parameters are required to describe a radially anisotropic (transversely isotropic) medium [*Love*, 1927]. Using the theory developed by *Takeuchi and Saito* [1972] and *Woodhouse and Dziewonski* [1984], one could attempt an inversion for all five elastic constants, $A$, $C$, $F$, $L$ and $N$ in the notation of *Love* [1927], but the data have only limited independent sensitivity to the first three parameters. Instead, *Ekström and Dziewonski* [1998] chose to model 3-D variations in $v_{SH} = (N/\rho)^{1/2}$ and $v_{SV} = (L/\rho)^{1/2}$. For simplicity, average isotropic sensitivity kernels for $\delta v_S$ calculated for anisotropic PREM were used in the inversion. The sensitivity of Rayleigh waves to independent variations in $v_{PH}$ and $v_{PV}$ were not considered in the analysis, but in the calculation of the sensitivity kernels an implicit variation in P-wave velocity such that $\delta v_P / v_P = 0.55\, \delta v_S / v_S$ was assumed.

In order to constrain the $v_{SH}$ and $v_{SV}$ velocities independently in the upper mantle, the data sets were divided into two groups: one with primary sensitivity to $v_{SH}$ and a second with primary sensitivity to $v_{SV}$. Travel time data were included in the $v_{SV}$ data set since all measured phases bottom in the lower mantle and have nearly vertical propagation paths in the upper mantle. Rayleigh wave dispersion and mantle and body wave data observed on the vertical and longitudinal components were also included in the $v_{SV}$ data set. Only fundamental mode Love wave dispersion and mantle wave data were included in the $v_{SH}$ data set.

The model coefficients (${}_kA_\ell^m$, ${}_kB_\ell^m$) were determined by formulating the inverse problem which minimizes the least-squares misfit between the observed data and corresponding model predictions. The data were inverted simultaneously for isotropic velocity perturbations in the lower mantle and independent $v_{SV}$ and $v_{SH}$ variations in the upper mantle. The inversion was damped towards a minimum model and smooth horizontal perturbations. The perturbations were also damped towards continuity across the 670 km discontinuity. No explicit radial damping was imposed in the inversion.

As discussed above, an important step in imaging the shallow mantle is the application of corrections for crustal structure. *Ekström and Dziewonski* [1998] corrected the Love and Rayleigh wave dispersion curves by subtracting the predicted effect of the crustal model CRUST-5.1 [*Mooney et al.*, 1998]. Travel time data are also corrected for this crustal model, as well as for topography and bathymetry. In addition, in order to account for the effects of azimuthal anisotropy, the Rayleigh wave dispersion data were corrected for the azimuthally varying terms of phase velocity before being included in the 3-D inversion.

It should be clear from this brief description that several approximations and assumptions were used in the construction of model S20A. In particular, the use of isotropic structural kernels could have a noticeable influence on the results. While it seems unlikely that the main features of S20A would change, it is likely that more details of the anisotropic structure of the upper mantle would be revealed by a more complete inversion.

### *6.2. A Section Through the Pacific Upper Mantle*

Plate 1 illustrates the most surprising finding from our anisotropic mantle model [*Ekström and Dziewonski*, 1998]. At 150 km depth below the center of the Pacific, $v_{SH}$ and $v_{SV}$ are significantly different from the PREM predictions, reflecting a large anomaly in radial anisotropy. In the rest of the world, the differences are much smaller, suggesting that the radial anisotropy in PREM [*Dziewonski and Anderson*, 1981] is a good global average.

Cross sections of the anisotropic and the isotropic heterogeneity of the Pacific upper mantle are shown in Plate 2. In the isotropic section, some familiar features are apparent. The subducting lithosphere beneath the Mariana Islands and South America is reflected in faster velocities. The spreading at the East Pacific Rise and behind the Mariana trench are seen as slow, shallow anomalies. The progressive cooling and thickening of the Pacific plate as it moves towards the northwest is also clear. A tongue of slow velocities extends from the East Pacific Rise towards the center of the Pacific, perhaps suggesting a connection with deep large-scale upwelling beneath the center of the plate.

The anisotropic cross section shows the very pronounced anomaly beneath the central Pacific in the depth range 125–200 km. The anisotropic structure is not very well resolved beneath 300 km, but it is intriguing that the regions in the transition zone in which $v_{SV}$ is greater than $v_{SH}$ are coincident with possible locations for vertical flow.

### *6.3. Shallow Mantle Structure Beneath Continents and Oceans*

The results of S20A show that anisotropy contributes significantly to the elastic heterogeneity in the mantle. This implies that isotropic velocity heterogeneity models which have been derived without consideration of anisotropy may have been biased. The isotropic velocity heterogeneity in the mantle is most likely related to variations in temperature and composition. From the model S20A we can calculate a Voigt average isotropic model which should represent this heterogeneity better than individual variations in $v_{SH}$ and $v_{SV}$:

$$\delta v_S^{\mathrm{Voigt}} = \frac{1}{3}(\delta v_{SH} + 2\delta v_{SV}). \tag{22}$$

Figure 6 shows average velocity profiles for various tectonic regions calculated from the isotropic portion of S20A, all referenced to the global average. Continents and oceans are seen to differ most from each other around 150 km depth, where the average difference is about 2%. Young oceans and stable continental interiors show the largest differences. It is interesting to note that the difference between stable continents and old oceans is essentially the same as that between all oceans and all continents. The profiles also suggest that the difference between continents and oceans may extend, at the ~1% level, to depths below 400 km. The presence of lithospheric roots of stable continental interiors is well established [*Jordan*, 1975]. If we choose a 1% fast $v_S$ limit as a seismological definition of a 'root', we conclude that the roots of the continents extend to an average depth of 250 km. Beneath some cratons, the 1% anomaly extends significantly deeper.

As pointed out by *Ekström and Dziewonski* [1998], the isotropic velocity anomaly $\delta v_S^{\mathrm{Voigt}}$ of S20A at 150 km depth correlates much better with the age of the sea

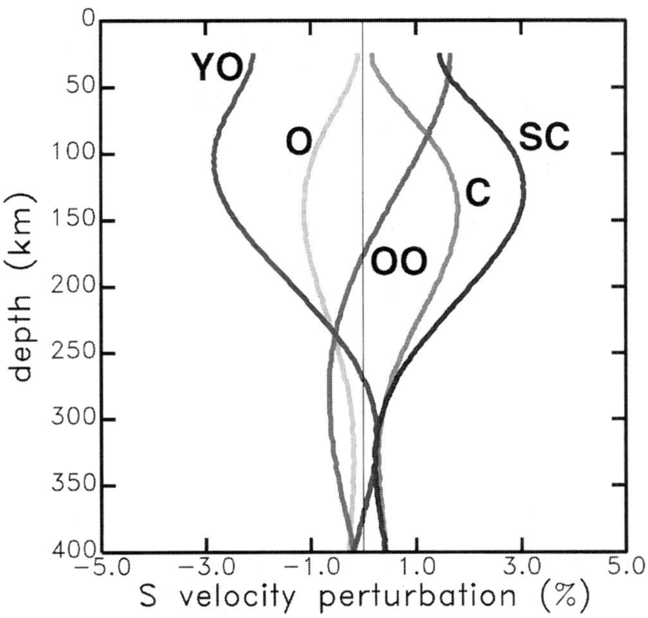


**Figure 6.** Average profiles of isotropic $S$ velocity perturbations for different tectonic areas. The profiles are referenced to the global average at each depth. O — ocean; OO — old ocean; YO — young ocean; C — continent; SC — stable continent.

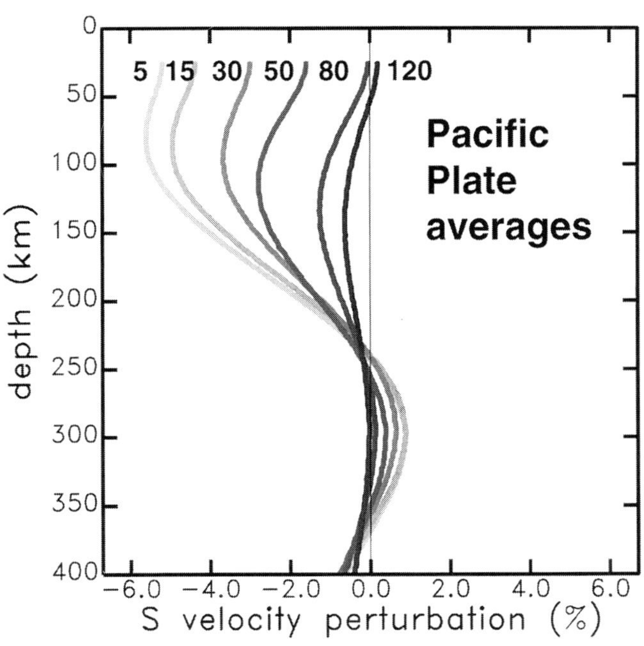


**Figure 7.** Average profiles of $S$ velocity perturbations in the Pacific mantle calculated for areas of different ages. The perturbations are calculated for the isotropic average model S20A(ISO) and have been references to the oldest sea floor (>140 Myr).

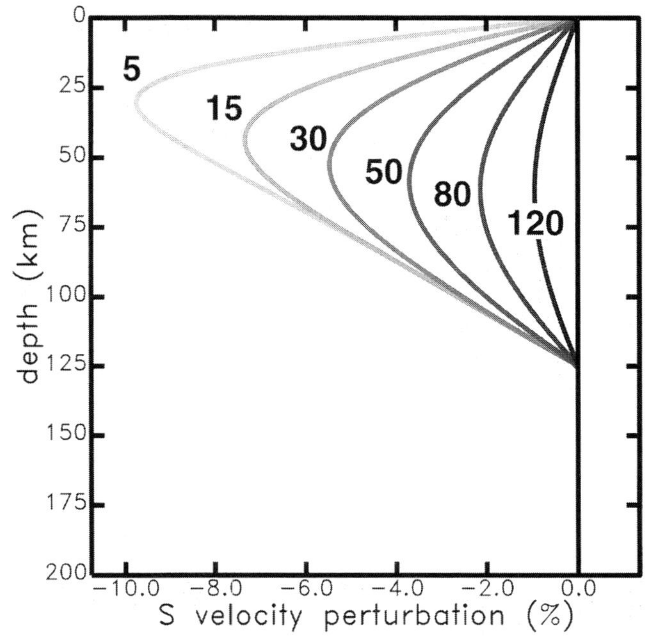


**Figure 8.** The predicted evolution of $S$ velocity with depth as a function of the age of the sea floor. The velocity perturbations have been referenced to a profile of 200 Myr age to be comparable with the observations. Note that no perturbations are predicted below 125 km depth, the assumed thickness of the plate. The parameters used in the calculation are: thickness of the plate, 125 km; bottom temperature, 1300°C; heat conductivity, $3.138 \times 10^{-5}$ K$^{-1}$; density, 3300 kg m$^{-3}$; heat capacity, 1.171 kJ kg$^{-1}$ K$^{-1}$; velocity derivative $(1/v_S)(\delta v_S/\delta T)$, $1.1 \times 10^{-4}$ K$^{-1}$.

floor than does $\delta v_{SV}$, which shows a pronounced slow velocity in the central Pacific. Figure 7 shows the average isotropic velocity profiles for different sea floor ages for the Pacific plate. A very clear age progression is seen in these profiles; the only major region in which the isotropic anomalies deviate substantially from the age progression is near the South Pacific Superswell [e.g., *McNutt*, 1998]. The surprising aspect of these profiles is the great depth extent of the variations. Figure 8 shows the predicted changes in $v_S$ based on a simple plate cooling model [*Parsons and Sclater*, 1977]. This plate model does not predict any seismic velocity variations beneath the base of the lithosphere at 125 km depth, while model S20A suggests variations to at least 225 km depth. While it is possible that the smooth polynomials used to parameterize the tomographic model may exaggerate the vertical extent of the velocity heterogeneity somewhat, it could not cause a doubling of the thickness of the anomaly. This result indicates that cooling extends below the plate, or that other mechanisms, such

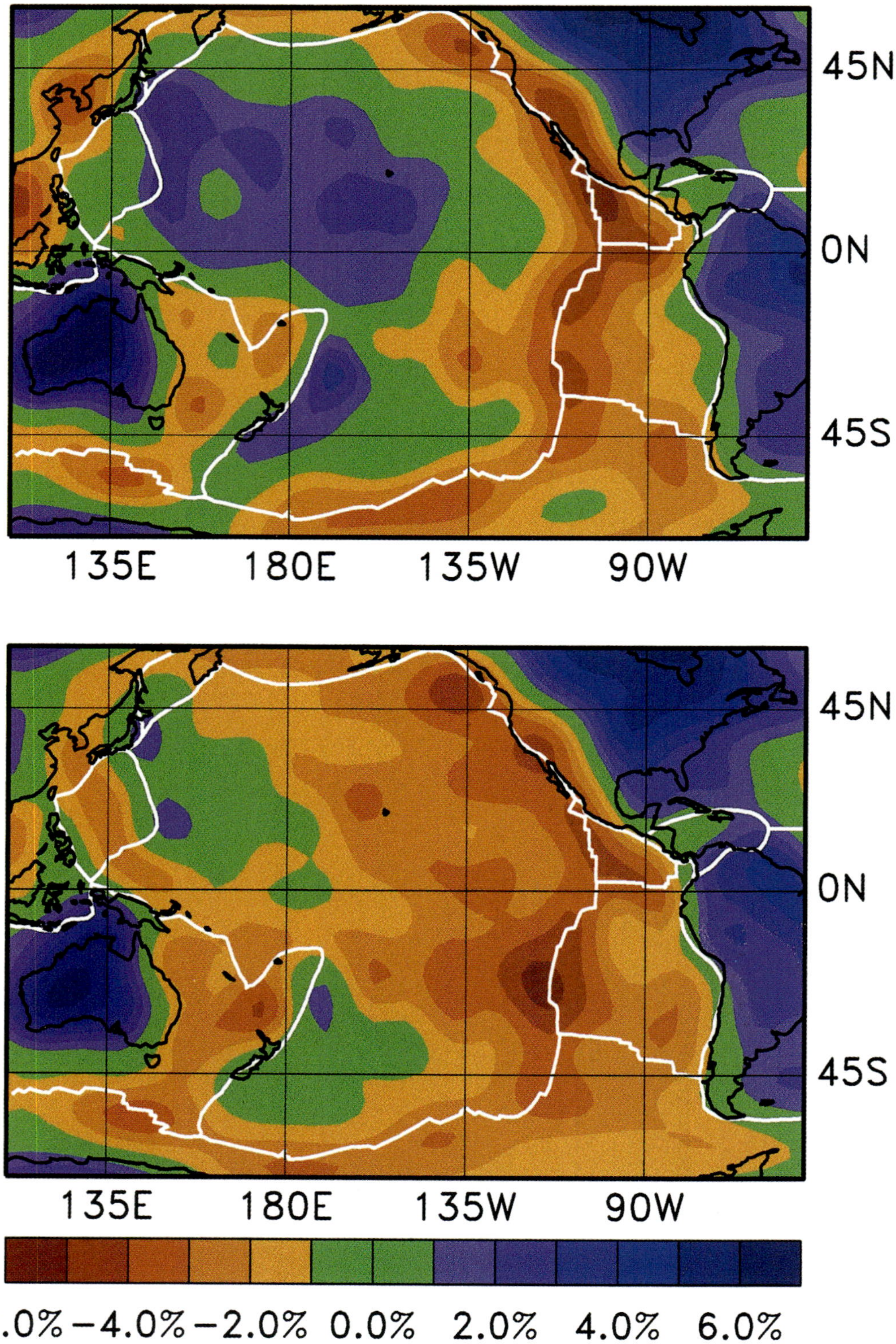


**Plate 1.** Maps showing velocity variations at 150 km depth for the Pacific. Top map shows $v_{SH}$ variations and bottom map shows $v_{SV}$ variations, both with respect to the values in the reference model PREM. In the central Pacific the difference between the two is greater than 5%.

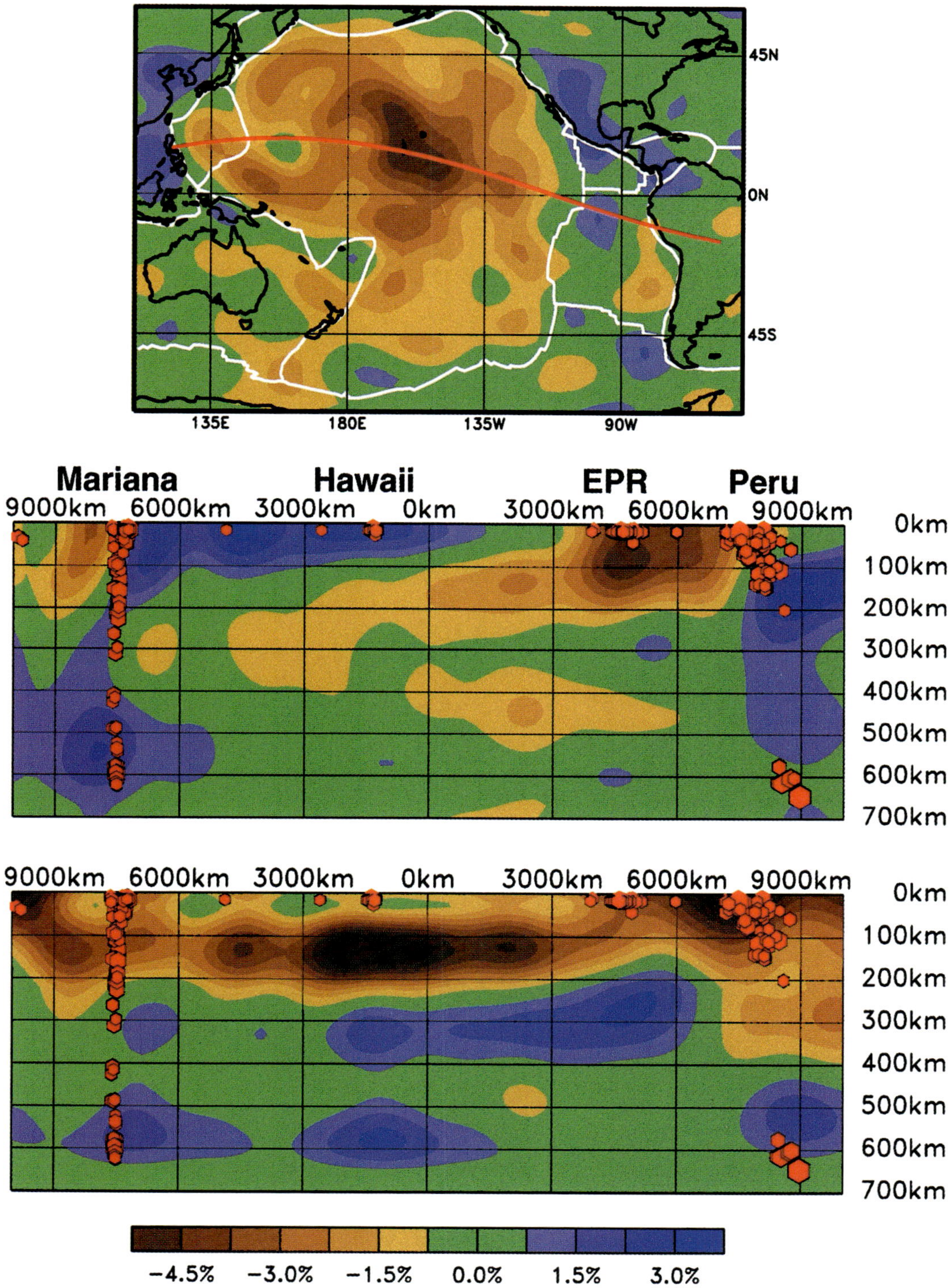


**Plate 2.** Cross sections through the model S20A. Top panel shows the radial anisotropy with respect to PREM at 150 km depth and the location of a cross section extending from the Philippines to Peru. The middle panel shows the isotropic velocity anomalies with respect to the global average for the top 700 km of the mantle. The seismicity is superimposed on the heterogeneity to illustrate the locations of the two subduction zones, the East Pacific Rise, and Hawaii. The bottom panel shows the full anisotropy, $(v_{SV} - v_{SH})/v_S$, including the radial anisotropy of PREM.

as the presence of partial melt, systematically affect the seismic velocities.

## 7. CONCLUSIONS

Recent measurements of Love and Rayleigh wave dispersion at intermediate periods have allowed a more detailed mapping of seismic velocities in the top few hundred kilometers of the mantle. The heterogeneity in this depth range is several times larger than that at other depths in the mantle, with the possible exception of the region above the core-mantle boundary. Results show that surface wave phase velocities cannot be explained simply by isotropic variations, but that anisotropy is large and variable in both the lithosphere and asthenosphere. The model S20A is a first attempt at interpreting the new surface wave data in terms of simple variations in radial anisotropy. More complete inversions, including 3-D variations in density, $v_P$, and azimuthal terms of anisotropy, are clearly possible.

*Acknowledgments.* I am grateful to Meredith Nettles, Jeannot Trampert and Rob van der Hilst for helpful comments on the manuscript. Model coefficients for the phase velocity maps of *Ekström et al.* [1997] and the 3-D velocity model of *Ekström and Dziewonski* [1998] are available on the World Wide Web: `http://www.seismology.harvard.edu/`. This work was funded by NSF grants EAR-92-19361 and EAR-98-05172 and AFOSR contract F49620-92-J-0392.

## REFERENCES

Agnew, D., J. Berger, R. Buland, W. Farrell, and F. Gilbert, International deployment of accelerometers: A network for very long period seismology, *Eos Trans. AGU*, *57*(4), 180–188, 1976.

Capon, J., Analysis of Rayleigh wave multipath propagation at LASA, *Bull. Seismol. Soc. Am.*, *60*, 1701–1731, 1970.

Chen, L., Booker, J. R., Jones, A. G., Wu, N., Unsworth, M. J., Wei, W., and Tan, H. Electrically conductive crust in southern Tibet from INDEPTH magnetotelluric surveying, *Science*, *274*, 1694–1696, 1996.

Dahlen, F. A., and J. Tromp, *Theoretical Global Seismology*, Princeton University Press, New Jersey, 1998.

Dorman, J., Seismic surface-wave data on the upper mantle, in *The Earth's Crust and Upper Mantle*, *Geophys. Monogr. Ser.*, vol. 13, edited by P. J. Hart, pp. 230–242, AGU, Washington, D. C., 1969.

Dziewonski, A. M., Upper mantle models from "pure path" dispersion data, *J. Geophys. Res.*, *76*, 2587–2601, 1971.

Dziewonski, A. M., and D. L. Anderson, Preliminary Reference Earth Model (PREM), *Phys. Earth Planet. Inter.*, *25*, 289–325, 1981.

Dziewonski, A. M., and J. M. Steim, Dispersion and attenuation of mantle waves through waveform inversion, *Geophys. J. R. Astron. Soc.*, *70*, 503–527, 1982.

Dziewonski, A. M., G. Ekström, and X.-F. Liu, Structure at the top and bottom of the mantle, in E. S. Husebye and A. M. Dainty (eds.), *Monitoring a Comprehensive Test Ban Treaty*, pp. 521–550, Kluwer Academic Publishers, Dordrecht, 1996.

Edmonds, A. R., *Angular Momentum in Quantum Mechanics*, Princeton Univ. Press, Princeton, N. J., 1960.

Ekström, G., J. Tromp, and E. W. F. Larson, Measurements and global models of surface wave propagation, *J. Geophys. Res.*, *102*, 8137–8157, 1997.

Ekström, G., and A. M. Dziewonski, Improved models of upper mantle *S* velocity structure (abstract), *Eos Trans. AGU*, *76(*46), Fall Meet. Suppl., F421, 1995.

Ekström, G., and A. M. Dziewonski, Imaging upper mantle S velocity structure: compatibility of different data sets (abstract), *Eos Trans. AGU*, *77(*46), Fall Meet. Suppl., F483, 1996.

Ekström, G., and A. M. Dziewonski, The unique anisotropy of the Pacific upper mantle, *Nature*, *394*, 168–172, 1998.

Evernden, J., Direction of approach of Rayleigh waves and related phenomena, II, *Bull. Seismol. Soc. Am.*, *44*, 159–184, 1954.

Forsyth, D. W., The early structural evolution and anisotropy of the oceanic upper mantle, *Geophys. J. R. Astron. Soc.*, *43*, 103–162, 1975.

Jordan, T. H., The continental tectosphere, *Rev. Geophys.*, *13*, 1–12, 1975.

Kanamori, H., Velocity and $Q$ of mantle waves, *Phys. Earth Planet. Inter.*, *2*, 259–275, 1970.

Kind, R., Ni, J. , Zhao, W., Wu, J., Yuan, X., Zhao, L., Sandvol, E., Reese, C., Nábělek, J., and, T. Hearn, Evidence from earthquake data for a partially molten crustal layer in southern Tibet, *Science*, *274*, 1692–1694, 1996.

Larson, E. W. F., J. Tromp, and G. Ekström, Effects of slight anisotropy on surface waves, *Geophys. J. Int.*, *132*, 654–666, 1998.

Laske, G., Global observations of off-great-circle propagation of long-period surface waves, *Geophys. J. Int.*, *123*, 245–259, 1995.

Laske, G., and G. Masters, Constraints on global phase velocity maps from long-period polarization data, *J. Geophys. Res.*, *101*, 16,059–16,075, 1996.

Lay, T., and H. Kanamori, Geometric effects of global lateral heterogeneity on long-period surface wave propagation, *J. Geophys. Res.*, *90*, 605–621, 1985.

Love, A. E. H., A treatise on the theory of elasticity, Cambridge University Press, 4th edition, 643 pp., 1927.

McNutt, M. K., Superswells, *Rev. Geophys.*, *36*, 211–244, 1998.

Masters, G., T. H. Jordan, P. G. Silver, and F. Gilbert, Aspherical Earth structure from fundamental spheroidal mode data, *Nature*, *298*, 609–613, 1982.

Montagner, J.-P., and T. Tanimoto, Global anisotropy in the upper mantle inferred from the regionalization of phase velocities, *J. Geophys. Res.*, *95*, 4794–4819, 1990.

Montagner, J.-P., and T. Tanimoto, Global upper mantle tomography of seismic velocities and anisotropies, *J. Geophys. Res.*, *96*, 20,337–20,351, 1991.

Mooney, W. D., G. Laske, and G. Masters, CRUST-5.1: A global crustal model at 5° × 5°, *J. Geophys. Res.*, *103*, 727–747, 1998.

Nakanishi, I., and D. L. Anderson, World-wide distribution of group velocity of mantle Rayleigh waves as determined by spherical harmonic inversion, *Bull. Seismol. Soc. Am.*, *72*, 1185–1194, 1982.

Nakanishi, I., and D. L. Anderson, Measurements of mantle wave velocities and inversion for lateral heterogeneity and anisotropy, 1, Analysis of great circle phase velocities, *J. Geophys. Res.*, *88*, 10,267–10,283, 1983.

Nakanishi, I., and D. L. Anderson, Measurements of mantle wave velocities and inversion for lateral heterogeneity and anisotropy, 2, Analysis by the single station method, *Geophys. J. R. Astron. Soc.*, *78*, 573–617, 1984.

Nataf, H.-C., I. Nakanishi, and D. L. Anderson, Measurements of mantle wave velocities and inversion for lateral heterogeneities and anisotropy, 3, Inversion, *J. Geophys. Res.*, *91*, 7261–7307, 1986.

Nishimura, C. E., and D. W. Forsyth, Rayleigh wave phase velocities in the Pacific with implications for azimuthal anisotropy and lateral heterogeneities, *Geophys. J. Int.*, *94*, 479–501, 1988.

Oliver, J., A summary of observed seismic surface wave dispersion, *Bull. Seismol. Soc. Am.*, *52*, 81–86, 1962.

Parsons, B., and J. G. Sclater, An analysis of the variation of ocean floor bathymetry and heat flow with age, *J. Geophys. Res.*, *82*, 803–827, 1977.

Peterson, J., H. M. Butler, L. G. Holcomb, and C. R. Hutt, The Seismic Research Observatory, *Bull. Seismol. Soc. Am.*, *66*, 2049–2068, 1976.

Smith, M. L., and F. A. Dahlen, The azimuthal dependence of Love and Rayleigh wave propagation in a slightly anisotropic medium, *J. Geophys. Res.*, *78*, 3321–3333, 1973.

Snieder, R., and G. Nolet, Linearized scattering of surface waves on a spherical Earth, *J. Geophys.*, *61*, 55–63, 1987.

Sobel, P. A., and D. H. von Seggern, Applications of surface-wave ray tracing, *Bull. Seismol. Soc. Am.*, *68*, 1359–1380, 1978.

Takeuchi, H., and M. Saito, Seismic surface waves, *Methods Comput. Phys.*, *11*, 217–295, 1972.

Tams, E., Über die Fortplanzungsgeschwindigkeit der seismischen Oberflächenwellen längs kontinentaler und ozeanischer Wege, *Centralbl. Mineral. Geol. Paläontol.*, no. 2, 44–52, 1921a.

Tams, E., Über die Fortplanzungsgeschwindigkeit der seismischen Oberflächenwellen längs kontinentaler und ozeanischer Wege, *Centralbl. Mineral. Geol. Paläontol.*, no. 3, 75–83, 1921b.

Tanimoto, T., The Backus-Gilbert approach to the 3-D structure in the upper mantle, I, Lateral variation of surface wave phase velocity with its error and resolution, *Geophys. J. R. Astron. Soc.*, *82*, 105–123, 1986.

Tanimoto, T., and D. L. Anderson, Lateral heterogeneity and azimuthal anisotropy of the upper mantle: Love and Rayleigh waves 100–250 s, *J. Geophys. Res.*, *90*, 1842–1858, 1985.

Toksöz, M. N., and D. L. Anderson, Phase velocities of long-period surface waves and structure of the upper mantle, 1, Great circle Love and Rayleigh wave data, *J. Geophys. Res.*, *71*, 1649–1658, 1966.

Trampert, J., and J. H. Woodhouse, Global phase velocity maps of Love and Rayleigh waves between 40 and 150 seconds, *Geophys. J. Int.*, *122*, 675–690, 1995.

Trampert, J., and J. H. Woodhouse, High resolution global phase velocity distributions, *Geophys. Res. Lett.*, *23*, 21–24, 1996.

Tromp, J., and F. A. Dahlen, Variational principles for surface wave propagation on a laterally heterogeneous Earth, II, Frequency-domain JWKB theory, *Geophys. J. Int.*, *109*, 599–619, 1992.

Tromp, J., and F. A. Dahlen, Maslov theory for surface wave propagation on a laterally heterogeneous Earth, *Geophys. J. Int.*, *115*, 512–528, 1993.

Wong, Y. K., Upper mantle heterogeneity from phase and amplitude data of mantle waves, Ph.D. thesis, Harvard Univ., Cambridge, Mass., 1989.

Woodhouse, J. H., and A. M. Dziewonski, Mapping the upper mantle: Three-dimensional modeling of Earth structure by inversion of seismic waveforms, *J. Geophys. Res.*, *89*, 5953–5986, 1984.

Woodhouse, J. H., and Y. K. Wong, Amplitude, phase and path anomalies of mantle waves, *Geophys. J. R. Astron. Soc.*, *87*, 753–773, 1986.

Wu, F. T., Mantle Rayleigh wave dispersion and tectonic provinces, *J. Geophys. Res.*, *77*, 6445–6453, 1972.

Zhang, Y.-S., and T. Lay, Global surface wave phase velocity variations, *J. Geophys. Res.*, *101*, 8415–8436, 1996.

---

G. Ekström, Department of Earth and Planetary Sciences, Harvard University, 20 Oxford Street, Cambridge, MA 02138. (e-mail: ekstrom@eps.harvard.edu)

# A Comparison Between Tomographic and Geodynamic Models of the Earth's Mantle

Charles Mégnin and Barbara Romanowicz

*UC Berkeley Seismological Laboratory, Berkeley, CA 94720*

We compute tomographic images of the short and long wavelength expansions of a 3-D spherical convection simulation to a resolution comparable to that of a recent seismic model of the whole mantle. The simulation incorporates realistic thermal effects associated with convection in the mantle, namely 20% bottom heating and phase transitions at 410 km and 670 km. A depth-dependent viscosity profile, and the history of plate motions starting in the mid-Mesozoic are also imposed. We show that tomography recovers the image of the convection simulation well, making it a useful tool to investigate physical parameters characteristic of mantle convection, in conjunction with geodynamic modeling. Our results indicate that the part of the temperature field corresponding to cold mantle heterogeneity in the convection simulation agrees relatively well with the fast velocities derived from tomography, but that current convection computations are lacking some key ingredients necessary to reproduce the character of upwellings correctly. Contingent upon the accuracy of the time dependence of the plate reconstruction, we place an upper bound of -4 $MPa.K^{-1}$ on the Clapeyron slope of the phase transition at 670 km. Comparisons between the images of the short and long wavelength model (developed up to degrees 32 and 12, respectively) provide insight into the spectral leakage in inversions of actual waveform data. We show that contamination of long-wavelength models by small-scale structure is present at all depths, increases with spherical harmonic degree, and has a peak magnitude of the order of 1% perturbation from *PREM*. Although this effect is small in absolute terms, it is significant on the scale of the models and casts doubts on the reliability of the smallest amplitude component of the heterogeneity pattern in long wavelength seismic images of the whole mantle.

## 1. INTRODUCTION

Three-dimensional images of the Earth's deep mantle obtained by seismic tomography provide us with increasingly accurate constraints on the large scale dynamics that drive plate motions. Since the earliest global tomographic models of the upper mantle [*Woodhouse & Dziewonski*, 1984; *Nataf et al.*, 1986], and even earlier regionalized models (e.g. [*Lévêque*, 1980; *Okal*, 1983]), it has been shown that the velocity structure in the uppermost mantle correlates with surface tectonics and that this correlation breaks down below about 250 km, with, in particular, a strong "degree 2" pattern in the transition zone [*Masters et al.*, 1982; *Souriau & Souriau*, 1983]. This pattern is shifted to the west with respect to the shallower signature at degree 2 of

The History and Dynamics of Global Plate Motions
Geophysical Monograph 121

plate tectonics, and is correlated with a degree 2 pattern in seismic attenuation [*Romanowicz et al.*, 1987; *Romanowicz*, 1990; *Montagner & Romanowicz*, 1993]. In the lowermost mantle, a large scale degree 2 pattern of geodynamical significance, correlated with that of the geoid, has also been found in the earliest studies [*Dziewonski et al.*, 1977; *Hager et al.*, 1985].

Over the years, as resolution in tomographic models has improved, greater details concerning the large scale flow in the mantle have been obtained. In particular, recent models with high resolution around subduction zones have allowed seismologists to image subducted lithospheric slabs through the upper mantle transition zone and into the lower mantle, and have shown a variety of situations, with slabs appearing to be lying flat at the upper-mantle/lower-mantle transition in some regions, and penetrating to large depths in the lower mantle in others. This has been shown both in regional studies (e.g. [*Spakman et al.*, 1989; *Zhou & Clayton*, 1990; *van der Hilst et al.*, 1991; *Fukao et al*, 1992; *van der Hilst & Seno*, 1993; *van der Hilst*, 1995]), and in whole mantle models (e.g. [*van der Hilst et al.*, 1997; *Grand*, 1994]). Because of a less favorable geometry in mid-oceanic areas, regions of upwellings, as expressed by low velocities in tomographic regions, are less well resolved, and, in particular, this has given rise to a vigorous debate regarding the continuity at depth of low velocities associated with mid-ocean ridges ( [*Zhang & Tanimoto*, 1992; *Su et al.*, 1992]).

While high resolution images increasingly will become available on a regional scale, global models of the whole mantle in which the structure is expressed as a truncated spherical harmonic expansion in the horizontal dimension provide useful constraints on the long wavelength characteristics of mantle circulation and how it relates to surface tectonics. Until recently, such models were limited to degree 12 (e.g. [*Su et al.*, 1994; *Li & Romanowicz*, 1996]), and are now progressively reaching higher degrees [*Masters et al.*, 1996; *Mégnin & Romanowicz*, 1997; *Liu & Dziewonski*, 1998; *Mégnin & Romanowicz*, 1999b], with the newest models showing good agreement in their low degrees (up to degree 8-12) with the previous generation. A common feature of these models is their prominent long wavelength component (*e.g.* [*Su & Dziewonski*, 1991]), with peaks at degree 2 in the transition zone and lowermost mantle, and predominance of degrees 1-8 in the upper mantle. These features provide strong constraints for numerical modeling of convection in the mantle, in that a variety of ingredients, such as level of heating from the bottom of the mantle, nature of the 670 km discontinuity, viscosity stratification, and boundary conditions at the top of the mantle, contribute to the spectral characteristics of the synthetic velocity models derived from numerical computations, which ultimately need to match the seismological results.

While good agreement between mantle models resulting from convection calculations and seismic tomography would go a long way towards improving our understanding of mantle dynamics, geodynamicists and seismologists have traditionally failed to produce compatible images, particularly the characteristic length scale of heterogeneity. Simple fluid dynamic models predict dominant wavelengths of the order of mantle depth [*Busse & Whitehead*, 1971]. This is in contradiction with both geoid observations (e.g. [*Lerch et al.*, 1982]) and images derived from seismic tomography (e.g. [*Tanimoto*, 1990; *Su et al.*, 1994; *Li & Romanowicz*, 1996; *Grand et al.*, 1997]) which find, as mentioned above, that the spectrum of heterogeneity in the mantle peaks at scales of 10,000 to 20,000 km. The disagreement remains with spherical axisymmetric convection simulations which predict a pattern of heterogeneity with a much broader spectrum (e.g. [*Machetel*, 1990]). The discrepancy was attributed to the presence of a filtering effect in the tomography and to computational limitations that did not permit the imaging of very fine scale mantle structure.

More recently, it was shown that, by introducing a depth dependent viscosity profile in the convection simulations [*Bunge et al.*, 1996], and more importantly by imposing plate motions [*Davies*, 1988; *Bunge & Richards*, 1996], a convective planform could be generated, characterized by the predominance of long wavelength circulation, which considerably improved the spectral agreement between geodynamic calculations and seismological models.

Although differences between seismic images of the mantle and convection calculations remain, the similarities now allow direct comparisons. In particular, the thermal effect of phase transitions on the convection planform in a realistic geometry produces spectrally localized features at transition zone depths. It is possible to use the convection models in conjunction with a realistic distribution of seismic data to estimate whether such features, if they exist, would be recovered in the inversion of real seismic data.

The spatial resolution of tomographic models and models resulting from convection calculations, however, are very different. Resolution in geodynamic models is controlled by the numerical mesh, (in the case of the convection model in this study, 10 million finite elements were used, corresponding to heterogeneity of the order of 50 km). In contrast, the resolution achieved by seismological models is limited by the filtering that the uneven data sampling, the regularization and the parametrization introduce in the inversion. For the purpose of comparison between seismological and numerical modelling, the recent "high resolution" models, as mentioned earlier, are accurate mostly in regions around subduction zones, so that long wavelength, global models are better suited for such comparisons. However, contamination of long wavelength seismic images by unmodeled small scale heterogeneity may result in a distorted model. Therefore, predicted seismic images of convection models must be computed by applying the same filtering process as has been applied to the real earth in the inversion of actual data. Such an approach allows to assess how a particular convection model would be "seen" through the blurred glass of global seismic tomography, and therefore to perform meaningful comparisons.

The issue of contamination of long wavelength tomographic models by small scale heterogeneity is central to the interpretation of seismic images in terms of mantle dynamics. By analyzing frequency shifts in the normal mode ${}_0S_{40}$, [*Snieder et al.*, 1991] have suggested that the presence of fine scale subduction zones and spreading ridge structures introduces appreciable spectral leakage into the even-degree expansion of the local frequency perturbation, up to spherical harmonic degree 6. However, the local frequency perturbations only represent the effects of the radially integrated earth structure. In the present study, we analyze how aliasing impacts the velocity models in different depth ranges using synthetic computations based on the combination of a recent numerical convection model [*Bunge et al.*, 1998] and the seismic sampling provided by a recent global tomographic study [*Li & Romanowicz*, 1996].

## 2. METHOD

The resolution of seismic models is generally tested by constructing synthetic input patterns such as 'checkerboards' and evaluating their recovery. In regional and global studies, more realistic synthetic structures, such as slabs, are also commonly evaluated (e.g. [*Spakman et al.*, 1989; *Fukao et al*, 1992; *van der Hilst et al.*, 1997]). [*Pulliam & Stark*, 1994] have also assessed the overall reliability of images from travel time tomography. Realistic numerical convection models provide the advantage of allowing to assess the resolution by seismic tomography of both large scale global features and localized heterogeneity, specific for example to subduction zones or hot upwellings, and to determine whether physical parameters such as the strength of phase transitions can be estimated.

Several studies have used tomographic sampling of convection simulations as an effective way to explore similarities between geodynamic and tomographic mantle models. [*Johnson et al.*, 1993] used the global distribution of available $SS-S$ and $ScS-S$ travel times to image spherical convection simulations of [*Tackley et al.*, 1994] and found general agreement between the spectra of the convection simulations and of the seismic models derived from these synthetic travel time data. [*Mégnin et al.*, 1997] argued that waveform tomography can successfully reproduce the spectral signature of various convection scenarii, each characteristic of a particular aspect of the physics of mantle convection. Their method consisted in applying the resolution matrix derived in the computation of the tomographic model *SAW12D* [*Li & Romanowicz*, 1996] to image eight convection calculations by [*Bunge et al.*, 1997]. Although convenient, this approach involves strong linear approximations and is not suitable for the analysis of short-wavelength contamination. An analysis of the aliasing of degrees 7-12 into a degree 6 model was attempted but the results proved strongly dependent on the choice of roughness penalty used in the modeling. It is important to note that such studies of convection calculations have so far focused on particular physical parameters corresponding to a dynamical mantle possibly very different from the actual earth. A new generation of 3-D spherical convection simulations of [*Bunge et al.*, 1998] that includes the principal thermal effects expected to be present in the mantle allows to make somewhat more meaningful comparisons between geodynamic and tomographic models, even though the currently available convection simulations only represent one step closer to a more realistic dynamic Earth. In the future, accurate modeling of mantle processes will require accounting for effects on the convective planform such as chemical and rheological stratification, partial melting, and temperature-dependent viscosity.

In this study, we consider the source-receiver distribution used in computing the whole mantle model *SAW12D*, consisting of 9,000 SH body wave seismograms low-pass filtered at 32s and 7,500 first and second orbit Love wavetrains, low-pass filtered at 80s. The body wave seismograms correspond to a collection of over 28,000 body phases, corresponding to $S$, $SS$, $S_3$,

Table 1. Body wave phases

| Body phase | Number of occurences |
|---|---|
| $S$ & $sS$ | 5,330 |
| $SS$ & $sSS$ | 6,578 |
| $S3$ & $sS3$ | 4,466 |
| $S4$ & $sS4$ | 2,354 |
| $ScS$ & $sScS$ | 4,226 |
| $ScS2$ & $sScS2$ | 3,098 |
| $ScS3$ & $sScS3$ | 132 |
| $S_{\text{diff}}$ & $sS_{\text{diff}}$ | 1,039 |
| Unidentified phases | 854 |
| Total | 28,077 |

$S_4$, $ScS$, $ScS_2$, $ScS_3$, $S_{\text{diff}}$ and the corresponding depth phases ($sS$, etc...). They are catalogued in Table 1. Modeling of waveforms with accurate body wave kernels [*Li & Tanimoto*, 1993; *Li & Romanowicz*, 1995] gives access to information contained in many phases, including those that arrive close together in time, thus providing improved sampling compared to single phase travel time delays.

We compute synthetic waveforms corresponding to the actual source-station configuration, in a predefined geodynamic model and invert them for up to spherical harmonic degree 12. The synthetic waveform data set and its subsequent inversion are computed using the 'nonlinear asymptotic coupling theory' (*NACT*) developed by [*Li & Romanowicz*, 1995]. This formalism takes into account the coupling of multiplets along and across mode branches, a prerequisite for the correct modeling of body waveforms [*Li & Tanimoto*, 1993]. Because of the zeroth order asymptotic expansion of the coupling terms, the sensitivity of the waveforms is restricted to the two-dimensional structure along the great-circle path [*Romanowicz*, 1987]. Resolution in the lowermost mantle is improved by a data selection scheme that isolates the different energy wavepackets of a given seismic trace [*Li & Romanowicz*, 1996]. This enhances the low amplitude phases, in particular the $S$ diffracted along the core-mantle boundary ($S_{\text{diff}}$) which constitutes the primary source of information on lowermost mantle heterogeneity [*Mégnin & Romanowicz*, 1999a].

All the models in this study are parametrized radially in Legendre polynomials up to degree 5 in the upper mantle and 7 in the lower mantle, and laterally in spherical harmonics, up to degree 12 for *SAW12D*, and up to degree 32 for the convection model.

### 2.1. *Temperature to velocity scaling*

In the absence of chemical heterogeneity, the shear velocity perturbation $\delta \ln V_s$ is obtained from the temperature field $\delta T$ from *GEMLAB 1* through the relation:

$$\delta \ln V_s = \frac{\alpha}{2}(1 - g)\,\delta T \qquad (1)$$

where $\alpha$ is the thermal expansivity $-\frac{\partial \ln \rho}{\partial T}|_P$, $g$ is the second Grüneisen parameter $\frac{\partial \ln \mu}{\partial \ln \rho}|_P$, $\mu$ the shear modulus, and P the pressure field.

In this study, we use as reference the thermal expansivity profiles of the spherically symmetric model *PREM* [*Dziewonski & Anderson*, 1981]. The parameter g is poorly constrained and we assign it the constant value of 4 throughout the mantle, which is within the range of estimates of [*Anderson, 1989*, p. 120]. Equation 1 thus reduces to

$$\delta \ln V_s(r,\theta,\phi) = \frac{\alpha(r)}{2}(1 - g)\,\delta T(r,\theta,\phi) \qquad (2)$$

### 2.2. *Aliasing of small-scale structure into long wavelength models*

We compute a set of seismograms for the input model truncated at degree 12 and invert for shear velocity perturbations from *PREM* up to degree 12. The resulting model is 'unaliased' in the sense that it is not contaminated by small scale heterogeneity (*i.e.* by structure at shorter wavelengths than degree 12). The difference between the unaliased and the input models results from the combination of several effects: 1) Gaps in spatial coverage; 2) Varying degree of sensitivity of the waveforms to the wavelength of structure (for instance very long period surface waves aren't sensitive to very small scale heterogeneity); 3) Inappropriate choice of tuning parameters and regularization in the inversion; 4) The parametrization, and, in particular, 5) possible artifacts due to separate vertical parametrization of the upper and lower mantle. In the present study, the same theory is used in the forward and the inverse problem. This analysis therefore cannot provide insight into the distortion caused by approximations in the formalism. The effect of modeling the waveforms with an asymptotic formalism is discussed in [*Clévédé et al.*, 1999] in the context of upper mantle tomography from long-period surface waveforms.

We compute a second set of synthetic seismograms with the same source-receiver distribution as the former, but this time through the short wavelength expansion of the model (degrees 0-32) and invert again for heterogeneity up to degree 12. In this respect, the procedure is analogous to the inversion of actual data as it allows us to examine possible aliasing of degree 13-32 structure into the degree 0-12 model. Hereafter, we refer to that model as the 'aliased' model.

We refer to the difference between the unaliased and the input model as the 'inherent' error in the model. In the absence of error arising from misjudgement in the selection of roughness penalty, the unaliased model is indeed the best possible model that can be derived with our methodology, given the data distribution, its frequency band and the model parametrization.

Although the effect of spectral leakage is clearly present on a given seismogram, its impact on velocity models may be modest because data averaging tends to remove incoherent structure. We estimate the amount of aliasing in tomographic images by subtracting the heterogeneity of the unaliased model from that of the aliased model. The resulting 3-D image then represents the spectral leakage of degrees 13-32 into the aliased model. This measure only represents an approximation of the aliasing in inversions of actual data, since the mantle clearly contains heterogeneity of finer scale than used for the purpose of this study. Although *GEMLAB 1* was developed up to degree 400, we truncated it to degree 32 due to computational limitations. The spectrum of *GEMLAB 1* is dominated by the long wavelengths, allowing us to model the first order effect of the contamination of the image from small-scale heterogeneity.

### 2.3. *Inversion procedure: tuning the regularization*

We use the same inversion procedure as that of [*Li & Romanowicz*, 1996]. A detailed description is given in that paper and we present here only the aspects relevant to this study:

The inverse problem is solved by making use of a stochastic formulation [*Jackson*, 1979] and the finite-dimensional model is obtained by the recursive application of the following Newton scheme to the Fréchet derivatives of the $L^2$-norm objective function [*Tarantola & Valette*, 1982]:

$$\begin{aligned} m_{i+1} &= m_i + \left(\mathbf{A}_i^T \mathbf{C_d}^{-1} \mathbf{A}_i + \mathbf{C_m}^{-1}\right)^{-1} \\ &\times \left(\mathbf{A}_i^T \mathbf{C_d}^{-1} \left(\mathbf{d} - \mathbf{u}(m_i)\right) - \mathbf{C_m}^{-1} m_i\right) \end{aligned} \quad (3)$$

Where $\mathbf{d}$ is the data, $\mathbf{u}(m_i)$ the synthetic seismogram evaluated for model $m_i$ and $\mathbf{A}_i$, the matrix of partial derivatives with respect to the parameters of model $m_i$.

The component of the model covariance matrix $\mathbf{C_m^{het}}$ corresponding to perturbations in lateral heterogeneity $dlnV_s$ is constructed according to:

$$\begin{aligned} \mathbf{m}^T \quad \left(\mathbf{C_m^{het}}\right)^{-1} \mathbf{m} &= \int_{\oplus} \eta_1 m^2 + \eta_2 \left(\frac{\partial m}{\partial r}\right)^2 \\ + \quad \eta_3 \left(\frac{\partial^2 m}{\partial r^2}\right)^2 &+ \eta_4 \mid \nabla_1 m \mid^2 d\oplus \end{aligned} \quad (4)$$

where $\oplus$ is the volume of the mantle, and $\nabla_1$ the surface gradient operator. A similar expression applies to the perturbations in the topography of mantle discontinuities. The selection of damping parameters $\eta_i$ is arbitrary. In this study, the problem of determining the optimal set of damping parameters is posed as an inverse problem in which we seek to identify the $\eta_i$ that will minimize the Euclidian distance between the output model $m_{i+1}$ and the target input geodynamic model. To minimize the number of iterations, we use as starting model $m_i$ the target model truncated to degree 8. The optimal set of model norm, lateral derivative, and first and second radial derivatives is computed with a simulated annealing algorithm coupled with a downhill simplex method [*Press et al.*,1994].

## 3. CONVECTION MODEL GEMLAB 1

We use as input model the time-dependent 3-D spherical convection calculation 'Geodynamical Model of Los Alamos and Berkeley' (*GEMLAB 1*) derived by [*Bunge et al.*, 1998] and characterized by:

- Imposition of the plate motion history starting in the mid-Mesozoic (119 Ma).
- An exothermic phase transition at 410 km with Clapeyron slope of $2MPa.K^{-1}$.
- An endothermic phase transition at 670 km with Clapeyron slope of $-4MPa.K^{-1}$.
- A rheology whereby the lithosphere and the lower mantle are 40 times more viscous than the upper mantle.
- 20% heating from the core.

This model is arguably the most realistic numerical models available to date, in that it predicts a spectrum of heterogeneity peaking at low degrees. However, we must keep in mind that it is, for the purpose of our study, only a convenient proxy for the real Earth. The strength of the Clapeyron slope of the two discontinuities is still an object of controversy, as is the magnitude of the viscosity stratification between upper and lower mantle. In addition the viscosity contrast imposed between lithosphere and asthenosphere is probably several orders of magnitude larger in the Earth than in *GEMLAB 1*.

### 3.1. *Resolution analysis*

Two sets of synthetic seismograms are computed from *GEMLAB 1*: using its degree 32 expansion (Figure 1) and using its degree 12 expansion (Figure 2). The resulting 'data sets' are inverted for heterogeneity up to degree 12, producing an aliased (Figure 3) and unaliased (Figure 4) model, respectively. Although a damping of amplitude is observed, maps at given depths do not display noticeable spatial distortion of the input structure. Spectral heterogeneity maps (*SHMs*) of the models are shown in Figure 5 and 6. The plots represent the rms amplitude of each model as a function of wavelength (horizontal axis) and depth (vertical axis). Horizontal lines are traced at 410, 670 and 2600 km depths. Since amplitudes in the mantle are dominated by the top and bottom boundary layers, little can be seen at other mantle depths. In Figure 6, we normalize the amplitudes of Figure 5 by the rms at each depth. Figures 5 & 6 (a) show the first 12 degrees of the input model *GEMLAB 1* and correspond to Figure 2. The uppermost mantle is characterized by a spectrum with amplitudes at all degrees and dominated by peaks at degrees 1, 4 and 5 which are most prominent between 100 km and 150 km depth. The signature of the phase transitions is observed at degrees 2 to 5, 7 and 10 between 400 km and 700 km depth. The heterogeneity pattern in most of the lower mantle is dominated by the long wavelengths. The D" region displays a broad spectrum, peaking at degrees 2-3 and showing energy up to degree 10.

Figures 5 & 6 (b) are the spectral representation of the aliased model in Figure 3, resulting from the inversion for up to degree 12 of the synthetics traced through the degree 32 model. In the uppermost mantle, the low degree spectral peaks are well recovered although a slight loss of amplitude and radial smearing of the degree 6 heterogeneity is observed. In the bottom of the mantle, the spectral peaks at degrees higher than 6 in the input model are noticeably attenuated.

Figures 5 & 6 (c) correspond to the model of Figure 4 and represent the unaliased model resulting from the inversion for up to degree 12 of the synthetics traced through the long wavelength (degree 12) model. Here again, a modest amount of radial smearing is observed in the amplitude peaks of the upper mantle. The spectral character of the input model is nonetheless well preserved.

Radial smearing in the upper mantle is also apparent in the rms profiles of Figure 5 (d). The radial character of the rapid amplitude excursions associated with the plates and the phase transitions in the input model (solid line) cannot be resolved perfectly with 32s body waves: vertical averaging is present in both the aliased (dotted line) and the unaliased (dashed line) models.

Figures 5 & 6 (e) and 7 represent the inherent error obtained by subtracting the model in Figure 2 (*GEMLAB 1*, expanded up to degree 12) from the model in Figure 4 (its unaliased version). The contamination appears strongest in the uppermost mantle and the transition zone where the strength of the amplitude peaks is underpredicted and the peaks are spread radially. The amplitude of the error is nonetheless modest when compared to the scale of the models: the spurious heterogeneity introduced never exceeds 1%.

In Figure 8 (top), we plotted the degree by degree euclidian distance between the unaliased and the long wavelength input models, relative to the former. The distance for spherical harmonic degree $s$ between two models $m$ and $p$, expanded radially in polynomials up to degree kmax is defined as:

$$D_s \equiv 100 \times \left( \frac{\sum_{k=0}^{kmax} \sum_{t=0}^{2*s+1} (m_k^{st} - p_k^{st})^2}{\sum_{k=0}^{kmax} \sum_{t=0}^{2*s+1} (m_k^{st})^2} \right)^{\frac{1}{2}} \quad (5)$$

In Figure 8, the thick and thin lines correspond to upper and lower mantle depths, respectively. The horizontal bars represent the total distance between the models. As expected, the relative error is greater in the lower mantle and increases with decreasing wavelength. The peak distance occurs at degree 12 where it represents 25% of the model amplitude in the upper mantle and 50% in the lower mantle. The average distance is 17% in the upper mantle and 32% in the lower mantle.

Figures 9 and 5 & 6 (f) represent the spectral leakage in the degree 12 model, obtained by subtracting the model in Figure 3 from the model in Figure 4. The small scale structure component of the model (degrees 13-32) maps preferentially at high degrees and in the transition zone. We also note that the effect of spectral leakage on the model exceeds that of the inherent error. The maps at different depths show spurious heterogeneity that exceeds 1% in the transition zone. Figure

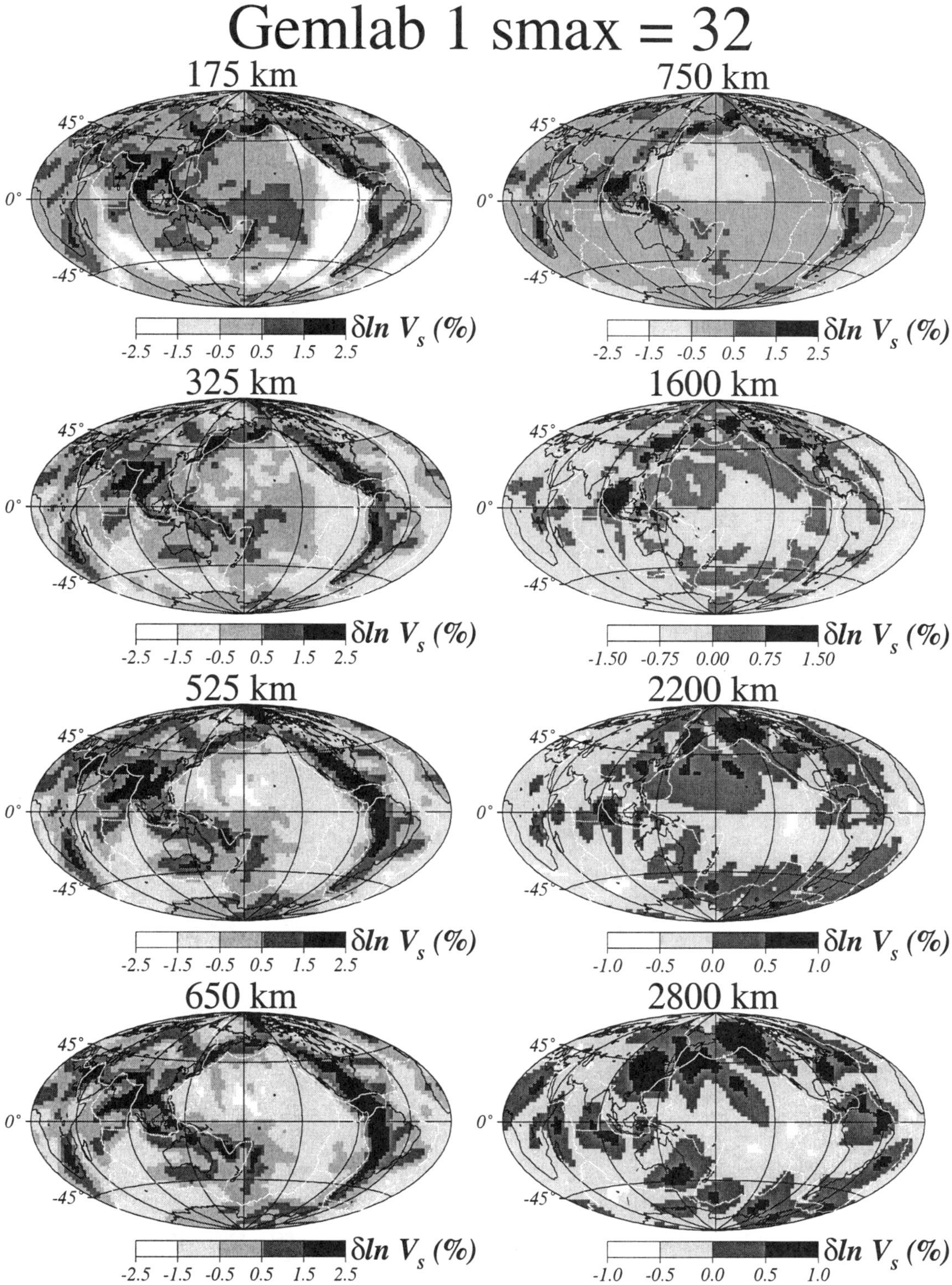


**Figure 1.** Convection model GEMLAB 1 expanded up to spherical harmonic degree 32.

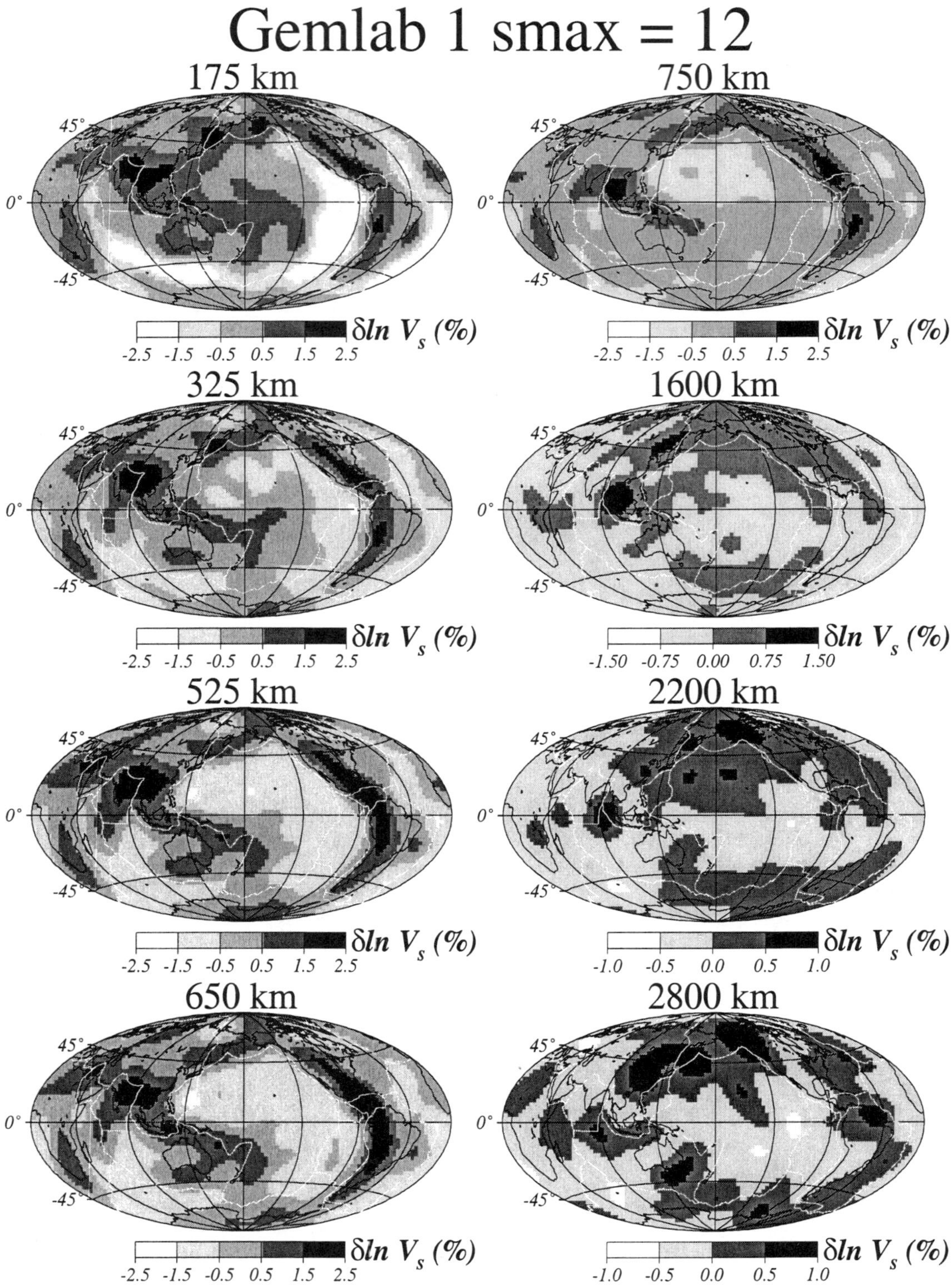


**Figure 2.** Convection model GEMLAB 1 expanded up to spherical harmonic degree 12.

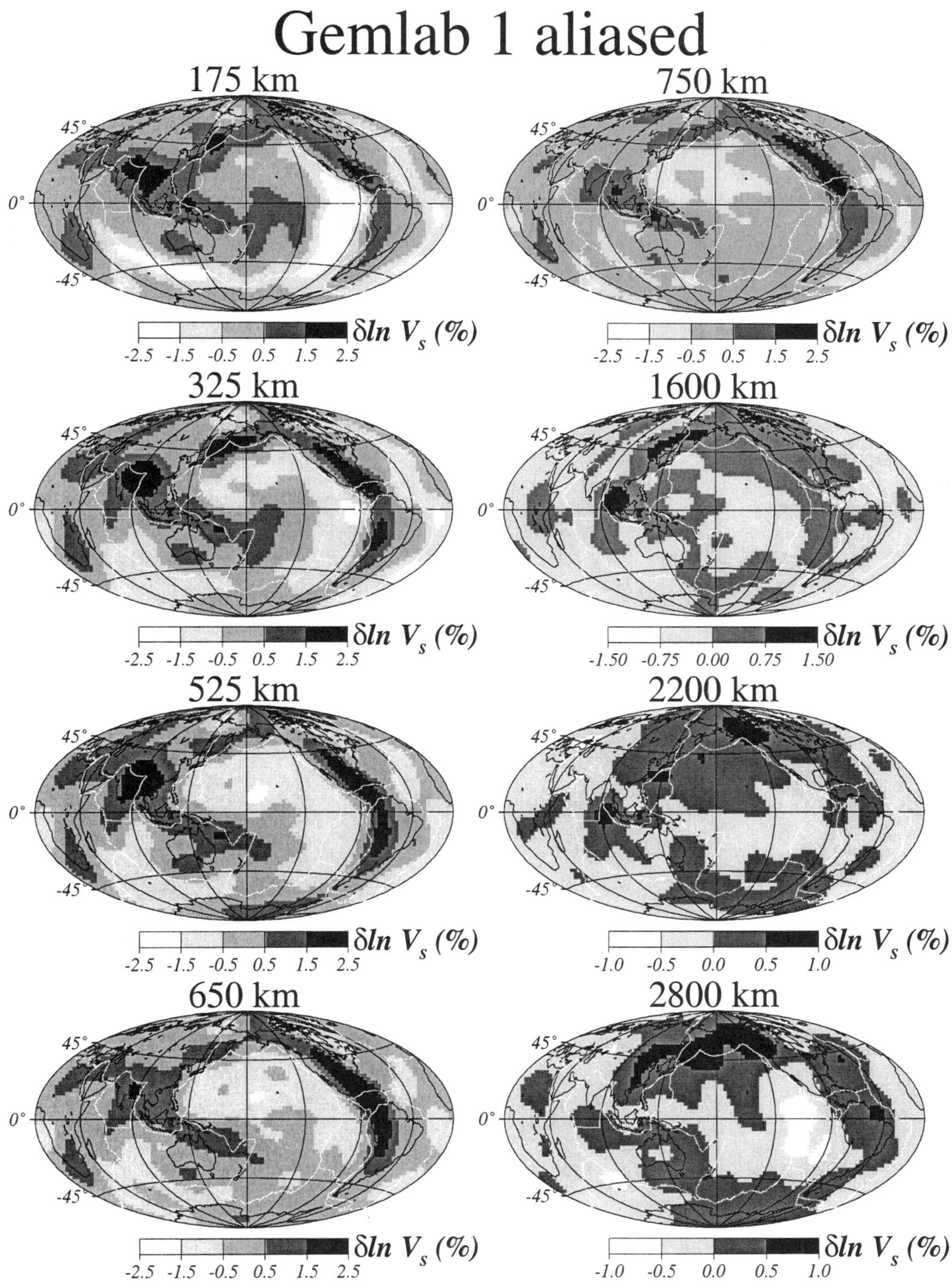


**Figure 3.** GEMLAB 1 aliased: Inversion of synthetics traced through the model in figure 1 for lateral heterogeneity up to degree 12.

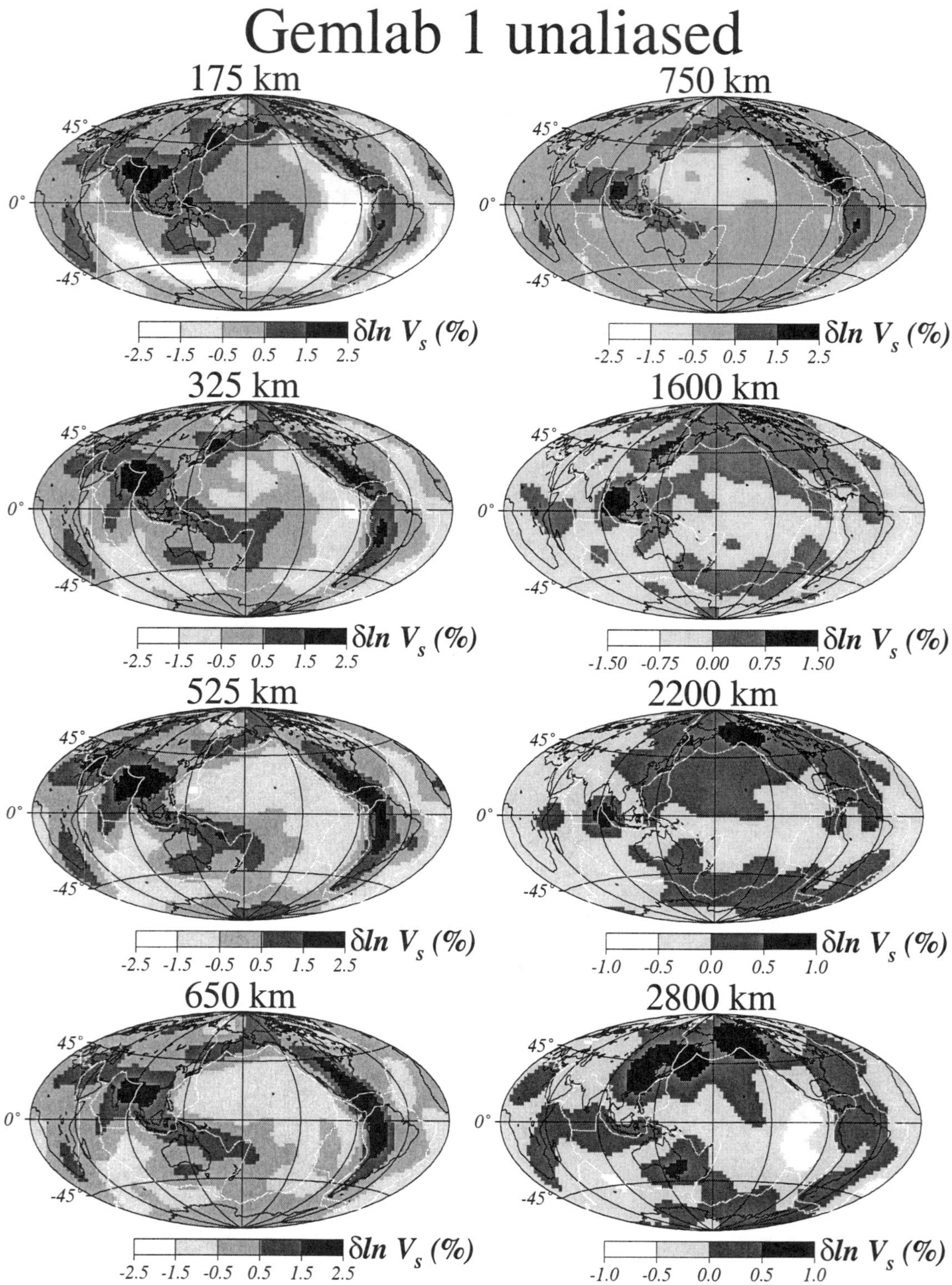


**Figure 4.** GEMLAB 1 unaliased: Inversion of synthetics traced through the model in figure 2 for lateral heterogeneity up to degree 12.

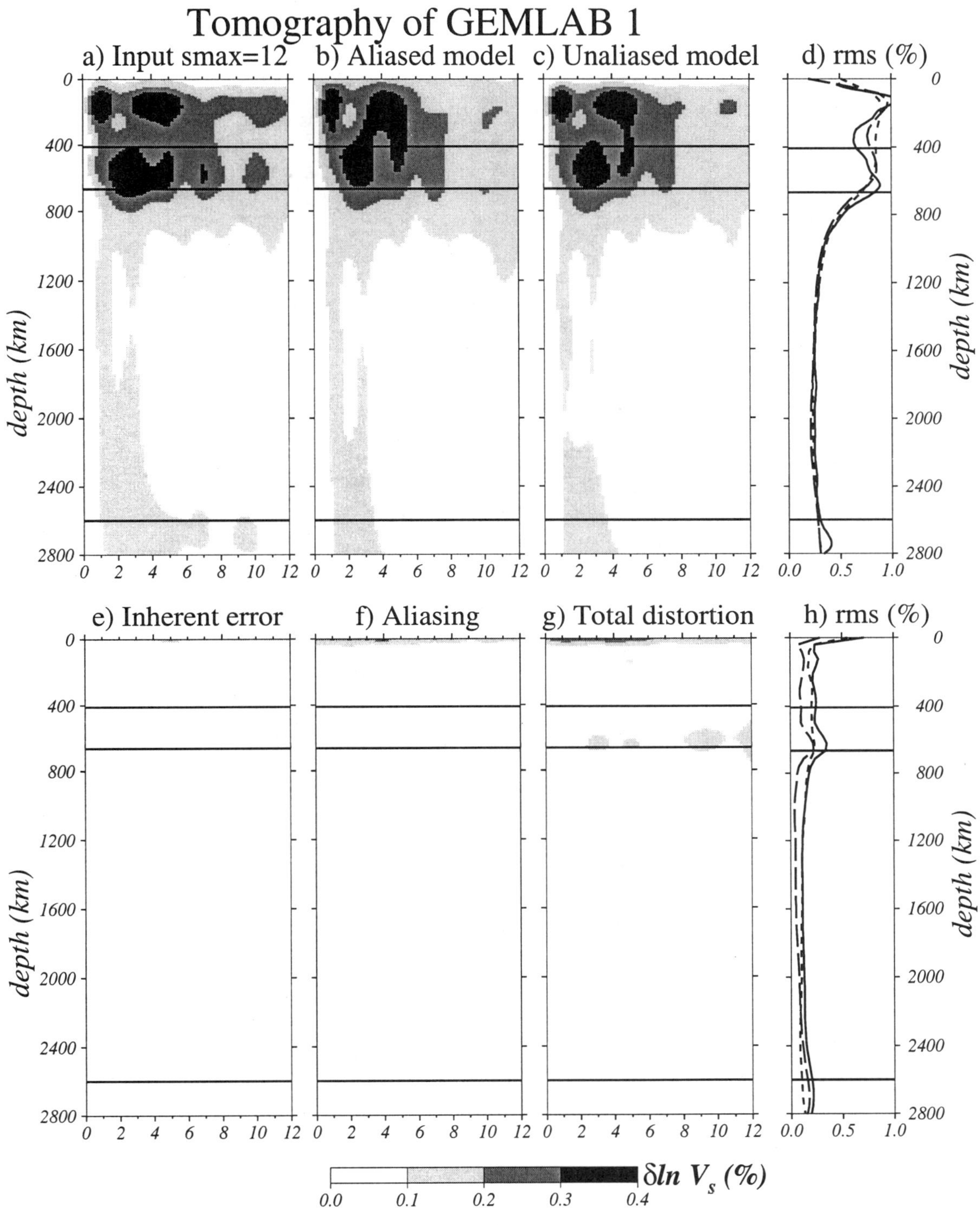


**Figure 5.** Spectral analysis of aliasing: a) Input model GEMLAB 1 (degrees 0-12). b) Inversion of synthetic data through GEMLAB 1 expanded up to degree 32. c) Inversion of synthetic data through GEMLAB 1 truncated at degree 12. d) rms profiles for the models in a) solid line -input model-, b) dotted line -aliased model-, c) dashed line -unaliased model-. e) Inherent error: difference between c) and a). f) Aliasing of degrees 13-32: difference between c) and b). g) Total distortion: difference between b) and a). h) rms profiles for e) dashed line -inherent error-, f) dotted line -aliasing-, and g) solid line -total distortion-.

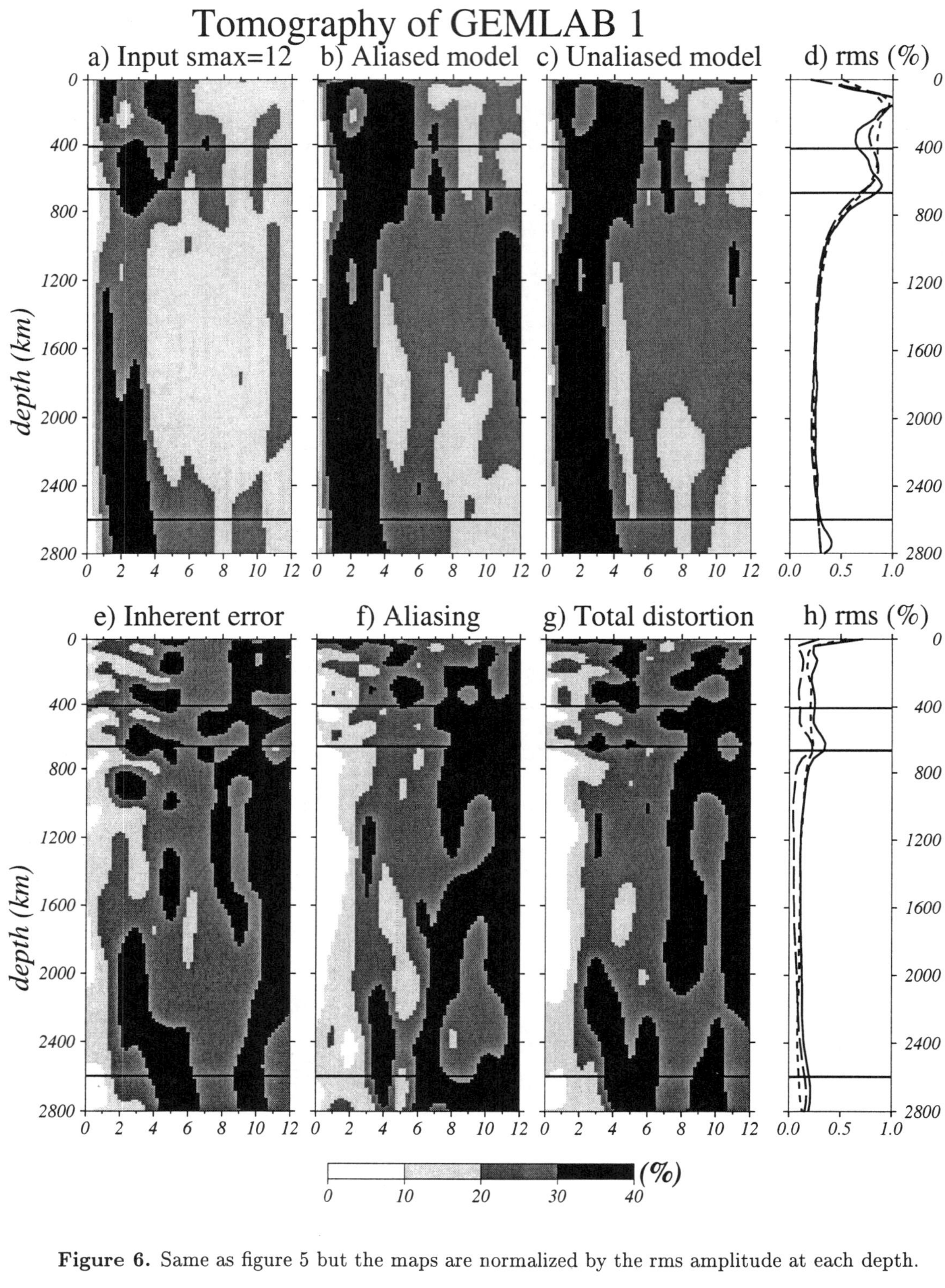


**Figure 6.** Same as figure 5 but the maps are normalized by the rms amplitude at each depth.

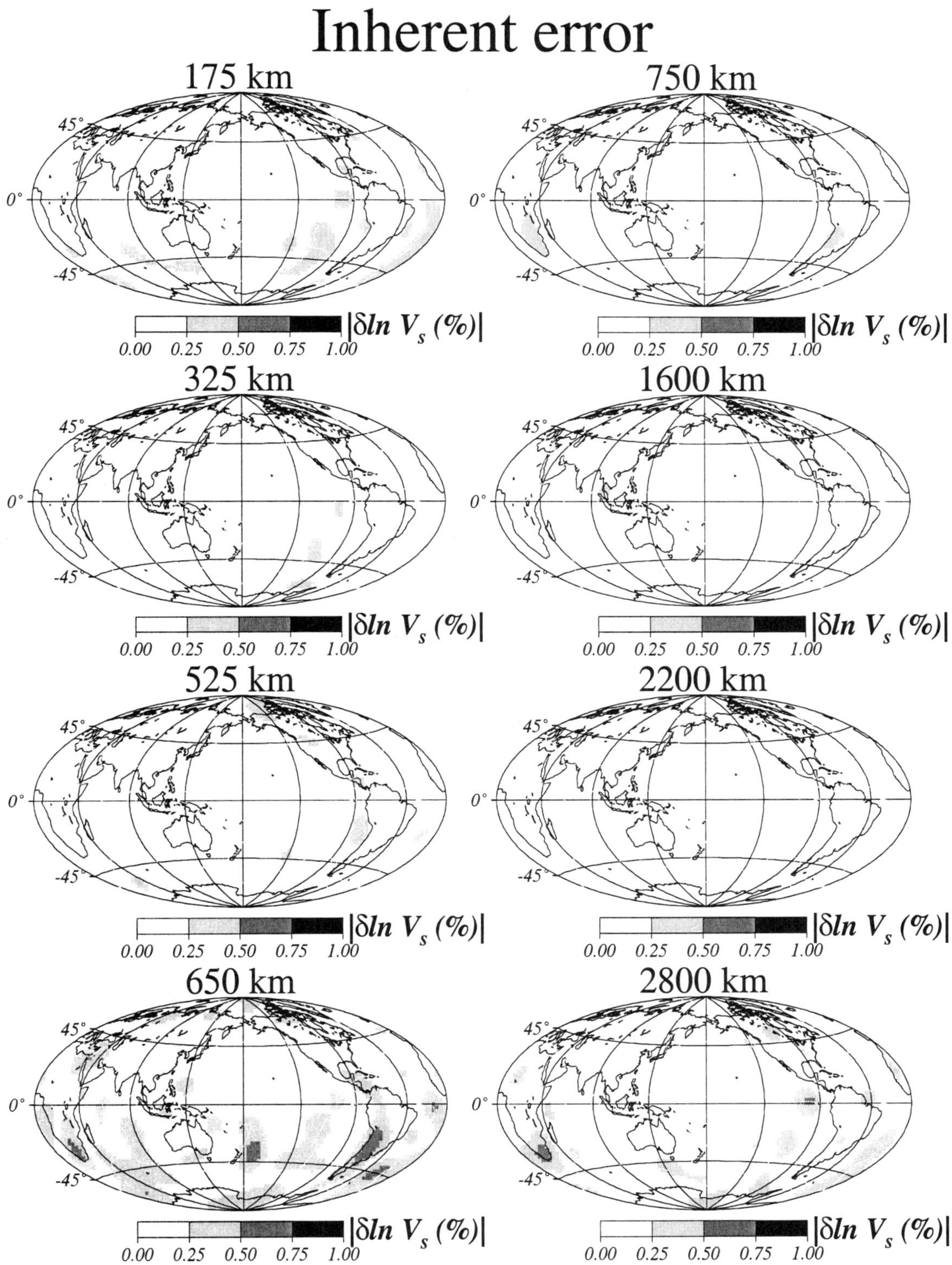


**Figure 7.** Spurious structure in the models caused by inherent error: absolute value of the difference between input model truncated at degree 12 and the unaliased model (Figures 5 & 6 e)).

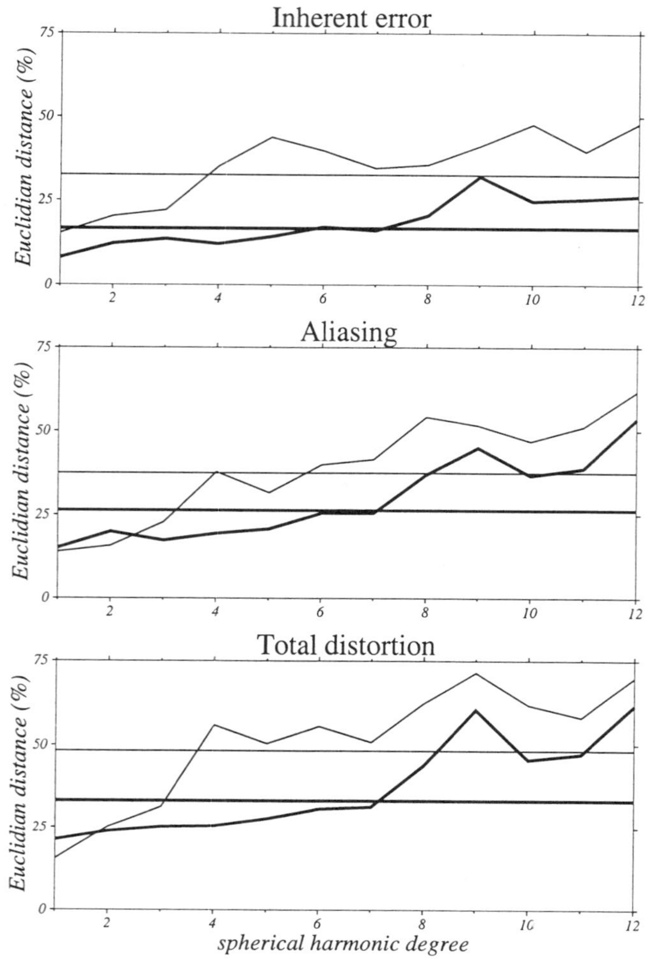


**Figure 8.** Top: Spherical harmonic degree by degree euclidian distance between unaliased and input model, relative to the unaliased model. Thick lines: upper mantle, thin lines: lower mantle. The horizontal lines represent the average distance. Middle: same as top, but for the difference between aliased and unaliased models. Bottom: Same, but for the difference between aliased and input model.

8 (middle) shows the dependence on wavelength of the spectral leakage, computed by taking the distance (5) between aliased and unaliased models. Here again, the peak distance occurs at degree 12, and is significantly stronger than the inherent error, representing 55% of model amplitude in the upper mantle and 70% in the lower mantle. The average distance is 28% in the upper mantle and 45% in the lower mantle.

Finally, Figures 5 & 6 (g) represent, in the absence of noise, the effects of the total error in the models, computed by subtracting model 1 from model 3. Radial smearing is noticeable at the surface and the amplitude of the error at high degrees near the 670 km discontinuity is comparable to that of the input structure (compare the peak at degree 10 in the transition zone in Figures 6 (a) and 6 (g)). Figure 8 (bottom) shows a distance between input and aliased models of 35% in the upper mantle and 50% in the lower mantle. As expected, the total error peaks at high degrees, where it represents 60% and 70% of model amplitude in the upper and the lower mantle respectively.

### *3.2. Comparison with tomography*

While *GEMLAB 1* only represents the current status of "realistic" convection computations, it is interesting to discuss how well its "aliased" version compares to a tomographic model obtained with real seismic data and the same geometry of paths and waveforms as used in the previous synthetic experiments. As mentioned earlier, the damping parameters used for the inversion of the aliased model are optimized for our source-receiver distribution and for waveforms that have sampled a mantle model containing short wavelengths. The prejudice they impose on the inversion is indeed that which best recovers degrees 0 through 12 of the synthetic input structure and therefore also represents the best we can do for the inversion of actual data for heterogeneity up to degree 12, under the assumption that recovering a model as close as possible to the convection model *GEMLAB 1* is a desirable goal.

Since model *SAW12D* was originally obtained using a different set of damping parameters, in order to compare more effectively the inversion results using actual data to the synthetic inversions performed here, we compute a new tomographic model based on our actual data set, using the optimized set of damping parameters determined in the calculation of the aliased *GEMLAB 1*. In this manner, we cannot completely confirm the reality of the features of the *GEMLAB 1* model that can be recovered through seismic tomography, but we can assess those features in the tomographic models that are significantly different from the predictions of the convection computation. An element of circularity is also present in calibrating smoothness constraints based on convection calculations made to match the tomography. To check whether this method could introduce features of the reference model where they do not exist, [*Mégnin et al.*, 1997] computed a set of synthetic seismograms in a convection model including a phase transition at 670 km, and derived a set of damping parameters adjusted to optimize the recovery of the input model. The inversion of this data set showed good restitution of the spectral footprint of the phase transition. A second set of synthetics was traced in another convection model

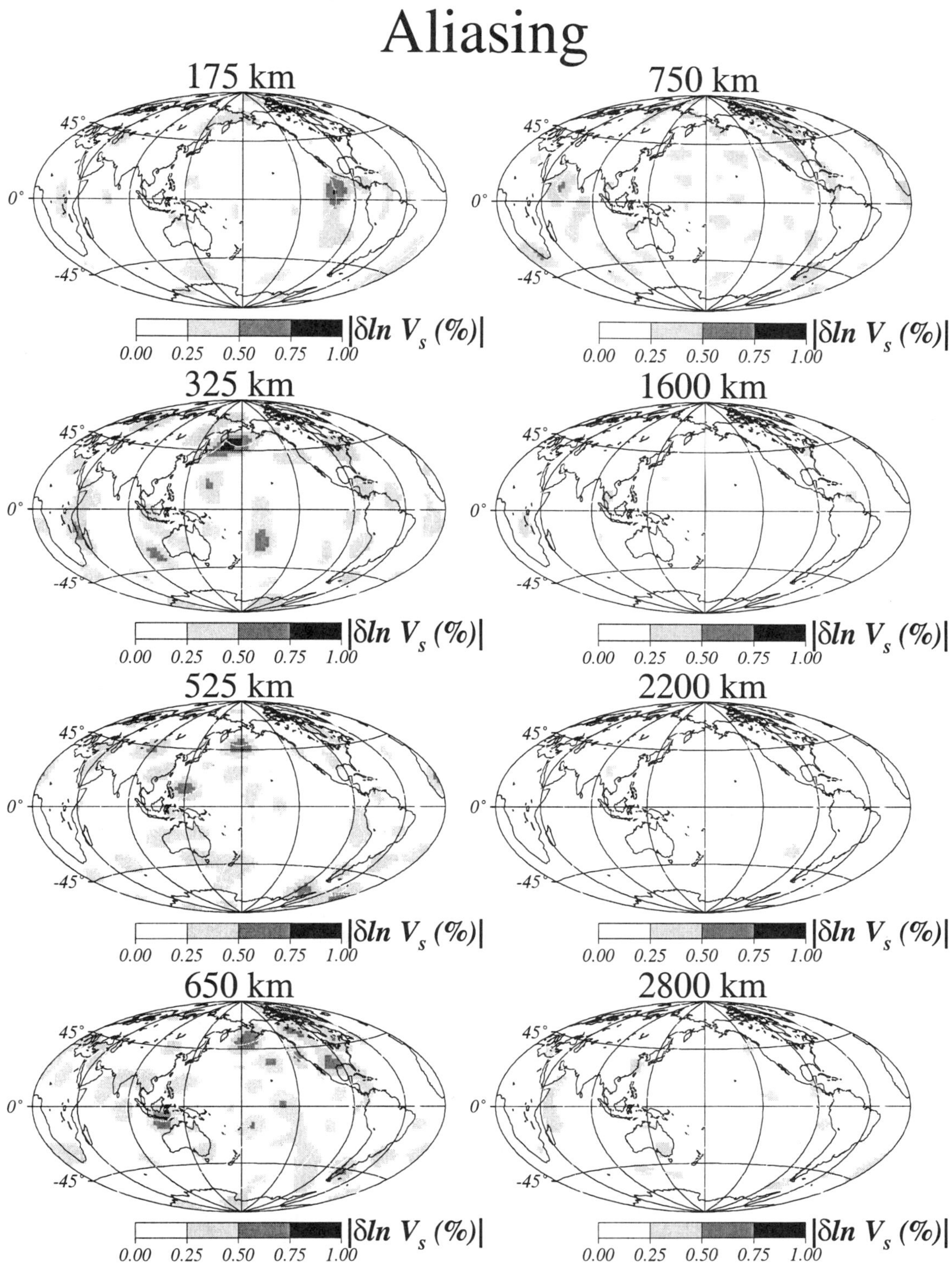


**Figure 9.** Spurious structure in the models caused by aliasing: absolute value of the difference between aliased and unaliased models (Figures 5 & 6 f))

which did not include the phase transition and was inverted with the same tuning parameters. The inversion of this new set of synthetics did not introduce the pattern associated with the phase transition in the control model, and hence suggests that there is no effective circularity in this methodology.

Figure 10 is a mantle model derived with damping parameters that best allow the recovery of the heterogeneity pattern of *GEMLAB 1*; it is readily compared to the aliased convection calculation of Figure 3. The models show both similarities and differences across the mantle. Notable differences are slab structures, which display more continuity in the convection simulation (*e.g.* under South America) In particular in the mid-mantle (1,600 km), the expectations from plate subduction show strong differences with the observations. Clearly, this discrepancy is not due to the tomography since continuity of slab structure is observed in the 'aliased' calculation. Large differences are also observed in D", where the convection calculation fails to reproduce the slow anomalies in the regions beneath the Central Pacific and Africa, probably due to the presence of chemical heterogeneity at the base of the mantle.

In Figure 11, we show a spectral comparison between the aliased model (left) and the image resulting from the optimized inversion of real data (center). The right of the plot shows the rms profiles for *SAW12* modified (solid line) and *GEMLAB 1* aliased (dashed line). The SHMs have been truncated at a minimum depth of 200 km because the convection model was not constructed with a realistic lithosphere.

An important similarity between the two models is in the transition zone where the spectral signature of the phase transitions in *GEMLAB 1*, although somewhat attenuated, is well reproduced in the tomographic pattern, up to degree 5. In both models, the heterogeneity at degrees 2 and 3 is largely unaffected by the phase transition and penetrates into the lower mantle down to 800 km depth. Convection simulations by [*Bunge et al.*, 1997; *Tackley et al.*, 1993] show that the introduction of a phase transition in the convection simulation increase the long wavelength amplitudes of the heterogeneity pattern in the transition zone. Based on the comparison between the two models, the convection calculation seems to overpredict the Clapeyron slope of the phase transition. Contingent upon the correct modeling of the time dependence of the subduction history, this observation suggests that the strength of the phase transition is less than $-4MPa.K^{-1}$.

Both models predict a strong pattern in the lower mantle at degree 2 and, in the D" region, at degrees 2 and 3. The tomography nonetheless shows considerably more complexity with heterogeneity at higher degrees, in particular in the lowermost mantle. In the lower mantle, the convection calculation shows a spectral peak at degree one which is not observed in the seismic images. Another important difference is the absence of the two large upwellings under Africa and the Pacific observed in all recent tomographic images of the whole mantle. This difference may be due to the presence of effects such as partial melting or other chemical anomalies in the lowermost mantle that have not been modeled in the convection simulation. It thus appears that, while current generation convection models seem to be relatively successful in modeling downwellings, future calculations must deal better with the characteristics of the upwellings as shown by tomography. This suggests that the introduction of yet unmodeled physical parameters such as rheological and chemical stratification (*e.g.* [*van der Hilst and Kárason*, 1999]), as well as partial melting and temperature-dependent viscosity may be necessary in order to bring better agreement between models and observations.

## 4. DISCUSSION AND CONCLUSION

In the present study, we used convection simulations to calibrate the recovery of heterogeneity characteristic of the mantle's convective pattern by seismic imaging techniques. We have applied a tomographic filter to results from 3-D spherical convection calculations that incorporate the history of plate motions starting in the Mid-Mesozoic, a viscosity contrast between upper and lower mantle, phase transitions at 410 km and 670 km, and 20% heating from the core. This model represents one of the most realistic convection computations currently available, in that it predicts a spectrum of heterogeneity peaked at low degrees in the lower mantle. The filtered convection model is compared to results derived from the inversion of actual waveform data, using an optimized set of damping parameters. A comparison between the two models allows to make the following inferences regarding the *GEMLAB 1* model:

1. The spectral peak at the low degrees in the pattern of heterogeneity is continuous from the transition zone to depths of about 800 km, across the 670 km discontinuity in both models.

2. In the lower mantle, the tomography shows a broader spectrum than is predicted by the convection simulation.

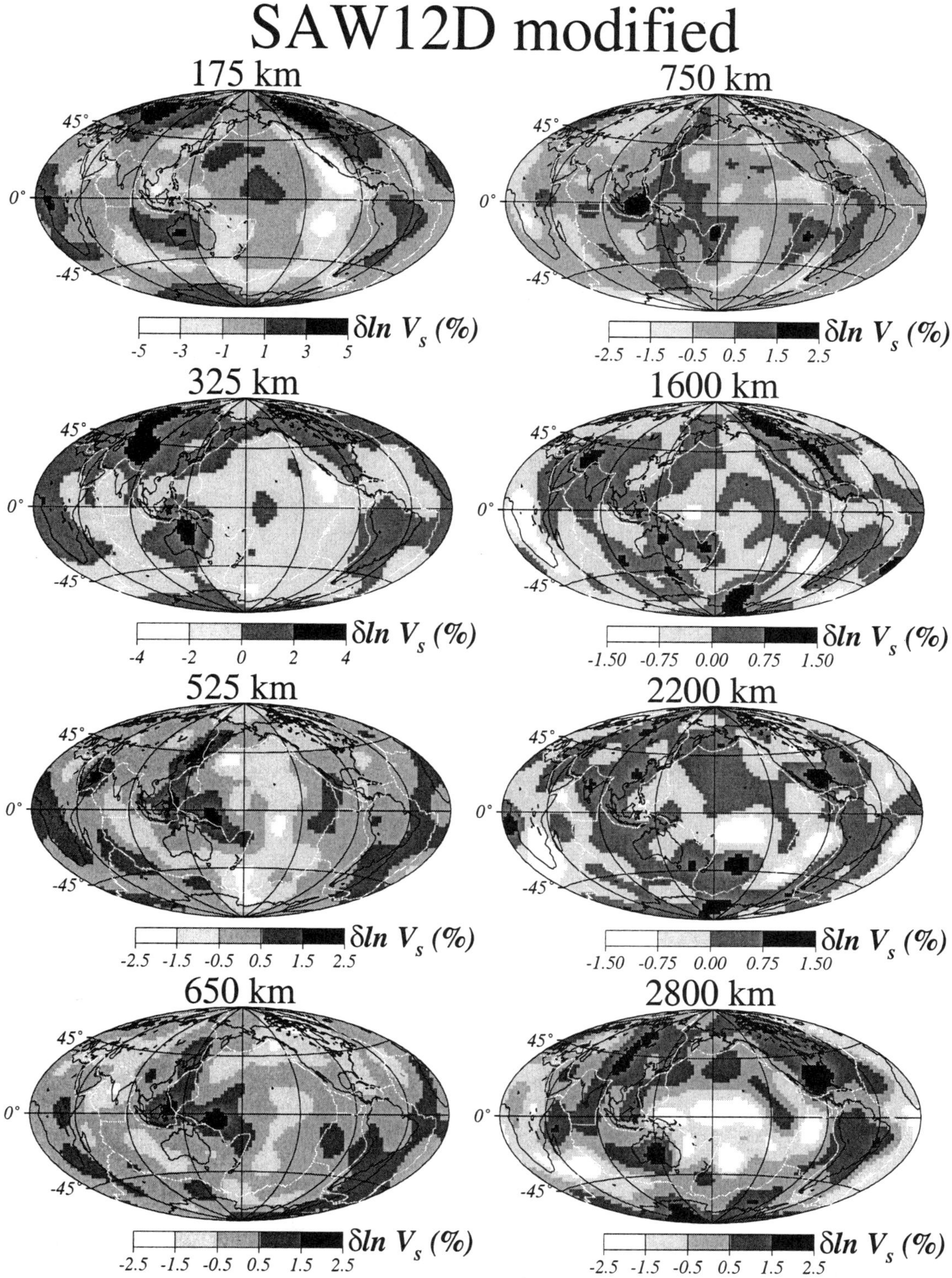


**Figure 10.** Tomographic model SAW12 modified.

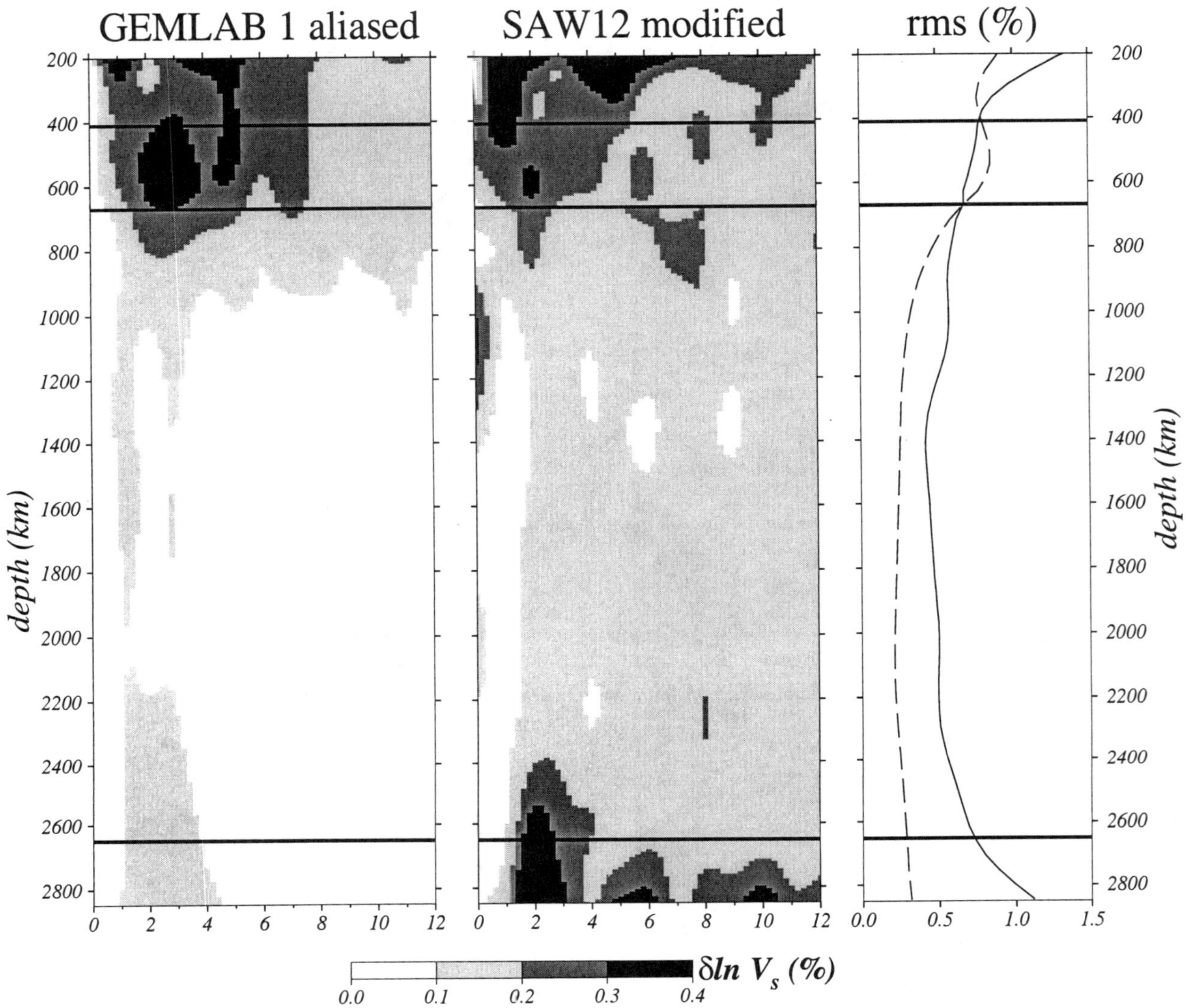


**Figure 11.** Spectral comparison between left: GEMLAB 1 aliased, and right: tomographic model SAW12 modified.

3. The large difference between the convection calculation and the tomography in the lowermost mantle indicates that the former must lack an important ingredient for correctly predicting the character of major upwellings. Part of the difference may be due to the constraints placed on the position of fossil slabs in the Mesozoic, but the thermal and chemical boundary conditions in the lowermost mantle do not yet seem to be well understood, and they must play an important role in the global dynamics.

4. Given that the convection calculations were made under assumption of a uniform mantle chemistry across the 670 km discontinuity, the images of transition zone structure are inconsistent with a Clapeyron slope larger than 4 $MPa.K^{-1}$. This conclusion is contingent upon the quality of the historical plate reconstruction.

The upper bound on the strength of the phase transition is also contingent on the choice of the second Grüneisen parameter $g$ (equation 2) that we use to convert the temperature field of the convection model into shear velocity perturbations. The pressure dependence of this parameter is poorly constrained experimentally. Nevertheless, for the radial thermal expansivity profile of *PREM*, the value of $4.0K^{-1}$ chosen here for $g$ produces a thermal field converted to velocity whose rms is in general agreement with that of of the tomography (see Figure 11, right).

Taking the difference between images of the long and short-wavelength expansions of the convection simulation, we quantify the effect of aliasing of small scale structure into long wavelength seismic images. We find that the amplitude of the spurious structure rarely exceeds 1% but is present at all depths, particularly at the shorter wavelengths. This effect is small when compared to the amplitude of the models in the upper mantle. It is nevertheless closer to the amplitude of the heterogeneity in most of the lower mantle (D" excepted) as seen in tomographic models.

*Acknowledgments.* We thank Catherine Thoraval, Hans-Peter Bunge and Philip B. Stark for helpful discusssions, and Rob van der Hilst, Mark Richards, and two anonymous reviewers for their comments. All figures were made with *GMT* [*Wessel & Smith*, 1991]. This study was supported by LANL-IGPP grant # 97-704. Berkeley Seismological Laboratory contribution 99-06.

## REFERENCES

D. L. Anderson, Theory of the Earth, Blackwell Scientific Publications, 1989.

Bunge, H.P., M.A. Richards, C. Lithgow-Bertelloni, J.R. Baumgardner, S. P. Grand and B. Romanowicz, Time scales and heterogeneous structure in geodynamic Earth models, *Science*, 280 No 5360, 91-95, April 3 1998.

Bunge, H.P., M.A. Richards, and J.R. Baumgardner, A sensitivity study of 3-D spherical mantle convection at $10^8$ Rayleigh number: Effects of depth-dependent viscosity, heating mode and an endothermic phase change, *J. Geophys. Res.*, 102, 11,991-12,007, 1997.

Bunge, H.P., M.A. Richards, and J. R. Baumgardner, Effect of depth-dependent viscosity on the planform of mantle convection, *Nature*, 379, 436-438, 1996.

Bunge, H.P. and M.A. Richards, The origin of long wavelength structure in mantle convection: effects of plate motions and viscosity stratification, *Geophys. Res. Lett.*, 23 No 21, 2,987-2,990, 1996.

Busse, F.H. and J.A. Whitehead, Instabilities of convection rolls in a high Prandtl number fluid, *J. Fluid Mech.*, 47, 305-320, 1971.

Clévédé, É., C. Mégnin, B. Romanowicz and P. Lognonné, Seismic waveform modeling in a three-dimensional Earth: asymptotic and non-asymptotic approaches, submitted to *Phys. Earth planet. Inter.*, 1999.

Davies, G.F., The role of the lithosphere in mantle convection, *J. Geophys. Res.*, 93, 10,451-10,466, 1988.

Dziewonski, A.M. and D.L. Anderson, Preliminary Reference Earth Model, *Phys. Earth planet. Inter.*, 25, 297-356, 1981.

Dziewonski, A.M., B.H. Hager & R.J. O'Connell, Large scale heterogeneities in the lower mantle, *J. Geophys. Res.*, 82, 1977.

Fukao, Y., M. Obayashi, H. Inoue, M. Nenbai, Subducting slabs stagnant in the mantle transition zone, *J. Geophys. Res.*, 97, 4,809-4,822, 1992.

Grand, S. P., R. D. van der Hilst and S. Widiyantoro, Global Seismic Tomography: A Snapshot of Convection in the Earth, *GSA Today*, 7 No 4, 1-7, 1997.

Grand, S.P., Mantle shear structure beneath the Americas and surrounding oceans, *J. Geophys. Res.*, 99, 11,591-11,621, 1994.

Hager, B.H., R.W. Clayton, M.A. Richards, R.P. Comer and A.M. Dziewonski, Lower mantle heterogeneity, dynamic topography and the geoid, *Nature*, 313, 541-545, 1985.

Jackson, D. D., The use of a priori data to resolve non-uniqueness in linear inversion, *Geophys. J. R. Astr. Soc.*, 57, 137-157, 1979.

Johnson S., G. Masters, P.J. Tackley and G.A. Glatzmaier, How well can we resolve a convecting Earth with seismic data ? *EOS Trans. AGU*, 74 No 43, Fall Meeting suppl.,80, (abstract) 1993.

Lerch, F.J., S.M. Klosko and G.B. Patel, A refined gravity model from LAGEOS (GEM-L2), *EOS Trans. AGU*, 63 No 74, Fall Meeting suppl., p. 906, 1982.

Lévêque, J.J., Regional upper mantle S velocity models from phase velocities of great-circle Rayleigh waves, *Geophys. J. R. Astr. Soc.*, 63, 23-43, 1980.

Li, X.D. and B. Romanowicz, Global mantle shear-velocity model developed using nonlinear asymptotic coupling theory, *J. Geophys. Res.* 101, 22,245-22,272, 1996.

Li, X.D. and B. Romanowicz, Comparison of global waveform inversions with and without considering cross-branch modal coupling, *Geophys. J. Int.*, 121, 695-709, 1995.

Li, X.D. and T. Tanimoto, Waveforms of long-period body waves in a slightly aspherical Earth model, *Geophys. J. R. Astr. Soc.*, 112, 92-102, 1993.

Liu, X.-F. and Dziewonski, A.M., Global analysis of shear wave velocity anomalies in the lower-most mantle, Gurnis M. editor, Core-mantle boundary region, American Geophysical Union, Washington, DC, 1998.

Machetel, P., Short-wavelength lower mantle seismic velocity anomalies, *Geophys. Res. Lett.* 17 No 8, 1,145-1,148, 1990.

Masters, G., S. Johnson, G. Laske, and H. Bolton, A shear-velocity model of the mantle, *Phil. Trans. R. Soc. London*, 354 1,385-1,410, 1996.

Masters, G., T.H. Jordan, P.G. Silver and F. Gilbert, Aspherical Earth structure from fundamental spheroidal mode data,*Nature*, 298, 609-613, 1982.

Mégnin, C. and B. Romanowicz, The effects of the theoretical formalism and data selection on mantle models derived from waveform tomography, *Geophys. J. Int.*, 138, 366-380, 1999.

Mégnin, C. and B. Romanowicz, A model of shear velocity in the mantle from the inversion of waveforms of body, surface and higher mode waveforms, submitted to *Geophys. J. Int.*, 1999.

Mégnin, C., H.-P. Bunge, B. Romanowicz and M. A. Richards, Imaging 3-D spherical convection models: What can seismic tomography tell us about mantle dynamics?, *Geophys. Res. Lett.*, 24 No 11, 1,299-1,302, 1997.

Mégnin, C. and B. Romanowicz, Improvements in waveform inversion techniques with applications to whole mantle structure, *EOS Trans. AGU* 78, p. F460 (abstract), 1997.

Montagner, J.P. and B. Romanowicz, Degrees 2, 4, 6 inferred from seismic tomography, *Geophys. Res. Lett.*, 20 No 7, 631-634, 1993.

Nataf, H.-C., I. Nakanishi & D.L. Anderson, Measurements of mantle wave velocities and inversion for lateral heterogeneities and anisotropy, *J. Geophys. Res.*, 91, 7,261-7,307, 1986.

Okal, E, Regional dispersion of first-order overtone Rayleigh waves, *Geophys. J. R. Astr. Soc.*, 72 No 2, 461-481, 1983.

Press, W.H., S.A. Teukolsky, W.T. Vetterling and B.P. Flannery, Numerical Recipes in C, 2nd ed., Cambridge University Press, 1994.

Pulliam, R.J. and P.B. Stark, Confidence regions for mantle heterogeneity, *J. Geophys. Res.*, 99, 6,931-6,943, 1994.

Romanowicz, B. The upper mantle degree two: constraints and inferences on attenuation tomographyfrom global mantle wave measurements, *J. Geophys. Res.*, 95, 11,051-11,071, 1990.

Romanowicz, B., Multiplet-multiplet coupling due to lateral heterogeneity: asymptotic effects on the amplitude and frequency of Earth's normal modes, *Geophys. J. R. Astr. Soc.*, 90, 75-100, 1987.

Romanowicz, B., G. Roult and T. Kohl, The upper mantle degree two pattern: constraints from GEOSCOPE fundamental spheroidal mode eigenfrequency and attenuation measurements, *Geophys. Res. Lett.*, 14 No. 12, 1219-1222, 1987.

Snieder, R., J. Beckers and F. Neele, The effect of small-scale structure on normal mode frequencies and global inversions, *J. Geophys. Res.*, 96, 501-515, 1991.

Souriau, A. & M. Souriau, Test of tectonic models by great circle Rayleigh waves, *Geophys. J. R. Astr. Soc.*, 73, 533-551, 1983.

Spakman, W., S. Stein, van der Hilst, R.D. and Wortel, R., Resolution experiments for NW Pacific subduction zone tomography, *Geophys. Res. Lett.*, 16 No 10, 1097-1100, 1989.

Su, W.-J.,R.L. Woodward and A. M. Dziewonski, Degree 12 modelof shear velocity heterogeneity in the mantle, *J. Geophys. Res.*, 99, 6,945-6,980, 1994.

Su, W.-J. Woodward, R.L. and A. M. Dziewonski, Deep origin of mid-ocean-ridge seismic velocity anomalies, *Nature*, 360, 149-152, 1992.

Su, W.-J. and A. M. Dziewonski, Predominance of long-wavelength heterogeneity in the mantle, *Nature*, 352, 121-126, 1991.

Tackley, P.J., D.J. Stevenson, G.A. Glatzmaier and G. Schubert, Effects of multiple phase transitions in a three-dimensional spherical model of convection in Earth's mantle, *J. Geophys. Res.*, 99, 15,877-15,901, 1994.

Tackley, P.J., D.J. Stevenson, G.A. Glatzmaier and G. Schubert, Effects of an endothermic phase transition at 670 km depth in a spherical model of convection in the Earth's mantle, *Nature*, 361, 699-704, 1993.

Tanimoto, T. Long-wavelength S-wave velocity structure throughout the mantle. *Geophys. J. Int.*, 100, 327-336, 1990.

Tarantola, A. and B. Valette, Generalized non-linear inverse problems solved using the least-squares criterion, *Rev. Geophys.*, 20, 219-232, 1982.

van der Hilst, R.D. and H. Kárason, Compositional heterogeneity in the bottom 1000 kilometers of Earth's mantle: toward a hybrid convection model, *Science*, 283, 1,885-1,888, 1999.

van der Hilst, R.D., S. Widiyantoro and E.R. Engdahl, Evidence of deep mantle circulation from global tomography, *Nature*, 386, 578-584, 1997.

van der Hilst, R.D., Complex morphology of subducted lithosphere in the mantle beneath the Tonga trench, *Nature*, 374, 154-157, 1995.

van der Hilst, R.D. and T. Seno, Effects of relative plate motion on the deep structure and penetration depth of slabs below the Izu-Bonin and Mariana arcs, *Earth Planet. Sci. Lett.*, 120, 395-407, 1993.

van der Hilst, R.D., E.R. Engdahl, W. Spakman and G. Nolet, Tomographic imaging of subducted lithosphere below northwest Pacific island arcs, *Nature*, 353, 37-43, 1991.

Wessel, P. & W.H.F. Smith, New version of the Generic Mapping Tool released, *EOS Trans. AGU*, 72 441, 445-446, 1991.

Woodhouse, J.H. & A.M. Dziewonski, Mapping the upper mantle: three-dimensional modeling of Earth structure by inversion of seismic waveforms, *J. Geophys. Res.*, 89, 5,953-5,986, 1984.

Zhang, Y.S. and T. Tanimoto, Ridges, hotspots and their interaction as observed in seismic velocity maps, *Nature*, 355, 45-49, 1992.

Zhou, H.W. & R.W. Clayton, P and S wave travel-time inversions for subducting slabs under the island arcs of the northwest Pacific, *J. Geophys. Res.* 95, 6,829-6,854, 1990.

C. Mégnin and B. Romanowicz, UC Berkeley Seismological Laboratory, 217 McCone Hall, Berkeley, CA 94720. (e-mail: pepe@seismo.berkeley.edu)

# Constraints on Mantle Convection From Seismic Tomography

Hrafnkell Kárason and Rob D. van der Hilst

*Department of Earth, Atmospheric, and Planetary Sciences,*
*Massachusetts Institute of Technology, Cambridge, Massachusetts*

Since the advent of global seismic tomography some 25 years ago, advances in technology, seismological theory, and data acquisition have allowed spectacular progress in our ability to image seismic heterogeneity in Earth's mantle. Using examples from some recently published tomographic models, we briefly review some concepts of seismic tomography, such as parameterization, and summarize how this class of imaging has contributed to our knowledge of the scale of mantle convection. With the presently available data seismologists can map the subduction of former oceanic lithosphere beneath most island arcs. Tomographic imaging has revealed that many slabs extend below the deepest earthquakes of the Wadati-Benioff zones and sink deep into the lower mantle, although intense deformation of flow trajectories has been detected in the upper mantle transition zone beneath several convergent margins (in particular in the western Pacific). We illustrate this with examples of our recent 3-dimensional model for mantle *P*-wavespeed, which is inferred from arrival time data of *P*, *pP*, *PKP*, and $P_{diff}$ waves. In combination with constraints from computational geodynamics the seismic images render mantle stratification at 660 km depth unlikely and show that surface plate motions are tied to large scale convective circulation. However, the ultimate fate of the slabs that penetrate into the lower mantle is still enigmatic and in some recent tomographic models the subduction related heterogeneity pattern vanishes at very large depth (2000 km or so). These observations have inspired models for mantle convection that have the potential for reconciling geophysical views and geochemical constraints, but much work remains to be done to establish the detailed pattern of mantle flow.

## INTRODUCTION

The relative motion of lithospheric plates at the surface of the Earth, which is directly related to natural hazards such as earthquakes and volcanoes and can cause long-term variations in climate, is the surface expression of slow, large scale deformation of rock in the deep interior of our planet. On a time scale of millions of years the mantle silicates flow in a process known as 'mantle convection', but on short time scales mantle material behaves as a solid. This viscoelastic behavior allows the propagation of longitudinal (*P*) and transverse (*S*) seismic waves, which speed up in "cold" downwellings and slow down in "hot" upwellings so that recordings of these waves can be used for the mapping of mantle flow. Our objective here is not to provide a rigorous review of seismic tomography but to highlight developments in seismic imaging pertinent to the scope of this special volume.

The History and Dynamics of Global Plate Motions
Geophysical Monograph 121

In the past half century, the scale of mantle convection, which relates critically to compositional stratification and the thermal and chemical evolution of our planet, has been one of the big puzzles in Earth Sciences. A major challenge has been to evaluate and reconcile the range of observations and constraints provided by different scientific disciplines [cf. *Albarède and Van der Hilst*, 1999]. Earth's heat budget (the balance between heat production and heat loss) and geochemical analyses of ocean floor basalts suggest that distinct mantle reservoirs have retained their identity for 2 billion years or more. One reservoir boundary is typically placed at 660 km depth, that is, between the upper and lower mantle. Seismic imaging and computational geodynamics indicate, however, that this interface is not an effective barrier to mantle flow and suggest that convective circulation occurs at a larger scale.

## THREE DECADES OF SEISMIC TOMOGRAPHY

Since the pioneering studies in the mid 1970's [see, for instance, *Julian and Sengupta*, 1973; *Sengupta and Toksöz*, 1976; *Aki et al.*, 1977; *Dziewonski et al.*, 1977; *contributions to the 1975 AGU Fall Meeting* (*EOS Trans. of the Am. Geophys. Un.*, 56, 393-396, 1975)], advances in technology (e.g., three to five orders of magnitude increase in computer processing speed, mass storage, and memory), inverse theory, and data quality and volume have vastly improved the tomographic imaging of Earth's deep interior structure. In a parallel development, our understanding of the images has improved dramatically because spectacular advances in computational geodynamics have facilitated the integration of the geological constraints on past plate motion at Earth's surface and the results of experimental and theoretical mineral physics with the snapshots of convection provided by seismic imaging.

### *Different inversion strategies*

For the imaging of global structure two methods have become popular, each with specific benefits and shortcomings. The first represents lateral variations in seismic properties by superposition of global basis functions, such as spherical harmonics. This class of tomography [for reviews see *Woodhouse and Dziewonski*, 1989; *Romanowicz*, 1991; *Montagner*, 1994; *Ritzwoller and Lavely*, 1995; *Ekström*, this volume; *Mégnin and Romanowicz*, this volume] is attractive for imaging structure at a long wavelength, $\lambda$, because the number of model parameters, which scales as $l^2$ — with harmonic degree $l$ inversely proportional to wavelength ($\lambda = 2\pi r/l$), is then small enough to resolve the coefficients by direct inversion of carefully selected and processed waveform data. Plate 1a depicts long wavelength variations in shear wavespeed at 1300 km depth according to the $l$=12 model of Su et al. [1994]; at this depth, the (half wavelength) resolution is 1350 km.

The large number of global basis functions required to describe structure at length scales of several hundred km or less, such as slabs of subducted lithosphere in the upper mantle, would prohibit direct inversion. Moreover, the coefficients can no longer be determined accurately owing to uneven data coverage on the relevant length scales, and artifacts can be introduced in regions of poor coverage [*Boschi and Dziewonski*, 1999]. It then becomes attractive to represent wavespeed variations by local basis functions, such as non-overlapping constant-slowness volumes (e.g., tetrahedrons, voronoi cells, rectangular blocks) or cubic splines or wavelets that interpolate between grid values. In this class of tomography, the large number of model parameters necessitates the use of iterative solvers such as *LSQR* [*Paige and Saunders*, 1982; *Nolet*, 1985] or *SIRT* [*Humphreys and Clayton*, 1988], with solution selection and resolution assessment less elegant than for direct inversions, but regularization (damping) can be used to avoid artifacts in regions of limited data coverage [*Spakman and Nolet*, 1988]. In the past decade the cell size used in regular grid inversions has decreased from about 6°x6° [*Inoue et al.*, 1990; *Pulliam et al.*, 1993] to 2°x2° [*Van der Hilst et al.*, 1997]. Plate 1b depicts *S* wavespeed variations in constant slowness blocks of 2.5°x2.5°x200km [*Grand, personal communication*, 1999; see *Grand et al.*, 1997], and Plate 1c depicts lateral variations in *P* wavespeed [*Kárason and Van der Hilst*, 2000] in 3°x3°x150km blocks. These map views reveal long linear features of faster-than-average wave propagation that reflect past episodes of plate convergence at Earth's surface. Even though the wavespeed variations may at first seem rather different, many structures in Plate 1a resemble the low-pass filtered version of those in Plates 1b and 1c, and if one considers the difference in nominal resolution the three models are, in fact, fairly consistent with each other in regions of adequate data coverage.

### *Flexible parameterization*

Uneven source and receiver distribution results in significant spatial variations in data coverage. Global basis functions offer no flexibility for regionalization but a local basis can be adapted to lateral variations in data coverage. Small blocks can be used in densely sampled regions without unnecessary overparameterization of poorly sampled regions [e.g., *Bijwaard et al.*, 1998]. Ideally such irregular grids reflect the spatial variation of resolution, but since this is difficult to quantify they have been designed on the basis of sampling [*Gudmundsson and Sambridge*, 1998;

*Bijwaard et al.*, 1998] or regional interest [*Abers and Roecker*, 1991; *Widiyantoro and Van der Hilst*, 1996]. Plate 2 illustrates the effect of different parameterizations.

*Data quality and coverage*

The success of any tomographic study depends critically on data quality. Most shear wave studies have been based on waveform data that were carefully selected and processed by individual investigators [e.g., *Woodhouse and Dziewonski*, 1984; *Masters et al.*, 1996; *Li and Romanowicz*, 1996; *Grand et al.*, 1997; *Van Heijst and Woodhouse*, 1999; *Ekström*, this volume; *Mégnin and Romanowicz*, this volume], but for global *P*-wave inversions such data sets are only now being constructed. With almost 15 million entries, the largest single data source available for tomography consists of the travel-time residuals processed and published by the International Seismological Centre (ISC). This data set is noisy but its size and redundancy allows the extraction of structural signal. Many researchers have processed the ISC data prior to inversion, but probably the most rigorous effort was made by *Engdahl et al.* [1998]. Using non-linear procedures for earthquake relocation and seismic phase re-identification they improved hypocenter parameters and travel time residuals for a large range of seismic phases. The striking agreement between images based either on travel times inferred from careful waveform processing (Plate 1b) or from routinely reported and processed phase arrivals (Plate 1c) demonstrates the value of the data processing by Engdahl and co-workers. It also demonstrates that — at least at the depth shown — compressional and shear wavespeed are highly correlated with each other.

Uneven data coverage continues to be one of the most persistent problems in travel time tomography, although the use of irregular grids can reduce some of its detrimental effects on imaging. For a given source-receiver distribution the sampling of Earth's structure can be improved by considering data and ray paths not only of direct *P* or *S* but of other phases as well. For *P*-wave imaging, seismologists have experimented successfully with depth phases (such as *pP*), which also helps constraining focal depth [*Engdahl et al.*, 1998], surface reflections (such as *PP* and *PPP*), and core reflected (*PcP*) and refracted (*PKP*) waves [cf. *Van der Hilst et al.*, 1991; *Van der Hilst and Engdahl*, 1991; *Obayashi and Fukao*, 1997; *Vasco and Johnson*, 1998; *Bolton and Masters*, 1998; *Van Heijst and Woodhouse*, 1999]. For *S* travel time tomography, arrival times of direct *S*, the multiple surface reflections such as *SS*, *SSS*, *SSSS*, core reflections and refractions *ScS*, *SKS*, and surface (Love) wave phase velocities have been used [cf., *Grand*, 1994; *Liu and Dziewonski*, 1998; *Widiyantoro et al.*, 1998]. Large gaps in data coverage remain, however. In regions where mantle structure cannot yet be resolved by body wave data one can impose zero perturbations from the reference model or use independent constraints on long wavelength variations, such as frequency shifts of Earth's free oscillations [*Kárason and Van der Hilst*, 1998; *Ishii and Tromp*, 1999].

*A new model for mantle* P-*wavespeed*

Because of our interest in the deep mantle [*Van der Hilst and Kárason*, 1999; *Kellogg et al.*, 1999], we made an effort to improve the sampling of lower mantle structure by incorporation of travel time data of core refracted *PKP* and diffracted $P_{\mathrm{diff}}$ waves [*Kárason and Van der Hilst*, 2000]. Figure 1 illustrates the ray geometry for these phases (NB. we used 3-D sensitivity kernels based on calculations by *Zhao et al.* [2000] for the back projection of the $P_{\mathrm{diff}}$ data but those are not shown here). In fact, we used differential times $PKP_{\mathrm{DF}}$-$PKP_{\mathrm{AB}}$, $PKP_{\mathrm{BC}}$-$PKP_{\mathrm{AB}}$, and $PKP_{\mathrm{DF}}$-$P_{\mathrm{diff}}$ to reduce effects of hypocenter mislocation and to facilitate the extraction of signal from structure in the deep mantle. Moreover, the requirement that several readings be made from a single record works as an excellent quality criterion [*Kárason and Van der Hilst*, 2000]. For these phases we used accurate travel times inferred from waveform cross correlation by *McSweeney* [1997] (1383 *PKP* differential times) and *Wysession* [1996] (542 $P_{\mathrm{diff}}$ differential times) as well as some 27,412 carefully selected and processed *PKP* differential data from the data base of Engdahl and co-workers. These data were used along with nearly 8 million *P* and *pP* data from the Engdahl catalog; see *Kárason and Van der Hilst* [2000] for details about the data integration. This augmented data set provides much better data coverage than the *P* data alone (Figure 2), in particular in the southern hemisphere.

Plate 3 illustrates how the lateral variation of *P*-wavespeed changes with increasing depth in the mantle. The signature of cratonic parts of ancient continents and the narrow linear anomalies associated with plate subduction characterize structure in the upper mantle. (We remark that in this model the thickness of continental lithosphere is not well resolved since for most continental receivers these the body waves used sample shallow mantle structure in only a small range of incident angles.) In the mid-mantle the amplitude of wavespeed variations is significantly smaller than in the upper mantle, but a pattern of long, tabular structural features emerges beneath the major convergent plate boundaries (see also Plate 1). The thickness of these tabular structures is much larger than the subducted slabs in the shallow mantle, which can perhaps be attributed to radial changes in viscosity [*Fischer et al.*, 1990; *Bunge et al.*, 1996]. Toward the base of the mantle

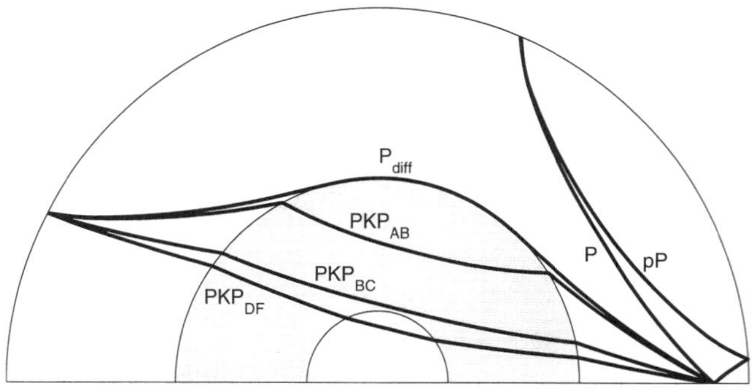


**Figure 1.** Ray geometry for the phases used in the *P*-wave study by Kárason and Van der Hilst [2000]. The core refracted *PKP* and diffracted $P_{\mathrm{diff}}$ waves provide better sampling of the deep mantle than *P* waves alone. We remark that instead of narrow rays we used 3-D sensitivity kernels to account for the fact that the $P_{\mathrm{diff}}$ travel times were measured from long period waveforms [*Wysession*, 1996]. The sensitivity kernels were based on mode summation [*Zhao et al.*, 1999].

the peak-to-peak amplitudes increase, albeit not as much as reported for shear wave perturbations [*Masters et al.*, 1996]. Figure 3 depicts an estimate of spatial resolution at selected depths. As expected from the uneven data coverage, the image recovery also reveals large geographical variations, but the changes in the character of heterogeneity from mid-mantle to the base of the mantle can not be attributed to sampling alone. The detail provided by the fine parameterization in regions of dense sampling is lost in the global maps of Plate 3 and Figure 3 but can be appreciated once we zoom into a particular region (Plate 2) or display structure by means of vertical sections across selected plate margins (Plate 4).

## RESULTS OF IMAGING PERTINENT TO THE ISSUE OF MANTLE CONVECTION

While many early results [*Dziewonski*, 1984; *Woodhouse and Dziewonski*, 1984] have proved to be robust, long wavelength models continue to be improved, and in recent studies the wavespeed variations are expanded up to $l$=24 [*Ekström*, this issue; *Mégnin and Romanowicz*, this issue]. This class of modeling does not, however, resolve trajectories of mantle flow in sufficient detail to determine unequivocally whether or not convection is stratified at 660 km. Indeed, its results can be used to argue either way [*Richards and Engebretson*, 1992; *Wen and Anderson*, 1995]. In the 1990's, travel time tomography with local basis functions has made several seminal contributions to our understanding of mantle convection because it has enabled the mapping of flow trajectories in unprecedented detail.

First, some slabs of subducted lithosphere penetrate into the lower mantle while others appear trapped in the upper mantle. This was first borne out by regional studies [e.g., *Zhou and Clayton*, 1990; *Van der Hilst et al.*, 1991; *Fukao et al.*, 1992] but has since been confirmed by other studies, including high resolution global inversions [*Widiyantoro*, 1997; *Bijwaard et al.*, 1998; *Kárason and Van der Hilst*, 1999]. Plate 4 provides examples of both styles of subduction. Slab deflection may occur beneath Izu Bonin and the southern Kuriles (Plate 4b,c), the Banda arc, and beneath the Thyrrenian Sea; deep slabs, sometimes severely deformed in the transition zone, have been detected beneath the Mariana, Tonga-Kermadec (Plate 4e), Sunda (4d), and northern Kurile arcs, the Philippines, the Aegean Sea (4a), and Central and South America (4f). Inspired by *Kincaid and Olson* [1987], *Van der Hilst and Seno* [1993] and *Van der Hilst* [1995] argued that the observed complexity does

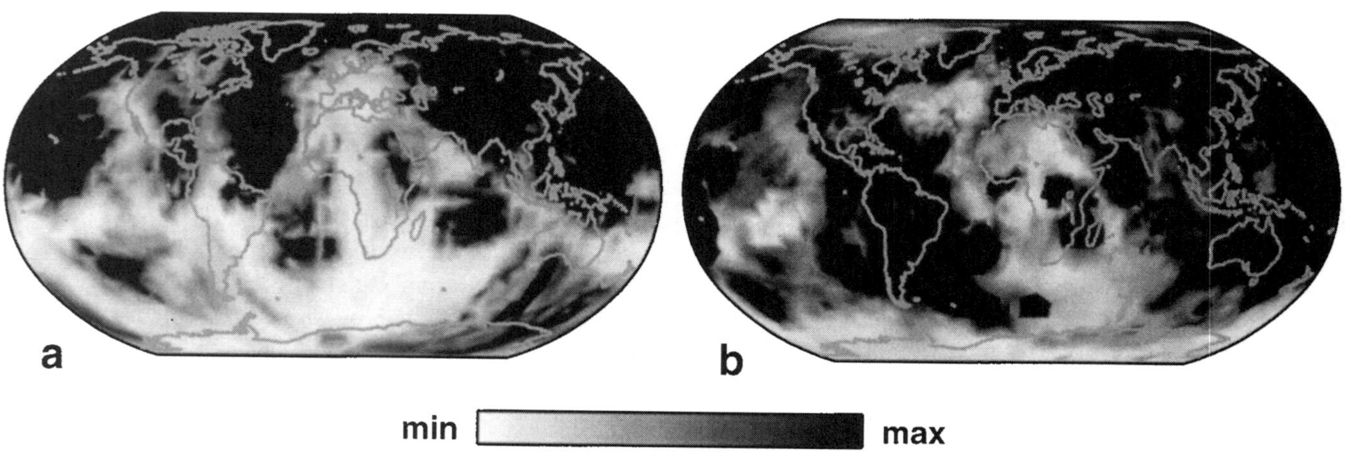


**Figure 2.** Improvement of sampling of deep mantle structure: (a) sampling by *P* and *pP* and (b) *PKP* and $P_{\mathrm{diff}}$. Shown here are the column sums of the sensitivity matrix **A** used by *Kárason and Van der Hilst* [2000].

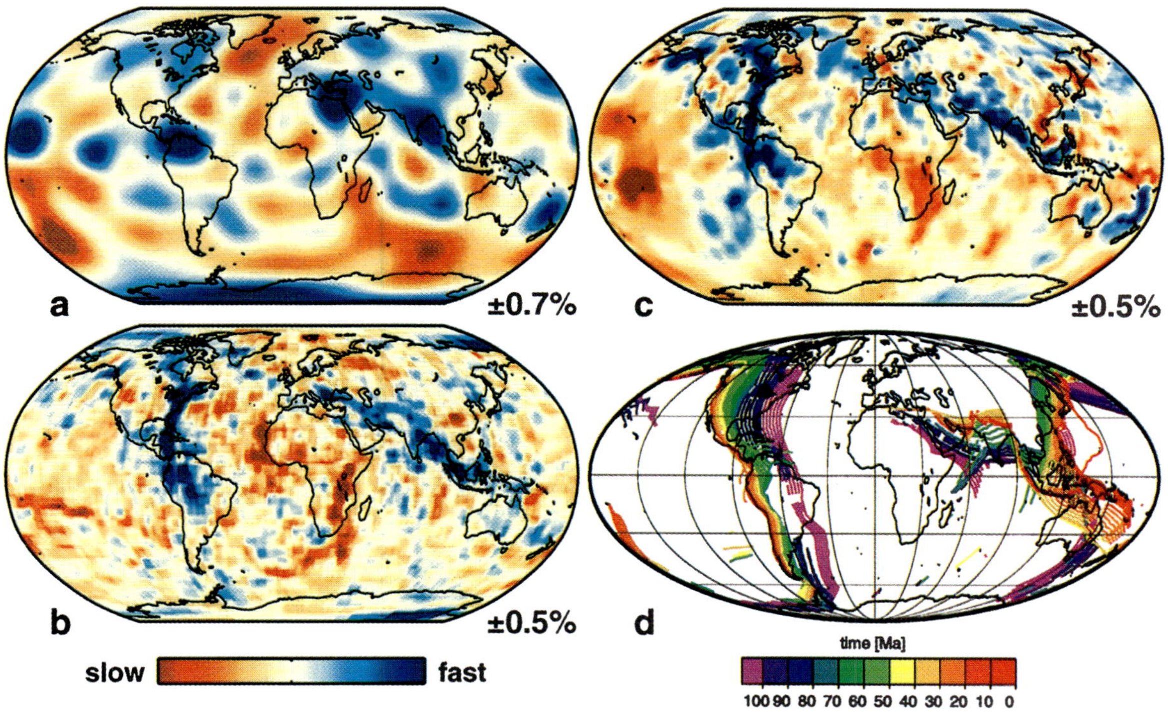


**Plate 1.** Robinson projections of lateral variation of seismic wavespeed at approximately 1300 km depth. (a) *S* wavespeed expanded on global basis functions (spherical harmonics up to degree and order 12) by *Su et al.* [1994]; (b) *S* wavespeed represented by regular blocks as local basis functions [*Grand, personal communication*, 1999; see *Grand et al.*, 1997]; (c) *P* wavespeed represented by blocks [Kárason and Van der Hilst, 2000]. The relative amplitude of the perturbations is given at the bottom right of each panel. (d) Mollweide projection of the history of plate convergence [cf. *Lithgow-Bertelloni and Richards*, 1998]; color indicates time interval in which subduction occurred.

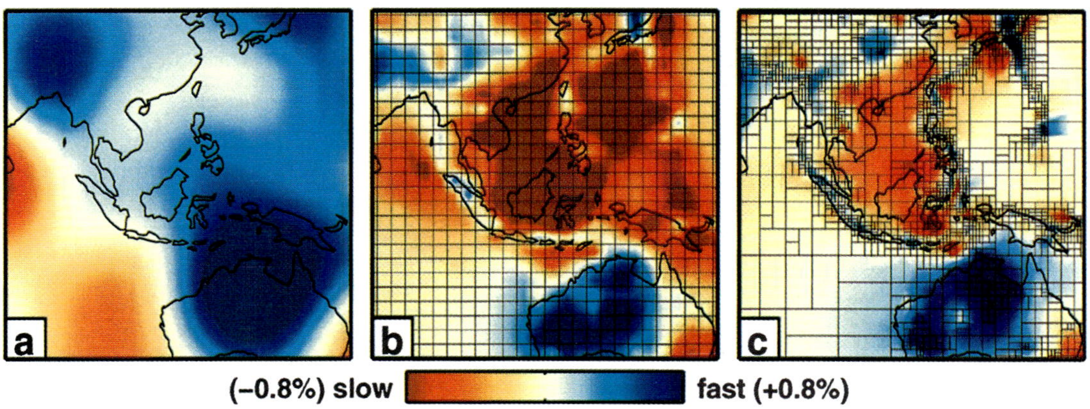


**Plate 2.** Lateral variation in seismic wavespeed at about 300 km depth beneath Australiasia according to different global models published in the last decade. (a) Spherical harmonics up to degree and order 12 [*Su et al.*, 1994]; (b) *P* wavespeed on a regular 3°x3° grid [*Kárason and Van der Hilst*, 2000]; (c) *P* wavespeed on an irregular grid [*Kárason and Van der Hilst*, 1999]. The image based on spherical harmonics is dominated by the high wavespeeds of the continental cratons in Australia and central Asia, and in the regular grid inversion the signature of the narrow slabs is overwhelmed by the low wavespeed in the back arc regions. Irregular gridding [e.g., *Bijwaard et al.*, 1998; *Kárason and Van der Hilst*, 1999] gives the most satisfactory rendering of the actual structure. Results similar to Plate 2c can be obtained by using small cells in a regular grid inversions [e.g., *Widiyantoro and Van der Hilst*, 1996] but for global inversions the number of model parameters would then become very large.

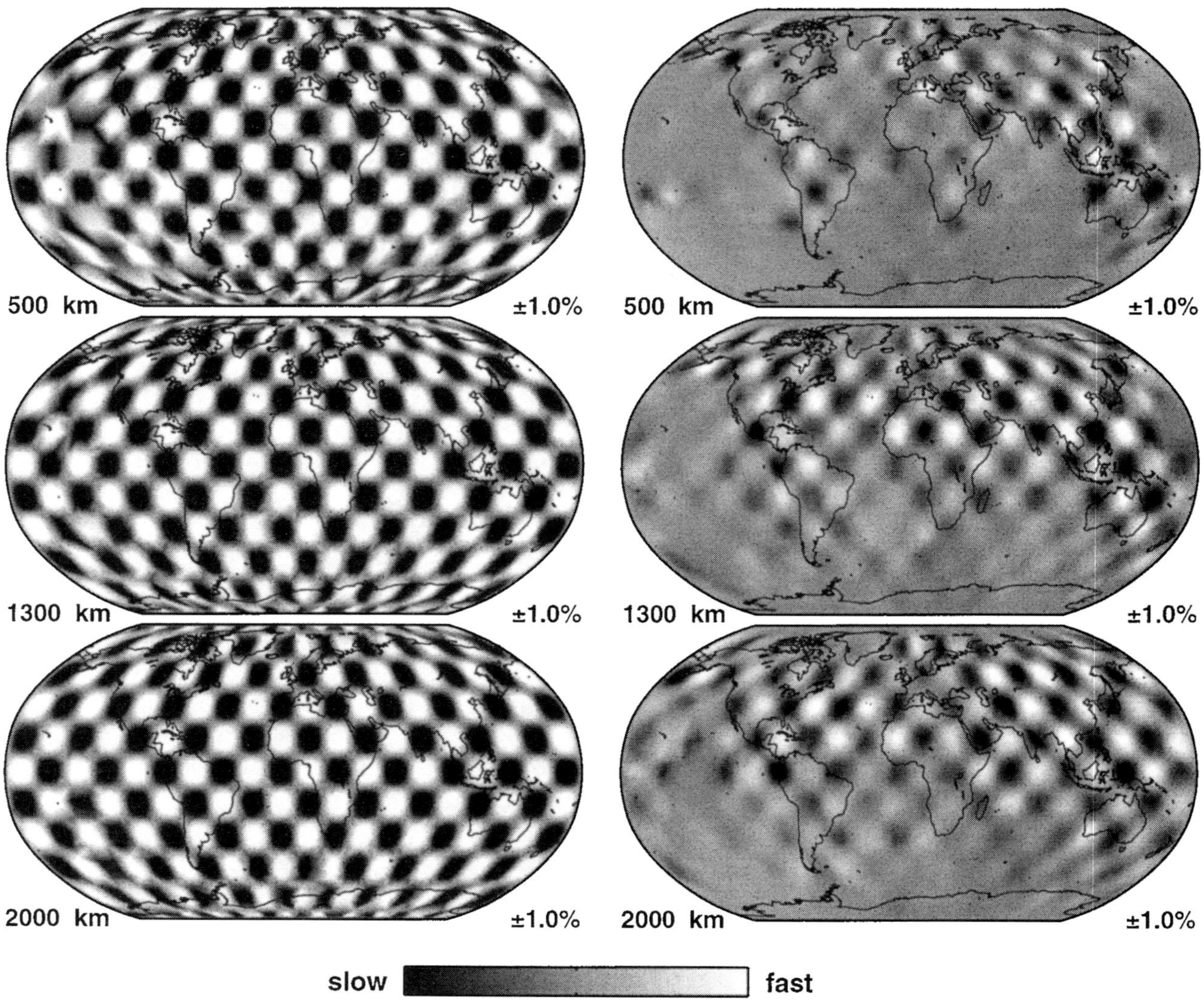


**Figure 3.** Estimate of spatial resolution at 500, 1300, and 2000 km depth. From the input model shown on the left we calculated synthetic travel time residuals by ray tracing and we then inverted them to test if (and where) we can resolve the know input pattern with the data coverage used. The input anomaly is ±2% in the center of each cell. This exercise shows that there are large regions in the mantle were resolution is poor owing to inadequate sampling, in particular in the upper mantle. Notice also that image recovery at 2000 km depth is equal to or better than at 1300 km, which indicates that the dramatic change in heterogeneity, compare, for instance, Plates 3d and e, is not simply related to sampling.

not imply stratification at 660 km depth but can be caused by interplay between relative plate motion (i.e., lateral trench migration) and the deformation of slabs when they encounter resistance (e.g., higher viscosity or a depressed phase boundary). Experimental [*Griffiths et al.*, 1995; *Guiltier-Frottier et al.*, 1995] and numerical [*Zhong and Gurnis*, 1995; *Davies*, 1995; *Christensen*, 1996] fluid dynamical experiments support this view, but selective weakening of the descending plate by grainsize reduction upon phase transformation may also contribute [*Riedel and Karato*, 1997]. The complexity caused by interaction of downwellings with the upper mantle transition zone persists to near 1000 km depth [*Van der Hilst and Kárason*, 1999] but the significance of this depth is not yet established and many slabs sink to even larger depth (Plate 4).

Second, independent *P* and *S* studies have begun to agree on structure as small as several hundred km [*Grand et al.*, 1997; *Van der Hilst et al.*, 1997] - a development that has increased the credibility of this class of imaging and may prove to be one of the milestones of tomography;

they revealed a relatively simple pattern of narrow high-wavespeed structures in the lower mantle beneath plate boundaries with a long history of subduction (Plate 1) and indicated that these deep structures often connect to seismogenic slabs in the upper mantle (Plate 4). Several slab structures disappear from tomographic view in the bottom 1000 km of the mantle (e.g., Plates 4a, d-f), but some fragments seem to connect to D" heterogeneity, for instance beneath eastern Asia and central America. Also the correlation between *P* and *S* images, which persists to large depth, may break down near the base of the mantle [*Grand et al.*, 1997; *Su and Dziewonski*, 1997; *Kennett et al.*, 1998].

Third, despite theoretical and practical difficulties significant progress has been made in the mapping of seismically slow anomalies, and several studies now suggest that mantle upwellings are continuous over a large depth range, supporting the view that 'plumes' originate (at a boundary layer) below the 660 km discontinuity [*Wolfe et al.*, 1997; *Bijwaard and Spakman*, 1999; *Goes et al.*, 1999; *Ritsema et al.*, 1999]. Image resolution — in particular in the radial direction — continues to form a formidable obstacle, however.

## SCALE OF MANTLE CONVECTION

These observations render untenable the conventional end-member views of either convective stratification at 660 km or undisturbed whole mantle flow. The 660 km discontinuity distorts mantle flow, occasionally resulting in local and transient layering, but many slabs penetrate to at least 1700 km depth in the mantle. The change in structure and heterogeneity spectrum between 1700 and 2300 km depth is probably real, but its origin is not yet known. It may point to stratification of some sort in the deep mantle [*Van der Hilst and Kárason*, 1999; *Kellogg et al.*, 1999], but it may also reflect changes in the nature of global plate motion in the distant past [*Richards and Engebretson*, 1992]. Alternatively, *Anderson* [1999] explains the slab-like lower mantle structures and their apparent continuity to upper mantle subduction zones by thermal coupling and coincidental alignment of structures in separately convecting upper and lower mantles. We take issue with this. It would be very fortuitous indeed if slowly changing structures in the sluggish lower mantle align with rapidly changing structures in the less viscous upper mantle. Thermal coupling can work [*Nataf*, 1988; *Čížková et al.*, 1999] but is too slow to explain the large depth of slab penetration beneath young convergent margins and is inconsistent with the constant dip angle inferred for several slabs. Moreover, if only heat is exchanged, thousands of kilometers of subducted lithosphere must have accumulated in the upper mantle beneath margins with a long subduction record, for which there is no observational evidence.

The geophysical evidence against stratification at 660 km depth is strong, and if isolated and seismically visible 'reservoirs' exist they are likely to reside at a larger depth. A mantle convection scenario that is then worth considering is one in which — apart from the upper and lower boundary layers (lithosphere and D", respectively) — three domains are identified. In the view postulated by *Kellogg et al.* [1999] and *Van der Hilst and Kárason* [1999], and discussed by *Albarède and Van der Hilst* [1999], undegassed and enriched material in the bottom 1000 km of the mantle has not (yet) mixed with the part of the mantle — the top 2000 km or so — that is involved in the recycling of slab material, has a relatively uniform major element composition, and represents the depleted and outgassed source of mid-ocean ridge basalts. The upper mantle transition zone (400-1000 km depth) divides the depleted part of the mantle in a well mixed, low viscosity domain and a deeper one with high viscosity and slower transients; here, mantle flow is distorted by viscosity stratification, effects of phase transitions in the mantle silicates, and changes in plate motion at Earth's surface.

For long term survival the deep domain must have a slightly higher intrinsic density than the overlying depleted mantle [*Tackley*, 1998; *Kellogg et al.*, 1999]. The interface between these domains would be close to isopycnic (i.e., compositional and thermal effects on buoyancy are in balance) and significant dynamic topography can develop with some slabs penetrating to near the CMB [*Kellogg et al.*, 1999]; this resembles the 'penetrative' convection proposed previously for stratification at 660 km depth [*Silver et al.*, 1988]. The anomalous deep mantle "layer" — if it indeed exists — could represent long-lived differences in composition or phase chemistry [*Van der Hilst and Kárason*, 1999]. Alternatively, in an evolutionary sense, it may not yet have been churned by subduction. It takes many tens of millions of years (or more) before changes in the plate configuration at Earth's surface affect the heterogeneity structure of the deep mantle. Moreover, as cold slabs sink, their negative buoyancy can be diminished (or enhanced) by phase reactions in the upper mantle transition zone and may be neutralized by compositional effects well above the very base of the mantle [*Kesson et al.*, 1998]. In this transient system recycling may involve only the upper and mid-mantle, and few subduction systems may have operated long enough to produce (sufficiently 'cold') slabs that reach the CMB and churn the deep layer. This can be tested by rigorous integration of plate reconstructions and tomographic images of mantle structure [*Besse, personal communication*, 1999] by means of numerical modeling [e.g., *Bunge et al.*, 1998].

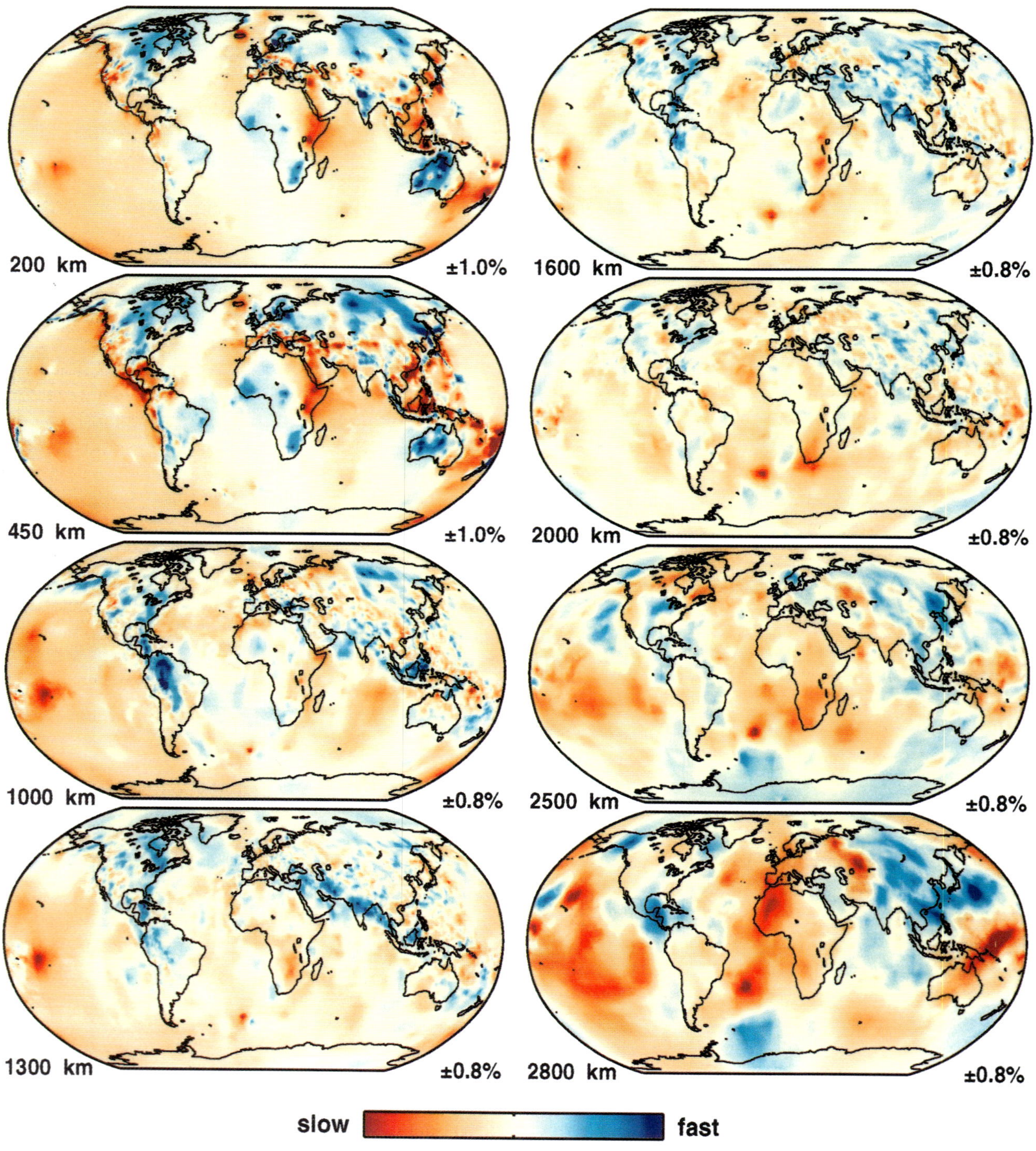


**Plate 3.** Robinson projections of the lateral variation in *P* wavespeed at several depths in Earth's mantle according to *Kárason and Van der Hilst* [1999]. This model is based on data from the seismic phases displayed in Figure 1, and thus benefits from improved sampling in the deep mantle (Figure 2), and the flexible parameterization illustrated in Plate 2 (although the effect of the irregular grid is hardly appreciated at this scale).

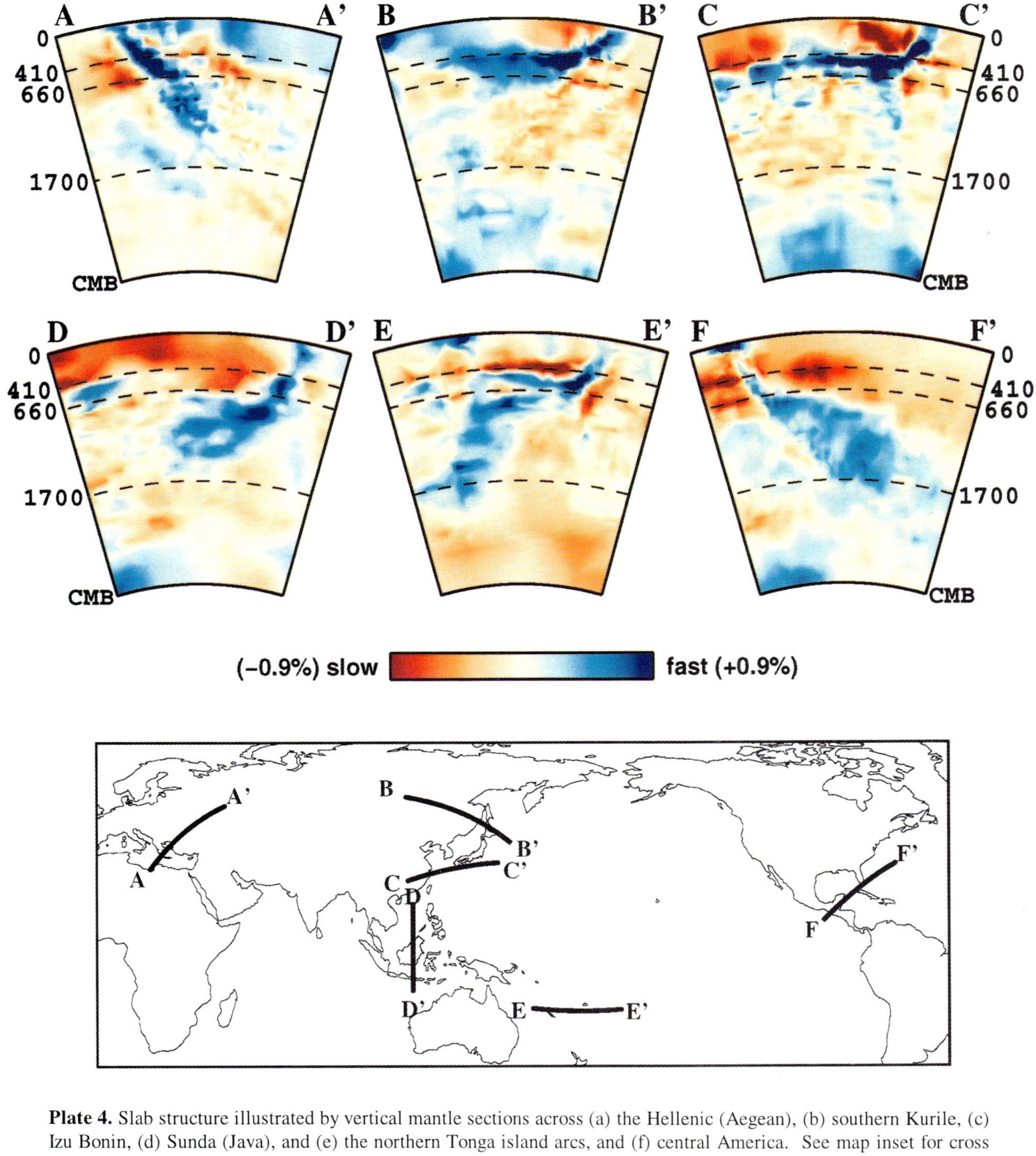


**Plate 4.** Slab structure illustrated by vertical mantle sections across (a) the Hellenic (Aegean), (b) southern Kurile, (c) Izu Bonin, (d) Sunda (Java), and (e) the northern Tonga island arcs, and (f) central America. See map inset for cross section locations.

## DIRECTIONS FOR FUTURE RESEARCH

Seismic tomography did not settle the debate about the scale of mantle convection just by demonstrating the deep penetration of many slabs, but it has moved research away from conventional — but inadequate — end-member convection models. Many aspects of mantle convection remain enigmatic [e.g., *Albarède and Van der Hilst*, 1999]. Seismic imaging will continue to play a central role in constraining the pattern of convective flow, but several issues must be sorted out. However dramatic the "red-and-blue" images, our actual understanding of them is unsatisfactory. We need better constraints on the amplitude of the changes in *P* and *S*-wavespeed, and their ratio $\delta ln V_S/\delta ln V_P$, and we need to integrate them with results from experimental and theoretical mineral physics to quantify changes in temperature, phase, and bulk composition. In addition, the trajectories of mantle convection must be delineated in even more detail in order to resolve outstanding issues, such as the ultimate fate of the slabs that penetrate across the upper mantle transition zone and the source and morphology of the return flow (plumes?). The latter requires a significant effort to improve data coverage beneath regions far away from seismically active plate boundaries. Seismic anisotropy can also be used as a tool for delineating flow trajectories [*Montagner*, 1998]. Continued imaging of mantle structure beneath convergent margins and better integration with the record of plate motions through computational geodynamic modeling will further our understanding of subduction and its partnership with mantle convection.

*Acknowledgments.* We thank Wei-jia Su and Steve Grand for providing us with their models, Li Zhao for the mode calculations that we used to approximate the sensitivity kernels for the $PKP_{DF}$-$P_{diff}$ travel time differentials, Bernard Steinberger for producing the diagram used as Plate 1d, and Mark Richards and Richard Gordon for constructive reviews. This research is supported by the National Science Foundation (EAR-9909492).

## REFERENCES

Abers, G.G., and S.W. Roecker, Deep structure of an arc-continent collision: Earthquake relocation and inversion for upper mantle *P* and *S* velocities beneath Papua New Guinea, *J. Geophys. Res.*, 96, 6379-6401, 1991.

Aki, K.A., A. Christoffersson, and E.S. Husebye, Determination of the three-dimensional seismic structure of the lithosphere, *Geophys. J. R. astron. Soc.*, 82, 277-296, 1977.

Albarède, F., and R.D. van der Hilst, New mantle convection model may reconcile conflicting evidence, *EOS Trans. Am. Geophys. Un.*, 45, 535-539, 1999.

Anderson, D.L., Tomography, Gutenberg, and Freud: are color cross sections really Rorschach tests? *EOS Trans. Am. Geophys. Un.*, 80, F719, 1999

Bijwaard, H. and W. Spakman, Tomographic evidence for a narrow whole mantle plume below Iceland, *Earth Planet. Sci. Lett.*, 166, 121-126, 1999.

Bijwaard, H., W. Spakman, and E.R. Engdahl, Closing the gap between regional and global travel time tomography, *J. Geophys. Res.*, 103, 30055-30078, 1998.

Boschi, L., and A.M. Dziewonski, High- and low resolution images of Earth's mantle: implication of different approaches to tomographic modeling, *J. Geophys. Res.*, 104, 25567-25594, 1999.

Bunge, H.-P., M.A. Richards, and J.R. Baumgardner, The effect of viscosity stratification on mantle convection, *Nature*, 379, 436-438, 1996.

Bunge, H.-P., Richards, M.A., Lithgow-Bertelloni, C., Baumgardner, J.R., Grand, S.P., and Romanowicz, B.A., Time scales and heterogeneous structure in geodynamic Earth models, *Science*, 280, 91-95, 1998.

Christensen, U.R., The influence of trench migration on slab penetration into the lower mantle, *Earth Planet. Sci. Lett.*, 140, 27-39, 1996.

Čížková, H., O. Čadek, A.P. van den Berg, and N.J. Vlaar, Can lower mantle slab-like anomalies be explained by thermal coupling between the upper and lower mantles? *Geophys. Res. Lett.*, 26, 1501-1504, 1999.

Davies, G.F., Penetration of plates and plumes through the mantle transition zone, *Earth Planet. Sci. Lett.*, 133, 507-516, 1995.

Dziewonski, A.M., B.H. Hager, and R.J. O'Connell, Large-scale heterogeneities in the lower mantle, *J. Geophys. Res.*, 82, 239-255, 1977.

Dziewonski, A.M., Mapping the lower mantle: determination of lateral heterogeneity up to degree and order 6, *J. Geophys. Res.*, 89, 5929-5952, 1984.

Engdahl, E.R., R.D. van der Hilst, and R. Buland, Global teleseismic earthquake relocation with improved travel times and procedures for depth determination, *Bull. Seism. Soc. Am.*, 88, 722-743, 1998.

Fischer, K., T.H. Jordan, and K.C. Creager, Seismic constraints on the morphology of deep slabs, *J. Geophys. Res.*, 93, 4773-4783, 1988.

Fukao, Y., M. Obayashi, H. Inoue, and M. Nenbai, Subducting slabs stagnant in the mantle transition zone, *J. Geophys. Res.*, 97, 4809-4822, 1992.

Goes, S., W. Spakman, and H. Bijwaard, A lower mantle source for central European volcanism, *Science*, 286, 1928-1931, 1999.

Grand, S.P., Mantle shear structure beneath the Americas and surrounding oceans, *J. Geophys. Res.* 99, 11,591-11,621, 1994.

Grand, S.P., R.D. van der Hilst, and S. Widiyantoro, Global seismic tomography: A snapshot of convection in the Earth, *Geol. Soc. Am. Today*, 7, No. 4, 1-7, 1997.

Griffiths, R.W., R.I. Hackney, and R.D. van der Hilst, A laboratory investigation of effects of trench migration on the descent of subducted slabs, *Earth Planet. Sci. Lett.*, 133, 1-17, 1995.

Guillou-Frottier, L., J. Buttles, and P. Olson, Laboratory experiments of the structure of subducted lithosphere, *Earth Planet. Sci. Lett.*, 133, 19-34, 1995.

Gudmundsson, O., and M. Sambridge, A regionalized upper mantle (RUM) seismic model, *J. Geophys. Res.*, 103, 7121-7130, 1998.

Humphreys, E., and R.W. Clayton, Adaption of backprojection tomography to seismic travel time problems, *J. Geophys. Res.*, 93, 1073-1085, 1988.

Inoue, H., Y. Fukao, K. Tanabe, and Y. Ogata, Whole mantle *P*-wave travel time tomography, *Phys. Earth Planet. Inter.*, 59, 294-328, 1990.

Ishii, M. and J. Tromp, Normal-mode and free-air gravity constraints on lateral variations in velocity and density of Earth's mantle, *Science*, 285, 1231-1236, 1999.

Julian, B.R., and M.K. Sengupta, Seismic travel time evidence for lateral inhomogeneity in the deep mantle, *Nature*, 242, 443-447, 1993.

Kárason, H., and R.D. van der Hilst, Improving seismic models of whole mantle *P*-wavespeed by the inclusion of data from differential times and normal modes, *EOS Trans. Am. Geoph. Un.*, 79, F656, 1998.

Kárason, H., and R.D. van der Hilst, New constraints on 3D variations in mantle *P* wavespeed, *EOS Trans. Am. Geoph. Un.*, 80, F731, 1999.

Kárason, H., and R.D. van der Hilst, Improving global tomography models of P-wavespeed I: Incorporation of differential times of refracted and diffracted core phases (*PKP, Pdiff*), *J. Geophys. Res.* (under review), 2000.

Kellogg, L.H., B.H. Hager, and R.D. van der Hilst, Compositional stratification in the deep mantle, *Science*, 283, 1881-1884, 1999.

Kennett, B.L.N., S. Widiyantoro, and R.D. van der Hilst, Joint seismic tomography for bulk-sound and shear wavespeed, *J. Geophys. Res.*, 103, 12469-12493, 1998.

Kesson, S.E., J.D. Fitz Gerald, and J.M. Shelley, Mineralogy and dynamics of a pyrolite lower mantle, *Nature,* 397, 252-255, 1998.

Kincaid, C. and P. Olson, An experimental study of subduction and slab migration, *J. Geophys. Res.*, 92, 13832-13840, 1987.

Li, X.-D. and B. Romanowicz, Global mantle shear-velocity model developed using nonlinear asymptotic coupling theory, *J. Geophys. Res.*, 101, 22245-22272, 1996.

Lithgow-Bertelloni, C. and M. Richards, The dynamics of Cenozoic and Mesozoic plate motions, *Rev. Geophys.*, 36, 27-78, 1998.

Liu, X.-F., and A.M. Dziewonski, Global analysis of shear wave velocity anomalies in the lower-most mantle, In: *Observational and theoretical constraints on the Core Mantle Boundary Region*, Gurnis, M., Wysession, M.E., Knittle, E., and Buffett, B.A. (Eds.), Geodynamics Series (American Geophysical Union), 28, 21-36, 1998.

Masters, G., S. Johnson, G. Laske, and H. Bolton, A shear-velocity model of the mantle, *Phil. Trans. R. Soc. Lond. A,* 354, 1385-1411, 1996.

McSweeney, T., *PhD thesis*, (supervisor K.C. Creager) University of Washington, Seattle, 1997.

Montagner, J.-P., Can seismic tomography tell us anything about convection in the mantle? *Rev. Geophys.*, 32, 115-137, 1994.

Montagner, J.-P., Where can seismic anisotropy be detected in Earth's mantle? In boundary layers ...., *Pure Appl. Geophys.*, 151, 223-156, 1998.

Nataf, H.-C., What is responsible for thermal coupling in layered convection? *J. Phys - Paris*, 49, 1707-1714, 1988.

Nolet, G., Solving or resolving inadequate and noisy tomographic system, *J. Comp. Phys.*, 61, 463-482, 1985.

Obayashi, M., and Y. Fukao, *P* and *PcP* travel time tomography for the core-mantle boundary, *J. Geophys. Res.*, 102, 17825-17842, 1997.

Paige, C. C., and M. A. Saunders, LSQR: an algorithm for sparse linear equations and sparse least squares, ACM Trans.Math.Soft., 8, 43-71 and 195-209, 1982.

Pulliam, R.J., Vasco, D.W. and Johnson, L.R., Tomographic inversions for mantle P-wave velocity structure based on the minimization of $l^2$ and $l^1$ norms of International Seismological Centre travel time residuals, *J. Geophys. Res.*, 98, 699-734, 1993.

Richards, M.A., and D.C. Engebretson, Large-scale mantle convection and the history of subduction, *Nature,* 355, 437-440, 1992.

Riedel, M.R., and S. Karato, Grain-size evolution in subducted oceanic lithosphere associated with the olivine-spinel transformation and its effects on rheology, *Earth Planet. Sci. Lett.*, 148, 27-43, 1997.

Ritsema, J., H.J. van Heijst, and J.H. Woodhouse, Complex shear wave velocity structure imaged beneath Africa and Iceland, *Science*, 286, 1925-1928, 1999.

Ritzwoller, M., and E.M. Lavely, 3-dimensional models of Earth's mantle, *Rev. Geophys.*, 33, 1-66, 1995.

Romanowicz, B., Seismic tomography of Earth's mantle, *Ann. Rev. Earth Planet. Sci.*, 19, 77-99, 1991.

Sengupta, M.K., and M.N. Toksöz, Three-dimensional model of seismic velocity variations in the earth's mantle, *Geophys. Res. Lett.*, 3, 84-86, 1976.

Silver, P.G., R.W. Carlson, and P. Olson, Deep slabs, geochemical heterogeneity, and the large-scale structure of mantle convection, *Ann. Rev. Earth Planet. Sci.,* 16, 477-541, 1988.

Spakman, W. and Nolet, G., Imaging algorithms, accuracy and resolution in delay time tomography, In: *Mathematical Geophysics: a Survey of Recent Developments in Seismology and Geodynamics,* pp. 155-188, ed. N.J. Vlaar *et al.*, Reidel, Dordrecht, 1988.

Su, W.-j., R.L. Woodward, and A.M. Dziewonski, Degree 12 model of shear velocity heterogeneity in the mantle, *J. Geophys. Res.,* 99, 6945-6981, 1994.

Su, W.-j., and A.M. Dziewonski, Simultaneous inversion for 3D variations in shear and bulk velocity in the mantle, *Phys. Earth Planet. Inter.,* 100, 135-156, 1997.

Van Heijst, H.J., and J.H. Woodhouse, Global high resolution phase velocity distributions of overtone and fundamental-mode surface waves determined by mode branch stripping, *Geophys. J. Int.*, 137, 601-620, 1999.

Van der Hilst, R.D., Complex morphology of subducted lithosphere in the mantle beneath the Tonga trench, *Nature*, 374, 154-157, 1995.

Van der Hilst, R.D., E.R. Engdahl, W. Spakman, and G. Nolet,

Tomographic imaging of subducted lithosphere below northwest Pacific island arcs, *Nature*, 353, 37-43, 1991.

Van der Hilst, R.D., and T. Seno, Effects of relative plate motion on the deep structure and penetration depth of slabs below the Izu-Bonin and Mariana island arcs, *Earth Planet. Sci. Lett.*, 120, 395-407, 1993.

Van der Hilst, R.D., S. Widiyantoro, and E.R. Engdahl, Evidence for deep mantle circulation from global tomography, *Nature*, 386, 578-584, 1997.

Van der Hilst, R.D., and H. Kárason, Compositional heterogeneity in the bottom 1000 km of Earth's mantle: towards a hybrid convection model, *Science*, 283, 1885-1888, 1999.

Vasco, D.W., and L.R. Johnson, Whole earth structure estimated from seismic arrival times, *J. Geophys. Res.*, 103, 2633-2671, 1998.

Wen, L.X., and D.L. Anderson, The fate of slabs inferred from seismic tomography and 130 million years of subduction, *Earth Planet. Sci. Lett.*, 133, 185-198, 1995.

Widiyantoro, S., *Ph.D. Thesis*, (supervisors B.L.N. Kennett and R.D. van der Hilst) Australian National University, Canberra, Australia, 1997.

Widiyantoro, S., and R.D. van der Hilst, Structure and evolution of lithospheric slab beneath the Sunda arc, Indonesia, *Science*, 271, 1566-1570, 1996.

Widiyantoro, S., Kennett, B.L.N., and Van der Hilst, R.D., Extending shear-wave tomography for the lower mantle using S and SKS arrival-time data, *Earth, Planets, and Space*, 50, 999-1012, 1998.

Widiyantoro, S., Kennett, B.L.N., and Van der Hilst, R.D., Seismic tomography with P and S data reveals lateral variations in the rigidity of slabs, *Earth Planet. Sci. Lett.*, 173, 91-100, 1999.

Wolfe, C.J., I.Th. Bjarnason, J.C. VanDecar, and S.C. Solomon, Seismic structure of the Iceland mantle plume, *Nature*, 385, 245-247, 1997.

Woodhouse, J. H., and A.M. Dziewonski, 1984, Mapping the upper mantle: Three-Dimensional modeling of earth structure by inversion of seismic waveforms, *J. Geophys. Res.*, 89, 5953-5986, 1994.

Woodhouse, J.H. and A.M. Dziewonski, Seismic modeling of the Earth's large scale structure, *Phil. Trans. R. Soc. A.*, 328, 291-308, 1989.

Wysession, M.E., Continents of the core, *Nature,* 381, 373-374, 1996.

Zhao, L., T.H. Jordan, and C. Chapman, Three-dimensional Fréchet differential kernels, *Geophys. J. Int.* (in press) 2000.

Zhong, S. and M. Gurnis, Mantle convection with plates and mobile, faulted plate margins, *Science*, 267, 838-843, 1995.

Zhou, H.-w, and R.W. Clayton, *P* and *S* wave travel time inversions for subducting slabs under the island arcs of the northwest Pacific, *J. Geophys. Res.*, 95, 6829-6851, 1990.

H. Kárason and R. D. van der Hilst, Department of Earth, Atmospheric and Planetary Sciences, Massachusetts Institute of Technology, Cambridge, MA, 02139, USA.
(emails: karason@mit.edu, hilst@mit.edu)

# Mantle convection and plate motion history: Toward general circulation models

Mark A. Richards

*Department of Geology and Geophysics, University of California, Berkeley, CA 94720*

Hans-Peter Bunge

*Department of Geological Sciences, Princeton University, Princeton, NJ 08544*

Carolina Lithgow-Bertelloni

*Department of Geological Sciences, University of Michigan, Ann Arbor, MI 48109*

Thermal boundary layer theory explains, to first order, the structure of oceanic lithosphere, i.e., depth and heat flow vs. age. For an internally heated mantle dominated by plate-scale flow, this theory also predicts that the main thermal heterogeneities in the mantle are generated by subducted oceanic lithosphere. This prediction is verified using Cenozoic and Mesozoic plate motion history to infer mantle heterogeneity structure related to subduction, leading to successful models for the observed geoid, present-day plate motions, and variations in seismic velocity in the mantle. These results suggest the feasibility of "general circulation models" for the mantle. Combining the motions of rheologically strong surface plates with a modest increase in mantle viscosity with depth results in convection dominated by downwellings at convergent plate margins, with little "small-scale" structure. Large-scale mantle heterogeneity structure responds to changes in plate motions on a time-scale of about $150 Myrs$, not coincidently, the same timescale for which oceanic plate motions can be determined from the magnetic isochron record. Applying Cenozoic and Mesozoic plate motion history to 3-D spherical convection models, we derive a general circulation model that is largely in accord with seismic tomography, with the exception of a major low-velocity structure beneath Africa. We also examine the evolution of core-mantle boundary structure and potential applications to variations in the character of geomagnetic reversals since the Cretaceous. Improvements

The History and Dynamics of Global Plate Motions
Geophysical Monograph 121

to the general circulation model should include better estimates of uncertainties in plate reconstructions, better treatment of rheology in the computational models, theoretical exploration of the "initial condition" problem, and treatment of chemical layering. All of these improvements appear feasible within the next decade.

## 1. INTRODUCTION

This paper is a summary of our efforts over the past decade to integrate the history of global plate motions into our understanding of mantle dynamics. Our work differs from "first principles" numerical modeling of mantle convection in that we have developed deterministic models for mantle convection that can be compared explicitly with a broad range of data, such as seismic tomography, the gravity field, and dynamic topography. A result of our effort has been the development of "general circulation models" for the plate/mantle system that span Mesozoic and Cenozoic time [*Bunge et al.*, 1998]. The purpose of this paper is to provide a concise explanation for the development of these fledgling models.

An important theme running through our effort is the need to clearly distinguish model inputs and outputs, since the two can easily be confused as models take on increasing sophistication. The distinction is not always obvious: For example, models *constrained* by the history of plate motions cannot be used to *explain* these motions, but such models can be used to examine the instantaneous force balance on the plates acting at a given instant of time, and they can also be tested by careful comparison with the gravity field or independent results from seismic imaging. One must be cautious if modeling results are (hopefully) to be used outside the narrow limits of one's own discipline (geodynamics), lest the various roles of constraining data become confused. The reader is forewarned!

The philosophy of our approach is illustrated in Figure 1. The oceanic lithosphere is a thermal boundary layer at the top of the convecting mantle. The motion of this boundary layer is, roughly, piecewise constant due to the strength and fracturing of lithospheric plates, with upwelling flow at spreading centers and downwelling flow at subduction zones. Neglecting other "small-scale" modes of convection, such as hotspot plumes or other sub-plate scale instabilities [see, e.g., *Davies*, 1988], the buoyancy forces driving this flow result from release of potential energy as the cooling lithosphere moves away from ridges and sinks at subduction zones.

Shortly following the development of the theory of plate tectonics [*Wilson*, 1965; *Morgan*, 1968; *McKenzie and Parker*, 1967], it was shown in a classic paper

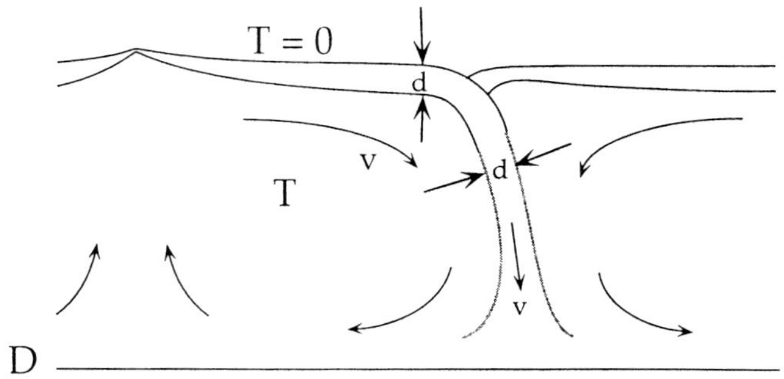


**Figure 1.** The relationship between the main thermal boundary layer (the lithosphere) and the dominant mode of mantle flow. Symbols are defined in the text.

by *Turcotte and Oxburgh* [1967] that a straightforward boundary layer model of mantle convection, similar to that illustrated in Figure 1, could explain the motion of plates, given the *Haskell* [1935] value for mantle viscosity, reasonable values for the thermal diffusivity and expansivity of mantle rock, and an estimate for the temperature at the base of the lithosphere. In a simplified formulation of this boundary layer theory, *Davies and Richards* [1992] showed that the buoyancy force of the sinking slab is much larger than that due to the thickening oceanic lithosphere. Balancing this force with viscous resistance to flow in the mantle, one obtains an estimate for the characteristic subduction velocity

$$v = D(g\rho\alpha T\sqrt{\kappa}/4\eta)^{2/3} \qquad (1)$$

or

$$v = gDd\rho\alpha T/(4\eta) \qquad (2)$$

where $D = 3x10^6 m$, $\rho = 4000 kg/m^3$, $\alpha = 2x10^{-5}/K$, $T = 1400^o$, $\kappa = 10^{-6} m^2/s$, and $\eta = 10^{22} Pa - s$, are, respectively, the depth, density, thermal expansivity, internal temperature, thermal diffusivity, and viscosity of the mantle. The characteristic thickness of the thermal lithosphere, $d$, is given by the square root of the product of the thermal diffusivity and the timescale for convective flow, $D/v$, so that

$$d = \sqrt{\kappa D/v} \qquad (3)$$

Using the numerical values listed above, we obtain an estimate for the characteristic subduction velocity of $v = 86mm/yr$, similar to observed plate velocities, and an estimate for the characteristic thermal boundary layer thickness of $d = 33km$, roughly consistent with the observed thickness of oceanic lithosphere. The thermal Rayleigh number, $Ra$ [see, e.g., *Turcotte and Schubert*, 1980, for a complete definition], representing the vigor of convection for this analysis is given by

$$Ra = \rho \alpha g T D^3 / \kappa \eta \approx 3x10^6 \quad (4)$$

or about 3000 times the critical $Ra$ for onset of thermal convection. This is probably an underestimate, because the more appropriate Rayleigh number based on internal heating is larger, and the viscosity of the upper mantle is probably a factor of 10-100 less than the mantle average estimate used. Thus the ratio of buoyancy forces to viscous forces is so large that convection is, *a priori*, expected to be time-dependent, or chaotic, regardless of the presence or absence of plates.

The boundary layer theory outlined above provides a first order explanation of the driving force of plate tectonics, as well as the thickening of the oceanic lithosphere with the square root of plate age. However, the theory also makes a clear prediction regarding the thermal heterogeneity structure of the sublithospheric mantle, namely, that it should be dominated by cold downwellings which are generated by subduction on roughly the timescale that it takes for a cold slab to sink to the bottom of the mantle. A great deal of this paper is devoted to demonstrating the validity and power of this simple idea.

We begin by demonstrating that the history of subduction, as mapped by plate tectonicists for Mesozoic and Cenozoic time, can explain the large-scale heterogeneity structure of the mantle as evidenced by the shape of the gravity field. We also confirm, as a consistency check, that this heterogeneity structure gives rise to instantaneous plate motions very much like those observed for the present. Next, we consider the basic fluid dynamics using large-scale computational models of mantle convection with plate motions imposed at the surface. These computational models are then applied to achieving a deeper understanding of the timescales for convective cycling in the mantle, the effects of key model parameters on thermal heterogeneity structure, and the evolution of the core-mantle boundary region.

## 2. MANTLE DYNAMICS AND PLATE TECTONICS

A rough correspondence between the locations of major geoid lows and Cretaceous subduction zones was pointed out by *Chase and Sprowl* [1983], and the complementary correspondence of the African - North American geoid high with the ancient Pangean supercontinent was noted by *Anderson* [1982]. The geoid, or shape of the Earth's gravitational equipotential surface (sea level in the oceans), is dominated by these broad highs and lows, and indeed the geoid spectrum is dominated by the longest wavelength structure at spherical harmonic degrees $L = 2 - 3$. *Richards and Engebretson* [1992] showed that the low-degree geoid was highly correlated with the time integrated history of subduction since mid-Mesozoic time ($200Ma$), and *Ricard et al.* [1993] developed a quantitative model for both the geoid and seismic heterogeneity by dropping subducted "slablets" into the mantle and letting them sink vertically at plate-like velocities. The implication of this work is that mantle heterogeneity structure is generated by active cold downwellings at subduction zones, with complementary passive upwelling regions, as expected for a convecting system dominated by internal heating and cooling of an upper boundary layer [e.g., *McKenzie et al.*, 1974].

We have constructed a model for mantle heterogeneity by mapping subduction zones in the hotspot reference frame since $200Ma$ [*Engebretson et al.*, 1985; 1992], calculating slab buoyancies from a thermal cooling model based on subducted plate age, and allowing the slabs to sink vertically in the mantle. Thermal diffusion of the slab temperature anomalies is neglected, because only lengthscales of thousands of kilometers are considered. ($L = 15$ corresponds to a horizontal wavelength of about $2700km$.) Likewise, horizontal advection of slabs is ignored, as the relatively slow motion of hotspots implies little horizontal transport in the time required for a slab to sink to the bottom of the mantle.

The slab sinking velocity is set to the convergence rate in the upper mantle, but the slabs are slowed by a factor of $ln\ \eta^*$, where $\eta^*$ is the ratio of upper to lower mantle viscosity. In the model shown, $\eta^*$ is 40, which yields a best fit to both the geoid and observed seismic velocity heterogeneity [e.g., *Su and Dziewonski*, 1991; *Li and Romanowicz*, 1997]. Thus the slabs slow down by a factor of about 4 ($ln$ 40=3.69) as they sink into the lower mantle, consistent with flow models for layered viscosity [*Gurnis and Davies*, 1986; *Richards*, 1991].

Figure 2 shows planforms of the slab model at various depths for the present-day and also at $56Myrs$, emphasizing that the model contains the element of time-evolution of mantle structure. Note that the large-scale structure changes little during Cenozoic time, as evidenced by the persistent circum-Pacific pattern of subducted slabs, and also by the fact that the predicted

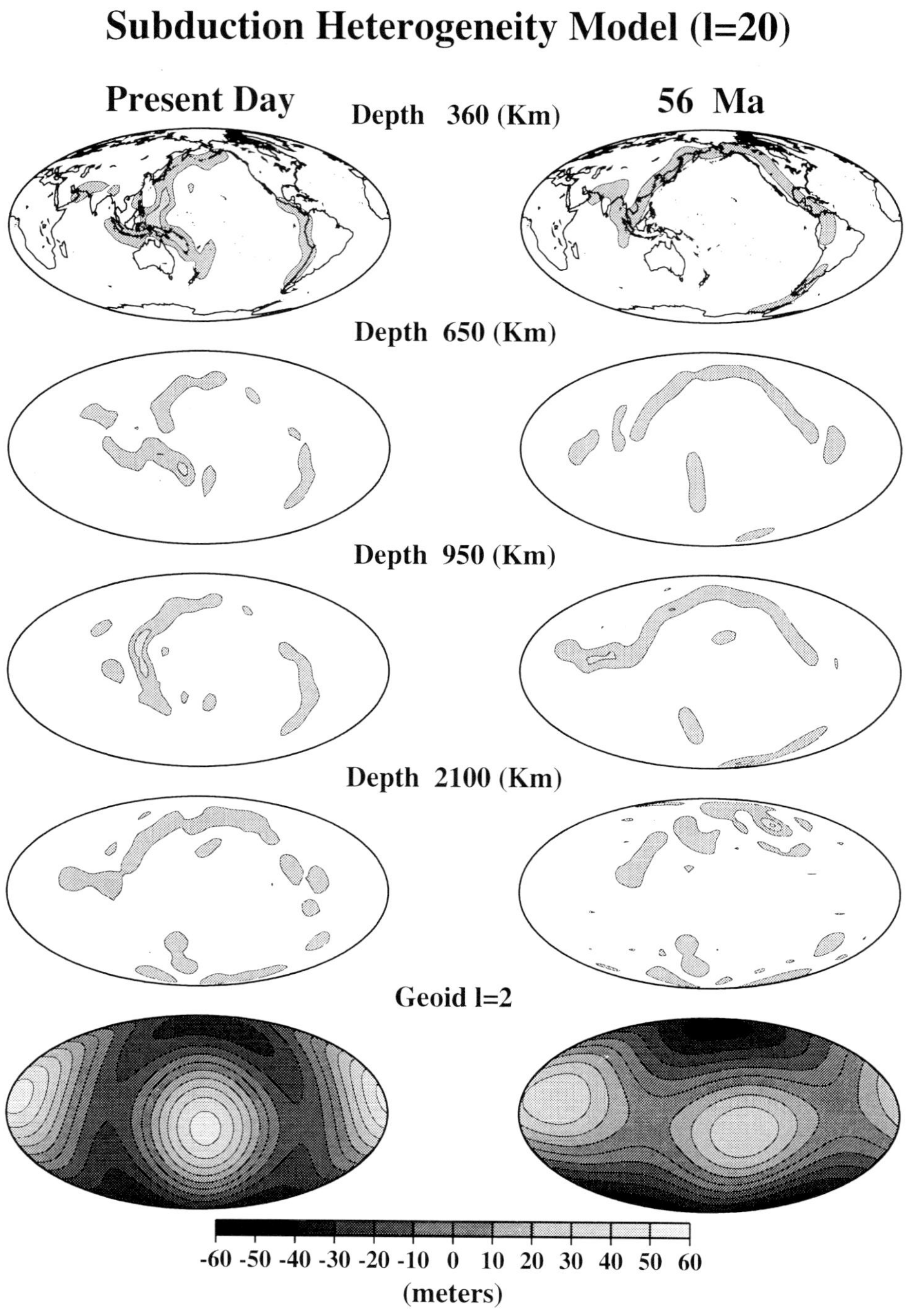


**Figure 2.** Mantle density heterogeneity model for the present-day Earth and at $56Ma$ at depths of 360, 650, 950, and $2100km$. Contoured regions represent areas of high density with respect to the surrounding mantle, in other words, the locations of the subducted slabs at that depth. The spatial and temporal changes in the location of subduction (from mostly the northern Pacific to the western Pacific) are well illustrated by the high densities in the northern regions at the bottom of the mantle and the high density regions in the western Pacific for the upper mantle. Subduction of the oceanic part of the Indian plate under Eurasia is also well illustrated. In the top panel the continental outlines in their present position have been superimposed for reference. The bottom panels show the harmonic degree 2 geoid at both times, illustrating how little the large-scale pattern of subduction (and mantle heterogeneity) have changed through Cenozoic time.

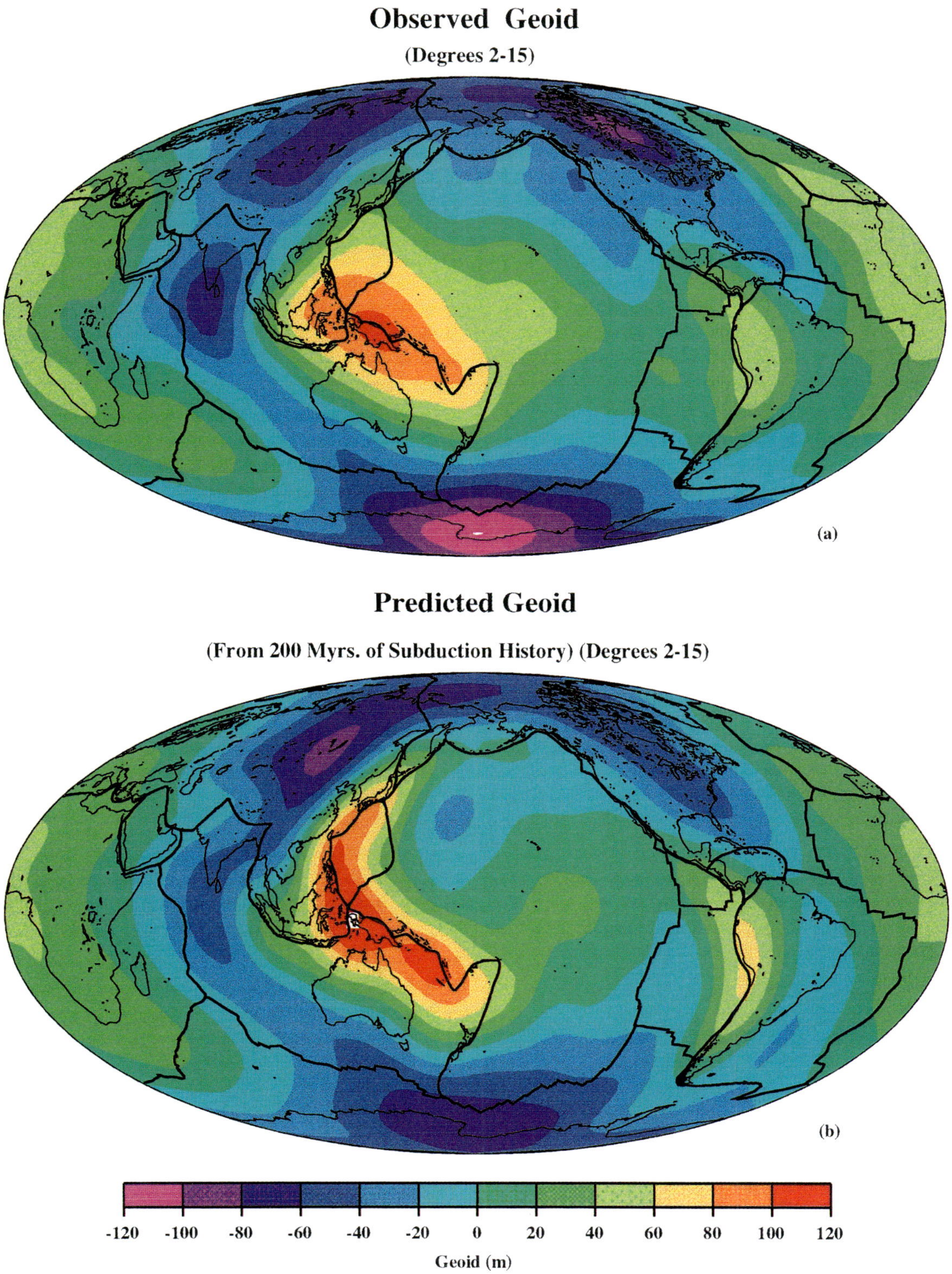


**Plate 1.** (a) Observed nonhydrostatic geoid and (b) synthetic geoid from Cenozoic and Mesozoic subduction history, using a slowing factor of 4 as slabs sink into the lower mantle and a factor of 40 increase in viscosity with depth. (See *Lithgow-Bertelloni and Richards* [1998] for more details of the model.)

degree 2 geoid (bottom panels of Figure 2) are little-changed over a $56Myr$ time interval.

Plate 1a shows the observed nonhydrostatic geoid up to harmonic degree 15, and Plate 1b shows the recent "slab model" geoid of *Lithgow-Bertelloni and Richards* [1998]. The presence of cold subducted slabs corresponds to high seismic velocity, and to geoid lows. The latter may seem anti-intuitive, but the geoid signal is compensated by downwarp of the Earth's surface and core-mantle boundary due to negative buoyancy (and *vice versa*) for models with modest viscosity contrasts between the upper and lower mantle [e.g., *Richards and Hager*, 1984]. The correlation between observed and model geoids is significant at greater than the 99% confidence level, and the success of this model is impressive given that there is really only *one* free parameter, namely $\eta^*$. It is particularly noteworthy that $\eta \approx 30 - 50$ is now considered to be entirely consistent with independent constraints on mantle viscosity structure from post-glacial rebound [*Mitrovica*, 1996]. Thus application of plate motion history data to the simple ideas illustrated in Figure 1 explains the main features of mantle heterogeneity structure and dynamics in terms of physical model parameters that are fairly well-constrained.

Having employed the history of plate motions to explain the major features of mantle dynamics, it is instructive to examine, in a quantitative way, the evolution of global plate motions. Although there are several ways to do this, we have found it most useful to consider separately the component of surface plate motions related to divergent and convergent boundaries (*poloidal* motion) and the component related to strike-slip, or transform, boundaries (*toroidal* motion). This essentially mathematical decomposition [see *Hager and O'Connell*, 1979] is useful, because internal buoyancy forces generate only poloidal flow in the absence of lateral viscosity variations, while the surface plates themselves are largely responsible for the toroidal component.

Figure 3a shows an analysis of poloidal/toroidal partitioning of surface plate motions through Cenozoic and Mesozoic time, based on reconstructions compiled by various authors [*Gordon and Jurdy*, 1986; *Engebretson et al.*, 1985; 1992; *Larson and Chase*, 1972; *Larson and Pitman*, 1972; *Scotese*, 1990; *Minster and Jordan*, 1978; *Gripp and Gordon*, 1991]. Error estimates are based on formal uncertainties in rotation poles for Cenozoic motions, while *ad hoc* error estimates for the Mesozoic are based on extremal bounds implied by alternative plate motion reconstructions. Toroidal motions appear relatively constant in time, and they are of amplitude 1/3 or less that of poloidal motions (see Figure 3b). (Note that numerous references in the literature to "equipartitioning" of toroidal and poloidal motions are therefore misleading.) By contrast, there is a large increase in poloidal motions as we go back toward the end of the Mesozoic ($65Ma$), and the fact that this increase is not mirrored in the toroidal component remains a mystery without obvious explanation [see *Lithgow-Bertelloni and Richards*, 1995, for a more complete analysis].

More important to the present discussion is the fact that overall plate motions, including global subduction rates, are almost a factor of two higher in the mid-Cretaceous than at the present time. As a result, the mid-mantle heterogeneities in the slab heterogeneity model that are most responsible for generating the observed geoid anomalies (and for explaining the large-scale seismic heterogeneity structure) are dominated by subducted slabs from Cretaceous time. Hence, the original observation of [*Chase and Sprowl*, 1983] regarding correspondence of the geoid and Cretaceous subduction patterns is readily understood in terms of plate tectonic history. This latter point emphasizes the crucial importance of accounting for the character of time-dependence in the plate/mantle convecting system: There are no "closed" convection cells of lasting duration in the mantle, and interpretation of mantle structure and dynamics requires a consideration of the history of plate motions.

As a consistency check on the model described above, we have modeled global plate motions driven by the slab-derived heterogeneity structure. The present-day plate geometry at the surface is specified, and the internal density structure is used in a straightforward flow calculation for the tractions on these plates, yielding instantaneous plate velocities. Following *Hager and O'Connell* [1981] and *Ricard and Vigny* [1989], the driving tractions computed analytically in a Newtonian viscous mantle are balanced by the tractions due to viscous resistance to plate motion. We again use a layered viscosity mantle of $\eta^* = 40$, although the derived plate motions do not prove very sensitive to the viscosity structure, but rather to the absolute viscosity [*Lithgow-Bertelloni and Richards*, 1998]. For an upper mantle viscosity of $1x10^{21} Pa - sec$ we derive global plate motions in the no-net rotation reference frame that are very similar to those observed (Figure 4a,b), explaining 96% of the horizontal divergence (poloidal) component and 66% of the radial vorticity (toroidal) component of the surface velocity field, and obtaining a global correlation coefficient between observed and predicted plate motions of 0.91. (By contrast, the best

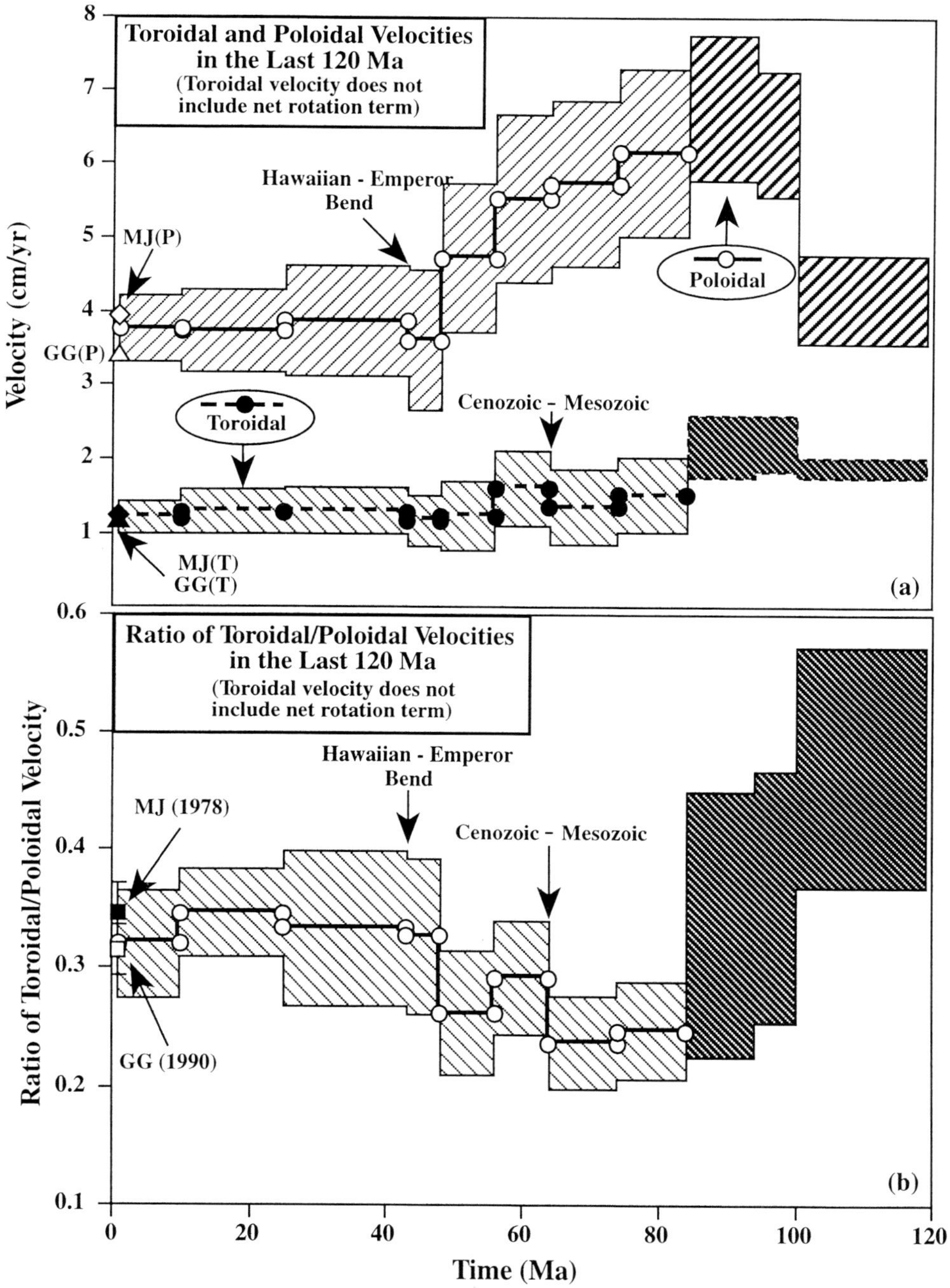


**Figure 3.** (a) Toroidal (solid circles and dashed line) and poloidal (open circles and solid line) velocities for the past $120 Myrs$, net rotation not included. The Cenozoic estimates are based on *Gordon and Jurdy* [1986], and the Mesozoic estimates are based on *Engebretson et al.* [1985; 1992]. The circles mark the beginning and end of each stage. The light hatched areas (slanted upward to the right for poloidal and to the left for toroidal) around the solid and dashed lines represent the $2\sigma$ confidence intervals on the velocities. The dark hatched areas represent the range (not formal uncertainties) of velocities from 84 to $119 Ma$, given two sets of rotation poles [*Engebretson et al.*, 1985; 1992]. The open and solid rhombohedra labeled MJ are the poloidal and toroidal velocity, respectively, for AM1 [*Minster and Jordon*, 1978]; the open and solid triangles labeled GG are the corresponding values for HS2-NUVEL1 [*Gripp and Gordon*, 1991]. The $2\sigma$ error bars are smaller than the symbols. (b) ratio of toroidal to poloidal velocities; net rotation not included. Hatching is as (a). Solid and open squares are the ratios for AM1-2 and HS2-NUVEL1, respectively.

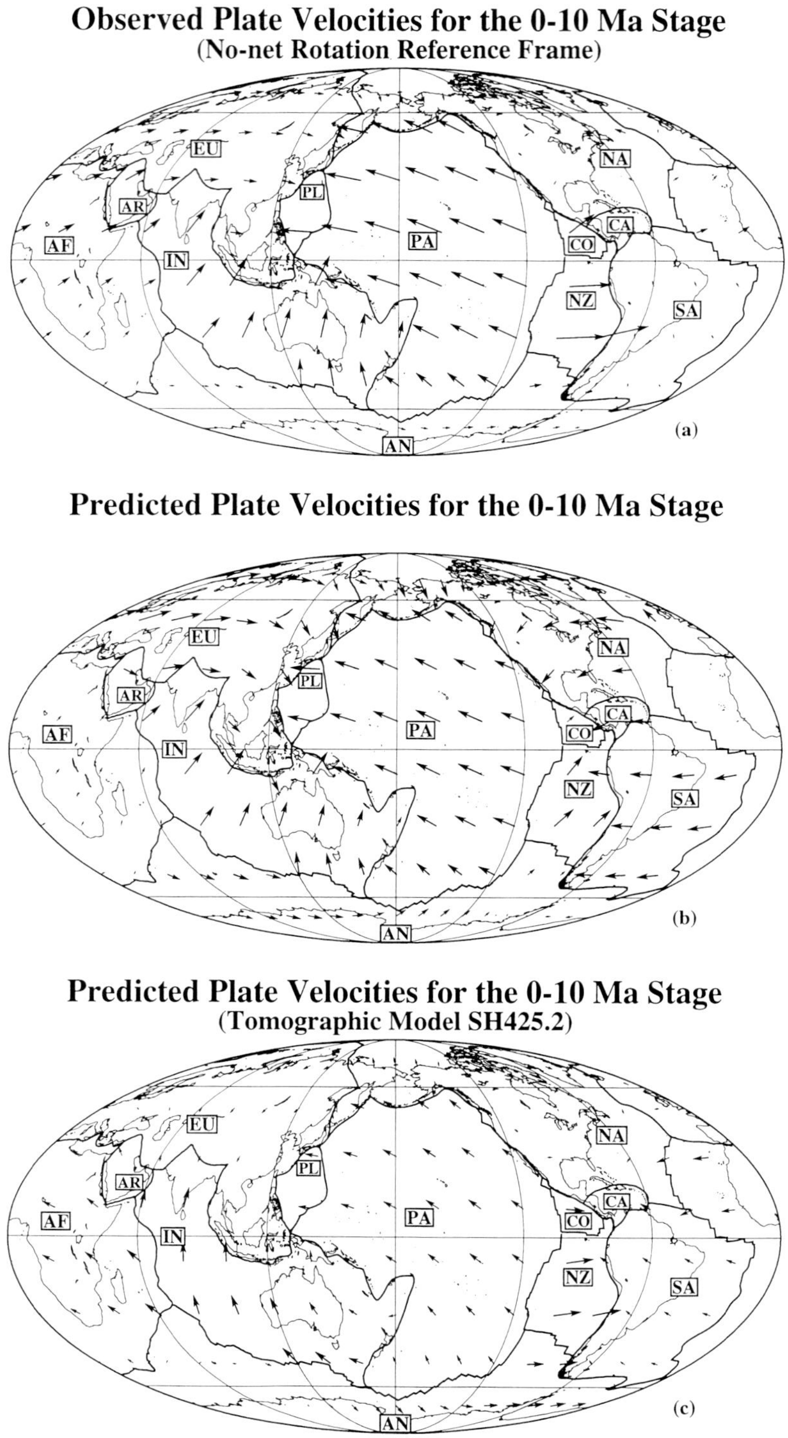


**Figure 4.** (a) Observed velocities in the no-net-rotation reference frame for 0-10$Ma$, (b) predicted velocities using our mantle heterogeneity model based on subduction history, and (c) predicted velocities using tomographic model SH425.2 [*Su and Dziewonski*, 1991]. Plate boundaries are indicated by solid black lines. Plate name abbreviations are indicated as well. The arrows indicate the direction of motion of each plate, and the length of the arrow is proportional to the magnitude of the plate velocity. For reference, the continents are also shown in their present position.

fitting plate motion model derived from similar fitting with a seismic tomography-based mantle density heterogeneity field (Figure 4c) yields a correlation coefficient of only 0.64, and explains only 66% of the horizontal divergence and 20% of the radial vorticity.) Thus the subduction history approach yields a model that is self-consistent, although not perfectly so, in connecting the generation of mantle buoyancy forces and the forces that drive plate motions.

The slab model for instantaneous global plate motions can also be applied to past plate motion reconstructions (with similar success) [*Lithgow-Bertelloni and Richards*, 1995], because the slab model predicts mantle heterogeneity throughout Mesozoic and Cenozoic time. It is important to note that such models cannot explain sudden changes in plate motions, such as may have occurred at Hawaiian-Emperor Bend time ($43Ma$), nor does any model based simply on continuous advection of density anomalies in the mantle [but see *Norton*, 1994, for an opposing view of this possible "non-event"]. On the other hand, an important feature of the subduction history approach is that aspects of the temporal evolution of the convecting system can be considered, whereas seismic imaging only applies to the present-day velocity heterogeneity of the mantle. It is this temporal aspect of our approach that we emphasize in the next section, where we apply a full thermal convection model to the ideas illustrated above.

## 3. CONVECTION MODELS CONSTRAINED BY PLATE MOTIONS

A major goal of geodynamics is to understand the role of plates in mantle convection, and this goal really consists of two parts: First, we need to understand why plates exist at all, and this is thought to be mainly an issue of appropriate rheological characterization [see, e.g., *Weinstein*, 1996; *Bercovici*, 1993; 1995; *Zhong et al.*, 1998]. Second, we need to understand the effect of plates on the underlying convecting system, keeping in mind of course that the plates form the upper boundary layer. Developing general circulation models for the mantle requires that we address the latter goal, i.e., that we clearly understand how convection is modulated by surface plates. As long as plate motion history data are used to constrain such models (which appears to us the only feasible route), then the rheological causes of plates remain a secondary issue. In fact, general circulation models should soon be able to address important questions regarding plate boundary rheology [*Baumgardner et al.*, 1998].

Unfortunately, there are a number of poorly determined physical parameters, or conditions, of mantle convection (other than plates) that may affect the nature of flow: compositional layering, phase transitions, heat source distribution, pressure and temperature dependence of viscosity, etc. In this paper, we arbitrarily ignore compositional layering, because it is very difficult to model at present, and because there is increasing geophysical evidence that flow associated with plate motions penetrates most of the depth of the mantle [e.g., *van der Hilst et al.*, 1997]. We also ignore lateral variations in viscosity, aside from the existence of lithospheric plate boundaries at the surface, because radial viscosity variations within the deep mantle appear much more influential [*Tackley*, 1996]. The effects of radial viscosity stratification, heat source distribution, and phase changes have been investigated systematically by *Bunge et al.* [1996; 1997], who were able to identify and distinguish the characteristics of flow controlled by these model parameters.

In general, high $Ra$ number convection produces horizontal scales for upwelling and downwelling flow structure that are comparable to, or less than, the depth of the convecting fluid, so that in the case of mantle convection, simple isoviscous convection models exhibit heterogeneity dominated by sub-plate-scale flow corresponding to wavelengths of order 1000-2000$km$, or spherical harmonic degrees 20-40 [*Bunge et al.*, 1996]. For Earth, where the large plates of dimension 5000-10,000$km$ represent the dominant pattern of flow, and where the geoid indicates an even redder heterogeneity spectrum [*Chase*, 1979; *Crough and Jurdy*, 1980], it is necessary to elucidate mechanisms that produce larger characteristic scale-lengths. One obvious mechanism is the strength of the plates themselves, which to some degree must inhibit small-scale downwellings, but this mechanism alone appears insufficient [*Bunge and Richards*, 1996].

Numerical modeling studies of mantle convection in 3-D spherical geometry and at high $Ra$ number have demonstrated that increasing viscosity with depth, bottom heating from the core, and a flow-inhibiting endothermic phase reaction (e.g., 670$km$ spinel - perovskite) all tend to redden the thermal heterogeneity spectrum [*Tackley et al.*, 1993; 1994; *Bunge et al.*, 1996; 1997]. We have found that the largest of these effects, at least for Earth's mantle, is an increase in viscosity with depth by about 1-2 orders of magnitude, which, for reasons not entirely clear, strongly promotes an upper surface convection planform of long, linear downwellings, even in the absence of "subducting" plates [*Bunge et al.*, 1996].

Plates 2a,b show the near-surface and mid-mantle thermal fields for convection at $Ra = 10^8$, with purely internal heating (no core heat), and with an increase in mantle viscosity by a factor of 20 through the transition zone (500-1000$km$ depth). Near the surface, flow is dominated by long, cold, linear downwelling features which thicken considerably at depth within the higher viscosity lower mantle. We must emphasize that the surface flow is *not* plate-like at all, since the complementary (passive) upwelling structures are broad and diffuse, rather than being concentrated at narrow ridges, and also since the model contains no physics that would give rise to transform faults.

When observed present-day plate motions are used to constrain the surface velocity field (Plates 2c,d), the long, linear downwelling structures naturally lock onto zones of plate convergence (subduction zones). (The present-day motions are imposed for many overturn times, so that flow has settled into quasi-steady-state at the time shown, "quasi" referring to the fact that flow at $10^8$ Rayleigh number remains time-dependent, regardless of plate motion constraints.) However, much subplate scale structure remains in the form of convective rolls [*Richter and Parsons*, 1975] beneath the plates at an amplitude that should be (and is not) clearly observable from geoid and bathymetry data [*Davies*, 1988]. This problem is largely eliminated by making the plates strong, which we accomplish by increasing the viscosity of the upper mantle thermal boundary layer (lithosphere) to a value equal that of the lower mantle (Plates 2e,f). This stabilizes the boundary layer to small-scale instabilities, and results in both upper- and mid-mantle flow that is completely dominated by concentrated cold thermal downwellings at subduction zones.

These models show that the combination of (1) a largely internally-heated mantle, (2) the motions of strong surface plates, and (3) a modest increase in viscosity with depth yields a mantle flow dominated by sinking of the upper boundary layer at convergence zones, with little other significant active flow structure. Condition (1) is consistent with geochemical evidence for the distribution of radioactive heat sources within the Earth [*Wasserburg et al.*, 1964], condition (2) is obvious in the post plate tectonics world, and condition (3) is consistent with a broad range of geodynamic constraints on mantle viscosity structure [*Mitrovica and Forte*, 1996]. Therefore the simple model of Plates 2e,f provides a starting point for integrating plate tectonics and mantle convection. In fact, this model should be considered a "first-cut" general circulation model, limited primarily by the fact that it includes only present-day plate motions.

The need to account for plate motion history is fairly obvious: Plate motions in general, and subduction zones in particular, are not stable in time. Thus, for example, the largely extinct Farallon - North American subduction zone, which is now dominated by transform and spreading motion along the San Andreas - northern East Pacific Rise system, is not evident in Plate 2f, whereas seismic tomography clearly shows the old Farallon plate as a prominent feature sinking through the lower mantle beneath North, Central, and South America [*Grand*, 1994; *Van der hilst et al.*, 1997].

## 4. EFFECTS OF PLATE MOTION CHANGES

Introducing plate motion history into convection models involves a conceptual and philosophical hurdle that may not at first be obvious: Namely, that we can no longer just impose a single regime of plate motions until convection achieves equilibrium conditions, but instead we must run a history of plate motions that requires some rational guess at unknown mantle "initial conditions" at some time in the rather distant past. There are at least two possible solutions to this problem: (1) Develop some mathematical procedure for projecting our fairly good knowledge of present-day mantle structure into the past, using plate motions as a constraint, i.e., essentially running the equations governing convection backward in time. (2) Impose a long-enough plate motion history in forward models so that the mantle "memory" of initial conditions is essentially lost.

The first of these two methods [*Bunge, et al.*, 1992] requires recasting the energy equation, which cannot be simply time-reversed due to the presence of a diffusion term, in "adjoint" form, and solving an iterative optimization problem [e.g., *Chavent et al.*, 1975]. This method remains under development by one of us (HPB). However, we have achieved reasonable success with the second of these methods, which we now describe.

We employ a fairly "realistic" convection model, described more completely in *Bunge et al.* [1997; 1998], that includes compressible flow, a constant-temperature core supplying about 18% of the mantle heat flow (the remainder being internal heating), positive and negative Clapeyron slope phase transitions at the 410$km$ and 670$km$ discontinuities, respectively, and a factor of 40 increase in mantle viscosity through the transition zone. In order to understand both the essence of the "initial condition" problem and its possible resolution, we consider a simple model experiment: We impose on the convection model the regime of plate motions mapped at 119$Ma$ (mid-Cretaceous) until quasi-steady-state is achieved. Then we suddenly impose the plate motion

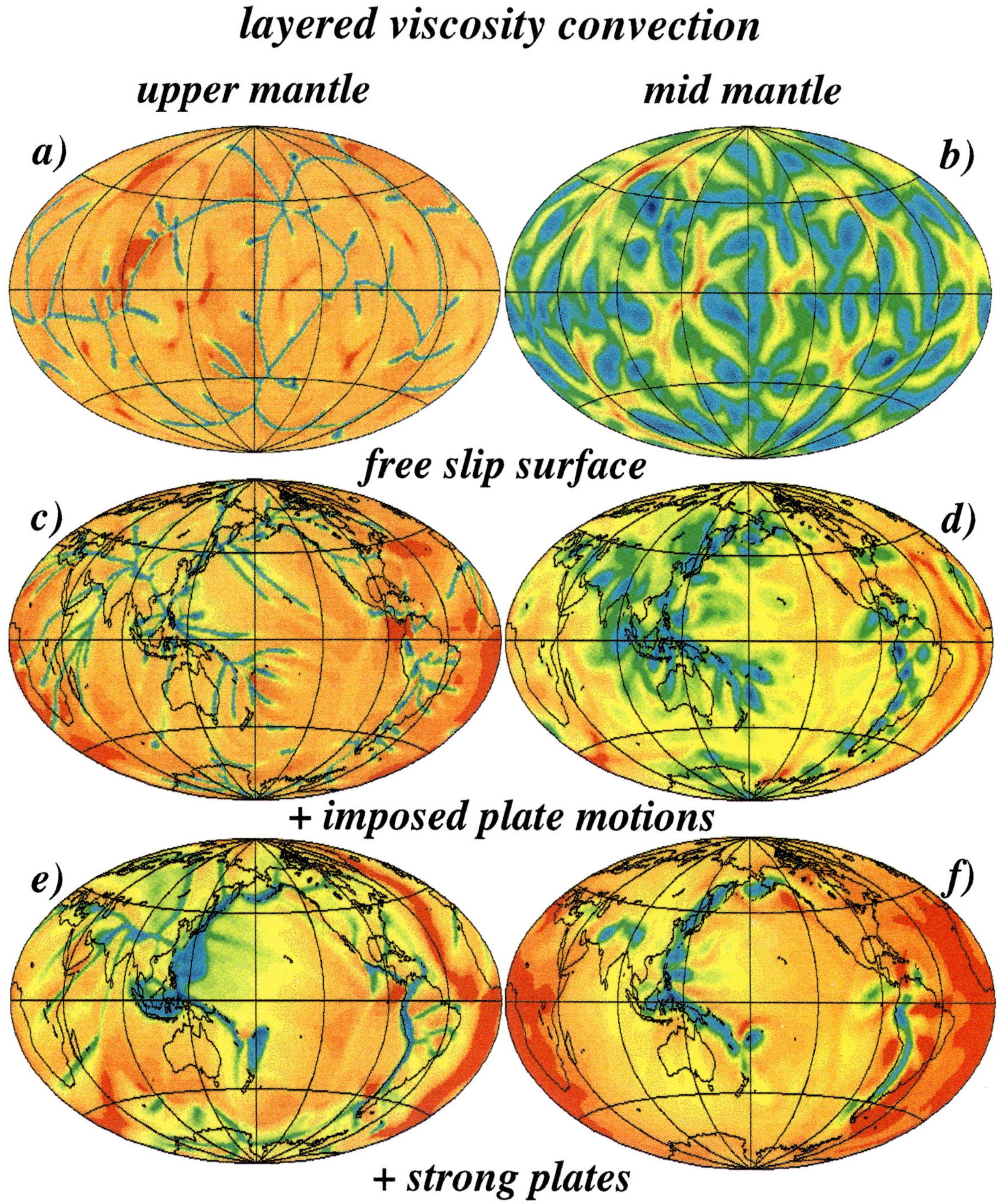


**Plate 2.** Temperature fields at $300 km$ depth (left panels) and $1500 km$ depth (right panels) for internally-heated convection with a viscosity contrast of 20 between the upper and lower mantle. In (a) and (b) the upper surface boundary condition is free-slip. In (c) and (d) the present-day plate motions are imposed as a velocity boundary condition, and in (e) and (f) the viscosity of the upper $300 km$ of the mantle (corresponding to the upper thermal boundary layer thickness) has been raised by a factor of 30. Blue is cold and red is hot. In (e) and (f) the pattern of convection has completely locked on to the imposed pattern of plate motions at the surface, and the temperature field is dominated by cold downwellings at subduction zones.

regime of the present-day, and observe the adjustment of the convecting system.

Figure 5 illustrates this adjustment by plotting the global correlation between the temperature field before the plate motion change and the temperature field for times afterward. The various curves show correlations for the entire mantle, the upper mantle, the lower mantle, and the core-mantle boundary (D") region. The correlations all begin at unity. Afterward, the whole mantle curve drops steadily until it levels off at a correlation of about 0.2 after an elapsed time of about $175Myrs$. The other curves behave similarly, with the most rapid adjustment occurring, not surprisingly, in the upper mantle, and the slowest in the D" layer. These results suggest that if we can map plate motions from about Middle Jurassic time forward, then general circulation models can be constructed for late Cenozoic time with relatively little knowledge of initial conditions.

The results of Figure 5 are not entirely surprising, but they are of fundamental importance to our overall goal, and worthy of additional comment. First, the adjustment time is comparable to the vertical transit time for these models, which is controlled by the radial viscosity structure. Second, this adjustment time is also comparable to the time over which global plate reconstructions are feasible. These reconstructions are limited by a complete lack of seafloor paleomagnetic data prior to the Jurassic, and this is really no coincidence: The oldest seafloor should represent, roughly, the vertical transit time in the mantle. Third, the adjustment time of about $175Myrs$ applies not only to the uppermost mantle, where plate-related flow is most directly felt, but also for the deep mantle and core-mantle boundary region, even with a significant heating component from the core! (In fact, core heating actually speeds up the adjustment process.) Thus our simple 2-stage model experiment addresses simultaneously a number of key questions concerning general circulation models, providing rather clear answers for the case of whole mantle flow. Obviously, these model experiments need to be repeated for layered flow models, where the adjustment times, at least for the lower mantle, may be considerably longer.

## 5. GENERAL CIRCULATION MODELS

We now proceed to implement all the tools developed above. We begin by running the plate-constrained model described in the previous section to quasi-steady-state with the $119Ma$ plate motion stage, and then impose the same sequence of plate motion reconstructions used in generating Figures 2-4 and Plate 1, ending the model run at present time. (We call this model GEMLAB, or Geodynamic Earth Model of Los Alamos and Berkeley; see *Bunge et al.*, 1998.) The lithosphere and lower mantle are a factor of 40 more viscous than the upper mantle. Although there are many facets of this model we could examine, we choose to concentrate here on comparisons with seismic tomography and on an analysis of the evolution of CMB structure.

Plates 3 and 4 show comparisons of two GEMLAB models with shear wave velocity structure at $1100km$ depth in the mantle, and just above the core-mantle boundary. GEMLAB2 differs from GEMLAB1 in that the phase transitions are ignored and core heating is removed. The model of *Grand et al.* [1997] is constructed from body shear wave tomography, with allowances for ray propagation in regionally heterogeneous structure, and is developed with local basis functions (cells). The model SAW12 of *Li and Romanowicz* [1996] is constructed from a completely independent data set of long-period transverse shear waves, surface waves, and normal modes, and employs global basis functions (spherical harmonics up to degree 12).

In the mid-mantle (Plate 3), the two seismic models differ significantly in character, with *Grand's* model dominated by small-scale high-velocity (blue) structures beneath active and extinct subduction zones, and *Li and Romanowicz's* being smoother and showing more complementary high-velocity structure. *Grand's* model is strikingly similar to that of *Van der Hilst et al.* [1997], which is constructed entirely from short-period compressional wave arrival times. All seismic models seem to agree on a pronounced low-velocity structure beneath southern Africa [see also *Su and Dziewonski*, 1991], although the local and global methods present very different views of the sharpness of this feature.

The GEMLAB models produce almost entirely cold downwelling structures concentrated beneath subduction zones. Interestingly, GEMLAB2 does a better job than GEMLAB1 in explaining the high-resolution body wave tomography, especially the extinct Farallon plate imaged beneath North and South America. The reason for this is that GEMLAB2 lacks an endothermic phase reaction at $670km$ depth, and allows easy penetration of cold slabs into the lower mantle. At this early stage of modeling maturity, we are somewhat reluctant to claim this result as evidence against a strong effect from the spinel-perovskite phase change at $670km$ depth. Rather it serves as an illustration of the "laboratory" nature of

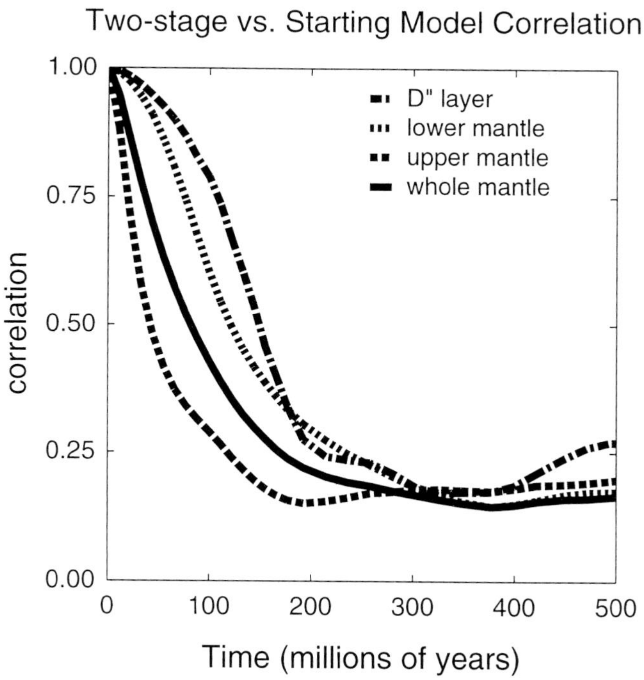


**Figure 5.** Global cross-correlation (see *Ricard et al.*, 1993, for a definition) between the evolving mantle convection model and the initial condition as a fuction of time. Correlations are shown separately for the upper, the lower, and the whole mantle, and for the core-mantle boundary region. A correlation of 1.0 means perfect correlation. The model evolves from perfect correlation at the beginning toward a final state corresponding to the present-day velocity boundary conditions. Most of the adjustment occurs within the first $150 Myrs$. The final correlation value of 0.2 reflects the correleation between the Mesozoic and present-day plate tectonic regime.

the GEMLAB models, and the strong sensitivity of details of mantle structure to model parameters that are not well-constrained *a priori*.

Near the core-mantle boundary (Plate 4), the nature of structure in the seismic models and in the GEMLAB1 model are substantially different. All the models are dominated by a large-scale, circum-Pacific pattern of high velocity or cold temperature anomalies. However, the seismic models indicate very high-amplitude low velocity features, especially beneath Africa and the southeastern Atlantic, that do not appear at all in the GEMLAB1 model, even though the latter contains a significant component of bottom heating. Thus it is fairly clear that these huge low velocity anomalies at the CMB cannot be described merely as "passive" upwelling structures complementary to the pattern of subduction, or as active upwelling structures modulated by the global pattern of subduction. Instead, they appear to demand a rather active lowermost mantle, perhaps involving a signature of chemical heterogeneity inferred from the high amplitudes of the velocity anomalies. In this case, the GEMLAB comparison serves to clarify a major discrepancy between seismic observation and the simple whole-mantle convection models developed in this paper. A major computational challenge is to develop models that include some degree of chemical layering, or a thermo-chemical boundary layer at the bottom of the mantle, in order to further elucidate the geodynamic significance of the seismic results.

## 6. APPLICATION TO CORE-MANTLE BOUNDARY EVOLUTION

A number of applications of these explicit Earth models are currently under development, including a time-history of intraplate stresses and dynamic topography during Cenozoic time, the main novelty being that we can begin to consider the temporal evolution of such observables. Here we consider another application that is directly related to the questions of the mantle "response time" developed in Figure 5. Namely, we consider the evolution of core-mantle boundary thermal structure.

With the advent of new computational models for the Earth's dynamo and magnetic reversals [*Glatzmaier and Roberts*, 1995], an immediate question arises as to how core dynamics are affected by interactions with the mantle. This coupling could be either thermal or topographic, both of which should be dominated by mantle convection, since the viscosity of the mantle is many orders of magnitude greater than that of the outer core. Recent dynamo modeling shows that the stability of the dynamo to reversals is sensitive to the large-scale shape of the thermal field imposed by the mantle, with harmonic degree 2 structure tending to stabilize the magnetic field [*Glatzmaier*, private comm.). Dynamo theory [*Merrill and McFadden*, 1988] suggests that odd harmonics might stabilize the field, while even harmonics might be destabilizing. Thus the transition from the mid-Cretaceous long-normal geomagnetic period (about 80-120$Ma$), during which few if any magnetic reversals occurred, to rapid reversals, as in the present era, might be related to major changes in the large-scale thermal structure of the core-mantle boundary.

Can the GEMLAB models predict such changes? Or, more specifically, can the GEMLAB models tell us anything reliable about the differences between core-mantle boundary structure in the mid-Cretaceous and at the present time?

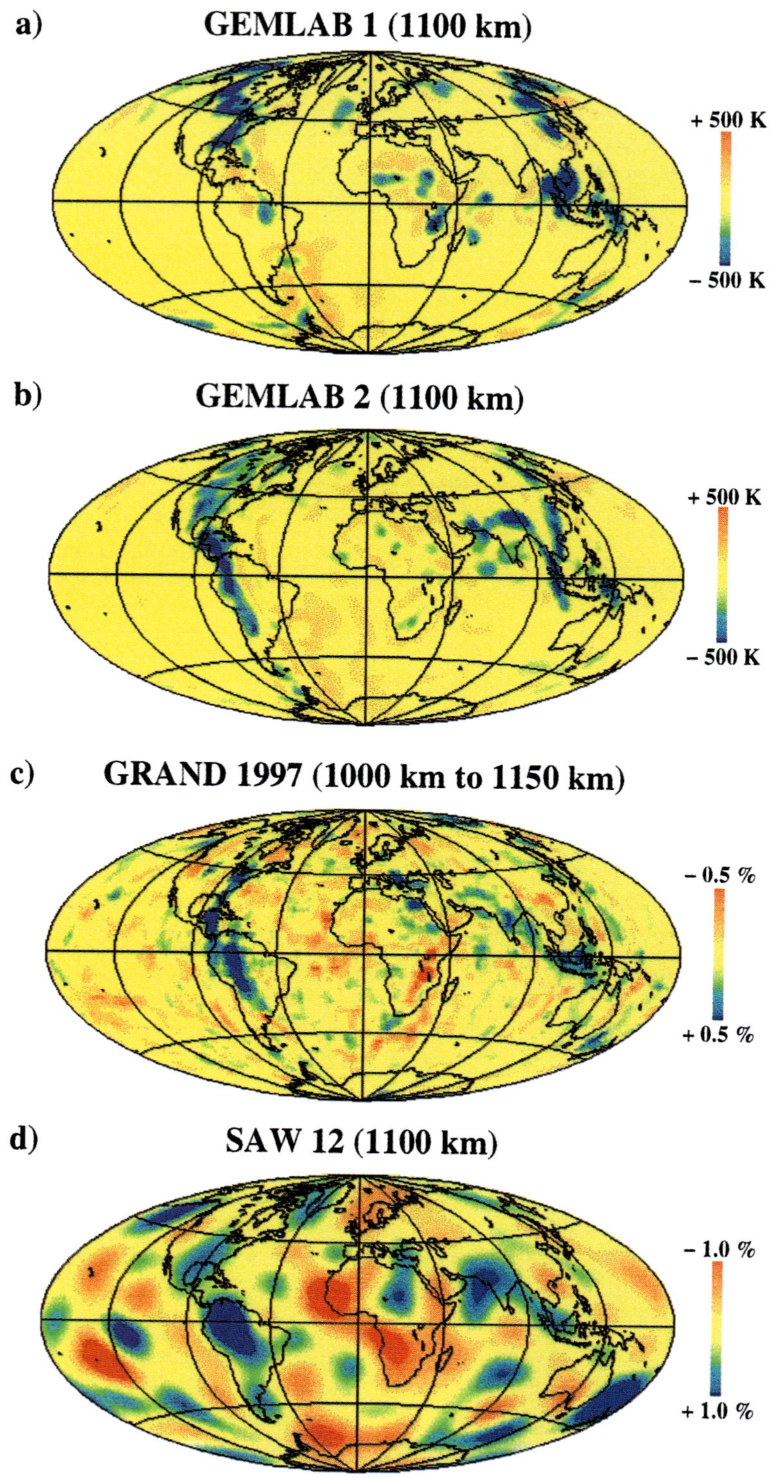


**Plate 3.** Maps of lateral heterogeneity (mean values removed) at a depth of 1100*km*. (a) The standard Earth model GEMLAB1 described in the text. (b) GEMLAB2 has no bottom heating or phase transitions. (c) Seismic shear wave model of *Grand et al.* [1997]. (d) Seismic shear wave model of *Li and Romanowicz* [1996]. For the convection model, blue is cold and red is hot. For the seismic models, blue is fast shear wave speed and red is slow.

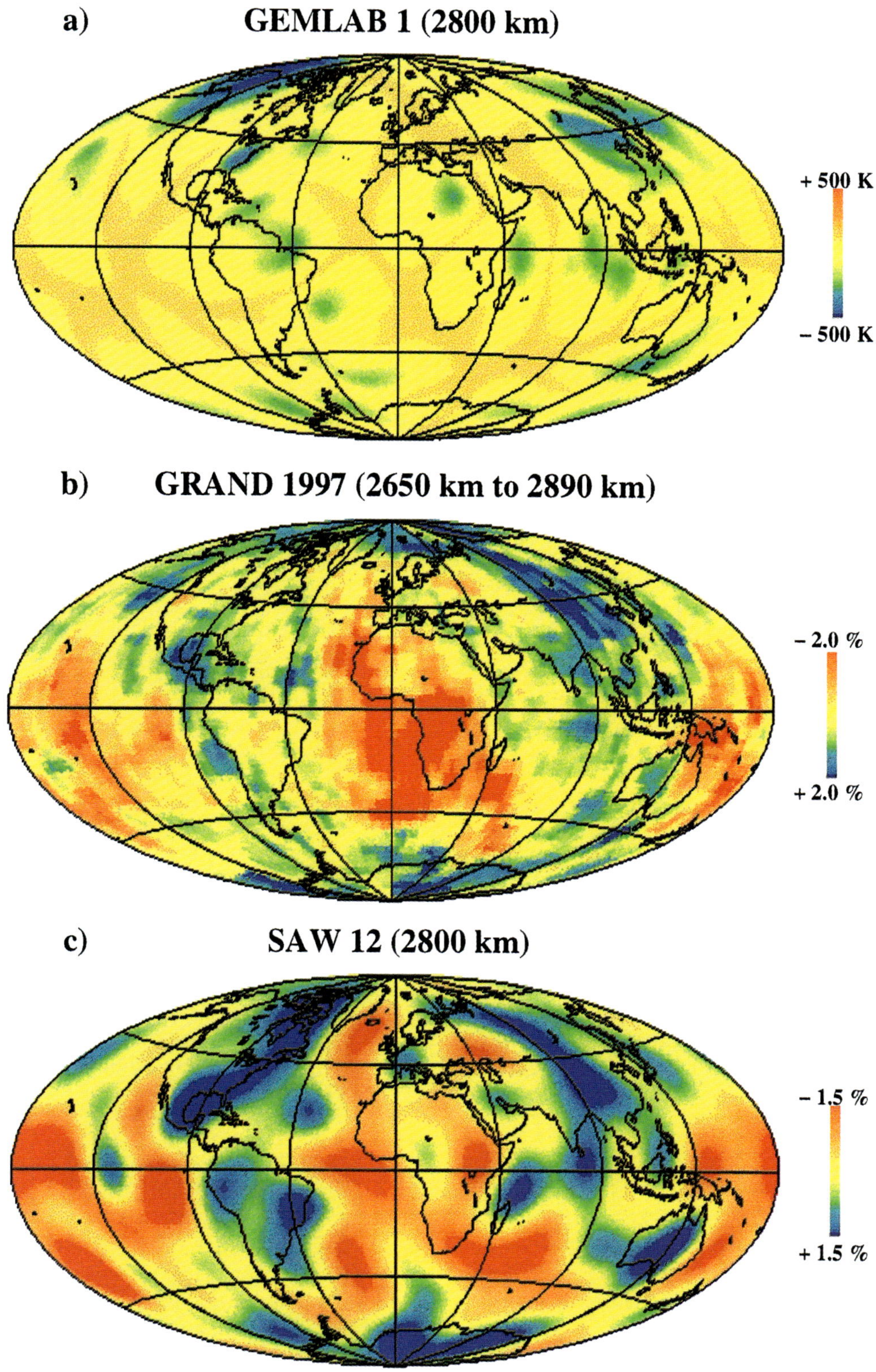


**Plate 4.** Same as maps of Figure 8 except that the depth is 2800*km*. (a) GEMLAB1, (b) *Grand et al.* [1997], and (c) *Li and Romanowicz* [1996].

Figure 5 shows that the core-mantle boundary response to changing plate motions occurs on a timescale of 100-150$Myrs$. Thus a reliable estimate of CMB structure in the mid-Cretaceous requires a good knowledge of Jurassic and Triassic plate motions, which we do not have at present, so the prognosis for assessing changes in CMB structure since the last long-normal event appears marginal, at best. However, some useful insight into the questions raised is provided by analysis of the temporal behavior of the CMB structure in the GEMLAB1 model.

Figure 6a shows a time-correlation plot for each harmonic degree component of the near-CMB temperature field from GEMLAB1. At each time, the thermal heterogeneity field in the bottom 50$km$ of the mantle is correlated with that at the beginning of the model run (similar to what was done in Figure 5), so that we begin with a perfect correlation of 1.0 at 100$Ma$. The lowest harmonics ($L = 1, 2, 4$ are shown) change very little over the 100$Myr$ model timespan. Only the higher harmonics ($L = 6, 8$ are shown), which are not be very well-determined given uncertainties in plate motion reconstructions, show significant changes. Thus it is clear that the GEMLAB models, in their current state of development, have little of a predictive nature to say about changing CMB conditions over the long timespan of interest regarding the Cretaceous long-normal period.

However, it must be noted that the relatively "static" nature of CMB thermal structure in Figure 6a is somewhat artificial, because for all time preceeding 100$Ma$, the model was run with a constant surface velocity field according to the 119-100$Ma$ reconstruction. What would happen if we had equally good knowledge of prior plate motions?

To address this question, we introduce a somewhat artificial model experiment: We begin a model run with present-day plate motions, run the plate motion history for Cenozoic and Mesozoic time backwards to the 119-110$Ma$ stage, and then forward again. Thus we produce a model that spans about 200$Myrs$ with varying surface plate motions representative of the Earth, but an obviously artificial history for the first 119$Myrs$. This model allows a somewhat better assessment of what we might learn of CMB history since Cretaceous time if we indeed had a better set of plate motion constraints prior to 119$Ma$.

Figure 6b shows correlations of CMB structure for the time-reversed model, with correlations computed relative to the 100$Ma$ time. Now, the correlations for the latter half of the model time more nearly represent the rate of change in CMB structure likely to occur in the Earth. (Note that the correlations prior to 100$Ma$ behave as in Figure 6a, only reversed, as expected.) Now we see, for example, that the $L = 2$ correlation drops to a value of about 0.8 in about 50$Myrs$, while the $L = 4$ correlation falls to about 0.4 in about the same amount of time. Thus the GEMLAB1-type model does produce significant changes in low-order CMB structure

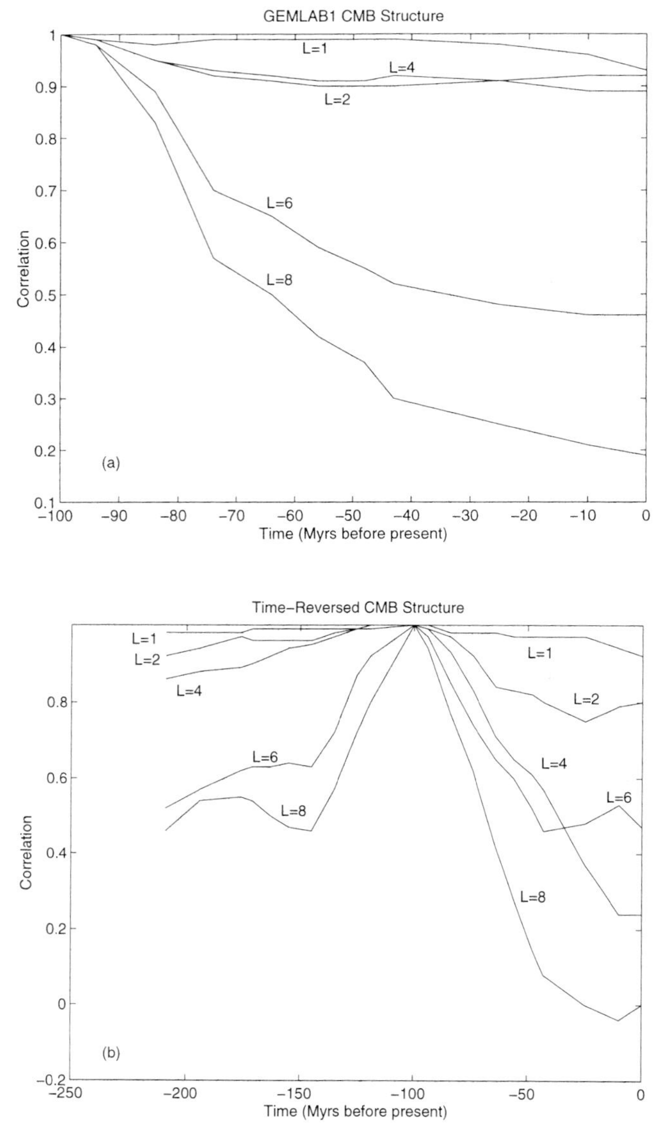


**Figure 6.** Correlation between the thermal structure near the core-mantle boundary in time and the structure at time 100$Ma$. In (a) the model is started by imposing the 119-100$Ma$ plate motions for a long time, and then running forward with subsequent reconstructions. In (b) the sequence is run in reverse and then forward, but correlations are still shown relative to structure at 100$Ma$.

on timescales relevant to the Cretaceous long-normal question. However, $L = 1$ appears quite persistent even in the time-reversed model, so that if this component exercises a stabilizing influence on the dynamo, it is possible that it changes little on the $100Myr$ timescale.

These model experiments are unsuccessful, at least so far, in providing a reliable estimate of what low-order changes may have occurred in CMB structure since the Cretaceous long-normal event. However, they do give an indication of how rapidly low-order structure may change in a reasonable whole-mantle convection model, and they also indicate that even rough information on global plate motions from earlier times (prior to $119Ma$), incorporated into the GEMLAB model, could help constrain the mantle boundary condition to the core dynamo from mid-Cretaceous time onward. It is also noteworthy that the essence of the problem is the "initial condition" conundrum referred to in the previous section. Hence the CMB evolution question might be addressed better following the development of a formal inverse theory in the GEMLAB models.

## 7. CONCLUSIONS

We have explored simple models for mantle convection, testing the idea that plate motion constraints are sufficient to give some realism. The basic model assumptions are that the mantle is isochemical and heated largely from within by radioactive decay, that there is a modest but significant increase in mantle viscosity with depth, that the motions of rheologically strong surface plates represent the main flow, and that the plate motion record is (just) sufficiently long to constrain the development of most mantle heterogeneity structure. These assumptions lead to a first-order model that satisfies some observations rather well, especially the geoid and seismic tomography. The models verify a relevant time-scale of about $150Myrs$ for the development of mantle heterogeneity structure in response to plate motions, so that initial conditions do not present an intractable problem for models with similarly long plate motions histories applied. The combination of plates and a high-viscosity lower mantle results in a mantle heterogeneity spectrum similar to that observed, with negligible small-scale structure. It is noteworthy that more rapid plate motions in the mid- late-Cretaceous have resulted in the main signature of mid-mantle heterogeneity structure associated with subducted slabs. On the other hand, the large low-velocity structure in the mid-mantle beneath Africa, and perhaps other structures of a similar nature, are not accounted for in our models, even with considerable degrees of bottom heating from the core [*Bunge et al.*, manuscript in preparation]. The implication is that there may be a major component of thermo-chemical convection in the deep mantle.

These models need to be improved in at least four ways. First, a better accounting of uncertainties in plate motions and plate geometries is needed, especially for Mesozoic time. Generally, rotation pole uncertainties are better-characterized, because they are derived directly from paleomagnetic data, but we have so far had to use "end-member" rotation poles prior to Cenozoic time to estimate extremal uncertainties. Uncertainties in plate boundaries are even more difficult to ascertain, and there has been little effort to do so. Generally, one is presented with best-guess plate boundaries, or, perhaps, alternative boundaries [e.g., *Engebretson et al.*, 1985]. A major collaborative effort between tectonicists and geodynamicists is needed for this somewhat unrewarding, but nevertheless important, task.

A second major improvement needed is better treatment of rheology throughout the mantle, especially that associated with plates and plate boundaries. There are many possible approaches to modeling plate rheology [see, e.g., *Bercovici et al.*, this volume], but the main shortcoming with regard to the type of contstrained models we have developed is a lack of one-sided subduction and stiff subducted slabs. For example, comparing the body S-wave tomography and GEMLAB models in Plate 4a, it is clear that we have "misplaced" the subducted Farallon slab about $1000km$ westward in the models, largely due to a lack of an appropriately shallow dip angle associated with that subduction. From the computational and physical standpoint, repairing such problems stands as a much more formidable task than other issues such as model resolution.

A third needed improvement, as discussed briefly above, is in the area of "initial conditions," or how one goes about estimating mantle structure at the beginning of model runs. Improving upon our *ad hoc* method will require considerable theoretical and computational development, beyond the scope of this discussion. Fortunately, such methods have already been employed in other areas of fluid dynamics, such as hydrology, oceanography, and meteorology [*Sun*, 1994; *Schroter*, 1988; *Talagrand and Courtier*, 1987], and the mathematical tools necessary are fairly well-developed.

Lastly, it will be important to address non-adiabatic density increases in the mantle, that is, chemical layering of the flow. Such layering is suggested from a variety of evidence, especially geochemical [see *Hofmann*, 1993, for a comprehensive and recent review], but also seismological [*Grand*, 1998], although the older paradigm of a mantle strictly layered at 670*km* depth has been largely abandoned. Instead, it appears likely that the mantle is partially or imperfectly layered, with thermal and chemical heterogeneities being of similar magnitude, and with significant radial mass transfer occurring at all levels. Treating such problems introduces two major difficulties for any general circulation models. First, treatment of chemical heterogeneities causes serious computational problems [e.g., *van Keken et al.*, 1997], and second, all considerations of relevant timescales must also be reconsidered, i.e., chemical layering will introduce a considerably longer lag between lower mantle structure and surface plate motions. Thus the first-order success of the models we have presented should be accepted with considerable caution, and the challenges lying ahead are considerable.

*Acknowledgments.* The authors thank Peter Olson and Clem Chase for helpful reviews. We also thank John Baumgardner, David Engebretson, and Yanick Ricard for collaborations on various aspects of this research over the past decade. This work was supported by grants to MAR from the National Science Foundation, the National Aeronautics and Space Administration, and the Institute for Geophysics and Planetary Physics at Los Alamos National Laboratory.

## REFERENCES

Anderson, D.L., Hotspots, polar wander, mesozoic convection, and the geoid, *Nature*, *297*, 391-393, 1982.

Baumgardner, J.R., M.A. Richards, W.-S. Yang, and C. Lithgow-Bertelloni, 3-D spherical models of plate motion with laterally varying rheology, *EOS Trans. AGU*, *79*, F911, 1998.

Bercovici, D., A simple model of plate generation from mantle flow, *Geophys. J. Int.*, *114*, 635-650, 1993.

Bercovici, D., A source-sink model of the generation of plate tectonics from non-Newtonian mantle flow, *J. Geophys. Res.*, *100*, 2013-2030, 1995.

Bunge, H.-P. and M.A. Richards, The backward problem of plate tectonics and mantle convection, *EOS Trans. AGU*, *73*, 281, 1992.

Bunge, H.-P. and M.A. Richards, The origin of long-wavelength structure in mantle convection, *Geophys. Res. Lett.*, *23*, 2987-2990, 1996.

Bunge, H.-P., M.A. Richards, and J.R. Baumgardner, Effect of depth-dependent viscosity on the planform of mantle convection, *Nature*, *379*, 436-438, 1996.

Bunge, H.-P., M.A. Richards, and J.R. Baumgardner, A sensitivity study of three-dimensional spherical mantle convection at $10^8$ Rayleigh number: Effects of depth-dependent viscosity, heating mode, and an endothermic phase change, *J. Geophys. Res.*, *102*, 11,991-12,007, 1997.

Bunge, H.-P., M.A. Richards, C. Lithgow-Bertelloni, J.R. Baumgardner, S.P. Grand, and B.A. Romanowicz, Time scales and heterogeneous structure in geodynamic Earth models, *Science*, *280*, 91-95, 1998.

Chase, C.G., Subduction, the geoid, and lower mantle convection, *Nature*, *282*, 464-468, 1979.

Chase, C.G. and D.R. Sprowl, The modern geoid and ancient plate boundaries, *Earth Planet. Sci. Lett.*, *62*, 314-320, 1983.

Chavent, G., M. Dupy, and P. Lemmonnier, History matching by use of optimal theory, *Soc. Pet. Eng. J.*, *15*, 74-86, 1975.

Crough, S.T. and D.M. Jurdy, Subducted lithosphere, hotspots, and the geoid, *Earth Planet. Sci. Lett.*, *48*, 15-22, 1980.

Davies, G.F., Ocean bathymetry and mantle convection, II. Small-scale flow, *J. Geophys. Res.*, *93*, 10,481-10,488, 1988.

Davies, G.F. and M.A. Richards, Mantle convection, *J. Geology*, *100*, 151-206, 1992.

Engebretson, D.C., A. Cox, and R.G. Gordon, Relative motions between oceanic and continental plates in the Pacific basin, *Geol. Soc. Amer. Special Paper*, *206*, 1985.

Engebretson, D.C., K.P. Kelly, H.J. Cashman, and M.A. Richards, 180 million years of subduction, *GSA Today*, *2*, 93-95, 1992.

Forte, A.M. and J.X. Mitrovica, New inferences of mantle viscosity from joint inversion of long-wavelength mantle convection and post-glacial rebound data, *Geophys. Res. Lett.*, *23*, 1147-1150, 1996.

Glatzmaier, G.A. and P.H. Roberts, A three-dimensional self-consistent computer simulation of a geomagnetic field reversal, *Nature*, *377*, 203-209, 1995.

Gordon, R.G. and D.M. Jurdy, Cenozoic global plate motions, *J. Geophys. Res.*, *91*, 12,389-12,406, 1986.

Grand, S.P., Mantle shear structure beneath the Americas and surrounding oceans, *J. Geophys. Res.*, *99*, 11,591-11,621, 1994.

Grand, S.P., Two barriers to flow in the deep mantle, *EOS Trans. AGU*, *79*, F586, 1998.

Grand, S.P., R.D. van der Hilst, and S. Widiyantoro, Global seismic tomography: A snapshot of convection in the Earth, *GSA Today*, *7*, 1-7, 1997.

Gripp, A.E. and R.G. Gordon, Current plate velocities relative to the hotspots incorporating the NUVEL-1 global plate motion model, *Geophys. Res. Lett.*, *17*, 1109-1112, 1990.

Hager, B.H. and R.J. O'Connell, Kinematic models of large-scale flow in the Earth's mantle, *J. Geophys. Res.*, *84*, 1031-1048, 1979.

Hager, B.H. and R.J. O'Connell, A simple global model of plate dynamics and mantle convection, *J. Geophys. Res.*, *86*, 4843-4867, 1981.

Haskell, N.A., The motion of a viscous fluid under a surface load, *Physics,*, *6*, 265-269, 1935.

Hofmann, A.W., Mantle geochemistry: The message from oceanic volcanism, *Nature*, *385*, 219-229, 1993.

Larson, R.L. and C.G. Chase, Late Mesozoic evolution of the western Pacific ocean, *Geol. Soc. Amer. Bull.*, *83*, 3627-3644, 1972.

Larson, R.L. and W.C. Pitman, World-wide correlation of Mesozoic magnetic anomalies and its implications, *Geol. Soc. Amer. Bull.*, *83*, 3645-3662, 1972.

Li, X. and B. Romanowicz, Global mantle shear-velocity model developed using nonlinear asymptotic coupling theory, *J. Geophys. Res.*, *101*, 22,245-22,272, 1996.

Lithgow-Bertelloni, C. and M.A. Richards, Cenozoic plate driving forces, *Geophys. Res. Lett.*, *22*, 1317-1320, 1995.

Lithgow-Bertelloni, C. and M.A. Richards, The dynamics of Cenozoic and Mesozoic plate motions, *Rev. Geophys.*, *36*, 27-78, 1998.

Lithgow-Bertelloni, C., M.A. Richards, Y. Ricard, R.J. O'Connell, and D.C. Engebretson, Toroidal-poloidal partitioning of plate motions since 120 Ma, *Geophys. Res. Lett.*, *20*, 375-378, 1993.

McKenzie, D.P. and R.L. Parker, The North Pacific: an example of tectonics on a sphere, *Nature*, *216*, 1276-1280, 1967.

McKenzie, D.P., J.M. Roberts, and N.O. Weiss, Convection in the Earth's mantle: Towards a numerical simulation, *J. Fluid Mech.*, *62*, 465-538, 1974.

Merrill, R.T. and P.L. McFadden, Secular variation and the origin of geomagnetic field reversals, *J. Geophys. Res.*, *93*, 11,589-11,597, 1988.

Minster, J.-B. and T.H. Jordan, Present-day plate motions, *J. Geophys. Res.*, *83*, 5331-5354, 1978.

Mitrovica, J.X., Haskell [1935] revisited, *J. Geophys. Res.*, *101*, 555-569, 1996.

Morgan, W.J., Rises, trenches, great faults, and crustal blocks, *J. Geophys. Res.*, *73*, 1959-1982, 1968.

Norton, I.O., Plate motions in the North Pacific: The 43 MA non-event, *Tectonics*, *14*, 1080-1094, 1995.

Ricard, Y., M.A. Richards, C. Lithgow-Bertelloni, and Y. Le Stunff, A geodynamic model of mantle density heterogeneity, *J. Geophys. Res.*, *98*, 21,895-21,909, 1993.

Ricard, Y. and C. Vigny, Mantle dynamics with induced plate tectonics, *J. Geophys. Res.*, *94*, 17,543-17,560, 1989.

Richards, M.A., Hotspots and the case for a high viscosity lower mantle, in *Glacial Isostasy, Sea Level, and Mantle Rheology*, R. Sabadini et al. (eds.), pp. 571-588, Kluwer Acad., Norwell, Mass., 1991.

Richards, M.A., D.C. Engebretson, Large-scale mantle convection and the history of subduction, *Nature*, *355*, 437-440, 1992.

Richards, M.A., and B.H. Hager, Geoid anomalies in a dynamic Earth, *J. Geophys. Res.*, *89*, 5987-6002, 1984.

Richter, F.M. and B. Parson, On the interaction of two scales of convection in the mantle, *J. Geophys. Res.*, *80*, 2529-2541, 1975.

Schroter, J., Driving of non-linear time-dependent ocean models by observation of transient tracers – a problem of constrained optimisation, in *Ocean Circulation Models: Combining Data and Dynamnics*, D.L.T. Anderson and J. Willebrand (eds.), Kluwer Acad. Publ., Dordrecht, 605pp., 1988. *89*, 5987-6002, 1984.

Scotese, C.R., Atlas of Phanerozoic plate tectonic reconstructions, *Paleomap Proj. Tech. Rep.*, *10-90-1*, 1990.

Su, W.-J. and A.M. Dziewonski, Predominance of long-wavelength heterogeneity in the mantle, *Nature*, *352*, 121-126, 1991.

Sun, N.-Z., *Inverse Problems in Groundwater Modeling*, Kluwer, Dordrecht, 337 pp., 1994.

Tackley, P.J., Effects of strongly variable viscosity on three-dimensional compressible convection in planetary mantles, *J. Geophys. Res.*, *101*, 3311-3332, 1996.

Tackley, P.J., D.J. Stevenson, G.A. Glatzmaier, and G. Schubert, Effects of an endothermic phase transition at 670 km depth on a spherical model of convection in Earth's mantle, *Nature*, *361*, 699-704, 1993.

Tackley, P.J., D.J. Stevenson, G.A. Glatzmaier, and G. Schubert, Effects of multiple phase transitions in three-dimensional spherical models of convection in Earth's mantle, *J. Geophys. Res.*, *99*, 15,887-15,901, 1994.

Talagrand, O. and P. Courtier, Variational assimilation of meteorological observations with adjoint vorticity equations. Part I, Theory, *Q.J.R. Meteorol. Soc.*, *113*, 1311-1328, 1987.

Turcotte, D.L. and E.R. Oxburgh, Finite amplitude convective cells and continental drift, *J. Fluid Mech.*, *28*, 29-42, 1967.

Turcotte, D.L. and G. Schubert, *Geodynamics: Applications of Continuum Physics to Geological Problems*, John Wiley and Sons, New York, 450pp., 1982.

van der Hilst, R., S. Widiyantoro, and R. Engdahl, Evidence for deep mantle circulation from global tomography, *Nature*, *386*, 578-584, 1997.

van Keken, P.E., S.D. King, H. Schmeling, U.R. Christensen, D. Neumeister, and M.-P. Doin, A comparison of methods for the modeling of thermomechanical convection, *J. Geophys. Res.*, *102*, 22,477-22,495, 1997.

Wasserburg, G.J., G.J.F. MacDonald, F. Hoyle, and W.A. Fowler, Relative contributions of uranium, thorium, and potassium to heat production in the Earth, *Science*, *143*, 465-467, 1964.

Weinstein, S.A., Thermal convection in a cylindrical annulus with a non-Newtonian outer surface, *PAGEOPH*, *146*, 551-572, 1996.

Wilson, J.T., A new class of faults and their bearing on continental drift, *Nature*, *207*, 343-347, 1965.

Zhong, S., M. Gurnis, and L. Moresi, Role of faults, nonlinear rheology, and viscosity structure in generating plates from instantaneous mantle flow models, *J. Geophys. Res.*, *103*, 15,255-15,268, 1998.

---

H.-P. Bunge, Department of Geological Sciences, Princeton University, Princeton, NJ 08544. (e-mail: bunge@geodynamics.princeton.edu)

C. Lithgow-Bertelloni, Department of Geological Sciences, University of Michigan, Ann Arbor, MI 48109. (e-mail: crlb@umich.edu)

M. Richards, Department of Geological Sciences, University of California, Berkeley, CA 94720. (e-mail: markr@seismo.berkeley.edu)

# Large Igneous Provinces and Plate Tectonics

Olav Eldholm

*Department of Geology, University of Oslo, Oslo, Norway*

Millard F. Coffin

*Institute for Geophysics, University of Texas at Austin, Austin, Texas*

Large Igenous Provinces (LIPs) constitute broad areas, $>10^5$ km$^2$, of mafic volcanic and plutonic rocks erupted over $\sim10^6$ yr. Compared with plate dimensions, LIPs are smaller, localized features; they form both in plate interiors and at plate boundaries. LIPs originate in the mantle, via mass and energy transfer which acts both independently of, and in conjunction with, the wide-ranging upwellling systems producing new oceanic crust by sea floor spreading along the mid-oceanic ridge system. Plate tectonic theory does not readily explain the massive magmatism, which is most commonly attributed to mantle plumes. Most LIPs form in extensional oceanic or continental plate tectonic settings, suggesting a relationship with thinned lithosphere. Deformed LIP complexes in intraplate continental settings suggest formation throughout most of Earth history. The post-150 Ma LIP record reveals both many events and large melt volumes from 135-85 Ma, and a distinct decline since 50 Ma. These trends may reflect variations in mantle circulation and have links to global environmental change. Following formation, oceanic LIPs may be carried laterally by the plates to regions of plate convergence. Subsequent accretion of major LIPs into continental crust contributes episodically to continental growth.

## INTRODUCTION

The plate tectonics paradigm provides an orderly and systematic mechanism to describe the temporal evolution of the Earth's crust and upper mantle which has proven to be highly successful. However, plate tectonic theory does not readily account for some crustal features and tectonic processes. An example is short-lived, voluminous emplacements of predominantly mafic extrusive and intrusive rocks by processes not directly linked to crustal production by steady-state sea floor spreading (Plate 1). We use the term crustal production for new crust formed in MOR and LIP settings, and accretion and arc magmatism for crustal formation in subduction settings. The spatial and temporal occurrences of these features, commonly termed Large Igneous Provinces (LIPs), suggest localized and episodic emplacement. Thus, these products of a dynamic, but apparently also unpredictable mantle, contrast greatly with large-scale, steady-state horizontal plate tectonic movements of the lithosphere and extensive zones of upwelling and subduction at plate boundaries.

The interplay and cause-and-effect relationship between plate tectonic processes and the more local areas of upwelling and massive melting are by no means simple. Nonetheless, how the core and mantle get rid of their internal energy determines all global tectonic processes except those of extraterrestrial origin. LIPs attest to past energy transfer to the

The History and Dynamics of Global Plate Motions
Geophysical Monograph 121

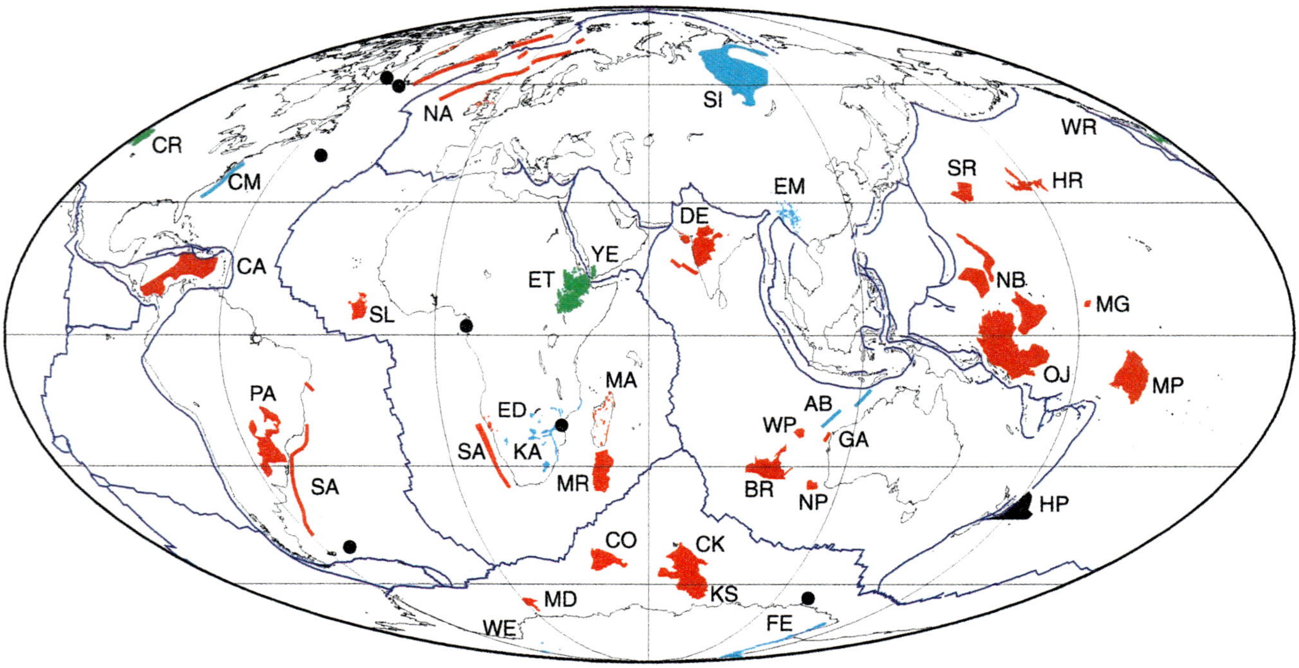


**Plate 1.** Global transient LIP distribution: light blue >150 Ma, red 150-50 Ma, green 50-0 Ma, black undated. Volcanic margin LIPs (lineations and circles) are mainly based on occurrences of seaward dipping reflections in seismic profiles. Current plate boundaries appear in dark blue.

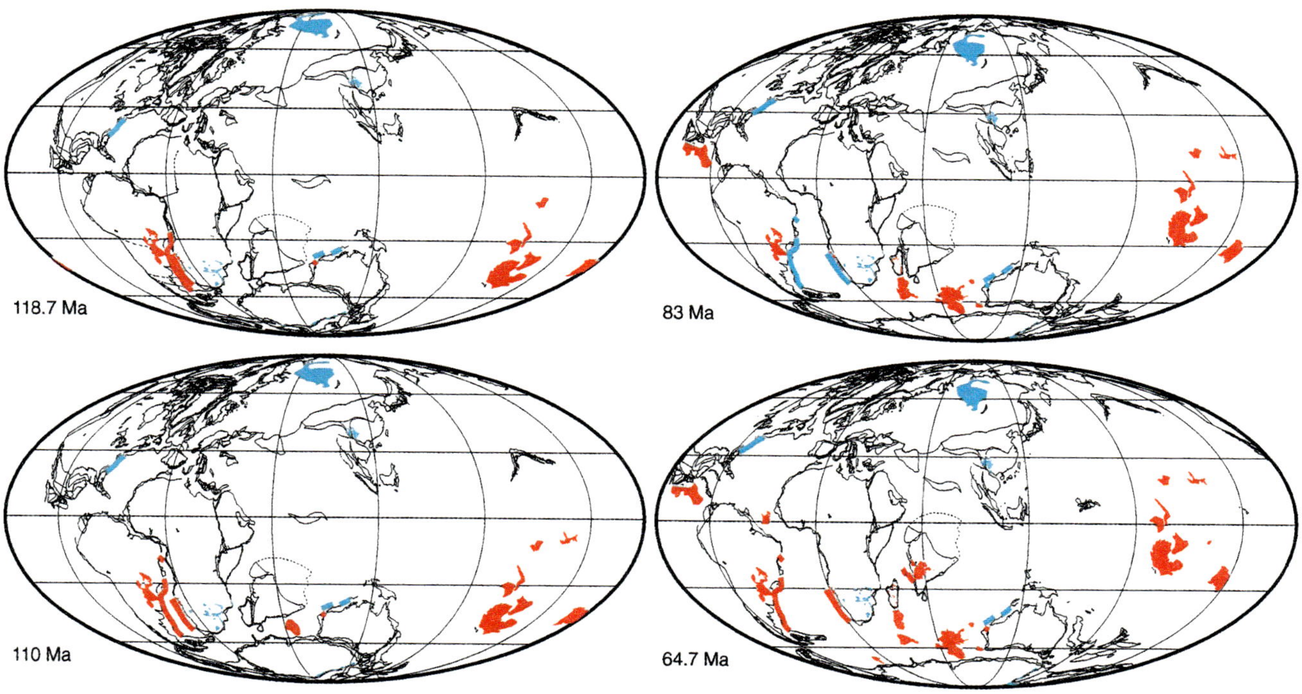


**Plate 2.** Plate tectonic reconstructions at 118.7, 110, 83, and 64.7 Ma [*Coffin et al.*, 1998]. The 110 Ma reconstruction is within the Cretaceous Normal Superchron. Color coding as in Plate 1.

Earth's surface in a mode different from the processes dominated by plate tectonics. Today, the mid-ocean ridge (MOR) system (Plate 1) accounts for >90% of the mass and energy transfer from mantle to crust [*Davies*, 1988; *Sleep*, 1990], whereas crustal production rates for the most massive LIPs may have exceeded the global MOR production rate over intervals of 2-3 m.y. or less. Throughout the ~150 m.y. for which significant crust exists in the oceans today, LIP production appears to have intermittently augmented sea floor spreading mafic crustal production, and at certain periods dominated the mantle mass and energy flux, contributing significantly to crustal growth and evolution.

LIPs are commonly associated with mantle plumes and their associated hotspots reflecting deep, narrow upwelling systems. However, mantle plumes have to date only been imaged convincingly in geophysical data beneath Iceland [*Wolfe et al.*, 1997; *Shen et al.*, 1998; *Bijwaard and Spakman*, 1999]. Although, the concept of mantle plumes and hotspots has been recognized for decades [*e.g., Wilson*, 1963; *Morgan*, 1971, 1981], it is only during the last ten years that LIPs have achieved widespread attention as a monitor of mantle dynamics and as a potential instigator of global change [*e.g., Mahoney and Coffin*, 1997]. To evaluate these interior and exterior causes and consequences, we need information about dimensions, structure, composition and crustal production rate for many LIPs. Unfortunately, the available database for most LIPs is indeed still very meager compared with most other geological provinces of similar size. Therefore, it is particularly important to be aware of the limitations in the actual data that form the basis for quantitative analysis and modeling.

We have previously summarized a number of LIP related observations and inferences [*Coffin and Eldholm*, 1994]. New regional geophysical surveys have since been reported from some provinces, and the quality of age dating has improved considerably. In this study we first, from an observational standpoint, classify LIPs and describe their plate tectonic settings, structure and dimensions, as well as distributions in space and time. Then, we discuss the main formation models, relations to mantle dynamics, and the history and fate of LIPs within the plate tectonic cycle. For potential environmental LIP implications we refer to summaries by *Rampino and Stothers* [1988], *Courtillot* [1990], *Larson* [1991a, 1995], *Coffin and Eldholm* [1993a, 1994], *Courtillot et al.* [1996], and *Kerr* [1998]. Note that the reference list highlights overview papers with illustrations adequate for non-specialists and for educational purposes.

## PLATE TECTONIC SETTINGS

Understanding the interaction of LIPs and plate tectonics, and their emplacement and deformational history, requires knowledge of the plate tectonic setting from LIP formation to destruction. Although LIPs are expressions of the most voluminous transient basaltic volcanism on Earth, most LIPs are characterized by initial relatively small-scale igneous activity prior to a main short-lived event of massive basaltic volcanism, plutonism, and overall crustal thickening. Submarine ridges and seamount groups show that the main event is commonly followed by less voluminous, persistent or sporadic volcanism, of variable intensity and duration [*e.g., Richards et al.*, 1989; *Duncan and Richards*, 1991; *White*, 1991]. Moreover, basalts from the Kerguelen and Ontong Java oceanic plateaus indicate that the main emplacement event was followed by another, smaller transient event 20-30 Ma later [*Bercovici and Mahoney*, 1994; *Pringle et al.*, 1997]. Because the main formation event represents production volume and rate maxima, it is most significant for plate tectonics, geodynamics and global change. We therefore focus on provinces formed during this transient period of greatly enhanced igneous activity (Plate 1), and subdivide LIPs into four categories:

- Continental flood basalt (CFB): Composite accumulations of horizontal or subhorizontal flows, erupted in rapid succession over large areas of continental lithosphere, primarily by fissure eruptions
- Oceanic Plateau (OP): Broad, largely flat-topped features isolated from continents, generally rising >2000 m above the adjacent deep ocean basin sea floor; the extrusive cover may be subaerial or submarine.
- Ocean Basin Flood Basalt (OB): Extensive submarine flow and sill units overlying older oceanic crust formed by sea floor spreading. A sediment sequence pre-dating the flood basalt may separate the two igneous units.
- Volcanic Passive Continental Margin (VM): Outer margin covered by thick extrusive complexes constructed during continental breakup and initial sea floor spreading. Commonly recognized by wedges of seaward dipping reflections in seismic records.

The literature abounds in classifications of passive continental margins [*e.g., Eldholm et al.*, 1995] and of features in the ocean basins based on morphology, crustal structure and geological setting, *e.g.*, oceanic, continental and marginal plateau, microcontinent, submarine ridge, seamount province, and thermal swell [*e.g., Nur and Ben-Avraham*, 1982; *Schubert and Sandwell*, 1989]. In a LIP context, however, oceanic plateau and volcanic margin are genetic terms based on their tectono-magmatic formation and relative size. Applying this definition, *Coffin and Eldholm* [1993b, 1994] compiled the global Mesozoic and Cenozoic LIP distribution including transient, as well as persistent, oceanic features and CFBs. Subsequently, we have compiled an inventory of LIP parameters (dimensions, ages, and rates) known for transient

**Table 1.** Mesozoic and Cenozoic transient LIPs (Plate 1). Original area, o; extrusive crustal component only, x; subducted part, s; accreted terrain, a; and age in bracket a probable second transient event.

| LIP | Category | Mean, age, Ma | Area $10^6 km^2$ | Volume $10^6 km^3$ |
|---|---|---|---|---|
| Siberian Traps - SI | CFB | 250 | 2.5(o) | 1.2(x) |
| Emeishan - EM | CFB | 248 | 0.3 | |
| Wrangellia - WR | Opa | 225 | | |
| Central Atlantic Magmatic Province – CM (SDRs only) | CFB, VM | 200 | 0.1 | 0.7 |
| Karoo, Ferrar, Weddell Sea – KA, FE, WE | CFB, VM | 184 | 3.1 | 1.5(x) |
| Hikurangi Plateau - HP | OP | ? | 0.7 | 2.7(s) |
| Argo Basin Margin - AB | VM | 155 | 0.1 | |
| Shatsky Rise - SR | OP | 147 | 0.2 | 2.5 |
| Magellan Rise - MG | OP | 145 | 0.5 | 1.8 |
| Gascoyne margin - GA | VM | 136 | | 0.8 |
| Parana, Etendeka, South Atlantic margins – PA, ED, SA | CFB, VM | 132 | 2.3(o) | 21.0 |
| Manihiki Plateau - MP | OP | 123 | 0.8 | 8.8 |
| Ontong Java Plateau - OJ | OP | 121 (90) | 1.9 | 44.4 |
| Nauru Basin - NB | OB | 111 | 1.8 | 0.9 |
| South Kerguelen - KS | OP, VM | 110 | | 6.0 |
| Naturaliste Plateau - NP | VM | 101 | | 1.2 |
| Hess Rise - HR | OP | 99 | 0.8 | 9.1 |
| Wallaby Plateau - WP | OP | 96 | 0.4 | 1.5 |
| Caribbean - CA | OP | 88 | 1.1 | 4.5 |
| Madagascar, Madagascar Ridge, Conrad Rise – MA, MR, CO | CFB, OP | 88 | 1.6 | 4.4 |
| Central Kerguelen, Broken Ridge – CK, BR | OP | 86 | | 9.1 |
| Sierra Leone Rise – SL | OP | ~73 | 0.9 | 2.5 |
| Maud Rise – MD | OP | ~73 | 0.2 | 1.2 |
| Deccan - DE | CFB/VM | 65 | 1.8(o) | 8.6 |
| NE North Atlantic - NA | CFB/VM | 54 | 1.2 | 9.9 |
| Yemen, Ethiopia – YE, ET | CFB | 30 | 0.8(o) | 0.5(x) |
| Columbia River - CR | CFB | 16 | 0.2 | 1.3 |

features (Table 1). The LIP parameters in this paper refer to this inventory, which will be published elsewhere.

In a plate tectonic framework, we first show the global LIP distribution with respect to the present-day plate configuration (Plate 1). Then, we reconstruct LIPs to Cenozoic and Cretaceous paleopositions using the PLATES [*Coffin et al.*, 1998] database (Plate 2). The present-day spatial distribution is relatively unsystematic, and transient LIPs occur in intraplate environments both within the continental and oceanic parts of plates. Their younger, persistent parts, on the other hand, record the movement of the lithosphere as it moves over a melt source in the asthenosphere. However, a strong link with divergent plate boundaries emerges when the transient events are placed in the plate tectonic setting when they occurred (Plate 2). From the Cenozoic and Cretaceous spatial LIP distributions we infer the following trends:

- LIPs are most dense in the west-southwest Pacific Ocean.
- Most LIPs form at, or in the vicinity of, a diverging plate boundary.
- Few CFBs exist in the continental lithosphere several hundreds of kilometers away from plate boundaries.

A consequence of these plate tectonic settings is that LIPs may have two types of crustal structures. A LIP formed off-axis, *i.e.*, intraplate, will sandwich and intrude the pre-existing crustal block; an on-axis feature, however, is entirely composed of crustal rocks produced during the transient event (Table 2, Figure 1). The voluminous outpouring of far-flowing basaltic lava during an off-axis LIP-forming event may, of course, affect a neighboring plate boundary; vice versa an on-axis event will flood and underplate areas of pre-existing crust. In particular, on-axis breakup volcanism may cover large areas of adjacent continental crust.

After its on- or off-axis formation, the LIP becomes part of the rigid plate, or plates, its movement and fate governed by the plate tectonic cycle. Oceanic plateaus seem to be carried passively by the plate and subside with the adjacent oceanic plate, maintaining the relative basement relief acquired during formation [*Coffin*, 1992]. There is little evidence for significant post-constructional deformation as the volcanic margin LIPs cools, subsides and becomes covered by sediment with time. In fact, recent work off Norway shows that thick and underplated crust on the volcanic margin is almost entirely aseimic, suggesting it is stronger than the surrounding crust [*Byrkjeland et al.*, 1999, in press]. On the other hand, some faulting characterizes the Ontong Java [*Kroenke*, 1972] and Kerguelen [*Houtz et al.*, 1977] plateaus. Eventually, an oceanic LIP may arrive at a region of plate convergence where it will deform by subduction and obduction. The impact on LIPs by plate destruction is documented by three oceanic plateaus: Wrangellia [*Richards et*

**Table 2.** Transient LIP categories and plate tectonic settings during their main formation event.

| Category | Plate tectonic setting | *Type locality* |
|---|---|---|
| Continental Flood Basalt | Off-axis | Siberian Traps |
| Oceanic Plateau | On-axis<br>Off-axis | Shatsky Rise<br>Ontong Java Plateau? |
| Volcanic Passive Margin | On-axis | NE North Atlantic |
| Ocean Basin Flood Basalt | Off-axis | Nauru Basin |

*al.*, 1991] and part of Caribbean [*e.g.*, *Kerr et al.*, 1997] forming accreted terranes in mountain belts; and by the accretion of part of the Ontong Java Plateau onto the Solomon Island Arc [*Hughes and Turner*, 1977; *Tejada et al.*, 1996].

Although most primary evidence for pre-Cretaceous LIPs may have been eroded or deformed by plate convergence, continental data indicate that large igneous events may have occurred repeatedly from Archean to Cretaceous time. For example, *Ernst and Buchan* [1997] have proposed three new Mesozoic and 14 pre-Mesozoic LIPs (Figure 2), and locations of paleoplumes, based on giant radiating dike swarms. Several volcanic sequences dating back at least to 2.7 Ga have been suggested to represent LIP analogs [*e.g.*, *Saunders et al.*, 1996], in particular parts of Archean greenstone belts have been proposed to be unsubducted oceanic plateaus [*Abouchami et al.*, 1990; *Puchtel et al.*, 1997] and the North Atlantic volcanic margin basalts resemble the ~2.0 Ga old Cape Smith fold belt volcanics in northern Quebec [*Saunders et al.*, 1996].

Spacecraft exploration of the solar system has revealed LIP-like features on Venus, Mars, the Moon, and possibly Mercury, and hot spot volcanism on Io. In view of the absence of evidence for global tectonism on these planetary surfaces, it has been suggested that a mode of mantle circulation and internal energy loss typical for LIPs is more prevalent throughout the solar system than the present steady-state plate tectonic processes dominating on Earth [*Head and Coffin*, 1997].

## CRUSTAL STRUCTURE AND COMPOSITION

Study of individual LIPs has expanded during the past few years and new data, particularly from marine geophysics, seismic tomography, geochemistry, petrology and geodynamic modeling, reveal important structural and compositional similarities [*e.g.*, *Mahoney and Coffin*, 1997]. Knowledge of the upper crustal structure based on regional coverage of high-quality seismic images has gradually improved on some volcanic margins and on the Kerguelen Plateau, but seismic reconnaissance surveys are still the main data source for most other provinces. The seismic images from the middle and lower crust are commonly poor and the geometry is mostly inferred from seismic velocity and/or gravity data, although high-quality velocity transects have been carried out on the Kerguelen Plateau and on volcanic margin segments in the North Atlantic Ocean.

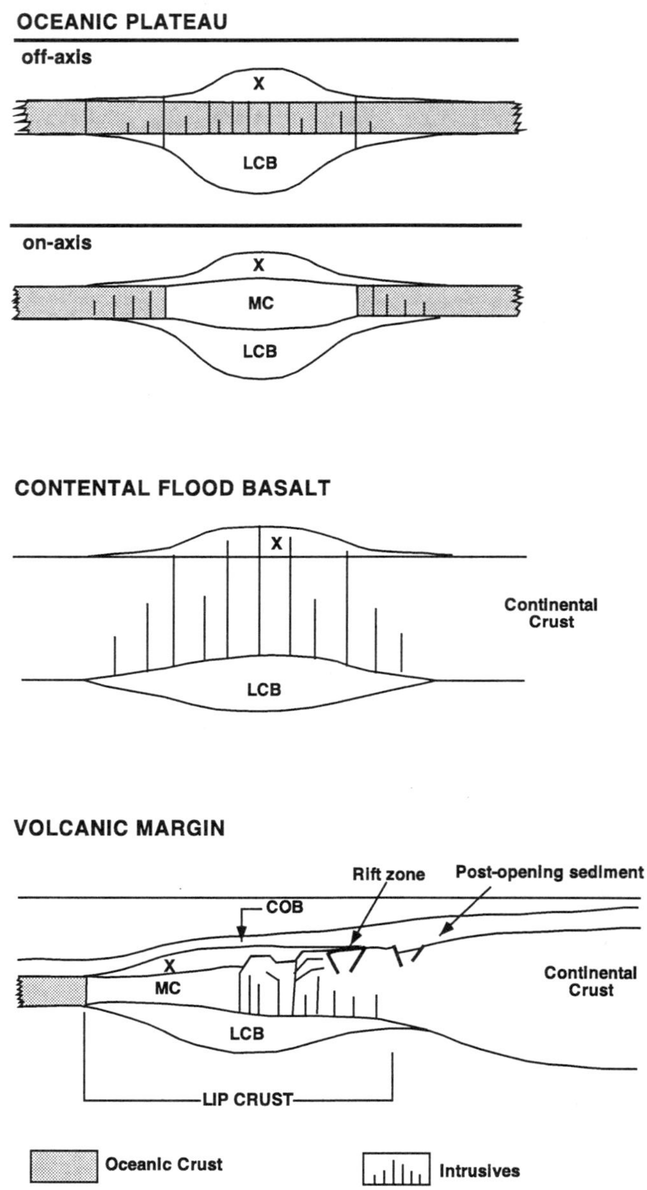


**Figure 1.** Schematic LIP plate tectonic settings and gross crustal structure. LIP crustal components are: extrusive cover (X), middle crust (MC) and lower crustal body (LCB).

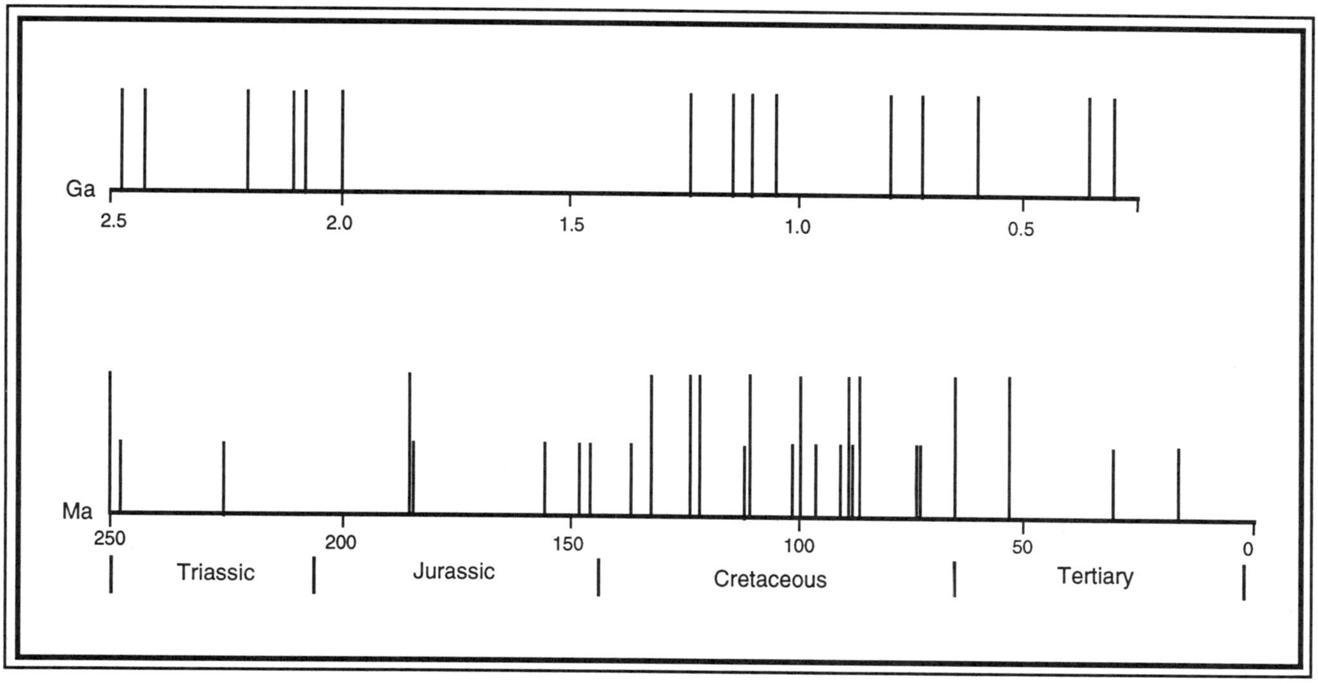


**Figure 2.** Temporal LIP distribution. Top: giant radiating and non-radiating dike swarms associated with the arrival of a plume head at the base of the continental lithosphere that can be dated with an uncertainty <250 m.y. compiled by *Ernst and Buchan* [1997]. Bottom: Mesozoic and Cenozoic LIPs (Table 1).

Because of their accessibility to sampling, CFBs and accreted parts of oceanic plateaus offer the most petrological, geochemical, and geochronological data. For other LIPs, the compositional and emplacement age information arrive from dredges and a limited number of shallow basement penetration drill holes, of which the deepest well has penetrated 914 m into extrusive rock (Table 3). It is revealing that the first multichannel seismic reflection transect on the Earth's largest igneous province, the Ontong Java Plateau, was acquired in 1998 [*Coffin et al.*, in prep.]. Moreover, only three drill holes have reached basement in this province, which covers an area equal to one-third of the contiguous United States, penetrating 9, 25 and 149 m of lavas, *i.e.*, sampling not more than ~0.5% of the total igneous crust. Thus, the available database relative to the size of the LIP constructions restrict us to addressing first-order questions in terms of global LIP comparisons and inferences.

Excluding ocean basin flood basalts, seismic crustal structure of different LIPs is grossly similar in three ways [*Coffin and Eldholm*, 1994; *Eldholm and Grue*, 1994]. One, the crystalline crust is much thicker than the 7.1-km-thick average oceanic crust [*White et al.*, 1992]. The thickest LIP crust, on the Ontong Java Plateau, is about 4.5 times as thick as oceanic crust [*Gladczenko et al.*, 1997]. Two, the seismic velocity distribution suggests a division into three main seismic units; and three, the lower unit comprises a high-velocity, high-density lower crustal body (LCB) (Figure 1). Unit velocities are 3.7-6.0 km/s for the upper, extrusive crustal unit reaching ~6.5 km/s at the base of the thickest constructions; on average 6.5-6.7 km/s for the middle crust; and 7.0-7.6 km/s for the LCB.

Seismic images on NE North Atlantic volcanic margins reveal typical wedges of seaward dipping reflections (SDRs) shown by drilling to be dominantly tholeiitic basalt flows and thin interbedded sediments [*Roberts et al.*, 1984; *Eldholm et al.*, 1987; *Larsen et al.*, 1994; *Duncan et al.*, 1995]. Similar images are reported from a number of volcanic margins world-wide (Plate 1) and from the Kerguelen Plateau [*Coffin et al.*, 1990; *Schaming and Rotstein*, 1990] and Caribbean ocean basin flood basalts [*Mauffret and Leroy*, 1997; *Diebold et al.*, 1999, in press], while only weak and local intrabasement reflections have yet been observed on the Ontong Java Plateau [*Hagen et al.*, 1993; *Coffin et al.*, in prep.]. Previously, most margins were considered non-volcanic but improved quality in seismic imaging has led to recognition of SDRs on many passive margin segments, and they appear to dominate in the Atlantic and Indian oceans. However, the magnitude of volcanism differs among these margins, and the conjugate margins in the NE North Atlantic and in the South Atlantic south of Walvis Ridge-Rio Grande Rise reveal

**Table 3.** Main compositional and age data base for offshore LIPs. TD, total depth; DSDP, Deep Sea Drilling Project; ODP, Ocean Drilling Program.

| Data sources | Remarks |
|---|---|
| Dredges | Exposed surface rock |
| Far-field drill holes- volcaniclastic horizons- sills and low-angle dikes | Mostly cuttings from industry holes—difficult to access |
| DSDP and ODP drill holes w/TD in basement —penetration <50 m | DSDP (spot coring) ODP (continuous coring) Not optimal for LIP studies |
| Dedicated LIP drillholes —basement penetration >50 m | |
| Leg (number of holes): | Max. basement penetration, m |
| 84 Hatton Bank VM (2) | 183 |
| 104 Vøring VM (1) | 914 |
| 120 Kerguelen OP (1) | 54 |
| 130 Ontong Java OP (1) | 149 |
| 152 Greenland VM (1) | 779 |
| 163 SW Greenland VM (2) | 131 |
| 183 Kerguelen-Broken Ridge OP/VM (7) | 232 |

particularly massive igneous activity during lithospheric breakup (Plate 1).

Seismic modeling based on core and downhole log data from NE North Atlantic margin drill sites shows that the dimensions and physical property distribution for most individual, and composite, lava units will not produce the observed intrabasement reflections. Therefore, the seismically anisotropic SDRs have been attributed to extensive, thick individual or composite flow units and to non-geologic seismic interference [*Planke and Eldholm*, 1994; *Planke and Cambray*, 1998]. If the extrusive complex is mainly constructed by horizontal and weakly dipping lavas, it may be difficult to image even in high-quality seismic data. The great variety in seismic character in extrusive LIP terranes reflects volume and rate of magma production, constructional environment, syn- and post-constructional deformation, and subsidence [*Eldholm et al*., 1995]. However, integrating physical volcanology data [*e.g*., *Self et al*., 1997], downhole logs, and seismic characteristics provides a volcanostratigraphic tool to determine emplacement environments during LIP construction [*Planke et al*., 1999, in press].

The abundant, predominantly subhorizontal tholeiitic lava flows in CFBs, accreted parts of Cretaceous oceanic plateaus, and in oceanic LIP drill cores record a first-order similarity in structure and composition of the extrusive LIP cover. The lavas are generally emplaced by fissure eruptions in subaerial or shallow marine environments. However, the apparent deep water environment, below the calcium carbonate compensation depth [*Sliter and Leckie*, 1993], of the Ontong Java Plateau lavas remains enigmatic. Initial eruption and intrusions are sometimes accompanied by subsidiary high-temperature melts such as picrites, and intermediate and felsic lavas and intrusives are commonly associated with initial and late emplacement periods [*e.g*., *Carlson*, 1991]. Except for CFBs, most LIP basalts show relatively little crustal contamination. The major element compositions of CFBs appear more diverse than MOR basalt [*Macdougall*, 1988], and the lavas have a "lithospheric signature" [*Lassiter and DePaolo*, 1997]. The LIPs basalt geochemistry and petrology indicate at least four different mantle source reservoirs suggesting lower to upper mantle sources [*Hart et al*., 1992], and it appears that the observed differences in chemistry and isotopes mostly relate to the amount of crustal contamination and the source compositions of the subcontinental lithosphere and asthenosphere [*e.g*., *Saunders et al*., 1992; *Coffin and Eldholm*, 1994].

The composition of the middle and lower LIP crust (Figure 1) clearly depends on the tectono-magmatic emplacement setting. In an intraplate setting, the middle crustal sandwich of older oceanic or continental compositions is expected to be strongly intruded in places. In terms of seismic velocities, the on-axis produced middle crust resembles a thick oceanic layer 3A [*Ewing and Houtz*, 1979], and may consist of a dike complex changing into gabbroic rock with depth [*Zehnder et al*., 1990].

The LCB on volcanic margins is part of the initial, thick oceanic crust, but continues some distance landward as an underplated body beneath extended and thinned continental crust (Figure 1) contributing considerably to crustal growth. A compositional mix of cumulate gabbro and ultramafics was proposed by *Eldholm and Grue* [1994], who also pointed out that the relatively similar depths to the top of separate LCBs could reflect post-emplacement metamorphism. The composition of the lower crust has been discussed by several investigators, based on predicted elevated mantle temperatures, observed LCB velocities and thicknesses, and petrological and geochemical constraints (see summaries in *Kelemen and Holbrook* [1995] and *Gladczenko et al*. [1997]). Suggested compositions include: high MgO melts [*White and McKenzie*, 1989], fractionation of a picritic melt into olivine-rich gabbroc cumulates [*Farnetani et al*., 1996; *Neal et al*., 1997], and oxide-rich gabbro lenses acting as high-velocity wave guides [*Fountain et al*., 1994]. Moreover, *Gladczenko et al*. [1997] evaluated the potential impact of secondary metamorphism on the thick Ontong Java Plateau crust, and concluded that the available data satisfy a lower crust produced either by ponded and fractionated primary melts or by secondary recrystallization. In addition, many volcanic margins and oceanic plateaus LIPs form at, or near, diverging plate boundaries leading to varying eruption and intrusion conditions within the same LIP. This suggests that the detailed composition of the lower LIP crust may have to await strategic deep drilling, and potential onshore analogs.

## DIMENSIONS AND PRODUCTION RATES

To evaluate the importance of LIPs as a window into boundary conditions for mantle circulation and melt potentials, as well as to global environmental change, we need to calculate crustal volumes and production rates. Globally, these parameters may be compared with the plate tectonic, or MOR, production volumes and rates to study LIP forming processes, and the Earth's total mafic crustal production budget through geological time.

It is important to keep in mind that the surface and subsurface expressions of LIPs, in most cases, yield minimum dimensions for the area and volume of the extrusive LIP cover, mainly due to syn- and post-constructional erosion which is particularly efficient in subaerial and shallow marine environments. The Deccan LIP, for example, is considered to have had a former extent three to four times greater than the present surface area of the Deccan Traps CFB [*e.g.*, *Todal and Eldholm*, 1999, in press]. In addition, the lower and middle crust are not well mapped in many LIPs, as discussed above. Depending on the availability and quality of seismic data crustal volumes may be calculated from: 1) global relief data, assuming a homogeneous crust and isostatic equilibrium [*Schubert and Sandwell*, 1989]; 2) the volume of the extrusive cover, calculating its isostatic compensation using representative densities for the three main crustal layers; 3) crustal geometries entirely from seismic surveys, or combined with gravity modeling of the lower crustal configuration. Airy-type compensation is assumed, and the methods are applicable to off-axis plate tectonic settings by accounting for the existing continental or oceanic crustal block. The crustal volumes of the <150 Ma old LIPs (Plate 1) are based on these procedures and a conservative approach (Table 1); thus, they are considered minima.

The quality of radiometric dates have greatly improved as older K-Ar values are being supeseded by more precise and reproducible $^{40}$Ar-$^{39}$Ar dates. The result is tighter age groupings, further supporting the short-lived nature of the main effusion event. In addition, biostratigraphic data, sea floor spreading type magnetic anomalies, and in particular magnetic polarity reversals recorded by the lavas, may further constrain the duration of the most intense volcanism. Another, less used but powerful method is far-field volcaniclastic horizons in sedimentary basins (Table 3) as demonstrated for the North Atlantic LIP (Figure 3). Compositionally and temporally, the extensive and well-dated Balder ash formation in northwest Europe corresponds to the main volcanic event constructing the onshore CFBs and the SDRs on the volcanic margins. Furthermore, less intense and compositionally variable ash beds on either side of Balder record the onset and waning of the breakup event [*Knox and Morton*, 1988; *Eldholm et al.*, 1989].

Because only limited portions of the transient features have yet been properly dated, the duration of the transient magmatism is not well known for many LIPs. Therefore, we employ a mean age for each transient LIP forming event (Table 1, Figure 2). Because LIP production rates critically depend both on reliable volumes and durations, available data permit only the most general conclusions:

- LIPs range in volume from ~1 to at least 45 x $10^6$ km$^3$.
- The duration of most transient emplacement events is <3 m.y., and probably <1 m.y. for some LIPs.
- The production rate of the largest LIP, Ontong Java Plateau, may have been greater than the 16.8 km$^3$/y present-day global MOR production rate of *Larson and Olson* [1991].
- Several larger LIPs yield production rates exceeding 25% of today's global MOR production rate.

## LIP FORMATION MODELS

The various LIP surface manifestations reflect a spatial and temporal record of magma production peaking during the transient phase. In terms of upper mantle circulation, commonly inferred from studies of the MOR system, the spatial and temporal LIP evolution offers an additional window to study the dynamic behavior of the Earth's mantle. The fact that LIPs, emplaced over intervals of the order of 1 m.y, are associated in many cases with persistent but far less voluminous magmatism over the subsequent 10-100 m.y., has led to their association with diapir-like mantle plumes providing a simple explanation for excess energy fluxes relative to the ambient mantle. The transient event marks the penetration of a massive melt anomaly through the lithosphere, *i.e.*, the plume "head", whereas subsequent persistent activity reflects the "tail"'s imprint on the lithosphere as it moves relative to the underlying plume [*Richards et al.*, 1989]. However, not all oceanic LIPs have obvious plume tails [*Coffin and Eldholm*, 1994], and they appear absent or poorly developed in intraplate continental settings. Although most agree that increased mantle temperatures set the stage for decompressional melting, the origin of LIPs via plumes is debated. Plume models include: initiation of a plume head-tail system by upper mantle plume flux [*White and McKenzie*, 1989; 1995]; arrival at the base of the lithosphere of a new plume from a thermal boundary layer (D") at the base of the mantle [*e.g.*, *Campbell and Griffiths*, 1990; *Hill*, 1991; *Davies*, 1997]; or plumes in a system of combined whole- and layered mantle

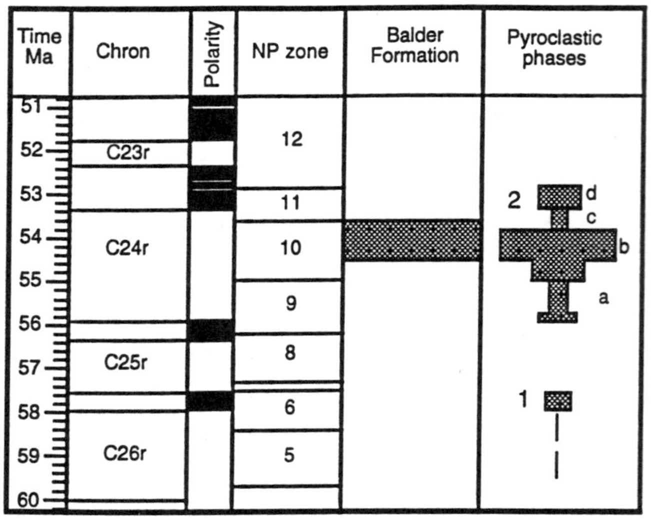


**Figure 3.** Far-field signature of the North Atlantic LIP as recorded by pyroclastic deposits in the North Sea sedimentary basin from *Knox and Morton* [1988]. Pyroclastic phase 2 represents breakup volcanism and phase 1 initial volcanism. Horizontal bars indicate relative intensity. Subphases 2b-c are compositionally similar to the SDRs drilled on the volcanic margins and correspond to the laminated mudstone of the Balder Formation which produce a widely recognizable far-field "ash marker" in seismic data. Note that the duration of the most intense breakup volcanism is <1 m.y. Timescale and calcareous nannoplankton (NP) zones from *Berggren et al.* [1995].

circulation [*Machetel and Weber*, 1991; *Tackley et al.*, 1993] where the endothermic phase change at a depth of ~660 km marks separation of the upper and lower mantle. Alternatively, *Anderson* [1994] and *King and Anderson* [1995] suggest that plumes are not required to produce flood basalts, preferring instead small-scale lateral mantle flow from thick to thin lithosphere across cratonic boundaries where extension is driven by plate tectonic forces. Nonetheless, *Coffin and Eldholm* [1994] inferred from calculation of total partial melt volumes that at least the greatest LIPs originate in the lower mantle (Figure 4), an inference consistent with evidence of significant, but limited input of lower mantle material into the upper mantle [*Hart et al.*, 1992; *Stein and Hofmann*, 1994].

Volumetric differences among LIPs (Table 1), implying a wide size range of mantle melt anomalies, and the spatial LIP distribution (Plate 1) indicate that three main factors determine whether a LIP is formed and its size. These are the source intensity, *i.e.*, the intensity of the melt anomaly which is governed by temperature, fluid content and fertility (basalt content) of the asthenosphere; the vulnerability of the lithosphere to melt penetration; and the plate speed over the source. A subsequent, persistent LIP tail depends on the same factors. Thus, we may make inferences about LIP formation and development independent of the cause of the melt anomaly; in fact, the variability in size (Table 1) appears to allow multi-scale mantle circulation.

Because large melt volumes are most easily explained by decompressional melting, thick lithosphere impedes the process [*Lassiter and DePaolo*, 1997]. Thus, the thick lithosphere beneath shield areas, for example, inhibits both the size of the melt anomaly and melt migration. In fact, observational data suggest that the melt of all but the most vigorous mantle upwelling either ponds at the base of the crust and/or intrudes (Figure 1), inducing crustal growth and regional uplift, or deflecting to regions of thinner lithosphere. Conversely, thin lithosphere enhances melting and is more conducive for rapid melt transfer to the surface. In particular, positive relief at the top of the asthenophere caused by previous or ongoing lithospheric rifting, or along diverging plate boundaries, will induce melting and/or act as a conduit or catchment for the melt [*Artyuskov et al.*, 1980; *Thompson and Gibson*, 1991; *Sleep*, 1997]. This formation concept is consistent with the plate tectonic settings of most LIPs, *i.e.*, they occur preferentially in regions conducive to voluminous decompressional melting such as thin oceanic lithosphere and continental lithosphere where accentuated relief reflects extension and thinning by rifting. Almost all post-Jurassic LIPs (Plate 2, Table 1) formed at or close to plate boundaries; even the prominent Siberian Platform CFB has been associated with extension and 8-10 km of crustal thinning [*Zonenshain et al.*, 1990]. This concept, however, does not depend on a cratonic boundary as implied by *King and Anderson* [1995]. Similarly, thin oceanic lithosphere and accessible pathways may explain the common association of oceanic LIPs with transform faults and triple junctions [*e.g.*, *McNutt et al.*, 1989; *Sager and Han*, 1993].

## THE MANTLE WINDOW

LIP sizes (Table 1) and emplacement rates (Figure 5) combined with their spatial, temporal and plate tectonic settings (Plates 1, 2; Figure 1) raise questions about how Earth's crust forms. Does it form in three different ways: plate tectonic MOR production, LIP production, and arc magmatism and accretion? If so, are the mantle driving processes independent or interrelated? These are fundamental questions related to how our planet has evolved.

To monitor mantle energy flux and dynamics over time, we need a representative frame of reference. The preponderance of LIPs on oceanic crust and at or near passive margins imply that these areas are the most reliable surface recorders of mantle melt anomalies. In spite of at least two major Jurassic LIPs, the Central Atlantic Magmatic Province (CAMP) [*Marzoli et al.,* 1999] and Karoo-Ferrar (Table 1), we do not

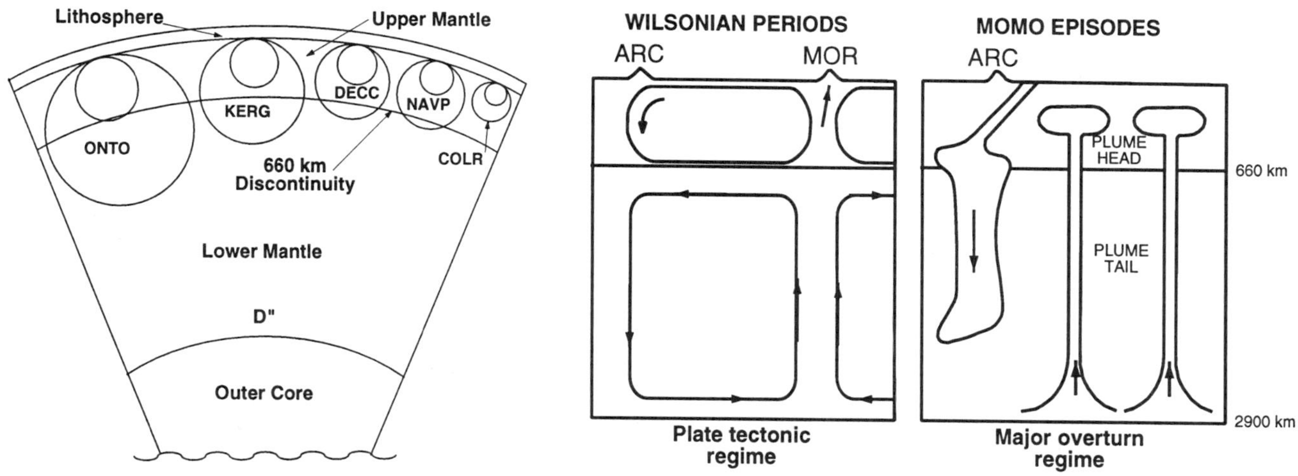


**Figure 4.** Left: LIP volume-based static estimates of asthenospheric melt volumes, assuming a simple spherical geometry and 5 and 30% partial melting [*Coffin and Eldholm*, 1994]. Right: schematic illustration of the two modes of mantle convection, Wilsonian or plate tectonic periods and MOMO episodes, proposed by *Stein and Hofmann* [1994].

include the pre-latest Jurassic record to avoid large biases from subducted crust. Assuming the same continent and ocean area distribution as at present, we retain a mantle window covering about 60% of the earth's surface over the past 150 m.y. Notwithstanding, the pre-Jurassic record (Figure 2) strongly suggests that LIP forming processes were active also prior to the Mesozoic era.

Using available crustal structure data, we quantify the volume of excess crust produced during the history of a mantle melt source relative to the steady-state MOR production of oceanic crust. This crustal addition comprises three igneous components emplaced during the initial phase, transient event, and persistent phase, respectively. The scattered and largely intrusive initial phase rocks are difficult to quantify volumetrically, but are considered relatively small. The contribution during the persistent phase is still small but significant from blocky submarine ridges, whereas seamounts chains are less important. For example, the Ninetyeast Ridge production rate is 0.24 $km^3$/yr for the 80-40 m.y. period [*Pringle et al.*, 1997], compared to 0.06 $km^3$/yr for the 60-40 m.y. Emperor, and 0.16-0.03 $km^3$/yr for the Hawaiian seamount chains, respectively [*White*, 1993].

To minimize the effect of the uncertainty in the duration of the transient event, we have calculated LIP volumes, and corresponding emplacement rates, over 5 and 15 m.y. intervals since 150 Ma (Figure 5). To quantify the total LIP production, we also include the excess crust produced along plume tails based on data from *Schubert and Sandwell* [1989], *White* [1993] and *Pringle et al.* [1997]. However, our mantle window does not provide a complete record of LIP production because part of the LIP crust has been removed or deformed by plate convergence. To account for subducted and obducted crust we introduce a simple "subduction correction" (Figure 5), assuming that all oceanic crust is recirculated to the mantle over 200 m.y. and a constant average LIP 5-m.y.-production volume over the same period. For comparison, we include the MOR production rates of *Larson* [1991b]. We conclude (Figure 5) that:

- A significant amount of new crust is produced by LIPs; however, the major volumetric contribution is restricted to the transient events.
- Cretaceous and Paleocene time was marked by vigorous LIP formation.
- Early Cretaceous time was dominated by the formation the Ontong Java Plateau.
- LIP frequency (Figures 5 and 7) and intensity have decreased dramatically since ~50 Ma.

The subduction correction underestimates the subducted crust due to the increased Cretaceous and Paleocene LIP production, further supporting the temporal production change over the past 150 m.y., which we stress is based on trends in the data rather than absolute volumes and rates. The LIP production rates are smaller than those of *Larson and Olson* [1991], who did not remove existing oceanic crust from oceanic plateaus and plume tails, but accounted for subduction by including twin features, *i.e.*, on–ridge formation, for western Pacific oceanic plateaus and ridges with the exception of Ontong Java Plateau and Line Islands.

MOR and LIP production appear to correlate positively over time; however, we note that the spreading rates during the long Cretaceous normal polarity superchron (CNS),

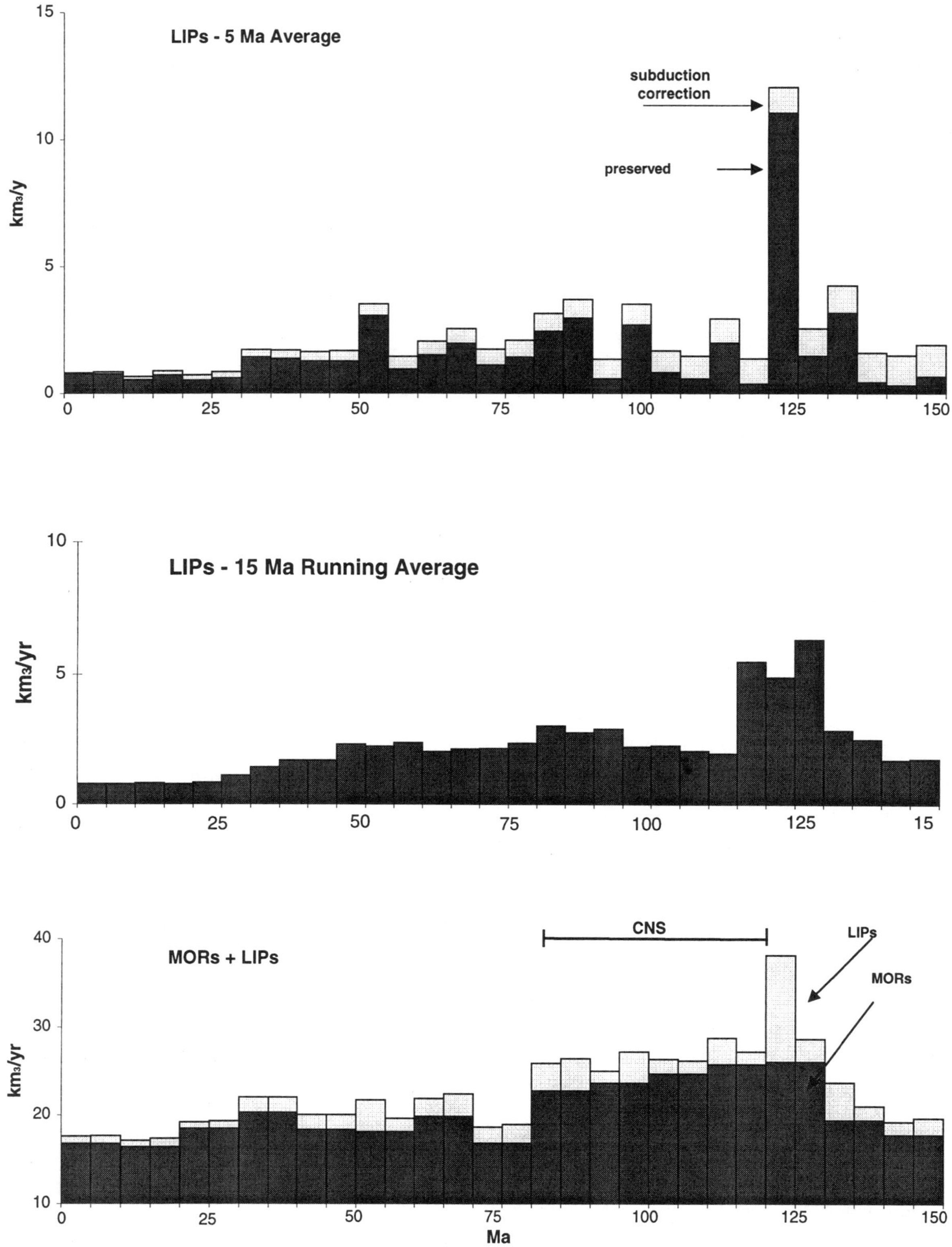


**Figure 5.** Crustal production rates. Top: rates averaged over 5 m.y. intervals for transient and persistent LIPs with subduction correction. Middle: subduction-corrected rates above smoothed by a 15 m.y. running aveage. Bottom: 5 m.y. average LIP and global MOR production rates [*Larson*, 1991b] adjusted to the *Gradstein et al.* [1995] and *Berggren et al.* [1995] time scales. CNS, Cretaceous Normal Superchron.

120.4-83.5 m.y. [*Gradstein et al.*, 1994] (Figure 5), are poorly constrained. Nonetheless, MOR crustal production has varied considerably less than LIP production, implying that LIP formation and sea floor spreading reflect fundamentally different mantle processes. This is also suggested by our calculations for the production of subduction corrected LIP crust corresponding to 6.5., 12.4, and 13.8% of the MOR crust produced during the 0-50, 50-100 and 100-150 Ma time periods.

Although some periodicity in LIP forming events has been suggested since 3 Ga [*Yale and Carpenter*, 1998], the LIP record over the past 150 m.y. does not provide strong evidence for periodicity. We rather interpret the main temporal trends in terms of the mantle convecting in two main modes: one feeds the MORs, the other feeds LIPs. We further believe that the LIP record prior to the Cretaceous period provides strong support that the two systems can be traced far back in earth history. The association of many pre-Cretaceous LIPs with accreted terranes and rift systems [*Saunders et al.*, 1996; *Ernst and Buchan*, 1997] indicates a relationship between major melt anomalies and thin lithosphere similar to that of post-Jurassic time. However, higher upwelling mantle temperatures may have caused some penetration of relatively thick continental lithosphere in a younger Earth. Conversely, the relative importance of LIP production may have diminished since early Eocene time (Figures 3 and 5). According to *Stein and Hofmann* [1994] the normal mode of plate tectonics prevails during so-called Wilsonian periods characterized by opening and closing of oceans and by separate upper and lower mantle mantle convection, with plumes originating mostly from near the base of the upper mantle. On the other hand, during MOMO (mantle overturn, major orogenies) episodes cold material accumulated near the base of the upper mantle descends into the lower mantle inducing multiple major plumes rising from near the core-mantle boundary to the surface where they may lead to formation of LIPs (Figure 4).

Many have pointed out that Cretaceous time was marked by especially high LIP activity (Figures 3 and 5), and that the Ontong Java Plateau resulted from a deep mantle plume [*e.g.*, *Larson*, 1991a, b], *i.e.*, reflecting a MOMO episode of whole mantle overturn (Figure 4). The temporal LIP distribution over the past 150 m.y. suggests that the 135-80 m.y. interval is active both in terms of number of events and in melt volume. This interval encompasses the CNS, which is characterized by warm climates, black shale (oceanic anoxic) events, biotic and oceanic isotope fluctuations and high sea levels [*e.g.*, *Larson*, 1991a; *Bralower et al.*, 1993, 1997]. On the other hand, the relatively high production levels continued into early Eocene time. We find diminishing activity since 50 Ma, a time period characterized by climatic deterioration, equally spectacular and important, posing key questions such as:

- Is the reference period long enough to detect reliable (*e.g.*, unaliased) temporal trends?
- What is the LIP reference level since 150 Ma?
- Does the increased Early Cretaceous activity level entirely reflect the Ontong Java Plateau melt anomaly, or superplume [*Larson*, 1991b, 1995]?
- Is there a link between LIP production and Cenozoic climatic change?

In other words, what is normal, and what is anomalous or reflects an evolutionary trend? The answers are not obvious, but one should bear in mind that the only known melt anomaly of a size that convincingly requires a MOMO event is the Ontong Java Plateau and its coeval and adjacent LIPs in the SW Pacific Ocean, *i.e.*, the Greater Ontong Java LIP of *Coffin and Eldholm* [1994]. Thus, a plausible message is that MOMO events may occur on a timescale of 100-200 Ma, whereas the archetypical LIP occurring on a 10-30 Ma timescale primarily reflects upper mantle circulation superimposed on the plate tectonic cycle of sea floor spreading and subduction. In fact, MOMO events may have occurred as far back as 2.7 Ga [*Nelson*, 1998]. Plume processes may have been more important in the past, perhaps related to higher heat production and lower viscosity in the mantle [*Fyfe* 1978; *Davies*, 1992]. Thus, it is tempting to speculate whether the diminishing activity reveals an evolutionary trend, perhaps linked to cooling of the Earth and the development of a more mature, and predictable mantle, in contrast to the model of *Allégre* [1997].

## LIPS, THE PLATE TECTONIC CYCLE, AND CONTINENTAL GROWTH

The thermal structure of the asthenosphere, mapped by seismic tomography, shows that the main mode of mantle convection is flow-controlled and scaled by plates. While upwelling is characteristic for the lower mantle, the upper mantle is dominated by subduction [*e.g.*, *Dziewonski and Woodhouse*, 1987; *Anderson et al.*, 1992; *Davies and Richards*, 1992; *Anderson*, 1998]. If and how LIP formation relates to plate tectonic processes commonly depends upon cause-and-effect assumptions for various types of mantle circulation models, and we believe some disagreement in the literature is partly semantic.

The spatial LIP distribution and their plate tectonic emplacement settings at or near diverging plate boundaries (Figures 1 and 2) suggest a direct spatial relationship between transient LIP formation and lithospheric extension, plate breakup, and oceanic crust formed by sea floor spreading.

This inter-relationship is apparent because lithospheric configurations allowing excess decompressional melting and vulnerability to melt migration, *i.e.*, those which facilitate LIP formation, are attributed to plate tectonic processes. Thus, the key question becomes what causes initial rifting in the vicinity of a LIP, far-field plate tectonic forces or the mantle thermal anomaly?

Depending on the lithospheric setting both mechanisms may be applicable. For example, a vigorous plume from the lower mantle may have the potential to initiate continental breakup [*e.g., Hill*, 1991]. However, the implication that melting may precede lithospheric extension for high plume temperatures and the <10 m.y. duration between lithospheric impact and melt generation, do not satisfy observational data from many LIPs [*Saunders et al.*, 1992]. On the other hand, a plume at the base of the lithosphere will only lead to large-scale melting if the lithosphere is extended [*White and McKenzie*, 1989], though it has been proposed that a stationary continental lithosphere may be weakened by thermal erosion and intrusion [*e.g., Kent*, 1991; *Davies*, 1994]. Studies on volcanic margins support the influence of lithospheric relief on the magnitude of melting and LIP formation. A 15-20 m.y., or longer, rift phase commonly precedes onset of sea floor spreading; in fact, volcanic margin LIP formation may also occur in lithosphere which has been extended, and thinned, during previous rift episodes [*e.g., Eldholm et al.*, 1995]. Hence, the rift system appears to capture the plume, providing conditions for enhanced melting. An example is the NE North Atlantic LIP developing within a previous regional Mesozoic rift system, where onset of renewed rifting occurred about 20 m.y. prior to breakup [*Skogseid et al.*, 1999, in press], whereas initial magmatism first appeared 12-13 m.y. later [*Saunders et al.*, 1997]. If not an instigator of plate tectonic events, the tectono-magmatic settings of LIPs imply that the combination of mantle melt anomalies and lithospheric vulnerability due to ongoing and/or previous extensional deformation may be considered an accelerator and modifier of plate tectonic processes.

The slowdown in LIP production and plume activity since 50 Ma may also have important plate tectonic consequences. If the 150 m.y. mantle window is representative, one may speculate that the impact of plume-driven mantle circulation on plate tectonic change has diminished with time.

Within the plate tectonic cycle, growth of continental crust is generally associated with plate convergence, primarily through arc magmatism and accretion. We have previously shown how LIPs contribute to growth of the oceanic crust, relative to normal steady-state production, and to continental crust by episodic magmatic pulses volumetrically expanding the crust by the processes of intrusion, effusion and underplating. Because LIPs move with the plate, they may have been important contributors to continental growth and stabilization over long time periods (Figure 2). Noting the apparent episodicity in continental crustal growth [*Moorbath*, 1978], a link with MOMO episodes (Figure 4) was proposed by *Stein and Hofmann* [1994] taking the Ontong Java Plateau as an example.

The importance of MOMO episodes for continental growth gains support from buoyancy analysis of crustal bodies expanded in thickness with respect to normal oceanic crust. *Cloos* [1993] has suggested that up to ~17-km-thick oceanic plateaus are inherently subductable, while a thickness of ~30 km is required to "jam" the subduction zone initiating changes in plate boundary geometry and mountain building by lateral accretion. Accordingly, of the present LIPs, only the Ontong Java Plateau has the potential of creating collisional orogenesis. However, parts of the 15-20-km-thick Caribbean LIP crust exist locally on the adjacent continent [*Mauffret and Leroy*, 1997; *Sinton et al.*, 1998]. Furthermore, accreted Ontong Java Plateau rocks appear on Malaita and adjacent islands [*Tejada et al.*, 1996; *Petterson et al.*, 1997] where subduction of probably <30-km-thick plateau flank crust along the North Solomon Trench was initiated as recently as 4 Ma [*Phinney et al.*, 1999, in press]. These observations suggest that <30-km-thick crust also subducts poorly. Thus, we concur that a LIP formed during a MOMO episode has the potential of inducing significant continental growth, delayed by the time it takes for the intervening oceanic lithosphere to become subducted. However, the same effect is also expected for other major LIPs, some of which may have been produced by plumes originating in the upper mantle, *i.e.*, during Wilsonian periods (Figure 4).

## SUMMARY – AND A LOOK AHEAD

Compared with with plate dimensions, LIPs are smaller and isolated features in plate interiors and at plate boundaries. Study of LIPs shows that they reflect mantle dynamics, mass and energy transfer that acts both independently of, and in conjunction with, the wide-ranging upwellling systems leading to partial melting and construction of new oceanic crust by sea floor spreading along the MOR system. Three principal factors determine whether a LIP will appear on the earths surface, and its size: 1) the source intensity, or the size of the melt anomaly, depending on temperature, fluid content and fertility of the asthenosphere; 2) the vulnerability of the lithosphere to melt penetration, largely a function of lithospheric dynamics; and 3) the plate speed over the source.

Relative to the record of global plate tectonic history in today's oceans and continental margins, most LIPs occur in oceanic or in continental breakup plate tectonic settings, suggesting a first-order relationship with thin lithosphere. The record of deformed LIP-associated rock complexes in presently intraplate continental settings, dating back ~3 Ga, indicates that LIPs formed throughout most of Earth history. LIP longevity is also consistent with observations on many planetary surfaces. What evolutionary trends the post-150 Ma distribution (Figure 5) contains is not entirely obvious, though much attention has been directed towards the increased number of events and large melt volumes in the 135-85 m.y. interval and coeval environmental change. On the other hand, the birth of only two relatively small LIPs, Ethiopia-Yemen and Columbia River (Plate 1) since 50 Ma may provide an equally important message.

During their formation LIPs provide an additional component of crustal production and growth relative to sea floor spreading. After formation, oceanic LIPs are carried laterally by the plates, eventually reaching a subduction zone. The subsequent accretion of major LIPs onto continental crust make these "old" LIPs episodic contributors to continental growth. However, the growth takes place over a far longer period that the LIP emplacement. In addition, formation of volcanic margins and continental flood basalts adds to continental growth. Another contribution arrives from melt anomalies beneath thick lithosphere leaving underplated bodies and intrusive complexes without a typical LIP manifestation at the surface.

LIP sizes and distributions reflect different source intensities, thus in principle allowing different source origins. Nonetheless, most evidence is compatible with mantle plumes as the primary instigator; the greatest LIPs most probably originate from near the base of the lower mantle. Regardless of mantle circulation models, the plume geometry and source depths contrast greatly with the relatively shallow upwelling systems associated with diverging plate boundaries and sea floor spreading. In fact, it appears that developing and developed plate boundary regions are particularly attractive for LIP formation by their increased melt potential and their ability to act as melt catchments and funnels. A consequence is that rifting may be initiated by far-field plate tectonic forces rather than by the plume itself.

Looking ahead, we have to keep in mind the "underdeveloped" state of the LIP database. We have shown that LIP evolution and development, together with plate tectonics, play key roles in our understanding of the dynamic processes shaping the earth. Moreover, LIP formation may contribute importantly to global environmental change. Still, our knowledge of many LIPs is not much better than that of the MORs prior to the general acceptance of the plate tectonics paradigm about 30 years ago [*Dick and Mevel*, 1997]. To obtain improved estimates of LIP dimensions, emplacement ages and production rates, which in turn provide boundary conditions for geodynamic modeling, we need large, interdisciplinary studies of key LIPs. These studies will include:

- Intensified geophysical surveying by highly sophisticated seismic techniques to obtain velocities and images of the entire crust. Because the available seismic LIP profiles are all two-dimensional while the point (volcanoes) and line sources (fissures) in extrusive LIP terrane create three-dimensional features, selective three-dimensional seismic surveys are required.
- Strategic scientific transect drilling, complemented by selected deep holes, to obtain compositional and age information, noting that volcanic margin drilling has proved the feasibility of drilling with good core recovery in LIP terrane.
- Increased emphasis on comparative studies, including both the relatively young and largely undeformed features, as well as accreted LIP terranes.

*Acknowledgements.* We thank the many colleagues who have provided valuable input to our LIP inventory on which this contribution is based, and to the UTIG PLATES Project and Lisa Gahagan for providing plate reconstructions. Support of the Research Council of Norway, the National Science Foundation, Joint Oceanographic Institutions/U.S. Science Support Program, the Ocean Drilling Program, and UTIG (P. Stoffa) is gratefully acknowledged. Gwen Watson and Jan Thompson ably supplied technical services. The University of Texas Institute for Geophysics contribution #1431.

## REFERENCES

Abouchami, W., M. Boher, A. Michard, and F. Allarede, A major 2.1 Ga event of mafic magmatism in West Africa: an early stage of crustal accretion, *J. Geophys. Res., 95*, 17605-17629, 1990.

Allégre, C.J., Limitation on the mass exchange between the upper and lower mantle: the evolving convection regime of the Earth, *Earth Planet Sci. Lett., 150*, 1-6, 1997.

Anderson, D.L., The sublithospheric mantle as the source of continental flood basalts; the case against the continental lithosphere and plume head reservoirs, *Earth Planet. Sci. Lett., 123*, 269-280, 1994.

Anderson, D.L, The scales of mantle convection. *Tectonophysics, 284*, 1-17, 1998.

Anderson, D.L., T. Tanimoto, and Y.-S. Chang, Plate tectonic and hotspots, *Science, 256*, 1645-1651, 1992.

Artyuskov, E.V., A.E. Shlesinger, and A.L. Yanshin, The origin of vertical crustal movements within lithospheric plates, *Geodyn. Ser., 1*, Amer. Geophys. Union, 37-51, 1980.

Bercovici, D., and J. Mahoney, Double flood basalt and plume head separation at the 660-km discontinuity, *Science, 266*, 1367-1369, 1994.

Berggren, W.A., D.V. Kent, C.C. Swisher, and M.-P. Aubrey, A revised Cenozoic geochronology and chronostratigraphy, *SEPM Spec. Publ., 54*, 129-212, 1995.

Bijwaard, H., and W. Spakman, Tomographic evidence for a narrow whole mantle plume below Iceland, *Earth Planet. Sci. Lett., 166*, 121-126, 1999.

Bralower T.J., P.D. Fullagar, C.K. Paull, G.S. Dwyer, and R.M. Leckie, Mid-Cretaceous strontium isotope stratigraphy of deep-sea sections, *Bull. Geol. Soc. Am., 109*, 1421-1442, 1997.

Bralower T.J., W.V. Sliter, M.A. Arthur, R.M. Leckie, D. Allard, and S.O. Schlanger, Dysoxic/anoxic episodes in the Aptian-Albian (Early Cretaceous), *Geophysical Monograph* 77, Amer. Geophys. Union, 5-137, 1993.

Byrkjeland, U., H. Bungum, and O. Eldholm, Seismotectonics of the Norwegian continental margin, *J. Geophys. Res.*, in press.

Campbell, I.H., and R.W. Griffiths, Implications of mantle plume structure for the evolution of flood basalts, *Earth Planet Sci. Lett., 99*, 79-83, 1990.

Carlson, R.W., Physical and chemical evidence on the cause and source characteristics of flood basalt volcanism, *Austral. J. Earth Sciences, 38*, 525-544, 1991.

Chen, Y., S.C. Solomon, I.T. Bjarnason, and C.L. Wolfe, Seismic evidence for a lower-mantle origin of the Iceland plume, *Nature, 395*, 62-65, 1998.

Cloos, M., Lithospheric buoyancy and collision orogenesis: subduction of oceanic plateaus, continental margins, island arcs, spreading ridges and seamounts, *Geol. Soc. Am. Bull., 105*, 715-737, 1993.

Coffin, M.F., Emplacement and subsidence of Indian Ocean plateaus and submarine ridges, Geophysical Monograph 70, Amer. Geophys. Union, 115-125, 1992.

*Coffin, M.F., and O. Eldholm, Large Igneous Provinces, *Sci. Am.*, 42-49, Oct. 1993a.

Coffin, M.F., and O. Eldholm, Scratching the surface: estimating dimensions of large igneous provinces, *Geology, 21*, 515-518, 1993b.

*Coffin, M.F., and O. Eldholm. Large igneous provinces: crustal structure, dimensions, and external consequences, *Rev. Geophys., 32*, 1-36, 1994.

Coffin, M.F., L.A. Lawver, and L.M. Gahagan, Atlas of Paleogeographic Reconstructions, PLATES Progress Report No. 215-0798, *Unversity of Texas Institute for Geophysics Technical Report 181*, 88 pp., 1998.

Coffin M.F., K. Mochizuki, M. Wiederspahn, T.P. Gladczenko, and O. Eldholm, in press. Structure and stratigraphy of the OJP-Nauru Basin Transect and north-central OJP: MCS and sonobuoy study, In: Taira A. (ed.), Cruise Report KH98-1, Leg 2 Cairns-Pohnpei.

Coffin, M.F., M. Munschy, J.B. Colwell, R. Schlich, H.L. Davies, and Z.-G. Li, Seismic stratigraphy of the Raggatt Basin, southern Kerguelen Plateau: tectonic and paleoceanographic implications, *Geol. Soc. Am. Bull., 102*, 563-579, 1990.

*Courtillot, V.E., A volcanic eruption. *Sci. Am.*, 85-92, Oct. 1990.

*Courtillot, V., J.J. Jaeger, Z. Yang, G. Feraud, and C. Hofmann, The influence of continental flood basalts on mass extinctions: where do we stand?, *Geol. Soc. Am. Spec. Paper, 307*, 513-525, 1996.

Davies, G.F., Ocean bathymetry and mantle convection, *J. Geophys. Res., 93*, 10467-10480, 1988.

Davies, G.F., On the emergence of plate tectonics, *Geology, 21*, 515-518, 1992.

Davies, G.F., Thermomechanical erosion of the lithosphere by mantle plumes, *J. Geophys. Res., 99*, 15709-15722, 1994.

*Davies, G.F., The mantle dynamical repertoire: plates, plumes, overturns and tectonic evolution. *J. Austral. Geol. Geophys., 17*, 93-99, 1997.

Davies, G.F., and M.A. Richards, Mantle convection, *J. Geol., 100*, 151-206, 1992.

Dick, H., and Mevel, C., The oceanic lithosphere & scientific drilling into the 21st century, Workshop Rept., JOI, Inc., Washington, DC, 89 pp., 1997.

Diebold, J., N. Driscoll, and EW-9501 Science Team, New insights on the formation of the Caribbean Basalt Province revealed by multichannel seismic images of volcanic structures in the Venezuelan Basin, In Caribbean Sedimentary Basins, P. Mann (ed.), Elsevier Basin of the World Ser., in press.

Duncan, R.A., and M.A. Richards, Hotspots, mantle plumes, flood basalts, and true polar wander, *Rev. Geophys., 29*, 31-50, 1991.

Duncan, R.A., H.C. Larsen, J.F. Allan, et al., *Proc. ODP, Init. Repts, 163*, College Station, TX (Ocean Drilling Program), 1995.

Dziewonski, A.M., and J.H. Woodhouse, Global images of the Earth's interior, *Science, 236*, 37-48, 1987.

Eldholm, O., and K. Grue, North Atlantic volcanic margins: dimensions and production rates, *J. Geophys. Res., 99*, 2955-2968, 1994.

Eldholm, O., J. Thiede, and E. Taylor, Evolution of the Vøring volcanic margin, *Proc. ODP, Sci. Res., 104*, 1033-1065, 1989.

Eldholm, O., J. Skogseid, S. Planke and T.P. Gladczenko, Volcanic margin concepts, In Banda E., M, Talwani and M. Torné, (eds.), Rifted Ocean-Continent Boundaries, NATO ASI Series Volume, Kluwer Acad. Publs., 1-16, 1995.

Eldholm, O., J. Thiede, E. Taylor, et al., 1987, *Proc. ODP, Init. Repts, 104*, College Station, TX (Ocean Drilling Program), 1987.

Ernst, R.E., and K.L. Buchan, Giant radiating dyke swarms: their use in identifying pre-Mesozoic large igneous provinces and mantle plumes, Geophysical Monograph 100, Amer. Geophys. Union, 297-333, 1997.

Ewing, J., and R. Houtz, Acoustic stratigraphy and structure of the oceanic crust, *Maurice Ewing Ser. 2, Amer. Geophys. Union*, 1-14, 1979.

Farnetani, C.G., M.A. Richards, and M.S. Ghiorso, Petrological models of magma evolution and deep crustal structure beneath hotspots and flood basalt provinces, *Earth Planet. Sci. Lett., 143*, 81-94, 1996.

Fountain, D.M., T.M. Boundy, H. Austrheim, and P. Rey, Eclogite-facies shear zones – deep crustal reflectors?, *Tectonophysics, 232*, 411-424, 1994.

Fyfe, W.S., Evolution of the Earth's crust: modern plate tectonics to ancient hot-spot tectonics?, *Chem. Geol.*, 23, 89-114, 1978.

Gladczenko, T.P., M.F. Coffin, and O. Eldholm, Crustal structure of the Ontong Java Plateau: modeling of new gravity and existing seismic data, *J. Geophys. Res.*, *99*, 22711-22729, 1997.

Gradstein F.M., F.P. Agterberg, J.G. Ogg, J. Hardenbol, P van Veen, J. Thierry, and Z. Huang, A Triassic, Jurassic and Cretaceous time scale, *SEPM Spec. Publ.*, *54*, 95-126, 1995.

Hagen, R.A., L.A. Mayer, D.C. Mosher, L.W. Kroenke, T.H. Shipley, and E.L. Winterer, Basement structure of the northern Ontong Java Plateau, *Proc. ODP, Sci. Results*, *130*, 23-31, 1993.

Hart, S.R., E.H. Hauri, L.A. Oschmann, and J.A. Whitehead, Mantle plumes and entrainment: isotopic evidence, *Science*, *256*, 517-520, 1992.

Head, J.W., and M.F. Coffin, Large igneous provinces: a planetary perspective, Geophysical Monograph 100, Amer. Geophys. Union, 411-438, 1997.

Hill, R.I., Starting plumes and continental break-up, *Earth Planet. Sci. Lett.*, *104, 398-416*, 1991.

Houtz, R.E., D.E. Hayes, and R.G. Markl, Kerguelen Plateau bathymetry, sediment distribution and crustal structure, *Mar. Geol.*, *25*, 95-130, 1977.

Hughes, G.W., and C.C. Turner, Upraised Pacific Ocean floor, southern Malaita, Solomon Islands, *Geol. Soc. Am. Bull.*, *88*, 412-424, 1977.

Kelemen, P.B., and W.S. Holbrook, Origin of thick, high-velocity igneous crust along the U.S. East Coast margin, *J. Geophys. Res.*, *100*, 10077-10094, 1995.

Kent, R.W., Lithospheric uplift in eastern Gondwana: evidence for for a long-lived mantle plume system?, *Geology*, *19*, 19-23, 1991.

Kerr, A.C., Oceanic plateau formation: a cause of mass extinction and black shale deposition around the Cenomanian-Turonian boundary, *J. Geol. Soc.*, *155*, 619-626, 1998.

Kerr, A.C., J. Tarney, G.F. Marriner, A. Nivia, and A.D. Saunders, The Caribbean-Colombian Cretaceous Igneous Province: the internal anatomy of an oceanic plateau, Geophysical Monograph 100, Amer. Geophys. Union, 123-144, 1997.

King, S.D., and D.L. Anderson, An alternative mechanism of flood basalt formation, *Earth Planet. Sci. Lett.*, *136*, 269-279, 1995.

Knox, R.W.O'B., and A.C. Morton, The record of early Tertiary N Atlantic volcanism in sediments of the North Sea Basin, *Geol. Soc. Spec. Publ.*, *39*, 407-419, 1988.

Kroenke, L.W., Geology of the Ontong Java Plateau, Pd. D. dissertation, 119pp. Univ. of Hawaii, Honolulu, 1972.

Larsen H.C., A.D. Saunders, P.D. Clift, et al., *Proc. ODP, Init. Repts*, *152*, College Station, TX (Ocean Drilling Program), 1994.

Larson, R.L., Geological consequences of superplumes, *Geology*, *19*, 963-966, 1991a.

Larson, R.L., Latest pulse on Earth: evidence for a mid-Cretaceous superplume, *Geology*, *19*, 547-550, 1991b.

*Larson, R.L., The mid-Cretaceous superplume episode, *Sci. Am.*, *272*, 66-70, Feb. 1995.

Larson, R.L., and P. Olson, Mantle plumes control magnetic reversal frequency, *Earth Planet. Sci. Lett.*, *107*, 437-447, 1991.

Lassiter, J.C., and D.J. DePaolo, Plume/lithosphere interaction in the generation of continental and oceanic flood basalts: chemical and isotopic constraints, Geophysical Monograph 100, Amer. Geophys. Union, 335-355, 1997.

Machetel, P., and P. Weber, Intermittent layered convection in a model mantle with an endothermic phase change at 670 km, *Nature*, *350*, 55-57, 1991.

Macdougall, J.D., Continental flood basalts and MORB: a brief discussion of similarities and differences in their petrogenesis, In *Continental Flood Basalts*, edited by J.D. Macdougall, pp. 331-341, Kluwer Academic, 1988.

*Mahoney, J.J, and M.F. Coffin (eds.), Large Igneous Provinces. Continental, oceanic, and planetary flood volcanism, *Geophysical Monograph*, *100*, Amer. Geophys. Union, 1997.

Marzoli, A., P.R. Renne, E.M. Piccirillo, M. Ernesto, G. Bellieni, and A. De Min, Extensive 200-million-year-old continental flood basalts of the Central Atlantic Magmatic Province, *Science*, *284*, 616-618, 1999.

Mauffret, A., and S. Leroy, Seismic stratigraphy and structure of the Caribbean igneous province, *Tectonophysics*, *283*, 61-104, 1997.

McNutt, M., K. Fischer, S. Kruse, and J. Natland, The origin of the Marquesas fracture zone ridge and its implications for the nature of hot spots, *Earth Planet. Sci. Lett.*, *91*, 381-393, 1989.

Moorbath, S., Age and isotope evidence for the evolution of continental crust, *Phil. Trans. R. Soc. A*, *288*, 401-413, 1978.

Morgan, W.J., Convection plumes in the lower mantle, *Nature*, *230*, 42-43, 1971.

Morgan, W.J., Hotspot tracks and the opening of the Atlantic and Indian oceans, in *The Sea*, Vol. 7, edited by C. Emiliani, pp. 443-487, Wiley, New York, 1981.

Neal, C.R., J.J. Mahoney, L.W. Kroenke, R.A. Duncan, and M.G. Petterson, The Ontong Java Plateau, Geophysical Monograph 100, Amer. Geophys. Union, 183-216, 1997.

Nelson, D.R., Granite-greenstone crust formation on the Archaean Earth: a consequence of two superimposed processes, *Earth Planet. Sci. Lett.*, *158*, 109-119, 1998.

Nur, A., and Z. Ben-Avraham, Oceanic plateaus, the fragmentation of continents, and mountain building, *J. Geophys. Res.*, *87*, 3644-3661, 1982.

Petterson, M.G. and 8 others, Structure and deformation of north and central Malaita, Solomon Islands: tectonic implications for the Ontong Java Plateau-Solomon Arc collision, and for the fate of oceanic plateaus, *Tectonophysics*, *283*, 1-33, 1997.

Phinney, E.J, P. Mann, M.F. Coffin, and T.H. Shipley, Sequence stratigraphy, structure and tectonic history of the southwestern Ontong Java Plateau adjacent to the North Solomon Trench and Solomon Island Arch, *J. Geophys. Res.*, in press.

Planke S., and H. Cambray, Seismic properties of flood basalts from Hole 917A downhole data, southeast Greenland volcanic margin, *Proc. ODP, Sci. Res.*, *152*: College Station, TX (Ocean Drilling Program), 453-462, 1998.

Planke S., and O. Eldholm, Seismic response and construction of seaward dipping wedges of flood basalts: Vøring volcanic margin, *J. Geophys. Res.*, *99*, 9263-9278, 1994.

Planke, S., P.A. Symonds, E. Alvestad, and J. Skogseid, Seismic volcanostratigraphy of large volume basaltic extrusives on rifted margins, *J. Geophys. Res.*, in press.

Pringle, M.S., M.F. Coffin, and M. Storey, Estimated melt production of the Kerguelen hot spot, *Eos, Trans. Amer. Geophys. Union*, *78*, F728.

Puchtel, I.S., A.W. Hofmann, K. Mezger, K.P. Jochum, A.A.

Shchipansky, and A.V. Samsonov, Oceanic plateau model for continental crustal growth in the Archaean: a case study from the Kostomuksa greenstone belt, NW Baltic Shield, *Earth Planet. Sci. Lett., 155*, 57-74, 1997.

Rampino M.R., and R.B. Stothers, Flood basalt volcanism during the past 250 million years, *Science, 241*, 663-668, 1988.

Richards, M.A., R.A. Duncan, and V.E. Courtillot, Flood basalts and hot spot tracks: plume heads and tails, *Science, 246*, 103-107, 1989.

Richards, M.A., D.L. Jones, R.A. Duncan, and D.J. DePaolo, 1991, A mantle plume initiation model for the formation of Wrangellia and other oceanic flood basalt plateaus, *Science, 254*, 263-267, 1991.

Roberts, D.G., D. Schnitker, et al., *Init. Rept. DSDP, 81*, US Gov. Printing Office, Washington DC, 1984.

Sager, W.W., and H.-C. Han, Rapid formation of the Shatsky Rise oceanic plateau inferred from its magnetic anomaly, *Nature, 354*, 610-613, 1993.

Saunders, A.D., M. Storey, R.W. Kent, and M.J. Norry, Consequences of plume-lithosphere interaction, *Geol. Soc. Spec. Publ., 68*, 41-60, 1992.

Saunders, A.D., J. Tarney, A.C. Kerr, and R.W. Kent, The formation and fate of large oceanic igneous provinces, *Lithos, 37*, 81-95, 1996.

Saunders, A.D., J.G. Fitton, A.C. Kerr, M.J. Norry, and R.W. Kent, The North Atlantic Igneous Province, Geophysical Monograph 100, Amer. Geophys. Union, 45-93, 1997.

Schaming, M., and Y. Rotstein, Basement reflections in the Kerguelen plateau, south Indian Ocean: indications for the structure and early history of the plateau, *Geol. Soc. Am. Bull., 102*, 580-592, 1990.

Schubert, G., and D. Sandwell, Crustal volumes of the continents and of oceanic and continental submarine plateaus, *Earth Planet. Sci. Lett, 92*, 234-246, 1989.

Self, S., T. Thordarson, and L. Keszthelyi, Emplacement of continental flood basalt lava flows, Geophysical Monograph 100, Amer. Geophys. Union, 381- 410, 1997.

Sinton, C.W., R.A. Duncan, M. Storey, J. Lewis and J.J. Estrada, An oceanic flood basalt province with the Caribbean plate, *Earth Planet. Sci. Lett., 155*, 221-235, 1998.

Skogseid J., S. Planke, J.I. Faleide, T. Pedersen, O. Eldholm, and F. Neverdal, NE Atlantic continental rifting and volcanic margin formation, *Geol. Soc. London Soc., Spec. Publ.*, in press.

Sleep, N.H., Hotspots and mantle plumes: some phenomenology, *J. Geophys. Res., 95*, 6715-6736, 1990.

Sleep, N.H., Lateral flow and ponding of starting plume material, *J. Geophys. Res., 102*, 10001-10012, 1997.

Sliter, W.V., and R.M. Leckie, Cretaceous planktonic foraminifers and depositional evironments from the Ontong Java Plateau with emphsis on sites 803 and 807, *Proc. ODP, Sci Results, 130*, 63-84, 1993.

Stein, M., and A.W. Hofmann, Mantle plumes and episodic crustal growth, *Nature, 372*, 63-68, 1994.

Tackley, P.J, D.J. Stevenson, G.A. Glatzmaier, and G. Schubert, Effects of an endothermic phase transition at 670 km depth in a spherical model of convection in the Earth's mantle, *Nature, 361*, 699-704, 1993.

Tejada, M.L.G., J.J. Mahoney, R.A. Duncan, and M.P. Hawkins, Age and geochemistry of basement and alkalic rocks of Malita and Santa Isabel, Solomon Islands, southern margin of Ontong Java Plateau, *J. Petrol., 37*, 361-394, 1996.

Thompson, R.N., and S.A. Gibson, Subcontinental mantle plumes, hotspots and pre-existing thinspots, *J. Geol. Soc. London, 148*, 973-977, 1991.

Todal, A., and O. Eldholm, Continental margin off western India and Deccan Large Igneous Province, *Mar. Geophys. Res.*, in press.

*White, R.S., Ancient floods of fire. *Natural History*, 51-61, April 1991.

White, R.S., Melt production rates in mantle plumes, *Phil. Trans. R.. Soc. Lond., A342*, 137-153, 1993.

White, R.S., and D. McKenzie, Magmatism at rift zones: The generation of volcanic continental margins and flood basalts, *J. Geophys. Res., 94*, 7685-7729, 1989.

White, R.S., and D. McKenzie, Mantle plumes and flood basalts, *J. Geophys. Res., 100*, 17543-17583, 1995.

White, R.S., D. McKenzie, and R.K., O'Nions, Oceanic crustal thickness from seismic measurements and rare earth element inversions, *J. Geophys. Res., 97*, 19683-19715, 1992.

Wilson, T., Mantle plumes and plate motions, *Tectonophysics, 19*, 149-164, 1963.

Wolfe, C.J., I.T. Bjarnason, J.C. VanDecar, and S.C. Solomon, Seismic structure of the Iceland mantle plume, *Nature, 385*, 245-247, 1997.

Zehnder, C.M., J.C. Mutter, and P. Buhl, Deep seismic and geochemical constraints on the nature of rift-induced magmatism during breakup of the North Atlantic, *Tectonophysics, 173*, 545-565, 1990.

Yale, L.B., and S.J. Carpenter, Large igneous provinces and giant dike swarms: proxies for supercontinent cyclicity and mantle convection, *Earth Planet. Sci. Lett., 163*, 109-122, 1998.

Zonenshain, L.P., M.I. Kuzmin, and L.M. Natapov, Geology of the USSR: a plate-tectonic synthesis, *Geodynamics Ser., 21*, Amer. Geophys. Union, 242 pp., 1990.

*overview papers for non-specialists and for educational purposes.

---

Millard F. Coffin, University of Texas Institute for Geophysics, 4412 Spicewood Springs Road, Bldg. 600, Austin, TX 78759-8500 USA; mikec@ig.utexas.edu

Olav Eldholm, Department of Geology, University of Oslo, POB 1047 Blindern, N-0316, Oslo, Norway; olav.eldholm@geologi.uio.no

# Recent progress on the plate motion relative to hotspots

Yasushi Harada

*Geographical Survey Institute, Tsukuba, Ibaraki, Japan*

Yozo Hamano

*Department of Earth and Planetary Physics, Faculty of Science, University of Tokyo, Japan*

In the history of modeling plate motion, stage poles (or Euler poles) and rotation angles are often used to describe the rotation of the plate relative to hotspots. As the model improves, more and more stages are required to fit observed data of locations and ages of seamounts that are created by the hotspots. To describe the plate motion with the stage poles however, one must assume that Euler poles are fixed to the hotspot reference frame for certain intervals of time. Generally speaking, Euler poles do not need to be fixed relative to hotspots, and without knowing the timing and the numbers of the stages, the accurate positions of the stage poles are never determined. Therefore, we presented a new method for calculating total rotation poles for a plate motion with respect to hotspots irrespective of whether the stage poles are fixed relative to hotspots or not. A new model for the motion of the Pacific plate obtained by this method provides a better fit to the positions and the ages of the observed hotspot tracks of the Pacific plate than the previous models do. Most of the previous models do not have errors assigned, but this suggests that the actual size of the errors are much larger than previously estimated, and that Euler pole may move more rapidly than formerly considered. The new method is also useful for investigating the validity of the assumptions of fixed hotspots and a rigid plate. From the fact that the new model can explain positions and ages of observed hotspot tracks that are not used to construct the model, the upper limits of mean motion as a result of the combination of the inter-hotspot motion and the intra-plate deformation in the Pacific plate is estimated to be about 2 to 3 mm/yr.

## 1. INTRODUCTION

The earth has moving rigid shells on the surface called plates, and the kinematics of the plate motion is referred to as plate tectonics, which is a very special characteristic of the planet. As a plate moves, volcanic chains are sometimes created on the plate by hot plumes that come from the mantle below. Wilson [1963, 1965] called the hot plumes hotspots. The hotspots have the peculiar feature that they stay still or move very slowly compared with the velocity of the plate although they are thought to be in the convecting mantle. The idea

The History and Dynamics of Global Plate Motions
Geophysical Monograph 121

of fixed hotspots and rigid plates is the most basic assumption for the modeling of the plate motions relative to the hotspots. This paper will focus on the recent progress about the way the plate motions are modeled, and the validity of the assumptions of the fixed hotspots and the rigid plates. For this purpose, we analyzed the Pacific plate motion with respect to hotspots, since the Pacific plate motion can be determined most accurately because of the size of the plate, the number of hotspots and the fast speed relative to them.

## 2. FORMER STUDIES ON THE DETERMINATION OF THE PLATE MOTION WITH RESPECT TO HOTSPOTS

Morgan [1971, 1972] deduced that hotspot positions are fixed relative to one another from the fact that the trends of the Pacific seamount chains (hotspot tracks) can be explained with two rotations of the plate at around two stage poles and angles (Table 1, thick lines in Fig. 1). The set of stage poles (rotation axis) and rotation angles gives principal parameters for the plate rotation relative to hotspots; often called "absolute motion". If hotspots are fixed relative to each other, it can be used as a convenient reference frame for the motion of a plate, and the motion can be fully described with positions of the rotation poles (also referred to as Euler poles) and the rotation angles. The position of the stage poles can be found by fitting a small circle to the trend of a hotspot track such as the track of Hawaii (Fig. 1). If the stage pole is found, the trends of the other hotspot tracks can be calculated and checked with the pole as shown in Fig. 1 (see Easter-Line, Cobb, Caroline, Macdonald, Louisville). As the knowledge of the locations and the ages of the seamounts increase, the disagreement between the modeled and the observed hotspot tracks come to be highlighted. The Louisville hotspot track [Lonsdale, 1988] of the southern Pacific, for instance, was poorly known at the time of the model of Morgan [1972], and the model failed to explain the hotspot track (Fig. 1). Other researchers, such as Duncan and Clague [1985] (the model is later revised by Watts *et al.* [1988]), Engebretson *et al.* [1985], Lonsdale [1988], Fleitout and Moriceau [1992], and Wessel and Kroenke [1997] tried more sets of stage poles and angles to fit every trend of the hotspot tracks of the same age (Table 1). Fleitout and Moriceau [1992] for example, used the satellite gravity anomaly [Sandwell and Smith, 1992a, 1992b] as well as the topography of the seamounts to find the stage poles, and succeeded in obtaining more sets of stage poles and angles that agree with the hotspot tracks of Hawaii, Easter-Line, Cobb, Caroline, Macdonald, and even Louisville (Table 1, thin lines in Fig. 1). Wessel and Kroenke [1997] used additional stage poles, but the positions of their stage poles are different from those of others. Fig. 2 shows the difference of the positions for the stage poles of the three models (Morgan [1972]: small square dots, Fleitout and Moriceau [1992]: thin line, Wessel and Kroenke [1997]:

Table 1. Stage poles of the previous models for the motion of the Pacific plate with respect to hotspots for post-100 Ma

| Age [Ma] | Lat.[deg.] | Long.[deg.] | Ang.[deg.] |
|---|---|---|---|
| | *Morgan, 1972* | | |
| 0-42 | 67 | 287 | 34 |
| 42-80 | 67 | 250 | 45 |
| | *Duncan and Clague, 1985* | | |
| 0-42 | 68.0 | 285.0 | 34.0 |
| 42-65 | 17.0 | 253.0 | 14.0 |
| 65-74 | 22.0 | 265.0 | 7.5 |
| 74-100 | 36.0 | 284.0 | 15.0 |
| | *Engebretson* et al. *1985* | | |
| 0-5 | 57 | 285 | 4.7 |
| 5-28 | 69 | 289 | 18.2 |
| 28-43 | 69 | 289 | 10.4 |
| 43-74 | 22 | 269 | 20.2 |
| 74-100 | 48 | 299 | 24 |
| | *Fleitout and Moriceau, 1992* | | |
| 0-18 | 72 | 282 | 14.4 |
| 18-24 | 77 | 301 | 5.4 |
| 24-43 | 54 | 302 | 14.8 |
| 43-66 | 11 | 255 | 11.3 |
| 66-78 | 26 | 252 | 7.3 |
| 78-92 | 41 | 276 | 17.8 |
| | *Wessel and Kroenke, 1997* | | |
| 0-3 | 25.00 | 333.00 | 3.6 |
| 3-15.1 | 71.00 | 272.50 | 13.2 |
| 15.1-19.9 | 67.26 | 353.10 | 4.0 |
| 19.9-43.1 | 57.17 | 297.52 | 14.4 |
| 43.1-54.2 | 16.19 | 261.79 | 6.8 |
| 54.2-65 | 9.67 | 248.59 | 6.1 |
| 65-74 | 22.00 | 265.00 | 7.5 |
| 74-100 | 18.90 | 319.50 | 13.4 |

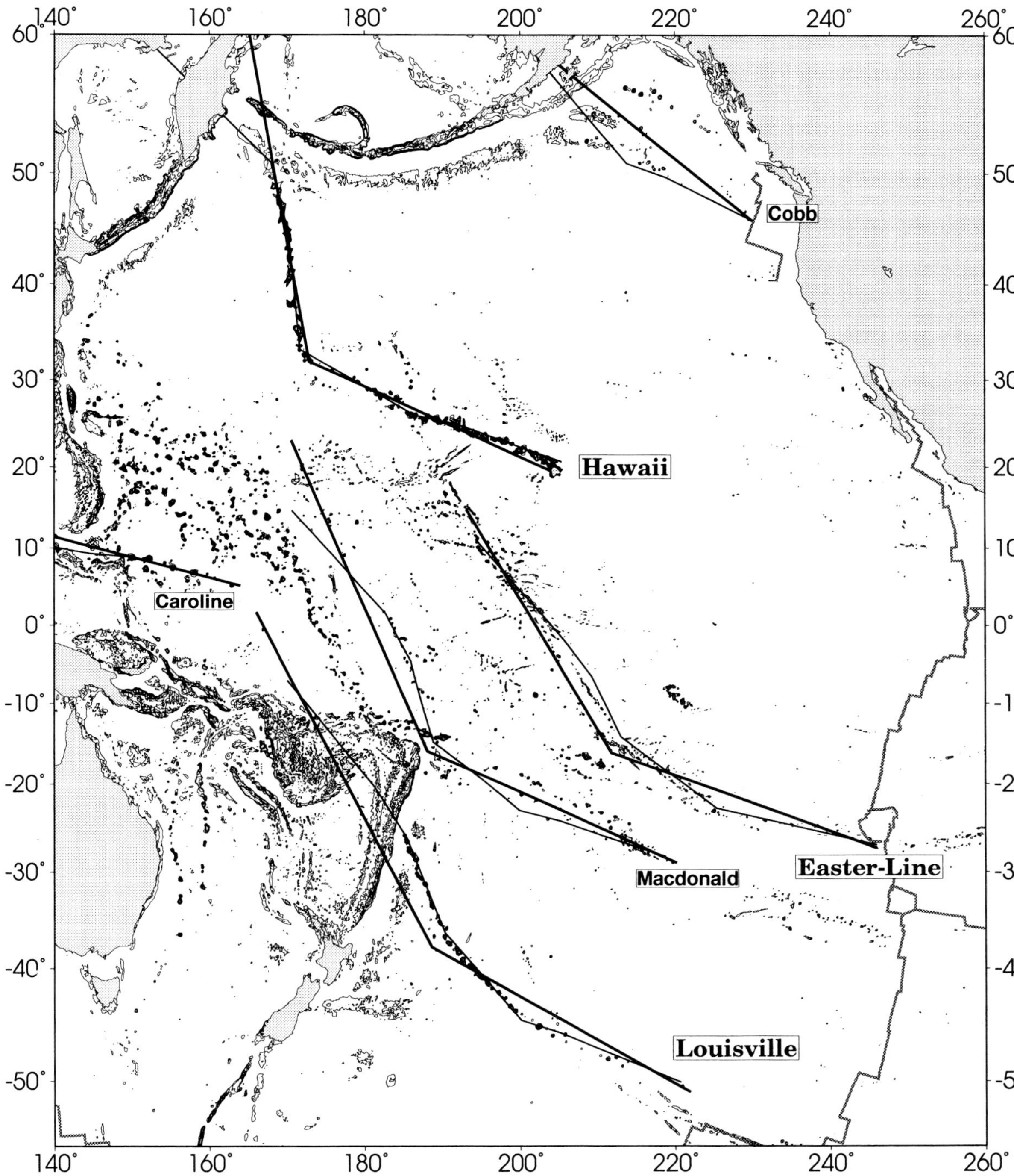


**Figure 1.** Two examples for the previous models of the hotspot tracks of the Pacific plate (Morgan [1972]: thick line, and Fleitout and Moriceau [1992]: thin line). The actual seamounts (hotspot tracks) are shown by the contour map of the gravity anomaly above 40 mgal. The parameters of these models are shown in Table 1. Note that the trend of the Louisville hotspot track in the south Pacific is not explained by Morgan's model. On the other hand, the model of Fleitout and Moriceau [1992] is successful in explaining the Louisville hotspot track by using more stage poles. The present position of the Louisville hotspot for both models is assumed to be at a small seamount located east of the youngest dated (0.5 Ma) seamount [Lonsdale, 1988].

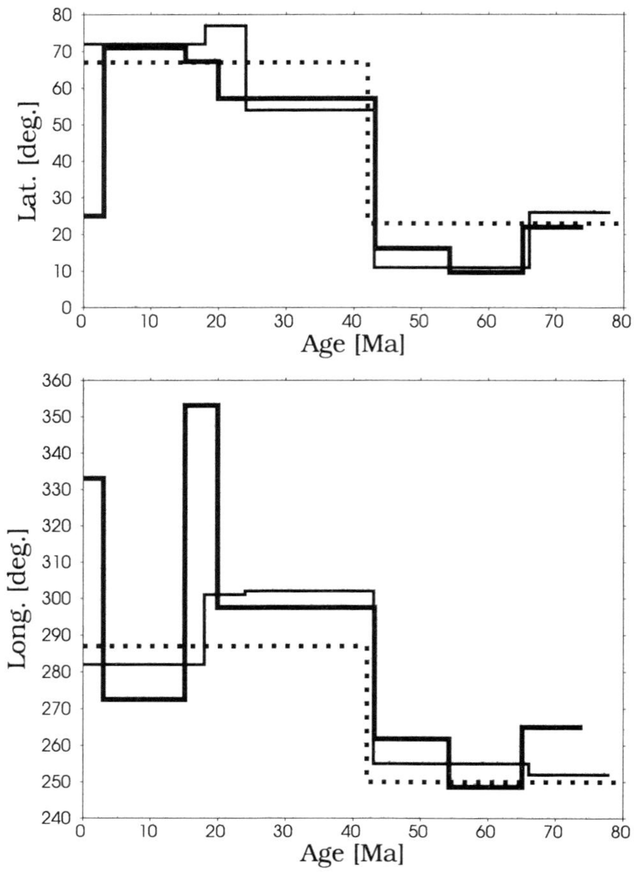


**Figure 2.** An example for the comparisons of stage poles of the previous models for the motion of the Pacific plate, Morgan [1972]: small square dots, Fleitout and Moriceau [1992]: thin line, Wessel and Kroenke [1997]: thick line. The parameters of these models are shown in Table 1. There is a clear tendency that the shorter the intervals of the stages, the larger the dispersion for the position of the poles become, but no errors of the poles are estimated. It is natural that the shorter the stages, the harder it is to estimate the uncertainties of the stage poles.

thick line). As the number of the stage poles increase, the positions of the poles change more radically. Although the errors of those poles are hard to estimate, it is probable that the size of the uncertainties are not decreasing as those models are revised, even if those give a better fit to the actual hotspot tracks.

### *2.1. The Previous Method of Obtaining Stage Poles*

Most previous models are based on a trial-and-error method of small-circle fitting : fitting a proper section of a hotspot track (e.g., Hawaiian seamounts) with a small circle and finding a stage pole of the small circle, and then changing the position of the pole a little so as to fit other hotspot tracks of the same age (Fig. 1). This method fails, however, if the section of a hotspot track is too short, since the uncertainty of the position of the pole of the small circle become large. The size of the uncertainty will not be reduced radically by adding the constraints of other hotspot tracks because of the limited number of contemporaneous hotspot tracks. This means that the more stage poles are assumed, the less reliable the model becomes. Owing to those problem, the errors of the poles for previous models are not estimated except by Molnar and Stock [1987]. In addition to the problem, the former methods have a difficulty in dealing with the timings and the numbers of the stages. Because of the sparsity and the ambiguity of the age data for the hotspot tracks, it is hard to determine the timings and the numbers of the stages accurately.

To avoid this situation, maybe one should discard the small-circle fitting method and the describing of plate motions with stage poles. It is because, generally, Euler poles are not necessarily fixed with respect to the hotspot reference frame. Therefore we need a new method of determining plate motion that does not assume the fixed Euler poles. It is more preferable if the new method can indicate the limitation for the assumptions of fixed hotspots and rigid plates. Since if those assumptions are false, the determination of the motion of the plate relative to hotspots is almost meaningless.

## 3. A NEW TECHNIQUE FOR OBTAINING PRECISE POSITIONS OF TOTAL ROTATION POLES

### *3.1. The Polygonal Finite Rotation Method*

A pole of the finite rotation of a polygon on a spherical surface can easily be calculated by the positions of the polygons before and after the rotation (see appendix). When hotspots are fixed with respect to each other, the shape of a polygon, which consists of coeval positions of the hotspots, does not change with age. If hotspots create continuous chains of seamounts, and present positions of the hotspots are known, those hotspot tracks can be regarded as the traces of the vertices of the polygon as a result of the plate motion. Therefore, to select as many discrete polygons as needed that are congruent with the present position polygon along each continuous region of the hotspot track, finite poles for the rotation from the present polygon to each polygon can be calculated almost continuously until no congruent polygon can be selected from the hotspot tracks (Fig. 3). This polygonal finite rotation method is powerful because of the following reasons. First, it can systematically calculate continuous poles for the total rotation without assuming the stillness of the Euler pole

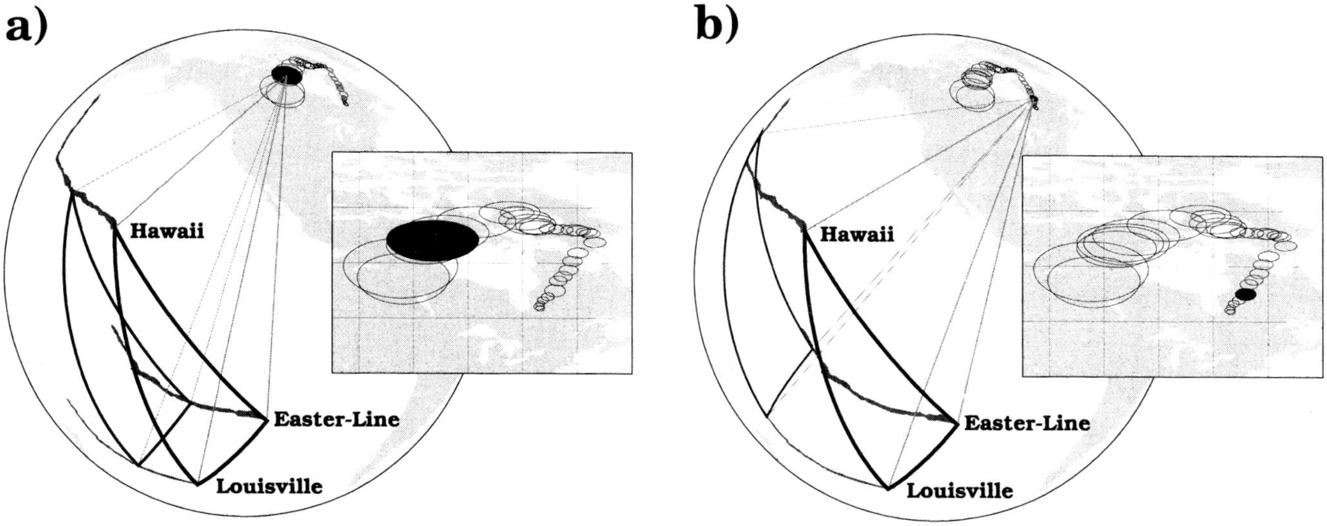


**Figure 3.** The almost continuous trace of the poles for total rotation of the Pacific plate can be calculated from more than two continuous hotspot tracks. Fixed hotspots and a rigid plate are strictly assumed here. First, from the satellite gravity anomaly, three continuous regions for the seamounts that are created by Hawaii, Louisville and Easter-Line hotspots are determined. Second, in each continuous region, three sets of points that construct the congruent polygons with the polygon of the present positions of three hotspots are selected discretely as many as needed until the polygons can be regarded as continuously distributed along each track. Third, the pole of the total rotation from the 0-age polygon to each congruent polygon can be calculated (see appendix). Owing to the uncertainties (up to 100 km) of the positions of the seamounts from which the vertices of the polygons are taken, the distribution for the pole positions has a scatter. The region for the pole positions can be scaled versus rotation angle, and for arbitrary rotation angle, the angular standard deviations within a 1° window of rotation angle are calculated as the uncertainties. In this case, the uncertainties are decreasing as the total rotation angle increase (compare black solid circles in **a** and **b**). This is partly effect of 2-D projection of the poles, and partly because of the inter-hotspot motion and the intra-plate deformation. Note that these uncertainties of this model not include the uncertainties caused by the inter-hotspot motion or the intra-plate deformation. Also note that the poles can be automatically calculated without knowing the ages of the hotspot tracks, therefore, these poles are determined totally independently from the ages except for the present positions of the hotspots. Poles for 0 - 8 Ma are omitted for clarity in this figure.

during a finite span of the age. Second, the poles of the finite rotation can be calculated solely from the geometry of the tracks if only the present positions of the hotspots are known. In the previous method, the age of the seamounts are required to select the contemporary sections of the tracks, whereas in this method, the ages of the seamounts can be compared after the calculation of the poles to see if the method is successful or not. Third, the poles of the finite rotation for the polygon on the spherical surface are calculated more accurately than the calculations of the stage poles if the numbers of the vertices and the size of the polygon are large. The errors of the stage poles are hard to estimate if the intervals of the stage are short, whereas in this method, the uncertainties of the pole positions can be estimated from the size of the region for the distribution of the pole positions (see the caption of Fig. 3). Fourth, the limitation of the fixed hotspots and a rigid plate can be deduced by the new method. If there are large inter-hotspot motion or intra-plate deformation, it is probable that no polygons whose shape is same as the zero age polygon can obtained from the each hotspot track. Only when the inter-hotspot motion and intra-plate deformation is not so large but gradually occurring in a direction, the rough quantity of those effects can be estimated from how the inconsistency of the model for the plate motion occurs.

In the present study, we used three hotspot tracks (Hawaii, Louisville, and Easter-Line) to determine the finite rotation of the Pacific plate. The location of the active or very young volcano of Loihi (18.92°N, 155.25°W) is used as the zero age point of the Hawaii hotspot. The present position of the Louisville hotspot is assumed to be at a small seamount (50.9°S, 138.1°W)

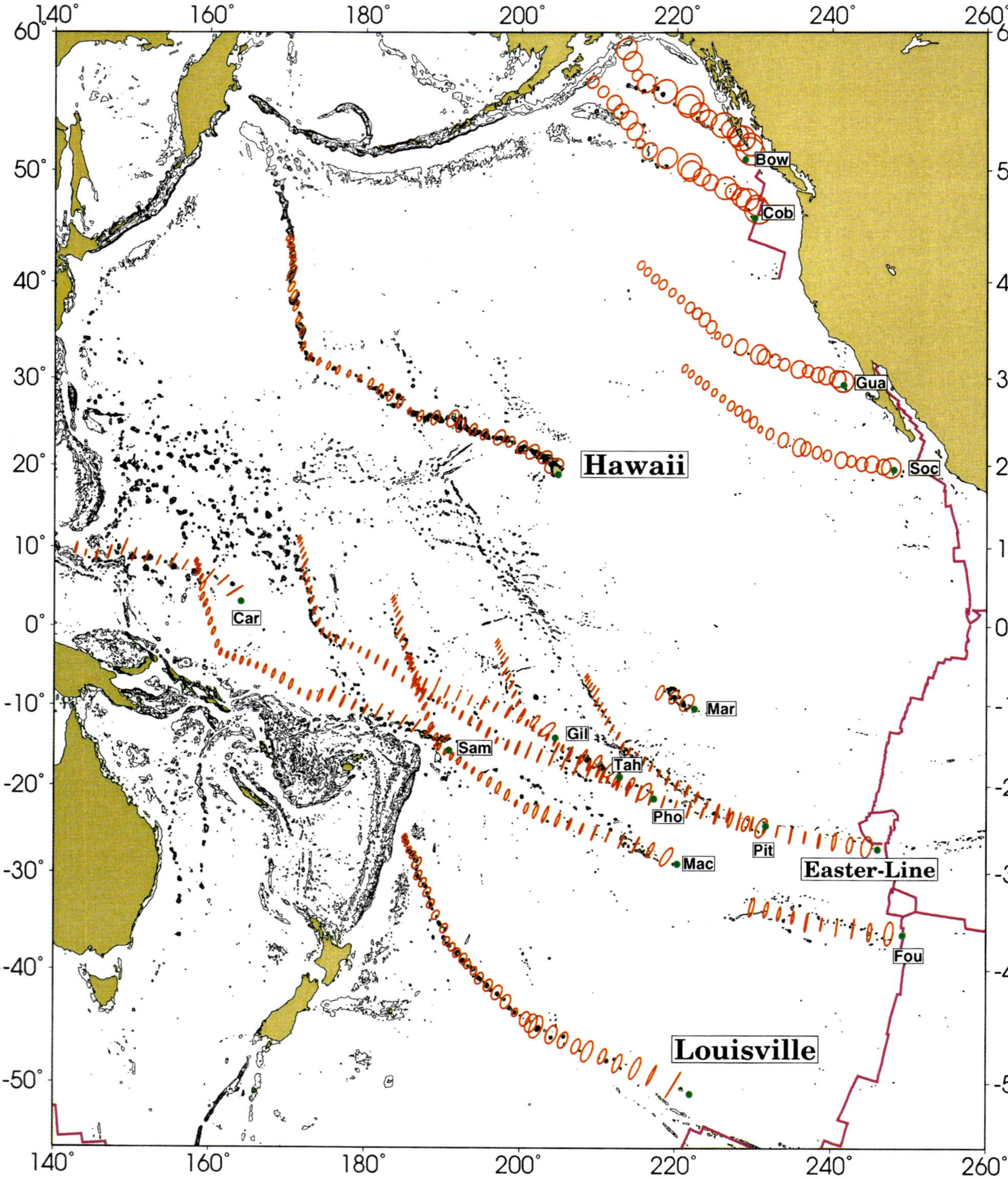


**Plate 1.** Theoretical hotspot tracks (red ellipses) calculated from the newly determined poles and angles. The size of the ellipses shows the estimated error of the positions for 2 Myr interval. We allowed large errors for younger poles (0 - 10 Ma) because the younger part of the Louisville track have only a few seamounts. Green points are the present positions of the hotspots (some are assumed in this study). Note that this result is based on the condition that fixed hotspots and a rigid plate are strictly assumed, that is, these ellipses do not include the uncertainties caused by the inter-hotspot motion and the intra-plate deformation. Therefore, the disagreements between modeled and observed tracks can be considered as results of the two effects (see text for details). The seamounts are shown by the contour map of the gravity anomaly above 40 mgal. The abbreviations are Bow: Bowie, Cob: Cobb, Car: Caroline, Gua: Guadalupe, Soc: Socorro, Mar: Marquesas, Pit: Pitcairn, Mac: Macdonald, Pho: Phoenix, Fou: Foundation Gil: Gilbert-Marshall, Tah: Tahiti, Sam: Samoa.

located east of the youngest dated (0.5 Ma) seamount [Lonsdale, 1988]. Both hotspot tracks are almost continuous chains in the gravity anomaly [Sandwell and Smith, 1992a, 1992b]. We added the Easter-Line hotspot track, which is also continuous in the gravity anomaly. The present position of the hotspot is not established yet, but it is assumed to be at the southern end of the Easter micro-plate in this analysis. The conditions of additional tracks make the dispersion of the pole positions narrower, because the constraints for the existences of the congruent polygons become harder.

## 4. RESULTS

### *4.1. The Geometrical Positions of the Hotspot Tracks*

The polygon (triangle in this case) was almost successfully traced as far as the western end of the Louisville track (the shortest track among the three), and the calculation was made to the total rotation angle of about 40° (Fig. 3). There is a little disagreement (about 100 km) among the three areas of tracks where the congruent polygon cannot be exactly traced at the oldest part (60 - 70 Ma). We allowed a band about 200 km wide along the Easter-Line track as a region of existence of the hotspot for this part in order to calculate the total poles. The regions obtained for the positions of the total rotation poles (Fig. 3) are not uniform in size but mainly become smaller as the rotation angle increase. This partly indicates that the inter-hotspot motion or the intra-plate deformation are occurring but not too large to apply the polygonal finite rotation method up to around the western end of the Louisville track. Without any information of age data except for the present positions, the theoretical hotspot tracks of the Pacific plate other than the three tracks can be calculated using the obtained poles and angles (Plate 1).

The calculated fine structure of the Hawaii, Louisville, and Easter-Line hotspot tracks in Plate 1 are naturally reproduced by the poles and angles. The theoretical hotspot tracks other than the three tracks have good agreements with the positions of the observed seamounts. The calculated Cobb hotspot track in the northeastern Pacific cover most of the Cobb-Eickelberg seamount chain. The present location of the hotspot is known as the Axial volcano at 46°N, 130°W [Desonie and Duncan 1990]. The calculated Bowie, Guadalupe, and Socorro hotspot tracks cover the observed seamounts, although the number of the seamounts in each track is small. The trace of the Caroline hotspot track is well-matched with the calculated track if we assume that the present location is at a small seamount which is seen in the satellite gravity anomaly (3.1°N, 164.0°E). The Marquesas and Tahiti hotspots have recent volcanoes, and the trends of the short seamount chain are explained by the modeled tracks. The older trends of the Gilbert-Marshall and Phoenix hotspot tracks are well explained by the calculated tracks if the present positions are assumed at 14.1°S, 155.5°W and 21.5°S, 142.8°W respectively. The present positions of the Samoa, Macdonald, and Pitcairn hotspots are relatively well known, and the calculated hotspot tracks explain the observed seamounts that are probably created by their respective hotspots. The Foundation seamounts [Mammerickx, 1992] are not an established hotspot track, but the calculated Foundation hotspot track largely covers the observed seamounts assuming the present position is on the ridge (36.6°S, 110.8°W). As above, most of the positions of the hotspot tracks in the Pacific plate are well explained by the new model.

### *4.2. The Ages of the Hotspot Tracks*

Independent of the agreement for the theoretical and observed positions of the hotspot tracks, the radiometric ages of the seamounts along the tracks are examined. If there are relative motions of hotspots which are parallel to the motion of the plate, inconsistency will be found in the age analysis. Based on the comparison between theoretical angles of the total rotation and observed radiometric ages of the hotspot tracks, the $^{40}Ar/^{39}Ar$ ages of the Hawaii, Louisville and Easter-Line hotspot tracks (Table 2) correspond with each other (Fig. 4). The $^{40}Ar/^{39}Ar$ ages of the Cobb hotspot track (Table 2) also shows a good agreement with the age curve obtained from the $^{40}Ar/^{39}Ar$ ages of the Hawaii and Louisville hotspot tracks. Almost all other radiometric ages of the Pacific hotspot tracks are K-Ar ages, and they tend to show a little younger age than the $^{40}Ar/^{39}Ar$ ages of the same samples (e.g., Schlanger *et al.* [1984]). The K-Ar ages of the other hotspot tracks are scattered near or under the age curve. The K-Ar ages of the Caroline hotspot track [Keating *et al.* 1984] agree with the modeled ages if the present location is assumed to be at the small seamount (3.1°N, 164.0°E). Some of the hotspot tracks such as Phoenix have no age data. Other tracks such as Guadalupe, Socorro, Samoa, Pitcairn, and Macdonald have only a few dated seamounts along the younger part of the tracks. From those age data, significant inconsistency in the new model could not be found.

Using the relationship of the angle of total rotation and age (Fig. 4), ages are assigned to the model. Overall, a new model of the motion of the Pacific plate relative to the hotspots is constructed with almost continuous poles of the finite rotation for the past 70 Myr.

Table 2. Total fusion ages of the seamounts of the Pacific hotspot tracks

| Name | Lat.[deg.] | Long.[deg.] | Age [Ma] | Ref.[a] |
|---|---|---|---|---|
| *Hawaiian hotspot track* | | | | |
| Lanai | 20.8 | -156.3 | 1.28 ± 0.04 | 1 |
| Laysan | 25.8 | -171.8 | 21.4 ± 0.6 | 2 |
| Laysan | 25.8 | -171.8 | 21.1 ± 0.6 | 2 |
| Laysan | 25.8 | -171.8 | 20.3 ± 0.3 | 2 |
| Laysan | 25.8 | -171.8 | 20.6 ± 0.3 | 2 |
| Necker | 25.3 | -172.0 | 23.4 ± 2.8 | 2 |
| Diakakuji | 32.7 | 172.0 | 43.4 ± 1.6 | 3 |
| Koko | 34.4 | 171.4 | 48.1 ± 0.8 | 4 |
| Ojin | 37.5 | 170.3 | 56.2 ± 1.7 | 5 |
| Ojin | 37.5 | 170.3 | 56.6 ± 1.7 | 5 |
| Ojin | 37.5 | 170.3 | 57.2 ± 1.7 | 5 |
| Jingu | 38.4 | 171.2 | 55.4 ± 0.9 | 6 |
| Nintoku | 41.2 | 170.2 | 55.6 ± 1.7 | 5 |
| Nintoku | 41.2 | 170.2 | 56.3 ± 1.7 | 5 |
| Suiko | 44.4 | 170.0 | 64.3 ± 1.9 | 5 |
| Suiko | 44.4 | 170.0 | 81.5 ± 9.1 | 5 |
| Suiko | 44.4 | 170.0 | 65.8 ± 3.5 | 5 |
| *Louisville hotspot track* | | | | |
| Unnamed | -50.43 | -139.15 | 0.5 ± 0.2 | 7 |
| Unnamed | -48.2 | -148.8 | 12.5 ± 0.4 | 7 |
| Valerie | -41.61 | -164.2 | 34.4 ± 0.6 | 7 |
| Valerie | -41.61 | -164.2 | 36.6 ± 0.6 | 7 |
| Unnamed | -38.33 | -167.73 | 45.5 ± 0.5 | 7 |
| Unnamed | -36.95 | -169.83 | 44.6 ± 0.5 | 7 |
| Unnamed | -33.95 | -171.19 | 53.4 ± 2.6 | 7 |
| Unnamed | -30.01 | -173.25 | 61.2 ± 0.9 | 7 |
| Unnamed | -27.28 | -174.21 | 66.7 ± 1.4 | 7 |
| Unnamed | -27.28 | -174.21 | 70.1 ± 1.4 | 7 |
| Unnamed | -25.53 | -175.04 | 65.9 ± 0.6 | 7 |
| Unnamed | -25.53 | -175.04 | 66.6 ± 0.6 | 7 |
| *Easter-Line hotspot track* | | | | |
| Unnamed | -15.0 | -149.0 | 41.8 ± 0.9 | 8 |
| Unnamed | -15.0 | -149.0 | 47.4 ± 0.9 | 8 |
| Unnamed | -9.0 | -150.4 | 70.5 ± 1.1 | 8 |
| Unnamed | -7.3 | -151.3 | 71.9 ± 1.4 | 8 |
| *Cobb hotspot track* | | | | |
| Cobb | 46.77 | -130.83 | 3.27 ± 0.3 | 9 |
| Gluttony | 47.14 | -131.46 | 1.55 ± 1.4 | 9 |
| Lust | 47.5 | -131.51 | 4.40 ± 1.07 | 9 |
| Sloth | 47.62 | -131.70 | 5.20 ± 0.32 | 9 |
| Warwick | 48.03 | -132.45 | 6.91 ± 0.3 | 9 |
| Unnamed | 48.28 | -133.22 | 7.73 ± 0.33 | 9 |
| Eickelberg | 48.29 | -133.09 | 9.03 ± 0.4 | 9 |
| Eickelberg | 48.53 | -133.1 | 7.05 ± 0.15 | 9 |
| Pathfinder | 50.6 | -143.2 | 23.5 ± 0.2 | 10 |
| Pathfinder | 50.6 | -143.2 | 23.8 ± 0.2 | 10 |
| Pathfinder | 50.6 | -143.2 | 22.9 ± 0.7 | 10 |
| Miller | 53.32 | -144.2 | 27.0 ± 1.6 | 10 |

Table 2. (continued)

| Name | Lat.[deg.] | Long.[deg.] | Age [Ma] | Ref.[a] |
|---|---|---|---|---|
| Miller | 53.32 | -144.2 | 23.2 ± 2.5 | 10 |
| Miller | 53.32 | -144.2 | 27.8 ± 2.2 | 10 |
| Murray | 53.6 | -148.3 | 25.6 ± 0.3 | 10 |
| Murray | 53.6 | -148.3 | 22.5 ± 2.0 | 10 |
| Murray | 53.6 | -148.3 | 24.8 ± 1.6 | 10 |
| Murray | 53.6 | -148.3 | 25.1 ± 1.2 | 10 |
| Murray | 53.6 | -148.3 | 25.6 ± 1.5 | 10 |
| Patton | 54.3 | -150.3 | 27.2 ± 2.9 | 10 |
| Patton | 54.3 | -150.3 | 26.1 ± 0.8 | 10 |
| Patton | 54.3 | -150.3 | 32.0 ± 2.7 | 10 |
| Patton | 54.3 | -150.3 | 30.2 ± 0.5 | 10 |

[a]1: Bonhommet *et al.* [1977], 2: Dalrymple *et al.* [1981], 3: Dalrymple & Clague [1976], 4: Clague & Dalrymple [1973], 5: Dalrymple *et al.* [1980], 6: Dalrymple & Garcia [1980], 7: Watts *et al.* [1988], 8: Schlanger *et al.* [1984], 9: Desonie & Duncan [1990], 10: Dalrymple *et al.* [1987]

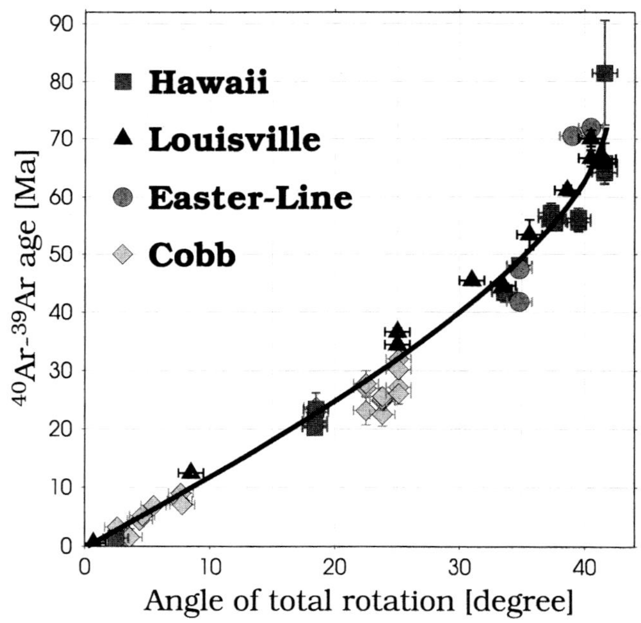


**Figure 4.** The observed radiometric ages of the hotspot tracks are plotted as a function of the angles of the total rotation of the Pacific plate (see Table 2). Only $^{40}Ar/^{39}Ar$ ages are used to preserve data quality. Note that although the age data are never used in calculating the angles, the ages of the four hotspot tracks are in quite good harmony with each other. The standard age curve is obtained by cubic-spline fitting to the age data. The age data still are not good enough to estimate the inter-hotspot motion that parallel to the plate motion.

Table 3. Poles of the total rotation of the Pacific plate with respect to hotspots

| Age [Ma] | Lat.[deg.] | Long.[deg.] | Ang.[deg.][a] | Err.[deg.][b] |
|---|---|---|---|---|
| 2 | 32.10 | 258.30 | 1.9254 | 38.4 |
| 4 | 40.90 | 255.20 | 3.6026 | 14.9 |
| 6 | 50.80 | 258.50 | 5.2678 | 10.9 |
| 8 | 54.20 | 255.70 | 6.9193 | 6.9 |
| 10 | 57.10 | 258.10 | 8.5555 | 4.6 |
| 12 | 59.00 | 257.60 | 10.174 | 5.4 |
| 14 | 63.10 | 262.30 | 11.774 | 3.3 |
| 16 | 64.00 | 263.60 | 13.353 | 3.2 |
| 18 | 65.30 | 265.90 | 14.909 | 3.3 |
| 20 | 64.00 | 263.50 | 16.441 | 3.7 |
| 22 | 67.40 | 271.50 | 17.946 | 2.8 |
| 24 | 69.10 | 277.60 | 19.424 | 2.0 |
| 26 | 69.10 | 279.00 | 20.871 | 1.1 |
| 28 | 67.80 | 280.40 | 22.287 | 1.8 |
| 30 | 67.00 | 281.50 | 23.669 | 1.8 |
| 32 | 66.70 | 282.30 | 25.016 | 1.4 |
| 34 | 66.40 | 282.70 | 26.326 | 1.3 |
| 36 | 66.20 | 285.30 | 27.597 | 0.9 |
| 38 | 66.10 | 287.00 | 28.828 | 0.9 |
| 40 | 66.00 | 289.50 | 30.016 | 1.0 |
| 42 | 65.90 | 290.20 | 31.160 | 0.9 |
| 44 | 65.70 | 291.70 | 32.257 | 0.8 |
| 46 | 65.40 | 292.30 | 33.307 | 0.8 |
| 48 | 63.80 | 293.20 | 34.307 | 0.9 |
| 50 | 61.90 | 290.10 | 35.255 | 0.9 |
| 52 | 60.40 | 289.10 | 36.151 | 0.9 |
| 54 | 58.80 | 288.40 | 36.991 | 0.9 |
| 56 | 57.60 | 287.40 | 37.775 | 0.9 |
| 58 | 56.80 | 286.50 | 38.499 | 0.9 |
| 60 | 55.00 | 285.90 | 39.164 | 1.1 |
| 62 | 53.90 | 284.60 | 39.765 | 0.7 |
| 64 | 53.30 | 283.70 | 40.303 | 0.6 |
| 66 | 52.60 | 283.20 | 40.775 | 0.5 |
| 68 | 52.10 | 283.00 | 41.179 | 0.6 |
| 70 | 51.80 | 283.20 | 41.515 | 0.5 |

[a]Total rotation angles.
[b]The errors of the poles are estimated as the angular standard deviations of the total rotation poles within every 1° window of the angle.

Table 4. Stage poles for 4 Myr intervals of this models

| Age [Ma] | Lat.[deg.] | Long.[deg.] | Ang.[deg.] |
|---|---|---|---|
| 4 | 67.76 | 258.88 | 3.51 |
| 8 | 68.50 | 266.30 | 3.33 |
| 12 | 73.62 | 305.96 | 3.38 |
| 16 | 64.02 | 263.07 | 3.08 |
| 20 | 59.05 | 360.68 | 3.79 |
| 24 | 57.54 | 289.35 | 2.92 |
| 28 | 56.58 | 288.96 | 2.78 |
| 32 | 57.30 | 303.40 | 2.64 |
| 36 | 54.48 | 319.66 | 2.56 |
| 40 | 57.05 | 310.35 | 2.29 |
| 44 | 35.50 | 293.65 | 2.34 |
| 48 | 16.98 | 254.01 | 2.99 |
| 52 | 14.42 | 261.15 | 2.46 |
| 56 | 9.06 | 257.82 | 2.27 |
| 60 | 14.47 | 241.84 | 1.84 |
| 64 | 13.19 | 255.44 | 1.24 |
| 68 | 17.68 | 284.08 | 0.40 |

## 5. DISCUSSION

### *5.1. Rigid Plates and Fixed Hotspots*

In the calculation above, we assume fixed hotspots and rigid plates. The limitations of the two assumptions can deduced from the disagreements between theoretical tracks calculated from the new model and observed tracks which is shown in Plate 1. There are some disagreements in the western ends of the Cobb and Bowie tracks in the northwestern Pacific, and a slight disagreement in the western ends of the Easter-Line track. Although it is difficult to distinguish between the effect of the inter-hotspot motion and that of the intra-plate deformation, if the disagreements between modeled and observed data are only found in areas near subduction zones, the cause of the disagreements can be considered not by the inter-hotspot motion but by the plate deformation. In the case of the Cobb and Bowie tracks, the deviation becomes large as the tracks reach the axis of the Aleutian trench, and the older end of the seamounts looks moved seaward by the regional deformation of the subducting plate. Therefore, it is probable that this deviation is caused by the subduction of the Pacific plate. The slight disagreement of the Easter-Line track may show that the confidence regions for the solution of the polygonal finite rotation method are about to fail, and probably no solution can be obtained from the hotspot tracks older than 70 Ma. If the maximum deviation of 200 km occurred by the constant inter-hotspot motion perpendicular to the plate motion for the past 70 Myr, the relative motion of the hotspots can be estimated to be about 3 mm/yr (in other words, this analysis is not sensitive to the relative motion of the hotspots less than 3 mm/yr). However, this case can also be explained by the plate deformation. The disagreement can considered to have occurred not at the Easter-Line track but at the Louisville track in the southern Pacific, when the model is corrected to fit the older part of the Easter-Line track more strongly. Then the corrected modeled

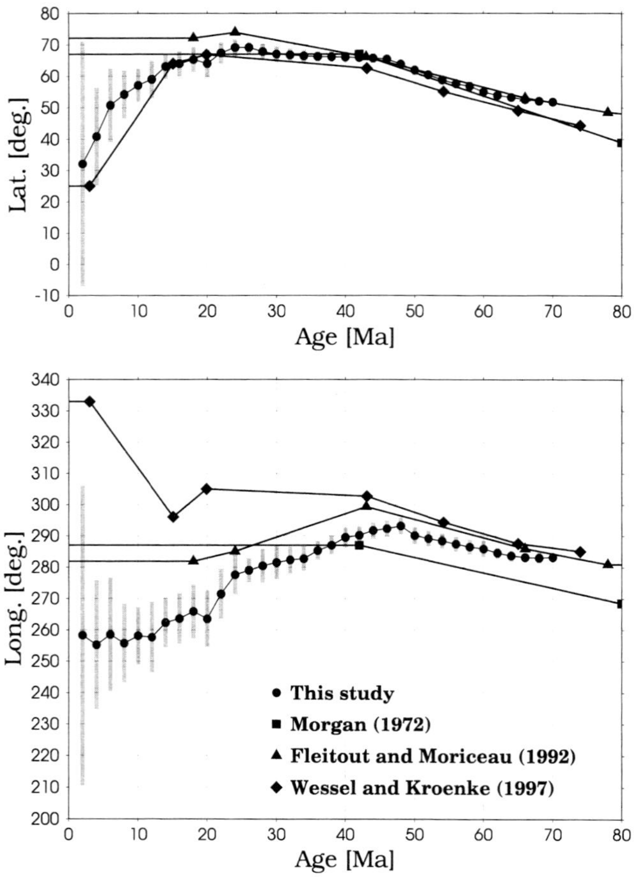


**Figure 5.** Comparisons of the total rotation poles for the Pacific plate determined by the present study: solid circle with gray error bar and others (Morgan [1972]: solid square, Fleitout and Moriceau [1992]: solid triangle, Wessel and Kroenke [1997]: solid diamond). The poles of the total rotation of the previous models are calculated from the stage poles (Table 1). Note that most of the poles of the previous models are outside the limit of the estimated error bars of this model. See text.

Louisville track will not agree with the oldest part of the track, but will drift a little to the west. Therefore the disagreement can be explained by the influence of the subduction of the southern part of the Pacific plate toward the Tonga trench.

If there are relative motions of hotspots which are parallel to the motion of the plate, the sensitivity depends on the quality of the age data of the seamounts. If the errors of the age data are about $\pm 10$ %, the uncertainties of the position of the 70 Ma seamounts are $\pm 700$ km when the Pacific plate moves at 100 mm/yr, the sensitivity can be roughly estimated as $\pm 10$ mm/yr if the motions are occurring in uniform directions.

One thing that needs further scrutiny in this analysis is the determinations of the present locations of the hotspots. If the assumed present positions of these hotspots are incorrect, some discrepancy between the modeled and the observed track may appear. Generally, it is difficult to find the present locations of hotspots, unless the seamounts of the hotspot tracks are continuously erupted like the Hawaii-Emperor seamount chain. Some hotspots made only several discrete seamounts for tens of million years (Guadalupe, Socorro), others ceased to create seamounts millions of years ago (Gilbert-Marshall, Phoenix), these hotspots do not put strong constraint on this model of the plate motion. If there is no recent volcanism, the present position of the hotspot has to be assumed by fitting the whole trend of the modeled track to several observed seamounts. The hot-spotting technique that Wessel and Kroenke [1997] developed is useful in this case.

### *5.2. Comparisons of the Euler Poles for Plate Motion*

Fig. 5 shows the difference of the positions of the poles for the total rotation. Some poles of the previous models are far away from the coeval error bars of the new model. This difference may indicates that the conventional assumption of the fixed Euler pole to the hotspot reference frame for certain intervals of time, is not acceptable for the precise determination of the Euler poles. In other words, the former estimations of the errors for the Euler poles may be too small. Molnar and Stock [1987], for instance, estimated the size of the error for the model of the Pacific plate motion by Duncan and Clague [1985] as the 95% confidence region of radius 200 km around the pole. But the differences for the coeval positions of poles between the previous models and the new model are significant and sometimes as large as 1500 km or larger (Fig. 5). Thus, it is probable that there was a considerable under-estimation of the errors of the former models.

From the total rotation poles, we calculated the stage poles for every 4 Myr interval to compare with the previous models (Fig. 6). The general profile is largely the same as the previous models, but the new stage poles have finer features and some peaks such as 18 Ma peak correspond with that of Wessel and Kroenke [1997], others such as 10 Ma peak do not. If the Euler pole moves for the time scale of shorter than 4 Myr, the comparison of the (Fig. 6) would be meaningless, but even though we calculate the stage poles for shorter interval, the errors of the stage poles become larger and the comparison become much more difficult. In this sense, comparing in total rotation poles is more preferable when the Euler pole is moving.

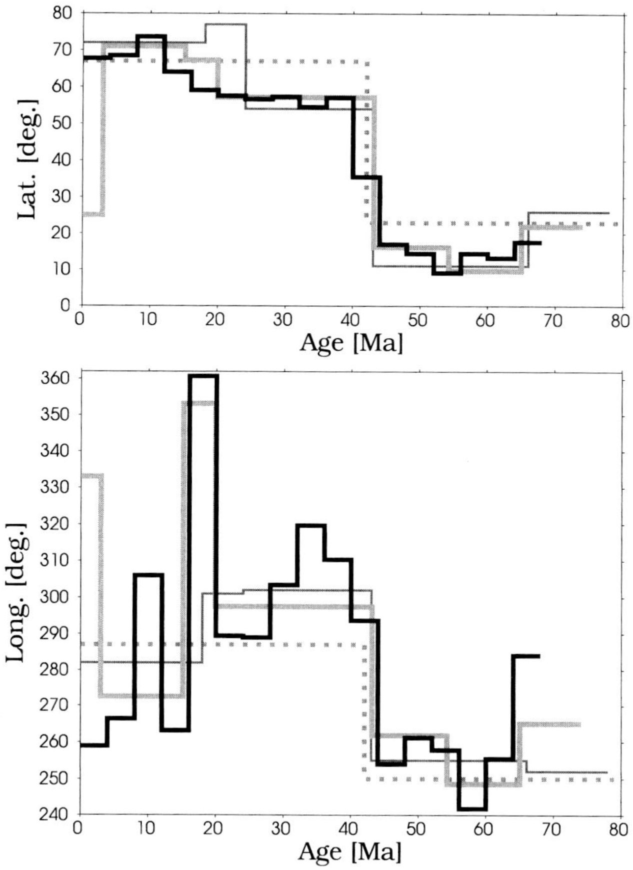


**Figure 6.** Comparisons of the stage poles of this study and the previous models for the motion of the Pacific plate. The stage poles for 4 Myr intervals of this study: thick black line, Morgan [1972]: small square gray dots, Fleitout and Moriceau [1992]: thin gray line, Wessel and Kroenke [1997]: thick gray line. Only mean values are shown. The typical errors of the stage poles for 4 Myr intervals are as large as 20° - 25°. Note that the longitudinal errors are large when the latitudes of the poles are high, therefore the errors of the stage poles after 40 Ma are much larger than those of before 40 Ma in this figure. The parameters of these models are shown in Table 1 and Table 4. See text.

## 6. CONCLUSION

A new method of calculating the continuous poles for the finite rotation of a plate made it possible to calculate the precise poles without assuming fixed Euler poles relative to hotspots for some span of time, and to estimate the limits of two assumptions of fixed hotspots and rigid plates. The model of the Pacific plate motion is successfully constructed by the method and almost no inconsistency is found between the positions and ages of the modeled and the observed hotspot tracks, including not only tracks used to construct the model but also the other tracks in the Pacific plate. This suggests that the inter-hotspot motion and the intra-plate deformation of the Pacific plate is small enough for the two assumptions to apply over the past 70 Myr. The difference for the positions of the Euler poles between the new model and the previous models also suggests that the plate motions are not steady with respect to the hotspot reference frame for the time scale of more than a few millions of years, and that the errors of the former models for the plate motion are much larger than previously estimated.

## APPENDIX : GEOMETRICAL EXPLANATION OF THE NEW METHOD

Fig. 7 shows the method for calculation of the poles of total (finite) rotation from two continuous hotspot tracks. A and B are zero age points on the Hawaii and the Louisville hotspot track respectively. Here we assume fixed hotspots and rigid plate, therefore arbitrary pairs (A' and B') on each track of the equal angular distance with arc AB is contemporaneous. A pole of finite rotation from A to A' is on the great circle ($C_1$) as a perpendicular bisector of arc AA', whereas a pole of B to B' is on the great circle ($C_2$) as a perpendicular bisector of arc BB'. Therefore the pole (P) of total rotation from arc AB to arc A'B' is determined from the intersection of the two great circles ($C_1$, $C_2$). The

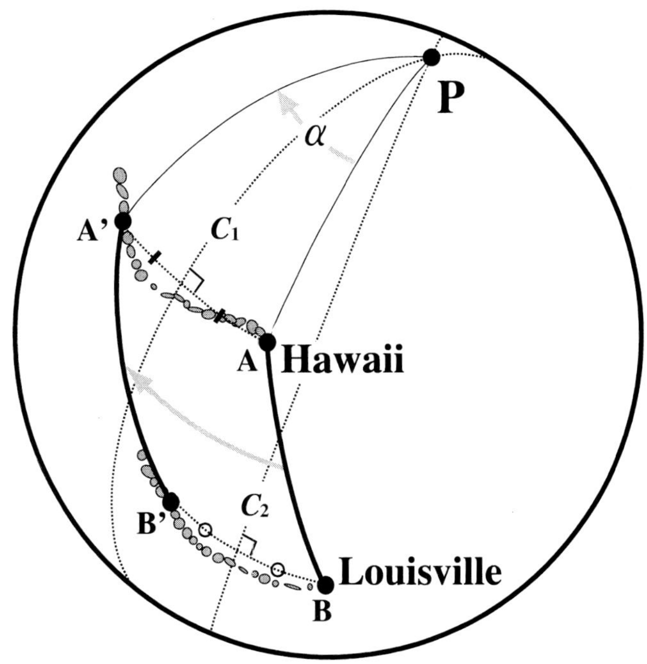


**Figure 7.** Diagram showing the new method of calculation for the poles of finite rotation of the plate with respect to hotspots. See appendix for details.

pair A'B' can be chosen along each track almost continuously up to the western end of Louisville hotspot track (about 70 Ma), so the poles are determined almost continuously. Additional continuous hotspot tracks such as Easter-Line, make the dispersion of the obtained poles smaller.

## REFERENCES

Bonhommet, N., Beeson, M. H., and Dalrymple, G. B., A contribution to the geochronology and petrology of the island of Lanai, Hawaii, *Geol. Soc. America Bull.*, *88*, 1282-1286, 1977.

Clague D. A. and Dalrymple, G. B., Age of Kōkō Seamount, Emperor Seamount chain, *Earth Planet. Sci. Lett.*, *17*, 411-415, 1973.

Dalrymple, G. B. and Clague D. A., Age of the Hawaiian-Emperor bend, *Earth Planet. Sci. Lett.*, *31*, 313-329, 1976.

Dalrymple, G. B. and Garcia, M. O., Age and chemistry of volcanic rocks dredged from Jingu Seamount, Emperor Seamount Chain, *Init. Repts. DSDP*, *55*, Washington (U.S. Govt Printing Office), 685-694, 1980.

Dalrymple, G. B., M. A. Lanphere, and D. A. Clague, Conventional and $^{40}Ar/^{39}Ar$ K-Ar ages of volcanic rocks from Ojin (Site 430), Nintoku (Site 432), and Suiko (Site 433) Seamounts and the chronology of volcanic propagation along the Hawaiian-Emperor Chain, *Init. Repts. DSDP*, *55*, Washington (U.S. Govt Printing Office), 659-676, 1980.

Dalrymple, G. B., Clague D. A., Garcia, M. O., and Bright, S. W., Petrology and K-Ar ages of dredged samples from Laysan Island and Northampton Bank volcanos, Hawaiian Ridge, and evolution of the Hawaiian-Emperor Chain, *Bull. Geol. Soc. Amer.*, Part II, *92*, 884-933, 1981.

Dalrymple, G. B., Clague D. A., and Vallier, T. L., $^{40}Ar/^{39}Ar$ petrology and tectonic significance of some seamounts in the gulf of Alaska, *in Seamounts, Islands and Atolls*, Washington, D.C.(AGU), 1987.

Desonie, D. N., and R. A. Duncan, The Cobb-Eickelberg seamount chain: Hotspot volcanism with mid-ocean Ridge basalt affinity, *J. Geophys. Res.*, *95*, 12697-12711, 1990.

Duncan, R. A. and D. A. Clague, Pacific plate motion recorded by linear volcanic chains, in *The ocean Basins and Margins*, pp.89-121, eds Nairn, A. E. M, Stehli, F. G. and Uyeda, S, Plenum Press, 1985.

Engebretson, D. C., Cox, A. and R. G. Gordon, Relative motions between oceanic and continental plates in the Pacific Basin, *Special Paper Geol. Soc. Am.*, *206*, 1-59, 1985.

Fleitout, L. and C. Moriceau, Short-wavelength geoid, bathymetry and the convective pattern beneath the Pacific Ocean, *Geophys. J. Int.*, *110*, 6-28, 1992.

Keating, B. H., D. P. Mattey, J. J. Naughton, D. Epp, A. Lazarewicz, and D. Schwank, Evidence for a hot spot origin of the Caroline islands, *J. Geophys. Res.*, *89*, 9937-9948, 1984.

Lonsdale, P., Geography and history of the Louisville hotspot chain in the Southwest Pacific, *J. Geophys. Res.*, *93*, 3078-3104, 1988.

Mammerickx, J., The Foundation seamounts: tectonic setting of a newly discovered seamount chain in the south Pacific, *Earth Planet. Sci. Lett.*, *113*, 293-306, 1992.

Molnar, P., and J. Stock, Relative motions of hotspots in the Pacific, Atlantic and Indian Oceans since late Cretaceous time, *Nature*, *327*, 587-591, 1987.

Morgan, W.J., Convection plumes in the lower mantle, *Nature*, *230*, 42-43, 1971.

Morgan, W.J., Plate motions and deep mantle convection, *Geol. Soc. Am. Mem.*, *123*, 7-22, 1972.

Sandwell, D. T., and W. H. F. Smith, Global marine gravity from ERS-1, Geosat and Seasat reveals new tectonic fabric, *Eos Trans AGU*, *73*, Fall Meet. Suppl., 133, 1992a.

Sandwell, D. T., and W. H. F. Smith, Global gravity, 1:45,000,000 world map, Scripps Inst. of Oceanogr., La Jolla, Calif., 1992b.

Schlanger, S. O., M. O. Garcia, B. H. Keating, J. J. Naughton, W. W. Sager, J. A. Haggerty, J. A. Philpotts, and R. A. Duncan, Geology and geochronology of the Line Islands, *J. Geophys. Res.*, *89*, 11261-11272, 1984.

Watts, A. B., J. K. Weissel, R. A. Duncan, and R. L. Larson, Origin of the Louisville Ridge and its relationship to the Eltanin Fracture Zone system, *J. Geophys. Res.*, *93*, 3051-3077, 1988.

Wessel, P. and L. Kroenke, A geometric technique for relocating hotspots and refining absolute plate motions, *Nature*, *387*, 365-369, 1997.

Wilson, J.T., A possible origin of the Hawaiian Islands, *Canad. J. Phys.*, *41*, 863-870, 1963.

Wilson, J.T., Evidence from ocean islands suggesting movements in the Earth, *Phil. Trans. Roy. Soc.*, *A*, *258*, 145-167, 1965.

Y. Harada, Crustal Deformation Research Division, Geography and Crustal Dynamics Research Center, Geographical Survey Institute, Kitasato 1, Tsukuba, Ibaraki, 305-0811, Japan. (e-mail: harada@gsi-mc.go.jp)

Y. Hamano, Department of Earth and Planetary Physics, Faculty of Science, University of Tokyo, Hongo 7-3-1, Bunkyo-ku, Tokyo, 113-0033, Japan. (e-mail: hamano@geoph.s.u-tokyo.ac.jp)

# Global Hotspot Reference Frames and Plate Motion

Ian O. Norton

*Exxon Exploration Company, Houston, Texas*

Most of the earth's new magmatic material is produced at volcanic centers associated with plate boundary processes. These are either arc volcanoes above subduction zones or linear spreading centers along divergent plate boundaries. Scattered about Earth's surface, however, are many volcanic centers that are not associated with plate boundary processes. These are usually found far from plate boundaries, and they continue to produce magma from a well-defined central location for many tens of millions of years. Geochemical studies of the products of these volcanic centers shows that their magma comes from deep within the mantle, below the tectonic plates that form the top 100 km of the earth. The term 'hotspot' has been coined to describe these volcanic features. If the tectonic plate on which the volcanic center is located moves relative to the underlying magma source, a trail of volcanic features can be produced on the plate, a 'hotspot track'. Probably the best-known example is associated with the island of Hawaii, with the Hawaii-Emperor seamount chain in the northern Pacific Ocean forming the 'track'. An extension of the hotspot concept is the proposal that all the global hotspots stay fixed relative to one another and thus provide a complete reference frame in which to describe plate motion. This would be incredibly useful, as no other reference frame exists in which to describe absolute plate motion through time. The hotspot reference frame was thought to provide this, and many published studies of plate motion make use of this reference frame. As studies of plate motions have advanced, however, it has become clear that the global hotspots do not stay fixed relative to each other. In this paper it is proposed that there are three global families of hotspots, a Pacific family, an Indo-Atlantic family and also Iceland, a major hotspot in terms of volcanic material but not fixed to the other two families. Hotspots within each of the first two families have remained relatively fixed for the past 80 million years.

## INTRODUCTION

The concept of hotspots was introduced by *Wilson* [1963, 1965] as an explanation for the existence of linear volcanic features in the world's ocean basins that are not associated with mid-ocean ridges. One of the best known examples of this type of feature is the Hawaii-Emperor island and seamount chain in the North Pacific (Figures 1 and 2). Wilson suggested that the islands and seamounts are produced by a deep mantle volcanic source that underlies the moving Pacific plate, producing a string of islands and seamounts 'arranged like plumes of smoke which are being carried down-wind from their source' [*Wilson*, 1963]. Since

The History and Dynamics of Global Plate Motions
Geophysical Monograph 121

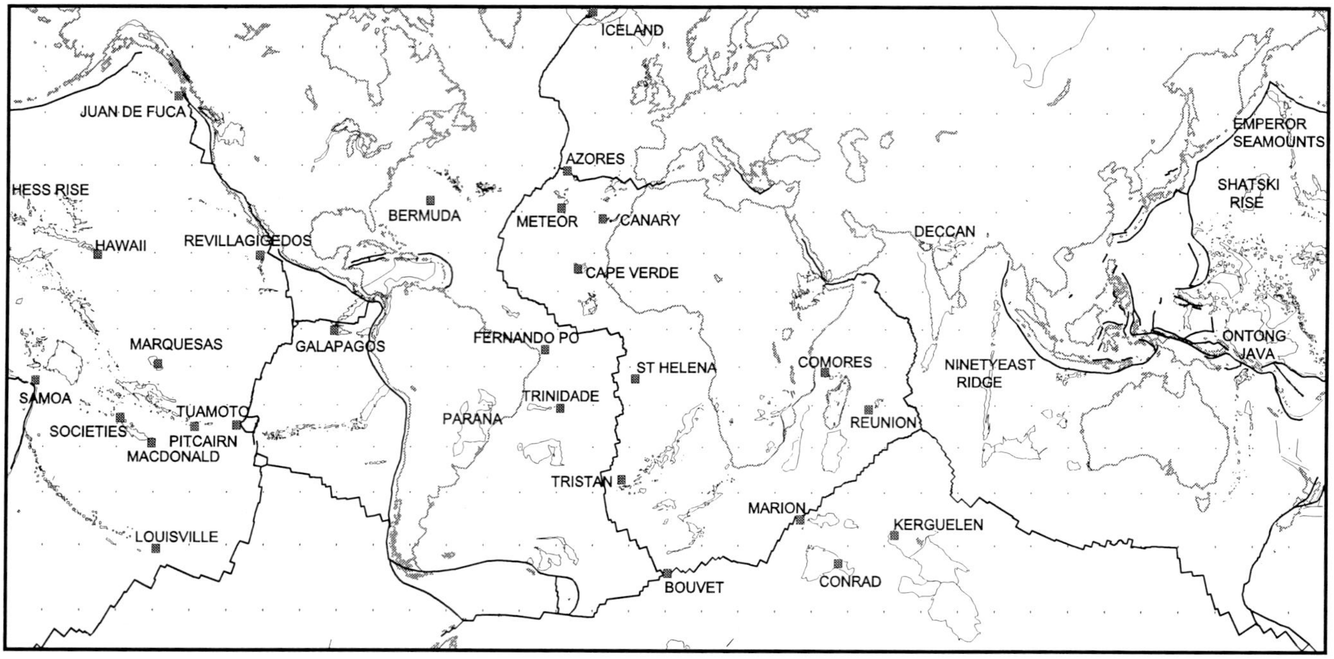


**Figure 1.** Global hotspots as proposed by *Wilson* [1965], *Morgan* [1972], *Duncan and Clague* [1985], *Lonsdale* [1988], *Duncan and Richards* [1981], *Morgan* [1983], *O'Connor and Duncan* [1990] and *Müller et al.* [1993]. Hotspot tracks are shown using the large igneous province compilation by *Coffin and Eldholm* [1993].

Wilson's seminal papers, better knowledge of volcano ages along the seamount chain has confirmed his model. Age progression along the Hawaii-Emperor seamount chain with distance from Hawaii is illustrated in Figure 3, from *Clague and Dalrymple* [1989]. This figure illustrates the rather constant speed of the Pacific plate relative to the Hawaii hotspot. *Morgan* [1972a] proposed a modifier to the hotspot model. He suggested that a hotspot is the surface manifestation of a deep mantle plume whose motion (if any) is independent of plate motion. He further suggested that, if we assume that the Hawaii hotspot has remained fixed in a mantle reference frame, the absolute velocity of the Pacific plate could be measured using mapped directions of motion from figures such as Figure 1, and the rate of motion from Figure 3. Morgan also extended this idea to all of the global hotspots, and proposed the assumption that hotspots remain fixed relative to each other and therefore provide a reference frame for measuring all plate motion. One implication of the fixed hotspot assumption in the North Pacific is that the 60 degree bend in the Hawaii-Emperor chain at 43 Ma should be the result of a sudden change in direction of motion of the Pacific plate. This is discussed later in this paper.

Early experimental tests of the fixed hotspot assumption using plate motion measurements tended to confirm that selected groups of hotspots could indeed be regarded as internally fixed [*Morga*n, 1971, 1972a, b], although *Morgan* [1972a] did suggest minor motion of 5 mm/yr. between Pacific hotspots. In contrast, *Burke et al.* [1973] compared plate reconstructions of Atlantic bordering continents with reconstructions using a fixed reference frame that included only Atlantic hotspots. They used (Figure 1) Meteor (they used the name Colorado Seamounts), Azores, Tristan, Discovery and Bouvet and found that these hotspots could be divided into two groups with about 1.8 cm/yr relative motion between the groups. The two groups were split between the North Atlantic hotspots Meteor and Azores and the three hotspots from the South Atlantic. *Morgan* [1981] took issue with *Burke et al.'s* [1973] selection of what defined the chosen hotspot tracks. He used 11 Atlantic hotspots (Table 1) and was able to show that these constituted a fixed reference frame for the past 200 million years. *Morgan* [1981] extended the analysis to the Indian Ocean hotspots (Table 1) and showed that these hotspots were also fixed to the Atlantic hotspots, although in a later paper confirming a fixed Atlantic hotspot reference frame, *Morgan* [1983] excluded the Indian Ocean hotspots from his analysis. The Atlantic and Indian Ocean hotspots were further analyzed by *Duncan and Richards* [1991] and *Müller et al.* [1993]. These authors concluded that the Indian

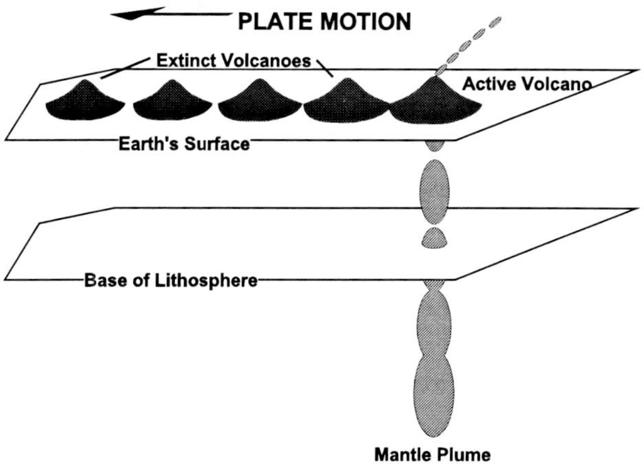


**Figure 2.** Cartoon illustration of a hotspot. A mantle plume penetrating the lithosphere produces a volcano on the earth's surface. The tectonic plate making up the lithosphere moves relative to the hotspot, so a chain of extinct volcanoes is left as a track of the hotspot.

and Atlantic Ocean hotspots did indeed constitute a fixed reference frame, but with the exception of an analysis described by *Morgan* [1981] which showed a lack of correlation between the Pacific and Indo-Atlantic hotspots, it is noteworthy that the Pacific hotspots (and sometimes Iceland) were excluded from these calculations.

Comparisons of the Indo-Atlantic and Pacific hotspots using plate circuit calculations have consistently shown relative motion of up to 20 mm/yr. [*Molnar and Atwater,* 1973; *Molnar and Francheteau,* 1975, *Morgan,* 1981; *Molnar and Stock,* 1987; *Norton,* 1995]. In fact, there is no published study of plate motions which supports the idea that the Pacific and Indo-Atlantic hotspots plus Iceland can be regarded as fixed to each other. These analyses are extended in this paper and, as will be explained, the global hotspots can be grouped into three 'families', each of which appear to have distinct relative motion characteristics. These groups fall into three oceanic domains: the Pacific, Indo-Atlantic and, in the far northern Atlantic, Iceland.

## PACIFIC OCEAN HOTSPOTS

The Pacific Ocean basin has several remarkably linear chains of seamounts and islands, all trending in an approximately northwest-southeast direction (Figure 4). In the central western Pacific there are also many oceanic plateaus and seamounts that do not follow any obvious trend. These features are mostly of Jurassic or Early Cretaceous age, however [*Duncan and Clague,* 1985], and will be excluded from this analysis. This is because we will be using plate circuit calculations to compare hotspot motions. These calculations require spreading ridge boundaries and the Pacific was only connected to the rest of the world via a spreading ridge from 90 Ma to present. The hotspot model has been very successful in explaining the general trend of the mid Cretaceous and younger features in the Pacific, especially for the Hawaii-Emperor chain (Figure 3). The problem has come in trying to explain all the features in terms of fixed hotspots. *Morgan* [1972a] suggested that Hawaii, MacDonald Seamount and Tuamoto (Sala y Gomez) are fixed hotspots and calculated trajectories for the hotspot tracks across the Pacific basin. With the bathymetric and age control data available at the time, *Morgan's* [1972a] assumption of three fixed hotspots seemed reasonable, although he did suggest minor inter-hotspot motions of about 5 mm/yr. to explain differences between his synthetic trajectories and locations of seamounts in the Central Pacific.

*Duncan and Clague* [1985] added seven more hotspots to those proposed by *Morgan* [1972a, b] (Table 1). They calculated synthetic trajectories for these hotspots, but did not show a detailed comparison between these trajectories and observed volcano ages or locations, especially in the complicated Central Pacific where they suggested several parallel and in some cases overlapping hotspot tracks. *Duncan and Clague* [1985] included the Louisville ridge in the far South Pacific in their analysis. Later detailed map-

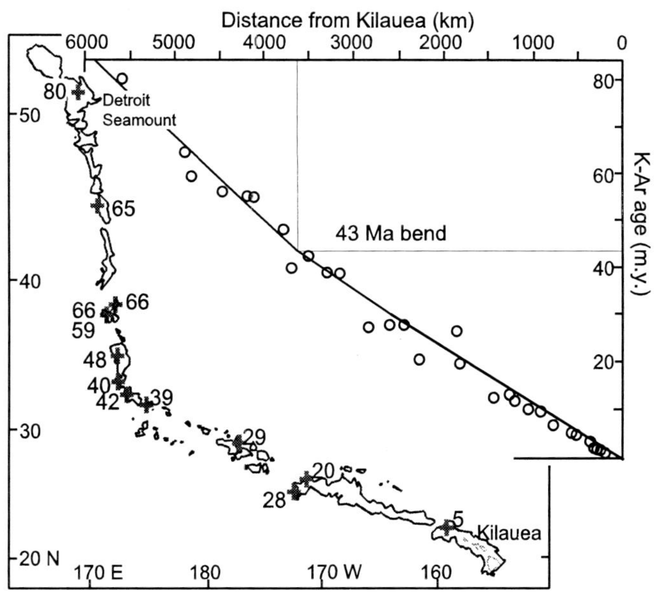


**Figure 3.** Plot of age versus distance from Hawaii for the Hawaii-Emperor seamount chain. Data from *Clague and Dalrymple,* [1989] and, for Detroit seamount, from *Tarduna and Gee* [1995].

**Table 1.** Pacific and Indo-Atlantic hotspots as suggested by different authors.

| **Pacific** | | | | |
|---|---|---|---|---|
| *Wilson* [1965] | *Morgan* [1972a] | *Duncan and Clague* [1985] | *Lonsdale* [1988] | *Duncan and Richards* [1991] |
| Hawaii | Hawaii | Hawaii | Hawaii | Hawaii |
| | MacDonald Seamount (Austral Islands chain) | Austral Seamounts chain | | |
| | Sala y Gomez-Tuamoto | Sala y Gomez-Tuamoto | | |
| | Juan de Fuca (JdF) | Kodiak-Bowie (=JdF?) | | |
| | | Cobb-Eickelberg (=JdF?) | | |
| | | Revillagigedos | | |
| | | Marquesas | | Marquesas |
| | | Pitcairn | | Pitcairn |
| | | Societies | | Societies |
| | | Caroline | | |
| | | Louisville Ridge | Louisville Ridge | Louisville Ridge |
| | | Samoa | | Samoa |
| | | | | Juan Fernandez |
| | | | | Galapagos |
| | | | | San Felix |
| | | | | Bowie |
| | | | | Cobb |
| | | | | Socerro |

ping of the Louisville ridge by *Watts et al.* [1988] and *Lonsdale* [1988a] improved knowledge of this feature. Age progression along the ridge, plus the occurrence of a '43 Ma bend' that could be correlated with the bend in the Hawaii-Emperor chain led both *Watts et al.* [1988] and *Lonsdale* [1988a] to suggest that the Hawaii and Louisville hotspots could be regarded as relatively fixed at least for the past 70 million years. This was an important result because the simplicity of the Louisville track (i.e. a linear, age-progressive feature like the Hawaii-Emperor chain) allowed rotation poles for the Pacific plate to a hotspot reference frame to be calculated, without the multi-hotspot complications of the Central Pacific.

Some of the wide variety in choice of what constitutes a viable hotspot and associated track comes from uncertainties in mapping of seamount locations and in age determinations of seamounts. Seamount mapping in the sixties and seventies (the early days of development of the hotspot model) relied entirely on ship-borne depth soundings, so vast areas of the world oceans did not have adequate coverage for accurate maps. This situation improved dramatically in the early eighties with the advent of satellite-based gravity observations. By using gravity anomalies produced by seamounts, *Craig and Sandwell* [1988] were able to use Seasat satellite gravity to detect many previously unmapped seamounts in their global compilation. Global coverage was not complete in the Seasat data, with between-track gaps of as much as 100 km [*Craig and Sandwell,* 1988].

The more recent global gravity derived from satellite missions has provided dense global coverage. An updated global seamount compilation has yet to be undertaken with these data, but *Wessel* [1997] has used the Geosat and ERS-1 data to produce a map of seamounts on the Pacific plate. This data is shown with rectangular symbols in Figure 4. The symbols are sized according to the vertical gravity gradient, which *Wessel* [1997] relates to seamount size. Also outlined in Figure 4 are large igneous provinces '(LIPS)' compiled by *Coffin and Eldholm* [1993 and 1994]. These volcanic features are thought to have formed over mantle plumes. Some of these plumes were apparently short-lived and produced only individual LIPS (e.g. the Ontong-Java Plateau), but others persisted through geologic time to produce linear trails on the moving plates, i.e. hotspot tracks. These latter are the hotspots we are concerned with in this paper. In the Pacific (Figure 4) the general northwest-southeast trend of seamounts which first led to the fixed hotspot hypothesis is apparent in the figure. Also shown are calculated tracks of four of the proposed Pacific hotspots listed in Table 1, Hawaii, MacDonald Seamount, Sala y Gomez and Louisville. The hotspot tracks were calculated using rotations to the Pacific plate as listed in Table 2. Following *Watts et al.* [1988] and *Lonsdale* [1988a], the rotations were derived to provide good matches to the Hawaii-

**Table 1 (continued).** Pacific and Indo-Atlantic hotspots as suggested by different authors.

| **Indo-Atlantic** | *Morgan* [1972a] | *Morgan* [1983] | *O'Connor and Duncan* [1990] | *Müller et al.* [1993] |
|---|---|---|---|---|
| | Tristan da Cunha | Tristan da Cunha | Tristan da Cunha | Tristan da Cunha |
| | Reunion | | Reunion | Reunion |
| | Bouvet Island | Bouvet Island | Bouvet Island | |
| | Ascension Island | Ascension Island | Ascension Island | |
| | Cape Verde Islands | Cape Verde Islands | Cape Verde Islands | |
| | Canary Islands | Canary Islands | Canary Islands | |
| | | Discovery Seamount | | |
| | | Vema Seamount | | |
| | | Trinidade Island | Martin Vas (=Trinidade) | |
| | | St Helena Island | St Helena Island | |
| | | Fernando Po Island | Fernando Po Island | |
| | | Guyana | | |
| | | Meteor | | Meteor |
| | | Bermuda | Bermuda | |
| | | Azores | Azores | |
| | | Madeira | | |
| | | Iceland | Iceland | |
| | | Labrador | | |
| | | Eifel | | |
| | | | | Conrad Rise |
| | | | Kerguelen | Kerguelen |
| | | | Tasman Seamount | Tasman Seamounts |
| | | | New England-Corner Seamounts | New England-Corner Seamounts |
| | | | Bermuda | |
| | | | Comores | |
| | | | Afar | |
| | | | Jan Mayen | |
| | | | Discovery Seamount | |
| | | | Shona Ridge | |
| | | | Marion Island | |
| | | | Crozet Island | |
| | | | East Australia | |
| | | | Tasman | |

Emperor and Louisville tracks. Modifications to these authors' rotation poles were made to take into account improved age control of the old end of the Emperor seamount chain [*Tarduno and Cottrell,* 1997]. Age data for Pacific seamounts and islands are shown along with the calculated tracks from Figure 4 in Figure 5. Age data come from compilations by *Duncan and Clague* [1985], *Craig and Sandwell* [1988], *Morgan* [1972a], *Watts et al.* [1988] and *Tarduno and Cottrell* [1997]. Not all known age data for the Hawaii-Emperor chain are shown, as the age progression along this chain is well-established *[Clague and Dalrymple,* 1989; see Figure 3]. As can be seen, the fit to these tracks is excellent. In contrast, as pointed out by *Morgan* [1972a, b], the fit to other linear trends of seamounts and islands in the Central Pacific is not nearly so good. Nevertheless, we will assume here that the Hawaii and Louisville hotspots do indeed provide a fixed hotspot reference frame for the Pacific plate. We assume that Central Pacific complications are due to multiple hotspots [*Duncan and Clague,* 1985] and/or interactions between hotspots, fracture zones and spreading ridges [*Epp,* 1984*; McNutt et al.,* 1989; *Ito et al.,* 1995]

## INDO-ATLANTIC HOTSPOTS

Indo-Atlantic hotspots are shown in Figure 6. This hotspot family is dominated by three hotspots: Tristan, Reunion and Kerguelen, each of which has produced very dis-

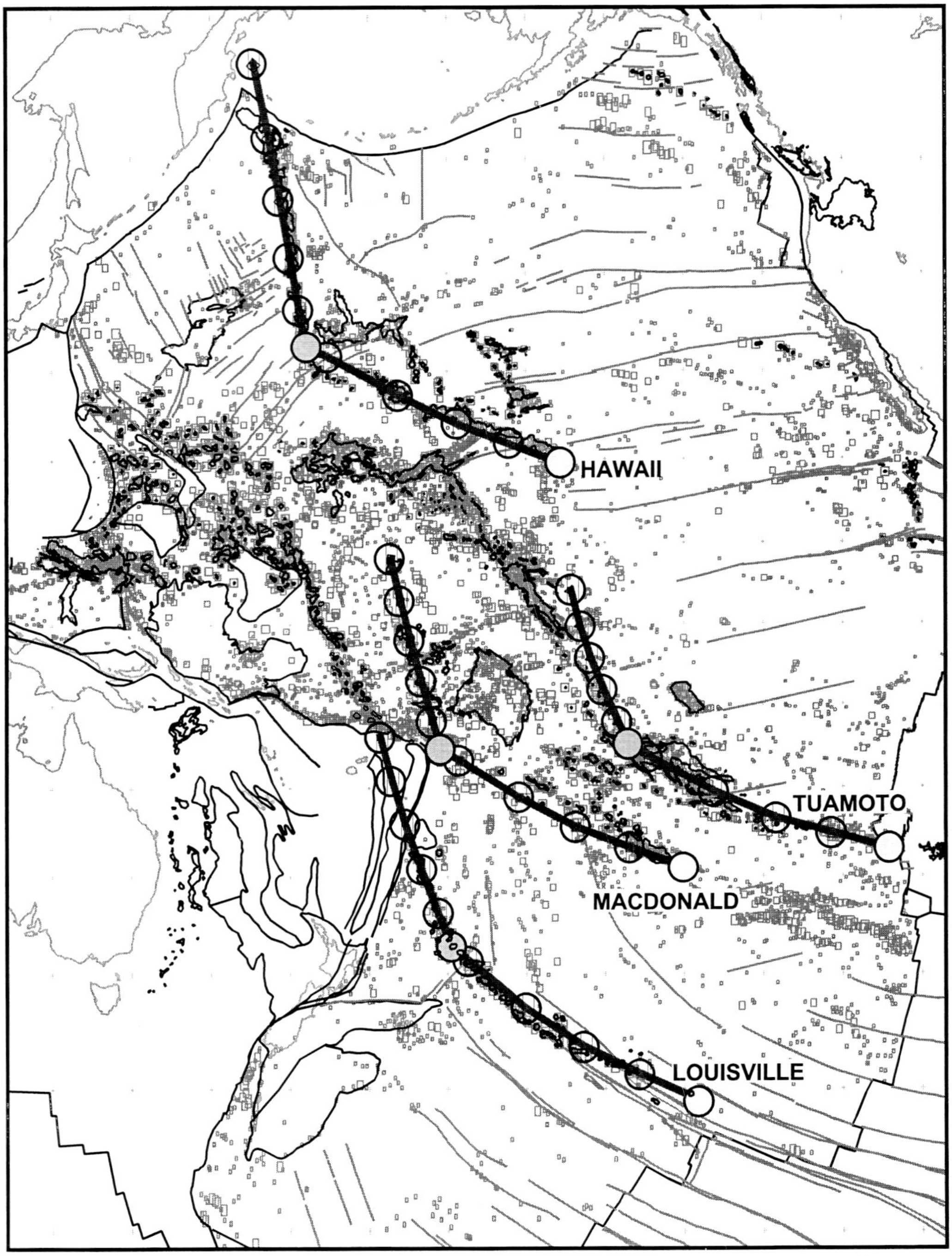


**Figure 4.** Hotspots on the Pacific plate. Rectangular symbols are seamounts as determined from satellite altimetry by *Wessel* [1997], sized according to assumed seamount size. Heavy outlines are hotspot tracks mapped by *Coffin and Eldholm* [1993] in their compilation of Large Igneous Provinces (LIPs). Closed circles denote the present location of the hotspots used in this study. Open circles are calculated positions of the hotspots every 10 million years, with the addition of the position at 43 Ma (grey circle). Present plate boundaries are shown in black; grey lines are fracture zones. The hotspot tracks are calculated using the following rotations: 0 to 25 Ma, 70.79°N, 71.77°W, -17.27°; 25 to 43 Ma, 59.85°N, 66.42°W, -16.33°; 43 to 81 Ma, 13.65°N, 99.70°W, -20.0°. Timing of 25 Ma for the first pole is as suggested by *Lonsdale* [1988], and the 81 Ma age is for the Detroit seamount as determined by *Tarduno and Cottrell* [1997].

**Table 2.** Rotation poles

| Magnetic Anomaly | Age (my) | Rotation Pole Latitude | Longitude | Angle | Reference |
|---|---|---|---|---|---|
| Kula to Pacific | | | | | |
| 24 | 52.9 | 0.00 | 0.00 | 0.00 | 1 |
| 25 | 56.1 | 14.48 | 100.76 | -1.32 | 2 |
| 26 | 57.7 | 14.48 | 100.76 | -2.09 | 2 |
| 27 | 61.1 | 14.48 | 100.76 | -3.23 | 2 |
| 28 | 63.0 | 14.48 | 100.76 | -3.67 | 2 |
| 29 | 64.4 | 14.48 | 100.76 | -4.32 | 2 |
| 30 | 66.6 | 14.48 | 100.76 | -6.04 | 2 |
| 32 | 72.2 | 14.48 | 104.23 | -8.66 | 2 |
| 33 | 79.1 | 14.48 | 104.23 | -11.03 | 2 |
| Farallon-Pacific | | | | | |
| 5 | 10.6 | 72.48 | 55.57 | -6.49 | 2 |
| 6 | 19.6 | 72.35 | 58.62 | -12.39 | 2 |
| 13 | 33.3 | 85.06 | 39.45 | -32.30 | 2 |
| 18 | 39.3 | 86.63 | 86.08 | -43.16 | 3 |
| 21 | 47.1 | 85.55 | 123.80 | -54.08 | 3 |
| 25 | 56.1 | 82.14 | 102.33 | -61.87 | 3 |
| 32 | 72.2 | 81.82 | 96.98 | -71.42 | 3 |
| 34 | 83.0 | 80.50 | 97.59 | -79.26 | 3 |
| Pacific-Antarctica | | | | | |
| A2a | 2.6 | 67.03 | -73.72 | 2.42 | 4 |
| A3A | 5.9 | 67.91 | -77.93 | 5.42 | 4 |
| A4a | 8.9 | 69.68 | -77.06 | 7.95 | 4 |
| A5a | 12.3 | 71.75 | -73.77 | 10.92 | 4 |
| A5d | 17.5 | 73.68 | -69.85 | 15.17 | 4 |
| A6c | 24.1 | 74.72 | -67.28 | 19.55 | 4 |
| A10 | 28.3 | 74.55 | -67.38 | 22.95 | 4 |
| A13 | 33.5 | 74.38 | -64.74 | 27.34 | 4 |
| A20 | 42.5 | 74.90 | -51.31 | 34.54 | 4 |
| A21 | 47.9 | 74.52 | -50.19 | 37.64 | 4 |
| A24 | 53.4 | 73.62 | -52.50 | 40.03 | 4 |
| A26 | 61.1 | 71.38 | -55.57 | 44.90 | 4 |
| A31 | 67.7 | 69.33 | -53.44 | 51.05 | 4 |
| A32 | 72.2 | 68.38 | -51.50 | 54.54 | 4 |
| 33O | 79.1 | 66.05 | -54.73 | 58.61 | 4 |
| A34 | 83.0 | 65.99 | -51.89 | 62.71 | 2 |
| Fit | 90.0 | 65.23 | -50.12 | 67.64 | 2 |
| Antarctica-Africa | | | | | |
| 5 | 10.6 | 11.47 | -43.84 | 1.59 | 2 |
| 6 | 19.6 | 6.41 | -37.16 | 2.91 | 2 |
| 13 | 33.3 | 9.44 | -39.90 | 5.59 | 2 |
| 20 | 43.2 | 12.21 | -41.52 | 7.82 | 2 |
| 21 | 47.1 | 12.80 | -42.07 | 8.77 | 2 |
| 23 | 51.3 | 10.38 | -43.41 | 9.72 | 2 |
| 25 | 56.1 | 6.72 | -41.90 | 10.36 | 2 |
| 28 | 63.0 | -1.89 | -37.94 | 11.25 | 2 |
| 32 | 72.2 | -5.76 | -39.08 | 13.85 | 2 |
| 33 | 76.4 | -3.76 | -39.75 | 15.49 | 2 |
| 34 | 83.0 | -0.98 | -40.08 | 17.79 | 5 |

**Table 2 (continued).** Rotation poles

| Magnetic Anomaly | Age (my) | Rotation Pole Latitude | Longitude | Angle | Reference |
|---|---|---|---|---|---|
| Africa-North America | | | | | |
| 2 | 1.9 | 80.44 | 56.37 | -0.37 | 6 |
| 3A | 5.9 | 82.40 | 43.10 | -1.44 | 6 |
| 5 | 10.6 | 80.12 | 50.80 | -2.52 | 6 |
| 6 | 19.6 | 79.57 | 37.84 | -5.29 | 6 |
| 13 | 33.3 | 76.41 | 7.61 | -9.78 | 6 |
| 20 | 43.2 | 76.52 | 10.05 | -13.05 | 6 |
| 21 | 47.1 | 77.92 | 15.59 | -14.26 | 6 |
| 25 | 56.1 | 81.91 | 16.62 | -17.61 | 6 |
| 30 | 66.6 | 82.51 | -0.63 | -20.96 | 6 |
| 32 | 72.2 | 81.35 | -9.15 | -22.87 | 6 |
| 33O | 79.1 | 78.30 | -18.35 | -27.06 | 6 |
| 34 | 83.0 | 75.80 | -22.81 | -29.93 | 6 |
| North America-Europe (=Asia) | | | | | |
| 5 | 10.6 | 65.38 | 133.58 | 2.44 | 7 |
| 6 | 19.6 | 68.92 | 136.74 | 4.97 | 7 |
| 13 | 33.3 | 65.64 | 136.95 | 7.51 | 7 |
| 20 | 43.2 | 66.69 | 137.48 | 9.87 | 7 |
| 21 | 47.1 | 67.19 | 136.85 | 10.91 | 7 |
| 24 | 52.9 | 63.96 | 142.79 | 13.15 | 7 |
| 25 | 56.1 | 63.14 | 141.66 | 14.22 | 7 |
| 30 | 66.6 | 64.84 | 143.96 | 16.95 | 7 |
| 33 | 79.1 | 66.17 | 147.74 | 19.00 | 7 |
| 34 | 83.0 | 66.54 | 148.91 | 19.70 | 7 |
| Iceland hotspot to Europe | | | | | |
| 5 | 10.6 | 68.07 | 138.20 | 0.23 | 8 |
| 6 | 19.6 | 67.84 | 144.00 | 2.25 | 8 |
| 13 | 33.3 | 58.10 | 150.70 | 2.70 | 8 |
| 21 | 47.1 | 32.34 | 137.72 | 4.00 | 8 |
| 25 | 56.1 | -43.98 | 153.67 | 10.52 | 8 |
| India-Africa | | | | | |
| 5 | 10.6 | 29.22 | 39.89 | -4.79 | 8 |
| 6 | 19.6 | 21.73 | 41.41 | -10.46 | 8 |
| 13 | 33.3 | 19.55 | 43.38 | -18.15 | 8 |
| 21 | 47.1 | 18.84 | 42.19 | -23.37 | 8 |
| 23 | 51.3 | 20.67 | 37.75 | -26.58 | 8 |
| 25 | 56.1 | 22.79 | 29.38 | -29.93 | 8 |
| 28 | 63.0 | 23.12 | 28.80 | -38.48 | 8 |
| 32 | 72.2 | 24.27 | 19.43 | -46.19 | 8 |
| 34 | 83.0 | 21.19 | 23.91 | -53.90 | 8 |
| Fit | 90.0 | 19.66 | 25.25 | -58.61 | 8 |
| South America to Africa | | | | | |
| 5 | 10.6 | 60.00 | -39.00 | 3.15 | 9 |
| 6 | 19.6 | 59.50 | -38.00 | 7.50 | 9 |
| 8 | 26.2 | 59.00 | -36.00 | 10.00 | 9 |
| 13 | 33.3 | 57.50 | -34.00 | 13.38 | 9 |
| 20 | 43.2 | 57.50 | -32.00 | 17.60 | 9 |
| 22 | 49.4 | 59.00 | -31.50 | 20.10 | 9 |
| 25 | 56.1 | 61.00 | -32.00 | 22.30 | 9 |
| 28 | 63.0 | 63.00 | -33.30 | 24.00 | 9 |
| 29 | 64.4 | 63.00 | -33.30 | 24.30 | 9 |

**Table 2 (continued).** Rotation poles

| Magnetic Anomaly | Age (my) | Rotation Pole Latitude | Longitude | Angle | Reference |
|---|---|---|---|---|---|
| 30 | 66.6 | 63.00 | -33.30 | 24.70 | 9 |
| 31 | 68.2 | 63.00 | -33.50 | 25.40 | 9 |
| 32 | 72.2 | 63.00 | -33.50 | 26.60 | 9 |
| 33 | 73.6 | 63.00 | -33.50 | 27.90 | 9 |
| 33R | 79.1 | 63.18 | -34.19 | 30.96 | 9 |
| 34 | 83.0 | 62.29 | -34.78 | 33.58 | 9 |
| Indo-Atlantic hotspots to Africa | | | | | |
| | 10.0 | 60.00 | -30.00 | 1.80 | 10 |
| | 20.0 | 51.50 | -44.40 | 4.25 | 10 |
| | 30.0 | 43.00 | -45.00 | 6.60 | 10 |
| | 40.0 | 38.60 | -41.80 | 9.00 | 10 |
| | 45.0 | 37.10 | -40.80 | 10.21 | 10 |
| | 50.0 | 33.00 | -40.80 | 12.02 | 10 |
| | 55.0 | 30.70 | -40.90 | 13.16 | 10 |
| | 60.0 | 29.90 | -42.00 | 14.18 | 10 |
| | 65.0 | 29.30 | -42.90 | 15.21 | 10 |
| | 70.0 | 25.30 | -40.20 | 16.68 | 10 |
| | 75.0 | 21.30 | -39.40 | 18.25 | 10 |
| | 80.0 | 17.90 | -38.80 | 19.90 | 10 |
| | 85.0 | 19.20 | -41.40 | 21.94 | 10 |
| | 90.0 | 19.40 | -41.90 | 23.31 | 10 |
| | 95.0 | 19.60 | -42.40 | 24.68 | 10 |
| | 100.0 | 18.90 | -41.40 | 25.35 | 10 |
| | 110.0 | 17.70 | -39.50 | 26.71 | 10 |
| | 120.0 | 18.90 | -39.70 | 27.47 | 10 |
| | 130.0 | 16.70 | -37.50 | 28.52 | 10 |
| Pacific hotpots to Pacific | | | | | |
| | 25.0 | 70.79 | -71.77 | -17.27 | 8 |
| | 43.0 | 65.65 | -66.66 | -33.44 | 8 |
| | 81.0 | 45.45 | -78.09 | -47.40 | 8 |
| | 90.0 | 42.09 | -79.14 | -51.50 | 8 |

References
1, *Byrne* [1979]; 2, *Norton* (95); 3, *Rosa and Molnar* [1988]; 4, *Cande et al.* [1995]; 5, *Patriat et al.* [1985]; 6, *Klitgord and Schouten* [1986]; 7, *Srivastava and Roest* [1989]; 8, this paper; 9, *Shaw and Cande* [1990]; 10, *Müller et al.* [1993].

tinct linear ridges on the sea floor. There are several other volcanic features that have been proposed as hotspots and associated tracks but, like the Pacific, different authors have chosen different members of the hotspot family with little explanation for the choices. Three different choices are summarized in Table 1. *Morgan* [1972a, b and 1983] and *O'Connor and Duncan* [1990] focused on the Atlantic hotspots and Reunion, showing how they formed a family without apparent internal relative motion. The Kerguelen-Ninetyeast Ridge hotspot was excluded from these calculations. *Royer et al.* [1991] used improved Indian Ocean relative motion data to show how the Kerguelen-Ninetyeast Ridge hotspot was not fixed to the Atlantic hotspots, but moved at about 1.4 cm/yr. from 84 to 36 Ma. *Müller et al.* [1993] recalculated the Indo-Atlantic hotspot paths (including the Ninetyeast Ridge) and came to the conclusion that they all formed a group without any internal relative motion. Figure 6 is a figure similar to one by *Müller et al.* [1993], showing the Indo-Atlantic hotspot motion. Here, I use the hotspot to Africa rotations from *Müller et al.* [1993] to illustrate relative motions between the hotspots and overlying plates. I have added some hotspots proposed by Morgan and Duncan and the Marion hotspot track as proposed *by Storey et al.* [1995]. I also use timing of which hotspots belonged to different plates as proposed by *Royer et al.* [1991] and *Müller et al.* [1993].

The Azores hotspot proposed by *Burke et al.* [1973] and *Morgan* [1983] is excluded, as the Azores are not created by hotspot volcanism, rather they are the result of compressional deformation along the Africa-Europe plate boundary [e.g. *Sartoni et al.,* 1994]. Figure 6 shows that the Indo-Atlantic hotspot family forms a group with remarkably little apparent internal relative motion, at least for the past 80 million years. The Tristan hotspot and, to some extent, the New England seamount chain are the only ones with well-defined tracks for times older than 80 Ma, so determination of relative hotspot motion becomes more difficult before 80 Ma. As in the Pacific, there are some minor mismatches between calculated and observed paths. The Ninetyeast Ridge mismatch reaches about 200 km at 70 Ma, but a problem with this hotspot is the difficulty of accurately defining its present position [*Morgan,* 1983; Müller et al., 1993]. This leads to uncertainties in defining its past track. Bouvet Island, a proposed hotspot in the South Atlantic, does not have any track emanating from its present position, suggesting that it may not in fact be a hotspot. I have included St. Helena in the figure to show that the band of seamounts trending away from the present island does indeed approximately follow the Indo-Atlantic hotspot path. Age progression along this feature has not been established, but present-day volcanism at the track's northeast end in Cameroon suggests a model more complicated than a single hotspot track for this feature [*Morgan,* 1981]

## ICELAND

Iceland is the present-day expression of a mantle plume that was the source of enormous volumes of lava erupted as Greenland separated from Europe 55 million years ago [*White and McKenzie,* 1989]. It is a classic example of volcanism associated with continental breakup. Initial volcan-

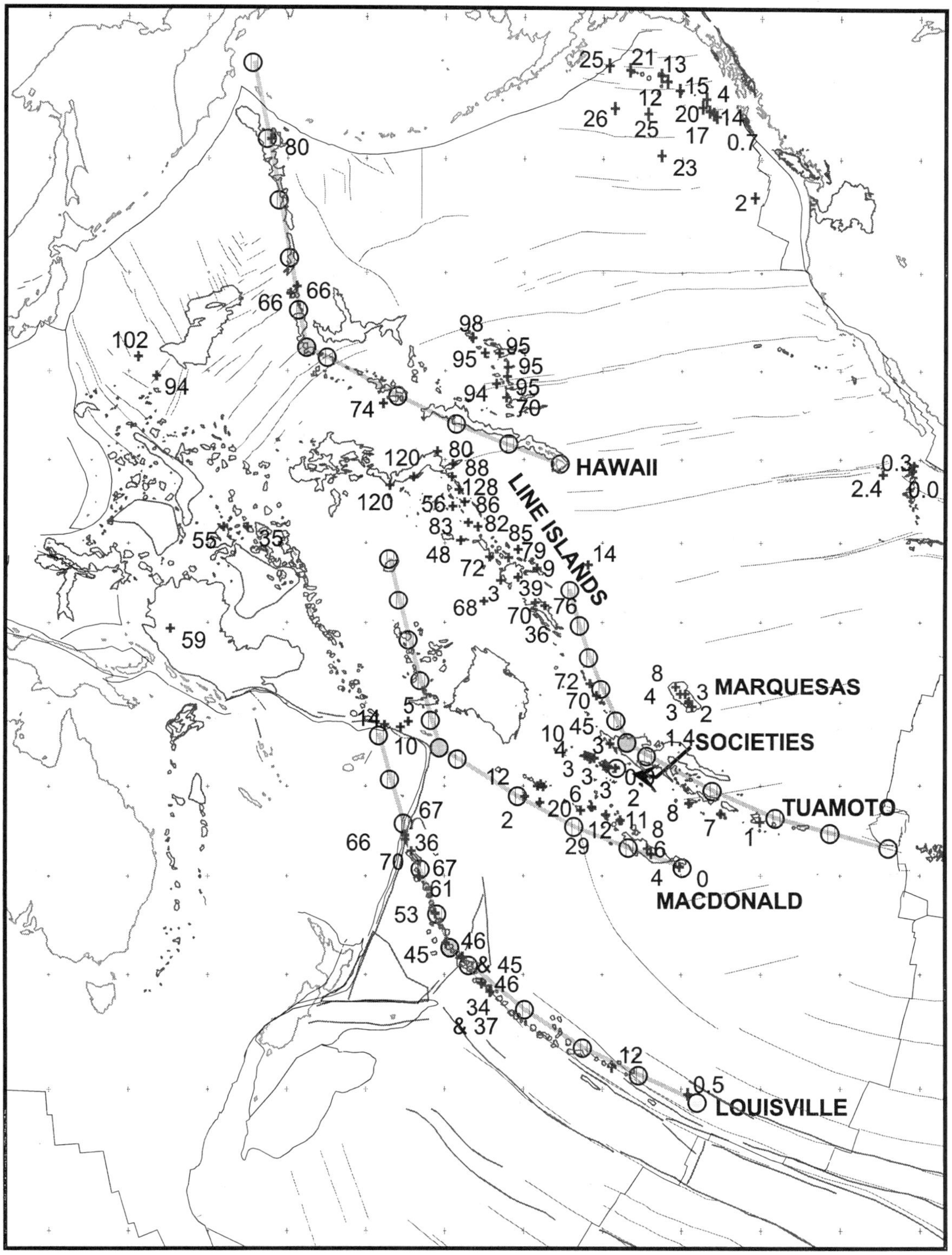


**Figure 5.** Calculated hotspot tracks from Figure 4 plus age data for seamounts and islands in the Pacific. For references see the text. The match of calculated tracks to the Hawaii-Emperor and Louisville tracks is excellent, but not so good for the Central Pacific tracks.

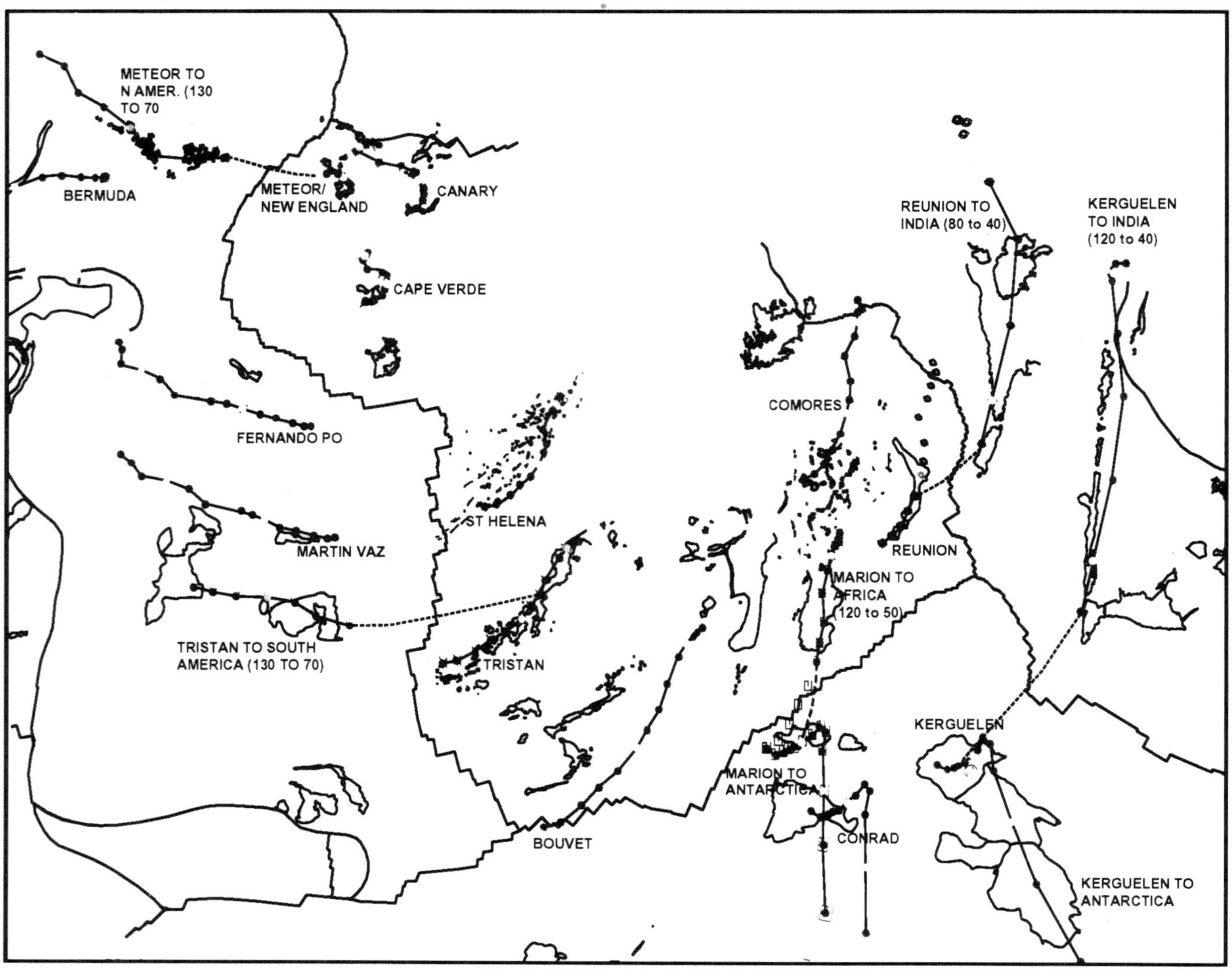


**Figure 6.** The Indo-Atlantic hotspots. Hotspot tracks are shown with symbols every 10 million years. Hotspot to Africa rotations are from *Müller et al.,* [1993]. Plate rotations to Africa are listed in Table 2. The fit of many of the tracks to mapped oceanic ridges and plateaus is excellent. These feature are from a compilation by *Coffin and Eldholm* [1993], with the addition of the Cameroon volcanic line (track of the St Helena hotspot). This latter feature is sometimes excluded from hotspot tracks because there is active volcanism at both ends [*Morgan,* 1983] and there is no age data on intervening seamounts.

ism formed the Skaergard Intrusion in Greenland and the North Atlantic Volcanic province in the Faeroes, northern Scotland and Ireland. Also formed were the volcanic features visible on seismic records as 'seaward dipping reflectors' on both margins. The hotspot trails created as Europe and Greenland separated are the Iceland-Faeroes Ridge on the European Plate and the Iceland-Greenland Ridge on the Greenland Plate. In Figure 7, these features are shown along with two calculated hotspot tracks, one for the Indo-Atlantic hotspots relative to Greenland and the other for the same hotspots relative to Europe. If Iceland has remained fixed to the Indo-Atlantic family for the past 55 million years, the tracks shown in Figure 7 should follow the apparent hotspot trails. The calculated trail relative to Greenland should follow the Iceland-Greenland Ridge and intersect the coast of Greenland at the Skaergard Intrusion at 55 Ma. The calculated trail relative to Europe should follow the Iceland-Faeroes Ridge to reach the Faeroes at 55 Ma. Neither of the tracks matches at all; the one to Europe, in particular, does not even come close. The track to Greenland reaches the coast at 30 Ma. This means that the Iceland hotspot did not remain fixed to the other Indo-Atlantic hotspots.

A similar calculation for Indo-Atlantic hotspots relative to Greenland was made by *Lawver et al.* [1993], which led them to propose that the Iceland hotspot of today is indeed fixed to the Indo-Atlantic hotspots, i.e. it is not the same one that created the North Atlantic volcanic province at 55 Ma. Figure 7 shows that this is not possible, because the track of Indo-Atlantic hotspots to Europe also mismatches.

## COMPARISON OF HOTSPOT FRAMES

So we are led to the conclusion that Iceland and the Indo-Atlantic hotspots have not remained fixed to each other. We will now compare the Indo-Atlantic and Pacific families. To calculate motion between them, we need to be able to calculate relative motion between the Pacific plate and the Indo-Atlantic plates - we need to use a plate circuit that includes only spreading ridge boundaries [*Atwater,* 1989]. Our Indo-Atlantic hotspot motions are relative to Africa, so to calculate them relative to the Pacific the plate circuit is Africa-East Antarctica-West Antarctica-Pacific. *Molnar and Stock* [1987] attempted this type of calculation. They found that rather large relative motion between the Indo-Atlantic and Pacific hotspots was required, but at the time there was some uncertainty about whether there was a hidden plate boundary in Antarctica or in the South Pacific that was not taken into account in the plate circuit calculations [*Stock and Molnar,* 1987]. This was a legitimate concern as the far southern Pacific was so poorly mapped at the time. *Stock and Molnar* [1987] proposed that there was an extra plate, the Bellinghausen plate, in the South Pacific, but even taking this extra plate into account did not explain mismatches between Indo-Atlantic and Pacific hotspots [*Molnar and Stock,* 1987]. In these papers Stock and Molnar relied on magnetic lineations and fracture zones mapped from ship borne data, which at the time was sparse, particularly on the Antarctic plate. Since these papers, knowledge of South Pacific tectonics has greatly improved as a result of mapping of fracture zones from satellite gravity data [*Sandwell and Smith,* 1997]. *Mayes et al.* [1990] used early versions of this satellite data to update mapping of fracture zones and magnetic lineations. Most recently, spatial resolution of the gravity data has greatly improved through the release of the Geosat data [*Sandwell and Smith,* 1997]. Re-mapping of magnetic lineations using the Geosat data to constrain fracture zones [*Norton,* 1995] and also with detailed bathymetric and new magnetic isochron mapping [*Cande et al.,* 1995] has demonstrated large changes in spreading direction that were not discernible using the scattered ship borne and earlier gravity data that was available previously. *Cande et al.* [1995] suggested that the Bellinghausen plate only existed as an independent plate prior to anomaly 27, 61 Ma. Thus this plate is unlikely to be a factor in Tertiary age reconstructions, especially at 43 Ma, the time of the bend in the Hawaii-Emperor chain.

The other place where a hidden Tertiary age plate boundary has been postulated is in Antarctica, specifically between West and East Antarctica. This boundary would affect plate circuit calculations as the Africa-Antarctica plate motion is calculated from the Southwest Indian Ridge between Africa and Antarctica, whereas Antarctica-Pacific data comes from the South Pacific, i.e. between the Pacific and West Antarctica. There is a fundamental difference between the crust of East and West Antarctica. East Antarctica is an Archean-cored continental plate, whereas West Antarctica has much thinner crust, most of which accreted to Gondwana during Paleozoic and Early Mesozoic time [*Dalziel and Elliot,* 1982]. Present-day tectonic activity within the Antarctic plate is limited to active volcanism in West Antarctica and, as represented by Mt Erebus, on the margin of East Antarctica [*LeMasurier and Thomson,* 1989]. Lack of seismicity within the Antarctic plate suggests no internal active plate boundaries, but whether there was an internal plate boundary during Tertiary time is less clear, partly because ice cover limits available geologic data [*Dalziel and Elliot,* 1982]. If an internal boundary did exist during Tertiary time, it presumably would have connected to the South Pacific spreading system on the Pacific side of Antarctica and to the Southwest Indian Ridge on the Atlantic-Indian side, through the Weddell Sea. No such boundary has ever been found, even using the latest satellite gravity data to map tectonic features in the Southern Oceans where ship borne surveying has been limited. A plate boundary is not apparent in the South Pacific [*Norton,* 1995]. Tectonic models of the Weddell Sea show that it formed during Mesozoic time and there is no apparent need to invoke any extra Tertiary age plate boundaries [*Livermore and Hunter,* 1996]. Thus we will assume that there is no internal Antarctic plate boundary and for the plate circuit connecting Africa to the Pacific use Africa-Antarctica-Pacific, but the issue of an internal Antarctic plate boundary is certainly the weak link in the plate circuit. If the mismatch between the Pacific and Indo-Atlantic hotspot reference frames [*Morgan,* 1981; *Molnar and Stock,* 1987; *Tarduno and Gee,* 1995; *Cande et al.,* 1995; *Norton,* 1995] is entirely due to a hidden plate boundary within Antarctica, the amount of motion needed within Antarctica is rather large. *Norton* [1995] showed that, to reconcile the

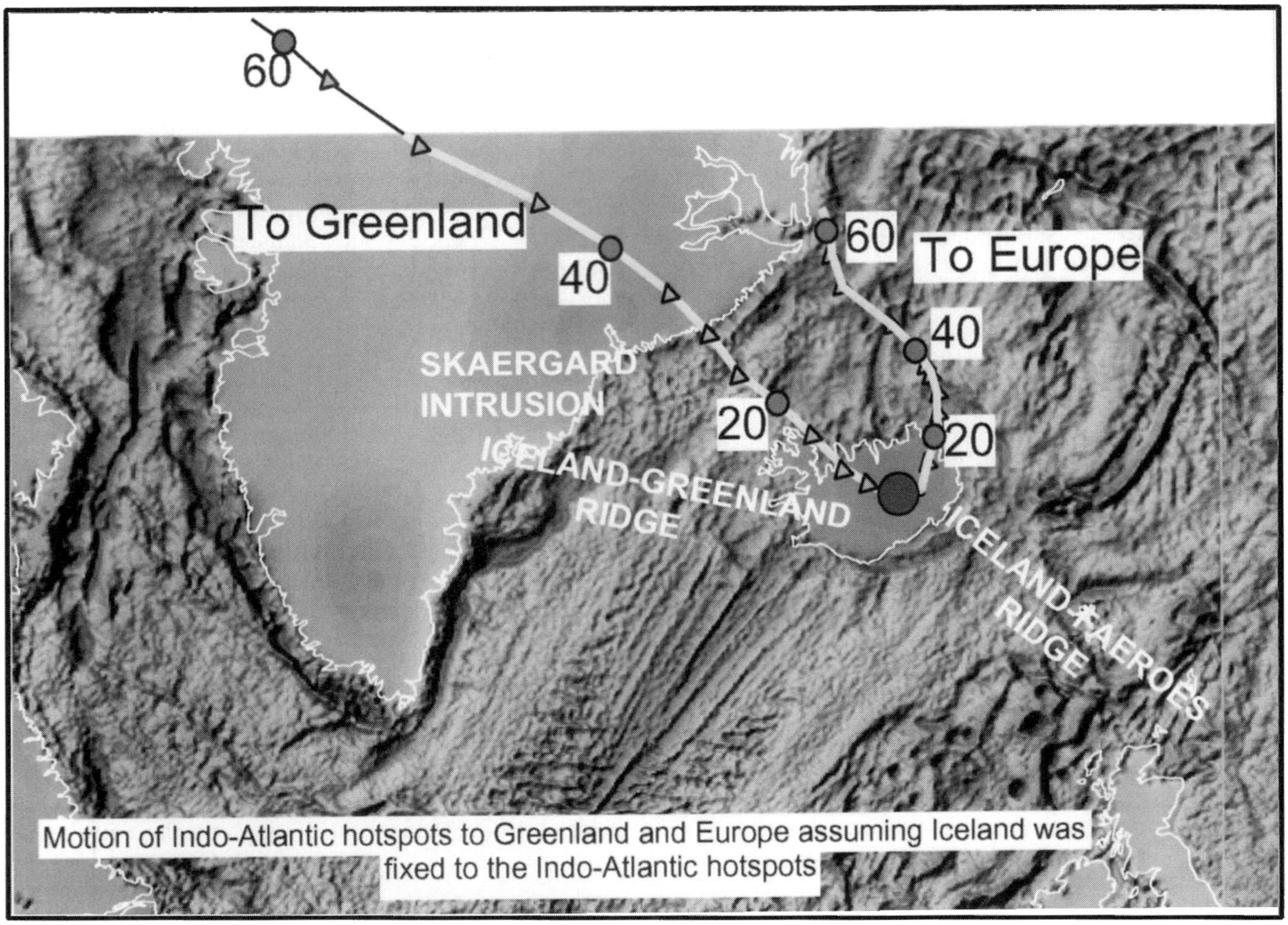


**Figure 7.** Motion of the Indo-Atlantic hotspot frame with respect to Europe and Greenland, with start point at the Iceland hotspot. Base map is Geosat gravity *Sandwell and Smith* [1997]. If the Iceland hotspot was fixed to the Indo-Atlantic family, the calculate track to Europe would follow the Iceland-Faeroes ridge and the track to Greenland would follow the Iceland-Greenland ridge. Both tracks would intersect the continental margins at 55 Ma. The large mismatch is interpreted to show that the Iceland hotspot has not remained fixed to the Indo-Atlantic frame.

Pacific and Indo-Atlantic hotspot frames with an internal Antarctic plate boundary would require 830 km of motion between West and East Antarctica since 63 Ma, a motion amount that certainly should have produced an observable record.

Using this plate circuit, Figure 8 shows the Indo-Atlantic hotspots relative to the Pacific. Here, the hotspots are shown at positions calculated every 10 million years (plus a point at 43 Ma), with start points at the same positions as shown in Figure 4. The plate circuit calculations are carried back as far as possible using only spreading ridge boundaries in the circuit. The South Pacific formed between 80 and 90 million years ago with rifting of New Zealand from Antarctica [e.g. *Stock and Molnar,* 1987]. Prior to this time the Pacific plate was separated from Antarctica by a subduction zone, so there is no way to use a plate circuit to calculate relative motion between the Pacific and Indo-Atlantic plates before 80 Ma. Another uncertainty in using plate circuit calculations is in the observational errors associated with mapping magnetic anomalies and fracture zones used in individual relative motion calculations around plate circuits. These errors are cumulative in plate circuit calculations. *Stock and Molnar* [1988] analyzed relative motion between the Pacific plates and North America. The longest plate circuit calculation they used was for Farallon to North América motion. This involves calculating four relative motions to complete the plate circuit: Farallon to Pacific, Pacific to Antarctica, Antarctica to Africa and Africa to

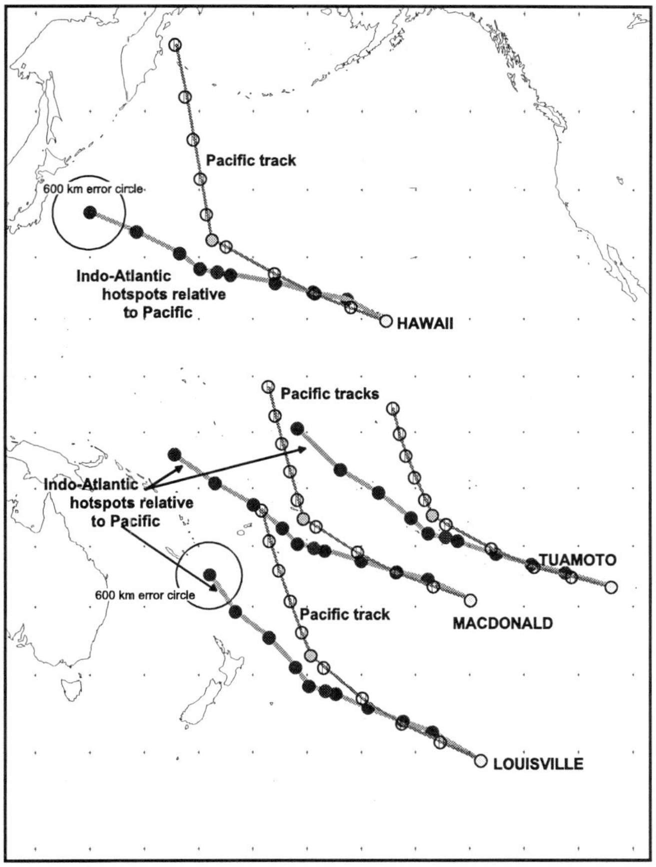


**Figure 8.** Indo-Atlantic hotspot frame relative to the Pacific for the past 90 million years, with symbols every 10 million years plus 43 Ma. The Pacific hotspot tracks from Figure 4 are shown with open circles. Closed circles are calculated Indo-Atlantic hotspots relative to the Pacific, calculated using the plate circuit Indo-Atlantic hotspot-Africa-Antarctica-Pacific. Error circles are discussed in the text. Pacific to Antarctic motion can only be determined from 90 Ma, as before that time the Pacific was separated from Antarctica by a subduction zone. If the Indo-Atlantic hotspots were fixed to Pacific hotspots, the two tracks would overlap.

North America. *Stock and Molnar* [1988] used reasonable estimates of the mapping uncertainties in each of these relative motion calculations and estimated that the total error was about ±600 km for early Tertiary positions of the Farallon plate relative to North America. They estimated that this error decreased to about ±140 km in the late Tertiary. The plate circuit calculation used in Figure 8 involves only three stages, Indo-Atlantic hotspots to Africa, Africa to Antarctica and Antarctica to Pacific, so the errors are likely to be less than for the four-stage Farallon to North America calculations of *Stock and Molnar* [1988]. But using an upper bound of ±600 km for an error estimate, a circle of radius 600 km is shown for the 80 Ma calculated positions of the Indo-Atlantic hotspots relative to Hawaii and Louisville. Even with this error range taken into account, there is a large mismatch between the Indo-Atlantic and Pacific hotspot tracks apparent in Figure 8.

It is readily apparent from Figure 8 that, even though there is agreement in trend between the Indo-Atlantic and Pacific hotspots for the past 20 million years, the two hotspot families have not remained fixed to each other for the past 80 million years. Averaged over the last 80 My, the rate of motion between the two hotspot reference frames is 27 ± 8 km/my (using the 600 km error circle as our error estimate). Note too that the 43 Ma bend is not repeated in the calculated Indo-Atlantic tracks. In Figure 9, the calculation is repeated in the opposite direction - the Pacific hotspot family motion is plotted relative to the Indo-Atlantic hotspots, with initial positions at the Indo-Atlantic hotspots. Again, it is apparent that the two hotspot families have not remained fixed to each other.

## THE 43 Ma EVENT

Another method of examining relative motions is shown in Figure 10. Here, with a start point at Hawaii, motion of every major plate plus the Indo-Atlantic and Iceland hotspots is plotted at reconstructed times every 5 million years. The Iceland track is from rotations (Table 2) that move the hotspot from its present position towards Europe along the Iceland-Faeroe Ridge and also towards Greenland along the Iceland-Greenland Ridge. Although the general trend of the Hawaii seamounts is followed by several of the paths for 43 Ma to present - including the other hotspots - the 43 bend in the Hawaii-Emperor seamount chain is not matched at all. Exceptions are the calculated track for North and South America. These show shapes that are tantalizingly close to the Hawaii-Emperor seamount chain, except that 'the bend' is at 47 Ma, not 43 Ma. In the fixed hotspot model, the 43 Ma bend in the Hawaii-Emperor seamount chain would be due to a change in motion direction of the Pacific plate. The magnitude of the implied change in motion direction, 60 degrees, should produce dramatic tectonic events ('43 Ma events') in Pacific-bordering plates. Absence of these events in Asia led *Norton* [1995] to look more closely at relative plate motions in the North Pacific. This discussion is extended here.

To investigate possible 43 Ma plate interactions, it is instructive to look at a plate reconstruction of the Pacific for 43 Ma. This is shown in Figure 11, a plate reconstruction of the Pacific and its bordering plates at 43 Ma. This reconstruction uses the plate reconstruction parameters in Table 2, plus western Pacific rotations after *Lee and*

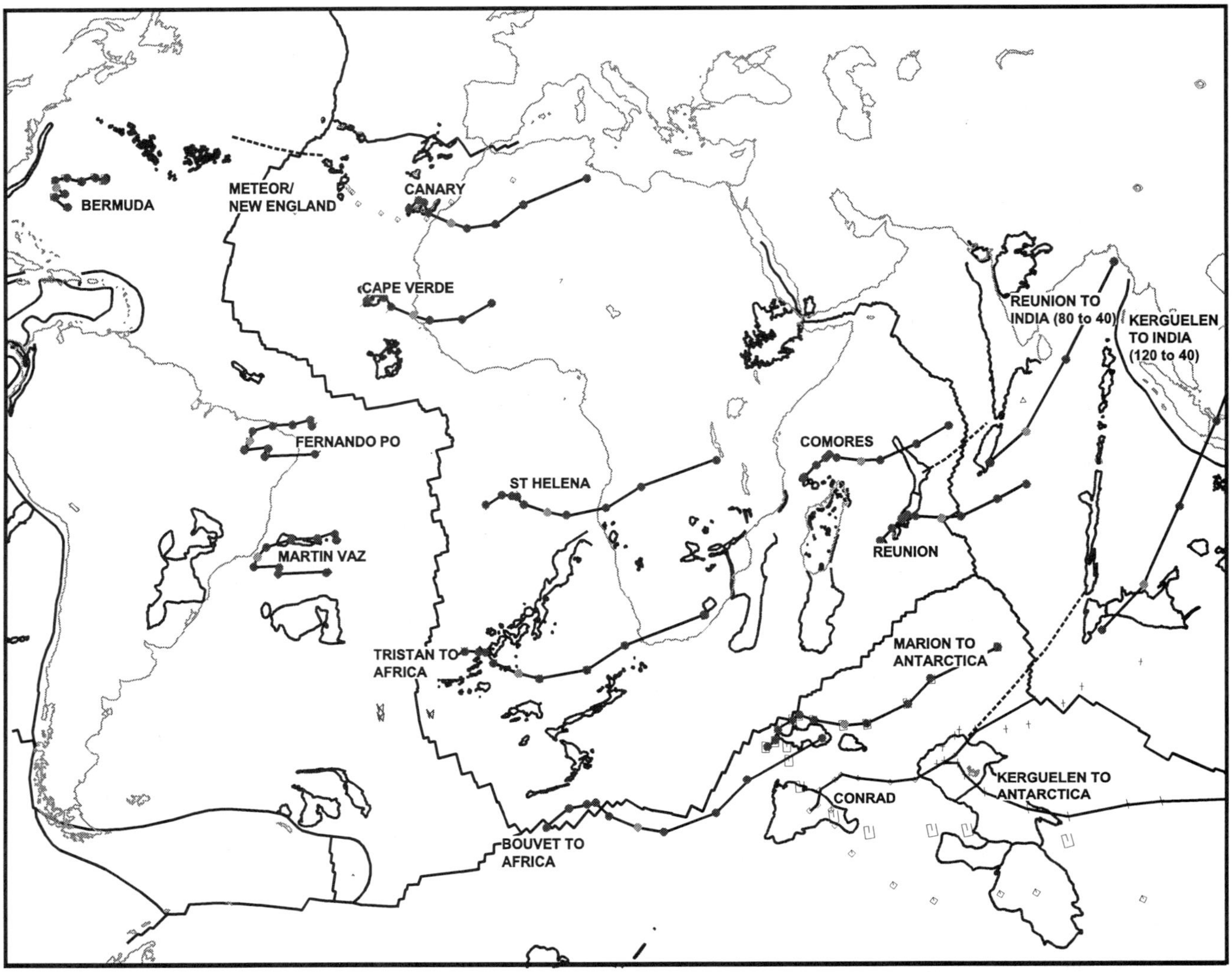


**Figure 9.** Similar to Figure 8, Pacific hotspot frame relative to the Indo-Atlantic hotspots for the past 90 million years. Compare to Figure 6. Indo-Atlantic hotspot tracks are shown with the LIPS data [*Coffin and Eldholm,* 1993]. If the Indo-Atlantic hotspots were fixed to Pacific hotspots, the calculated Pacific tracks would overlap the bathymetric hotspot tracks.

*Lawver* [1992]. Cross hatched areas denote the large areas of oceanic plates that have been subducted since 43 Ma. Plate boundaries within these areas are of course speculative, but follow the trends proposed by *Engebretson et al.* [1985]. It is also possible that there were plates existing at 43 Ma that have entirely disappeared. An example of this may be the area between the Philippine plates (NP and SP) and Asia, shown cross hatched and labeled 'Extra plate?' in figure 11. Another area of unknown tectonics lies between the Pacific plate and the Aleutian arc. Here, it is not known whether the Kula or Pacific plates were subducting beneath the Aleutian arc at 43 Ma. East-west magnetic lineations east of the Emperor seamounts young to the north. These anomalies were produced by spreading between the Pacific and Kula (now vanished) plates (see review in *Atwater,* [1989]). The youngest of these magnetic lineations is number 23, 51 million years old, but it is not known when spreading between the Kula and Pacific plates stopped and the Kula plate became part of the Pacific plate. *Norton* [1995] followed *Byrne* [1979] in assuming that the spreading direction change between 54 and 49 Ma that is mapped in the NE Pacific [*Atwater,* 1989] coincided with cessation

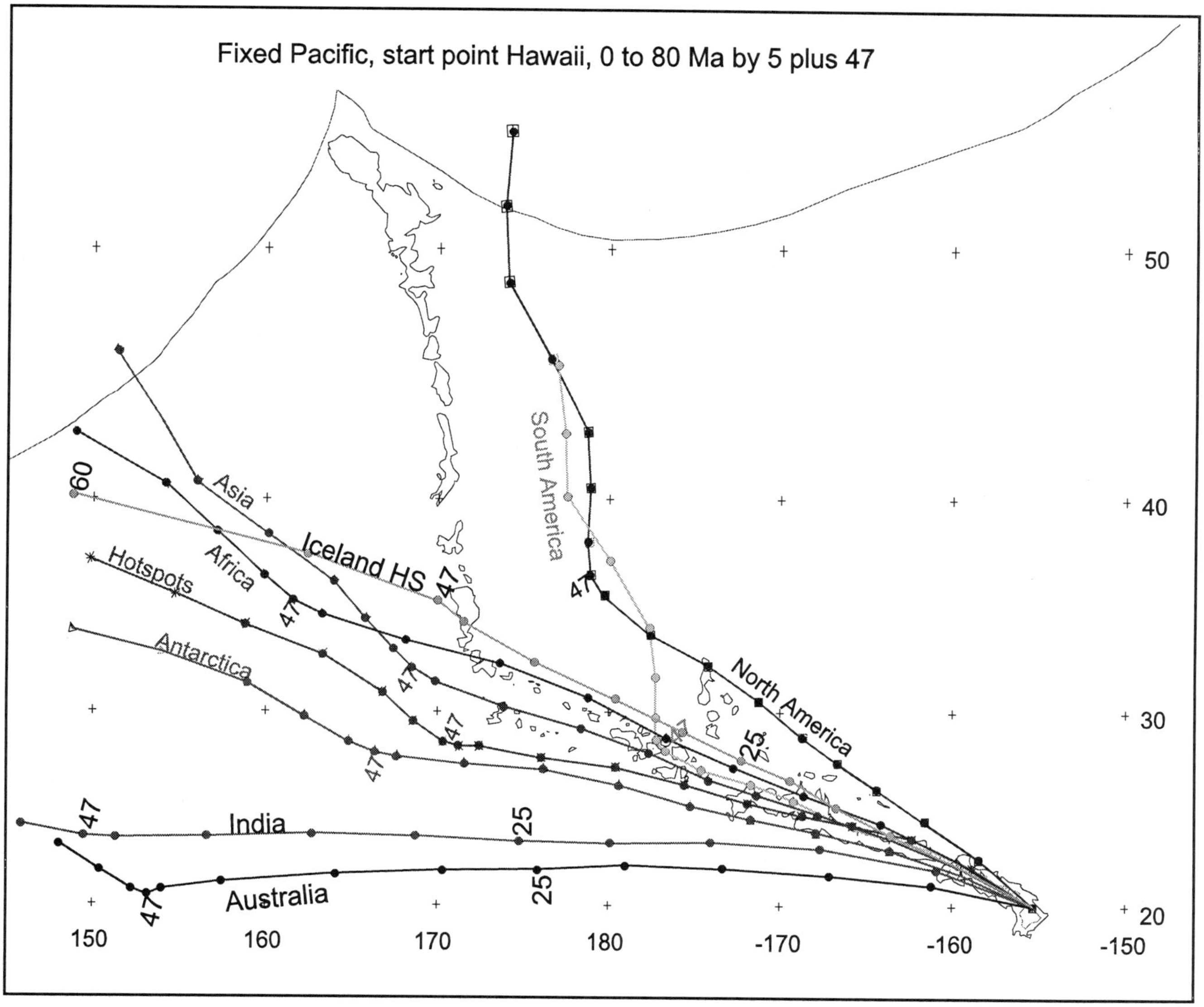


**Figure 10.** Calculation of motion of various plates relative to the Pacific. Each track starts at Hawaii and follows the motion of a point attached to the following plates: North and South America, Asia, Africa, Antarctica, India and Australia. Also shown are tracks of the Iceland and Indo-Atlantic hotspots. If any of these global plates or hotspots was fixed relative to the Hawaii hotspot, its calculated track would follow the Hawaii-Emperor seamount chain. Only North America and, to some extent, South America seem to be even close, but the significance of this is not clear. This is discussed further in the text.

of Kula-Pacific spreading, but this is an assumption. *Lonsdale* [1988b] mapped some magnetic lineations east of the Emperor seamounts that he interpreted as being produced by Kula-Pacific spreading that lasted until 42 Ma.

Figure 11 shows a possible plate boundary extending north from the known Pacific plate to the Aleutian Arc. This shows a possible Kula-Pacific boundary. If the Kula plate still existed as a separate plate at 43 Ma then the Pacific plate was not touching North America - the Kula, Farallon and Vancouver plates bordered western North America [*Stock and Molnar,* 1988; *Atwater,* 1989]. If the Kula plate was already part of the Pacific plate, then the Pacific could have been touching North America, depending on where the northern end of the Vancouver-Pacific spreading center is drawn. Figure 10 shows that there was a large change in direction of relative motion between the Pacific

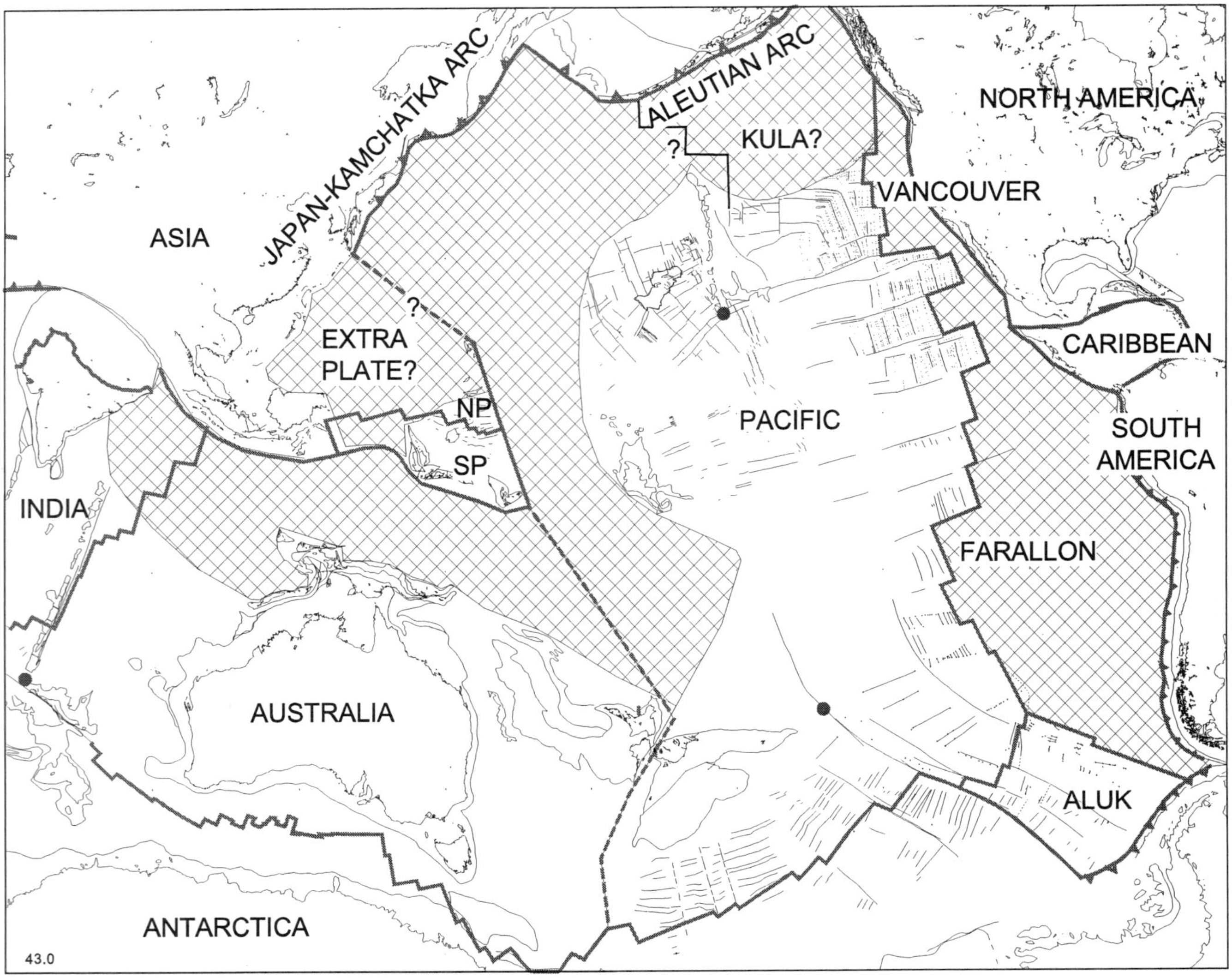


**Figure 11.** Plate reconstruction for the Pacific and its bordering plates at 43 Ma with North America fixed. Cross hatched areas denote oceanic crust that has been subducted since 43 Ma. NP = North Philippine plate, SP = South Philippine plate.

and North America at 47 Ma; if the Pacific was indeed touching North America at this time, there should be a record of this event in present-day southeast Alaska.

In the northwest Pacific, between the Pacific plate and Asia in Figure 11, there is no evidence for missing Tertiary age plates, so the simplest interpretation is that the Pacific was subducting beneath Asia. As shown in Figure 10, there was no abrupt change in Asia-Pacific motion at 43 Ma. Thus it is not necessary to suggest that the Kamchatka-Japan subduction system suffered any major reorganization associated with the 43 Ma event. This is why *Norton* [1995] suggested that there was no major 43 Ma tectonic event associated with an abrupt change in direction of the Pacific plate at 43 Ma - even though there was an abrupt change (at least at 47, not 43) relative to North America, this would only have any tectonic significance if there was a common plate boundary. As discussed above, such a boundary may exist in southeast Alaska, depending on interpretation of Kula plate history. It will be interesting to see whether future refinements in plate motion models and also in age control of the 43 Ma bend in the Hawaii-Emperor seamount show that the 47 Ma bend in calculated plate motion of

North America shown in Figure 10 does actually coincide with the Hawaii-Emperor bend and also possibly with demise of the Kula plate.

## CONCLUSIONS

In this review it has been shown that there are at least three hotspot 'families' that have very little internal relative motion - the Pacific (80 million years), Indo-Atlantic (130 million years) and Iceland hotspot families (55 million years). Relative motion between the families is rather large and makes the notion of a fixed global hotspot reference frame untenable. Note that the question of an internal Antarctic plate boundary, while perhaps not fully resolved and therefore a source of uncertainty in comparisons of the Pacific and Indo-Atlantic hotspots, still does not help with explaining the difference in motion between Iceland and the Indo-Atlantic hotspots. Why these three hotspot groups should form is not clear. The Pacific is separated from the other two by subduction zones which may provide mantle 'levees' separating different mantle convection regimes, but this does not explain why Iceland and the rest of the Indo-Atlantic hotspots are not relatively fixed. *Duncan and Richards* [1991] showed a figure (their Figure 1) that grouped the global hotspots into Pacific and Indo-Atlantic domains. They showed that the two groups correlate with mapped geoid residuals. Perhaps this observation can provide a starting point for models to explain how the two major groups can exist, although Iceland could still challenge the assumptions.

A final point about hotspots and oceanic volcanic features: all the hotspot tracks used in this study are located on oceanic crust. The LIPS identified by *Coffin and Eldholm* [1993] include some on continental crust, but these do not form well-defined tracks. This difference between oceanic and continental hotspot manifestation has been related to crustal thickness [*Morgan* 1983]. But there are several oceanic volcanic edifices that are not obviously related to hotspot tracks. These are usually isolated plateaus such as the Shatski and Hess rises in the North Pacific (Figure 1). The Shatski Rise, in particular, is of interest. Although it is possible that it formed on a short-lived hotspot, the fact that is located along the track of a triple junction is remarkable. *Aslanian et al.* [1994] noted this correlation, and suggested that plateau formation was related to excess volcanism at the triple junction. This triple junction would have undergone some major reorganization when the Kula plate formed about 90 Ma [*Atwater,* 1989]. Exactly how this happened is not known because of the difficulty of mapping age of ocean crust in the Cretaceous quiet zone. Could the locus of excess volcanism associated with the triple junction have evolved into the Hawaii hotspot at 90 Ma when the Kula plate formed? I leave this as an unanswered question.

*Acknowledgments.* I thank Mike Coffin for the use of his LIPS data file, Dave Sandwell for discussions on seamount data, Paul Wessel for access to his seamount data file and especially Richard Gordon for careful reviews of this paper. Tectonic data (present day plate boundaries, fracture zones and locations of mapped magnetic isochrons) were compiled at the University of Texas Institute for Geophysics (PLATES project). I thank Exxon Exploration for permission to release this paper.

## REFERENCES

Acton, G.D., and R.G. Gordon, Paleomagnetic tests for Pacific plate reconstructions and implications for motion between hotspots, *Science, 263,* 1246-1254, 1994.

Aslanian, D., M. Sahabi, J.-L. Olivet, J. Goslin, L. Gilg-Capar, M. Maia, Recq, P. Beuzart, L. Geli, Genesis of topographic anomalies in the oceans, part 1: the role of triple junctions, EOS Trans. AGU 75, 321, 1994.

Atwater, T., Plate tectonic history of the northeast Pacific and western North America, in *The Geology of North America,* vol. N, *The Eastern Pacific Ocean and Hawaii,* edited by E.L. Winterer, D.M. Hussong, and R.W. Decker, pp. 21-72, Geol. Soc. of Am., Boulder, Colo., 1989.

Besse, J. and V. Courtillot, Revised and synthetic Apparent Polar Wander paths of the African, Eurasian, North American and Indian plates, and True Polar Wander since 200 Ma, *J. Geophys. Res., 96,* 4,029-4,050, 1991.

Burke, K., W.S.F. Kidd, J.T. Wilson, Relative and latitudinal motion of Atlantic hotspots, *Nature, 245,* 133-137, 1973.

Byrne, T., Late Paleocene demise of the Kula-Pacific spreading center, *Geology, 7,* 341-344, 1979.

Cande, S.C., C.A. Raymond, J. Stock and W.F. Haxby, Geophysics of the Pitman fracture zone and Pacific-Antarctic plate motions during the Cenozoic, *Science, 270,* 947-953, 1995.

Clague, D.A., and G.B Dalrymple, Tectonics, geochronology and origin of the Hawaii-Emperor Chain, in *The Geology of North America,* vol. N, *The Eastern Pacific Ocean and Hawaii,* edited by E.L. Winterer, D.M. Hussong, and R.W. Decker, pp. 188-217, Geol. Soc. of Am., Boulder, Colo., 1989.

Coffin, M.F. and O. Eldholm, Large igneous provinces, *Scientific American,* 42-49, Oct. 1993.

Coffin, M.F. and O. Eldholm, Large Igneous Provinces: crustal structure, dimensions, and external consequences. *Reviews of Geophysics,32,* 1-36, 1994.

Craig, C.H. and D.T. Sandwell, Global distribution of seamounts from Seasat profiles, *J. Geophys. Res., 93,* 10,408-10,420, 1988.

Dalziel, I.W.D. and D.H. Elliot, West Antarctica: problem child of Gondwana, *Tectonics, 1,* 3-19, 1982.

Duncan, R.A., and D.A. Clague, Pacific plate motion recorded by linear volcanic chains, in *The Ocean Basins and Margins,* vol. 7A, *The Pacific Ocean*, edited by A.E.M. Nairn, F.G. Stehli, and S. Uyeda, pp. 89-121, Plenum, New York, 1985.

Duncan, R.A., and M.A. Richards, Hotspots, mantle plumes, flood basalts and true polar wander, *Rev. Geophys., 29,* 31-50, 1991.

Engebretson, C.D., A. Cox and R.G. Gordon, Relative motions between oceanic and continental plates in the Pacific Basin, *Geol. Soc. Amer. Spec. Pap. 206,* 59 pp., 1985.

Epp, D., Possible perturbations to hotspot traces and implications for the origin and structure of the Line Islands, *J. Geophys. Res., 89,* 11,237-11,286, 1984.

Harrison, C.G.A. and T. Lindh, Comparison between the hot spot and geomagnetic field reference frame, *Nature, 300,* 251-252, 1982.

Ito, G., M.K. McNutt and R.L. Gibson, Crustal structure of the Tuamotu Plateau, 15°S, and implications for its origin, *J. Geophys. Res., 100,* 8,097-8,114, 1995.

Klitgord, K.D., and H. Schouten, Plate kinematics of the central Atlantic, in *The Geology of North America,* vol. M, *The Western North Atlantic Region*, edited by P.R. Vogt and B.E. Tucholke, pp. 351-378, Geol. Soc. of Am., Boulder, Colo., 1986.

Lawver, L.A. and R.D. Müller, Iceland hotspot track, *Geology, 22,* 311-314, 1994.

Lee, T.Y. and L.A. Lawver, Cenozoic plate reconstructions of Southeast Asia, *Tectonophys., 251,* 149-180, 1995.

LeMasurier, W.E. and J.W. Thomson, Volcanoes of the Antarctic Plate and Southern Oceans, *Antarctic Res. Ser., 48,* pp 487, AGU, Washington, D.C., 1989.

Livermore, R.A. and R.J. Hunter, Mesozoic spreading history of the southern Weddell Sea, in *Weddell Sea tectonics and Gondwana break-up,* Storey, B.C., King, E.C. and R.A. Livermore (eds.), Geological Society Special Publication 108, pp 227-241, 1996.

Lonsdale, P., Geography and history of the Louisville hotspot chain in the Southwest Pacific, *J. Geophys. Res., 93,* 3078-3104, 1988a.

Lonsdale, P., Paleogene history of the Kula plate: Offshore evidence and onshore implications, *Geol. Soc. Am. Bull., 100,* 733-754, 1988b.

Mayes, C.L., L.A. Lawver and D.T. Sandwell, Tectonic history and new isochron chart of the South Pacific, *J. Geophys. Res., 95,* 8,543-8,567, 1990.

McNutt, M.K, K. Fischer, S. Kruse and J. Natland, The origin of the Marquesas fracture zone ridge and its implications for the nature of hot spots, *Earth Planet. Sci. Lett., 91,* 381-393, 1989.

Molnar, P. and T. Atwater, Relative motions of hotspots in the mantle, *Nature, 246,* 288-291, 1973.

Molnar, P. and J. Francheteau, The relative motion of hotspots in the Atlantic and Indian Oceans during the Cenozoic, *Geophys. J. R. Astron. Soc., 43,* 763-774, 1975.

Molnar, P. and J. Stock, Relative motions of hotspots in the Pacific, Atlantic and Indian oceans since Late Cretaceous time, *Nature, 327,* 587-591, 1987.

Morgan, W.J., Convection plumes in the Lower Mantle, *Nature, 230,* 42-43, 1971.

Morgan, W.J., Plate motions and deep mantle convection, in *Studies in Earth and Space Sciences (Hess Volume),* edited by R. Shagam et al., *Mem. Geol. Soc. Am. 132,* pp. 7-122, 1972a.

Morgan, W.J., Deep mantle convection and mantle plumes, *Amer. Assoc. Petrol. Geol. Bull., 56,* 203-213, 1972b.

Morgan, W.J., Hotspot tracks and the opening of the Atlantic and Indian Oceans, in *The Sea,* vol. 7, *The Oceanic Lithosphere,* edited by C. Emiliani, pp. 443-475, Wiley-Interscience, New York, 1981.

Morgan, W.J., Hotspot tracks and the early rifting of the Atlantic, *Tectonophysics, 94,* 123-139, 1983.

Müller, R.D., J.-Y. Royer, and L.A. Lawver, Evidence for hotspot motion in the Late Cretaceous/Early Tertiary, *Geol. Soc. Am. Abstr. Programs, 23,* A318, 1991.

Müller, R.D., J.-Y. Royer, and L.A. Lawver, Revised plate motions relative to the hotspots from combined Atlantic and Indian Ocean hotspot tracks, *Geology, 21,* 275-278, 1993.

Norton, I.O., Plate motions in the North Pacific: the 43 Ma non-event, *Tectonics, 14,* 1080-1094, 1995.

O'Connor, J.M. and R.A. Duncan, Evolution of the Walvis Ridge - Rio Grande Rise hot spot system: implications for African and South American plate motions over plumes, *J. Geophys. Res., 95,* 17,475-17,502, 1990.

Patriat, P., J. Ségoufin, J. Goslin, and P. Beuzart, Relative positions of Africa and Antarctica in the Upper Cretaceous: Evidence for non-stationary behaviour of fracture zones, *Earth Planet. Sci. Lett., 75,* 204-214, 1985.

Rosa, J.W.C., and P. Molnar, Uncertainties in reconstructions of the Pacific, Farallon, Vancouver and Kula plates and constraints on the rigidity of the Pacific and Farallon (and Vancouver) plates between 72 and 34 Ma, *J. Geophys. Res., 93,* 2997-3008, 1988.

Royer, J.-Y., J.W. Peirce and J.K. Weissel, Tectonic constraints on hotspot formation of the Ninetyeast Ridge, in Proceedings of the Ocean Drilling Program, Scientific Results, 121-122, 1991.

Sandwell, D.T. and W.H.F. Smith, Marine gravity anomaly from Geosat and ERS-1 satellite altimetry, *J. Geophys. Res., 102,* 10.039-10,054, 1997.

Sartoni, R., L. Torelli, N. Zitellini, D. Peis and E. Lodolo, Eastern segment of the Azore-Gibralter line (central-eastern Atlantic): an oceanic plate boundary with diffuse compressional deformation, *Geology, 22,* 555-558, 1994.

Shaw, P.R. and S.C. Cande, High-resolution inversion for South Atlantic plate kinematics using joint altimeter and magnetic anomaly data, *J. Geophys. Res., 95,* 2625-2644, 1990.

Simkin, T., L. Siebert and E. Venzke, Smithsonian Global Volcanism Program, http://www.volcano.si.edu/gvp/volcano/regions.htm, 1999.

Srivastava, S.P., and W.R. Roest, Seafloor spreading history II-IV, in *East Coast Basin Atlas Series: Labrador Sea,* edited by J.S. Bell, Map sheets L17-2 - L17-6, Atl. Geosc. Cent., Geol. Surv. of Can., Dartmouth, N.S., 1989.

Stock, J. and P. Molnar, Revised history of early Tertiary plate motion in the south-west Pacific, *Nature, 325,* 495-499, 1987

Stock, J., and P. Molnar, Uncertainties and implications of the Late Cretaceous and Tertiary position of North America relative to the Farallon, Kula and Pacific plates, *Tectonics, 7,* 1339-1384, 1988.

Storey, M.T., J.J. Mahoney, A.D. Saunders, R.A. Duncan, S.P. Kelley and M.F. Coffin, Timing of hotspot-related volcanism and the breakup of Madagascar and India, *Science, 267,* 852-855, 1997.

Tarduno, J.A. and J. Gee, Large-scale motion between Pacific and Atlantic hotspots, *Nature, 378,* 477-480, 1995.

Tarduno, J.A. and R.D. Cottrell, Paleomagnetic evidence for motion of the Hawaiian hotspot during formation of the Emperor seamounts, *Earth Planet. Sci. Lett.,* 171-180, 1997.

Watts, A.B., J.K. Weissel, R.A. Duncan, and R.L. Larson, Origin of the Louisville Ridge and its relationship to the Eltanin fracture zone system, *J. Geophys. Res., 93,* 3051-3077, 1988.

Wessel, P., Sizes and ages of seamounts using remote sensing: implications for intraplate volcanism, *Science, 277,* 802-805, 1997.

White, R. and D. McKenzie, Magmatism at rift zones: the generation of volcanic continental margins and flood basalts, *J. Geophys. Res., 94,* 7,685-7,729, 1989.

Wilson, J.T., A possible origin of the Hawaiian Islands, *Can. J. Physics, 41,* 863-870, 1963.

Wilson, J.T., Convection currents and continental drift, *Phil. Trans. R. Soc. London, 258A,* 1965.

Ian O. Norton, Exxon Exploration Company, P.O. Box 4778, Houston, Texas 77210-4778. (ian.o.norton@exxon.sprint.com).

# Fast Paleogene Motion of the Pacific Hotspots From Revised Global Plate Circuit Constraints

Carol A. Raymond

*Jet Propulsion Laboratory, California Institute of Technology, Pasadena, CA*

Joann M. Stock

*Seismological Laboratory, California Institute of Technology, Pasadena, CA*

Steven C. Cande

*Scripps Institution of Oceanography, University of California, San Diego, CA*

Major improvements in Late Cretaceous-early Tertiary Pacific-Antarctica plate reconstructions, and new East-West Antarctica rotations, allow a more definitive test of the relative motion between hotspots using global plate circuit reconstructions with quantitative uncertainties. The hotspot reconstructions, using an updated Pacific-hotspot kinematic model, display significant misfits of observed and reconstructed hotspot tracks in the Pacific and Indian Oceans. The misfits imply motions of 5-80 mm/yr throughout the Cenozoic between the African-Indian hotspot group and the Hawaiian hotspot. Previously recognized misfits between reconstructed Pacific plate paleomagnetic poles and those of other plates might be accounted for within the age uncertainty of the paleomagnetic poles, and non-dipole field contributions. We conclude that the derived motion of the Hawaiian hotspot relative to the Indo-Atlantic hotspots between 61 Ma and present is a robust result. Thus, the Pacific hotspot reference frame cannot be considered as fixed relative to the deep mantle. The bend in the Hawaiian-Emperor Seamount chain at 43 Ma resulted from a speedup in the absolute motion of the Pacific plate in a westward direction during a period of southward migration of the hotspot. The relationship between the hotspot motion and plate motion at Hawaii suggests two possible scenarios: an entrainment of the volcanic sources in the asthenosphere beneath the rapidly moving plate while the hotspot source drifted in a plate-driven counterflow deeper within the mantle, or drift of the hotspot source which was independent of the plate motion, but responded to common forces, producing synchronous changes in hotspot and plate motion during the early Tertiary.

## INTRODUCTION

For more than two decades, research has been carried out to determine the absolute motions of the Earth's tectonic plates. This work is motivated by the desire to investigate the fundamental driving forces of the plates, and motion of

The History and Dynamics of Global Plate Motions
Geophysical Monograph 121

the deep mantle. Three main reference frames have emerged for defining past and present absolute plate motions: the hotspot, paleomagnetic, and mean-lithosphere frames (c.f. Jurdy, 1990). In the hotspot reference frame, first introduced by Morgan (1971), the motions of individual plates containing well-dated, progressive linear volcanic island chains, attributed to a set of fixed sources impinging on the bottom of a moving plate, are determined from the apparent motion of the plate over the sources (mantle plumes, or hotspots) (Minster and Jordan, 1978; Morgan, 1981; Jurdy and Gordon, 1984; Gordon and Jurdy, 1986). The motions of plates with strong expressions of time-progressive volcanism ("hotspot tracks") are used to determine the motions of the other plates via plate circuits. In the work reported here, the plates for which motion relative to hotspots is well-determined are the Pacific plate (Lonsdale, 1988), the African plate and the Indian plate (Müller et al., 1993). Hotspot tracks on other plates are not considered robust enough to define their motion relative to their hotspots. To construct models of absolute plate motions in a hotspot reference frame, global plate reconstructions, or plate circuits, are used to connect every mobile plate to the fixed hotspots. These plate circuits include errors due to the quality of the data used to resolve relative plate motions.

Another important reference frame for examining absolute plate motions is the no-net-rotation, or mean-lithosphere, reference frame. In this treatment, the mean angular motion of all the tectonic plates is determined, then subtracted, from each individual plate's motion to arrive at the motions relative to this framework (Jurdy and Van der Voo, 1974; Jurdy, 1990). This reference frame can be defined based solely on the plate geometry and relative motions. A related reference frame, defined by Solomon and Sleep (1974), is the no-net-torque reference frame, in which the torques exerted by subducting slabs and ridges are either held in balance (uniform drag beneath the plates), or are balanced by non-uniform drag beneath the plates. Solomon and Sleep (1974) derive a set of plate motion models under different assumptions of the drag force, and no motion relative to the mantle, that do not differ greatly from absolute motion models derived under the fixed hotspot assumption. Some of these models result in a net rotation of the lithosphere.

The paleomagnetic reference frame is defined by the axial geocentric dipole (AGD) hypothesis, whereby all paleomagnetic poles are assumed to be representative of the ancient spin axis, because the Earth's field is assumed to be perfectly dipolar and the dipole axis to be collocated with the spin axis. This reference frame constrains only two of the three parameters of the plate relative to the spin axis (paleolatitude and paleoazimuth), leaving paleolongitude undetermined. This reference frame is the only means to establish the absolute plate motions in pre-Mesozoic time, and serves as an important alternative reference frame for Mesozoic and Cenozoic time. Differences in motions of the plates within the paleomagnetic and hotspot reference frames have been cited as evidence of true polar wander, which refers to a motion of the entire solid earth relative to the spin axis. Therefore, the data sets which may be used to examine the kinematics, and thus dynamics, of plate motions through time are the relative motions of the plates, their motions relative to the hotspots, and paleomagnetic data which indicate the ancient position of the geomagnetic pole (paleopole) for a specific locality and time.

Significant revisions in relative plate motion models in the Southern Ocean, enabled by new and better fracture zone trends revealed by Geosat Geodetic Mission (GM) altimetry data (Sandwell and Smith, 1992), combined with significant new magnetic anomaly data sets, motivate a reanalysis of the global plate circuit connecting the Pacific plate to the rest of the world. Cande et al (1995) used dense magnetic anomaly data collected in a flowline corridor straddling the Pitman Fracture Zone on the southern Pacific-Antarctic Ridge, combined with the first new magnetic anomaly data in decades on the Marie Byrd land margin of Antarctica, to derive new Pacific-Antarctic finite rotations with smaller uncertainties. These new reconstructions differ significantly from earlier reanalyses of Stock and Molnar (1987), and Mayes et al. (1992), and define the plate kinematic evolution of the South Pacific basin with significantly higher resolution and confidence, although the pole path bears remarkable similarity to the original analysis of Molnar et al (1975).

These data have been combined with updated reconstructions of the Indian Ocean (Royer et al., 1988; Royer and Sandwell, 1989; Royer and Chang, 1991, Royer et al., 1997) to construct a new set of global plate circuits in order to test the hypothesis of fixed hotspots. The longevity and distinct geochemical character of hotspot magmas suggest that they tap regions of long isolated mantle (e.g. Anderson, 1995). Thus, the issue of the fixity of hotspots has important implications for the source region of the hot upwellings and the chemical heterogeneity of the mantle, as well as in defining a reference frame for evaluating plate driving forces.

It has long been recognized that the absolute motion of the Pacific plate derived under the assumption of fixed hotspots is inconsistent with independent evidence of the paleolatitudinal position of the Pacific plate through the Tertiary (Gordon and Cox, 1980a; Jurdy and Gordon, 1984; Acton and Gordon, 1994). Paleolatitude evidence from the distribution of equatorial sediment facies (van Andel, 1974; Suarez and Molnar, 1980; Gordon and Cape, 1981); paleomagnetic inclinations of seafloor rocks and sediments (Gordon, 1982); and paleomagnetic poles derived from inversions of seamount magnetic anomalies (Harrison et al., 1975), shape analysis of seafloor spreading anomalies (Schouten and Cande, 1976; Cande, 1976; Acton and Gordon, 1991; Petronotis et al., 1992; 1994), analyses of drilled basalts (Kono, 1980), and all of the above combined with sediment facies (e.g. Gordon and Cox, 1980b; Gordon, 1982), consistently indicate that the northward motion of the Pacific plate relative to the Earth's rotational (spin) axis was slower than the northward motion of the plate relative to the Hawaiian hotspot.

High-quality, independent data sets (sediment facies, seamount paleopoles, inclinations of drilled basalts and sediments), documenting the paleolatitudinal motion of the other non-Pacific plates, are not as complete. Nonetheless, Morgan (1981) proposed that motion occurred between the global hotspot reference frame and the spin axis since 60 Ma, based on the misfit of paleomagnetic poles from globally distributed sites to the present north pole when reconstructed using absolute motions defined in the hotspot reference frame. The general lack of resolution of paleomagnetic poles, together with the possibility that they are corrupted by several types of systematic errors (remagnetizations, overprints and non-dipole components, e.g. Epp et al., 1983) leaves some ambiguity in the interpretation of the difference in motion of the Pacific plate relative to the hotspot and spin axis reference frames. We test the view that the hotspot tracks do not define a fixed reference frame, and interpret the results of our global plate circuit tests in terms of the rate of relative motion between the sources of the hotspot tracks, and the implications for paleomagnetic reconstructions.

Previous studies have come to conflicting conclusions concerning the fixity of hotspots. Gordon and Cox (1980a), Gordon and Jurdy (1986) and Acton and Gordon (1994) concluded from global plate circuit tests that the misfit of hotspot reconstructions, and the discordance between Pacific plate paleomagnetic poles and coeval poles reconstructed from the other continents, were consistent with the joint hypotheses of fixed hotspots and an unaccounted-for plate boundary in the South Pacific. The existence of a "missing" or unrecognized plate boundary between the North Pacific and East Antarctica was demonstrated by the extremely poor fit of the reconstructed hotspot tracks, indicating unrealistically large hotspot drift rates, as well as the non-closure of the local Pacific-Antarctic-Australia plate circuit in the early Tertiary, indicating motion between East and West Antarctica. These studies all favored a boundary between the North and South Pacific, active from early to mid Tertiary time, because of geometric constraints that limit the amount of motion that could be accommodated at an East-West Antarctic boundary.

Stock and Molnar (1987) examined the Pacific-Antarctic link of the global plate circuit and proposed that the Pacific-Antarctic finite reconstructions derived by Molnar et al. (1975) for Late Cretaceous and early Tertiary time actually described the motion between Antarctica and an oceanic plate separate from the Antarctic plate, which they named the Bellingshausen plate. They calculated Pacific-Antarctic rotations using only the early Tertiary Pacific plate seafloor spreading data, under the assumption of symmetric spreading. These rotations improved the agreement between the African hotspot frame and the Hawaiian track, particularly in reproducing the bend near 43 Ma. Molnar and Stock (1987) used this revised plate motion circuit, including quantitative uncertainties in the plate rotations, to argue for relative motions on the order of 10-20 mm/yr between the Hawaiian hotspot source and the sources of major hotspot tracks in all the other ocean basins. Jurdy (1990) offered more evidence for these relative motions based on the divergence, in the mean-lithosphere reference frame, of the Pacific hotspots from the paleomagnetic (spin axis) and African hotspot reference frames through Cenozoic time.

## GLOBAL PLATE CIRCUIT TESTS OF HOTSPOT FIXITY

The Pacific-Antarctic reconstructions of Cande et al. (1995) allow a far more rigorous test of the errors in the plate circuit and the fixity of hotspots. Global plate circuits exploit the concept of plate rigidity to predict the relative motion of a pair of plates by adding the relative motions of a series of intervening plates, which define the circuit. They are useful for estimating the history of relative motion between pairs of plates that share a convergent or transcurrent plate boundary, rather than a divergent spreading axis. Uncertainties in plate rotation parameters can be propagated through the circuit to accurately estimate the uncertainty in the derived rotation parameters; however, ignorance of a boundary in the circuit will invalidate the result. The new Pac-Ant reconstructions are inserted into plate circuits connecting the Pacific hotspots to the African and Indian hotspots, and vice-versa. The first link in the circuits is to reconstruct the starting plate (African, Pacific) to its past position relative to assumed fixed hotspot sources beneath it. We adopted the model of Müller et al. (1993) of the absolute motion of the Indian and African plates relative to their hotspot sources; this model combines hotspot tracks on the Indian, Antarctic and African plates with relative plate motion data to derive a consistent absolute plate motion model under the assumption that the hotspots beneath these plates are stationary relative to each other. The predictions of their model are consistent with the observed hotspot tracks in the Indian and Atlantic Oceans, indicating that the hotspot tracks define a single, consistent reference frame in this sector. It has been recognized for some time that the Kerguelen hotspot track is inconsistent with the joint hypotheses of fixed Indo-Atlantic hotspots and a hotspot location at Kerguelen Island. The solution has been to assume the Kerguelen source resides at the western edge of the northern Kerguelen plateau, beneath Skiff Bank (Duncan and Richards, 1991). However, a recent seismic survey shows a thick sediment cover in the presumed hotspot location (D. Müller, pers. comm.). It is likely that the source of the Kerguelen track lies near Kerguelen Island, indicating that the Kerguelen hotspot source is not fixed relative to the other Indo-Atlantic hotspot sources. Our reconstructions use the far west location for the Kerguelen source, which is most consistent with the fixed-hotspot assumption.

*A Revised Pacific-Hotspot Motion Model*

We use the Pacific-hotspot model of Lonsdale (1988) as a starting point for our reconstructions. Lonsdale's model is strongly constrained by data from the Louisville Ridge in the South Pacific, as well as data from the Hawaiian track. We made several modifications to Lonsdale's model to improve the fit to the better-constrained Hawaiian track without violating the fit to the Louisville track. First, we added a change in motion at chron 3 (5 Ma), corresponding to the change in Pacific plate absolute motion proposed by Engebretson et al. (1985). This is a simplification of the change in motion on the Pacific-Antarctic Ridge at chron 3A (5.9 Ma) observed by Cande et al. (1995). We then adjusted the pole and the rate of motion between 25 Ma and 43 Ma, and increased the rate between 43 Ma and 70 Ma, to achieve an excellent fit to both the position and ages of the Hawaiian volcanoes (Plate 1a). The resulting fit to the Louisville seamount chain (Plate 1b) is also good; however, a new zero-age position of the Louisville hotspot is required to register the predicted and observed tracks. The new Pacific-hotspot rotations are given in Table 1. Our inferred zero-age position for the Louisville hotspot source is at 51.5°S, 141.0°W. There is no evidence (i.e., recent volcanism or seismicity) to support the inferred position of the Louisville source; however, it lies halfway between Lonsdale's position and the position inferred by Wessel and Kroenke (1997) using the "hot-spotting" technique (53.5°S, 141.12°W).

We tested this new Pacific-hotspot model against the Duncan and Richards (1991) model and found little difference in the reconstructions; the conclusions we draw would be valid with either model.

*Revised Plate Circuits*

We used plate circuits to transfer the reference frame of the African/Indian hotspots to the Pacific basin, and calculated predicted tracks of the Hawaiian and Louisville hotspots. Similarly, we transferred the new Pacific-hotspot reference frame to the Indian Ocean and calculated predicted tracks of the Reunion and Kerguelen hotspots. The plate circuits used are listed in Table 2. A 100 km circular uncertainty in the position of the hotspot source was assumed to derive the uncertainties for the reference frame rotations. Traversing the circuit in both directions, using independently derived hotspot reference frame rotations for different plates, allows us to evaluate any systematic error which may be associated with these hotspot reference frame models. The detailed age progression and tight geometrical constraints of the Hawaiian track allow a more robust evaluation of the fit of the African/Indian hotspot model to the Pacific basin data. For the same reasons, the Pacific motion relative to the hotspots is arguably better determined than the model of the Indian plate motion relative to its hotspots.

The circuits listed in Table 2 comprise the revised Pacific-Antarctic rotations (Cande et al., 1995) and revised reconstructions of the Indian Ocean that incorporate Geosat constraints (Royer and Chang, 1991; Royer and Sandwell, 1989), as well as copious recent aeromagnetic data at chron 5 (Royer et al., 1997). Quantitative uncertainties estimated during the rotation fitting procedure are available for all the Pac-Ant reconstructions, and for Indian Ocean reconstructions at chrons 5, 6 and 13. Antarctica-India reconstructions for chrons older than 13 (Royer and Sandwell, 1989) did not include uncertainties; for these we estimated partial uncertainty rotations based on paleo-plate boundary geometry and conservative data uncertainties. The reconstructed tracks are referenced to the traditional location of the Hawaiian hotspot (19.6°N, 204.5°E), our inferred position of the Louisville hotspot (51.5°S, 141.0°W), an accepted position of the Reunion hotspot (21.2°S, 55.6°E) and the position of the Kerguelen hotspot used by Duncan and Richards (1991) and Müller et al. (1993), which lies at the western edge of the Kerguelen plateau near Skiff Bank (49°S, 63°E).

The new marine geophysical data from the South Pacific have documented a fragmented ridge system in the Pacific-Antarctic basin prior to chron 27 (~61 Ma). There appear to have been two or more ridges separating plates on the Antarctic side of the basin from the Pacific plate (Cande et al., 1995), as was suggested by Stock and Molnar (1987) when they proposed the Bellingshausen plate hypothesis. A reconstruction of the basin at chron 27 is shown in Figure 1, which was published by Heinemann et al. (1999). This reconstruction shows the geometry of the three plates (Marie Byrd Land, Bellingshausen, and Aluk) on the Antarctic margin. The boundary between the Bellingshausen plate and the Antarctic plate is known to be divergent or transtensional between the ridge and the margin, but its character upon reaching the continental margin at ~245°E is unknown, and here presumed to be a subduction zone between 245°E and 270°E. The eastern boundary of the Bellingshausen plate may have been a convergent or transcurrent boundary with the Aluk plate, based on deformation of the seafloor near Peter I Island seen in satellite and shipboard gravity, and multichannel seismic data (Gohl et al., 1997).

Previous Pacific-Antarctic reconstructions have relied heavily on data from the more well studied northern end of the basin, which we refer to as Pacific-Bellingshausen (Pac-Bell), after Stock and Molnar (1987). We conclude that the previously poorly studied southern part of the basin

**Table 1.** Finite Rotations for Pacific-Hotspot motion relative to the Pacific plate. Positive rotations are counterclockwise. Ages from Cande and Kent (1995).

| Chron | Age (Ma) | Latitude | Longitude | Angle |
|---|---|---|---|---|
| c3 | 5.0 | -56.50 | 104.90 | 4.70 |
| c5 | 10.83 | -65.98 | 108.75 | 9.24 |
| c6 | 21.16 | -70.45 | 111.79 | 17.43 |
| --- | 25.0 | -71.20 | 112.44 | 20.49 |
| c13 | 33.3 | -69.14 | 117.41 | 25.44 |
| c18 | 40.22 | -67.88 | 119.88 | 29.60 |
| --- | 43.0 | -67.46 | 120.64 | 31.28 |
| c20 | 43.87 | -66.87 | 119.71 | 31.43 |
| c21 | 47.86 | -64.25 | 116.15 | 32.60 |
| c24 | 53.25 | -60.86 | 112.54 | 34.14 |
| c28 | 63.3 | -55.19 | 108.08 | 37.47 |
| c31 | 68.68 | -52.51 | 106.42 | 39.39 |
| c33 | 78.78 | -48.08 | 104.10 | 43.22 |

(spreading between Campbell Plateau and Marie Byrd Land) is a more likely candidate for Pacific-Antarctic spreading; thus our reconstructions differ significantly from previous ones. The Indian Ocean reconstructions also must take into account changing plate boundary geometry. For the reconstructions older than chron 13 we reconstructed Antarctica directly to India; however, for younger times, the deformation of the greater Indian plate necessitates linking the circuit from Antarctica to Africa to India. Deformation of the greater Indian (Indo-Australian) plate since chron 5 has been documented (Royer et al., 1997), and is included in our circuit calculations.

### *East-West Antarctic Motion*

Many previous investigators have concluded that the lack of closure of early Tertiary global plate circuits, and the misfit of reconstructed hotspot tracks derived from them, could be explained by an unaccounted-for plate boundary between East and West Antarctica. Cenozoic extension along the Transantarctic Mountain front is evidence for such a boundary (Cooper et al., 1991; Wilson, 1995); however, strong constraints on the amount and timing of the extension are lacking. Gordon and Cox (1980a) tested the amount of motion that would be required to close the circuit if all motion occurred on an East-West Antarctic boundary. The resulting overlap of West Antarctica with South America in their model caused them to reject the possibility that East-West Antarctic motion could explain the misfit on the hotspot reconstructions.

Rotation parameters describing motion between East and West Antarctica have been derived from new magnetic anomaly and fracture zone data in the conjugate South Tasman Sea and in the Adare Trough in the northern Ross Sea, Antarctica (Cande, S. C., J. M. Stock, D. Müller and T. Ishihara, Cenozoic motion between East and West Antarctica, submitted to Nature). These new data allow a three-plate solution (with uncertainties) for Australia-East Antarctica-West Antarctica motions at chron 13o; the magnetic anomalies record motion between chron 8o and chron 20o (26 Ma to 44 Ma). We have performed global plate circuit reconstructions with and without inclusion of the motion between East and West Antarctica to illustrate how this previously poorly constrained leg of the circuit contributes to the agreement between the observed and reconstructed tracks.

The new marine geophysical data from the Adare Trough indicate significant motion began between East and West Antarctica at around 44 Ma. It is possible that, prior to that time, motion between East and West Antarctica was taken up at another boundary, possibly along the Transantarctic Mountains, which might affect the plate circuit through Antarctica. The main phase of extension within the Ross Embayment is thought to have occurred between 105-85 Ma based on geological evidence from the Transantarctic Mountains (Wilson, 1995) and Ross Sea (Cooper et al., 1991). Lawver and Gahagan (1994) reject the possibility of a significant amount of extension within the Ross Embayment during the early Tertiary, based on the fit between Marie Byrd Land and the Campbell Plateau offshore New Zealand at 86 Ma. Therefore, any significant amount of early Tertiary extension would have occurred west of Marie Byrd Land and the eastern Ross Sea, within the western Ross Sea. Cooper et al. (1991) propose a hiatus in deformation within the western Ross Sea between the formation of grabens in the Cretaceous and Eocene uplift of the Transantarctic Mountains, however, the ages of major unconformities within the Ross Sea are poorly resolved. The only possible evidence of a boundary in the vicinity of the Ross Sea, prior to spreading at the Adare Trough, is a maximum of 50 km of extension within the Iselin Trough, orthogonal to the Adare Trough, just east of the Iselin Bank. Such extension would have likely occurred between approximately 62 Ma and 53 Ma. We conclude that we have captured the bulk of the motion between East and West Antarctica during the Tertiary by closing the Adare Trough. Only a small amount of further extension might have occurred, and this would have a negligible impact on the conclusions drawn from the reconstructions.

**Table 2.** Plate Circuit Parameters

| | Reference | Chrons |
|---|---|---|
| hs-pac | This paper | 5, 6, 13, 18, 21, 24, 25, 27, 28, 31 |
| pac-ant | Cande et al., 1995 | 5, 6, 13, 18, 21, 24, 25, 27, 28, 31, |
| want-eant | Cande et al., submitted | 13, 20 |
| ant-afr-ind | Royer & Chang, 1991; Royer et al., 1997 | 5, 6, 13 |
| ant-aus(s. ind)-ind | Royer & Sandwell, 1989; Royer et al., 1988; Royer et al., 1997 | 18, 21, 24, 28, 31 |

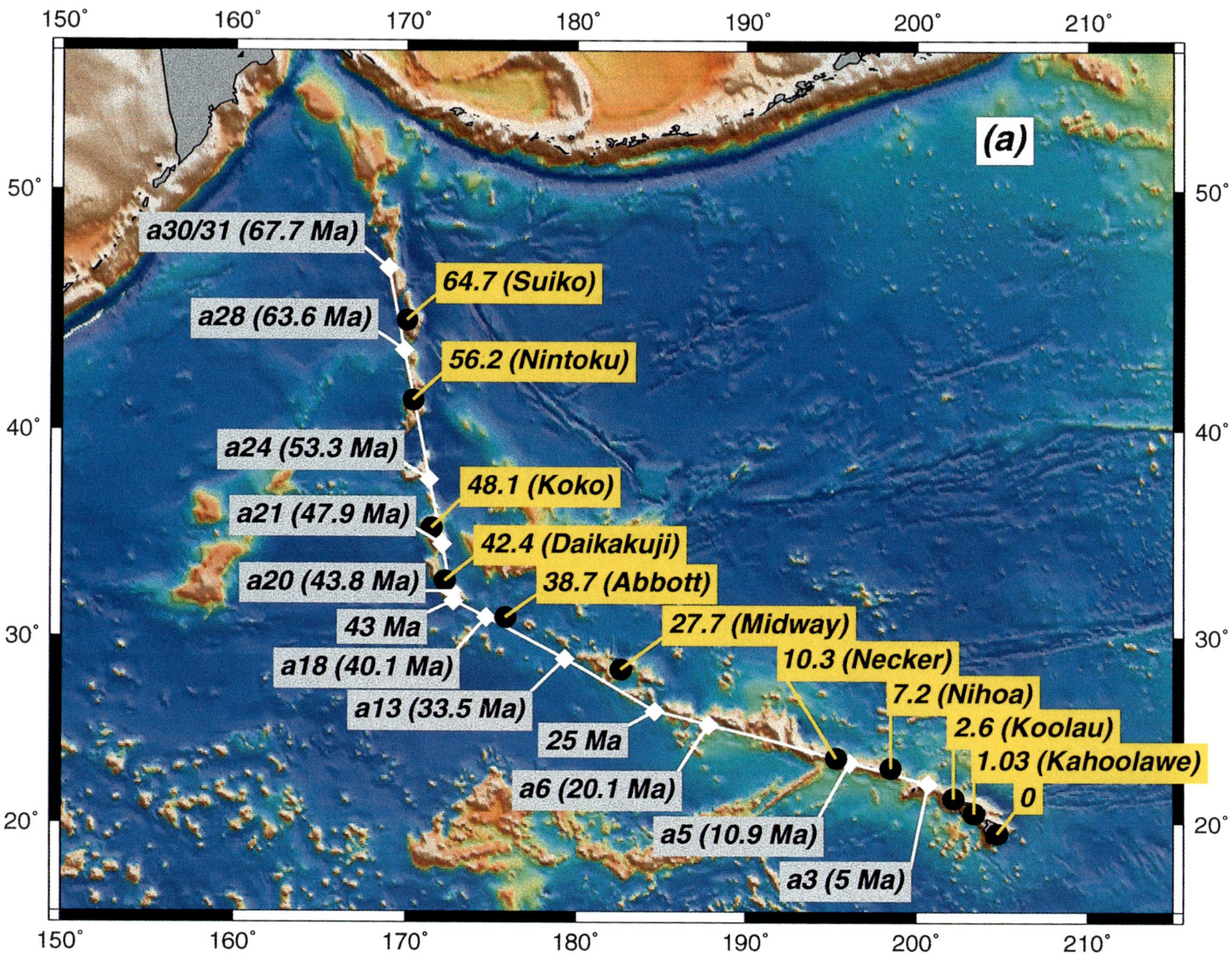


**Plate 1**. Revised Pacific-hotspot motion model shown as white line compared with the (a) Hawaiian-Emperor chain in the North Pacific, and (b) the Louisville seamount chain in the South Pacific. The annotated white diamonds give chrons and ages (from Cande and Kent, 1995) for the reconstruction. Annotated black dots give locations and ages of well-dated seamounts and islands along the chains (Clague and Dalrymple, 1989; Clague and Jarrard, 1973; Lonsdale, 1988). Image shows bathymetry from the ETOPO5 data set (NGDC, 1988).

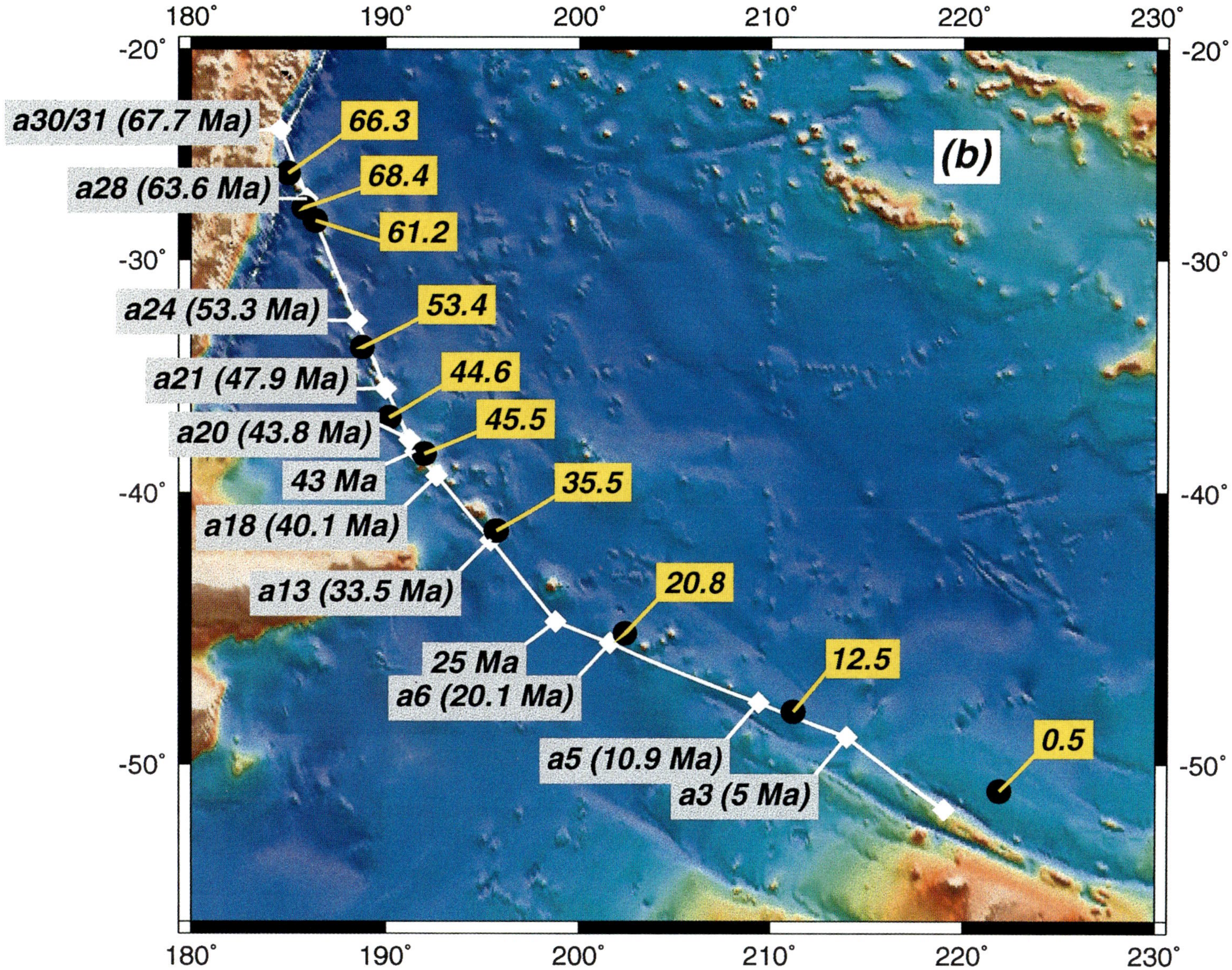


**Plate 1.** Continued.

## RESULTS

Tracks of the Pacific hotspots reconstructed in the Indo-Atlantic reference frame to the Hawaiian-Emperor and Louisville chains, and the Indo-Atlantic hotspots reconstructed in the Pacific reference frame to the Chagos-Laccadive/Reunion and Kerguelen/Ninety East Ridge tracks are shown in Plate 2. Both sets of reconstructions clearly demonstrate a misfit that systematically increases with age between the reconstructed and observed tracks, regardless of the starting hotspot model. The most striking element of this misfit is the failure of the reconstructed track to reproduce the sharp bend in the Hawaiian-Emperor chain. Inclusion of East-West Antarctic motion improves the fit only slightly.

A subtler, yet persistent age misfit between the observed and reconstructed hotspot tracks is most clearly expressed for the well-dated Hawaiian chain. All the reconstructed points in the Pacific fall on seamounts that are younger than predicted, and the magnitude of the age discrepancy increases with the age. The opposite effect is observed for the reconstructed Indian points, although the misfit is not as clear because of the larger uncertainties.

### *Motion of the Hawaiian Hotspot in the Indo-Atlantic Reference Frame*

The motion of the Hawaiian hotspot source relative to the Indo-Atlantic hotspot reference frame for intervals bounded by our well-resolved reconstructions is shown in Figure 2 (a, b). The drift rate was calculated as latitudinal and longitudinal displacement over time since 68 Ma to facilitate comparison to paleomagnetically-determined paleolatitude. The results show rapid westward drift of the hotspot source (~73-80 mm/yr) during the Paleocene (69-56 Ma), with a smaller southward component (~33-44 mm/yr). Our results for the oldest interval, 68-61 Ma, however, would be invalid if Campbell Plateau was moving with respect to the northern Pacific plate prior to the reorganization at 61 Ma. The drift changes to a southward motion at ~44 mm/yr during the interval from 56 to 43 m.y, slowing to ~33 mm/yr and becoming directed southwest between 43 and 33 Ma. A slow, dominantly eastward and southward drift persists to the present at rates of 5-20 mm/yr. These rates are remarkable in that the oldest rates (68-33 Ma) are many times faster than previous plate circuit estimates.

The recent drift rates (40 Ma to present) generally agree with the predicted southeast drift for the Hawaiian and Louisville sources (12 mm/yr and 9 mm/yr, respectively) derived by Steinberger and O'Connell (1998). These authors perform a numerical simulation of a 3-D flow field driven by plate motion to solve for hotspot migration in a no-net-rotation reference frame. These authors also predict a similar (12 mm/yr) southeastward drift of the Kerguelen hotspot source since 40 Ma, but verv little motion of the Reunion hotspot (2 mm/yr).

### *Comparison to Motion of the Hawaiian Hotspot in the Spin Axis Reference Frame*

The derived rates support the fast southward drift deduced by Tarduno and Cottrell (1997) using paleomagnetic data on Late Cretaceous seamounts in the Emperor chain. These authors attribute the latitudinal discrepancy between the paleolatitudes of Detroit seamount (81 Ma) and Suiko seamount (65 Ma) and the present latitude of the Hawaiian hotspot to southward motion of the Hawaiian hotspot. Although they consider true polar wander (TPW) as a possible explanation for the Suiko-Hawaii latitudinal discrepancy, they reject TPW as a possible cause of the Detroit paleolatitude discrepancy because continental paleomagnetic data predict less, not more, TPW during this interval. They suggest southward motion of the sources of the seamounts, relative to a spin axis reference frame, of 30-50 mm/yr between 81 and 43 Ma, with possible rates as high as 64 mm/yr $\pm$ 43 mm/yr. These rates agree well with our estimates of southward motion of the hotspot source, relative to the Indo-Atlantic hotspot reference frame, of 44 mm/yr, 33 mm/yr, and 42 mm/yr for the intervals 68-61 Ma, 61-56 Ma, and 56-43 Ma, respectively.

An independent estimate of southward drift of the Hawaiian hotspot source comes from a paleomagnetic pole determined by a comprehensive analysis of the skewness of anomaly 25r (Petronotis et al., 1994). By comparing their skewness pole to other paleomagnetic and sediment facies data, Petronotis et al. (1994) conclude that the Pacific plate was nearly stationary relative to the spin axis during the Eocene. They estimate a southward drift of the hotspot source of 52 $\pm$ 24 mm/yr between 57 and 39 Ma, corresponding to our southward drift estimate of 42 mm/yr between 56 and 43 Ma. An analysis of the skewness of anomaly 32 further suggests that the rapid southward hotspot drift began earlier, with drift rates of 64 mm/yr between 73 and 57 Ma (Petronotis and Gordon, 1999).

Thus, a somewhat coherent picture of the latitudinal drift of the Hawaiian hotspot source is emerging from several lines of evidence summarized above. The favored scenario is for fast southward drift of the hotspot source (roughly 40-50 mm/yr) during the Paleocene and Eocene. Only very minor southward drift occurred post-late Eocene.

Quantitative uncertainties for the rotation parameters used in the global plate circuit yield error ellipses for the position of the hotspot track predicted by the reconstructions. The step of calculating the hotspot source drift at Hawaii from measuring the displacement of the observed and reconstructed track does not easily yield quantitative errors. The errors due to the uncertainty in the reconstructed positions can be visualized from the ellipses on the reconstructed points. However, because we are comparing the

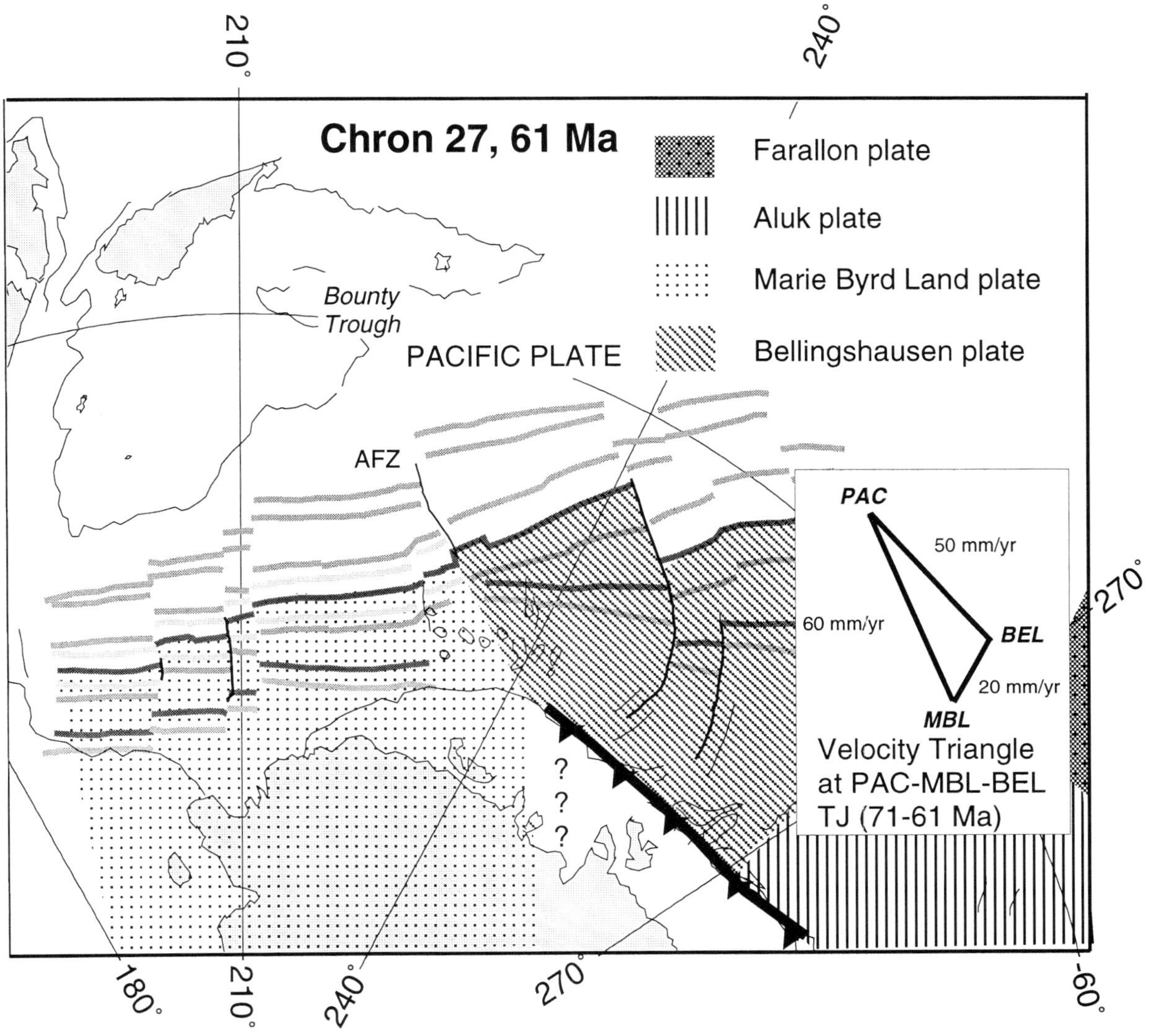


**Figure 1.** Reconstruction of the South Pacific Basin at chron 27 (61 Ma) showing the locations of the Pacific, Bellingshausen, Marie Byrd Land, Aluk and Farallon plates (from Heinemann et al., 1999). Schematic magnetic lineations are shown to illustrate the evidence for the separate Bellingshausen and Marie Byrd Land plates on the Antarctic margin. The Campbell Plateau is shown in the upper left corner, attached to the reconstructed North Island of New Zealand, and the location of the Bounty Trough within the Campbell Plateau is labeled.

ages of the anomalies with the ages of the dated volcanic edifices, the uncertainty in the age of both types of data contributes to the error in the derived drift rates. It is difficult to quantify this error, but we have observed that the drift rate results are sometimes quite sensitive to the ages assigned to the anomalies, since they are being compared to fixed ages for the volcanoes. Also, solving for the drift with a very narrow time window, such as for an individual anomaly, results in noise in the derived rates. The rather smooth nature of the variation in drift rate that we show for the selected intervals indicates this noise has been suppressed, and errors caused by the uncertainty in the magnetic chron ages have been reduced.

## DISCUSSION

Before the misfits in the hotspot reconstructions can be interpreted unequivocally as evidence for inter-hotspot motion, we must consider the possibility that a systematic error remains in the plate circuit. It seems unlikely that errors in the reconstructions of plates to their respective hotspots contribute significantly to the misfit because we see similar misfits using two completely independent starting models (one for Indo-Atlantic hotspots and one for Pacific hotspots). We have included a new East-West Antarctica rotation with quantitative uncertainties that is constrained by seafloor data, thereby eliminating that link

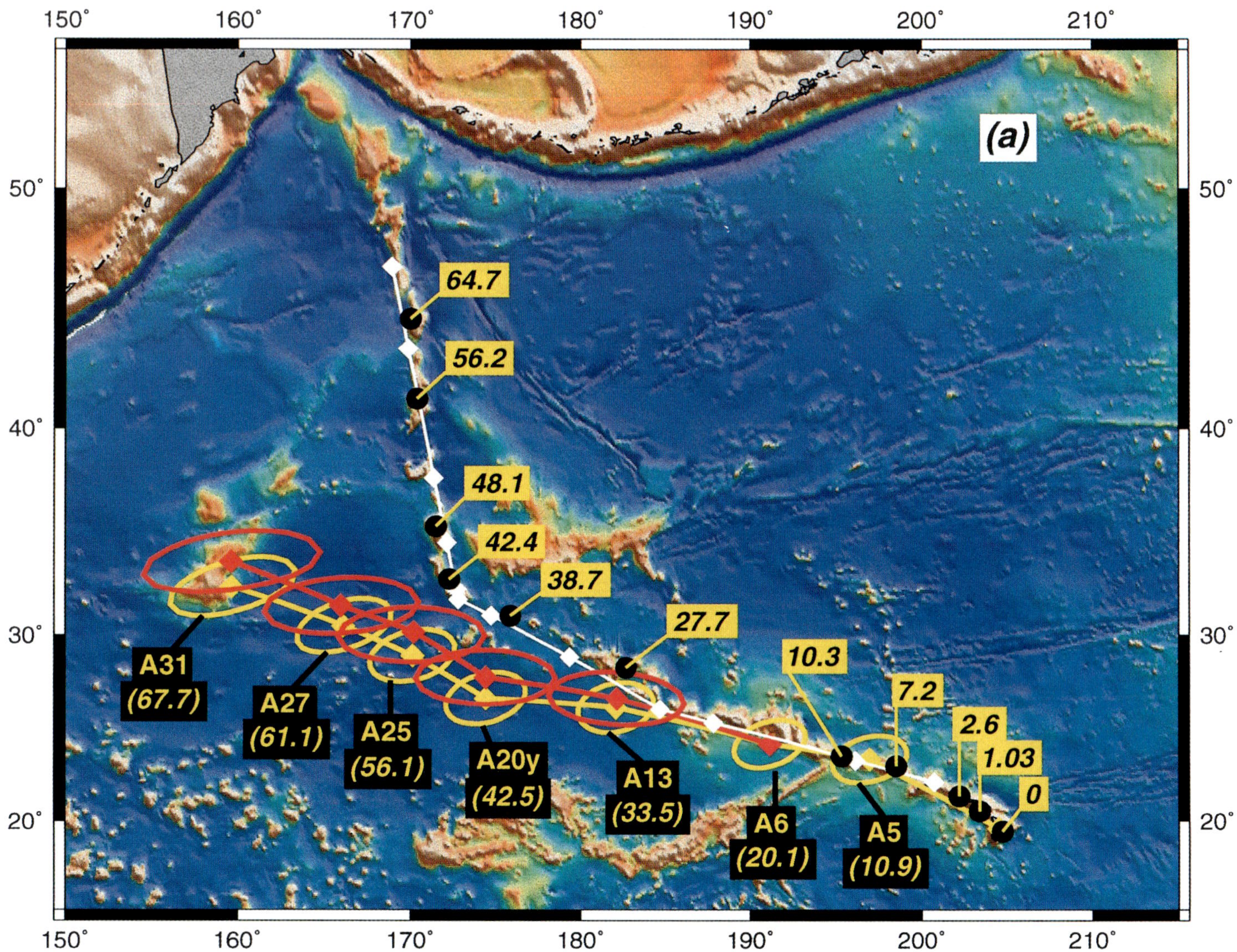


**Plate 2.** Reconstructions of Pacific-hotspot motion in the Indo-Atlantic reference frame to (a) the Hawaiian-Emperor chain and (b) the Louisville chain, and (c) reconstructions of Indo-Atlantic hotspots in the Pacific frame to Reunion and Kerguelen, superimposed on ETOPO5 bathymetry image. Reconstructed tracks shown in yellow with quantitative plate reconstruction error ellipses; chrons and ages as shown. Red track includes East-West Antarctic motion.

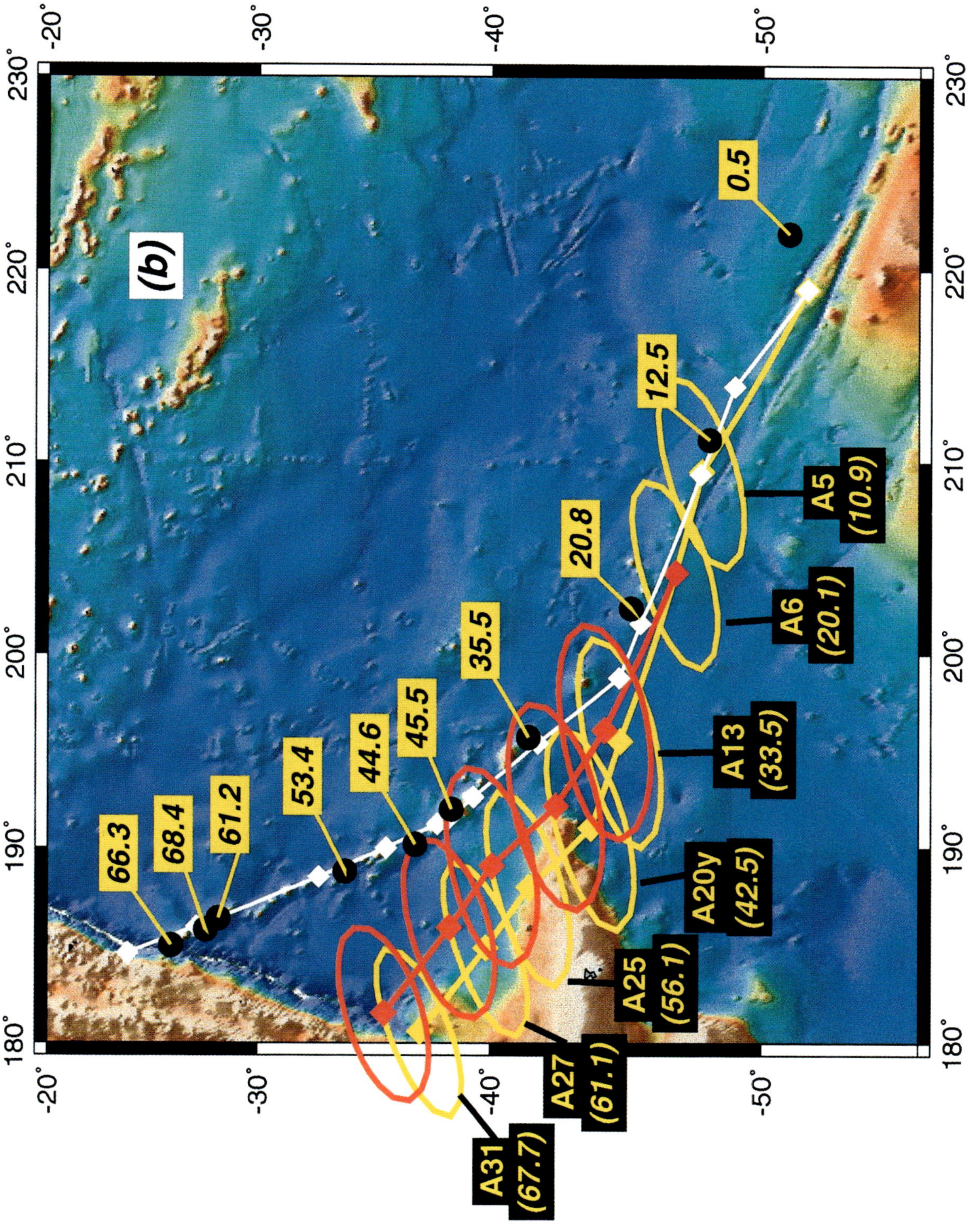


**Plate 2.** Continued.

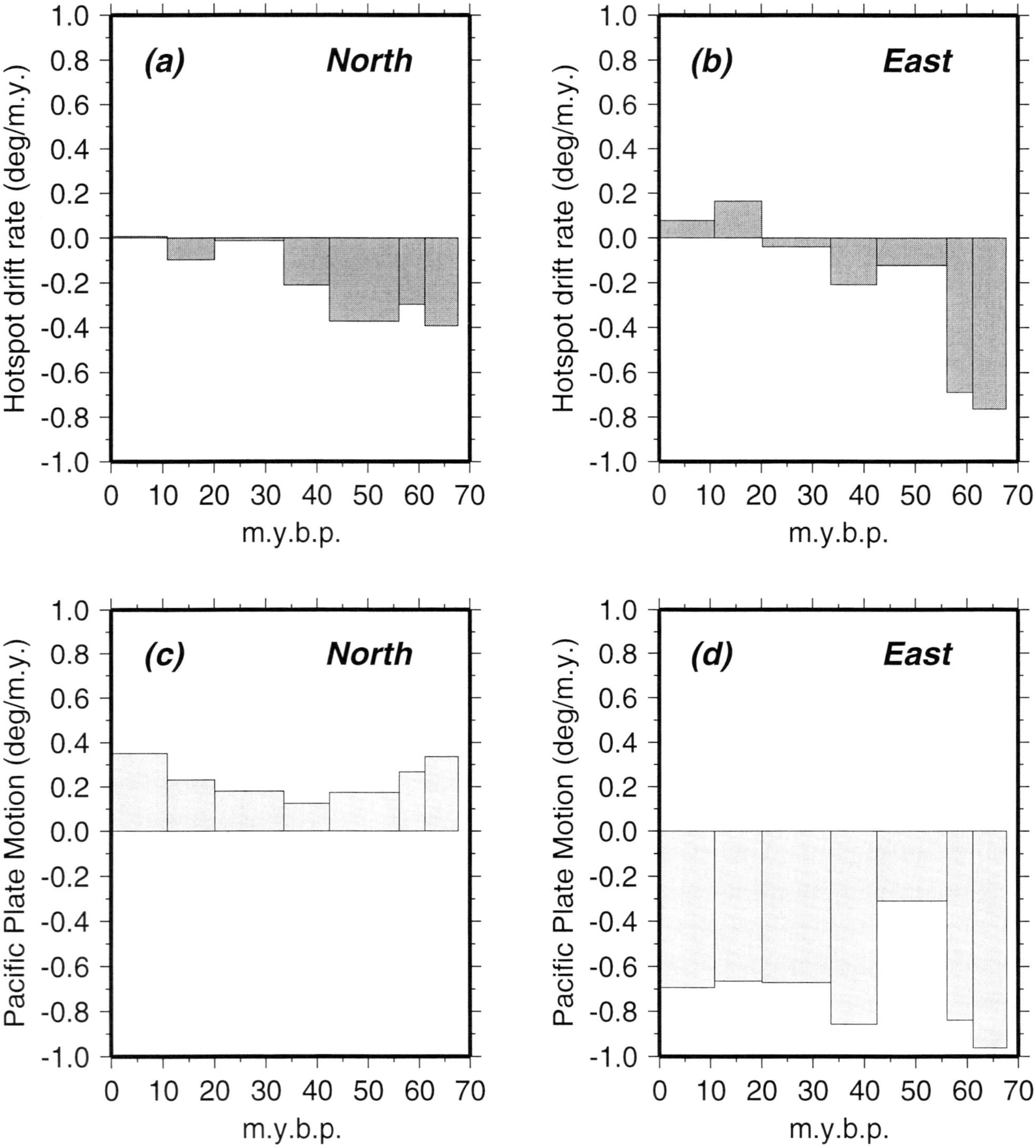


**Figure 2.** Inferred drift rate of the Hawaiian hotspot versus time in latitude (a) and longitude (b) compared to Pacific plate motion in the Indo-Atlantic fixed-hotspot reference frame, calculated at the location of the Hawaiian hotspot, in latitude (c) and longitude (d) versus time. (a) hotspot drift rate north in deg./m.y.; (b) hotspot drift rate east in deg./m.y.; (c) Pacific plate motion north in deg./m.y.; (d) Pacific plate motion east in deg./m.y.

as a major source of error. It is possible that the misfit in the reconstructions in early Tertiary time are caused by an unrecognized plate boundary within the Pacific plate, as proposed by Gordon and Cox (1980a), Gordon and Jurdy (1986) and Acton and Gordon (1994), such as between the Campbell Plateau and the north Pacific, which would then invalidate our Pac-Ant reconstructions. Detailed geophysical observations from the south Pacific are needed to address this possibility. Although an unrecognized plate boundary is possible during the early Tertiary, evidence for such a boundary of mid to late Tertiary age, where copious constraints on plate reconstructions exist, is lacking, yet we still observe significant misfits up to the present.

The main argument for an unrecognized plate boundary remaining in the circuit comes from global plate circuit reconstructions of paleomagnetic data, as were done most recently by Acton and Gordon (1994). While that study included rigorous treatment of the errors in the paleomagnetic data and the reconstructions, there are several possible additional systematic errors in the paleomagnetic data that could contribute to the discrepancies observed in their reconstructions. One possible source of error is non-dipole field components. Acton and Gordon (1994) recognize that a 4% geocentric axial quadropole component would contribute an additional $2^{\circ}$ of uncertainty to their reconstructions. The systematic shift of the high latitude poles (Greenland, Antarctica) away from the Pacific pole at 58 Ma suggests that eccentricity of the dipole field might be contributing to the misfit. Another unmodeled error is contributed by the age uncertainty in the derived poles. For instance, the reconstructed position of the 65 Ma Indian paleopole moves $2.5^{\circ}$ closer to the Pacific 65 Ma pole when reconstructed at 67 Ma (the older edge of the tightly-constrained age uncertainty on the Indian pole) because the Pacific plate is moving northward at that time. Consideration of these additional sources of error casts some doubt that the paleomagnetic reconstructions provide a compelling argument for a problem in the plate circuit.

*Implications for Mantle Dynamics*

Our new reconstructions imply that the motion of the Hawaiian hotspot source, relative to the Indo-Atlantic hotspots (Fig 2b) changed significantly at 56 Ma, when the direction of motion shifted from ~$245^{\circ}$ to $198^{\circ}$, and motion slowed from 80 to 44 mm/yr. No dramatic change in motion of the hotspot source is observed at the time of the Hawaiian-Emperor bend at 43 Ma. The motion of the Pacific plate in the Indo-Atlantic hotspot reference frame also shows no dramatic change in direction at 56 or 43 Ma. The remarkable bend in the Hawaiian-Emperor chain at 43 Ma is thus a consequence of a significant increase in the speed of westward plate motion (57 mm/yr), decoupling of the westward hotspot source motion from the plate motion, and a slight slowing of the southward hotspot drift (Fig 2), as viewed in the Indo-Atlantic fixed-hotspot reference frame.

An intriguing relationship is observed between the motion of the Hawaiian hotspot and that of the Pacific plate in the Indo-Atlantic reference frame. The hotspot track implies strong westward drift of its source, in the same direction as the motion of the plate between 68-56 Ma. The westward drift of the hotspot source slows dramatically at 56 Ma in concert with a change in plate speed, and then the source continues to drift slowly west between 56 and 33 Ma. The latitudinal drift of the hotspot is opposite to the plate motion between 68 and 33 Ma, as was previously noted by Petronotis and Jurdy (1990), and of similar magnitude. A possible explanation for these observations is that the asthenosphere was coupled to the fast-moving plate and entrained the magmatic source of the hotspot track in a linear shear flow. At the same time, the deeper sub-asthenospheric source of the hotspot track was drifting south, most likely in a plate-driven return flow (Chase, 1979; Olson, 1987; Duncan and Richards, 1991; Morgan and Smith, 1992). Previous models which predict that plumes may become advected in mantle flow and their conduits may bend considerably from vertical, support this inference (Richards and Griffiths, 1988; Duncan and Richards, 1991; Steinberger and O'Connell, 1998). The rates shown in Fig 2 (a) and (b) were derived from examining the surface expression of the hotspot, which represents the integrated motion of the source beneath the surface. Therefore we can only speculate on how the source conduit might be moving at different levels within the mantle.

Although we cannot resolve varying motion of the hotspot source with depth, we can still infer such motions based on the observations. If the plate and the hotspot source were indeed coupled during the period from ~68-43 Ma, as indicated by the strong correlation between westward plate motion and westward hotspot source motion, then the source of the Hawaiian hotspot (below the asthenosphere) may have actually drifted south faster than the rates calculated in Fig 2 (a). If the sub-asthenospheric hotspot conduit had its top sheared northward by northward plate motion, then the position of the surface expression of the hotspot would lie north of the source position at great depth. Assuming that the top of the source conduit moved north with the plate yields a (maximum) southward hotspot source drift rate of 60-80 mm/yr between 68 and 43 Ma.

The interpretation of entrainment of the hotspot source beneath the plate is problematic, however, in that the correlation of westward rates of motion decreases between 56 and 43 Ma during a period of slow plate motion, but does not increase when the plate resumes moving rapidly westward post-43 Ma. If the entrainment did occur, a lower sub-lithospheric viscosity post-56 Ma may have enabled rapid changes in plate motion (Ratcliff et al., 1998), and also decoupled the plate motion and hotspot drift post-43 Ma

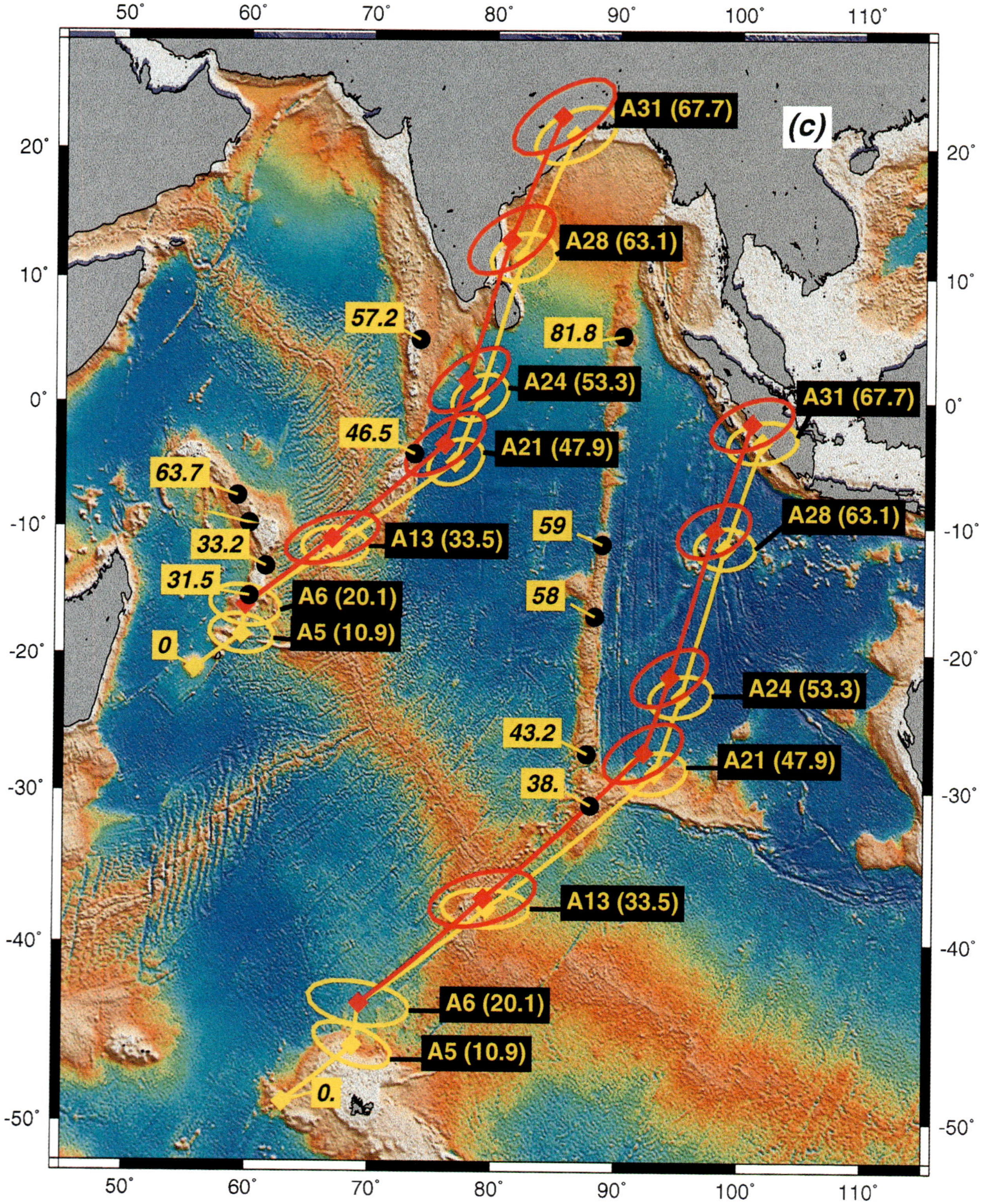


**Plate 2.** Continued.

(Steinberger and O'Connell, 1998). Post-33 Ma, the hotspot drift is weak (5-20 mm/yr), and slightly anti-correlated to plate motion, in general agreement with the prediction of Steinberger and O'Connell (1998).

To summarize, the evidence for synchronous changes in plate motion and hotspot drift between 68 and 43 Ma suggests that the motion of the plate may be controlling the motion of the hotspot source, including shearing of the source conduit beneath the plate and drift in a plate-driven return flow deeper within the mantle. Conversely, the motion of both the hotspot source and the plate may have been controlled by other events, such as changes in mantle flow patterns due to the approach and eventual destruction of ridges along the North American margin during the early Tertiary, and initiation of subduction in the western Pacific.

The fast Paleogene drift rates provide some constraints on the viscosity structure of the mantle beneath the Pacific during that time. The modeling results of Steinberger and O'Connell (1998) indicate that if the source conduit was indeed entrained in sub-lithospheric flow and/or sub-asthenospheric flow, then the viscosity should have been high enough in the upper mantle to allow advection of the density anomalies, but low enough to avoid overtilting of plume conduits. Steinberger and O'Connell (1998) suggest a viscosity higher than $3 \times 10^{19}$ Pa s and lower than $10^{21}$ Pa s to allow for the fast hotspot motion we infer within the upper mantle.

## CONCLUSIONS

Reconstructions of the Pacific hotspots in the Indo-Atlantic reference frame, and vice-versa, using updated plate rotations with quantitative uncertainties show significant and persistent misfit of the observed and reconstructed hotspot tracks. This misfit implies differential motion between the Pacific and Indo-Atlantic hotspot groups. Müller et al (1993) show no resolvable motion within the Indo-Atlantic hotspot group, except for a constrained fit to the Kerguelen track discussed previously. Our revised Pacific-hotspot model fits both Louisville and Hawaii, indicating little relative motion between Pacific hotspots, and suggesting that the misfit in the reconstructions is caused by motion of the Pacific hotspot group relative to the Indo-Atlantic group, although small relative motions within each group cannot be excluded. The plate circuit we used is still in question prior to 61 Ma because of remaining uncertainty about the plate kinematics of the southwest Pacific basin; however, the existence of a misfit throughout the Tertiary is incontrovertible.

The plate circuit test indicates drift of the Hawaiian hotspot relative to Indo-Atlantic hotspots of with rates as high as 80 mm/yr. Latitudinal drift of the Hawaiian hotspot relative to the Indo-Atlantic hotspots agrees with drift relative to the spin axis (Gordon and Cox, 1980a; Gordon and Cape, 1981; Petronotis et al., 1994; Tarduno and Cottrell, 1997). We conclude, because of possible systematic errors and scarcity of data, that paleomagnetic evidence for coherent motion between the spin axis and a fixed hotspot reference frame is less compelling than the evidence for southward drift of the Hawaiian hotspot relative to both the spin axis and the other hotspots, a conclusion that was also reached by Jurdy (1990).

The relationship between the Hawaiian hotspot track and the Pacific plate motion, referenced to the Indo-Atlantic hotspots and observed at Hawaii, suggests a coupling between plate motion and drift of the hotspot source at shallow levels between 68 and 43 Ma, with coeval fast southward drift of the source at a deeper level. These observations may be interpreted as indicating the influence of the plate motion on the motion of the hotspot source, or alternatively, as showing the influence of mantle dynamics on both the hotspot and the plate motion. Further modeling is needed to refine the estimates of hotspot drift and absolute plate motion for the Pacific plate to develop models which link the kinematics and dynamics.

*Acknowledgments.* This work benefited from discussions with many people, especially Richard Gordon, Bernhard Steinberger, Todd Ratcliff, Dietmar Mueller, Chuck DeMets and Gary Acton. Reviews by B. Steinberger and an anonymous reviewer improved the manuscript. GMT software was used to produce the figures (Wessel and Smith, 1991). This research was supported by NSF-OPP-93-17318. Part of the work was performed at the Jet Propulsion Laboratory, California Institute of Technology, under contract to the National Aeronautics and Space Administration. California Institute of Technology, Division of Geological and Planetary Sciences, Contribution number 8688.

## REFERENCES

Acton, G. D. and R. G. Gordon, A 65 Ma paleomagnetic plate for the Pacific plate from the skewness of magnetic anomalies 27r-31, *Geophys. J. Int., 106*, 407-420, 1991.

Acton, G. D. and R. G. Gordon, Paleomagnetic tests of Pacific plate reconstructions and implications for motion between hotspots, *Science, 263*, 1246-1254, 1994.

Anderson, D. L., Lithosphere, asthenosphere and perisphere, *Rev. Geophys., 33*, 125-149, 1995.

Cande, S. C., A pelaeomagnetic pole from Late Cretaceous marine magnetic anomalies in the Pacific, *Geophys. J. R. Astron. Soc., 44*, 547-566, 1976.

Cande, S. C., C. A. Raymond, J. Stock and W. F. Haxby, Geophysics of the Pitman fracture zone and Pacific-Antarctic plate motions during the Cenozoic, *Science, 270*, 947-953, 1995.

Cande, S. C. and D. V. Kent, Revised calibration of the geomagnetic polarity timescale for the Late Cretaceous and Cenozoic, *J. Geophys. Res., 100*, 6093-6095, 1995.

Chase, C. G., Asthenospheric counterflow: a kinematic model, *Geophys. J. R. Astron. Soc., 56*, 1-18, 1979.

Clague, D. A. and G. B. Dalrymple, Tectonics, geochronology and origin of the Hawaiian-Emperor chain, in *The Eastern*

*Pacific Ocean and Hawaii*, vol. N of *The Geology of North America*, edited by E. L. Winterer, D. M. Hussong, R. W. Decker, *Geological Society of America*, Boulder. CO, pp. 188-217, 1989.

Clague, D. A. and R. D. Jarrard, Tertiary pacific plate motion deduced from the Hawaiian-Emperor chain, *GSA Bull. 84,* 1135-1154, 1973.

Cooper, A. K., F. J. Davey and K. Hinz, Crustal extension and origin of sedimentary basins beneath the Ross Sea and Ross Ice Shelf, Antarctica, in *Geological Evolution of Antarctica*, edited by M. R. A. Thomson, J. A. Crame and J. W. Thomson, pp. 285-291, Cambridge University Press, New York, 1991.

Duncan, R. and M. A. Richards, Hotspots, mantle plumes, flood basalts and true polar wander, *Rev. Geophys., 29,* 31-50,1991.

Engebretson, D. C., A. Cox and R. G. Gordon, Relative motions between oceanic and continental plates in the Pacific Basin, *Geol. Soc. Am. Spec. Pap. 206, 59 pp., 1985.*

Epp, D., W. W. Sager, F. Theyer and S. R. Hammond, Hotspot-spin axis motion or magnetic far-sided effect?, *Nature, 303,* 318-320, 1983.

Gohl, K., F. Nitsche and H. Miller, Seismic and gravity data reveal Tertiary interplate subduction in the Bellingshausen Sea, southeast Pacific, *Geology, 25*, 371-374, 1997.

Gordon, R. G. and C. D. Cape, Cenozoic latitudinal shift of the Hawaiian hotspot and its implications for true polar wander, *Earth Plan. Sci. Lett., 55*, 37-47, 1981.

Gordon, R. G. and A. Cox, Paleomagnetic test of the early Tertiary plate circuit between the Pacific basin plates and the Indian plate, *J. Geophys. Res., 85,* 6534-6546, 1980a.

Gordon, R. G. and A. Cox, Calculating paleomagnetic poles for oceanic plates, *Geophys. J. R. Astron. Soc., 63*, 619-640, 1980b.

Gordon, R. G., The late Maastrictian paleomagnetic pole of the Pacific plate, *Geophys. J. R. Astron. Soc., 70*, 129-140, 1982.

Gordon, R. G. and D. Jurdy, Cenozoic global plate motions, *J. Geophys. Res., 91,* 12,389-12,406, 1986.

Harrison, C. G. A., R. D. Jarrard, V. Vacquier and R. L. Larson, Palaeomagnetism of Cretaceous Pacific seamounts, *Geophys. J. R. Astron. Soc., 42,* 859-882, 1975.

Heinemann, J., J. Stock, R. Clayton, K. Hafner, S. Cande and C. Raymond, Constraints on the proposed Marie Byrd Land-Bellingshausen plate boundary from seismic reflection data, *J. Geophys. Res., in press, 1999.*

Jurdy, D. M. and R. Van der Voo, A method for the separation of true polar wander and continental drift, including results for the last 55 m.y., *J. Geophys. Res., 79*, 2945-2952, 1974.

Jurdy, D. M. and R. G. Gordon, Global plate motions relative to the hot spots 64 to 56 Ma, *J. Geophys. Res., 89*, 9927-9936, 1984.

Jurdy, D. M., Reference frames for plate tectonics and uncertainties, *Tectonophysics, 182*, 373-382, 1990.

Kono, M., Paleomagnetism of DSDP leg 55 basalts and implications for the tectonics of the Pacific plate, *Initial Rep. Deep Sea Drill. Proj., 55*, 737-752, 1980.

Lawver, L. A. and Gahagan, L., Constraints on timing of extension in the Ross Sea region, *Terra Ant., 1*, 545-552,1994.

Lonsdale, P., Geography and history of the Louisville hotspot chain in the southwest Pacific, *J. Geophys. Res., 93,* 3078-3104, 1988.

Mayes, C. L., L. A. Lawver and D. T. Sandwell, Tectonic history and new isochron chart of the South Pacific, *J. Geophys. Res., 95,* 8543-8567, 1992.

Minster, J. B. and T. H. Jordan, Present-day plate motions, *J. Geophys. Res., 83,* 5331-5354, 1978.

Molnar, P., T. Atwater, J. Mammerickx and S. M. Smith, Magnetic anomalies, bathymetry and tectonic evolution of the South Pacific since the Late Cretaceous, *Geophys. J. R. Astron. Soc., 49,* 383-420, 1975.

Molnar, P. and J. Stock, Relative motions of hotspots in the Pacific, Atlantic, and Indian oceans since Late Cretaceous time, *Nature, 327*, 587-591, 1987.

Morgan, W. J., Convection plumes in the lower mantle, *Nature, 230*, 42-43, 1971.

Morgan, J., Hotspot tracks and the opening of the Atlantic and Indian oceans, in *Oceanic Lithosphere, (The Sea, vol. 7),* edited by C. Emiliani, pp.443-487, Wiley-Interscience, New York, 1981.

Morgan, J. P. and W. H. F. Smith, Flattening of the sea-floor depth-age curve as a response to asthenopsheric flow, *Nature, 359*, 524-527, 1992.

Müller, R. D., J-Y. Royer and L. A. Lawver, Revised plate motion relative to the hotspots from combined Atlantic and Indian Ocean hotspot tracks, *Geology, 21*, 275-278, 1993.

National Geophysical Data Center, ETOPO-5 bathymetry/topography data, *Data Announcement 88-MGG-02*, NOAA, U.S. Dept. Commerce, Boulder, CO, 1988.

Olson, P., Hot spots, swells and mantle plumes, *I:n* Ryan, M.P., ed., *Magma transport and Storage*, John Wiley & Sons Ltd., pp. 33-51, 1987.

Petronotis, K. E., and D. M. Jurdy, Pacific plate reconstructions and uncertainties, *Tectonophysics, 182*, 383-391, 1990.

Petronotis, K. E., R. G. Gordon and G. D. Acton, Determining paleomagnetic poles and anomalous skewness from marine magnetic anomaly skewness data from a single plate, *Geophys. J. Int., 118*, 1992.

Petronotis, K. E., R. G. Gordon and G. D. Acton, A 57 Ma paleomagnetic pole determined from a skewness analysis of crossings of marine magnetic anomaly 25R, *Geophys. J. Int., 118*, 529-554, 1994.

Petronotis, K. E. and R. G. Gordon, A Maastrichtian paleomagnetic pole for the Pacific plate determined from a skewness analysis of marine magnetic anomaly *32, Geophys. J. Intl., 139, 227-247*, 1999.

Ratcliff, J. T., D. Bercovici, G. Schubert and L. Kroenke, Mantle plume heads and the initiation of plate tectonic reorganizations, *Earth Plan. Sci. Lett., 156*, 195-207, 1998.

Richards, M. A. and R. W. Griffiths, Deflection of plumes by mantle shear flow: experimental results and a simple theory, *Geophys. J. R. Astron. Soc., 94,* 367-376, 1988.

Royer, J-Y. and D. T. Sandwell, Evolution of the eastern Indian Ocean since the Late Cretaceous- constraints from Geosat altimetry, *J. Geophys. Res., 94,* 13,755-13,782, 1989.

Royer, J-Y., P. Patriat, H. W. Bergh and C. Scotese, Evolution of the southwest Indian ridge from the Late Cretaceous (anomaly 34) to the middle Eocene (anomaly 20), *Tectonophysics, 155*, 235-260, 1988.

Royer, J.-Y. and T. Chang, Evidence for relative motions between the Indian and Australian plates during the last 20 Myr from plate tectonic reconstructions: Implications for the deformation of the Indo-Australian plate, *J. Geophys. Res., 96,* 11,779-11,802, 1991.

Royer, J-Y., R. G. Gordon, C. DeMets and P. R. Vogt, New limits on the motion between India and Australia since chron 5 (11 Ma) and implications for lithospheric deformation in the equatorial Indian Ocean, *Geophys. J. Int., 129*, 41-74, 1997.

Sandwell, D. T and W. H. F. Smith, Global marine gravity from ERS-1, Geosat and Seasat reveals new tectonic fabric, *EOS Trans. Amer. Geophys. Union, 73*, 133, 1992.

Schouten, H. and S. C. Cande, Palaeomagnetic poles from marine magnetic anomalies, *Geophys. J. R. Astron. Soc., 44,* 567-575, 1976.

Solomon, S.C. and N. H. Sleep, Some simple physical models for absolute plate motions, *J. Geophys. Res., 79*, 2557-2567, 1974.

Steinberger, B. and R. J. O'Connell, Advection of plumes in mantle flow: implications for hotspot motion, mantle viscosity and plume distribution, *Geophys. J. Int., 132*, 412-434, 1998.

Stock, J. and P. Molnar, Revised history of early Tertiary plate motion in the southwest Pacific, *Nature, 325*, 495-499, 1987.

Suarez, E. and P. Molnar, Paleomagnetic data and pelagic sediment facies and the motion of the Pacific plate relative to the spin axis since the Late Cretaceous, *J. Geophys. Res., 85,* 5257-5280, 1980.

Tarduno, J. A. and R. D. Cottrell, Paleomagnetic evidence for motion of the Hawaiian hotspot during formation of the Emperor Seamounts, *Earth Plan. Sci. Lett., 153*, 171-180, 1997.

van Andel, T. H., Cenozoic migration of the Pacific plate, northward shift of the axis of deposition, and paleobathymetry of the central equatorial Pacific, *Geology, 1,* 507-510, 1974.

Wessel, P. and W. H. F. Smith, Free software helps map and display data, *EOS Trans. Amer. Geophys. Union, 72*, 441-445, 1991.

Wessel, P., and L. Kroenke, A geometric technique for relocating hotspots and refining absolute plate motions, *Nature, 387,* 365-369, 1997.

Wilson, T. J., Cenozoic transtension along the Transantarctic Mountains-West Antarctic rift boundary, southern Victoria Land, Antarctica, *Tectonics, 14*, 531-545, 1995.

---

Steven C. Cande, Scripps Institution of Oceanography, La Jolla, CA 92093-0215, Carol A. Raymond, Jet Propulsion Laboratory, Mail Stop 183-501, 4800 Oak Grove Drive, Pasadena, CA 91109-8099, Joann M. Stock, Seismological Laboratory, Mail Stop 252-21, California Institute of Technology, Pasadena, CA 91125.

# Effects of Mantle Flow on Hotspot Motion

Bernhard Steinberger

*Institut für Meteorologie und Geophysik, Johann Wolfgang Goethe-Universität, Frankfurt am Main, Germany*

Richard J. O'Connell

*Department of Earth and Planetary Sciences, Harvard University, Cambridge, Massachusetts*

The motion of hotspots in large–scale mantle flow is discussed. The concept of mantle plumes and the experiments and observations on which it is based are reviewed. Results that support hotspot motion (from experiments, plate reconstructions and paleolatitude data) are contrasted with observations on which the concept of hotspot fixity is based (mainly geometry and age progression of hotspot tracks). A numerical model of hotspot motion in large-scale mantle flow is introduced. It is shown how hotspot motion may approximately represent flow at mid–mantle depth, particularly plate return flow. Such a model can explain some aspects of the observed hotspot distribution, yields predictions on the origin and shape of plume conduits and gives constraints on mantle viscosity structure. Some results for the motion of individual hotspots are shown, including coherent motion toward the southeast for Hawaii and Louisville, westward motion for Easter, and southward motion for Kerguelen. For Hawaii and Louisville, the calculated motion fits well the observed geometry and age progression of hotspot tracks. Furthermore, calculations can explain observed relative motions of Pacific and African hotspots back to 43 Ma without invoking any additional plate boundary in the Pacific-Antarctic region. Paleomagnetic and modelling results of polar motion are reviewed and combined with models of hotspot motion; resulting predictions of paleolatitudes are compared with observations for the Pacific plate. Calculated hotspot motion in combination with polar motion can explain any latitudinal shift of Pacific hotspots back to about 70-80 Ma.

## 1. INTRODUCTION

Hotspots have long been regarded as fixed and are therefore used as reference points for plate motion. Nevertheless, hotspot motion is expected from dynamical mantle flow models. There is as well observational evidence for hotspot motion. In this paper, we review arguments for and against hotspot motion, with emphasis on a qualitative discussion of the behavior of plume conduits in mantle flow in section 3. We then introduce our numerical model of plumes in mantle flow. This model was developed in order to contribute to the discussion about hotspot motion from a geodynamic perspective. Our results are in accordance with predictions arising

The History and Dynamics of Global Plate Motions
Geophysical Monograph 121

from the qualitative discussion. Our proposed model includes moderate motion of hotspots as well as polar motion and reconciles many observations.

## 2. EXPERIMENTAL AND OBSERVATIONAL EVIDENCE FOR MANTLE PLUMES

We begin by reviewing experiments and observations that support the existence of mantle plumes: Since the groundbreaking works of *Wilson* [1963], *Morgan* [1971,1972] and others it has been recognized that hotspots are caused by sources of heat – presumably upwellings from deep in the mantle – that are essentially stationary i.e. they move much more slowly than the tectonic plates. This insight was mainly based on the observation that active hotspot volcanoes are frequently associated with chains of islands and seamounts that get older with increasing distance from the active hotspot and that the age progressions can often be explained by a rigid plate moving over fixed hotspots. As an example, Figure 1 shows topography and age progression of the Reunion hotspot track on the African and Indian plate. This hotspot simultaneously left tracks on both plates for a period of time, while it presumably was close to the plate boundary. The age progression is different on both plates, as the plates moved at different speed relative to the hotspot.

Subsequent laboratory experiments showed that "If the upwelling material has less viscosity than the surrounding material, the structure ... eventually ascends as a spherical pocket of fluid fed by a pipe" [*Whitehead and Luther,* 1975]. This kind of upwelling became the standard model of a "mantle plume"; examples are shown in Figure 2. The exact shape and development of plumes depends on the viscosity contrast inside and outside the plume, whether the plume continues to be fed from the boundary layer, and to what extent material of the surrounding mantle is entrained. Detailed treatments are given by *Olson and Singer* [1985] and *Griffiths and Campbell* [1990]. When scaled to Earth dimensions, the experiments suggest a plumehead radius of a few hundred km, whereas the trailing conduit has a radius of a few tens of km at most; the ascent time of the plumehead through the mantle is of the order of 100 Ma, with large uncertainties. Obviously, the experiments are a simplification of plumes in the real Earth, which are expected to rise from the time-dependent boundary layer of a system convecting at high Rayleigh number. Nevertheless, the concept that plumeheads in the mantle establish rather narrow conduits is widely accepted. Those conduits may subsequently be affected by large-scale flow, as detailed in the next section.

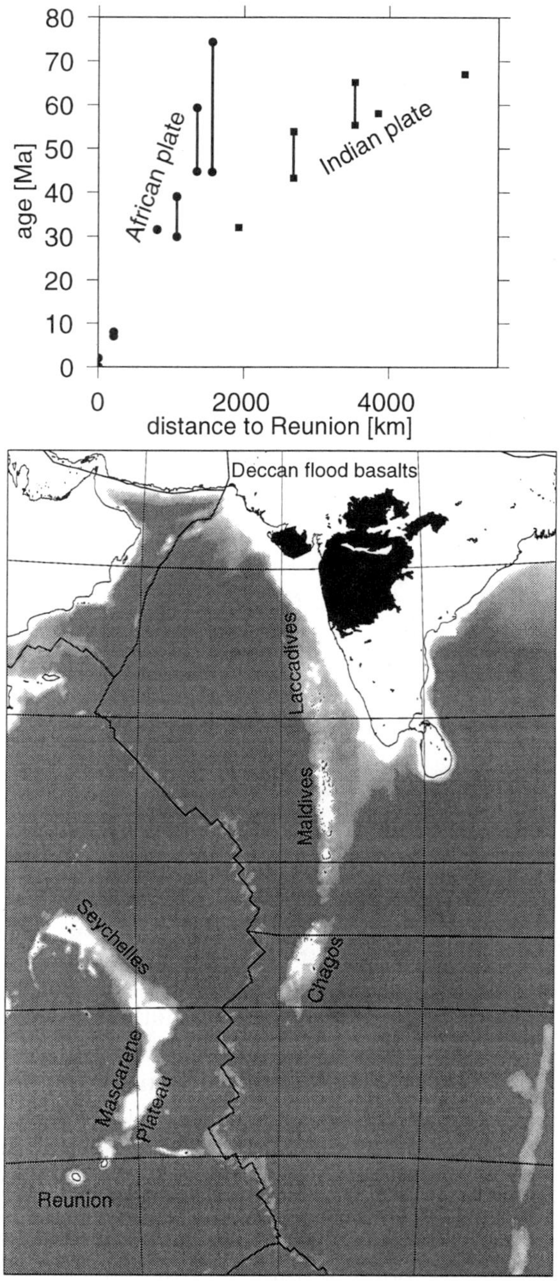


**Figure 1.** The Reunion hotspot track. The distribution of Tertiary extrusives in India (Deccan traps) is shown in black, according to *Choubert and Faure-Muret* [1976]. Topography is shown in greyshades between -3000 and -1000 m, in white above -1000 m, and in uniform grey below -3000 m. This figure and all other maps were made using GMT graphics [*Wessel and Smith,* 1995], with topography from ETOPO5 data [*National Geophys. Data Center,* 1988]. Age data of sea floor basalts along the track (shown as circles for African plate, squares for Indian plate) are from *Duncan and Hargraves* [1990] and references herein, and compiled by *Müller et al.* [1993].

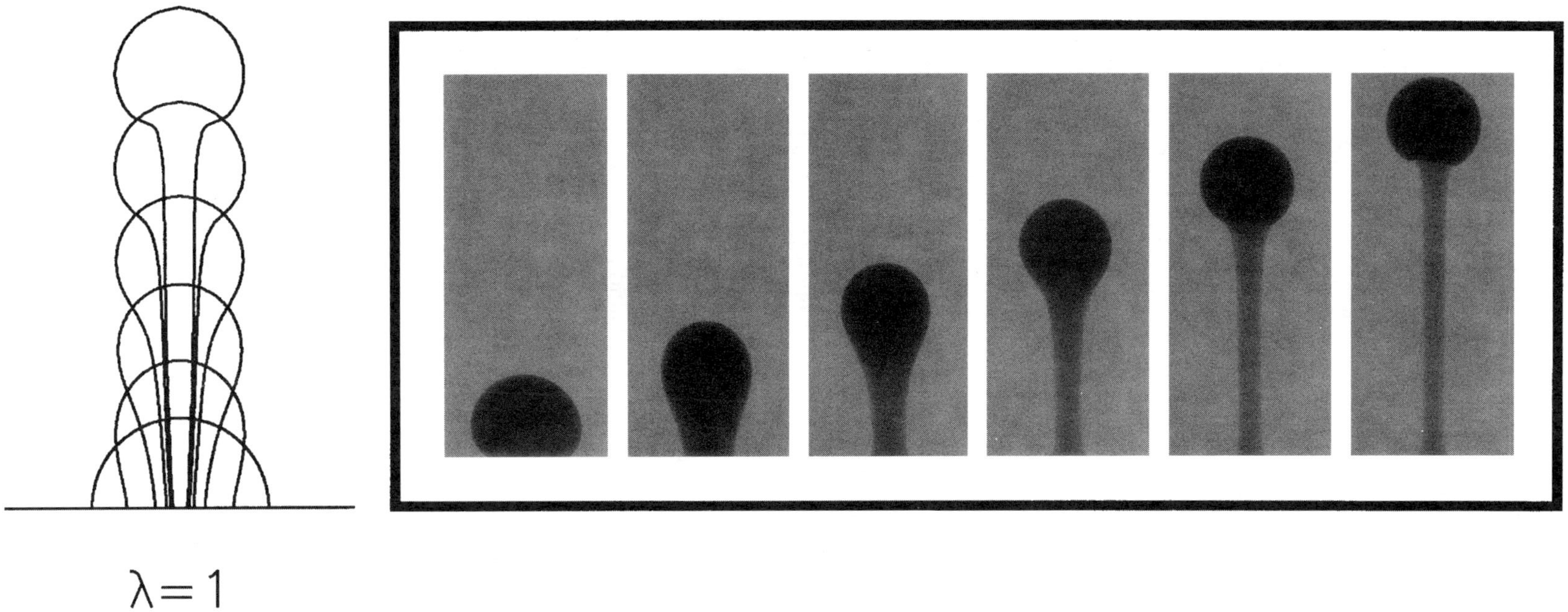


**Figure 2.** Experimental and numerical realization of a plume; figures are reproduced from *Manga et al.* [1993]. These results were obtained for a constant volume plume and the same viscosity $\eta$ inside and outside the conduit at low Reynolds number. Total time of the experiment is $42\eta/(\Delta\rho g r)$, where $\Delta\rho$ is density contrast, $g$ is gravity, $r$ is plumehead radius. This can be readily scaled to parameters expected for the Earth's mantle. Results look similar for constant feeding rate, except that the plumehead grows with time [*Olson and Singer*, 1985; *Griffiths and Campbell*, 1990], hence its rise speed will not approach a limit; the ratio of conduit to head diameter becomes smaller as the ratio of viscosity inside to outside the conduit decreases [*Whitehead and Luther*, 1975].

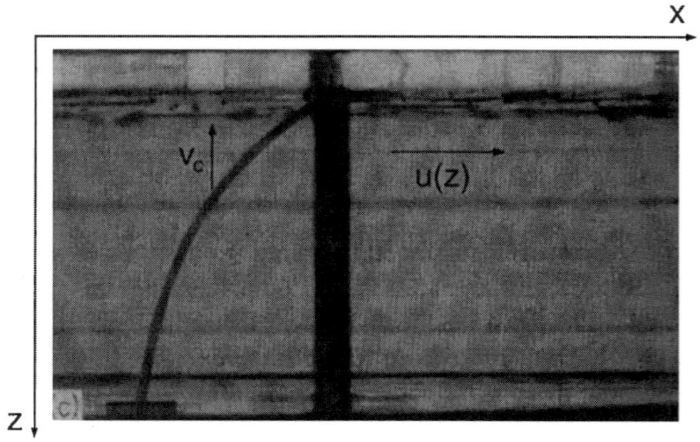


**Figure 3.** Experimental realization of a plume tilted in large-scale flow. Figure after *Richards and Griffiths* [1988].

The notion that plumes consist of "heads" and "tails" (conduits) is also supported by observations: Many plume tracks – corresponding to the motion of a plate over the tail – begin with extensive flood basalts – corresponding to the eruption of the plumehead. The clearest example is the association of the Deccan flood basalts with the Reunion hotspot track (Figure 1). The subject is discussed more extensively by *Richards et al.* [1989].

## 3. DISTORTION OF A PLUME CONDUIT IN SHEAR FLOW

*Richards and Griffiths* [1988] simulated the tilt of a plume in a simple experiment (Figure 3): The plume conduit is advected (horizontally) in large-scale flow (here: horizontal flow speed $u(z)$), but conduit elements also rise buoyantly (here: buoyant "Stokes" rising speed $v_c$ of the conduit through the surrounding viscous mantle – which is not to be confused with the rising speed of material through the conduit); total velocity of each conduit element is simply the vector sum of ambient flow velocity $\mathbf{u}$ and rising velocity $\mathbf{v}_c$. In the case of this experiment (stationary, fixed plume source; horizontal simple shear flow; constant viscosity of fluid outside conduit) a parabolic conduit shape results, as can be seen. The situation in the real Earth is more complicated: the source of the conduit is not fixed in space, but moves with the flow; mantle flow is not only horizontal; buoyant rising speed varies with depth; plumes are probably thermal features such that there is no sharp distinction between plume and surroundings. Nevertheless, the simple concept that motion of each conduit element can be calculated as vector sum $\mathbf{u} + \mathbf{v}_c$ may still be approximately valid for the Earth and is therefore used in our models.

Based on this simple concept we will now qualitatively discuss the behavior of hotspot plumes in large-scale mantle flow, beginning with the variables that determine $v_c$. The same effects as discussed here will also appear in our numerical model. Ambient mantle flow $\mathbf{u}$ will be further discussed in section 6.

A modified Stokes' formula for the rising speed of a plume conduit is

$$v_c = \frac{0.54 \Delta\rho g r_c^2}{\eta_{out}}, \qquad (1)$$

where $\Delta\rho$ is the density contrast between plume and surroundings, $g$ is gravity, $r_c$ is conduit radius and $\eta_{out}$ is the viscosity of the surrounding mantle. Obviously this equation is a rather crude approximation: Neither the effects of boundaries (surface and CMB) and viscosity interfaces nor any time dependence (such as for example discussed by *Sleep* [1992]) are considered. A more accurate treatment is however also prevented since many of the model parameters are not well constrained.

The numerical value 0.54 has been determined experimentally by *Richards and Griffiths* [1988] for a "chemical" plume (i.e. with different materials inside and outside the conduit), but may be different for a thermal plume conduit, which is not materially distinct from the surrounding mantle and hence may also rise by thermal entrainment of surrounding mantle [*Richards and Griffiths,* 1989].

$\Delta\rho$ is due to the temperature contrast between plume and surroundings. Estimates for plume excess temperature $\Delta T$ are of the order 200 to 300 K [e.g., *Schilling,* 1991]. For representative values $\Delta T = 300$ K, thermal expansivity $\alpha = 3 \cdot 10^{-5}\mathrm{K}^{-1}$ and mantle density $\rho_m = 3300\mathrm{kg\ m}^{-3}$ we obtain $\Delta\rho = 30\mathrm{kg\ m}^{-3}$. Whereas $\rho_m$ and $\Delta T$ increase with depth (see *Albers and Christensen* [1996] for calculations of the depth dependence of $\Delta T$), $\alpha$ is expected to decrease to values of less than $10^{-5}K^{-1}$ in the lower part of the lower mantle [*Chopelas and Boehler,* 1989], where therefore somewhat smaller values of $\Delta\rho$ are expected.

$\eta_{out}$ is usually inferred from models of postglacial rebound and the geoid. None of these methods can constrain details of mantle viscosity structure so that considerable uncertainties remain. Models of postglacial rebound give some kind of viscosity "average" for about the upper half of the mantle, with resolving power decreasing with depth. This "average" was determined to $10^{21}$ Pas in the classic paper by *Haskell* [1935], and a recent re-analysis by *Mitrovica* [1996] obtains a similar result. Some depth dependence can also be resolved, with results indicating a substantial increase of viscosity with depth [*Mitrovica*, 1996; *Lambeck and Johnston*, 1998]. For example, the preferred viscosity structure of *Lambeck and Johnston* [1998] features an increase from $4 \cdot 10^{20}$ Pas above the transition zone to $10^{22}$ Pas or greater in the lower mantle. Models of the geoid based on mantle density heterogeneities in a dynamic Earth were developed by *Ricard et al.* [1984] and *Richards and Hager* [1984]. They do not constrain absolute viscosity, but also show an increase of viscosity with depth. The limitations of this method have been discussed by *Thoraval and Richards* [1997], indicating that a robust feature of many models again is a substantial viscosity increase with depth (at least a factor of 30). A joint model of postglacial rebound and the geoid by *Mitrovica and Forte* [1997], also indicates a similar viscosity increase. The viscosity structure used here in most cases ($4 \cdot 10^{20}$ Pas between 100 and 400 km, $10^{21}$ Pas in the transition zone, and an increase in the lower mantle in four steps of 200 km from $10^{22}$ Pas to a maximum value of $4 \cdot 10^{22}$ Pas below 1270 km and above D", i.e. a viscosity increase with depth by a factor 100, henceforth referred to as "viscosity model A") is in accordance with the evidence cited above, however our models of hotspot motion favor an even stronger viscosity increase [*Steinberger and O'Connell*, 1998], and we will also show some results for the preferred viscosity structure of that paper (here referred to as "viscosity model B"). Similar to the approach presented here, a high-viscosity lower mantle has been previously inferred from models of hotspot motion [*Richards*, 1991].

$r_c$ can be estimated from scaling laboratory experiments to Earth dimensions to be a few tens of km or less (as mentioned above). Similar values in the upper mantle can also be inferred from the width of most hotspot tracks (of the order of 100 km or less) and associated gravity anomalies [e.g., *Morgan*, 1972]. $r_c$ might be somewhat larger in the lower mantle, as will be argued below. Also the fact that most hotspot plumes have not been observed by seismic tomography indicates narrow conduits. Seismic observations have been reported for the Iceland plume, indicating $r_c \sim 150$ km in the upper mantle [*Wolfe et al.*, 1997] and $r_c < 250$ km in the lower mantle [*Bijwaard and Spakman*, 1999], however this might be broader than usual, as this is the first tomographic evidence reported for a whole mantle plume.

Another estimate of $r_c$ can be made from Poiseuille's formula for flow in a pipe, which can be written

$$r_c = \left(\frac{8B\eta_{in}}{\pi\Delta\rho dp/dz}\right)^{\frac{1}{4}} \tag{2}$$

where $B$ is the anomalous mass flux of the plume conduit (defined as volume flux times difference of density inside and outside the conduit), $\eta_{in}$ is the viscosity inside the conduit and $dp/dz$ is the non-hydrostatic pressure gradient driving flow through the conduit. $\eta_{in}$ and $dp/dz$ are not well constrained, but at least eqn. 2 indicates how $r_c$ is expected to vary with anomalous mass flux and depth, and it gives an estimate in accordance with the above:

The anomalous mass flux $B$ has been determined for various hotspots based on the magnitudes of associated topographic swells [*Davies*, 1988; *Sleep*, 1990]. For plumes close to ridges, the length of the geochemical anomaly gives an additional constraint [*Schilling*, 1991]. Results indicate anomalous mass fluxes between about $0.3 \cdot 10^3$ kg/s for the smallest plumes and $8 \cdot 10^3$ kg/s for the largest one (Hawaii). Differences between various estimates are considerable, with uncertainties largest for the smallest hotspots.

Because of the higher temperature, $\eta_{in}$ is expected to be considerably less than $\eta_{out}$. If the temperature is adiabatic both inside and outside the conduit, the ratio $\eta_{in}/\eta_{out}$ may be constant, but due to thermal entrainment, $\eta_{in}/\eta_{out}$ is expected to increase with decreasing depth. Taking $\eta_{in} = 10^{19}$ Pas as a reasonable guess, and using the hydrostatic approximation $dp/dz = \Delta\rho g$ with $\Delta\rho = 30$ kg m$^{-3}$ as estimated above, we get $r_c \approx 40$ km for a representative value $B = 10^3$ kg/s, resp. $r_c \approx 70$ km for the Hawaiian plume flux $B = 8.3 \cdot 10^3$ kg/s (Table 1), in accordance with the above estimates. The actual value of $dp/dz$ might be considerably smaller [*Sleep*, 1992], but $\eta_{in}$ might also be smaller. Conduits of the largest plumes should be about a factor $(8/0.3)^{1/4} \sim 2$ thicker than for the smallest plumes, and for a viscosity increase of a factor 100, $r_c$ will increase with depth by a factor $(100)^{1/4} \sim 3$ or less.

In the upper mantle, the observed sharpness of the bend in the Hawaiian-Emperor-Chain gives an important constraint on the rising speed. As pointed out by *Griffiths and Richards* [1989], the deflection of the Hawaiian plume in the uppermost mantle (a layer of low viscosity below the lithosphere) is likely not more than 200 km. The reason is also explained in Figure 4. Using numerical values for plate motion $u_0 \approx 10$ cm/yr and a layer thickness of $d \approx 300$ km (estimated thickness of the mantle layer of lowest viscosity, as discussed above), this implies a rising speed of at least about 7.5 cm/yr, and correspondingly less for smaller plumes. In any case, a few cm/yr is probably a reasonable estimate for a conduit rising speed in the uppermost viscosity layer, corresponding to a horizontal deflection of typically a few hundred km or less and a typical risetime of the

Table 1. Parameters used for the hotspots shown. Ages are equal to age of associated flood basalt, where available (Tristan, Reunion, Kerguelen, Louisville), otherwise inferred from the length of the track (Samoa, Tahiti), or reasonable guesses are made, where the end of the track has been subducted (Hawaii, Easter). Anomalous mass fluxes $B$ are based on *Davies* [1988], *Sleep* [1990], *Schilling* [1991] and *Davies* [1992].

| Hotspot | Age, Ma | $B[10^3\text{kg/s}]$ |
|---|---|---|
| Easter | 100, 60 | 2.1 |
| Hawaii | 100 | 8.3 |
| Kerguelen | 117 | 0.9 |
| Louisville | 120 | 3.0 |
| Reunion | 67 | 1.4 |
| Samoa | 14 | 1.6 |
| Tahiti | 5 | 4.5 |
| Tristan | 125 | 1.0 |

conduit of the order of 10 Ma or less. For comparison, with the above estimates $r_c = 70$ km, $\Delta\rho = 30\text{kg m}^{-3}$ and $\eta_{out} = 4 \cdot 10^{20}$ Pas, eqn. 1 gives $v_c \approx 6$ cm/yr for the Hawaiian plume in the upper mantle. This shows that the above estimates are almost in accordance, but rising speed cannot be lower than estimated above (such as would be the case for a significantly thinner conduit).

Although Figure 4, which reiterates the argument made by *Griffiths and Richards* [1989], is a simplification, it summarizes the effects that we expect flow in the "low-viscosity" uppermost mantle to have on conduit shape and hotspot motion. We can classify three possible kinds of behavior (although in reality they will not be sharply distinct):

1. As the plumehead presumably rises comparatively fast, it will establish a conduit that is closer to vertical. In an initial stage (lasting a time equal to the rise time of the conduit through the layer of low viscosity, about 10 Ma or less) the conduit will get tilted, and the hotspot will move (a few 100 km, at most) in the direction of the plate, until the shape remains stationary

2. If the plate changes its motion, the conduit will adjust its shape, and the hotspot will move in the direction of the difference of plate motions before and after the change (again, a few 100 km at most, lasting about 10 Ma or less).

3. During other times, the conduit shape will remain stationary in the uppermost mantle. The observed length of hotspot tracks indicates typical hotspot lifetimes of $t_l = 100$Ma or more, and plate reconstructions show that plate motions also tend to remain constant over long periods of time, such that we expect this to be the usual case. In the simple case shown in Figure 4, this would correspond to a fixed hotspot. More realistically, the exit point of the plume from the lower mantle will also move, and the hotspot surface motion will approximately represent the motion of this point; the stationary shape corresponds to the conduit shown in Figure 3, with the difference that the plume source moves in our model.

Motion of that exit point (which is probably not sharply defined in the Earth as viscosity increase with depth may be gradual, but nevertheless provides a useful concept) may be due to two reasons: Firstly, it is due

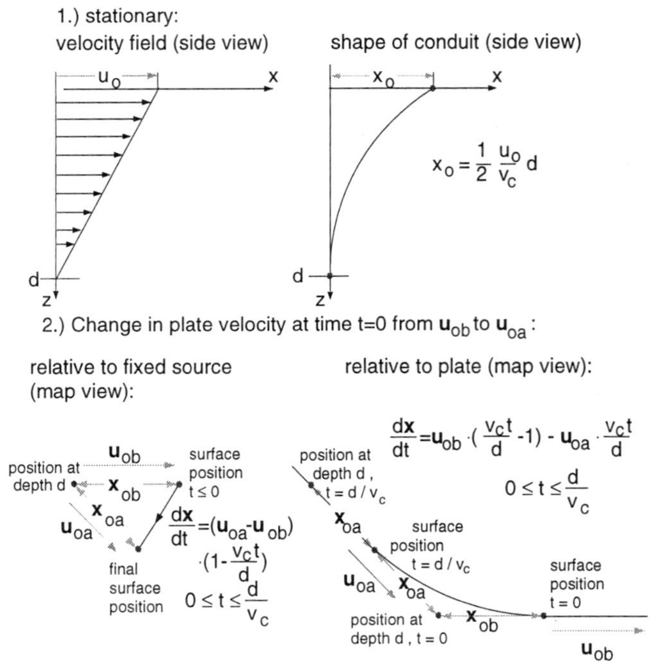


**Figure 4.** Plumes deflection in the upper mantle and curvature of hotspot track for change in plate velocity – a very simple example, after *Griffiths and Richards* [1989]: A given plate velocity, a constant exit point of the plume from the lower mantle and a simple shear flow in the upper mantle is assumed (top left). This gives rise to a parabolic conduit shape in the stationary case (top right). If the plate motion changes, the shape of the conduit will adjust to the new plate motion. During a time of adjustment, the hotspot surface position will move relative to the fixed source in a direction parallel to the difference of plate motion before and after the change (bottom left). Relative to the plate, this means that even if the plate motion changes discontinuously, the hotspot track does not have a sharp bend, but is curved with a radius of magnitude similar to the deflection of the conduit (bottom right).

to ambient mantle flow at that depth. Hence most of the time hotspot surface motion has a component representing flow at mid-mantle depth (that is, where the transition from "low" to "high" viscosity, hence from "fast" to "slow" conduit rising speed occurs). This flow is frequently toward ridges, opposite to plate motions and coherent under one plate. Secondly, if the conduit is already tilted below that depth and $v_c$ is not negligible, another component of motion is due to conduit elements rising from different locations. Such a tilt is caused by shearing of the conduit due to variations of flow velocity with depth, and the corresponding component of hotspot motion represents the change of flow velocity with depth. Therefore, as a general tendency, with increasing age of a hotspot, its motion will represent flow at increasing depth: This is even exactly true in the simple case of an initially vertical conduit distorted in a time-independent horizontal flow field with no lateral variations: In this case, the surface motion at a time $t$ after the conduit was vertical is equal to the flow speed at the depth from which the conduit element rose during time $t$, as can be easily shown.

In order to estimate the depth levels which thus may be represented in hotspot surface motion, we have to estimate $v_c$ in the lower mantle: for a constant conduit radius $r_c$ it should be approximately $v_c \sim 1/\eta_{out}$ (i.e. decrease by about a factor 100 in the model used here). If $\eta_{in}/\eta_{out}$ is constant, it follows from eqns. 1 and 2 $v_c \sim 1/\sqrt{\eta_{out}}$ (i.e. decrease by about a factor 10 in the model used here). As we have shown, the increase of $\eta_{in}$ with depth is likely less than the increase of $\eta_{out}$, hence $v_c$ in most of the lower mantle will probably be less than 10 % of the upper mantle value, i.e. not more than a few mm/yr. For example, for $r_c = 100$ km (more than what would be expected from scaling of lab experiments, but less than tomographic images below Iceland would indicate) $\eta_{out} = 4 \cdot 10^{22}$ Pas and other values as before, we obtain $v_c \approx 1.3$ mm/yr. With a representative hotspot lifetime of $t_l = 100$ Ma, this indicates, that hotspot surface motion probably represents flow in the upper part of the lower mantle most of the time, and that flow in the lowermost mantle has little effect on hotspot motion.

$v_c$ will not only decrease with depth in absolute terms, but also relative to ambient mantle flow speeds, as those do not decrease as strongly with depth: Numerical calculations of mantle convection with depth-dependent viscosity by *Gurnis* [1986] indicate that root mean square flow speeds decrease $\sim 1/\ln(\eta_{out})$ (i.e. by a factor of about 1/4.6 for a 100-fold viscosity increase), and our flow calculations give similar results, with typical flow speeds in the lower mantle of the order of 1 cm/yr. Therefore conduit motion in the lower mantle should be dominated by advection; the source regions of the plumes should move with the horizontal flow at the top of D" – frequently toward large-scale upwellings in the lower mantle. The lines of thought presented in this section are also illustrated in Figure 5. The mantle flow field that hence causes hotspot motion and conduit distortion is discussed in section 6.

## 4. EXTINCTION OF PLUME CONDUITS

In another laboratory experiment, *Whitehead* [1982] showed that plume conduits break up into separate drops if they become sufficiently distorted ($\gtrsim 60°$ from vertical – see Figure 6). However this result was again obtained for a chemical plume. *Richards and Griffiths* [1989] find that in thermal plumes the instability is surpressed by thermal entrainment, and that a plume conduit may be deflected to almost horizontal orientation without any instability. These results are for $r_c$ = 25km and $\eta_{out} = 10^{21}$ Pa s when scaled to the mantle, therefore they are probably relevant at least for smaller plumes. A strongly tilted conduit may however be wiped out due to thermal entrainment effects (also demonstrated by *Richards and Griffiths*, [1989]) even without an instability developing.

An instability may not develop either during the lifetime $t_l$, if the viscosity of the surrounding mantle is sufficiently high. A simple estimate for that condition is that buoyant rising of one conduit diameter takes longer than $t_l$, i.e.

$$2r_c/v_c > t_l \tag{3}$$

With the above estimates we find that this may well be the case in (at least the lower part of) the lower mantle.

Finally, even if an instability develops and a conduit breaks up into drops in the uppermost mantle, the conduit beneath may remain intact if the drops rise fast enough such that enough material and heat is carried through the conduit to prevent extinction. We used a simple scaling argument to estimate

$$v_c \gtrsim 2l\kappa/r_c^2, \tag{4}$$

[*Steinberger and O'Connell,* 1998] where $l$ is conduit length and $\kappa$ is thermal diffusivity, which has been found to be of the order of $10^{-6}$m$^2$s$^{-1}$ for mantle materials [e.g., *Osako and Ito,* 1991].

In summary, a strong tilt therefore favors destruction of plume conduits. However extinction may also occur without a strong tilt, due to thermal entrainment, lack of material supply, etc. On the other hand conduits may survive despite a strong tilt.

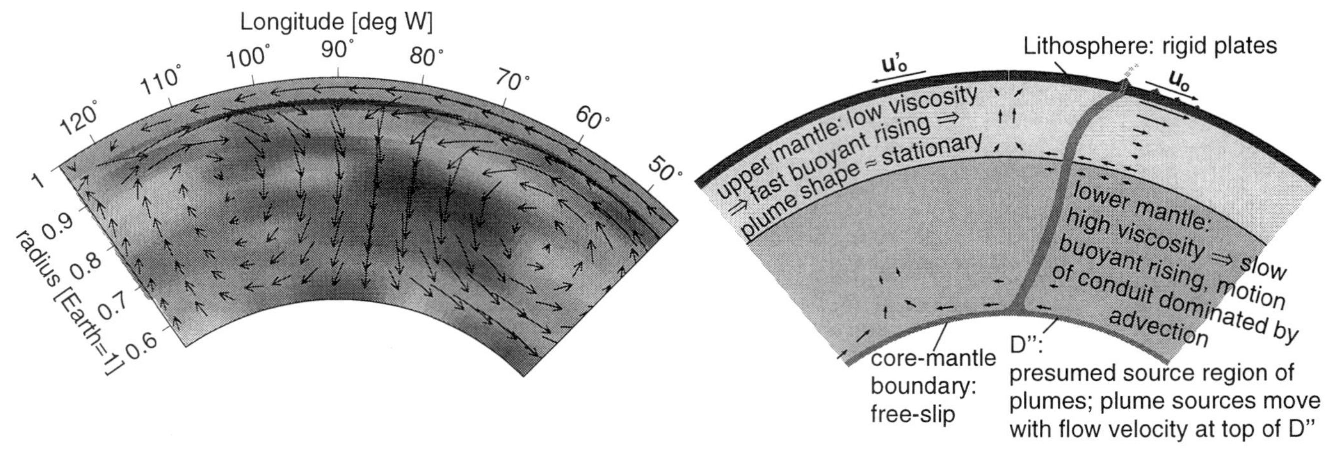


**Figure 5.** Left: Density anomalies beneath North America (model 3, i.e. inferred from Grand's tomographic model); cross section at 44.6 deg N; spherical harmonic expansion up to degree and order 32. Dark colors represent positive density anomalies (corresponding to fast anomalies in Grand's model), light colors represent negative anomalies (normalized to the maximum anomaly occurring in that depth layer). Also shown are arrows representing the E-W and vertical components of the induced present–day flow as calculated with the method of *Hager and O'Connell* [1979,1981] for viscosity model A. Length of arrows corresponds to displacement that would result from 50 Ma of constant flow. Right: Sketch showing how mantle flow may cause motion of hotspots and plume sources.

## 5. HOTSPOT MOTION VERSUS FIXITY – EVIDENCE FROM HOTSPOT TRACKS, PLATE RECONSTRUCTIONS AND PALEOMAGNETISM

The concept of hotspot fixity was first introduced after it was shown "that the Hawaiian-Emperor, Tuamotu-Line and Austral-Gilbert-Marshall island chains can be generated by the motion of a rigid Pacific plate rotating over three fixed hotspots" [*Morgan,* 1971]. More recent sampling shows that the age progression along the Tuamotu-Line and Austral-Gilbert-Marshall island chains is substantially more complicated than that expected from a single hotspot [*Jarrard and Clague,* 1977; *Duncan and Clague,* 1985; *Pringle and Duncan,* 1995; *McNutt et al.,* 1997]. Figure 7 however shows that geometry and age progression of several other hotspot tracks, in particular Hawaiian–Emperor and Louisville, can still be explained well by the fixed hotspot hypothesis. Of course, this is not a proof for hotspot fixity, since obviously hotspots moving coherently (i.e. without changing their distances) can explain the data equally well.

On the African hemisphere, the geometry and age progression of most hotspot tracks is not nearly as clear as for Hawaii and Louisville. *Müller et al.* [1993] showed that the fixed hotspot hypothesis is also compatible with the tracks there.

When looking at African and Pacific hotspot tracks combined, it becomes less clear whether hotspot fixity is in accord with observations. *Cande et al.* [1995] show a significant difference between the predicted Hawaiian hotspot track, assuming fixity relative to African hotspots, and the observed Hawaiian-Emperor chain. Their result is reproduced in Figure 8. The relative motion of Pacific and African hotspots, especially with regard to plate reconstructions in the Pacific-Antarctic region is also the subject of a paper by *Raymond et al.* [1999] in this volume. Another interpretation by *Harada* [1997] suggests that the Pacific and African hotspots are fixed within the uncertainties.

Another method to determine hotspot motion uses paleolatitude data that are inferred from the inclination of the characteristic remanent magnetization of drilled

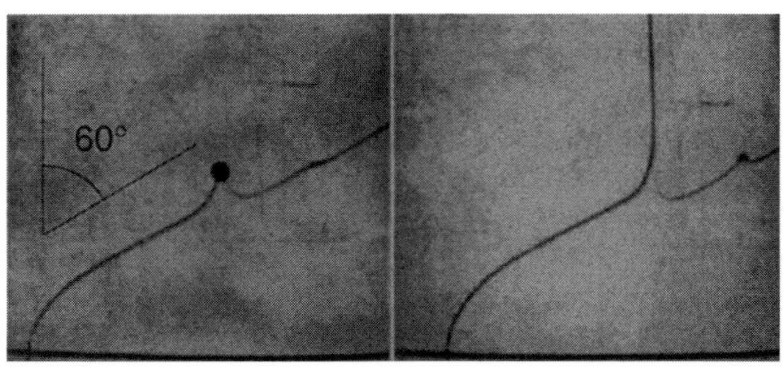


**Figure 6.** Laboratory experiment showing the breakup of a "chemical" plume conduit due to tilting. Figure after *Whitehead* [1982].

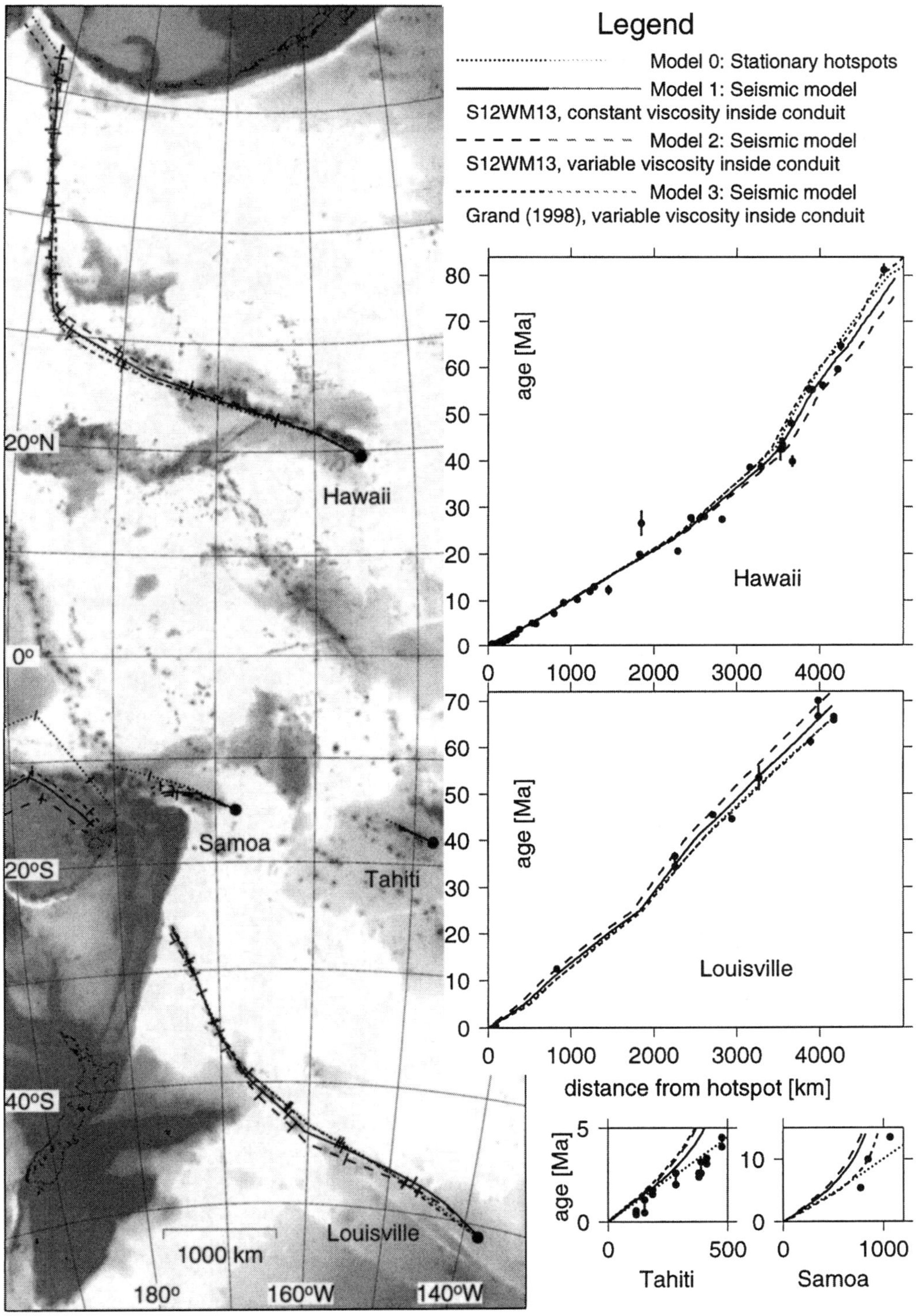


**Figure 7.** Hotspot tracks on the Pacific plate – observed and calculated for fixed and moving plumes (viscosity model A; three cases). Age data (shown as dots) for Hawaiian hotspot compiled by *Clague and Dalrymple* [1989] and from *Tarduno and Cottrell* [1997], for Louisville hotspot from *Watts et al.* [1988], for Tahiti and Samoa compiled by *Duncan and Clague* [1985]. Topography is shown in greyshades between -5000 and -2000 m, in uniform grey above -2000 m, and in white below -5000 m. Results for model B are also very similar.

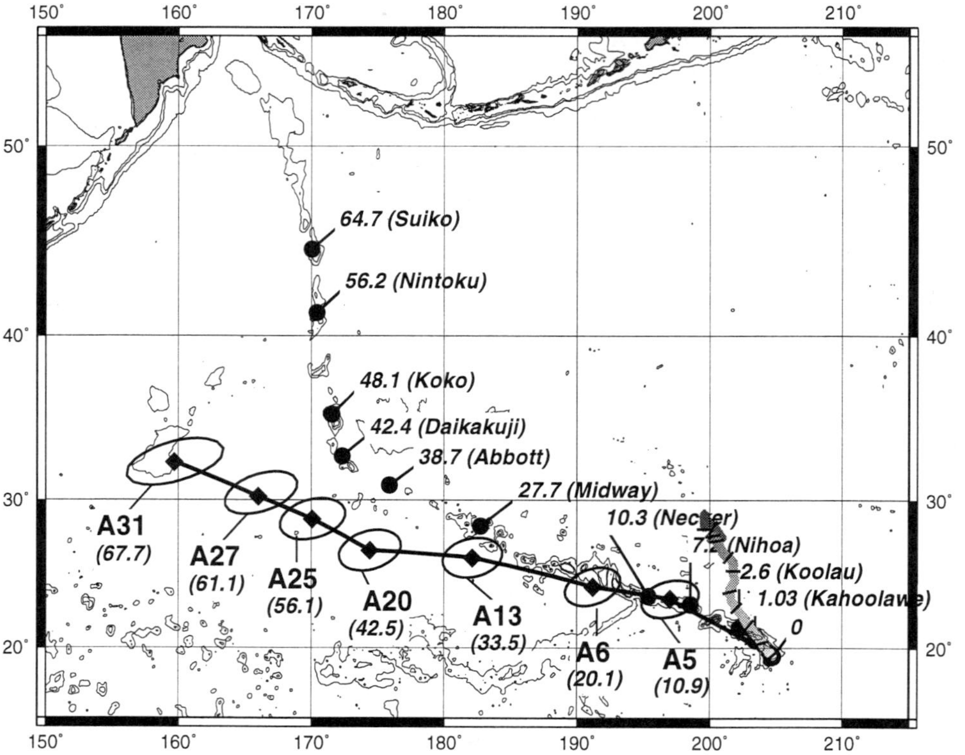


**Figure 8.** Prediction of the Hawaiian hotspot track from plate reconstructions assuming fixity relative to Indo-Atlantic hotspots. This prediction is different from the dotted line (assuming hotspot fixity) in Figure 7, because it is calculated for a given African absolute plate motion and a given relative motion Pacific vs. Africa, whereas in Figure 7 we use the best-fitting Pacific plate motion. Also shown is the motion of the Hawaiian hotspot (in the mean mantle reference frame of our flow calculations) as predicted from our model B, with greytones corresponding to age and tickmark interval 10 Ma. Figure modified after *Cande et al.* [1995].

basalts and sediment cores, using the axial dipole hypothesis: If all hotspots and the rotation axis were fixed relative to each other, the paleolatitude of hotspots should be the same as the present latitude for all times. This method can be extended to data other than from hotspot tracks: For a given model of absolute plate motion, based on hotspot tracks, the paleolatitude of any point on the plate can be predicted for all times, and compared with the observed paleolatitude. Using this method, *Sager and Bleil* [1987] find a "latitudinal shift of Pacific hotspots during the late Cretaceous and early Tertiary". In the top left panel of Figure 9 we redo their analysis, using our own model for the best fitting plate motion assuming fixed hotspots, and essentially reproduce their finding. Their results are however largely based on sediment cores and seamount paleomagnetism, which are less reliable than basalt cores. *Tarduno and Cottrell* [1997] have used data from basalt cores and find similar results, giving "evidence for motion of the Hawaiian hotspot during formation of the Emperor seamounts". The drawback here is that inclination data from basalt cores are far fewer in number.

## 6. A MODEL OF PLUMES IN LARGE–SCALE MANTLE FLOW

Since geodynamic models indicate that hotspots are unlikely to be fixed relative to one another, (as outlined in section 3) and the observational evidence for hotspot fixity versus mobility remains inconclusive (as shown in section 5) we have developed a numerical model of plumes in large–scale mantle flow. This enables us to make predictions of hotspot motion for various models of mantle flow and viscosity structure.

The details of the modelling procedure have been described elsewhere [*Steinberger and O'Connell,* 1998], so here we give only a brief overview with emphasis on understanding the physical processes: In a first step, a large–scale mantle flow field is calculated with the method of *Hager and O'Connell* [1979, 1981]. Using a spherical harmonic expansion, this method solves the

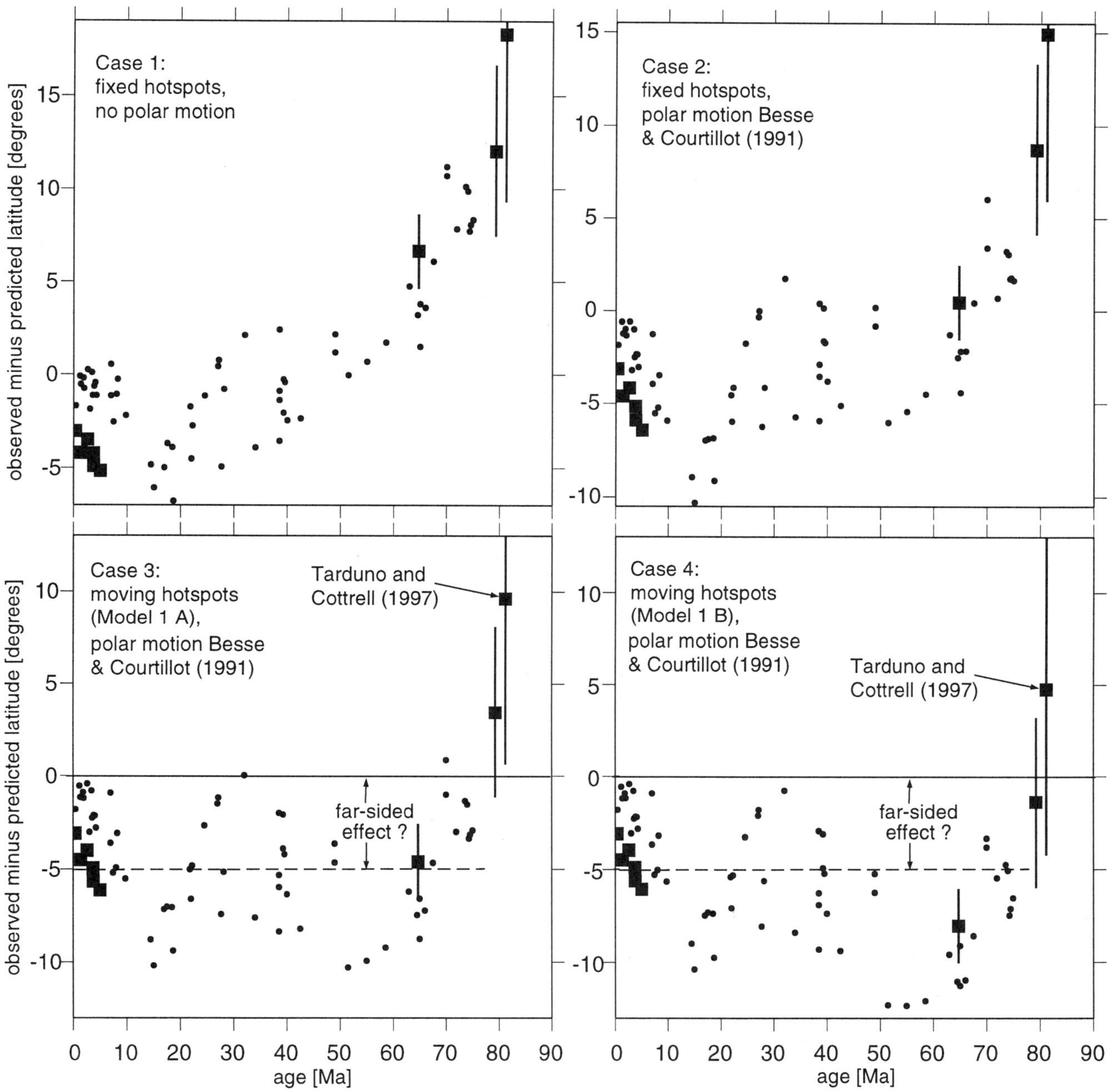


**Figure 9.** Latitudinal shift of Pacific hotspots – four cases. Observed paleolatitude data of samples on Pacific plate are from *Epp et al.* [1983], *Sager* [1984], *Sager and Bleil* [1987], *Tarduno and Gee* [1995], *Tarduno and Cottrell* [1997]. Larger squares represent more reliable data points (mostly on Hawaiian–Emperor chain). Similar to *Sager* [1984] we consider data points that are based on the paleomagnetism of extrusives, with a sufficiently large number of samples ($N \geq 18$), to be more reliable. For clarity, error bars (representing 95% confidence intervals) are only shown for more reliable data with ages $\geq$ 10 Ma. In the top left panel, predicted paleolatitudes are calculated by rotating samples from their drilling location back to the paleo-location where they were emplaced, using the measured age and a model of Pacific plate motion that optimizes the fit of the Hawaiian and Louisville hotspot tracks, assuming hotspot fixity. In the top right panel, the motion of the pole in the "fixed hotspot" reference frame is taken into account: In this case, the angular distance from the predicted paleo-location to the paleo-pole at the same age according to the *Besse and Courtillot* [1991] curve is the predicted paleo-latitude. In the two bottom panels, we use instead models of Pacific plate motion that optimize the fit of the Hawaiian and Louisville hotspot tracks, taking into account the calculated hotspot motion for two of our models (see Table 3).

equations of motion for a viscous mantle rheology, with viscosity varying with radius only. Because of the high viscosity, inertial forces are neglected.

Internal driving forces arise from mantle density heterogeneities $\delta\rho$ that are inferred from models of seismic s-wave speed anomaly $\delta v_s$. In most of the flow calculations shown here, we use either S12WM13 (by *Su et al.* [1994] – for our models 1 and 2), or the most recent model of S. Grand (retrieved via anonymous ftp and similar to *Grand et al.* [1997] – for our model 3). For viscosity model A, we use conversion factors $(\delta\rho/\rho)\,/\,(\delta v_s/v_s) = 0.2$ for S12WM13 resp. 0.4 for Grand's model, and only consider velocity heterogeneities below 220 km, in order to at least partly exclude effects of compositional continental roots. Both values are close to what has been inferred from geoid modelling [e.g., *Forte et al.,* 1993] and from laboratory experiments in combination with theoretical arguments [*Karato,* 1993]. A higher value is used for Grand's model because of the lower amplitudes of that model. For viscosity model B, we use conversion factors 0.4 for Grand's model and 0.2 for all other models at all depths, as in our previous paper, unless stated otherwise. The boundary conditions on the Earth's surface are taken to be the observed time–dependent global plate motions [*Gordon and Jurdy,* 1986; *Lithgow-Bertelloni and Richards,* 1998].

Mantle flow in our model is time-dependent for two reasons: First, plate motions change with time as the plate geometry changes and the velocities change; this changes the plate motion boundary condition that is applied to the model. Secondly, the flow field carries the internal density anomalies to new positions. We use the flow field to advect the density field backward in time from its present position for 68 Ma. This introduces some uncertainty in the model, since the advection is approximate. Nevertheless, the errors that accumulate over 68 My due to neglecting diffusion, viscous dissipation, adiabatic heating and cooling effects, are estimated to be relatively small. We have given a more detailed justification for the advection model in our previous work [*Steinberger and O'Connell,* 1998, appendix A2]. A constant density field is used before 68 Ma.

Figure 5 shows, as an example, a cross section below North America through model 3, as well as an inferred flow field. Note that both the density model and the calculated flow field are fully three–dimensional; we are however only able to plot the vertical and east-west component of the flow field. The dark structure in the center of the figure has been interpreted as the remains of the subducted Farallon plate, which are now in the lower mantle. As should be expected, this presumably cold and dense material drives a large–scale downward flow in our calculation. S12WM13 does not have as high resolution as more recent models such as Grand's. It does, however, provide a global model, which is needed to calculate a global flow field. In addition, the calculation of the flow provides an effective low–pass filter that averages the smaller scale features in the density field. Hence the lower resolution seismic models may still provide a sufficiently accurate flow field, especially in oceanic regions where more detailed tomographic results are not available. As an example, Figure 10 shows an inward flow at the base of the mantle towards an upwelling under southern Africa that corresponds to a large-scale negative seismic anomaly in this region. If Grand's model is used instead, the horizontal flow at the base of the mantle in this region is still dominated by this inward flow. Flow in the upper mantle is, to a larger degree, related to plate motions. Figures 10 and 11 show a flow at the top of the lower mantle that is frequently toward ridges i.e. has a "return flow" component opposite to plate motion. However, mantle flow in the depth range shown tends to be much slower than plate motions.

We then insert a plume conduit in the calculated large scale flow. We choose a vertical conduit as the initial condition, corresponding to the assumption that the plumehead that establishes the conduit (Figure 2) rises rapidly and is thus not significantly advected laterally. This assumption is discussed in more detail in *Steinberger* [1999]; here we just state that this assumption seems the most reasonable, because many hotspots are assumed rather old, such that the large-scale flow through which the plumehead has risen is poorly known, and also because an initially tilted conduit would further complicate the model by introducing more parameters. If a flood basalt is associated with the hotspot, we choose its age as the initial time. For example, for the Reunion hotspot we start with a vertical conduit 67 Ma ago (see Figure 1). All the ages used are listed in Table 1. The conduit is then advected and distorted in the flow field, whereby the velocity of each conduit element is a superposition of ambient mantle flow and vertical buoyant rising, as discussed in section 3.

Obviously, the separation of mantle flow in "plume flow" and "large scale flow" is somewhat artificial. For example, neither the flow into nor out of the conduit are considered. The separation is however justified since (a) plume conduits are expected to be thin features, as argued above, hence we expect their presence and

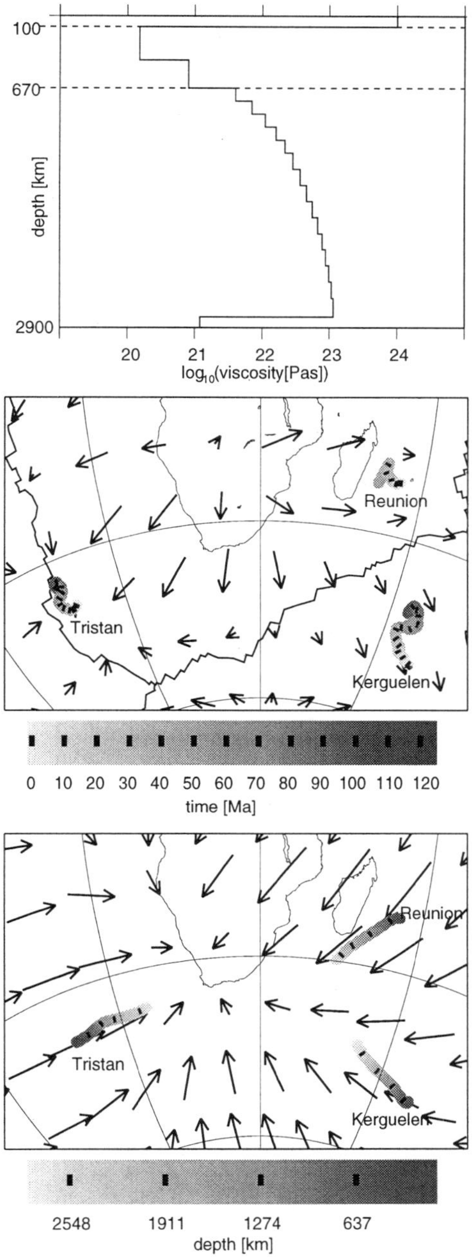


**Figure 10.** Calculated hotspot motion (mid-panel) and conduit shape (lower panel) for Tristan, Reunion and Kerguelen hotspots, for the viscosity model B shown. Arrows represent present–day flow at depths 800 km (mid-panel), resp. 2700 km (lower panel), with the arrow length corresponding to the total motion that would result from 68 Ma of constant flow. In the mid-panel, the motion of hotspots is plotted in the reference frame of the flow calculations (a mean mantle reference frame), with different shades of grey corresponding to different times, and tickmarks every 10 Ma. This hotspot motion is not to be confused with hotspot tracks on plates. In the lower panel, different shades of grey correspond to calculated conduit locations at different depths, with a tickmark interval of 0.1 Earth radii.

buoyancy does not significantly alter large-scale mantle flow, and (b) heat flow of plumes, hence their material flux is estimated to be substantially smaller than large-scale mantle flow: *Davies,* [1988] estimates that plumes supply less than 10 % of the total heat out of the mantle. Therefore we do not expect that inflow and outflow of plumes alters the large scale mantle flow field significantly either.

Based on eqns. (1) and (2), and in accordance with the discussion in section 3 we will use plume rising speeds as specified in Table 2, unless noted otherwise. Obviously these models are somewhat arbitrary. We therefore use several different models to show that results do not critically depend on exact values of $v_c$ chosen (as was already argued in section 3 – see also *Steinberger* [1999] where results for more hotspots are shown).

Figure 10 is an example of our calculations, which illustrates what is described above: In the middle panel, the recent surface motion of the hotspots tends to be similar in direction and magnitude to the horizontal flow in the mid–mantle (here shown at depth 800 km). Whereas Tristan and Reunion show very little motion during the past 40 Ma, the motion of Kerguelen is quite significant, in a south–easterly direction. This coincides with *Morgan* [1981], who finds Kerguelen to be the least fixed hotspot in the Atlantic and Indian Oceans.

The lower panel shows the calculated shape of three conduits. The bases of the conduits are advected with the flow toward the large upwelling under South Africa. Since the flow in the mid and upper mantle is in quite different directions, there is a significant tilt of the conduits, with the base of the conduits closer to the upwelling under South Africa. Corresponding results for viscosity model A are shown in *Steinberger* [1999] and are very similar.

## 7. MOTION OF HOTSPOTS IN THE PACIFIC

Another example is shown in Figure 11. Here calculations are not for actual hotspots; instead the initial hotspot locations are on a ten degree grid. The motion of a hotspot is only shown if the conduit has not been tilted more than 60° at depths between 670 and 1070 km. The figure thus illustrates the regions where plumes, if present, could have survived and given rise to long–lived hotspots on the surface. We restrict ourselves to this depth range, as instabilities will develop slowly at greater depths, and will not necessarily lead to extinction at shallower depths, according to estimates 3 and 4. The "pattern" of surviving plumes calculated does not critically depend on this depth range chosen. Figure 11 can therefore help to explain the

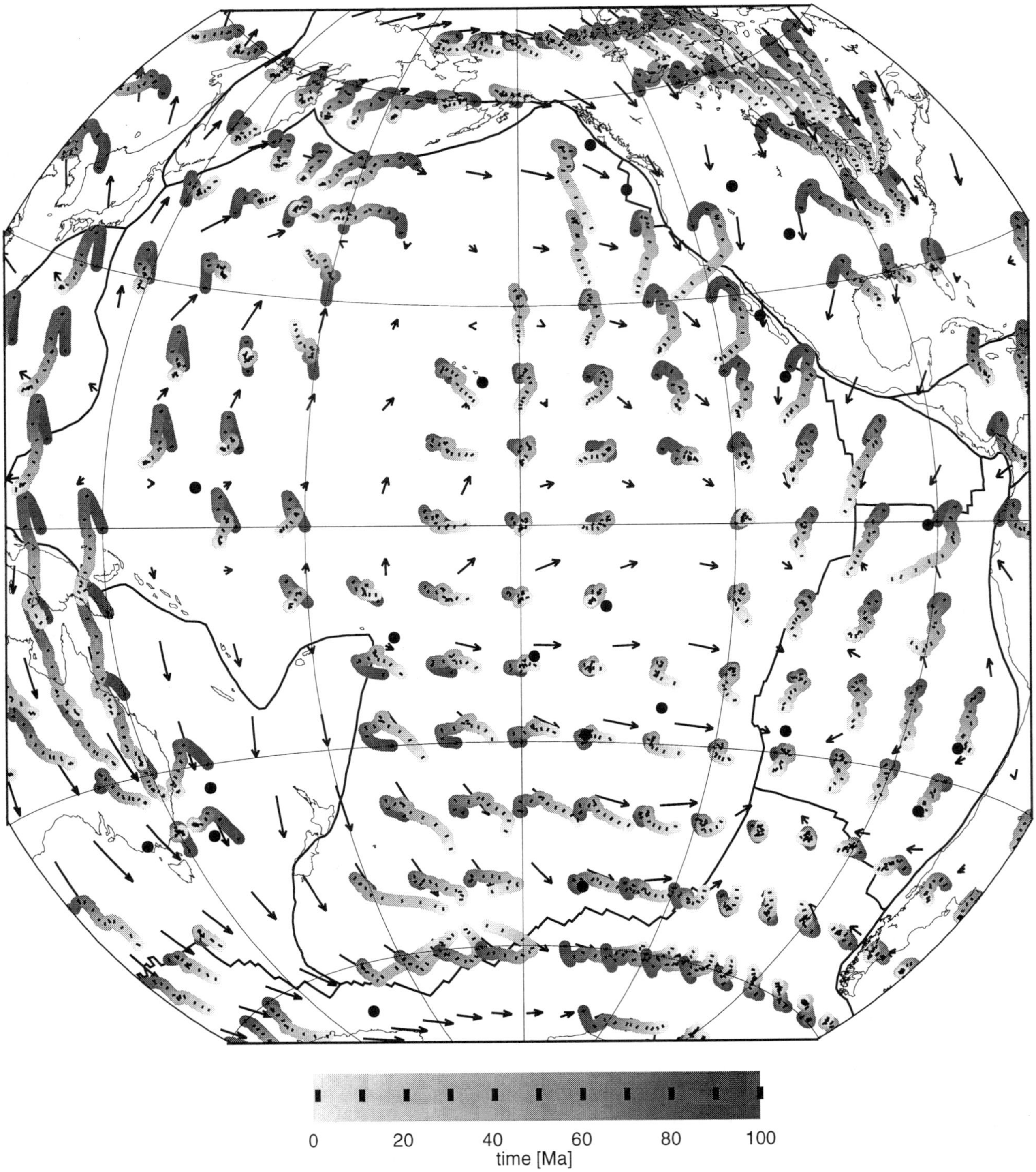


**Figure 11.** Motion and distribution of hypothetical hotspots (initially located on a 10-degree-grid) in the Pacific: Calculations are for model 1 A, with $v_c = 4\text{cm/yr}$ for $\eta_{out} = 10^{21}$ Pas (i.e. 10 cm/yr below the lithosphere), corresponding to a conduit radius of 88 km according to eqn. 1. Motion is only plotted for plumes tilted $< 60^\circ$ at depths between 670 and 1070 km. Initial surface positions of plumes are on a grid, with eruption of plumeheads 100 Ma ago. Arrows represent present–day flow at depth 1000 km, with the arrow length corresponding to the total motion in 50 Ma. Locations of actual hotspots are shown as black dots. Again, hotspot motion is plotted in the mean mantle reference frame.

observed hotspot distribution: In the regions of recent subduction and fast seismic anomalies, with the associated downward flow (see Figure 5), plume conduits frequently get tilted more than 60°, and the hotspots are more likely to become extinct. This agrees with the observation that hotspots cluster in the Pacific and around Africa, where tomography shows large–scale negative seismic anomalies in the lower mantle, hence large scale upwelling is inferred. The fact that Yellowstone is in a region of recent subduction is no contradiction to this, since its assumed age (equal to the Columbia River Basalts) is much less than 100 Ma: For this age (15 Ma) our models predict a Yellowstone conduit tilt of less than 60 degrees.

Results shown in Figure 11 are not specific for this particular viscosity structure – e.g. they are very similar for viscosity structure B [*Steinberger and O'Connell,* 1998, Figure 7]: both figures also show a coherent hotspot motion in a ~ south-easterly direction over much of the Pacific plate, including the locations of Hawaii and Louisville. This motion corresponds to a largely coherent flow in the mid–mantle opposite to plate motion. During the first ≈ 10 Ma, hotspot motion tends to be in a different direction, as explained in section 3. For the actual parameters of the Hawaiian hotspot, the motion for model B is included in Figure 8. Results for actual hotspots are also shown in *Steinberger* [1999] for model A and in Figure 12 for a greater number of models: A coherent motion of Hawaii and Louisville remains a common feature of most models. As mentioned above, coherent hotspot motion yields tracks that are compatible with the assumption of hotspot fixity; therefore our result can help to reconcile the findings of *Morgan* [1971] and others with a convecting mantle. This is further illustrated in Figure 7: By choosing a different "best fitting" Pacific plate motion (Table 3), the calculated hotspot motions can fit the observed tracks and age progression about as well as assuming hotspot fixity. We note that especially for Tahiti the models with a moving hotspot do not fit the age progression as well as assuming hotspot fixity. This misfit occurs because in our model, during the first few Ma, the initially vertical conduit is being tilted towards a steady state, as explained above. There are no indications for such an effect in the observed age progression. This probably shows that, if volcanism in the Tahiti chain is due to a hotspot plume in the sense of the model discussed here, the tilt of its conduit did not significantly change during the past ~ 5 Ma, in contrast to our model.

Table 2. Plume rising speeds (column 3) and corresponding variations (if any) of viscosity inside conduit (column 4) for a given model (column 1) and viscosity range outside conduit (column 2). $B_0 = 10^3$ kg/s and $\eta_0 = 10^{21}$ Pas are used; anomalous mass fluxes $B$ are listed in Table 1. Model A requires larger values $v_c \cdot \eta_{out}$ than model B, because of the higher viscosity below the lithosphere. A higher ratio $\eta_{in}/\eta_{out}$ is used in the layer of lowest viscosity in models 2 and 3 A, because that layer may be at least partly be fed by plumes [*Phipps Morgan et al.,* 1995], hence its viscosity may be closer to plume viscosity at the same depth.

| Model | $\eta_{out}$ | $v_c$ [cm/yr] | $\eta_{in}$ |
|---|---|---|---|
| 1 A | all | $1.72 \cdot \sqrt{B/B_0} \cdot \eta_0/\eta_{out}$ | const. |
| 2/3 A | $\geq 10^{21}$ Pas | $1.36 \cdot \sqrt{B/B_0} \cdot \sqrt{\eta_0/\eta_{out}}$ | $\sim \eta_{out}$ |
| 2/3 A | $\leq 10^{21}$ Pas | $1.36 \cdot \sqrt{B/B_0} \cdot \eta_0/\eta_{out}$ | const. |
| B | all | $0.86 \cdot \sqrt{B/B_0} \cdot \eta_0/\eta_{out}$ | const. |

Below the Nazca plate the flow in the mid–mantle is in a very different direction than that under the Pacific, especially for model B [*Steinberger and O'Connell,* 1998, Figure 7]. We therefore expect a motion of the Easter hotspot relative to Hawaii and Louisville of up to several cm/year. Figure 13 shows that all models with hotspot motion, especially model B, fit the observed age progression better than the fixed hotspot model. Moreover, for model B the calculated hotspot motion yields a hotspot location that is, for the assumed plate boundary, actually on the Nazca plate for longer periods of time, thus further helping to explain the observed features (Nazca and Sala y Gomez ridges) that have been associated with the Easter hotspot. Predicted tracks considerably differ among each other, and none of them fits observations too well; detailed reconstructions of this track are difficult for several reasons: The present location of the Easter hotspot is uncertain; the speed of the Pacific plate between 0 and 25 Ma might be overestimated in our models (see Table 3); a hotspot location on the Pacific plate may also lead to features on the Nazca plate (and vice versa) owing to plume-ridge interaction; there have been ridge jumps that may lead to kinks in the ridge and segments with reversed age progression, but the past location of the ridge and the timing of the ridge jumps are poorly known. A more detailed investigation of this hotspot track might actually help to improve plate reconstructions in this region. For example, the ridge topography around $85°W$, $25°S$ might indicate such a kink. When interpreted that way, it would mean that the ridge jumped from east to west

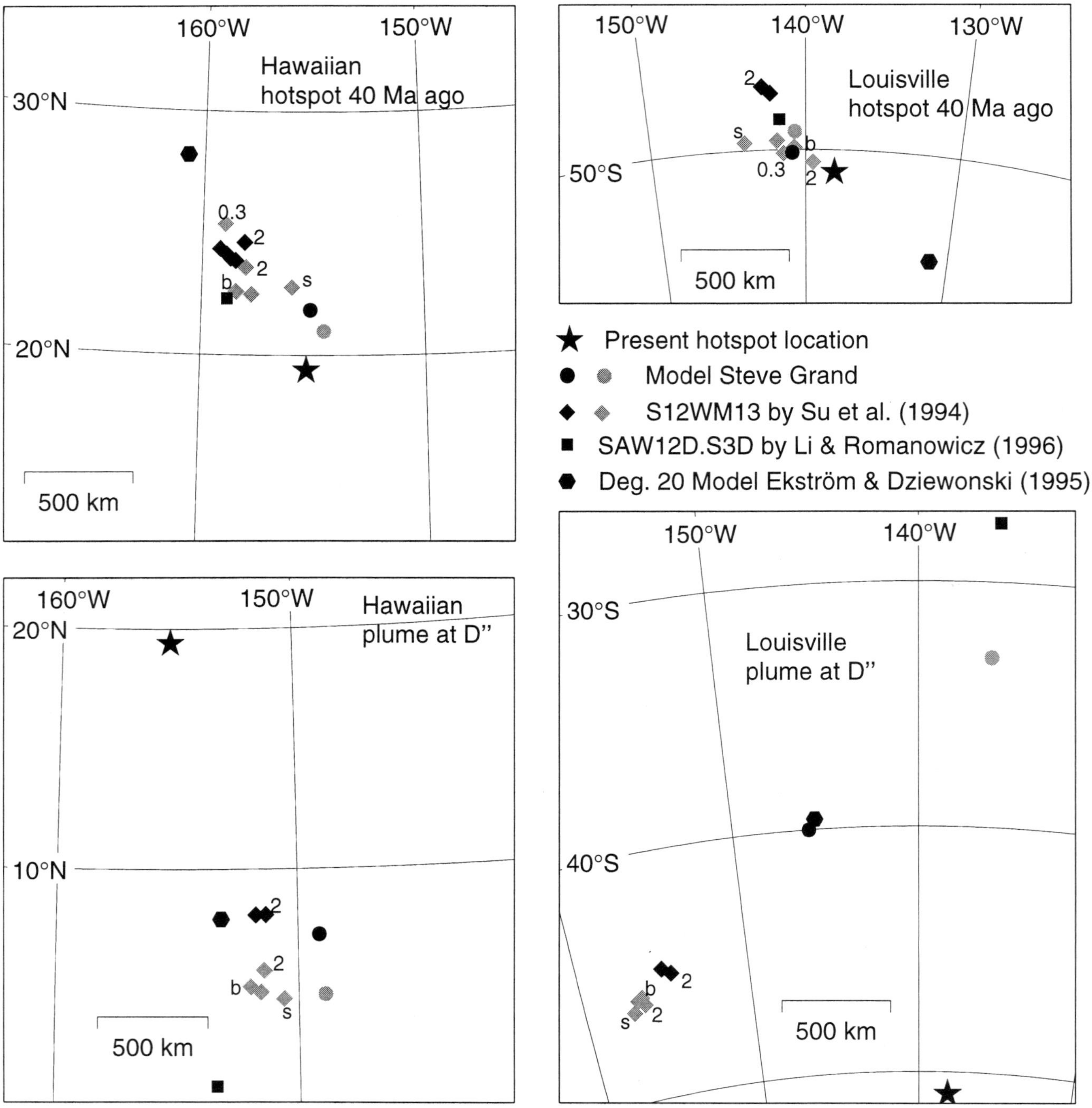


**Figure 12.** Summary of results for Hawaiian and Louisville hotspot for a greater number of models. Calculated hotspot surface locations 40 Ma ago (in mean mantle reference frame) and predicted locations of base of conduit at D" are shown. Results are shown for the viscosity models A (grey symbols) and B (black symbols). Grey symbols marked with a number "2" are for Model 2 (Table 2), whereas for black symbols marked with "2", $v_c = 0.86\ \mathrm{cm/yr} \cdot \sqrt{B/B_0} \cdot \sqrt{\eta_0/\eta_{out}}$ is used. For the data point marked with "0.3", a constant conversion factor $(\delta\rho/\rho)\,/\,(\delta v_s/v_s) = 0.3$ is used, otherwise conversion is done as specified in the text. The top left panel contains four black diamonds for assumed hotspot ages 100, 110, 120 and 130 Ma. Grey diamonds marked with the letter "b" (for big) and "s" (for small) correspond to buoyant rising speed increased resp. reduced by 50 %. Density anomalies are advected for 68 Ma in all cases except for black circles. Presumed present locations of Hawaiian and Louisville hotspots are shown as stars.

Table 3. Table of "absolute" Pacific plate rotations in a mean mantle reference frame. These were re-determined for the time intervals given in the first column (in Ma) for fixed hotspots (Model 0) and four models of moving hotspots. Our method of determining "best-fitting" plate motions is described in *Steinberger* [1999]. For each time interval, a constant rotation rate vector is assumed. Vectors are given in spherical coordinates (latitude[deg],longitude[deg],magnitude[deg/Ma]). Before 80 Ma, plate motions were not re-determined but adopted from *Lithgow-Bertelloni and Richards* [1998].

| | Model 0 | | | Model 1 A | | | Model 2 A | | | Model 3 A | | | Model B | | |
|---|---|---|---|---|---|---|---|---|---|---|---|---|---|---|---|
| t [Ma] | lat | lon | mag | lat | lon | mag | lat | lon | mag | lat | lon | mag | lat | lon | mag |
| 0– 5 | -64 | 99 | 1.00 | -64 | 101 | 0.94 | -65 | 97 | 0.93 | -65 | 90 | 1.03 | -64 | 107 | 0.91 |
| 5–25 | -73 | 122 | 0.87 | -77 | 110 | 0.82 | -77 | 97 | 0.84 | -74 | 133 | 0.86 | -79 | 92 | 0.80 |
| 25–43 | -59 | 122 | 0.58 | -55 | 156 | 0.51 | -57 | 157 | 0.51 | -63 | 151 | 0.57 | -56 | 168 | 0.48 |
| 43–80 | -8 | 105 | 0.59 | 1 | 104 | 0.55 | 1 | 101 | 0.56 | -7 | 104 | 0.54 | -7 | 99 | 0.49 |

of the hotspot around 22 Ma ago, and the hotspot has remained beneath the Nazca plate since then. The observed age progression (indicating a steeper slope in the age vs. distance plot ≈ 20 Ma ago) might even result from a change of plume tilt from toward the west (due to Pacific plate motion) to toward the east (due to Nazca plate motion), as discussed above. For comparison, in the plate boundary model used here, a ridge jump over the plume occurs 10 Ma ago, and the predicted age vs. distance curves have a steeper slope following that ridge jump. More age data would however be required to support such an interpretation.

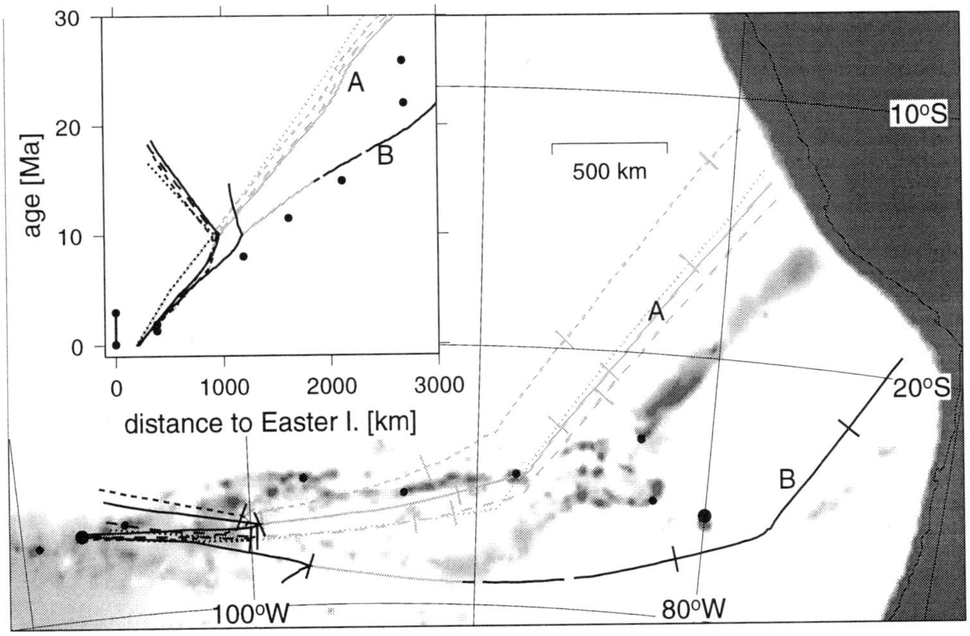


**Figure 13.** Easter hotspot track and age progression on the Nazca plate – Data [*O'Connor et al.*, 1995, and references herein] (shown as dots) and calculations for fixed and moving hotspots (four cases – legend see Figure 7); two solid lines are for viscosity models A and B as indicated. Assumed hotspot ages are 100 Ma for model A and 60 Ma for model B; if a significantly older age is assumed for model B, the calculated plume conduit is severely distorted. Location of the Easter hotspot is assumed half-way between Easter and Sala y Gomez Islands, as inferred from extrapolating age data on Sala y Gomez Ridge. Following previous suggestions [*O'Connor et al.*, 1995; *Morgan*, 1978], it is assumed that Easter Island is due to channeling of plume material toward the ridge. Black lines show the calculated hotspot track for the model of past plate motions and geometry used [*Gordon and Jurdy*, 1986]; segments with reversed age progression occur when the calculated hotspot location is on a part of the Pacific plate that subsequently gets transferred to the Nazca plate according to the model. Grey lines are calculated assuming the hotspot is always located on the Nazca plate regardless of any model of plate boundaries.

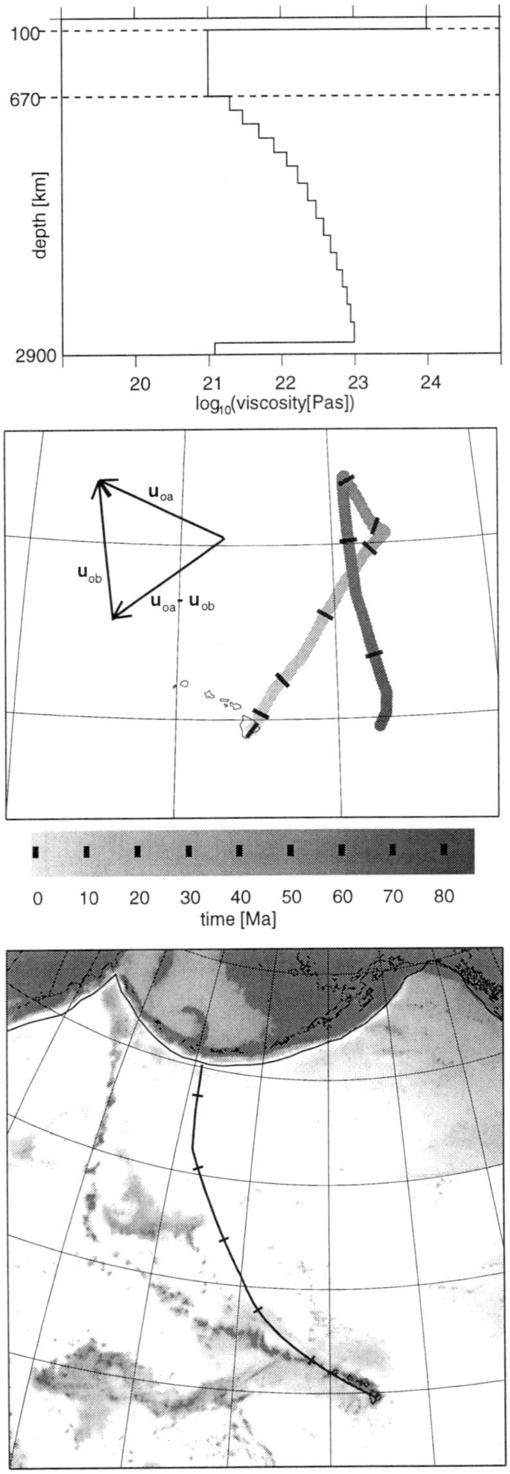


**Figure 14.** Calculated hotspot motion and track for Hawaiian plume, for viscosity model C (top panel). Predicted hotspot motion (in mean mantle reference frame) is shown in mid-panel, along with arrows indictating plate motions before and after 43 Ma, and the difference. The bottom panel shows the predicted hotspot track on the Pacific plate.

So far we have concentrated on model results that correspond well with observations. We conclude this section by showing a model that does not fit the observations, thus constraining the range of acceptable models. Figure 14 shows a result for a viscosity structure, similar to model B, but without the channel of low viscosity beneath the lithosphere, and with $v_c$ as in Table 2 for model B. The bottom panel shows that this model cannot reproduce the observed sharp bend of the Hawaiian-Emperor chain; rather it exhibits a fairly wide arc. In this model (C) the buoyant rising speed of the plume in the uppermost viscosity layer below the lithosphere is much less, and hence the deflection $x_0$ much larger, than in the previous models A and B. As we already saw for the simple model in Figure 4, the mid-panel of Figure 14 shows the hotspot moving approximately parallel to $\mathbf{u}_{0a} - \mathbf{u}_{0b}$ during a time following the change in plate velocity from $\mathbf{u}_{0b}$ to $\mathbf{u}_{0a}$. According to Figure 4 and Table 2, we should expect a horizontal plume deflection $x_0 \approx 90 - 150$km in the uppermost viscosity layer for models A and B, whereas $x_0 \approx 1150$km for model C, and a similar radius of curvature of the bend. Both results roughly agree with the respective numerical results shown in Figures 7 and 14, but only the first one is compatible with the observed sharp bend.

The results for model C thus indicate that a layer of low viscosity underlying the lithosphere is necessary to explain the observed hotspot tracks. While details of the mantle viscosity structure may not be constrained by these models, we can also conclude that a relatively high viscosity in at least parts of the lower mantle is required, simply in order to keep the mantle flow speeds low enough to achieve relative hotspot motions in accordance with observations. Examples were shown in *Steinberger and O'Connell* [1998].

## 8. IMPLICATIONS OF HOTSPOT MOTION FOR PLATE TECTONICS

In Figure 8, the calculated motion of the Hawaiian hotspot is shown together with the discrepancy between the observed Hawaiian hotspot track and the prediction assuming fixity relative to Indo-Atlantic hotspots, according to *Cande et al.* [1995]. Back to 43 Ma, the time of the bend of the Hawaiian–Emperor chain, the calculated hotspot motion explains the difference between the observed track and the prediction of *Cande et al.* [1995]. Before that however, the calculated motion of the hotspot is not sufficient to explain the discrepancy. Figure 12 shows that – independent of which tomographic model, viscosity structure, etc. is chosen – many models can significantly reduce that difference

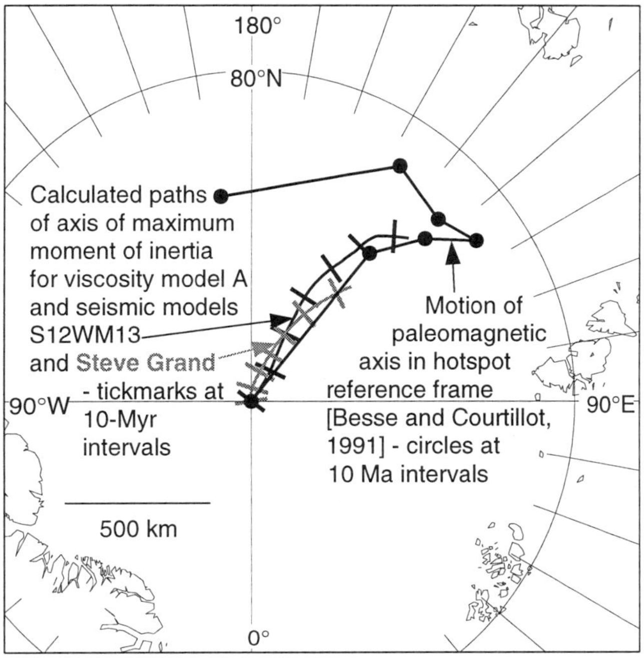


**Figure 15.** True polar wander – comparison of a curve based on paleomagnetic results with model calculations of changes of the principal axis of inertia due to advection of mantle density heterogeneities. The latter are inferred from the tomographic models S12WM13 [*Su et al.*, 1994] and from Grand's latest model (as in models 1/2 and 3), for viscosity model A, otherwise the model is the same as in *Steinberger and O'Connell* [1997]. Results for viscosity model B are considerably different; more generally, only viscosity models that are compatible with geoid modelling tend to yield true polar wander curves in agreement with paleomagnetic evidence.

during the last 43 Ma. None of the models, though, even approximately explains the much larger misfit before 43 Ma.

In Figure 9, the observed minus predicted latitudes on the Pacific plate are plotted under four different assumptions. As mentioned above, in the first case, which takes into account neither polar motion nor hotspot motion, a significant shift with time occurs. This shift is greatly reduced if polar motion, according to the curve of *Besse and Courtillot* [1991] is taken into account. That curve is inferred from continental paleomagnetic data, and is therefore independent of data from the Pacific plate. While details remain uncertain, a polar motion of 5 to 10 degrees from $\sim$ 160° E toward the present pole during the past $\sim$ 50 Ma is a common feature of several models of true polar wander [e.g., *Andrews,* 1985; *Livermore et al.,* 1984]. It is also supported by numerical models of polar motion based on calculated changes of the degree two geoid caused by advection of mantle density heterogeneities which we performed [*Steinberger and O'Connell,* 1997]. Figure 15 shows results for viscosity model A and the two tomographic models mainly used here: directions and magnitudes during the past $\approx$ 50 Ma are similar to the models based on paleomagnetism. Similar calculations, based on subduction history, were also recently performed by *Richards et al.* [1997].

In the lower panels of Figure 9, hotspot motion is also taken into account, and the best-fitting Pacific plate motion is re–determined. Because of the largely coherent hotspot motion opposite to plate motion, these models yield slower plate motions with a smaller northward component than the model with fixed hotspots. Additionally, the motion of the African plate is re–determined based on the calculated motion of the Reunion and Tristan hotspots shown in Figure 10. By considering the difference between our re–calculated motion of the African plate and its motion in the hotspot reference frame that was used by *Besse and Courtillot* [1991], their curve is converted into the same mean-mantle reference frame in which our calculations are done. The effects of this step are small (especially between 0 and 40 Ma) however, because of the small hotspot motion calculated during that time.

When these effects are included, there is essentially no trend in observed minus predicted paleolatitudes between 0 and 70-80 Ma. Figure 12 shows that the amounts of hotspot motion in the two models included in Figure 9 are fairly typical for a larger number of models, which will therefore all approximately remove any trend during that period. However there is a significant systematic offset from zero. The horizontal line through the data points is drawn by hand. An offset is even apparent in data from lava flows on the Hawaiian Island of ages 0 to 5 Ma. These data points are averaged over a large number of samples such that short term secular variations should be averaged out, and are considered more reliable [*Sager,* 1984]. It would be difficult to explain this offset of several degrees during the past 5 Ma by a hotspot motion unaccounted for in our model, especially since the Hawaiian track during the past 5 Ma gives no indication for a significant hotspot motion. The offset is therefore very likely due to other reasons and we conclude that the removal of any trend indicates that our models describe hotspot motion adequately, at least back to $\sim 70 - 80$ Ma ago.

The data point for Detroit Seamount (81 Ma) by *Tarduno and Cottrell* [1997] however is offset significantly from all other points. To determine whether this is due to hotspot motion more rapid than in our model, an episode of rapid true polar wander not accounted for in the *Besse and Courtillot* [1991] curve, or long-lived non-dipolar components in the Earth's magnetic field will clearly require more data. The offset in paleolatitude in Figure 9 may reflect a feature of the field that is persistent over relatively long times [*Johnson and Constable,* 1998].

## 9. DISCUSSION AND OUTLOOK

The purpose of this paper was to discuss the problem of hotspot motion from a geodynamic viewpoint. This problem is motivated by observations – some of them support the idea of hotspot fixity, others seem to require significant motions of hotspots. Following a general introduction to the current ideas about plumes and hotspots, we therefore reviewed these observations. We then presented a numerical model that is intended to explain most of the observations. In this context, the introduction also served to justify why this particular numerical approach was taken. Numerical results are shown here for only a few representative hotspots, where effects can be seen well, and/or results have important implications regarding plate motions. A more systematic treatment of all known hotspots is given in *Steinberger* [1999].

A coherent motion of Hawaiian and Louisville hotspots of the order of 1 cm/yr and a southward component of Hawaiian hotspot motion of similar magnitude during the Cenocoic is feature of most models tested. Our results therefore explain the apparent misfit between observed and predicted Hawaiian hotspot track that has been reported by *Cande et al.* [1995] only back to 43 Ma. Where the remaining misfit comes from and to what extent unrecognized plate boundaries can change the results is the subject of ongoing research [*Raymond et al.*, this issue]. In combination with polar wander, paleomagnetic data can however be explained back to about 70-80 Ma. This result does not depend on the particular polar wander curve chosen, as a polar motion of the order of 5 degrees away from the Pacific during the Cenocoic is a common feature of several paleomagnetic true polar wander curves as well as our predictions from the advection of mantle density heterogeneities. 5 degrees from polar motion plus about 7 to 8 degrees from hotspot motion are sufficient to explain the previously reported trend in observed minus predicted paleomagnetic data back to about 70-80 Ma. Our results do not support the hypothesis of a particularly rapid motion of the Hawaiian hotspot before 43 Ma, as proposed by *Norton* [1995], therefore a rather abrupt change in absolute plate motion at around 43 Ma for the Pacific plate is required, accompanied by corresponding adjustments in the motions of surrounding plates. Our results also give predictions of the shape of plume conduits: Plume source locations tend to be displaced relative to corresponding surface hotspots, approximately towards the center of large-scale upwellings under Southern Africa and the South Pacific (Figures 10 and 12).

We have shown that not all models for "reasonable" parameters (i.e. in agreement with other evidence) give results in agreement with observations. Therefore, our models can help to further constrain model parameters such as viscosity structure. Comparison of various models introduced here and in our other papers, that are in agreement with observations, shows that we can make reasonably robust predictions of direction and magnitude of hotspot motion and of the tilt of plume conduits. In general, predictions do not depend on particular model parameters chosen, but are similar for a great number of models. Because of model uncertainties, obviously no detailed predictions can be made at present. The validity of our predictions and hence of the proposed underlying physical mechanism is supported by the fact that our model can help to explain a number of previously unexplained observations, particularly for more recent times. The examples shown here focussed on the Hawaiian, Louisville and Easter hotspots and hence plate motions in the Pacific region. In other instances, significant misfits remain. We anticipate that, due to better model constraints, we will be able to make more accurate predictions of hotspot motion in the future, and that, in combination with better constraints on plate reconstructions and a greater number of paleomagnetic data, remaining misfits can be resolved.

*Acknowledgments.* We thank Steve Grand for sharing his latest, unpublished model via anonymous ftp. Some of the calculations were done by Maria Antretter. Based on comments by Carol Raymond, Mark Richards and an anonymous reviewer, the initial manuscript was profoundly rewritten. Consistency of viscosity structure A with models of the geoid was checked using a code written by Svetlana Panasyuk.

## REFERENCES

Albers, M., and U. R. Christensen, The excess temperature of plumes rising from the core-mantle boundary, *Geophys. Res. Lett.*, *23*, 3567-3570, 1996.

Andrews, J. A., True polar wander: An analysis of cenozoic and mesozoic paleomagnetic poles, *J. Geophys. Res.*, *90*, 7737-7750, 1985.

Besse, J., and V. Courtillot, Revised and synthetic apparent polar wander paths of the African, Eurasian, North American and Indian plates, and true polar wander since 200 Ma, *J. Geophys. Res.*, *96*, 4029-4050, 1991.

Bijwaard, H., and W. Spakman, Tomographic evidence for a

narrow whole mantle plume below Iceland, *Earth Planet. Sci. Lett., 166*, 121-126, 1999.

Cande, S. C., C. A. Raymond, J. Stock, and W. F. Haxby, Geophysics of the Pitman Fracture Zone and Pacific-Antarctic plate motions during the Cenozoic, *Science, 270*, 947-953, 1995.

Chopelas, A., and R. Boehler, Thermal expansion measurements at very high pressure, systematics, and a case for a chemically homogeneous mantle, *Geophys. Res. Lett., 16*, 1347-1350, 1989.

Choubert, G., and A. Faure-Muret, Atlas geologique du monde, *Commission de la carte geologique du monde*, Unesco, Paris, 1976.

Clague, D. A., and G. B. Dalrymple, Tectonics, geochronology, and origin of the Hawaiian-Emperor volcanic chain, in *The Geology of North America*, Vol. N, *The Eastern Pacific Ocean and Hawaii*, edited by E. L. Winterer, D. M. Hussong, and R. W. Decker, pp. 188-217, GSA, Boulder, CO., 1989.

Davies, G. F., Ocean bathymetry and mantle convection, 1. Large-scale flow and hotspots, *J. Geophys. Res., 93*, 10467-10480, 1988.

Davies, G. F., Temporal variation of the Hawaiian plume flux, *Earth Planet. Sci. Lett., 113*, 277-286, 1992.

Duncan, R. A., and D. A. Clague, Pacific plate motion recorded by linear volcanic chains, in *The Ocean Basins and Margins*, vol. 7a, edited by A. E. M. Nairn, F. G. Stehli, and S. Uyeda, pp. 89-121, Plenum, New York, 1985.

Duncan, R. A., and R. B. Hargraves, $^{40}Ar/^{39}Ar$ geochronology of basement rocks from the Mascarene Plateau, the Chagos Bank, and the Maldive Ridge, *Proc. Ocean Drill. Program Sci. Results, 115*, 43-51, 1990.

Ekström, G., and A. M. Dziewonski, Improved models of upper mantle S velocity structure (abstract), *Eos Trans. AGU, 76, Fall Meet. Suppl.*, F421, 1995.

Epp, D., W. W. Sager, F. Theyer, and S. R. Hammond, Hotspot-spin axis motion or magnetic far-sided effect? *Nature, 303*, 318-320, 1983.

Forte, A. M., A. M. Dziewonski, and R. L. Woodward, Aspherical structure of the mantle, tectonic plate motions, nonhydrostatic geoid, and topography of the core-mantle boundary, in *Dynamics of the Earth's Deep Interior and Earth Rotation, Geophys. Mon. Ser.*, Vol. 72, edited by J.-L. Le Mouël, D. E. Smylie, and T. Herring, pp. 135-166, AGU, Washington, DC., 1993.

Gordon, R. G., and D. Jurdy, Cenocoic global plate motions, *J. Geophys. Res., 91*, 12389-12406, 1986.

Grand, S. P., R. D. Van der Hilst, and S. Widiyantoro, Global seismic tomography: A snapshot of convection in the Earth, *GSA Today, 7*, 1-7, 1997.

Griffiths, R. W., and I. H. Campbell, Stirring and structure in mantle plumes, *Earth Planet. Sci. Lett., 99*, 66-78, 1990.

Griffiths, R. W., and M. A. Richards, The adjustment of mantle plumes to changes in plate motion, *Geophys. Res. Lett., 16*, 437-440, 1989.

Gurnis, M., Convective mixing in the Earth's mantle, Ph. D. Thesis, Aust. Nat. Univ., Canberra, 1986.

Hager, B. H., and R. J. O'Connell, Kinematic models of large-scale flow in the Earth's mantle, *J. Geophys. Res., 84*, 1031-1048, 1979.

Hager, B. H., and R. J. O'Connell, A simple global model of plate dynamics and mantle convection, *J. Geophys. Res., 86*, 4843-4867, 1981.

Harada, Y., New accurate models of the plate motions relative to the hotspots and the cause of the discrepancy in the global plate-motion circuit (abstract), *Eos Trans. AGU, 78, Fall Meet. Suppl.*, F721, 1997.

Haskell, N. A., The motion of a fluid under a surface load, 1, *Physics, 6*, 265-269, 1935.

Jarrard, R. D., and D. A. Clague, Implications of Pacific island and seamount ages for the origin of volcanic chains, *Rev. Geophys. Space Phys., 15*, 57-76, 1977.

Johnson, C. L., and C. G. Constable, Persistently anomalous Pacific geomagnetic fields, *Geophys. Res. Lett., 25*, 1011-1014, 1998.

Karato, S., Importance of anelasticity in the interpretation of seismic tomography, *Geophys. Res. Lett., 20*, 1623-1626, 1993.

Lambeck, K., and P. Johnston, The viscosity of the mantle: evidence from analyses of glacial-rebound phenomena, in *The Earth's Mantle*, edited by I. Jackson, pp. 461-502, Cambridge University Press, Cambridge, 1998.

Li, X.-D. and B. Romanowicz, Global mantle shear velocity model developed using nonlinear asymptotic coupling theory, *J. Geophys. Res., 101*, 22245-22272, 1996.

Lithgow-Bertelloni, C., and M. A. Richards, The dynamics of Cenozoic and Mesozoic plate motions, *Rev. Geophys., 36*, 27-78, 1998.

Livermore, R. A., F. J. Vine, and A. G. Smith, Plate motions and the geomagnetic field - II. Jurassic to Tertiary, *Geophys. J. R. astr. Soc., 79*, 939-961, 1984.

Manga, M., H. A. Stone, and R. J. O'Connell, The interaction of plume heads with compositional discontinuities in the Earth's mantle, *J. Geophys. Res., 98*, 19979-19990, 1993.

McNutt, M. K., D. W. Caress, J. Reynolds, K. A. Jordahl, and R. A. Duncan, Failure of plume theory to explain mid-plate volcanism in the southern Austral islands, *Nature, 389*, 479-482, 1997.

Mitrovica, J. X., Haskell [1935] revisited, *J. Geophys. Res., 101*, 555–569, 1996.

Mitrovica, J. X., and A. M. Forte, Radial profile of mantle viscosity: Results from the joint inversion of convection and postglacial rebound observables, *J. Geophys. Res., 102*, 2751–2769, 1997.

Morgan, W. J., Convection plumes in the lower mantle, *Nature, 230*, 42-43, 1971.

Morgan, W. J., Deep mantle convection plumes and plate motions, *Am. Assoc. Pet. Geol. Bull., 56*, 203-213, 1972.

Morgan, W. J., Rodriguez, Darwin, Amsterdam,.., a second type of hotspot island, *J. Geophys. Res., 83*, 5355-5360, 1978.

Morgan, W. J., Hotspot tracks and the opening of the Atlantic and Indian Oceans, in *The Sea*, Vol. 7, *The oceanic lithosphere*, edited by C. Emiliani, pp. 443-487, Wiley, New York, 1981.

Müller, R. D., J.-Y. Royer, and L. A. Lawver, Revised plate motions relative to the hotspots from combined Atlantic and Indian Ocean hotspot tracks, *Geology, 21*, 275-278, 1993.

National Geophys. Data Center, ETOPO-5 bathymetry/topography data, *Data Announc. 88-MGG-02*, Natl. Oceanic

& Atmos. Admin., U.S. Dep. Commer., Boulder, Colo., 1988.

Norton, I. O., Plate motions in the North Pacific: The 43 Ma nonevent, *Tectonics, 14*, 1080-1094, 1995.

O'Connor, J. M., P. Stoffers, and M. O. McWilliams, Time-space mapping of Easter Chain volcanism, *Earth Planet. Sci. Lett., 136*, 197-212, 1995.

Olson, P., and H. Singer, Creeping plumes, *J. Fluid Mech., 158*, 511-531, 1985.

Osako, M., and E. Ito, Thermal diffusivity of $MgSiO_3$ perovskite, *Geophys. Res. Lett., 18*, 239-242, 1991.

Phipps Morgan, J., W. J. Morgan, Y.-S. Zhang, and W. H. F. Smith, Observational hints for a plume–fed, suboceanic asthenosphere and its role in mantle convection, *J. Geophys. Res., 100*, 12753-12767, 1995.

Pringle, M. S., and R. A. Duncan, Radiometric ages of basement lavas recovered at Loen, Wodejebato, MIT and Takuyo-Daisan Guyots, northwestern Pacific Ocean, *Proc. Ocean Drill. Program Sci. Results, 144*, 547-560, 1995.

Raymond, C. A., J. M. Stock, and S. C. Cande, Fast Paleogene motion of the Pacific hotspots from revised global plate circuit constraints, in *The History and Dynamics of Global Plate Motions*, Geophys. Mon. Ser., edited by M. A. Richards, R. G. Gordon, and R. D. van der Hilst, AGU, Washington, DC., 1999.

Ricard, Y., L. Fleitout, and C. Froidevaux, Geoid heights and lithospheric stresses for a dynamic Earth, *Ann. Geophys., 2*, 267-286, 1984.

Richards, M. A., Hotspots and the case for a high viscosity lower mantle, in *Glacial Isostasy, Sea-Level and Mantle Rheology*, edited by R. Sabadini, K. Lambeck, and E. Boschi, pp. 571-587, Kluwer Academic Publishers, Dordrecht, 1991.

Richards, M. A., R. A. Duncan, and V. E. Courtillot, Flood basalts and hot spot tracks: Plume heads and tails, *Science, 246*, 103-107, 1989.

Richards, M. A., and R. W. Griffiths, Deflection of plumes by mantle shear flow: experimental results and a simple theory, *Geophys. J., 94*, 367-376, 1988.

Richards, M. A., and R. W. Griffiths, Thermal entrainment by deflected mantle plumes, *Nature, 342*, 900-902, 1989.

Richards, M. A., and B. H. Hager, Geoid anomalies in a dynamic Earth, *J. Geophys. Res., 89*, 5987-6002, 1984.

Richards, M. A., Y. Ricard, C. Lithgow-Bertelloni, G. Spada, and R. Sabadini, An explanation for Earth's long-term rotational stability, *Science, 275*, 372-375, 1997.

Sager, W. W., Paleomagnetism of Abbott Seamount and implications for the latitudinal drift of the Hawaiian hot spot, *J. Geophys. Res., 89*, 6271-6284, 1984.

Sager, W. W., and U. Bleil, Latitudinal shift of Pacific hotspots during the late Cretaceous and early Tertiary, *Nature, 326*, 488-490, 1987.

Schilling, J.-G., Fluxes and excess temperatures of mantle plumes inferred from their interaction with migrating mid-ocean ridges, *Nature, 352*, 397-403, 1991.

Sleep, N., Hotspots and mantle plumes: Some phenomenology, *J. Geophys. Res., 95*, 6715-6736, 1990.

Sleep, N., Time dependence of mantle plumes: Some simple theory, *J. Geophys. Res., 97*, 20007-20019, 1992.

Steinberger, B., Plumes in a convecting mantle: Models and observations for individual hotspots, *J. Geophys. Res.*, in press, 1999.

Steinberger, B., and R. J. O'Connell, Changes of the Earth's rotation axis owing to advection of mantle density heterogeneities, *Nature, 387*, 169-173, 1997.

Steinberger, B., and R. J. O'Connell, Advection of plumes in mantle flow: implications for hotspot motion, mantle viscosity and plume distribution, *Geophys. J. Int., 132*, 412-434, 1998.

Su, W.-J., R. L. Woodward, and A. M. Dziewonski, Degree 12 model of shear velocity heterogeneity in the mantle, *J. Geophys. Res., 99*, 6945-6980, 1994.

Tarduno, J. A., and R. D. Cottrell, Paleomagnetic evidence for motion of the Hawaiian hotspot during formation of the Emperor seamounts, *Earth Planet. Sci. Lett., 153*, 171-180, 1997.

Tarduno, J. A., and J. Gee, Large-scale motion between Pacific and Atlantic hotspots, *Nature, 378*, 477-480, 1995.

Thoraval, C., and M. A. Richards, The geoid constraint in global geodynamics: viscosity structure, mantle heterogeneity models and boundary conditions, *Geophys. J. Int., 131*, 1-8, 1997.

Watts, A. B., J. K. Weissel, R. A. Duncan, and R. L. Larson, Origin of the Louisville Ridge and its relationship to the Eltanin Fracture Zone System, *J. Geophys. Res., 93*, 3051-3077, 1988.

Wessel, P., and W. H. F. Smith, New version of the Generic Mapping Tools released, *Eos Trans. AGU, 76*, 329, 1995.

Whitehead, J. A., Instabilities of fluid conduits in a flowing earth - are plates lubricated by the asthenosphere? *Geophys. J. R. astr. Soc., 70*, 415-433, 1982.

Whitehead, J. A., and D. S. Luther, Dynamics of laboratory diapir and plume models, *J. Geophys. Res., 80*, 705-717, 1975.

Wilson, J. T., A possible origin of the Hawaiian islands, *Can. J. Phys., 41*, 863-870, 1963.

Wolfe, C. J., I. T. Bjarnason, J. C. VanDecar, and S. C. Solomon, Seismic structure of the Iceland mantle plume, *Nature, 385*, 245-247, 1997.

---

B. Steinberger, Institut für Meteorologie und Geophysik, Johann Wolfgang Goethe-Universität, Feldbergstr. 47, 60323 Frankfurt am Main, Germany. (e-mail: steinber@geophysik.uni-frankfurt.de)

R. J. O'Connell, Department of Earth and Planetary Sciences, Harvard University, 20 Oxford Street, Cambridge, MA 02138. (e-mail: oconnell@geophysics.harvard.edu)